D0086001
DISCARD

Understanding the HUMAN BODY

Phil Tate, D.A.
(Biological Education)

Instructor of Anatomy and Physiology
Phoenix College
Maricopa Community College District
Phoenix, Arizona

Rod. R. Seeley, Ph.D.

Professor of Physiology
Idaho State University
Pocatello, Idaho

Trent D. Stephens, Ph.D.

Professor of Anatomy and Embryology
Idaho State University
Pocatello, Idaho

St. Louis Baltimore Boston Chicago London Madrid Philadelphia Sydney Toronto

Editor-in Chief: James M. Smith
Editor: Robert J. Callanan
Developmental Editor: Laura J. Edwards
Project Manager: Gayle May Morris
Production Editors: Deborah L. Vogel/ Mary Drone
Manufacturing Supervisor: Betty Richmond
Designer: Susan Lane

Credits for all materials used by permission appear before the index.

Printed in the United States of America
Composition by Graphic World, Inc.
Printing/binding by Von Hoffmann Press, Inc.

Mosby–Year Book, Inc.
11830 Westline Industrial Drive
St. Louis, Missouri 63146

Library of Congress Cataloging in Publication Data

Tate, Philip.
Understanding the human body / Philip Tate, Rod R. Seeley, Trent D. Stephens.
p. cm.
Includes index.
ISBN 0-8016-7197-3
1. Human physiology. 2. Human anatomy. I. Seeley, Rod R.
II. Stephens, Trent D. III. Title.
QP36.T38 1993
612—dc20 93-21618
CIP

93 94 95 96 97 98 / 9 8 7 6 5 4 3 2 1

Preface

"YOU CAN'T SEE THE FOREST FOR THE TREES."

This saying has special relevance to the study of anatomy and physiology. There are so many trees in a forest, so much detailed information, that one can miss an appreciation of the forest itself, that is, the whole organism and the interactions of all its parts.

In anatomy and physiology courses, typically there are attempts to present as much material as possible. All the important information must be covered! Remember, however, that one definition of cover is to hide or conceal. So much detail can be presented that the main concepts are missed! There is a need for a book that presents the minimal amount of information. Such an approach, however, has its own dangers. The content can be so reduced that only individual pieces of information (trees) remain. There no longer is an entire, complex organism (forest) to be appreciated.

A balance between detail and the "big picture" can be achieved by considering how students learn. An approach commonly used by students is to list, define, and describe every detail of the information in an attempt to remember everything! Unfortunately, isolated facts are only remembered for a short time, usually until shortly after a test is taken. A better approach is to first *understand* the material by forming a conceptual framework that identifies and ties together the important points of a topic. In this text, explanations start with simple, easy-to-understand facts and are developed in a logical sequence. Students can see the "big picture" without being overwhelmed with details. Once the information is understood and the story makes sense, the details of the information naturally fall into place and are more easily remembered.

Remembering and understanding, however, are only the first steps in learning. They are important steps, and without them the next step is not possible. But there is a next step—developing the ability to use the understood material to solve problems. No book, no matter how much information it contains, can provide the answers to all possible questions. Students must develop the ability to apply what they know, analyze information, and synthesize solutions.

We believe problem-based learning can be developed through example and practice. By seeing how problems are solved, students gain insight into how to go about solving problems on their own. It gives them a "grab bag" of mental tricks to use for solving problems. By practicing, students develop and improve their mental skills. Eventually these skills become a natural and expected part of their approach to learning. Students develop a sense of confidence and accomplishment as they discover that they can indeed learn and solve problems.

The title of this text, *Understanding the Human Body*, was carefully chosen, because the term "understanding" has much to do with the philosophy of this book. Understanding is the bridge between remembering information and using information to solve problems. Information is not presented for its own sake, but for the understanding it provides. When a minimal number of the pieces of a puzzle are correctly arranged, the pattern of the puzzle can be seen. And with an understanding of the pattern, it is possible to fill in the blank spots, that is, use the information to solve problems.

Conceptual themes

The major conceptual themes emphasized throughout the text are the *relationship between structure and function* and the *regulation of homeostasis*. Just as the structure of a hammer makes it well-suited for the function of pounding nails, the forms of specific cells, tissues, and organs within the body allow them to perform specific functions effectively. For example, muscle cells contain proteins that make contraction possible, and bone cells surround themselves with a mineralized matrix that provides strength and support. Knowledge of structural and functional relationships makes it easier to understand anatomy and physiology and greatly enhances appreciation for the subject.

Homeostasis, the maintenance of an internal environment within an acceptably narrow range of values, is necessary for the survival of the human body. For instance, if the blood delivers inadequate amounts of oxygen to body cells, the heart and respiration rates increase until oxygen delivery to the body cells becomes adequate. The emphasis of this book is the regulatory mechanisms that normally maintain homeostasis. However, because failure of these mechanisms also illustrates how they work, pathologies resulting in dysfunction, disease, and possibly death also are presented.

A unique feature of this text is its consideration of microorganisms. Chapter 21 describes the basic biology of microorganisms with an emphasis on bacteria and viruses. Understanding how microorganisms grow and reproduce leads to an appreciation of disease symptoms, treatment strategies, and methods of preventing disease.

Relevant clinical examples

Clinical information should never be an end in itself. In some texts, mere clinical description or medical terminology represents a significant portion of the material. This text provides clinical examples to both promote interest and demonstrate relevance, but clinical information is used primarily to illustrate the application of basic knowledge. The ability to apply knowledge is a skill that goes beyond mere acquaintance with either clinical or basic anatomy and physiology content.

Analysis of practical problems

At best, some anatomy and physiology texts include a few "thought" questions that, for the most part, involve a restatement or a summary of content. Yet once students understand the material well enough to state it in their own words, it only seems logical for them to proceed to the next step—that is, to apply the knowledge to hypothetical situations. This text features two sets of problem-based learning questions in every chapter, *Predict Questions* and *Concept Review Questions* (to be highlighted in more detail later in this preface). Answers and explanations for the Predict Questions are provided at the back of the text. The explanations illustrate the methods used to solve problems and provide a model for the development of problem-based learning. The acquisition of such skills is necessary for a complete understanding of anatomy and physiology, it is fun, and it makes it possible for the student to deal with the many problems that occur as a part of professional and everyday life.

Illustration program

The statement, "A single picture is worth a thousand words," is especially true in anatomy and physiology. Structural and functional relationships become immediately apparent in the illustrations in this text. To maximize effectiveness, illustrations have been placed as close as possible to the narrative where they are cited, and special attention has been devoted to the figure legends, which summarize or emphasize the important features of each illustration. The illustrations also have been designed to be nonintimidating and aesthetically pleasing, features that encourage the student to spend time with the illustrations for maximum learning and pleasure. All the artwork in this textbook is in full color, making the illustrations attractive and emphasizing the important structures. In addition to the illustrations, photographs bring a dimension of realism to the text. In many cases, photographs are accompanied by line drawings that emphasize important features of the photograph.

Learning aids

The text must be an effective teaching tool. Because students learn best in different ways, a variety of teaching and learning aids are provided. This enables students to organize the material in their minds, determine the main points, and evaluate the progress of their learning.

Objectives. Each chapter begins with a series of *Objectives* that emphasize the important facts, topics, and concepts to be covered. The objectives are learning goals that focus attention on the material and tasks to be mastered by reading the chapter.

Vocabulary Aids. Learning anatomy and physiology is, in many ways, like learning a new language. To communicate effectively, a basic terminology, dealing with important or commonly used facts and concepts, must be mastered. At the beginning of each chapter are the *Key Terms*, a list of some of the more important new words to be learned along with their definitions. Throughout the text, these and additional important terms are presented in **boldface print.** When pronunciation of the word is complex, a *Pronunciation Key* is presented. Often simply being able to pronounce a word correctly is the key to remembering it. The *Glossary*, which collects the most important terms into one location for easy reference, also has a pronunciation key.

Understanding Essays. A unique feature of this text are the understanding essays. They are a more detailed examination of an important topic introduced in the main body of the text, providing a "taste" of what could be learned about a topic if it were examined in more detail. However, the intent is not to merely present more information. Instead, the understanding essay clarifies concepts by presenting an example of a concept in action. Furthermore, once the concept is better understood, the student is asked to solve a problem relating to the concept.

Some of the topics dealt with in the understanding essays include clinically important disorders such as diabetes mellitus, acidosis and alkalosis, cancer, and burns. Other topics emphasize fundamental processes such as osmosis, pH, action potentials, pain, antibody activity, and defecation. Structure and function are emphasized in the understanding essays that describe the vertebral column, muscle fibers, and the coronary circulation. The understanding essays also consider topical issues such as starvation and obesity, prevention of pregnancy, the human genome, and genetic recombination.

Boxed Essays. The boxed essays primarily emphasize pathologies of the human body. After an understanding of the normal operation of the body is presented in the text, the boxed essays survey representative diseases that disrupt homeostasis. In contrast to many anatomy and physiology texts, the boxed essays also describe diseases caused by microorganisms. By placing the essay material "aside," basic understanding is not obscured by the details of pathologies, and a better contrast between normal and abnormal conditions is achieved. A consideration of pathology adds relevance and interest that makes the material more meaningful.

Predict Questions. The Predict questions require the application of concepts. When reading a text it is very easy to become a "passive" learner; everything seems very clear to passive learners until they attempt to use the information. The predict question converts the "passive" learner into an "active" learner who must use new information to solve a problem. The answer to this kind of question is not a mere restatement of fact, but rather a prediction, an analysis of the data, the synthesis of an experiment, or the evaluation and weighing of the important variables of the problem. For example, "Given a stimulus, predict how a system will respond." Or, "Given a clinical condition, explain why the observed symptoms occurred." Predict questions are practice problems that help to develop the skills necessary to answer the Concept Review questions at the end of the chapter. In this regard, not only are possible answers given for the predict questions, but explanations that demonstrate the process of problem-solving are provided in Appendix B.

Chapter Summary. The chapter Summary is an outline that briefly states the important facts and concepts presented in the chapter. It provides a perspective of the "big picture" and can be used as a preview or review of the chapter.

Content Review. The Content Review questions are another method used in this text to turn the "passive" learner into an "active" learner. The questions systematically cover the content and require students to summarize and restate the content in their own words.

Concept Review. Following the mastery of the content review and therefore the chapter content, the Concept Review questions require the application of that content to new situations. These are not essay questions that involve the restatement or summarization of chapter content. Instead, they provide additional practice in problem-based learning and promote the development and acquisition of problem-solving skills. In addition, problem-based learning activities improve long-term retention of information because once the information is understood and used, it "belongs" to the student.

Chapter Test. At the end of each chapter are the Chapter Test questions. They are matching and fill-in-the-blank questions that provide students with the opportunity to test their knowledge. Answers are provided for all the questions in Appendix C.

Appendices. Appendix A is a table of common disorders caused by pathogens. The diseases of each major organ system are listed with a brief description of the disease. Appendix B contains the answers to the Predict questions and Appendix C contains the answers to Chapter Test questions.

Supplements

Any textbook can be used alone, but thoughtfully developed supplements increase its effectiveness for both student and instructor because the supplements are designed to support the pedagogical model developed in the text.

Study Guide. The study guide by Philip Tate and James Kennedy of Phoenix College supports the problem-based learning approach of the text. It introduces the student to the content of anatomy and physiology using matching, labeling, and completion exercises. A Mastery Learning Activity consisting of multiple-choice questions emphasizes comprehension of the material, evaluates progress, and prepares the students for classroom testing. In addition, a Final Challenges section consisting of essay questions provides practice with questions similar to the Predict and Concept Review questions of the textbook. Answers are given for all exercises, and explanations are furnished for the Mastery Learning Activity and the Final Challenges. The study guide provides the reinforcement and practice so essential for the student's success in the course.

Instructor's Manual. For the instructor, this ancillary will be an invaluable teaching resource. Each chapter contains a brief synopsis, lecture outline, suggested topics for class discussion, and a list of selected audiovisual resources. A test-bank of over 800 questions is also included in the manual. In addition, transparency masters of selected illustrations in *Understanding the Human Body* are included to enhance visual learning.

Computerized Test Bank. The test questions included in the instructor's manual are also available on a computerized test generation system compatible with IBM, Macintosh, and Apple computers. This system has many features that make it easy for the instructor to design tests and quizzes. All questions are categorized by subject matter and level of difficulty.

Transparency Acetates. A set of 60 full-color transparency acetates—with large, easy-to-read labels—is available to qualified adopters of the text for use as a teaching aid. These transparencies, which emphasize the major anatomical structures and physiological processes covered in the textbook, have been selected from illustrations in the text and provide a common vehicle for communication between the lecturer and the student.

Human Body Systems Software. Human body systems software helps beginning students further explore the human body. Individual modules introduce each of the eleven body systems. Each module contains an introduction, a tutorial with practice review questions, practical applications, and a final quiz. The software runs on IBM-compatible computers.

Human Cadaver Dissection Video. This 60-minute video takes the student through a dissection of the musculature of the human body as well as the internal organs of the thorax and abdomen. The video provides a detailed explanation of the dissection procedure in vivid close-up with clear, precise commentary.

Acknowledgments

The efforts of many people are required to produce a modern textbook. It is difficult to adequately acknowledge the contributions of all the people who have played a role. The encouragement and emotional support of our families were essential for the completion of this project.

We wish to express our gratitude to the present and past staff of Mosby for their steadfast help and encouragement. It has clearly been more than a vocation to them. Laura Edwards, Amy Winston, and Robert Callanan have worked with us in an untiring fashion to bring this work to completion. Their effort and contributions, as well as the efforts of the many others who have influenced the design and production of this text, are greatly appreciated.

We also thank the team of artists who have contributed to the text. Their attention to detail and their artistic contributions have made the text an attractive as well as an effective teaching tool.

We sincerely thank the reviewers. This book was conscientiously reviewed by people who not only have experience in the health fields, but who are exceptional teachers, as well. Their constructive comments and suggestions have added substantially to the quality of the text.

Philip Tate
Rod Seeley
Trent Stephens

Reviewers

Marguerite Brayton-Crokus, RN, *Cumberland Vocational Technical School*
Gayle F. Clark, BS ED, BSN, *Northern Tier Career Center*
Ginny A. Cohrs, RN, *Alexandria Technical College*
Barbara J. Cole, MS ED, *Phoenix College*
Judy A. Donaldson, RN, *North Seattle Community College*
Anthony G. Futcher, PhD, *Columbia Union College*
Saundra B. Porter, BSN, *Conway School of Practical Nursing*
Janet Rawlings, MA, *Antelope Valley College*
Jane Pelster Rosen, RN, BSN, *Blair Junior College*
Ann Senisi Scott, MA, *Nassau Tech BOCES*
Rita J. Schwieterman, BSN, *Dayton School of Practical Nursing*
Linda Strause, PhD, *University of California—San Diego*
Eileen M. Williams, RN, *Nassau Technological Center*

Contents in Brief

Contents

CHAPTER 1

Introduction to the Human Body

KEY TERMS

anatomical position
Position in which a person is standing erect with the feet forward, arms hanging to the sides, and the palms of the hands facing forward.

anatomy
Scientific discipline that investigates the structure of the body.

frontal plane
Plane running vertically through the body and separating it into anterior and posterior portions.

homeostasis
Existence and maintenance of a relatively constant environment within the body with respect to functions and the composition of fluids.

negative feedback
Mechanism by which any change from an ideal normal value is made smaller or is resisted.

physiology
Scientific discipline that deals with the processes or functions of living things.

positive feedback
Mechanism by which any change from an ideal normal value is made greater.

sagittal plane
Plane running vertically through the body and dividing it into right and left parts.

transverse plane
Plane running horizontally through the body and dividing it into superior and inferior parts; a horizontal cross section.

FEATURES

OBJECTIVES

After reading this chapter you should be able to:

1. Explain the importance of understanding the relationship between structure and function.
2. Define anatomy and physiology.
3. Describe the seven structural levels of organization of the body and give the major characteristics of each level.
4. Define homeostasis and explain why it is important.
5. Define negative feedback and positive feedback and describe their relationship to homeostasis.
6. Define the directional terms for the human body and use them to locate specific body structures.
7. Describe the three major planes of the body.
8. Name the regions of the body and describe the subdivisions of each region.
9. Describe the major trunk cavities.

HUMAN ANATOMY AND PHYSIOLOGY is the study of the structure and function of the human body. Knowledge of its structure and function makes it possible to understand how the body responds to a stimulus. For example, eating a candy bar results in an increase in blood sugar (the stimulus). Knowledge of the pancreas allows one to predict that the pancreas will secrete insulin (the response), which increases the movement of sugar from the blood into cells. Consequently blood sugar levels decrease back toward a normal range of values. Knowledge of structure and function also provides the basis for understanding disease. In one type of diabetes mellitus, for example, the pancreas does not secrete adequate amounts of insulin. Even though blood sugar levels increase, without adequate insulin there is not enough movement of sugar into cells. Therefore cells are deprived of a needed source of energy and they malfunction.

The study of anatomy and physiology is essential for those who plan a career in the health sciences, because a sound knowledge of structure and function is necessary for health professionals to perform their duties adequately. Knowledge of anatomy and physiology is also beneficial to the nonprofessional. This background improves your ability to evaluate physiological activities, understand recommended treatments, critically evaluate advertisements and reports in the popular literature, and interact with health professionals.

ANATOMY AND PHYSIOLOGY

Anatomy (ă-nat´o-me) is the scientific study of the body's structure. For example, the shape and size of the body's bones can be described. Anatomy includes a wide range of studies, such as the form of structures, their microscopic organization, and the relationship between the structure of a body part and its function. Just as a hammer's structure makes it well-suited for pounding nails, the structure of a body part allows it to perform its functions effectively. For example, bones provide strength and support because bone cells surround themselves with a hard, mineralized substance. Understanding the relationship between structure and function makes it easier to understand and appreciate anatomy.

Physiology (fiz´e-ol´o-je) is the scientific study of the processes or functions of living things. It is important in physiology to recognize structures as changing rather than unchanging. The major goals of physiology are understanding and predicting the body's responses to stimuli and understanding how the body maintains conditions within a narrow range of values in the presence of a continually changing environment. For example, in a hot environment the body responds by producing sweat, which helps cool the body and maintains normal body temperature.

STRUCTURAL AND FUNCTIONAL ORGANIZATION

The body can be studied at seven structural levels: chemical, organelle, cellular, tissue, organ, organ system, and organism (Figure 1-1).

The structural and functional characteristics of all organisms are determined by their chemical makeup. The **chemical** level of organization involves interactions between atoms and their combinations into molecules. The function of a molecule is related intimately to its structure. For example, collagen molecules are strong, ropelike fibers that give skin structural strength and flexibility. With old age, the structure of collagen changes and the skin becomes fragile and is torn more easily. A brief overview of chemistry is presented in Chapter 2.

An **organelle** is a structure contained within a cell that performs one or more specific functions. For example, the nucleus is an organelle containing the cell's hereditary information. Organelles are discussed in Chapter 3.

Cells are the basic living units of all plants and animals. Although cell types differ in their structure and function, they have many characteristics in common. Knowledge of these characteristics and their variations is essential to a basic understanding of anatomy and physiology. The cell is discussed in Chapter 3.

A group of cells with similar structure and function plus

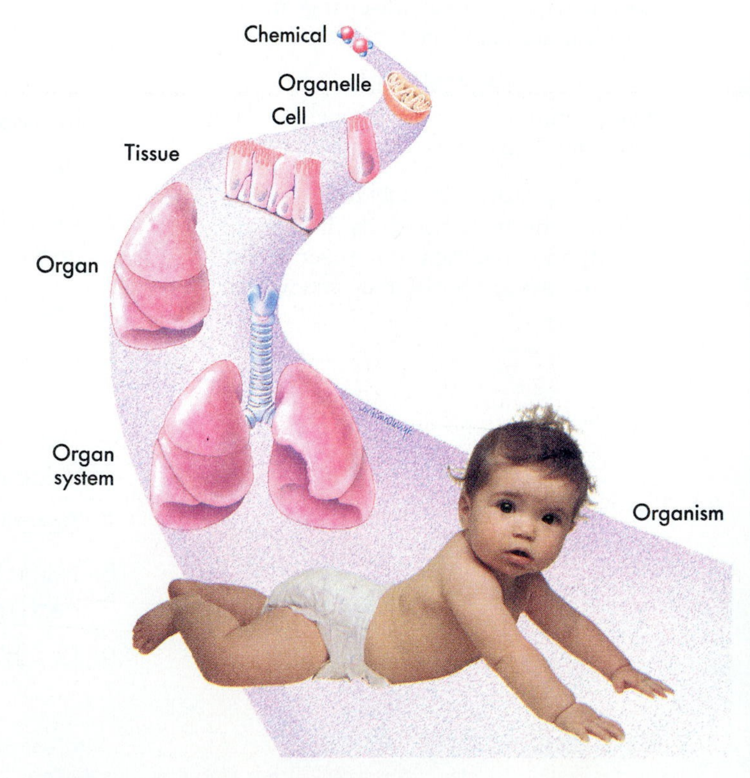

Figure 1-1 ● **Levels of organization.**
The seven levels of organization are the chemical, organelle, cellular, tissue, organ, organ system, and organism.

the extracellular substances located between them is a **tissue.** The many tissues that make up the body are classified into four primary tissue types: epithelial, connective, muscle, and nervous. Tissues are discussed in Chapter 4.

Organs are composed of two or more tissue types that together perform one or more common functions. The skin, stomach, eye, and heart are examples of organs.

An **organ system** is a group of organs classified as a unit because of a common function or set of functions. In this text the body is considered to have 11 major organ systems: the integumentary, skeletal, muscular, nervous, endocrine, cardiovascular, lymphatic, respiratory, digestive, urinary, and reproductive systems (Table 1-1).

An **organism** is any living thing considered as a whole, whether composed of one cell or many. The human organism is a complex of mutually dependent organ systems.

Table 1-1 Organ systems of the body

System	Major components	Functions
Integumentary	Skin, hair, nails, and sweat glands	Protects, regulates temperature, prevents water loss, and produces vitamin D precursors
Skeletal	Bones, associated cartilage, and joints	Protects, supports, and allows body movement, produces blood cells and stores minerals
Muscular	Muscles attached to the skeleton	Produces body movement, maintains posture, and produces body heat

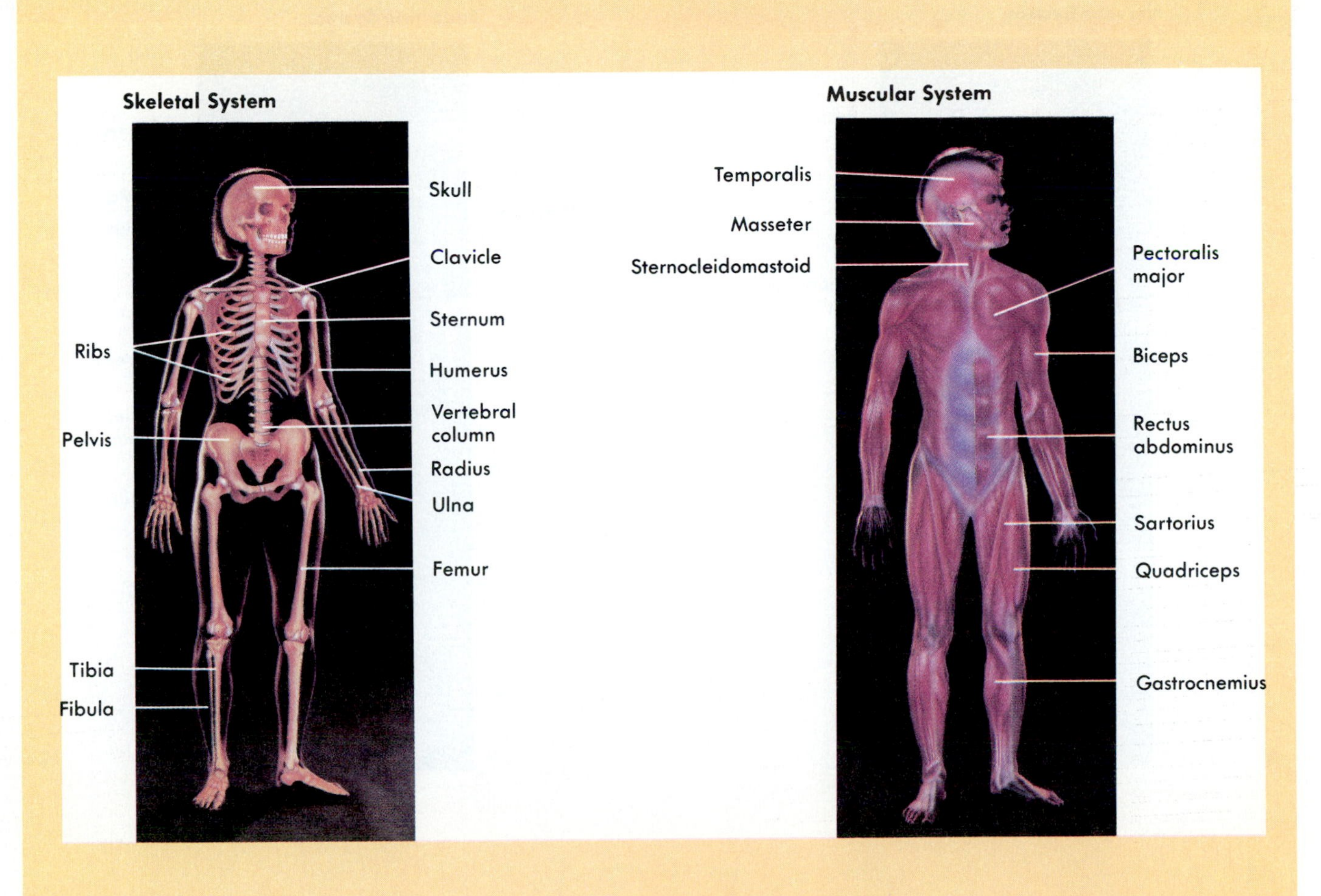

Table 1-1 Organ systems of the body—cont'd

System	Major components	Functions
Nervous	Brain, spinal cord, nerves, and sensory receptors	A major regulatory system: detects sensation, controls movements, controls physiological and intellectual functions
Endocrine	Endocrine glands such as the pituitary, thyroid, and adrenal glands	A major regulatory system: participates in the regulation of metabolism, reproduction, and many other functions
Cardiovascular	Heart, blood vessels, and blood	Transports nutrients, waste products, gases, and hormones throughout the body; plays a role in the immune response and the regulation of body temperature
Lymphatic	Lymph vessels, lymph nodes, and other lymph organs	Removes foreign substances from the blood and lymph, combats disease, maintains tissue fluid balance, and absorbs fats
Respiratory	Lungs and respiratory passages	Exchanges gases (oxygen and carbon dioxide) between the blood and the air and regulates blood pH
Digestive	Mouth, esophagus, stomach, intestines, and accessory structures	Performs the mechanical and chemical processes of digestion, absorption of nutrients, and elimination of wastes

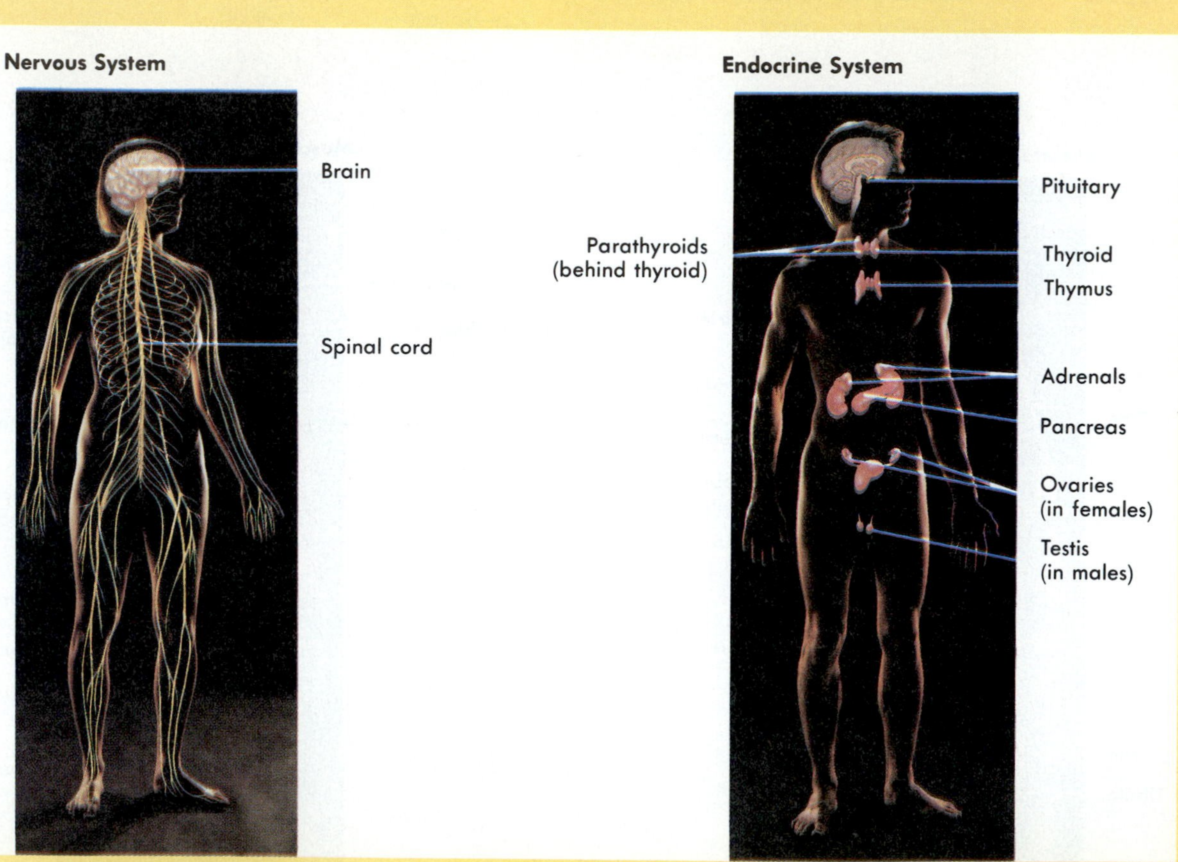

Cardiovascular System

Carotid artery
Jugular vein
Brachial artery
Superior vena cava
Pulmonary artery
Heart
Aorta
Inferior vena cava
Femoral artery and vein

Lymphatic System

Right lymphatic duct
Thoracic duct
Thymus gland
Spleen
Lymph node

Respiratory System

Pharynx
Trachea
Nasal cavity
Oral cavity
Larynx
Bronchus
Lungs

Digestive System

Pharynx
Gallbladder
Large intestine
Rectum
Salivary gland
Esophagus
Liver
Stomach
Small intestine

Continued.

Table 1-1 Organ systems of the body—cont'd

System	Major components	Functions
Urinary	Kidneys, urinary bladder, and the ducts that carry urine	Removes waste products from the circulatory system; regulates blood pH, ion balance, and water balance
Reproductive	Gonads, accessory structures, and genitals of males and females	Performs the processes of reproduction and controls sexual functions and behaviors

Urinary System

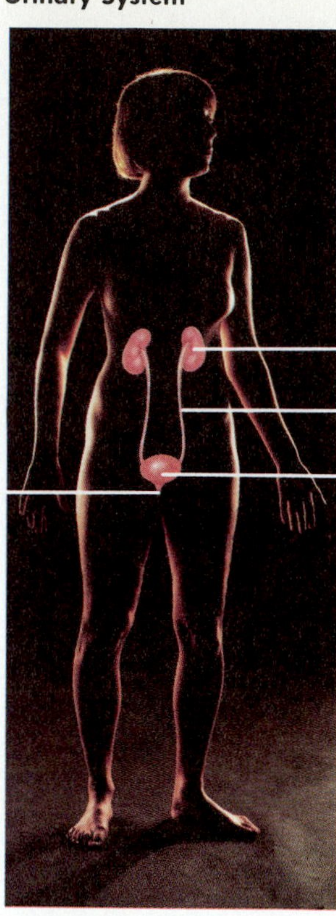

Reproductive System—Male

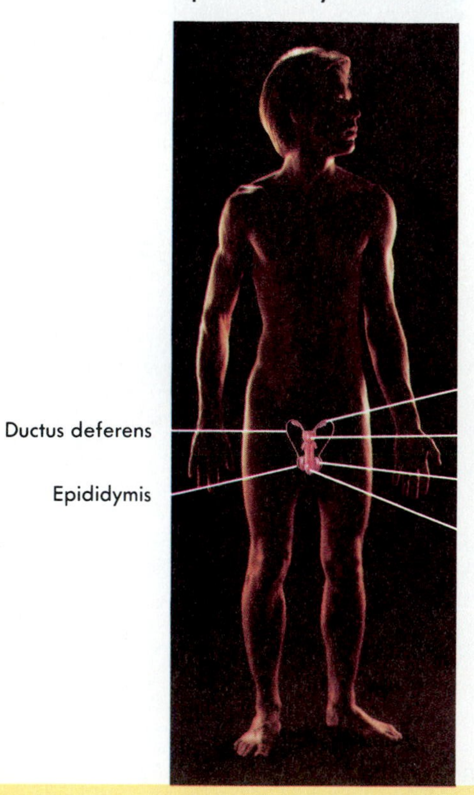

Reproductive System—Female

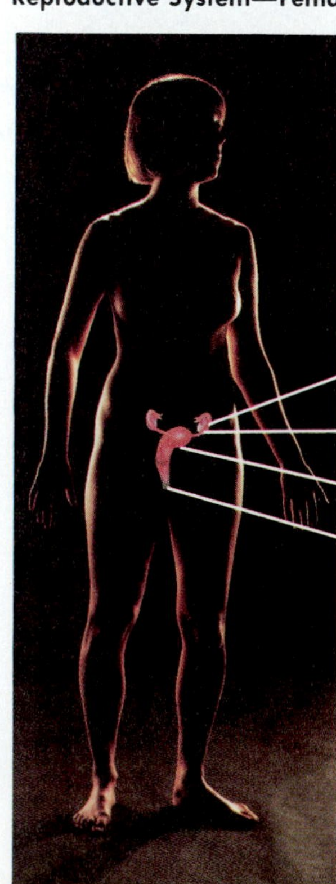

HOMEOSTASIS

Homeostasis (ho´me-o-sta´sis) is the existence and maintenance of a relatively constant environment within the body. Each cell of the body is surrounded by a small amount of fluid, and the normal function of that cell depends on the maintenance of its fluid environment within a narrow range of conditions, including volume, temperature, and chemical content. If the fluid surrounding cells deviates from homeostasis, the cells do not function normally and can even die. Disruption of homeostasis results in disease and possibly death.

The organ systems help control the cellular environment so that it remains relatively constant. For example, the digestive, respiratory, circulatory, and urinary systems function together so that each cell in the body receives adequate oxygen and nutrients and so that waste products do not accumulate to a toxic level.

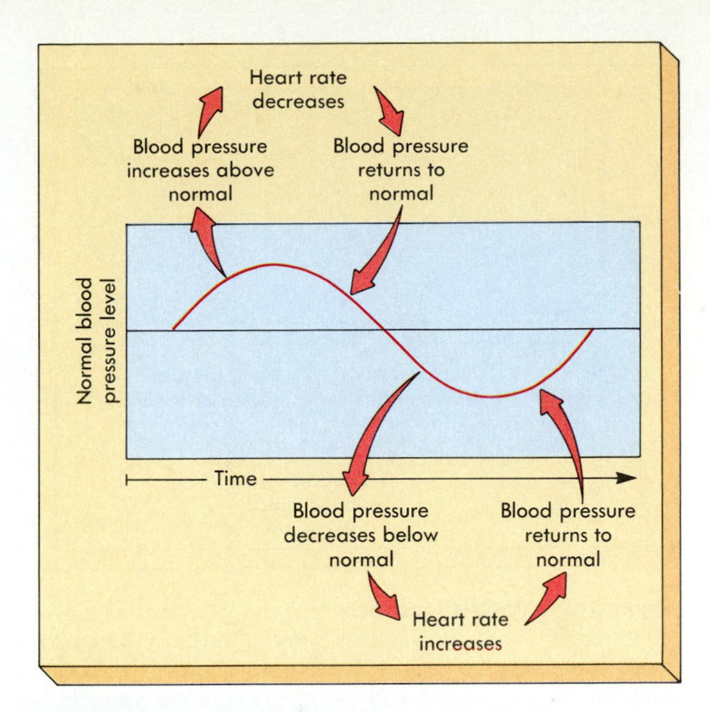

Figure 1-2 ● Negative feedback.
An increase in blood pressure initiates regulatory changes that cause heart rate to decrease, resulting in decreased blood pressure. Conversely, a decrease in blood pressure causes heart rate to increase, resulting in increased blood pressure.

Negative feedback

The systems of the body are regulated by **negative-feedback** mechanisms that function to maintain homeostasis. "Negative" means that any deviation from a normal value is made smaller or is resisted. Negative feedback does not prevent variation but maintains variation within a normal range.

The maintenance of normal blood pressure is an example of a negative-feedback mechanism (Figure 1-2). Blood pressure depends on contraction (beating) of the heart. Normal blood pressure is important because it is responsible for moving blood from the heart to tissues. The blood supplies the tissues with oxygen and nutrients and removes waste products. Thus normal blood pressure is required to ensure that tissue homeostasis is maintained. If blood pressure decreases slightly from its normal value, negative-feedback mechanisms increase heart rate, causing blood pressure to increase toward its normal value. If blood pressure increases slightly above its normal value, negative-feedback mechanisms decrease heart rate, causing blood pressure to decrease toward its normal value. As a result, blood pressure constantly rises and falls around the normal value, establishing a normal range of values for blood pressure.

1 Donating a pint of blood reduces blood volume, which results in a decrease in blood pressure (just as air pressure in a balloon decreases as air is let out of the balloon). What effect would donating blood have on heart rate? What happens if a negative-feedback mechanism does not return the value of some parameter such as blood pressure to its normal range?*

Positive feedback

Positive-feedback responses are not homeostatic and are rare in healthy individuals. "Positive" implies that when a deviation from a normal value occurs, the response of the system is to make the deviation even greater. Therefore positive feedback usually creates a "vicious cycle" leading away from homeostasis and, in some cases, results in death.

Inadequate delivery of blood to cardiac (heart) muscle is an example of positive feedback. Contraction of cardiac muscle generates blood pressure and moves blood. For normal contractions to occur, cardiac muscle is provided with an adequate blood supply through a system of blood vessels on the outside of the heart. In effect, the heart pumps blood to itself. Just as with other tissues, blood pressure must be maintained to ensure adequate delivery of blood to cardiac muscle. Following extreme blood loss, blood pressure decreases to the point that there is inadequate delivery of blood to cardiac muscle. As a result, cardiac muscle homeostasis is disrupted and cardiac muscle does not function normally. The heart pumps less blood, which causes the blood pressure to drop even further. The additional decrease in blood pressure causes less blood delivery to cardiac muscle, and the heart pumps even less blood, which again decreases the blood pressure (Figure 1-3). The process continues until the blood pressure is too low to sustain the cardiac muscle, the heart stops beating, and death results.

Following a moderate amount of blood loss, for example donating a pint of blood, negative-feedback mechanisms result in an increase in heart rate, which restores blood pressure. If the blood loss is severe, however, negative-feedback mechanisms may not be able to maintain homeostasis, and

*Answers to predict questions appear in Appendix B at the back of the book.

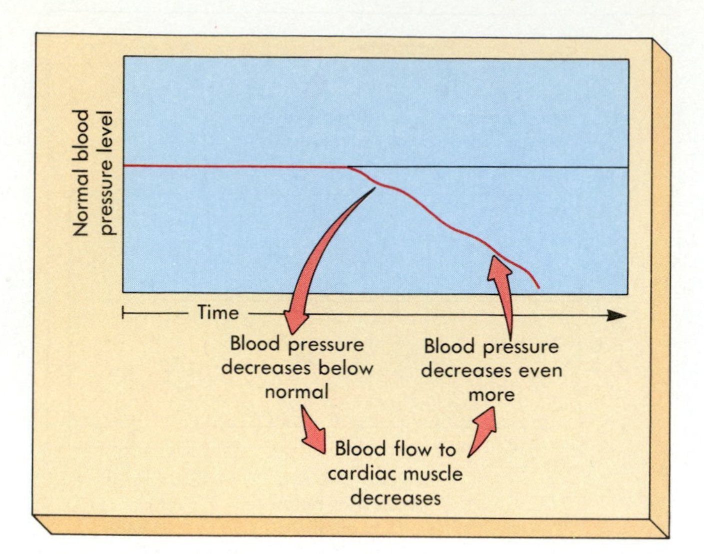

Figure 1-3 ● Positive feedback.
A decrease in blood pressure causes decreased blood flow to the heart. In this case the heart is unable to pump enough blood to maintain blood pressure and blood pressure decreases even more.

the positive-feedback effect of an ever-decreasing blood pressure can develop. This example illustrates a basic principle. Many disease states result from failure of negative-feedback mechanisms to maintain homeostasis. Medical therapy seeks to overcome illness by aiding the negative-feedback process. One example is a blood transfusion that reverses a constantly decreasing blood pressure and thus restores homeostasis.

A few positive-feedback mechanisms do operate in the body under normal conditions, but in all cases they are eventually limited in some way. Birth is an example of a normally occurring positive-feedback mechanism. Near the end of pregnancy, the uterus is stretched by the baby's large size and weight. This stretching, especially around the opening of the uterus, stimulates contractions of the uterine muscles. The uterine contractions push the baby against the opening of the uterus, stretching it further. This stimulates additional contractions that result in additional stretching. This positive-feedback sequence ends only when the baby is delivered from the uterus and the stretching stimulus is eliminated.

2 Is the sensation of thirst associated with a negative- or a positive-feedback mechanism? Explain. (Hint: What is being regulated when one becomes thirsty?)

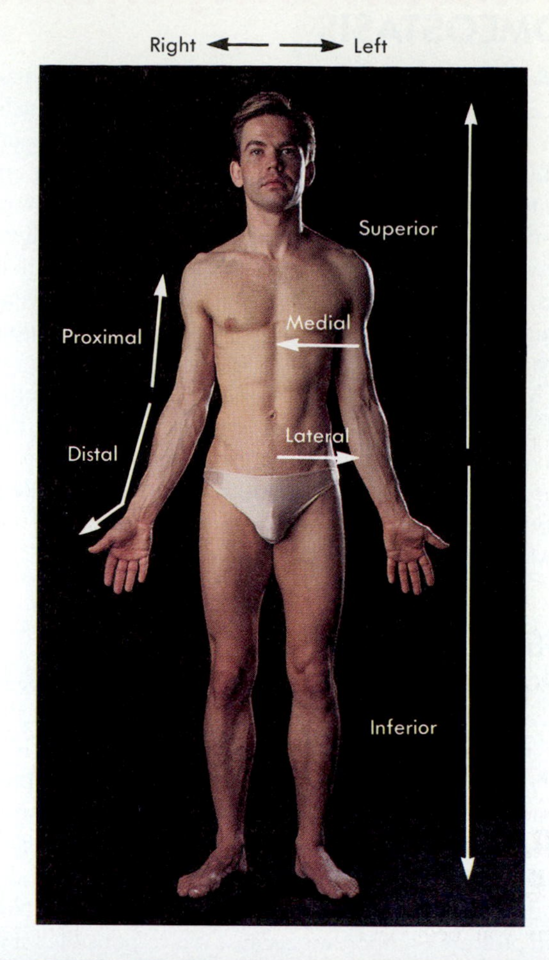

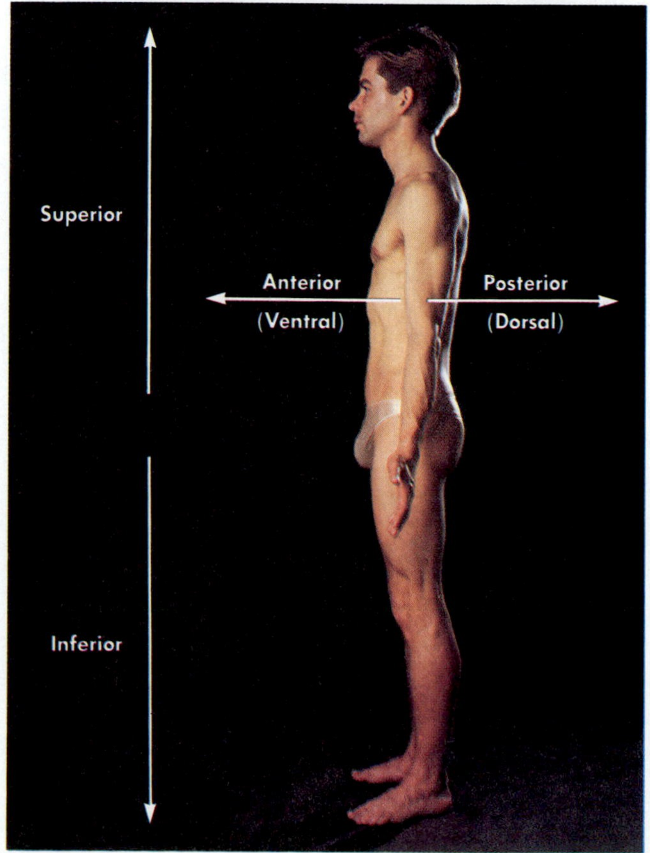

Figure 1-4 ● Directional terms.
A, Directional terms from the front. *Arrows* point in the indicated direction. **B,** Directional terms from the side. *Arrows* point in the indicated direction.

Historical notes

ANATOMY

For much of the history of anatomy, public sentiment made it difficult for anatomists to obtain human bodies for dissection. In the early 1800s, the benefits of human dissection for training physicians had become very apparent, and the need for cadavers had increased beyond the ability to acquire them legally. Thus arose the resurrectionists, or body snatchers. For a fee, and no questions asked, they removed bodies from graves and provided them to medical schools. Because the bodies were not easy to obtain and were not always in the best condition, two enterprising men named William Burke and William Hare went one step farther. They murdered 17 people and sold their bodies to a medical school. When discovered, Hare testified against Burke and went free. Burke was convicted, hanged, and then publicly dissected. Discovery of Burke's activities so outraged the public that sensible laws regulating the acquisition of cadavers were soon passed, and this dark chapter in the history of anatomy was closed.

PHYSIOLOGY

The idea that the body maintains a balance (homeostasis) can be traced back to ancient Greece. It was believed that the body supported four juices: the red juice of blood, the yellow juice of bile, the white juice secreted from the nose and lungs, and the black juice in the pancreas. Disease was thought to be caused by an excess of one of the juices, and normally the body would attempt to heal itself by expelling the excess juice. For example, mucus would be expelled from the nose of a person with a cold. This concept of body juices led to the practice of bloodletting to restore the body's normal balance of juices. Tragically, in the eighteenth and nineteenth centuries bloodletting went to extremes. When bloodletting did not result in improvement of the patient, it was taken as evidence that not enough blood had been removed to restore a healthy balance of the body's juices. The obvious solution was to let more blood, undoubtedly causing many deaths. Eventually the failure of this approach became obvious and the practice was abandoned. Fortunately we now understand better how the body maintains homeostasis.

TERMINOLOGY AND THE BODY PLAN

When you begin studying anatomy and physiology, the number of new words can seem overwhelming. In many ways it is like learning a new language. It is important to learn this terminology. Each term has a precise meaning and once you have learned it, you will be able to communicate effectively with others.

Directional terms

Directional terms refer to the body in the anatomical position regardless of its actual position. The **anatomical position** refers to a person standing erect with the feet forward, arms hanging to the sides, and the palms of the hands facing forward (Figure 1-4). In human anatomy, above is replaced by **superior,** below by **inferior,** front by **anterior** (or **ventral),** and back by **posterior** (or **dorsal).** Directional terms

Table 1-2 Directional terms for humans

Term	Definition	Example
Right	Toward the body's right side	The right ear
Left	Toward the body's left side	The left eye
Inferior	A structure below another	The nose is inferior to the forehead
Superior	A structure above another	The mouth is superior to the chin
Anterior (ventral)	The front of the body	The navel is anterior (ventral) to the spine
Posterior (dorsal)	The back of the body	The spine is posterior (dorsal) to the breastbone
Proximal	Closer to the point of attachment to the body than another structure	The elbow is proximal to the wrist
Distal	Farther from the point of attachment to the body than another structure	The wrist is distal to the elbow
Lateral	Away from the midline of the body	The nipple is lateral to the breastbone
Medial	Toward the middle or midline of the body	The bridge of the nose is medial to the eye
Superficial	Toward or on the surface	The skin is superficial to muscle
Deep	Away from the surface, internal	The lungs are deep to the ribs

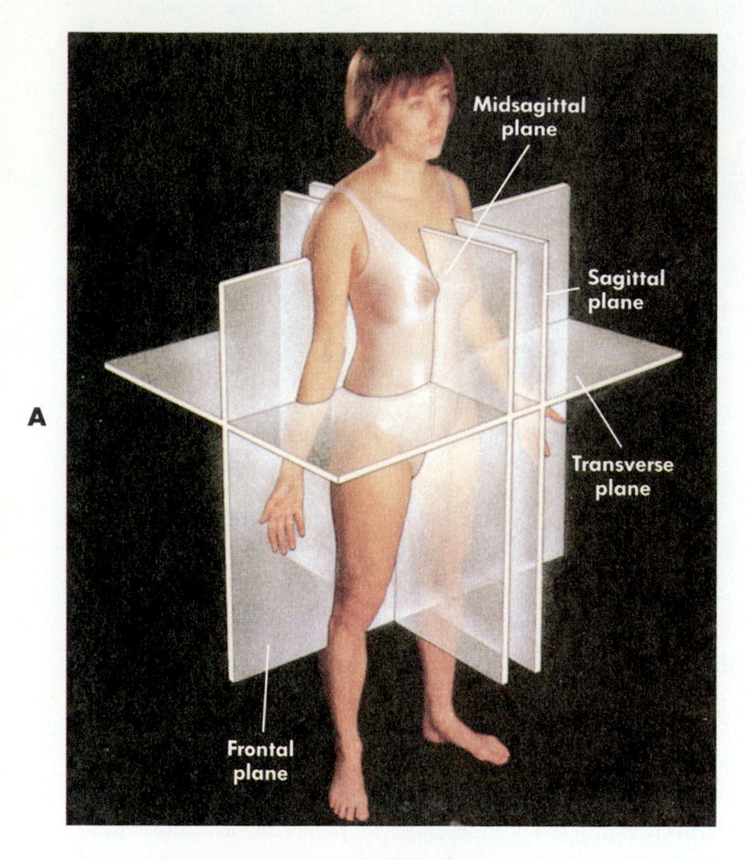

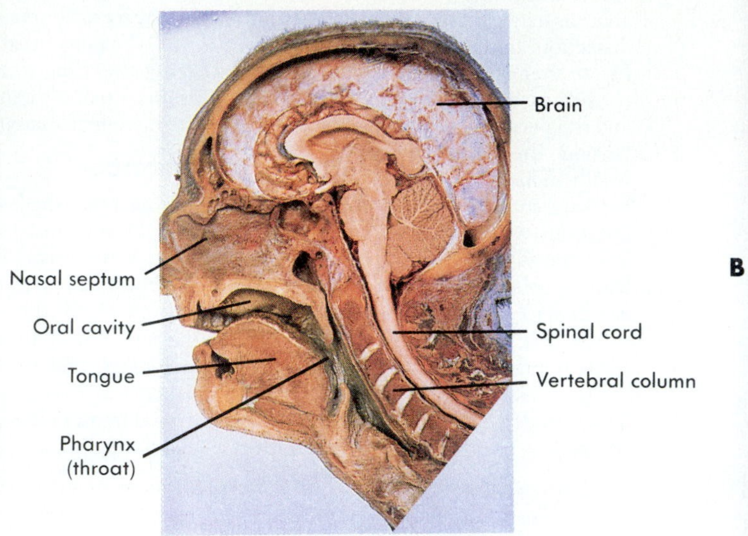

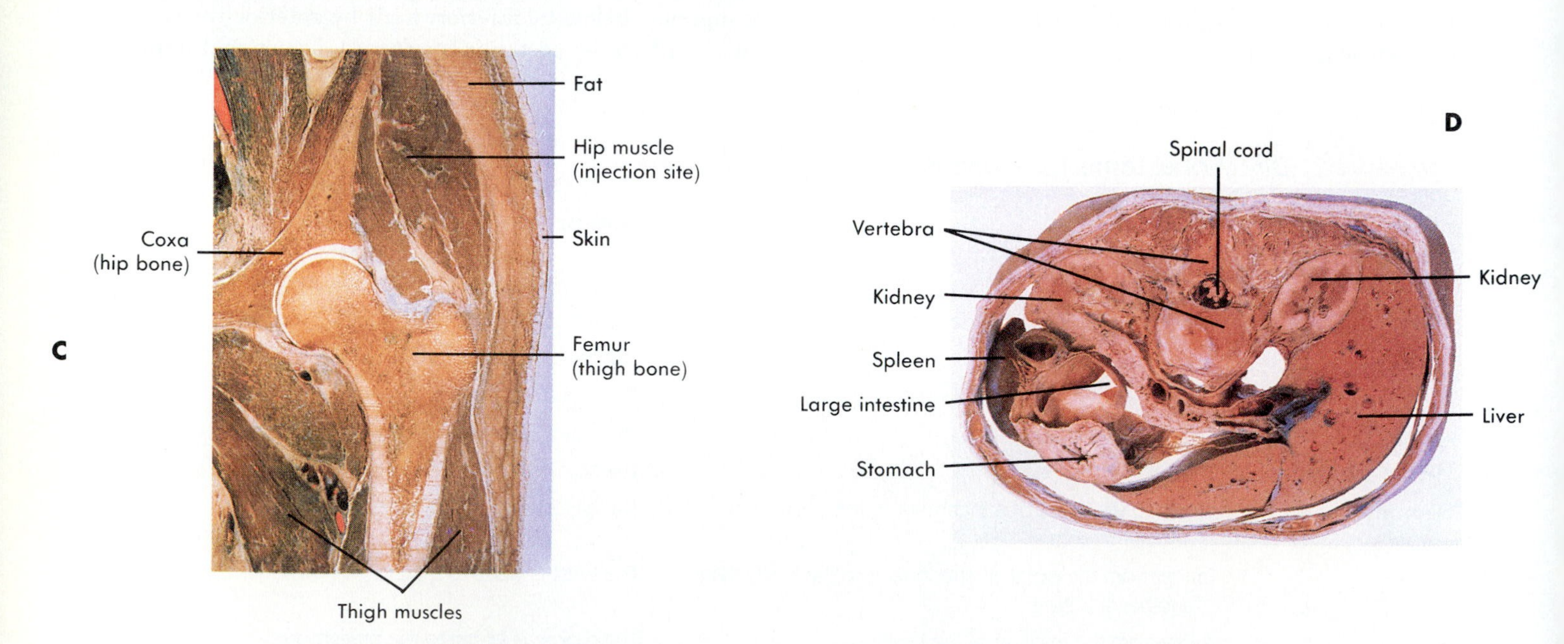

Figure 1-5 ● Planes of section.
A, The whole body. Planes are indicated by "glass" sheets. **B,** Midsagittal section of the head. **C,** Frontal section through the left hip. **D,** Transverse section through the abdomen.

are used to describe the position of structures in relation to other structures or body parts. For example, the neck is superior to the chest, but it is inferior to the head. Important directional terms are presented in Table 1-2 and are illustrated in Figure 1-4.

3 Provide the correct directional term for the following statement. When a man is standing on his head, his nose is ___________ to his mouth.

Planes

At times it is useful to discuss the body in reference to a series of planes (imaginary flat surfaces) passing through it (Figure 1-5). Sectioning the body is a way to "look inside" and observe the body's structures. A **sagittal** (saj´ĭ-tal) plane runs vertically through the body and separates it into right and left portions. If the plane divides the body into equal right and left halves, it is a **midsagittal** (mid´saj´ĭ-tal) plane. A **transverse** plane runs horizontally through the body and divides it into superior and inferior parts. A **frontal** plane runs vertically through the body and divides it into anterior and posterior portions.

Body regions

The body consists of the head, neck, trunk, upper limbs, and lower limbs. Specific names are given to different regions of the body (Figure 1-6). Some terms are commonly misused and should be noted. In anatomy, the **arm** does not refer to the entire upper limb, but to that region of the upper limb between the shoulder and the elbow. In a similar fashion, the **leg** refers to that portion of the lower limb between the knee and the ankle.

The trunk can be divided into three parts: the **thorax** is the chest, the **abdomen** is the region between the thorax

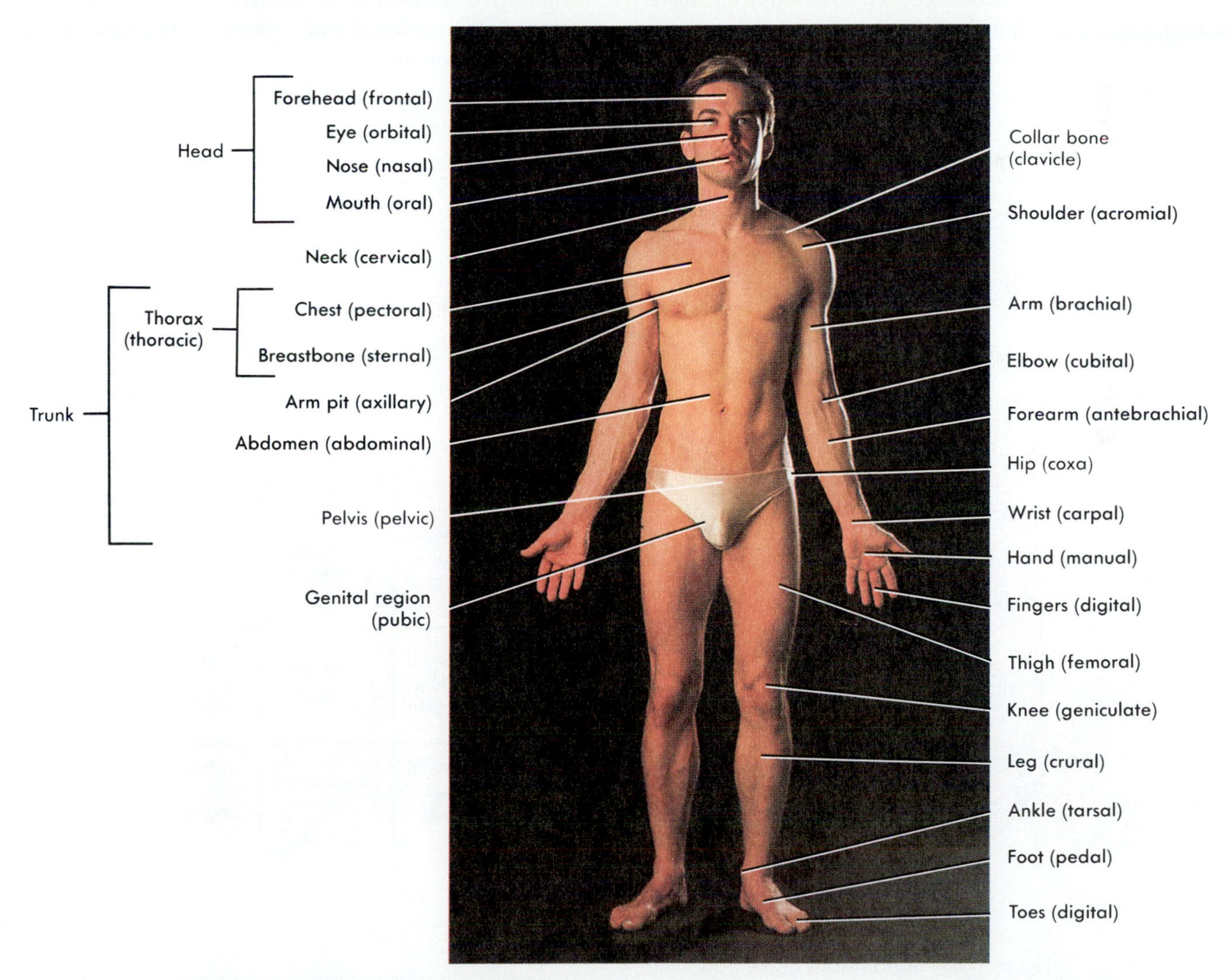

Figure 1-6 ● Body regions and structures.
The body consists of the head, neck, trunk, and limbs. The common names and anatomical names (in parentheses) are indicated for some parts of the body.

and pelvis, and the **pelvis** is the inferior end of the trunk associated with the hips. The abdomen often is subdivided superficially into **quadrants** (four areas) by two imaginary lines—one horizontal and one vertical—that intersect at the navel (Figure 1-7, A). The quadrants formed are the right-upper, right-lower, left-upper, and left-lower quadrants. Another method of subdividing the abdomen results in nine **regions,** which are formed by four imaginary lines—two horizontal and two vertical. These four lines create an imaginary tic-tac-toe figure on the abdomen (Figure 1-7, *B*). The quadrants or regions are commonly used by clinicians as reference points for locating the underlying organs. For example, the pain of an acute appendicitis is usually located in the right-lower quadrant.

Body cavities

The body contains many cavities, such as the nasal, cranial, and abdominal cavities. Some of these cavities open to the outside of the body, and some do not. Undergraduate anatomy and physiology textbooks sometimes describe a dorsal cavity, in which the brain and spinal cord are found, and a ventral body cavity that contains all the trunk cavities. The concept of a dorsal cavity is not described in standard works on anatomy and therefore is not emphasized here. Discussion in this chapter is limited to the major trunk cavities that do not open to the outside.

The trunk contains three large cavities: the thoracic cavity, the abdominal cavity, and the pelvic cavity (Figure 1-8). The **thoracic cavity** is surrounded by the rib cage and is separated from the abdominal cavity by the muscular diaphragm. The thoracic cavity is divided into two parts by a median structure called the **mediastinum** (me´de-as-ti´-num). The mediastinum is a partition containing the heart, thymus gland, trachea, esophagus, and other structures. The lungs are located on either side of the mediastinum.

The **abdominal cavity** is bounded primarily by the abdominal muscles and contains the stomach, intestines, liver, spleen, pancreas, and kidneys. The **pelvic cavity** is a small space enclosed by the bones of the pelvis and contains the urinary bladder, part of the large intestine, and the internal reproductive organs. The abdominal and pelvic cavities are not physically separated and sometimes are called the **abdominopelvic cavity.**

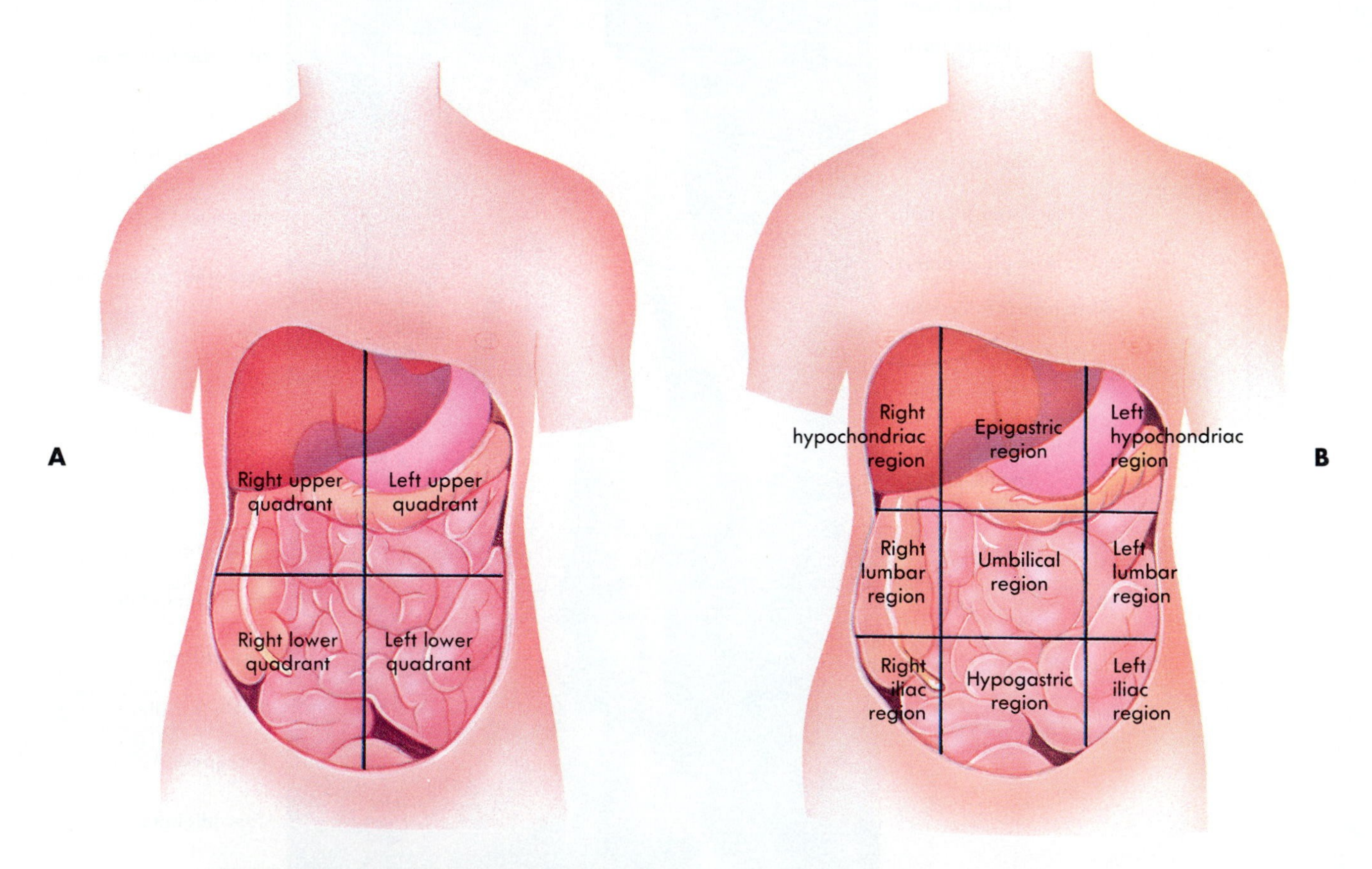

Figure 1-7 ● Subdivisions of the abdomen.
Lines are superimposed over internal organs to demonstrate the relationship of the organs to the subdivisions. **A,** Abdominal quadrants subdivide the abdomen into four areas. **B,** Abdominal regions subdivide the abdomen into nine areas

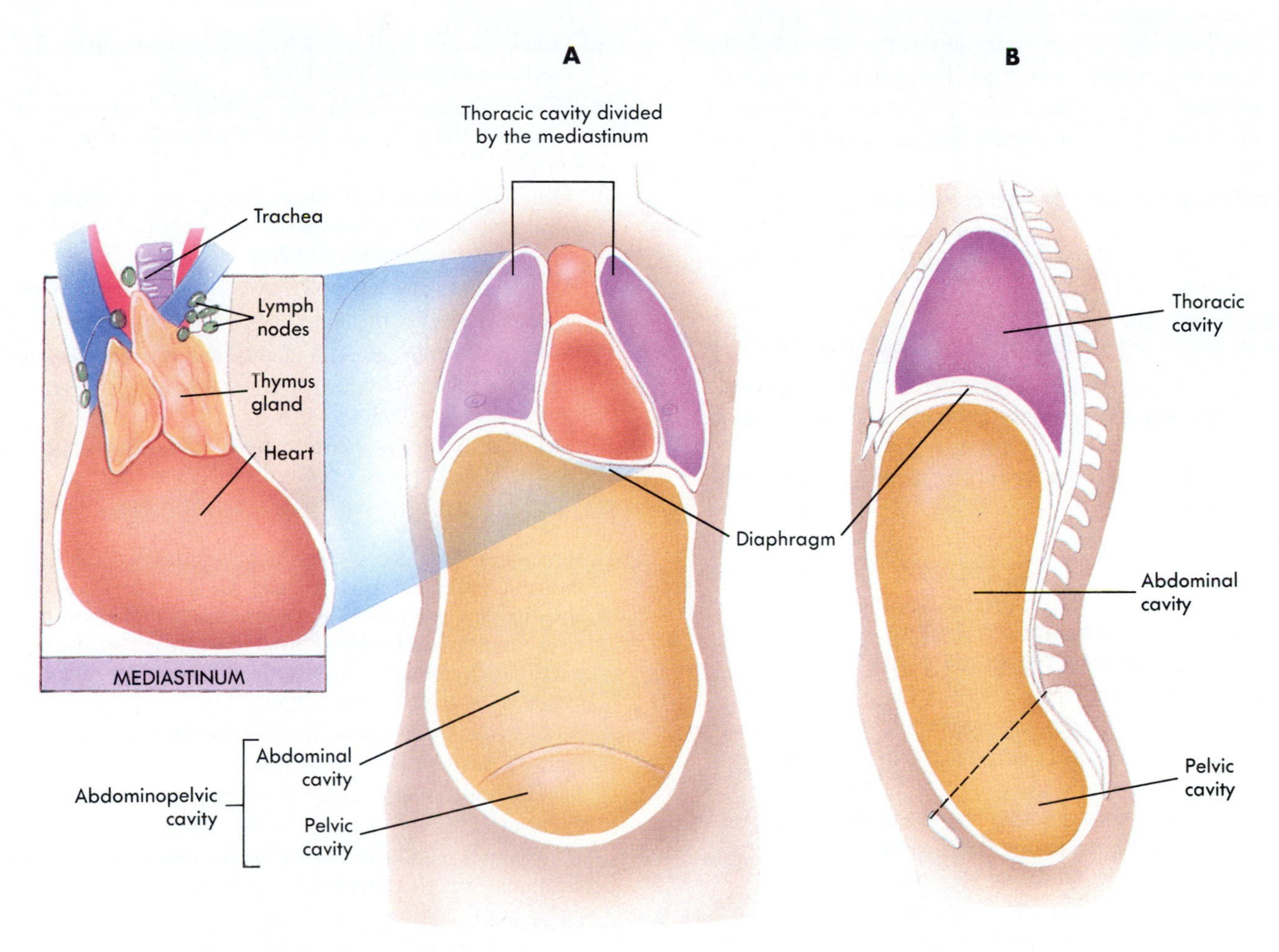

Figure 1-8 ● Trunk cavities.
A, Anterior view showing the major trunk cavities. The diaphragm separates the thoracic cavity from the abdominal cavity. The mediastinum, which includes the heart, divides the thoracic cavity. **B,** Sagittal view of trunk cavities. Imaginary plane shows the division between the abdominal and pelvic cavities. The mediastinum has been removed to show the thoracic cavity.

SUMMARY

A knowledge of anatomy and physiology can be used to predict the body's responses to stimuli when healthy or diseased.

ANATOMY AND PHYSIOLOGY

Anatomy is the study of the body's structures. Physiology is the study of the body's processes and functions.

STRUCTURAL AND FUNCTIONAL ORGANIZATION

The human body can be organized into seven levels: chemical (atoms and molecules), organelle (small structures within cells), cell (basic living unit of plants and animals), tissue (groups of similar cells), organ (two or more tissues), organ system (groups of organs), and organism.

The 11 organ systems are the integumentary, skeletal, muscular, nervous, endocrine, cardiovascular, lymphatic, respiratory, digestive, urinary, and reproductive systems (see Table 1-1).

HOMEOSTASIS

Homeostasis is the condition in which body functions, fluids, and other factors of the internal environment are maintained within a range of values suitable to support life.

Negative-feedback mechanisms operate to maintain homeostasis.

Positive-feedback mechanisms increase deviations from normal. Although a few positive-feedback mechanisms normally exist in the body, most positive-feedback mechanisms are harmful.

TERMINOLOGY AND THE BODY PLAN

Directional terms

A human standing erect with the feet forward, arms hanging to the sides, and palms facing forward is in the anatomical position.

Directional terms always refer to the anatomical position regardless of the body's actual position (see Table 1-2).

Planes

A sagittal plane divides the body into left and right parts, a transverse plane divides the body into superior and inferior parts, and a frontal plane divides the body into anterior and posterior parts.

Body regions

The body consists of the head, neck, trunk, and limbs.

The abdomen can be divided superficially into four quadrants or nine regions that are useful for locating internal organs or describing the location of a pain.

Body cavities

The thoracic cavity is bounded by the ribs and the diaphragm. The mediastinum divides the thoracic cavity into two parts.

The abdominal cavity is bounded by the diaphragm and the abdominal muscles.

The pelvic cavity is surrounded by the pelvic bones.

CONTENT REVIEW

1. Define anatomy and physiology.
2. List seven structural levels at which the body can be studied.
3. Define tissue. What are the four primary tissue types?
4. Define organ and organ system. What are the 11 organ systems of the body?
5. What does the term homeostasis mean? If a deviation from homeostasis occurs, what kind of mechanism restores homeostasis?
6. Define positive feedback. Why are positive-feedback mechanisms generally harmful?
8. What is the anatomical position? Why is it important to remember the anatomical position when using directional terms?
9. Define the following directional terms: superior, inferior, anterior (ventral), posterior (dorsal), proximal, distal, lateral, medial, superficial, and deep.
10. Define sagittal, midsagittal, transverse, and frontal planes.
11. Name the parts of the upper and lower limbs.
12. Describe the four-quadrant method and the nine-region method of subdividing the abdominal region. What is the purpose of these methods?
13. Define the following cavities: thoracic, abdominal, pelvic, and abdominopelvic. What is the mediastinum?

CONCEPT REVIEW

1. A man has lost blood as a result of a gunshot wound. Even though bleeding has been stopped, his blood pressure is low and dropping and his heart rate is elevated. Following a blood transfusion, his blood pressure increased and his heart rate decreased. Which of the following statements are consistent with these observations?
   A. Negative-feedback mechanisms can be inadequate without medical intervention.
   B. The transfusion interrupted a positive-feedback mechanism.
   C. The increased heart rate before the transfusion is an example of positive feedback.
   D. A and B.
   E. All of the above
2. During physical exercise, respiration rate increases. Two students are discussing the mechanisms involved: Student A claims they are positive feedback, and Student B claims they are negative feedback. Do you agree with Student A or Student B, and why?
3. Complete the following statements, using the correct directional terms for a human being.
   A. The navel is _____ to the nose.
   B. The heart is _____ to the breastbone (sternum).
   C. The ankle is _____ to the knee.
   D. The ear is _____ to the eye.
4. During pregnancy, which would increase more in size, the mother's abdominal or pelvic cavity? Explain.

CHAPTER TEST

Answers can be found in Appendix C

MATCHING For each statement in column A select the correct answer in column B (an answer may be used once, more than once, or not at all).

A	B
1. Subunit of a cell that performs specific functions.	anterior
2. Group of cells with similar structure and functions.	frontal
3. Two or more tissues.	inferior
4. Means "above."	lateral
5. Toward the back of the body.	medial
6. Toward the side of the body.	organ
7. Divides the body into left and right parts.	organelle
8. Divides the body into superior and inferior parts.	posterior
	sagittal
	superior
	tissue
	transverse

FILL-IN-THE BLANK Complete each statement by providing the missing word or words.

1. The maintenance of a relatively constant environment within the body is called ___________.
2. The discipline that attempts to predict the body's responses to stimuli and understand how the body maintains homeostasis is ___________.
3. ___________ mechanisms maintain or return a parameter such as blood pressure to its normal value.
4. The ___________ refers to a person standing erect with the feet forward, arms hanging to the sides, and the palms of the hands facing forward.
5. The toes are ___________ to the ankle.
6. The breastbone is ___________ to the nipple.
7. The ___________ cavity is bounded inferiorly by the diaphragm.
8. The ___________ divides the thoracic cavity into left and right parts.
9. The ___________ cavity is enclosed by bone and contains the urinary bladder.

CHAPTER

2

Chemistry, Matter and Life

OBJECTIVES

After reading this chapter you should be able to:

1. Name the subatomic particles of an atom and describe how they are organized.
2. Interpret a chemical formula.
3. Describe two types of chemical bonds.
4. Explain how ATP is used by cells to store and release energy.
5. Describe the pH scale and its relationship to acidity and alkalinity.
6. Explain why buffers are important.
7. List the properties of water that make it important for living organisms.
8. Describe four important types of organic molecules and their functions.
9. Describe how enzymes work.

KEY TERMS

acid
Any substance that is a proton donor; any substance that releases hydrogen ions.

atom
Smallest particle of an element that retains the properties of that element; composed of neutrons, protons, and electrons.

base
Any substance that is a proton acceptor; any substance that binds to hydrogen ions.

buffer
A chemical that resists changes in pH when either an acid or a base is added to a solution containing the buffer.

covalent bond
Chemical bond that is formed when two atoms share one or more pairs of electrons.

enzyme
A protein molecule that increases the rate of a chemical reaction without being permanently altered.

ion
Atom or group of atoms carrying an electrical charge due to loss or gain of one or more electrons.

ionic bond
Chemical bond that is formed when one atom loses an electron and another atom accepts that electron.

molecule
Two or more atoms of the same or different type joined by a chemical bond.

FEATURES

CHEMISTRY is the scientific study of the composition and the structure of substances and the many reactions they undergo. A basic knowledge of chemical principles is essential for understanding anatomy and physiology. For example, the physiological processes of digestion, muscle contraction, and generation of nerve impulses can be described in chemical terms. In addition, many abnormal conditions and their treatments can be explained in chemical terms.

BASIC CHEMISTRY

Matter is anything that occupies space. An **element** is matter composed of atoms of only one kind. **Atoms** (at´omz) are the smallest particles of an element that retain the properties of that element. For example, oxygen is an element composed of only oxygen atoms and carbon is an element composed of only carbon atoms.

The structure of atoms

Atoms are made up of **neutrons, protons,** and **electrons.** Neutrons have no electrical charge, protons have a positive electrical charge, and electrons have a negative electrical charge. The number of protons in an atom is equal to the number of electrons. Consequently, atoms are electrically neutral, with neither a positive nor a negative charge.

Protons and neutrons are organized in atoms to form a central **nucleus,** and the electrons move around the nucleus in regions called **orbitals** (Figure 2-1). These three-dimensional orbitals are often represented as a series of concentric circles around the nucleus.

Different elements have different numbers of protons. Elements commonly found in the human body are listed in Table 2-1. Note that the symbols used in the table can be used to refer to elements or to individual atoms.

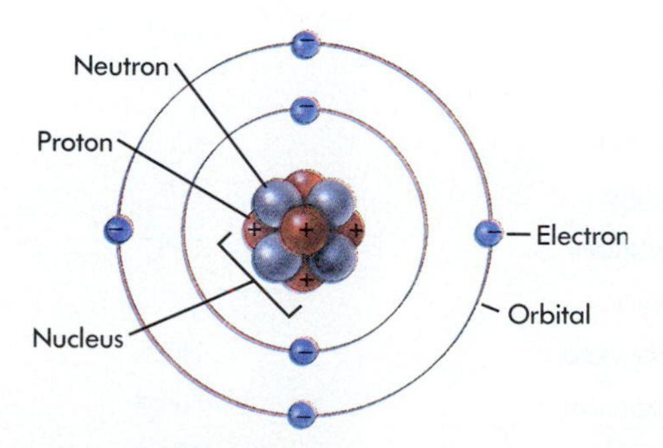

Figure 2-1 ● Model of an atom.
The nucleus contains neutrons, which have no charge, and positively charged protons. The concentric circles are the orbitals, in which the negatively charged electrons are located.

Table 2-1 ● Some common elements

Element	Symbol	Number of protons in element	Amount in human body by weight (%)
Hydrogen	H	1	9.5
Carbon	C	6	18.5
Nitrogen	N	7	3.3
Oxygen	O	8	65.0
Sodium	Na	11	0.2
Phosphorus	P	15	1.0
Sulfur	S	16	0.3
Chlorine	Cl	17	0.2
Potassium	K	19	0.4
Calcium	Ca	20	1.5
Iron	Fe	26	Trace
Iodine	I	53	Trace

Electrons and chemical bonds

Much of an atom's chemical behavior is determined by the electrons in its outermost orbitals. **Chemical bonds** are formed when the outermost electrons are transferred or shared between atoms. The resulting combination of atoms is called a **molecule.** If a molecule has two or more different kinds of atoms, it may be referred to as a **compound.** A molecule can be represented by a **chemical formula,** which consists of the symbols of the atoms in the molecule plus a subscript denoting the number of each type of atom. For example, the chemical formula for glucose (a sugar) is $C_6H_{12}O_6$. Thus glucose has 6 carbon, 12 hydrogen, and 6 oxygen atoms.

Ionic bonds

An **ionic** (i´on-ik) **bond** results when one atom loses an electron and another atom accepts that electron. For example, sodium (Na) can lose an electron that can be accepted by chlorine (Cl). The molecule that is formed is sodium chloride (NaCl), or table salt (Figure 2-2).

Atoms are electrically neutral because they have an equal number of protons and electrons. After an atom donates an electron, it has one more proton than it has electrons and is positively charged. After an atom accepts a donated electron, it has one more electron than it has protons and is negatively charged. These charged atoms are called **ions** (i´ons). Positively and negatively charged ions remain close together because oppositely charged ions are attracted to each other. The bond that results from this attraction is an ionic bond.

Ions are denoted by using the symbol of the atom from which the ion was formed. The charge of the ion is indicated by a superscripted plus ($^+$) or minus ($^-$) sign. For example, the sodium ion is Na^+ and the chloride ion is Cl^-. If more than one electron has been lost or gained, a number is used

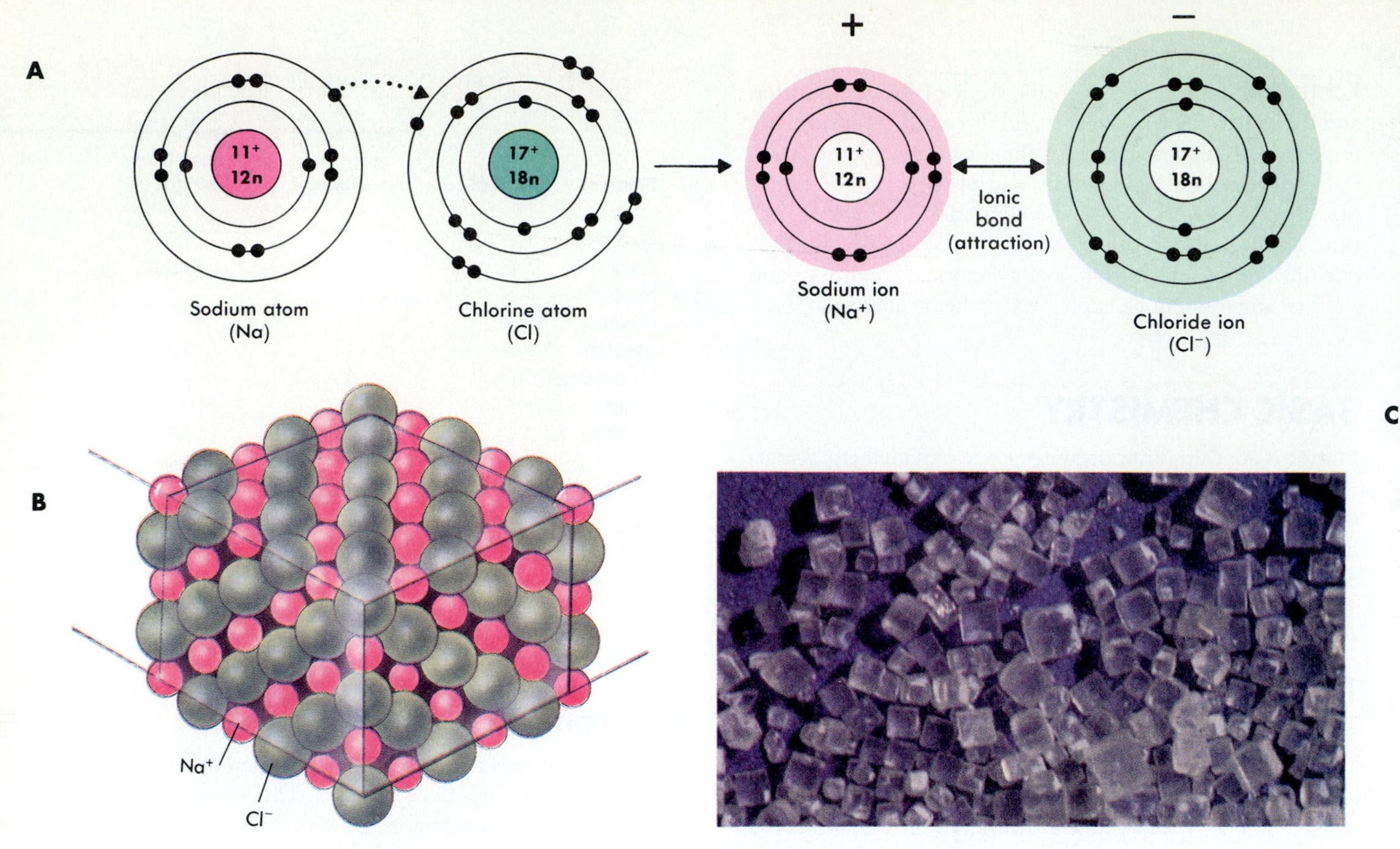

Figure 2-2 ● Ionic bond.
A, Sodium loses an electron, chlorine gains an electron, and they combine through an ionic bond to form sodium chloride (table salt). **B** and **C,** The sodium chloride is organized into crystals with a cube shape.

with the plus or minus sign. Thus Ca^{2+} is a calcium ion formed by the loss of two electrons. Table 2-2 lists some of the important ions found in the human body.

1 If an iron (Fe) atom lost three electrons, what would be the charge of the resulting ion? Write the symbol for this ion.*

Substances that produce ions when dissolved are sometimes referred to as **electrolytes** (e-lek´tro-lītz) because ions can conduct electrical current when they are in solution. An electrocardiogram (ECG) is a recording of the electrical currents produced by heart muscle. These currents can be detected by electrodes on the surface of the body because the ions in the body's fluids conduct the electrical currents.

Covalent bonds

A **covalent bond** results when two atoms share one or more pairs of electrons. Two hydrogen atoms, for example, can share their electrons to form a hydrogen molecule (Figure 2-3, A). Hydrogen atoms also can share electrons with an oxygen atom to form a water molecule (Figure 2-3, *B*). A carbon atom can share four of its electrons with other atoms, forming four covalent bonds (Figure 2-3, C). Carbon atoms

*Answers to predict questions appear in Appendix B at the back of the book.

Table 2-2 ● Important ions

Ion	Symbol
Calcium	Ca^{2+}
Sodium	Na^{+}
Potassium	K^{+}
Hydrogen	H^{+}
Hydroxide	OH^{-}
Chloride	Cl^{-}
Bicarbonate	HCO_3^{-}
Ammonium	NH_4^{+}
Phosphate	PO_4^{3-}
Iron	Fe^{2+}
Magnesium	Mg^{2+}

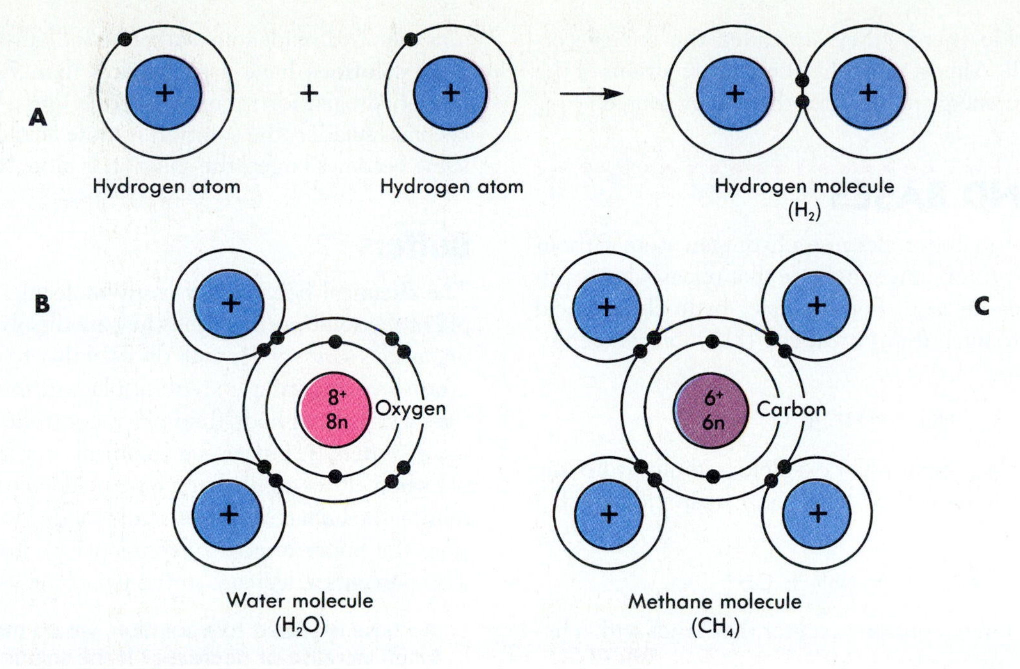

Figure 2-3 ● Covalent bond.
A, Each hydrogen atom has a single electron. The hydrogen atoms form a covalent bond and become a hydrogen molecule when the two electrons are shared between the two hydrogen atoms. **B,** A water molecule consists of two hydrogen atoms and an oxygen atom. Each hydrogen atom shares its electron with the oxygen atom, and the oxygen atom shares two of its electrons with the hydrogen atoms. **C,** A methane molecule consists of four hydrogen atoms and a carbon atom. Each hydrogen atom shares its electron with the carbon atom and the carbon atom shares four electrons with the hydrogen atoms.

are major components of most molecules in the human body. The large variety of molecules containing carbon results from the covalent bonds formed between carbon atoms, and between carbon atoms and hydrogen, oxygen, and nitrogen atoms.

CHEMICAL REACTIONS

A **chemical reaction** is the process by which atoms or molecules interact to form or break chemical bonds. The atoms or molecules present before the chemical reaction occurs are the **reactants** and those produced by the chemical reaction are the **products.** For example, the reactants sodium and chlorine combine to form the product sodium chloride.

Chemical reactions are important because of the products they form and the energy changes they produce. Energy exists in chemical bonds as stored energy. If the products of a chemical reaction contain less stored energy than the reactants, then energy is released. Most of the energy is released as heat. The body temperature of humans is maintained by heat produced in this fashion. The rest of the energy is used to synthesize (form) new molecules or to drive processes such as muscle contraction.

An example of a reaction that releases energy is the breakdown of the molecule adenosine triphosphate (ATP) to adenosine diphosphate (ADP) and a phosphate group (P). The phosphate group is attached to the ADP molecule by a phosphate bond (stored energy). When the bond between ADP and the phosphate group is broken, energy is released. Part of the energy is lost as heat and part is used by cells.

ATP ⟶ ADP + P + Heat +
Energy (for synthesis, muscle contractions, etc.)

2 Why does body temperature increase during exercise? (Hint: muscle contractions require energy.)

If the products of a chemical reaction contain more energy than the reactants, then the reaction requires the input of energy from another source. The energy released during the breakdown of food molecules is the energy source for this kind of reaction in the body. The energy from food molecules is used to synthesize molecules such as ATP, fats, and proteins. For example, the energy from food molecules is used to form a phosphate bond that attaches a phosphate group to ADP.

ADP + P +
Energy (from the breakdown of food molecules) ⟶ ATP

ATP molecules store energy that is released when the phosphate bond of ATP is broken. Because ATP molecules can store and provide energy, they are called the energy currency of the cell. Almost all of the chemical reactions of the cell that require energy use ATP as the energy source.

ACIDS AND BASES

An **acid** is a proton donor. Because a hydrogen atom without its electron is a proton, any substance that releases hydrogen ions in water is an acid. For example, hydrochloric acid (HCl) in the stomach forms hydrogen (H^+) ions and chloride (Cl^-) ions.

$$HCl \longrightarrow H^+ + Cl^-$$

A **base** is a proton acceptor. For example, sodium hydroxide (NaOH) forms sodium (Na^+) ions and hydroxide (OH^-) ions.

$$NaOH \longrightarrow Na^+ + OH^-$$

The hydroxide ion is a proton acceptor that binds with a hydrogen ion to form water (H_2O).

$$OH^- + H^+ \longrightarrow H_2O$$

A reaction between an acid and a base forms a **salt.** For example, hydrochloric acid combines with sodium hydroxide to form the salt sodium chloride.

$$HCl + NaOH \longrightarrow NaCl + H_2O$$

The pH scale

The **pH scale** (Figure 2-4), which ranges from 0 to 14, indicates the hydrogen ion concentration of a solution. Pure water is defined as a neutral solution. A **neutral solution** has an equal number of hydrogen ions and hydroxide ions and has a pH of 7. Solutions with a pH less than 7 are **acidic solutions,** and they have a greater concentration of hydrogen ions than hydroxide ions. **Alkaline** (al´kah-līn) **solutions** or **basic solutions** have a pH greater than 7, and they have fewer hydrogen ions than hydroxide ions. As the pH value becomes smaller, the solution is more acidic, and as the pH value becomes larger, the solution is more basic.

Buffers

The chemical behavior of many molecules changes as the pH of the solution in which they are dissolved changes. An organism's survival depends on its ability to maintain homeostasis by regulating body fluid pH within a narrow range. One way normal body fluid pH is controlled is through the use of buffers. A **buffer** is a chemical that resists changes in pH when either an acid or a base is added to a solution containing the buffer. When an acid is added to a buffered solution, the buffer removes hydrogen ions from the solution. This prevents a decrease in the pH of the solution.

3 If a base is added to a solution, would the pH of the solution increase or decrease? If the solution was buffered, what response from the buffer would prevent the change in pH?

WATER

A molecule of **water** consists of one atom of oxygen joined by covalent bonds to two atoms of hydrogen. Water has many useful properties for living organisms.

1. Many substances dissolve in water. Blood transports nutrients, gases, and waste products within the body. When ionic substances dissolve in water, the positive and negative ions separate or **dissociate** (Figure 2-5). The water molecules surround the positive and negative ions, keeping them in solution. When the ions

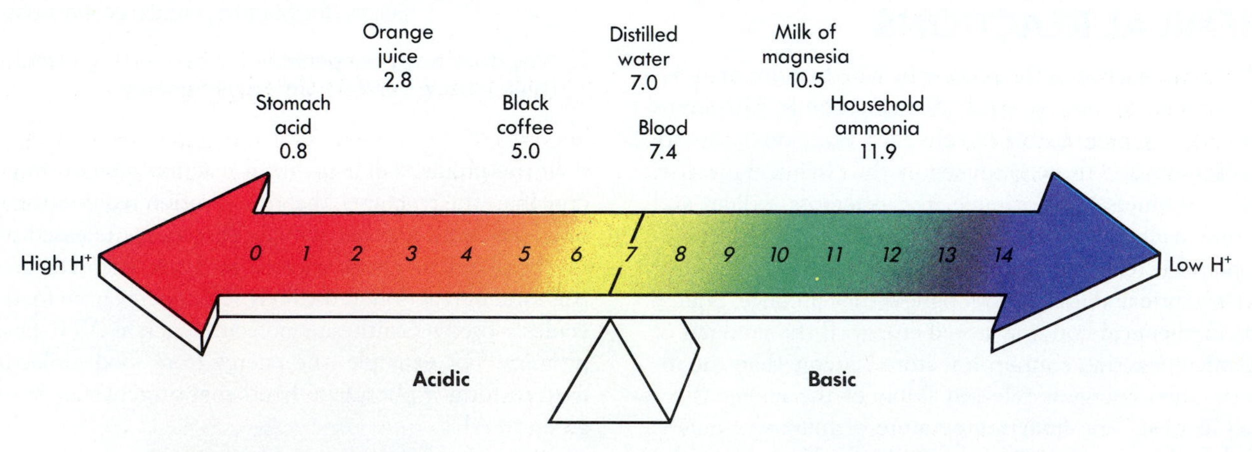

Figure 2-4 ● The pH scale.
A pH of seven is considered to be neutral, so the scale is depicted as balancing at that point. Values to the left (below 7) are acidic. The lower the value, the more acidic the solution. Values to the right (above 7) are basic or alkaline. The higher the value, the more basic the solution. Representative fluids and their pH values are listed above the Figure.

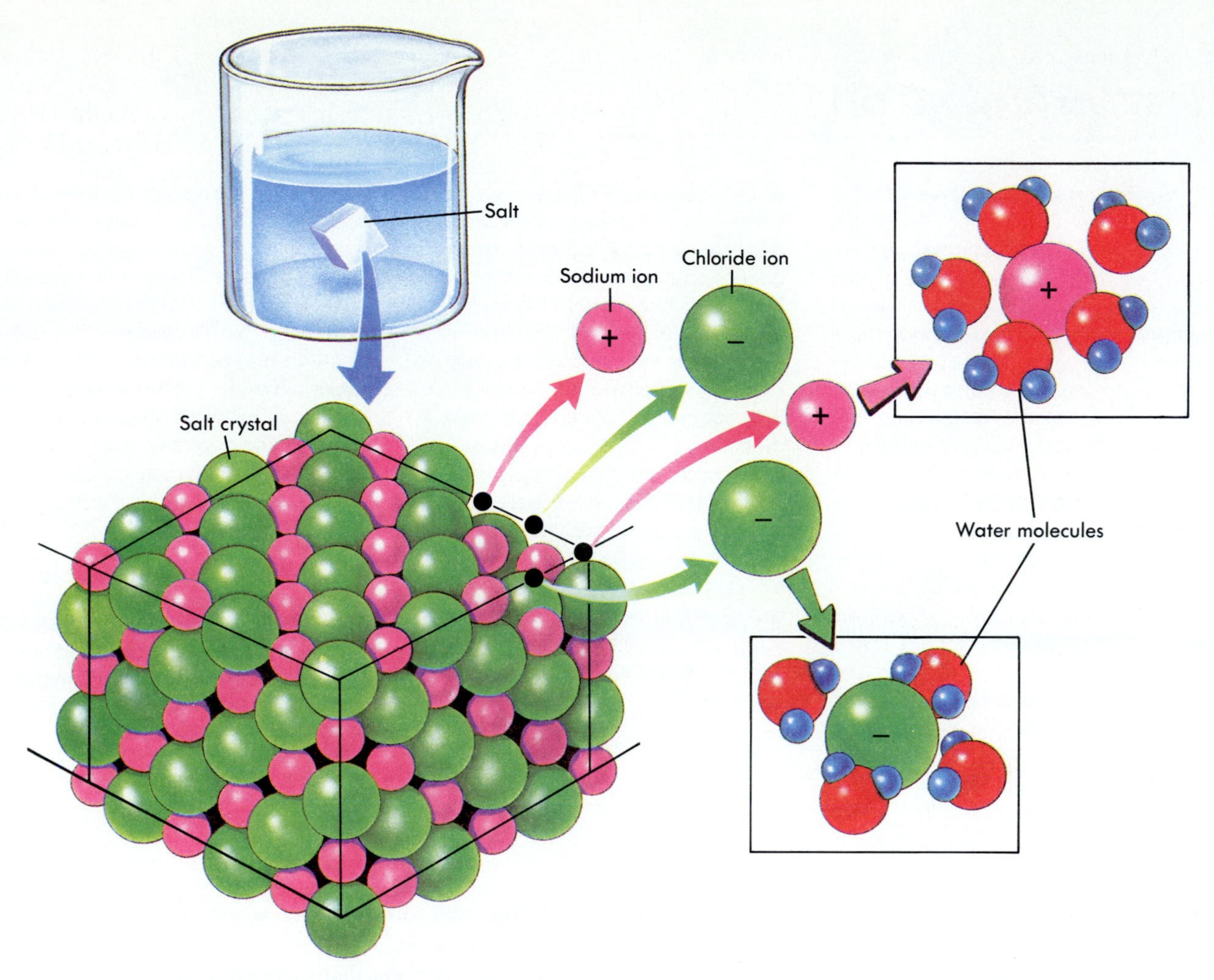

Figure 2-5 ● Dissociation.
Table salt dissolves in water when the individual sodium and chloride ions of salt each are surrounded by water molecules.

are in solution they can react with other molecules, so this property of water makes possible many of the body's chemical reactions.

2. Water is a necessary reactant or product in many chemical reactions. For example, during the digestion of food large molecules and water react to form smaller molecules.
3. Water is an effective lubricant. For example, tears protect the eye's surface from the rubbing of the eyelids.
4. Water can absorb large amounts of heat and remain at a stable temperature. Blood, which is mostly water, can transfer heat effectively from deep within the body to the body's surface. Blood is warmed deep in the body and then flows to the surface where the heat is released. In addition, water evaporation in the form of sweat carries large amounts of heat away from the body.

ORGANIC MOLECULES

Organic molecules are those that contain carbon. **Inorganic** molecules are all other molecules. An exception is carbon dioxide, which is traditionally considered an inorganic molecule. Important large organic molecules in humans are carbohydrates, lipids, proteins, and nucleic acids (Table 2-3).

Carbohydrates

The smallest **carbohydrates** are **monosaccharides** (mon-o-sak´ă-rīdz, one sugar) or simple sugars. Glucose (blood sugar) and fructose (fruit sugar) are important carbohydrate energy sources for many of the body's cells. Larger carbohydrates are formed by chemically binding monosaccharides. For this reason, monosaccharides are considered the building blocks of carbohydrates. **Disaccharides** (di-sak´ă-rīdz,

Understanding pH

The symbol pH stands for the power (p) of hydrogen ion (H^+) concentration. The power is a factor of 10, which means that a change in the pH of a solution by one pH unit represents a tenfold change in the hydrogen ion concentration. For example, a solution of pH 6 has 10 times the number of hydrogen ions as a solution with a pH of 7. Thus small changes in pH represent large changes in hydrogen ion concentration.

The normal pH range for human blood is 7.35 to 7.45. Many of the enzymes in the body are sensitive to small changes in pH. Enzymes are proteins that control how fast or slow chemical reactions occur within the body (see discussion of proteins in this chapter). If pH deviates from its normal range, enzymes do not function as well, and enzymes can even become nonfunctional. Consequently, the chemical reactions that are controlled by enzymes can be inhibited with harmful results for normal cell function. The condition of **acidosis** (as-ĭ-do´sis) results if blood pH drops below 7.35. The nervous system becomes depressed and the individual becomes disoriented and possibly comatose. **Alkalosis** (al´kah-lo´sis) results if blood pH rises above 7.45. The nervous system becomes overexcitable, and the individual may be extremely nervous or have convulsions. Both acidosis and alkalosis can result in death.

The body can closely regulate blood pH through buffers, the kidneys, and the respiratory system (see Chapter 18). For example, the respiratory system regulates blood pH by controlling the levels of carbon dioxide in the blood. Carbon dioxide (CO_2) can combine with water (H_2O) to form a hydrogen ion (H^+) and a bicarbonate ion (HCO_3^-).

$$CO_2 + H_2O \longleftrightarrow H^+ + HCO_3^-$$

This is a reversible reaction (indicated by a double-ended arrow). If carbon dioxide is added to blood it combines with water, the amount of hydrogen ions increases, and pH decreases. If carbon dioxide is removed from blood, hydrogen ions combine with bicarbonate ions to form more carbon dioxide. This decreases the amount of hydrogen ions and pH increases. Consequently, changing the amount of carbon dioxide in the blood affects hydrogen ion concentration and blood pH.

4 Emotional excitement can cause some individuals to hyperventilate. The rapid breathing removes carbon dioxide from the blood more rapidly than normal. Would a hyperventilating person develop acidosis or alkalosis? Explain.

two sugars) are formed when two monosaccharides join. For example, glucose and fructose combine to form sucrose (table sugar). **Polysaccharides** (pol-e-sak´ă-rīdz, many sugars) consist of many monosaccharides bound in long chains. Glycogen, or animal starch, is a polysaccharide of glucose. When cells containing glycogen need energy, the glycogen is broken down into individual glucose molecules that can be used as energy sources. Plant starch, also a polysaccharide of glucose, can be ingested and broken down into glucose.

Lipids

Fats, phospholipids, and **steroids** are examples of **lipids** (see Table 2-3). Fats are important energy storage molecules that also give the body padding and insulation. The building blocks of fats are **glycerol** (glis´er-ol) and **fatty acids. Triglycerides** are the most common type of fat molecule, and they have three fatty acids bound to a glycerol molecule.

Proteins

The building blocks of **proteins** are **amino** (ah-me´no) **acids.** There are 20 basic types of amino acids. Humans can synthesize 12 of these from simple organic molecules, but the remaining eight "essential amino acids" must be obtained in the diet. Although there are only 20 amino acids, hundreds to thousands of amino acids can combine in many different ways to form proteins with unique structural and functional features.

Proteins perform many important functions. For example, structural proteins provide the framework for many of the body's tissues, muscles contain proteins that are responsible for muscle contraction, and enzymes regulate the rate of chemical reactions.

At normal body temperatures, most chemical reactions would take place too slowly to sustain life if it were not for the body's enzymes. **Enzymes** (en´zīmz) increase the rate at which a chemical reaction proceeds by bringing the reactants together. The chemical reaction does not permanently change the enzyme or decrease the amount of the enzyme.

The three-dimensional shape of an enzyme is critical for its normal function. According to the **lock and key model** of enzyme action, the shape of an enzyme and the reactants allows the enzyme to bind easily to the reactants. Because the enzyme and the reactants must fit together, enzymes are very specific for the reactions they control, and each enzyme controls only one chemical reaction. After the reaction takes place, the enzyme is released and can be used again (Figure 2-6).

The ability of enzymes to function depends on their shape. If the bonds that maintain a protein's shape are broken, the protein becomes nonfunctional. This change in shape is called **denaturation,** and it can be caused by abnormally high temperatures or changes in pH of the body's fluids.

Table 2-3 ● Important organic molecules and their functions

Molecule	Elements	Building Blocks	Function	Examples
Carbohydrate	C, H, O	Monosaccharides	Energy	Monosaccharides can be used as energy sources. Glycogen (polysaccharide) is an energy storage molecule.
Lipid	C, H, O (P, N in some)	Glycerol and fatty acids (for fats)	Energy	Fats can be stored and broken down later for energy; per unit of weight fats yield twice as much energy as carbohydrates.
			Structure	Phospholipids and cholesterol are important components of cell membranes.
			Regulation	Steroid hormones regulate many physiological processes, e.g., estrogen and testosterone are responsible for many of the differences between males and females.
Protein	C, H, O, N (S in most)	Amino acids	Regulation	Enzymes control the rate of chemical reactions. Hormones regulate many physiological processes, e.g., insulin affects glucose transport into cells.
			Structure	Collagen fibers form a structural framework in many parts of the body.
			Energy	Proteins can be broken down for energy; per unit of weight they yield the same energy as carbohydrates.
			Contraction	Actin and myosin in muscle are responsible for muscle contraction.
			Transport	Hemoglobin transports oxygen in the blood.
Nucleic acid	C, H, O, N, P	Nucleotides	Regulation	DNA directs the activities of the cell.
			Heredity	Genes are pieces of DNA that can be passed from one generation to the next generation.
			Protein synthesis	RNA is involved in protein synthesis.

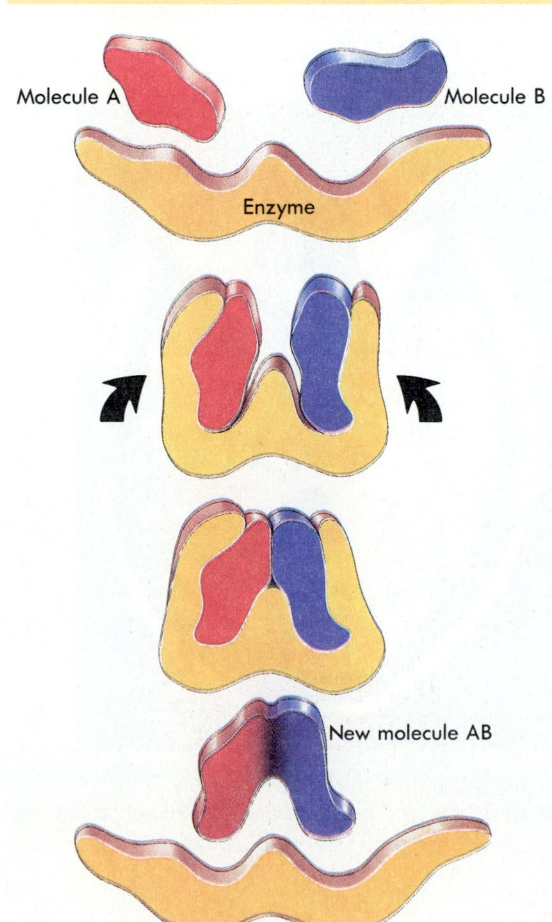

Figure 2-6 ● Enzyme action.
The enzyme brings the two reacting molecules together. This is possible because the reacting molecules "fit" the shape of the enzyme (lock and key model). After the reaction the unaltered enzyme can be used again.

The chemical events of the body are regulated primarily by mechanisms that control either the concentration or activity of enzymes. The rate at which enzymes are produced in cells, or whether the enzymes are in an active or inactive form, determines the rate of many chemical reactions.

Nucleic acids

The building blocks of **nucleic** (nu-kle´ik) **acids** are **nucleotides** (nu´-kle-o-tīdz). Each nucleotide consists of a sugar, a phosphate group, and an organic base. In **deoxyribonucleic** (de-ox´sĭ-ri´bo-nu-kle´ik) **acid (DNA)** the nucleotides contain the sugar deoxyribose. The millions of nucleotides in a DNA molecule form two strands that coil around each other to form a double helix. DNA is the genetic (hereditary) material of the cell, and it is responsible for controlling cell activities. **Ribonucleic** (ri´bo-nu-kle´ik) **acid (RNA)** is a single strand of nucleotides that contains the sugar ribose. Three different types of RNA are involved in protein synthesis (see Chapter 3).

Clinical applications of atomic particles

Protons, neutrons, and electrons are responsible for the chemical properties of atoms. They also are involved with other properties useful in a clinical setting. Understanding these properties has led to the development of methods for examining the inside of the body.

Isotopes (i´so-tōpz) are two or more forms of the same element that have the same number of protons and electrons, but different numbers of neutrons. For example, hydrogen has no neutrons, and its isotope deuterium has one. Water made with deuterium is called heavy water because of the weight of the "extra" neutron. Because isotopes have the same number of electrons, they are similar in their chemical behavior. The nuclei of some isotopes are stable and do not change. Radioactive isotopes, however, have unstable nuclei that lose neutrons or protons. Several different kinds of radiation can be produced when neutrons and protons, or the products formed by their breakdown, are ejected from the nucleus of the isotope.

The radiation given off by some radioactive isotopes can penetrate and destroy tissues. Dividing cells are more sensitive to radiation than nondividing cells. Radiation is used to treat cancerous (malignant) tumors because cancer cells are continually dividing. If the treatment is effective, the cancerous cells are killed with tolerable destruction of healthy tissue.

Radioactive isotopes also are used in diagnosis. The radiation can be detected, and the movement of the atoms throughout the body can be traced. For example, the thyroid gland normally transports iodine into the gland and uses the iodine in the formation of thyroid hormones. Radioactive iodine can be used to determine if iodine transport is normal.

Radiation can be produced in ways other than changing the nucleus of atoms. **X-rays** are radiation formed when electrons lose energy by moving from a higher energy orbital to a lower energy orbital. X-rays are used in examination of bones to determine if they are broken and in examination of teeth to see if they have caries (cavities). Mammograms are low-energy x-rays of the breast that can be used to detect tumors because the tumors are slightly more dense than normal tissue.

Computers are used to analyze a series of x-rays, each made at a slightly different body location. The picture of each x-ray "slice" through the body is assembled by the computer to form a three-dimensional image. A **computerized axial tomography (CAT) scan** is an example of this technique (Figure 2-A). CAT scans are used extensively to detect tumors and other abnormalities in the body.

Magnetic resonance imaging (MRI) is another method for looking into the body (Figure 2-B). The patient is placed in a very powerful magnetic field that aligns the hydrogen nuclei. Radiowaves given off by the hydrogen nuclei then are used by a computer to make an image of the body. Because MRI affects hydrogen, it is very effective for visualizing soft tissues that contain a lot of water. MRI technology is used to detect tumors and other abnormalities in the body.

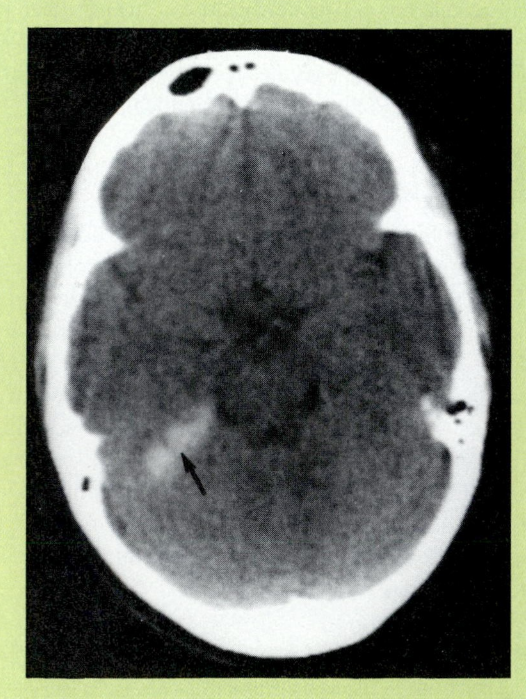

Figure 2-A CAT Scan.
CAT scan of patient with a cerebral hemorrhage *(arrow).*

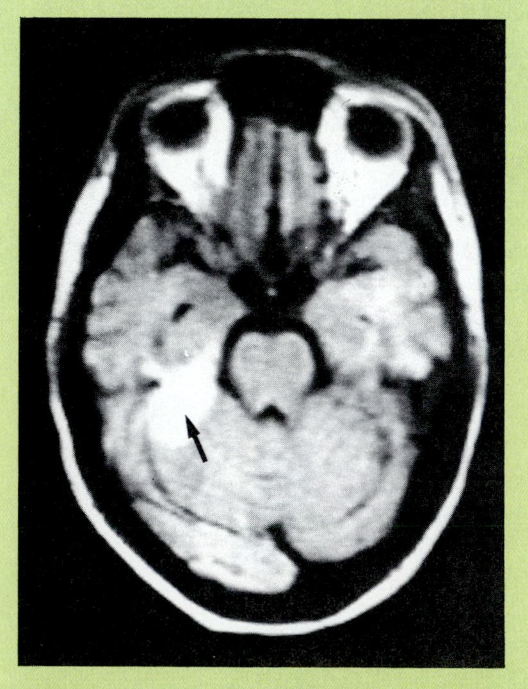

Figure 2-B MRI.
MRI of the same patient with a cerebral hemorrhage *(arrow).*

SUMMARY

Chemistry is the study of the composition and structure of substances and the reactions they undergo.

BASIC CHEMISTRY

The atom is the smallest unit of matter that is unable to be altered by chemical means.

An element is matter composed of one kind of atom.

The structure of atoms

Atoms consist of neutrons, positively charged protons, and negatively charged electrons.

Electrons are found in orbitals around the nucleus. The nucleus consists of protons and neutrons.

Each element has a unique number of protons.

Electrons and chemical bonds

A molecule is two or more atoms joined by chemical bonds.

An ionic bond results when an electron is transferred from one atom to another atom.

A covalent bond results when a pair of electrons are shared between atoms.

CHEMICAL REACTIONS

Reactants are atoms or molecules present before a chemical reaction and products are the atoms or molecules produced by a chemical reaction.

Chemical bonds are stored energy. The energy in the phosphate bond of ATP can be released as energy for cell use and as heat.

ACIDS AND BASES

Acids are proton (hydrogen ion) donors and bases are proton acceptors.

The pH scale

A neutral solution has an equal number of hydrogen ions and hydroxide ions and has a pH of 7.

An acidic solution has more hydrogen ions than hydroxide ions and has a pH of less than 7.

A basic solution has less hydrogen ions than hydroxide ions and has a pH greater than 7.

Buffers

Chemicals that resist changes in pH are buffers.

WATER

Water dissolves many substances, is necessary in many chemical reactions, is an effective lubricant, and transfers heat within the body.

Substances held together by ionic bonds dissociate in water to form ions.

ORGANIC MOLECULES

Organic molecules contain carbon (carbon dioxide is an exception), inorganic molecules do not.

Carbohydrates

Carbohydrates give the body energy. Monosaccharides are the building blocks that form more complex carbohydrates such as disaccharides and polysaccharides.

Lipids

Lipids provide energy (fats), are structural components (phospholipids), and regulate physiological processes (steroids). The building blocks of fats are glycerol and fatty acids.

Proteins

Proteins regulate chemical reactions (enzymes), are structural components, and cause muscle contraction. The building blocks of proteins are amino acids.

Enzymes speed up chemical reactions without being altered permanently. Enzymes are specific and bind to reactants according to the lock and key model.

Denaturation of proteins changes their shape and makes them nonfunctional.

Nucleic acids

Nucleic acids include DNA, the genetic material, and RNA, which is involved in protein synthesis. The building blocks of nucleic acids are nucleotides.

CONTENT REVIEW

1. Define chemistry. Why is an understanding of chemistry important?
2. Diagram the structure of an atom and label the parts. Compare the charges of the subatomic particles.
3. Define chemical bond, molecule, and compound. What is a chemical formula?
4. Distinguish between ionic and covalent bonds. Define an ion.
5. Define a chemical reaction.
6. Where does the energy to form ATP come from? What happens to the energy in ATP when ATP is broken down to ADP?
7. What is an acid, base, and salt? Describe the pH scale.
8. What is a buffer, and why are buffers important?
9. List four functions that water performs in the human body.
10. Define an organic and an inorganic molecule.
11. Name the four major types of organic molecules, give a function for each, and name the building blocks of each.
12. Describe the action of enzymes in terms of the lock and key model. What is denaturation?

CONCEPT REVIEW

1. If an atom of iodine (I) gained an electron, what would be the charge of the resulting ion? Write the symbol for this ion.
2. In terms of energy transfer in chemical reactions, explain why eating food is necessary for increasing muscle mass.
3. In a person with asthma there is decreased gas exchange between the lungs and the air. Consequently, there is a build-up of carbon dioxide in the blood. Would a person with asthma develop acidosis or alkalosis? Explain.
4. The kidneys are powerful regulators of blood pH. The kidneys accomplish pH regulation by excreting hydrogen ions from the blood into urine, which is eliminated from the body. If a patient has acidosis, would you expect his urine to be more or less acidic than normal? Explain.

CHAPTER TEST

Answers can be found in Appendix C

MATCHING For each statement in column A select the correct answer in column B (an answer may be used once, more than once, or not at all).

A	B
1. Anything that occupies space.	acid
2. Matter composed of atoms of only one kind.	amino acid
3. Subatomic particle with a negative charge.	base
4. Two or more atoms chemically bound together.	DNA
5. A charged atom.	electron
6. A proton (hydrogen ion) acceptor.	element
7. Solution with a pH of 4.	enzyme
8. General term for molecules containing carbon.	fat
9. Building block of carbohydrates.	inorganic
10. Energy storage molecule that also forms body padding.	ion
11. Building block of proteins.	matter
12. Protein that increases the rate at which chemical reactions proceed.	molecule
13. The genetic material.	monosaccharide
	neutron
	organic
	proton

FILL-IN-THE-BLANK Complete each statement by providing the missing word or words.

1. The scientific study of the composition and the structure of substances is called ___________.
2. The smallest particle of an element that retains the properties of that element is an ___________.
3. Within an atom, neutrons and protons are located in the ___________.
4. When one atom loses an electron and another atom accepts that electron an ___________ bond can be formed.
5. A solution that has more hydrogen ions than hydroxide ions is called ___________.
6. A chemical that resists changes in pH when either an acid or base is added to a solution is a ___________.
7. Positive and negative ions in water separate or ___________.
8. The building blocks of fats are ___________ and ___________.
9. According to the ___________ of enzyme action, the shapes of the enzyme and the reactants allow the enzyme to bind easily to the reactants.
10. ___________ is a change in shape that causes proteins (enzymes) to become nonfunctional.

CHAPTER

3

Cell Structures and Their Functions

KEY TERMS

active transport
Movement of materials across the plasma membrane that requires a carrier molecule and ATP.

cytoplasm
Cellular material surrounding the nucleus.

diffusion
Tendency for molecules to move from an area of higher concentration to an area of lower concentration in solution.

DNA
Deoxyribonucleic acid; the genetic material of cells, which directs the production of proteins.

mitochondrion
pl. mitochondria
Small, bean-shaped or rod-shaped structure in the cytoplasm that is a site of ATP production.

mitosis
Division of the nucleus that results in two daughter cells with exactly the same number and type of chromosomes as the parent cell.

nucleus
pl. nuclei
Cell organelle containing most of the cell's genetic material (DNA); controls the activities of the cell.

osmosis
Diffusion of water through a selectively permeable membrane from a less concentrated solution (less dissolved substances, more water) to a more concentrated solution (more dissolved substances, less water).

plasma membrane
Outermost component of the cell, surrounding and binding the rest of the cell contents; the cell membrane.

ribosome
Small, spherical, cytoplasmic organelle where protein synthesis occurs.

FEATURES

OBJECTIVES

After reading this chapter you should be able to:

1. Describe the structure and functions of the plasma membrane, nucleus, and nucleolus.
2. Compare the structure and function of rough and smooth endoplasmic reticulum.
3. Describe the roles of the Golgi apparatuses and secretory vesicles in secretion.
4. Explain the role of lysosomes in digesting material taken into cells.
5. Describe the structure and function of mitochondria.
6. Compare the structure and function of cilia, flagella, and microvilli.
7. Explain how substances cross the plasma membrane by passive and active transport processes.
8. Describe the process of protein synthesis.
9. Explain what is accomplished during mitosis and differentiation.

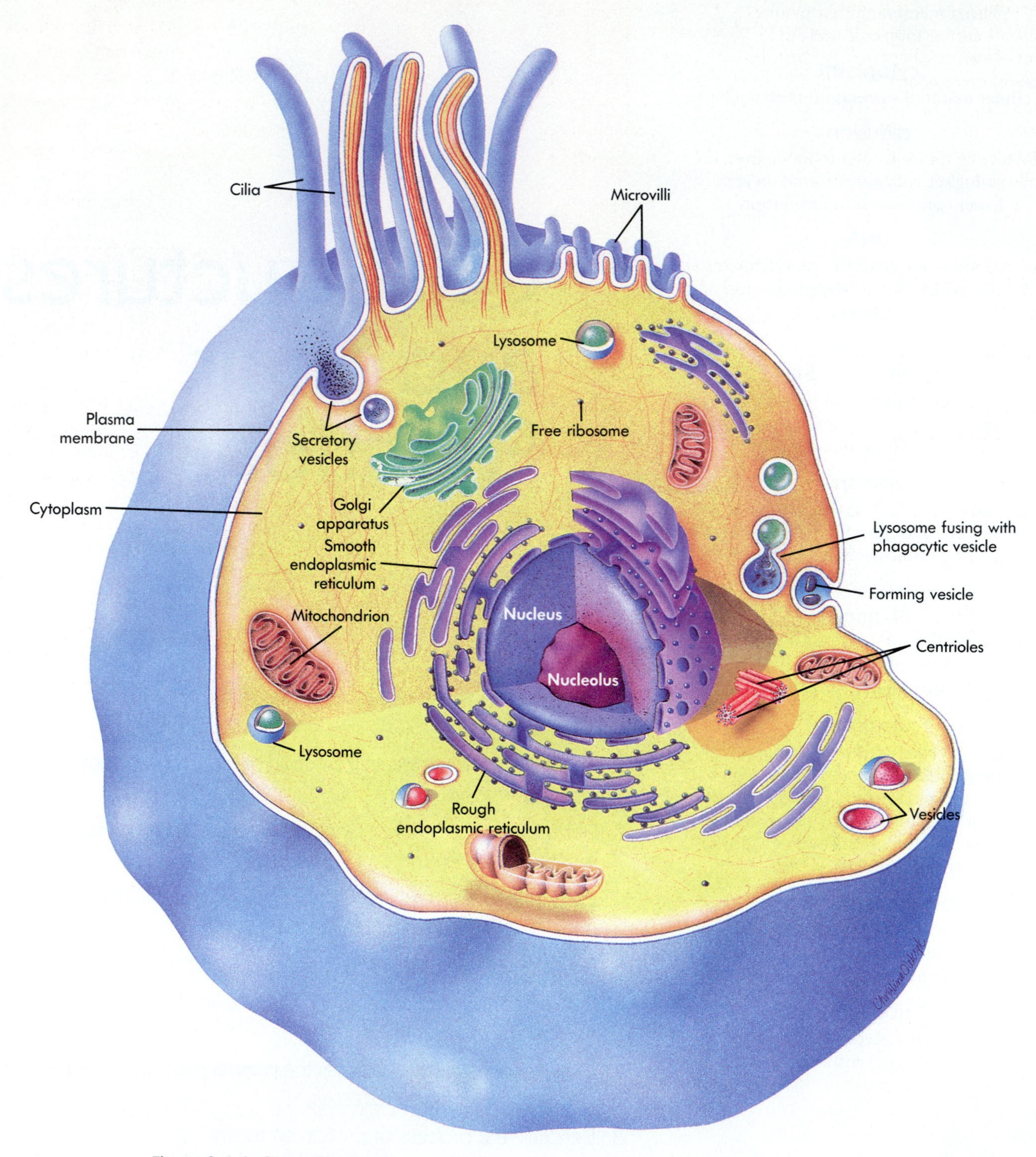

Figure 3-1 ● Generalized cell.
The major organelles of a typical cell are shown. No single cell contains all organelle types. In addition, some cells may contain many organelles of one type and another cell may contain very few.

THE CELL is the basic living unit of all organisms. The simplest organisms consist of a single cell, whereas humans are composed of trillions of cells. The study of cells is an important link between the study of chemistry in Chapter 2 and tissues in Chapter 4. A knowledge of chemistry makes it possible to understand cells because cells are composed of molecules that are responsible for many of the characteristics of cells. Cells, in turn, interact to form the tissues of the body. This chapter considers the structure of cells and how cells perform the activities necessary for life.

CELL STRUCTURE AND FUNCTION

Each cell is a highly organized unit. Within cells there are specialized structures called **organelles** (or´gă-nelz), which perform specific functions (Figure 3-1). At some time in their existence each cell contains a **nucleus** (nu´kle-us), an organelle containing the cell's genetic material, which controls the cell. The living material surrounding the nucleus is called **cytoplasm** (si´to-plazm), which contains many other types of organelles. The cytoplasm is enclosed by the **plasma** (plaz´mah) **membrane.**

The number and type of organelles within each cell determine the cell's specific structure and functions (Table 3-1). For example, cells secreting large amounts of protein contain well-developed organelles that synthesize and secrete protein, whereas muscle cells have organelles that enable the cells to contract.

Plasma membrane

The plasma membrane, or **cell membrane,** is the outermost component of a cell. The plasma membrane encloses the cytoplasm and forms a boundary between material inside the cell and material outside the cell. Substances outside the cell are called **extracellular** substances, and substances inside the cell are called **intracellular** substances. In addition to enclosing and supporting the cell contents, the plasma membrane is a selective barrier that determines what moves into and out of the cell. Many of the organelles within cells also have membranes similar in structure and function to the plasma membrane.

The major molecules that make up the plasma membrane are phospholipids and proteins (Figure 3-2). The phospholipids form a double layer of molecules with other lipids such as cholesterol interspersed among them. The lipids form a non-rigid framework that separates the intracellular fluid from the extracellular fluid.

Protein molecules are on the inner and the outer surfaces of the plasma membrane or extend across the plasma membrane. The proteins function as membrane channels, carrier molecules, receptor molecules, enzymes, or structural supports in the membrane. Membrane channels and carrier molecules are involved with the movement of substances through the plasma membrane (see membrane transport in this chapter). **Receptor molecules** are part of an intercellular communication system that enables coordination of the activities of cells. For example, a nerve cell can release a chemical messenger that moves to a muscle cell and binds to its receptor. The binding acts as a signal that triggers a response such as contraction of the muscle cell.

Table 3-1 • Organelles and their functions

Organelle	Function
Plasma membrane	Outer boundary of cell that regulates the movement of substances into and out of the cell; part of an intercellular communication system
Nucleus	Contains genetic material (DNA) that regulates cell activities by controlling protein synthesis
Nucleolus	Site of ribosome synthesis in the nucleus
Ribosome	Site of protein synthesis
Endoplasmic reticulum (ER)	Rough ER (with ribosomes) is a site of protien synthesis; smooth ER (without ribosomes) is a site of lipid synthesis
Golgi apparatus	Modifies protein structure and packages proteins in secretory vesicles
Secretory vesicle	Carries material produced in the cell to the plasma membrane through which the materials are released to the outside of the cell
Lysosome	Contains enzymes that digest materials taken into the cell
Mitochondria	Sites of ATP synthesis
Cilia	Move substances over the surface of cells
Flagella	Move sperm cells
Microvilli	Increase the surface area of cells, promoting absorption and secretion of materials

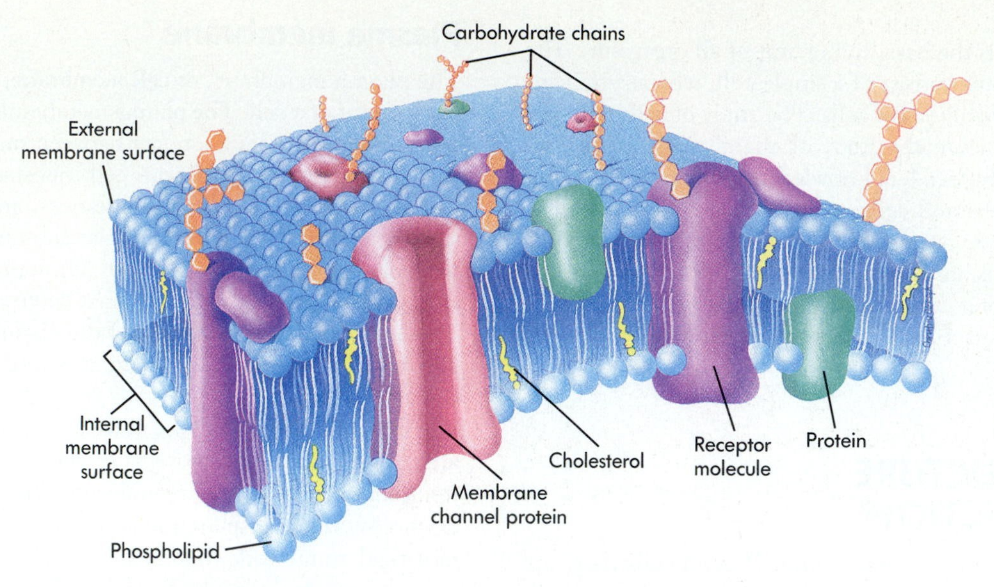

Figure 3-2 ● **The plasma membrane.**
The plasma membrane is composed of a double layer of phospholipid molecules with cholesterol molecules. Proteins, usually with attached carbohydrates, are on the surface of the phospholipid layer or extend across the phospholipid layer. Proteins can form membrane channels, carrier molecules, or receptor molecules.

Nucleus

The nucleus is a large organelle usually located near the center of the cell (see Figure 3-1). The nucleus contains the cell's genetic material **(DNA),** which determines the structure and function of each cell and of the individual (see protein synthesis in this chapter). All cells of the body have a nucleus at some point in their life cycle, although some cells, such as red blood cells, lose their nuclei as they mature. Other cells, such as skeletal muscle cells, contain more than one nucleus.

The nucleus is surrounded by two membranes with a narrow space between them. At many points on the surface of the nucleus, the inner and outer membranes come together to form **nuclear pores** (Figure 3-3). A **nucleolus** (nu-kle´o-lus) is a rounded, dense, well-defined nuclear body with no surrounding membrane (see Figure 3-1). Within the nucleolus the subunits of ribosomes are manufactured. The subunits move through the nuclear pores and join together in the cytoplasm to form functional ribosomes (see ribosomes in this chapter).

Cytoplasm

Cytoplasm is the cellular material outside the nucleus but inside the plasma membrane. Cytoplasm is approximately half fluid, in which chemical reactions take place, and half organelles, which are cellular subunits specialized to perform specific functions.

Ribosomes

Ribosomes (ri´bo-sōmz) are the organelles where proteins are produced (see protein synthesis in this chapter). Ribosomes can be found free in the cytoplasm or associated with a membrane called the endoplasmic reticulum (see Figure 3-1).

Rough and smooth endoplasmic reticulum

The **endoplasmic reticulum** (en´do-plaz´mik re-tik´u-lum) **(ER)** is a series of membranes that extend from the outer nuclear membrane into the cytoplasm (see Figure 3-3). **Rough endoplasmic reticulum** is ER with ribosomes attached to it. Because rough ER has numerous ribosomes, a large amount of rough ER in a cell indicates that it synthesizes protein. On the other hand, ER without ribosomes is called **smooth endoplasmic reticulum.** Smooth ER is a site for lipid synthesis in cells.

Golgi apparatus, secretory vesicles, and lysosomes

The **Golgi** (gol´je) **apparatus** consists of closely packed stacks of curved, membrane-bound sacs (see Figure 3-3). It concentrates, modifies, and packages materials. For example, proteins produced at the ribosomes enter the Golgi apparatus from the endoplasmic reticulum. The Golgi apparatus concentrates and, in some cases, chemically modifies the proteins by attaching carbohydrate or lipid molecules to them. The proteins then are packaged into **secretory vesicles** (ves´ĭ-klz), which pinch off from the margins of the Golgi apparatus. The secretory vesicles, which are small membrane-bound sacs, move to the surface of the cell. Their membranes then fuse with the plasma membrane, and the contents of the vesicle are released to the outside. The Golgi apparatuses are present in larger numbers and are most

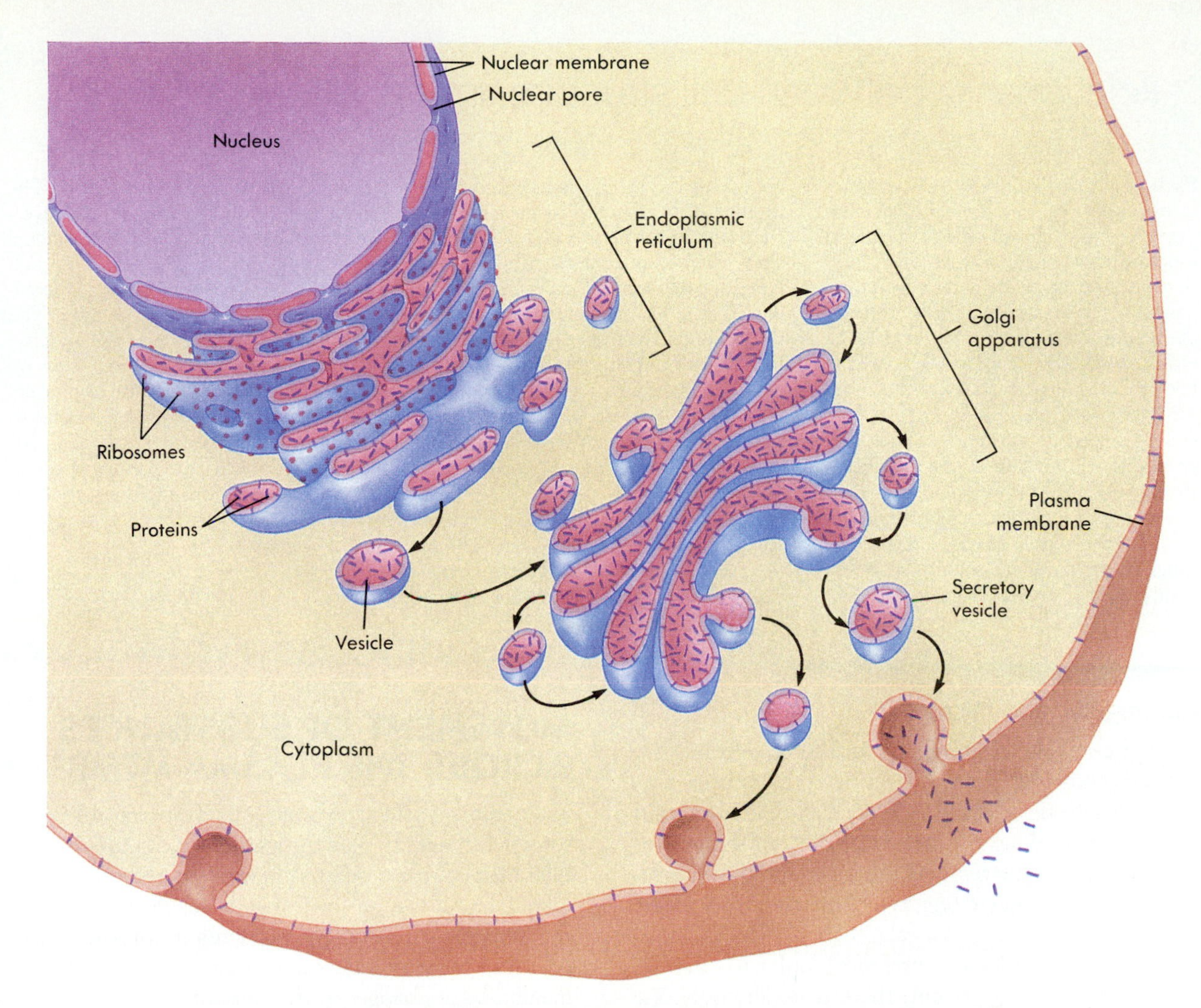

Figure 3-3 ● **The nuclear membrane, endoplasmic reticulum, and Golgi apparatus.** An inner and an outer membrane surround the nucleus. There are several nuclear pores where the inner and outer nuclear membranes become fused. The outer membrane is continuous with the endoplasmic reticulum (ER). Rough ER has ribosomes attached to its membrane and smooth ER has no ribosomes attached to it. The Golgi apparatus receives proteins from the rough ER and lipids from the smooth ER. The proteins and lipids are concentrated, modified, and packaged into secretory vesicles that move to the plasma membrane and release their contents to the outside of the cell.

highly developed in cells that secrete protein, such as the cells of the salivary glands or the pancreas.

Lysosomes (li´so-sōmz) are membrane-bound vesicles formed from the Golgi apparatus. They contain a variety of enzymes (proteins) that function as intracellular digestive systems (see Figure 3-1). The enzymes cause materials taken up by cells to break apart. For example, white blood cells take up bacteria and the enzymes within lysosomes destroy the bacteria. Also, when tissues are damaged, ruptured lysosomes within the damaged cells release their enzymes and digest both healthy and damaged cells. The released enzymes are responsible for part of the resulting inflammation (see Chapter 4).

Mitochondria

Mitochondria (mi´to-kon´dre-ah) are small bean-shaped or rod-shaped organelles with inner and outer membranes separated by a space (see Figure 3-1). Mitochondria are the major sites of adenosine triphosphate (ATP) production within cells. ATP is the major energy source for most chemical reactions within the cell, and cells with a large energy requirement have more mitochondria than cells with a lower energy requirement. When muscles enlarge as a result of exercise, for example, the number of mitochondria within the muscle cells increases and provides the additional ATP required for muscle contraction.

Relationship between cell structure and function

Each cell is well-adapted for the functions it performs, and the abundance of organelles in each cell reflects the cell's function. For example, epithelial cells that line the larger-diameter respiratory passages secrete mucus and transport the mucus toward the throat, where it is either swallowed or expelled from the body by coughing. Particles of dust and other debris suspended in the air become trapped in the mucus. Thus the production and transport of mucus from the respiratory passages function to keep the respiratory passages clean. Cells of the respiratory passages have abundant rough endoplasmic reticulum, Golgi apparatuses, and secretory vesicles. Cilia are on the surface of the respiratory epithelial cells. The ribosomes on the rough endoplasmic reticulum are the sites where the proteins, which are a major component of mucus, are produced. The Golgi apparatuses package the proteins and other components of mucus into secretory vesicles, which move to the surface of the epithelial cells. Their contents are released onto the surface of the epithelial cells. Cilia then propel the mucus toward the throat.

In people who smoke, the prolonged exposure of the respiratory epithelium to the irritation of tobacco smoke causes the respiratory epithelial cells to change in structure and function. The cells may flatten and form several layers of epithelial cells. The flat epithelial cells no longer contain abundant rough endoplasmic reticulum, Golgi apparatuses, secretory vesicles, or cilia. The cells are adapted to protect the underlying cells from irritation, but they no longer function to clean the respiratory passages. Extensive replacement of normal epithelial cells in respiratory passages correlates with chronic inflammation of the respiratory passages (bronchitis), which often exists in people who smoke.

Cilia, flagella, and microvilli

Cilia (sil´e-ah) project from the surface of cells and are capable of moving (see Figure 3-1). When present, they vary in number from one to thousands per cell. Cilia function to move materials along the surface of the cell. For example, cilia line part of the reproductive tract and their movement transports the oocyte (egg cell) from the ovary to the uterus.

Flagella (flă-jel´ah) have a structure similar to cilia but are much longer. When present, there is usually only one flagellum per cell. For example, each sperm cell has one flagellum, which functions to propel the sperm cell.

Microvilli (mi´kro-vil´i) are cylindrical extensions of the plasma membrane, but they do not move (see Figure 3-1). Microvilli are numerous on cells that have them, and they function to increase the surface area of cells. Microvilli are abundant on the surface of cells that line the intestine, kidney, and other areas where absorption is important.

Organelle interactions

To understand how a cell functions, the interactions between the organelles must be considered. For example, the transport of many food molecules into the cell by the plasma membrane requires ATP and plasma membrane proteins. The ATP is produced by mitochondria. Production of plasma membrane proteins requires amino acids transported into the cell by the plasma membrane. Instructions provided by the nucleus cause the amino acids to be combined at ribosomes to form proteins. Thus a picture of mutual interdependence of organelles emerges when the whole cell is examined. The next topics, movement of substances across the plasma membrane, protein synthesis, and cell division, illustrate the interactions of organelles that result in a functioning cell.

MOVEMENT OF SUBSTANCES ACROSS THE PLASMA MEMBRANE

The plasma membrane is **selectively permeable,** allowing some substances to pass through it but not others. Intracellular material has a different composition than extracellular material, and the survival of cells depends on maintaining the difference. In addition, nutrients continually must enter cells and waste products must exit. Rupture of the plasma membrane or changes in the amount or kinds of substances that pass through the membrane can lead to cell death.

Substances move across the plasma membrane by passive or active transport processes. **Passive transport** processes do not require energy derived from ATP, whereas **active transport** processes use energy from ATP.

Passive transport processes

The passive transport processes important to cells are diffusion, osmosis, and filtration.

Diffusion

Diffusion is the tendency for a substance to move from an area of higher concentration of that substance to an area of lower concentration of that substance. An example of diffusion is the movement of perfume molecules from an open bottle of perfume (where there is a higher concentration of perfume molecules) throughout a room (where there is a lower concentration of perfume molecules).

Diffusion has two important characteristics. First, the greater the difference in concentration, the faster a substance diffuses from the higher to the lower concentration. By analogy, the higher a hill, the greater is the height difference from the top to the bottom of the hill, and the steeper

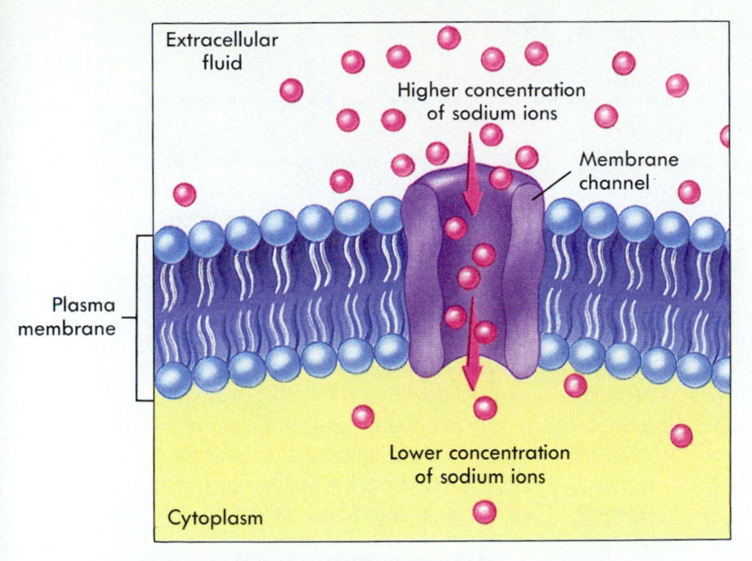

Figure 3-4 ● **Diffusion through membrane channels.** Substances that are small enough to pass through membrane channels can diffuse into and out of cells. Diffusion is from areas of higher to lower concentration.

is the slope of the hill. A car coasts down the slope of a steep hill faster than down a hill with a slope that is not so steep. Second, diffusion ends when the substance is evenly distributed, that is, when there is no longer a concentration difference. If a road has no slope, that is there is no high and low height difference, a car does not coast.

Substances diffuse through the plasma membrane in two ways. **Membrane channels** are large protein molecules that extend from one surface of the plasma membrane to the other (Figure 3-4). Ions and molecules that are smaller than the diameter of the membrane channel, such as sodium and potassium ions, can diffuse through the channels. Substances larger than the membrane channels usually cannot diffuse through the plasma membrane. However, some substances, such as water, carbon dioxide, oxygen, and urea, can bypass the membrane channels and diffuse directly through the phospholipid layers of the plasma membrane. In addition, glucose can diffuse through the phospholipid layers when assisted by certain protein molecules within the plasma membrane.

1 Urea is a toxic waste produced inside cells. It diffuses from cells into the blood and is eliminated from the body by the kidneys. What would happen to the intracellular and extracellular concentration of urea if the kidneys stopped functioning?*

Osmosis

Osmosis (os-mo´sis) is the diffusion of water across a selectively permeable membrane, such as the plasma membrane, from a region of higher water concentration to a region of lower water concentration. Water molecules can pass through the phospholipid layers of the plasma membrane. Osmosis is important to cells because large volume changes caused by water movement can disrupt normal cell functions.

Filtration

Filtration is the movement of fluid through a partition containing small holes. The fluid movement results from the force or weight of the fluid pushing against the partition. The fluid and substances small enough to pass through the holes move through the partition, but substances larger than the holes do not pass through the partition. For example, in a car, oil but not dirt particles pass through an oil filter. In the body, filtration occurs in the kidneys as a step in urine production. Blood pressure moves fluid from the blood through a partition, or filtration membrane. Water, ions, and small molecules pass through the filtration membrane to form urine, whereas larger substances such as proteins and blood cells remain in the blood.

Active transport processes

Active processes require the expenditure of energy. Specifically, adenosine triphosphate (ATP) breaks down to adenosine diphosphate (ADP) and a phosphate (P) with the release of energy (see Chapter 2). The energy can be used for transporting substances across the plasma membrane.

$$\text{ATP} \longrightarrow \text{ADP} + \text{P} + \text{Energy (for transport)}$$

Active processes include active transport and vesicle transport.

Active transport

Active transport is the movement of substances across the plasma membrane from a lower to a higher concentration. Movement from a lower to a higher concentration is possible because active transport uses the energy from ATP. This is like using the energy from gasoline to drive a car up a hill. Consequently, active transport accumulates substances on one side of the plasma membrane at concentrations many times greater than those on the other side. An example of active transport is the movement of amino acids from the small intestine into the blood.

Active transport involves **carrier molecules,** which are proteins that extend from one side of the plasma membrane to the other. After a molecule to be transported binds to a carrier molecule on one side of the membrane, the three-dimensional shape of the carrier molecule changes, and the transported molecule is moved to the opposite side of the plasma membrane. The transported molecule then is released by the carrier molecule, which then resumes its original shape and is available to transport another molecule (Figure 3-5).

*Answers to predict questions appear in Appendix B at the back of the book.

Understanding osmosis

Osmosis is the diffusion of water across the plasma membrane from a region of higher to lower concentration of water. For example, if there is a higher concentration of water inside the cell than outside the cell, water will move out of the cell. To predict whether water will move into or out of a cell, one needs to know the concentration of water. The concentration of a solution is not expressed in terms of water. Instead, the concentration of a solution is expressed in terms of the substances dissolved in the water. For example, if sugar solution A is twice as concentrated as sugar solution B, then solution A has twice as much sugar as solution B. As the concentration of a solution increases, the amount of water proportionately decreases. For example, a very concentrated sugar solution has very little water compared with a very dilute sugar solution, which is almost all water. Consequently, water diffuses from the less concentrated solution (fewer dissolved substances but more water molecules) into the more concentrated solution (more dissolved substances but fewer water molecules).

When water moves into cells they swell, and when water moves out of cells they shrink. Three terms describe solutions in which cells swell, remain unchanged, or shrink. When a cell is placed in a **hypotonic** (hi′po-ton′ik) solution, the solution has a lower concentration of dissolved substances and a higher concentration of water than the cytoplasm of the cell. Water moves by osmosis into the cell, causing it to swell. If the cell swells enough it can rupture, a process called **lysis** (li′sis; see Figure 3-A, *A*). When a cell is immersed in an **isotonic** (i′so-ton′ik) solution, the concentration of dissolved substances and water are the same on both sides of the plasma membrane. Therefore, the cell neither shrinks nor swells (Figure 3-A, *B*). When a cell is immersed in a **hypertonic** (hi′per-ton′ik) solution, the solution has a higher concentration of dissolved substances and a lower concentration of water than the cytoplasm of the cell. Water moves by osmosis from the cell into the hypertonic solution, resulting in cell shrinkage or **crenation** (kre-na′shun) (see Figure 3-A, *C*). Solutions injected into the circulatory system or tissues must be isotonic because swelling or crenation disrupts the normal function of cells and can lead to cell death.

2 In diabetes mellitus resulting from insufficient insulin production, glucose cannot enter cells. Instead, the glucose accumulates in the blood. Would you expect the cells of a person suffering from untreated diabetes mellitus to swell, shrink, or stay the same size? Explain.

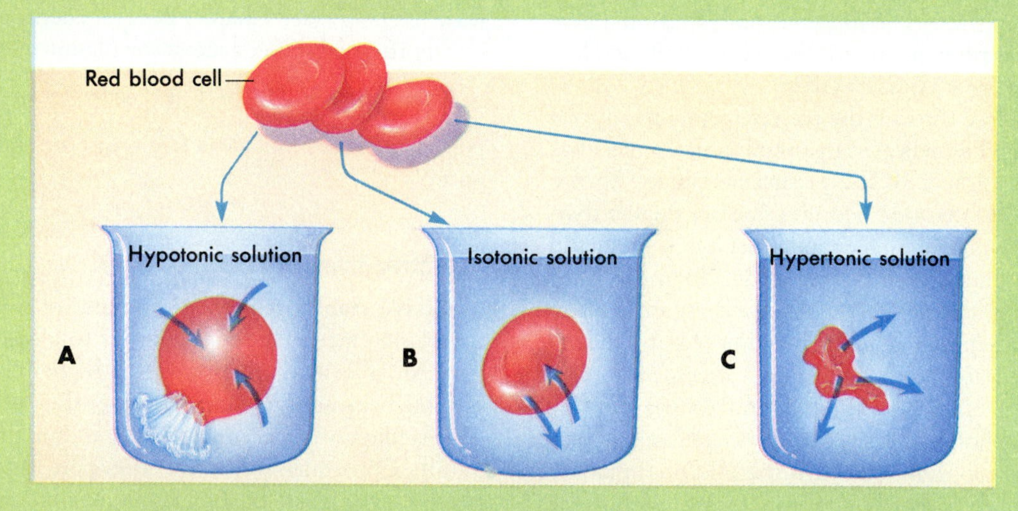

Figure 3-A Effects of hypotonic, isotonic, and hypertonic solutions on red blood cells. **A,** Hypotonic solutions result in swelling and finally, lysis of cells. **B,** Isotonic solutions result in normal-shaped cells. **C,** Hypertonic solutions result in crenation (shrinkage) of the cell.

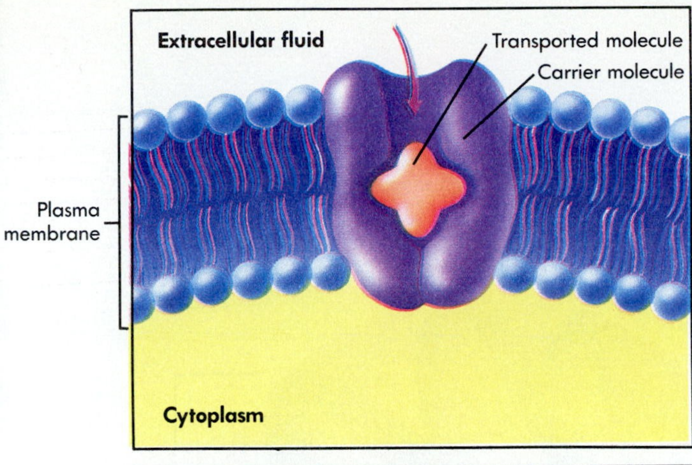

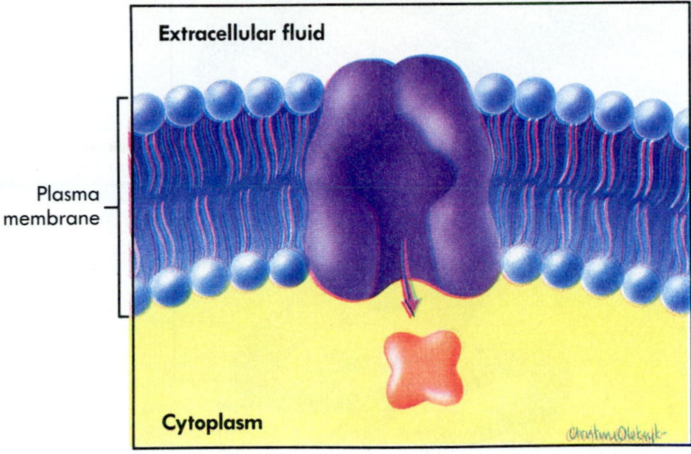

Figure 3-5 ● **Transport by a carrier molecule.**
The carrier molecule is a protein that extends across the plasma membrane. The carrier molecule binds with a molecule on one side of the plasma membrane, changes shape, and releases the molecule on the other side of the plasma membrane.

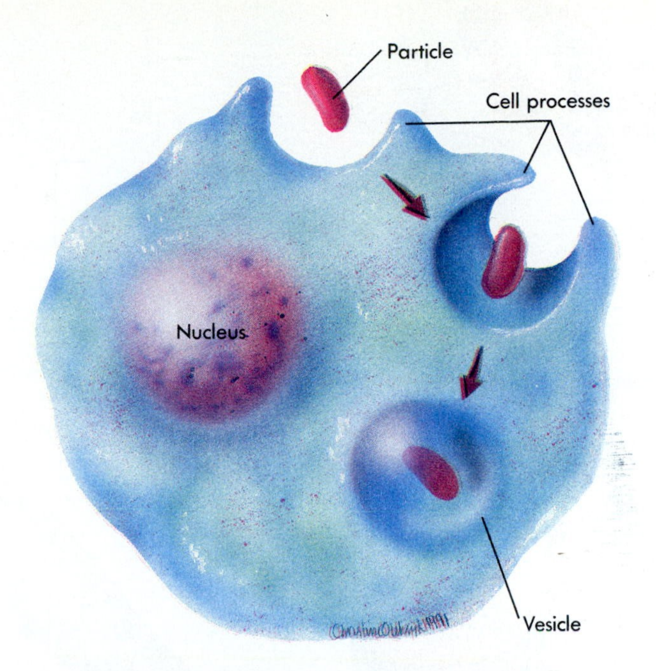

Figure 3-6 ● **Phagocytosis.**
Cell processes extend from the cell and surround the particle to be taken into the cell by phagocytosis. The cell processes fuse to form a vesicle that contains the particle. The vesicle is inside the cell.

Vesicle transport

A **vesicle** is a membrane-bound droplet found within the cytoplasm of a cell. Vesicles can be used by cells to move materials into and out of the cell. For example, **phagocytosis** (fag´o-si-to´sis), which means cell eating, is the movement of solid particles into the cell by the formation of a vesicle. A portion of the plasma membrane wraps around a particle and fuses so that the particle is surrounded by the membrane. That portion of the membrane then "pinches off" so the particle, surrounded by a membrane, is within the cytoplasm of the cell, and the plasma membrane is left intact (Figure 3-6). White blood cells and some other cell types phagocytize bacteria, cell debris, and foreign particles. Phagocytosis is an important means by which white blood cells take up and destroy harmful substances that have entered the body.

Pinocytosis (pin´o-si-to´sis) means cell drinking. This type of transport is distinguished from phagocytosis in that much smaller vesicles are formed and they contain liquid rather than particles. Pinocytosis is a common transport mechanism and occurs in certain kidney cells, epithelial cells of the intestine, liver cells, and cells that line capillaries.

In some cells, secretions accumulate within vesicles. These secretory vesicles then move to the plasma membrane where the vesicle membrane fuses with the plasma membrane, and the content of the vesicle is eliminated from the cell. This process is called **exocytosis** (eks-o-si-to´sis) (see Figure 3-3). Secretion of digestive enzymes by the pancreas, mucus by the salivary glands, and milk from the mammary glands are examples of exocytosis. In many respects exocytosis is similar to phagocytosis and pinocytosis but occurs in the opposite direction. All these processes require energy in the form of ATP and are therefore active processes.

3 List the organelles that would be common in cells that (A) use active transport to move large amounts of glucose into the cells, and (B) phagocytize foreign substances. Explain the function of each organelle you list.

PROTEIN SYNTHESIS

DNA directs the production of proteins within cells. The proteins produced in a cell function as enzymes regulating chemical reactions or as structural components inside and outside of cells. The ability of DNA to direct protein synthesis allows it to control the activities and the structure of cells and, therefore, the structural and functional characteristics of the entire organism. Whether an individual has blue eyes, brown hair, or other inherited traits is determined ultimately by DNA.

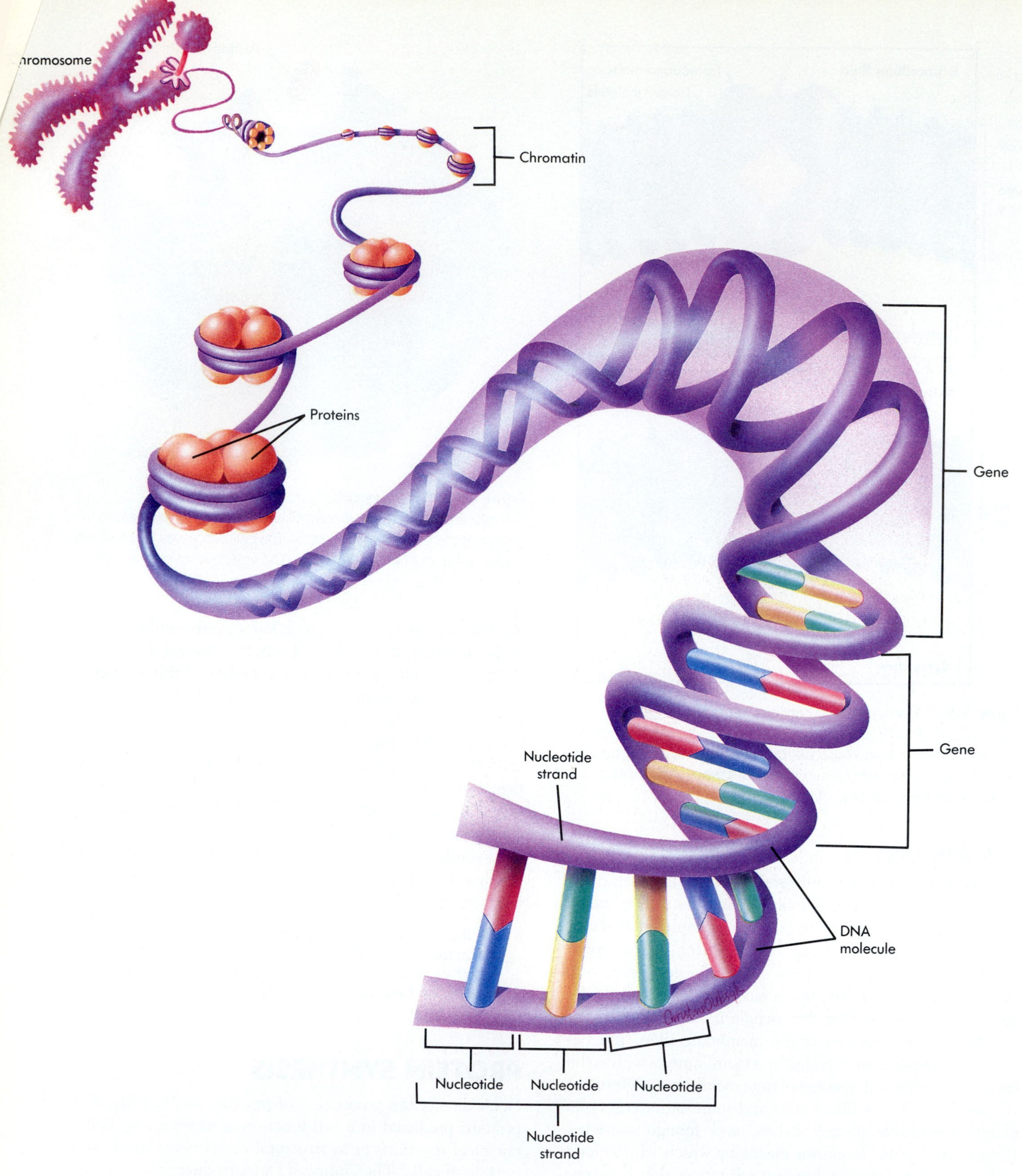

Figure 3-7 ● Structure of DNA.
Nucleotides join to form two strands of nucleotides. The strands are connected to form a twisted ladderlike structure, which is the DNA molecule. A sequence of nucleotides form a segment of the DNA molecule called a gene. Usually the DNA molecule, which is associated with some proteins, is stretched out, resembling a string of beads, and is called chromatin. During cell division, however, the chromatin condenses to form bodies called chromosomes.

A DNA molecule consists of nucleotides joined together to form two nucleotide strands (Figure 3-7). The two strands are connected and resemble a ladder that is twisted around its long axis. The nucleotides function as chemical letters that form chemical words. A **gene** is a sequence of nucleotides that is a chemical set of instructions (words) for making a specific protein. Each DNA molecule contains many different genes.

Recall from Chapter 2 that proteins consist of amino acids. The unique structural and functional characteristics of different proteins is determined by the kinds, numbers, and arrangement of their amino acids. Each gene in DNA is responsible for the production of a specific protein because each word in the chemical set of instructions of a gene represents a specific amino acid.

The production of proteins from genes in DNA involves two steps: transcription and translation, which can be illustrated with an analogy. Suppose a cook wants a recipe that is found only in a reference book in the library. Because the book cannot be checked out, the cook makes a copy, or **transcription,** of the recipe. Later, in the kitchen, the information contained in the copied recipe is used to prepare a meal. The changing of something from one form to another (from recipe to meal) is called **translation.**

In terms of this analogy, DNA (the reference book) contains many genes (recipes) for making different proteins (meals). DNA, however, is too large a molecule to pass through the nuclear pores to go to the ribosomes (kitchen) where the proteins (the meal) are prepared. Just as the reference book stays in the library, DNA remains in the nucleus. Therefore through transcription (Figure 3-8) the cell makes a copy of the information (words) in DNA necessary to make a particular protein. The copy, which is called **messenger RNA (mRNA),** travels from the nucleus to the ri-

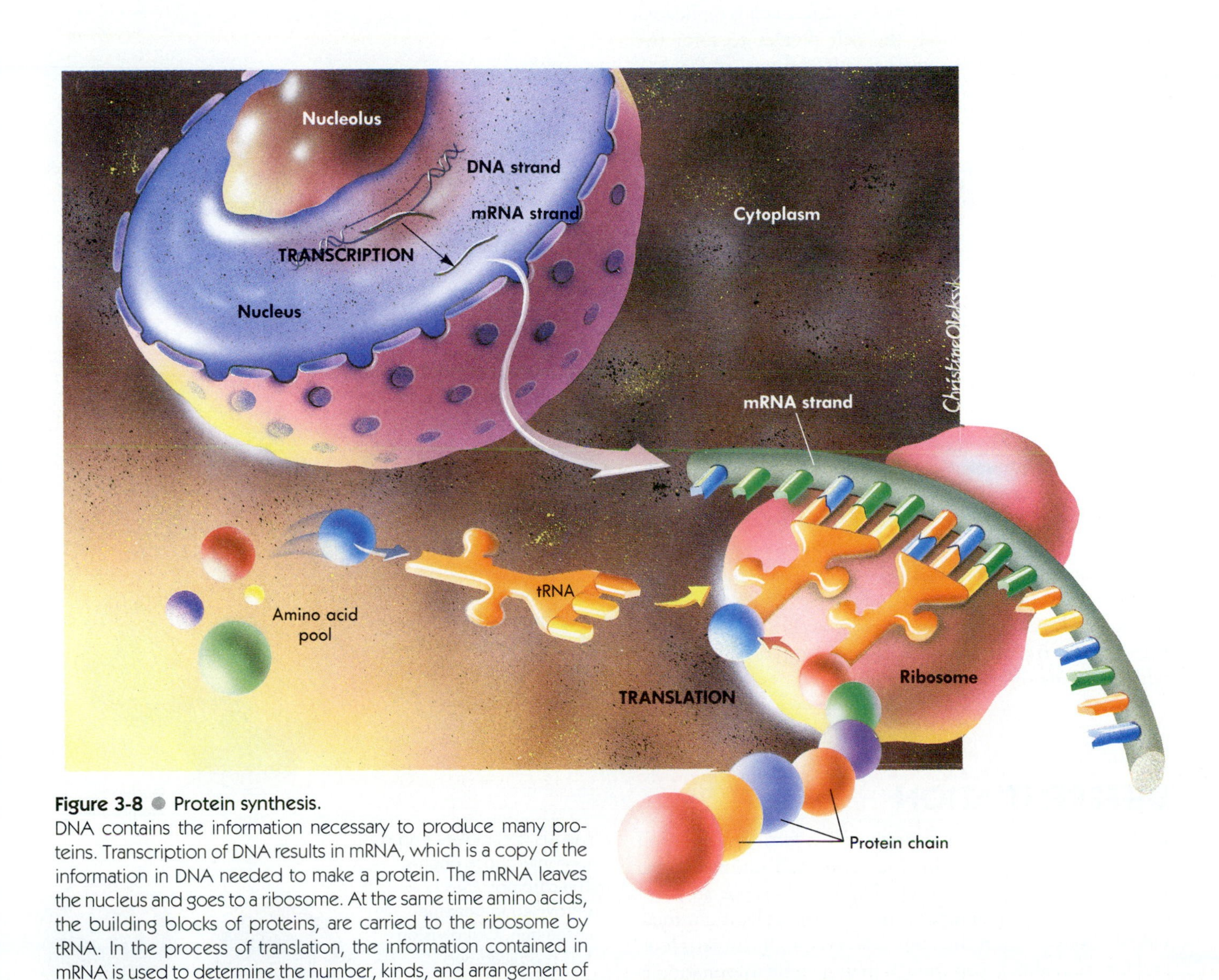

Figure 3-8 ● Protein synthesis.
DNA contains the information necessary to produce many proteins. Transcription of DNA results in mRNA, which is a copy of the information in DNA needed to make a protein. The mRNA leaves the nucleus and goes to a ribosome. At the same time amino acids, the building blocks of proteins, are carried to the ribosome by tRNA. In the process of translation, the information contained in mRNA is used to determine the number, kinds, and arrangement of amino acids in the protein.

bosomes in the cytoplasm where the information (words) in the copy is used to construct a protein, that is, translation. Of course, to turn a recipe into a meal the actual ingredients are needed. The ingredients necessary to synthesize a protein are amino acids. Specialized transport molecules, called **transfer RNA (tRNA),** carry the amino acids to the ribosome (see Figure 3-8).

In summary, the synthesis of proteins involves transcription, making a copy of part of the information in DNA (a gene), and translation, converting that copied information into a protein.

CELL DIVISION

All cells of the body, except those that give rise to reproductive cells (see Chapter 19), divide by **mitosis** (mi-to´sis). First, the DNA (genetic material) within a cell is replicated, or duplicated, and second, the cell divides to form two daughter cells with the same amount and type of DNA as the parent cell. Because DNA determines the structure and function of cells, the daughter cells have the same structure and perform the same functions as the parent cell. Growth and repair depend on mitosis. New cells added to the body as it grows and new cells that replace old or injured cells are produced by mitosis.

The period between active cell division is called **interphase** (Figure 3-9). During interphase, DNA is spread throughout the nucleus as thin threads called **chromatin** (kro´mah-tin) (see Figure 3-7). DNA is replicated during interphase and the cell has two complete sets of genetic material.

Mitosis, which follows interphase, results in the separation of the two sets of genetic material into two cells (see Figure 3-9). Mitosis is divided into four stages: prophase, metaphase, anaphase, and telophase. The chromatin coils up to form dense bodies called **chromosomes** (kro´mo-sōmz). Because the DNA was duplicated during interphase there is a duplicated set of chromosomes. The duplicated chromosomes line up along the center of the cell, the duplicated chromosomes separate, and an identical set of chromosomes moves toward the opposite ends of the cell. The cytoplasm of the cell then divides to form two cells, each of which has one set of chromosomes.

DIFFERENTIATION

Each individual begins as a single cell, a fertilized egg. Cell division produces two cells, four cells, and ultimately the trillions of cells that make up the body. Because all these cells were formed through mitosis, all the cells in an individual's body contain the same DNA. Not all the cells look and function alike, even though their genetic information is

Interphase: the time between cell division. DNA is found as thin strands of chromatin in the nucleus.

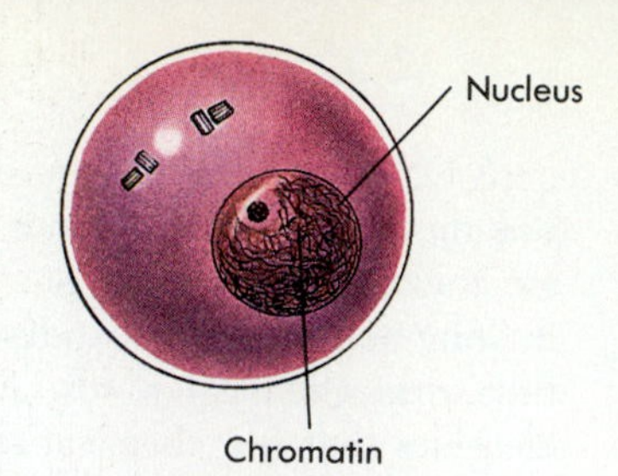

Prophase: the chromatin condenses into chromosomes. Organelles called *centrioles* (sen´tre-olz) move to opposite ends of the cell, establishing the direction in which the chromosomes will move.

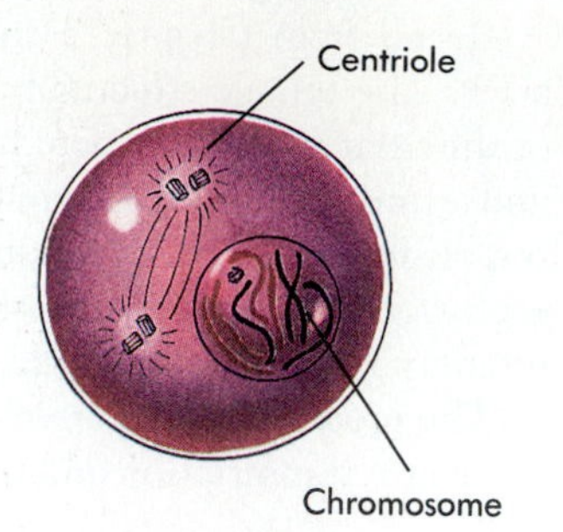

Metaphase: the chromosomes align in the center of the cell in association with protein fibers called spindle fibers.

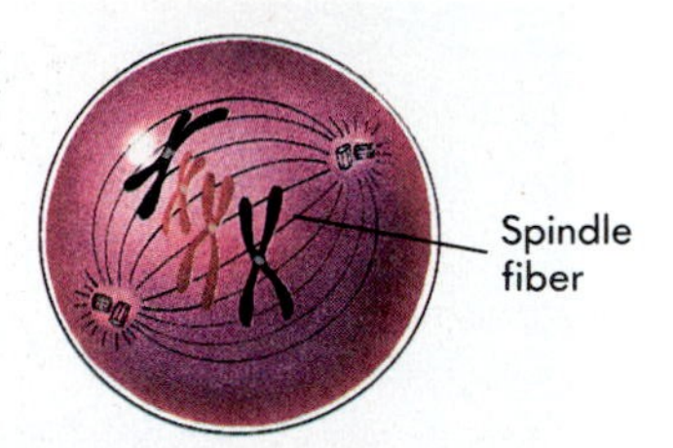

Anaphase: the chromosomes, assisted by the spindle fibers, move toward the centrioles at each end of the cell.

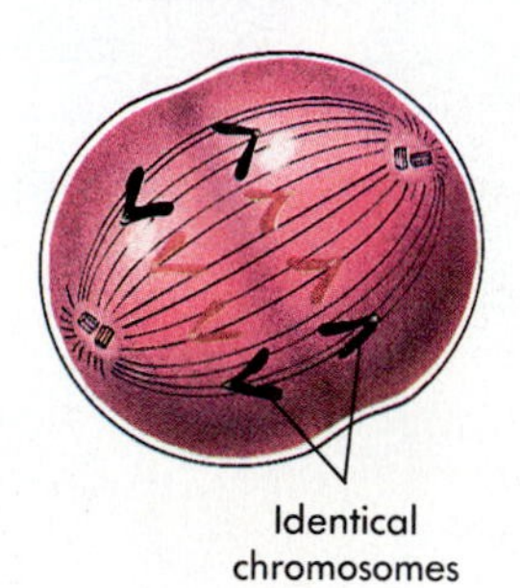

Telophase: the nuclear membrane reforms and the cytoplasm begins to divide into two cells.

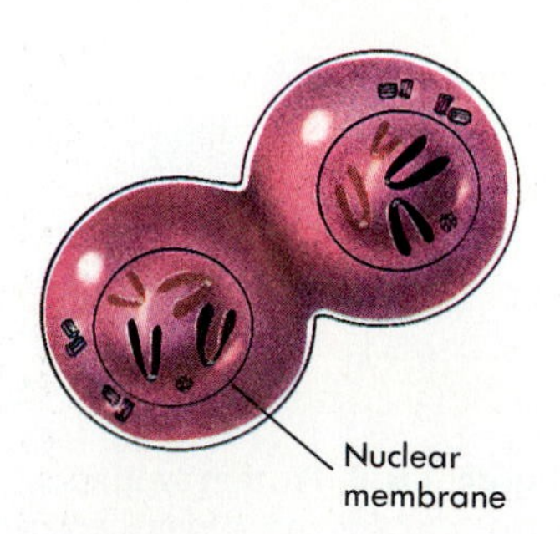

Mitosis is complete and a new interphase begins. The chromosomes have unraveled to become chromatin. Cell division has produced two daughter cells.

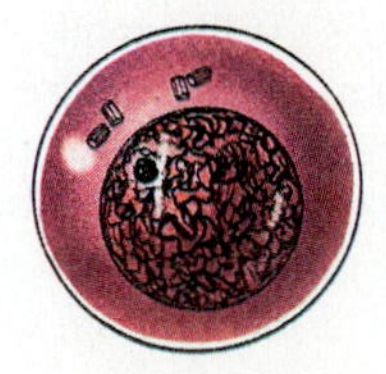
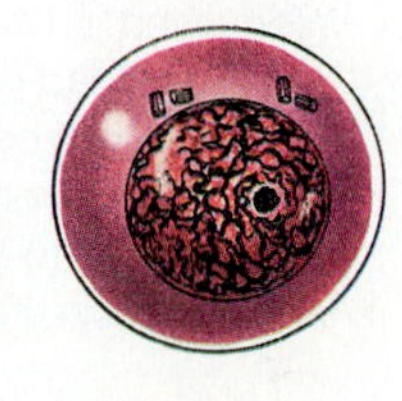

Figure 3-9 ● Cell division.

identical. Bone cells, for example, do not look like or function as fat cells or red blood cells.

The process by which cells develop specialized structures and functions is called **differentiation.** As cell numbers increase during development, cells become different from each other and give rise to the various cell types. As a result of differentiation, some portions of a cell's DNA are active while other portions of its DNA are inactive. The active and inactive sections of DNA differ with each cell type. The portion of DNA that is responsible for the structure and function of a bone cell is a different portion of DNA from that responsible for the structure and function of a fat cell. We do not fully understand the mechanisms that determine which portions of DNA are active in any one cell type, but it is this differentiation that produces the many cell types that function together to make a person.

SUMMARY

CELL STRUCTURE AND FUNCTION

Cells are highly organized units composed of living material. The nucleus contains genetic material, and cytoplasm is living material outside the nucleus.

Plasma membrane

The plasma membrane forms the outer boundary of the cell. It determines what enters and leaves the cell.

The plasma membrane is composed of a double layer of phospholipid molecules with protein molecules. The proteins function as membrane channels, carrier molecules, receptor molecules, enzymes, and structural components of the membrane.

Nucleus

The nucleus is enclosed by two separate membranes with nuclear pores.

DNA inside the nucleus is the genetic material of the cell and controls the activities of the cell.

The nucleolus is the site of production of ribosomal subunits.

Cytoplasm

Ribosomes are the site of protein synthesis.

Rough endoplasmic reticulum has attached ribosomes and is a major site of protein synthesis.

Smooth endoplasmic reticulum does not have ribosomes attached and is a major site of lipid synthesis.

The Golgi apparatus is a series of closely packed membrane sacs that function to concentrate and package into secretory vesicles the lipids and proteins produced by the endoplasmic reticulum.

Secretory vesicles are membrane-bound sacs that carry substances from the Golgi apparatuses to the plasma membrane, where the vesicle contents are released.

Membrane-bound sacs containing enzymes are called lysosomes. Within the cell the lysosomes break down ingested material.

Mitochondria are the major sites of ATP production.

Cilia move substances over the surface of cells, flagella move sperm cells, and microvilli increase the surface area of cells, which aids in absorption.

Organelle interactions

The interactions between organelles must be considered to fully understand cell function.

MOVEMENT OF SUBSTANCES ACROSS THE PLASMA MEMBRANE

The plasma membrane is selectively permeable, allowing some substances to pass through it but not others.

Passive transport processes

Diffusion is the movement of a substance from an area of higher concentration to an area of lower concentration.

Ions and small molecules can diffuse through membrane channels. Water, carbon dioxide, oxygen, urea, and glucose can bypass the membrane channels.

Osmosis is the diffusion of water across a selectively permeable membrane.

Filtration is the passage of a fluid through a membrane as a result of the force or weight of the fluid pushing against the membrane. Some of the substances in the fluid do not pass through the membrane.

Active transport processes

Active transport requires ATP and uses a carrier molecule to move substances from a lower to a higher concentration.

Phagocytosis is the movement of solid material into cells by the formation of a vesicle.

Pinocytosis is similar to phagocytosis, except that the material ingested is a liquid.

Exocytosis is the secretion of materials from cells by vesicle formation.

PROTEIN SYNTHESIS

Cell activity is regulated by enzymes (proteins), and DNA controls protein production.

A gene is a sequence of nucleotides in DNA.

Transcription is a copy (mRNA) of the information in a gene. Translation uses the information in the copy (mRNA) to produce proteins at the ribosomes.

CELL DIVISION

Cell division occurs by mitosis in all tissues except those that produce reproductive cells; mitosis produces new cells for growth and tissue repair.

DNA replicates during interphase, the time between cell division.

Mitosis results in the separation of the duplicated DNA into two cells.

DIFFERENTIATION

Differentiation, the process by which cells develop specialized structures and functions, results from the selective activation and inactivation of DNA.

CONTENT REVIEW

1. Define cytoplasm and cell organelle.
2. Describe the structure of the plasma membrane. What functions does it perform?
3. Describe the structure of the nucleus and nuclear pores. What is the function of the DNA and nucleolus found within the nucleus?
4. What is the function of ribosomes?
5. What is endoplasmic reticulum? Compare the structure and functions of rough and smooth endoplasmic reticulum.
6. Describe the Golgi apparatus, and state its function.
7. Where are secretory vesicles produced? What are their contents, and how are they released?
8. What is the function of lysosomes?
9. Describe the structure and function of mitochondria.
10. Describe the structure and function of cilia, flagella, and microvilli.
11. Distinguish between passive and active transport processes. List examples for each type of transport process.
12. Define diffusion. How does concentration difference affect how fast a substance diffuses? Under what conditions does diffusion stop?
13. In what ways can a substance diffuse through the plasma membrane?
14. Define osmosis. Why is osmosis important to cell function?
15. Define filtration. Give an example of filtration in the body.
16. Define active transport. Describe the use of ATP and carrier molecules in active transport.
17. Describe phagocytosis, pinocytosis, and exocytosis. What do they accomplish?
18. Describe how proteins are synthesized and how DNA determines the structure of proteins.
19. Describe what happens during interphase and mitosis. What kind of cells undergo mitosis?
20. Even though cells of the body have the same DNA, the cells are structurally and functionally different. How is this possible?

CONCEPT REVIEW

1. Suppose that a cell has the following characteristics:
   - many mitochondria
   - well-developed rough endoplasmic reticulum
   - well-developed Golgi apparatuses
   - numerous vesicles
   - many microvilli at the cell surface

   Predict the major function of the cell. Explain how each characteristic supports your prediction.
2. Secretory vesicles fuse with the plasma membrane to release their contents to the outside of the cell. In this process the membrane of the secretory vesicle becomes part of the plasma membrane. Because small pieces of membrane are continually added to the plasma membrane, one would expect the plasma membrane to become larger and larger as secretion continues. The plasma membrane, however, stays the same size. Explain how the plasma membrane stays the same size.
3. A man's body was found floating in the salt water of San Francisco Bay, which is about the same concentration as body fluids. When seen during an autopsy, the cells in his lung tissues were clearly swollen. Choose the most logical conclusion.

   A. He probably drowned in the bay.

   B. He may have been murdered elsewhere.

   C. He did not drown.
4. Patients with kidney failure can be kept alive by a dialysis machine, which removes toxic waste products from the blood. In a dialysis machine, blood flows past one side of a dialysis membrane and dialysis fluid flows on the other side of the membrane. The dialysis membrane is selectively permeable. Small molecules such as ions, glucose, and urea, can pass through the dialysis membrane but larger molecules such as proteins cannot. If you wanted to use a dialysis machine to remove from blood only the toxic waste product urea, what could you use for the dialysis fluid?

   A. A solution that is isotonic and contains only protein.

   B. A solution that is isotonic and contains the same concentration of substances as blood, except for having no urea in it.

   C. Distilled water.

   D. Blood.
5. In sickle cell anemia a protein inside red blood cells does not function normally. Consequently, the red blood cells become sickle shaped and plug up small blood vessels. It is known that sickle cell anemia is hereditary and results from changing one nucleotide for a different nucleotide within the gene that is responsible for producing the protein. Explain how this change results in an abnormally functioning protein.

CHAPTER TEST

Answers can be found in Appendix C

MATCHING 1 For each statement in column A select the correct answer in column B (an answer may be used once, more than once, or not at all).

A	B
1. Outer boundary of the cell.	cilia
2. Substances outside of cells.	extracellular
3. Contains the cell's genetic material (DNA).	Golgi apparatus
4. Site of protein synthesis.	intracellular
5. Site of lipid synthesis.	lysosome
6. Packages proteins and lipids for secretion.	microvilli
7. Contains intracellular digestive enzymes.	mitochondria
8. Site of ATP production.	nucleus
9. Move materials along the surface of cells.	plasma membrane
10. Increase cell surface area.	rough endoplasmic reticulum
	smooth endoplasmic reticulum

MATCHING 2 For each statement in column A select the correct answer in column B (an answer may be used once, more than once, or not at all).

A	B
1. Nucleotides in DNA that contain instructions for making a protein.	chromatin
2. The process of making a copy of the information in DNA.	chromosome
3. Carries copied information from DNA to ribosomes.	gene
4. Carries amino acids to ribosomes.	interphase
5. Process by which cells are produced for growth and repair.	mitosis
6. Time during which DNA replicates.	mRNA
7. DNA dispersed throughout the nucleus as thin threads.	transcription
8. DNA coiled to form dense bodies.	translation
	tRNA

FILL-IN-THE BLANK Complete each statement by providing the missing word or words.

1. A membrane that allows some substances to pass through it but not others is said to be __________.
2. The three passive transport processes important to cells are __________, __________, and __________.
3. Substances such as sodium and potassium ions enter or leave cells by diffusing through __________.
4. Osmosis is the diffusion of __________ through a selectively permeable membrane.
5. The movement of liquid through a partition due to the force or weight of the liquid pushing against the partition is called __________.
6. The process requiring ATP that uses a carrier molecule to move substances from lower to higher concentrations is called __________.
7. __________ is the ingestion of solid particles by the formation of a vesicle.
8. The movement of materials out of cells by the formation of a vesicle is called __________.

CHAPTER

4

Tissues, Glands, and Membranes

KEY TERMS

connective tissue
Major tissue type consisting of cells surrounded by large amounts of extracellular material; functions to hold other tissues together and provides a supporting framework for the body.

epithelial tissue
Major tissue type consisting of cells with a basement membrane, little extracellular material, and no blood vessels; covers the free surfaces of the body and forms glands.

gland
A single cell or a multicellular structure that secretes substances into the blood, into a cavity, or onto a surface.

histology
Scientific discipline that studies the structure of cells, tissues, and organs in relation to their function.

mucous membrane
Thin sheet consisting of epithelium and connective tissue that lines cavities that open to the outside of the body; many contain mucous glands, which secrete mucus.

muscle tissue
Major tissue type consisting of cells with the ability to contract; includes skeletal, cardiac, and smooth muscle.

nervous tissue
Major tissue type consisting of neurons, which have the ability to conduct action potentials,
and neuroglia, which are support cells.

serous membrane
Thin sheet consisting of epithelial and connective tissue that lines cavities not opening to the outside of the body; does not contain glands but does secrete serous fluid.

tissue
A collection of cells with similar structure and function, and the substances between the cells.

OBJECTIVES

After reading this chapter you should be able to:

1. List the characteristics of epithelial tissue.
2. Classify and give an example of the major types of epithelium.
3. Explain the functional significance for epithelium of the following: cell layers and cell shapes.
4. Define and categorize glands.
5. Describe the basis for classifying connective tissue and give examples of each major type.
6. Name the three types of muscle tissue and list their functions.
7. State the functions of nervous tissue and describe a neuron.
8. List the structural and functional characteristics of mucous membranes and serous membranes.

FEATURES

A TISSUE is a group of cells with similar structure and function and the material between the cells. **Histology** (his-tol´o-je) is the study of tissues and their function. Knowledge of tissue structure and function is necessary to understand how individual cells are organized in tissues, and how tissues are organized to form organs, organ systems, and the complete organism. There is a relationship between the structure of each tissue type and its function, and between the tissues in an organ and the organ's function.

There are four major tissue types: epithelial, connective, muscular, and nervous tissue. This chapter describes these tissue types and the basic characteristics of glands and membranes. Glands and membranes are composed mainly of epithelial and connective tissue, and they are common components of many of the organ systems described in later chapters.

EPITHELIAL TISSUE

Epithelium (ep-ĭ-the´le-um; plural: epithelia) covers surfaces of the body or forms glands. Surfaces of the body include the skin and the lining of cavities such as the digestive tract and blood vessels. Epithelium consists almost entirely of cells with little extracellular material between them. Most epithelia have a **free surface,** which is not in contact with other cells, and the epithelial cells usually rest on a **basement membrane,** which attaches the epithelial cells to underlying tissues (Figure 4-1). Epithelium does not have blood vessels, so gases and nutrients that reach the epithelium must diffuse across the basement membrane from underlying tissues.

Classification

Epithelia are named according to the number of cell layers and the shape of the cells in the tissue. **Simple epithelium** consists of a single layer of cells, and **stratified epithelium** consists of more than one layer of epithelial cells with some epithelial cells sitting on top of other epithelial cells. Categories of epithelium based on cell shape are **squamous** (skwa´mus; flat), **cuboidal** (cubelike), and **columnar** (tall and thin). In most cases, an epithelium is given two names such as simple squamous, simple columnar, or stratified squamous.

Simple squamous epithelium is a single layer of thin, flat cells (Figure 4-2, *A*). Because substances easily pass through this thin layer of tissue, it often is found where diffusion takes place. For example, the respiratory passages end as small sacs called alveoli. The alveoli consist of simple squamous epithelium that allows oxygen from the air to diffuse into the body, and for carbon dioxide to diffuse out of the body into the air.

Simple cuboidal epithelium is a single layer of cubelike cells (Figure 4-2, *B*) and **simple columnar epithelium** is a single layer of cells that are tall and thin (Figure 4-2, *C*). These cells have a greater volume than simple squamous cells and contain more cell organelles, which enable them to perform complex functions. In kidney tubules, for example, simple cuboidal epithelium excretes waste products from the body into the urine, and reabsorbs useful materials from the urine back into the body. The simple columnar epithelium of the small intestine produces and releases digestive enzymes that help complete the process of digesting food. The columnar cells then absorb the digested foods.

Text continued on p. 48.

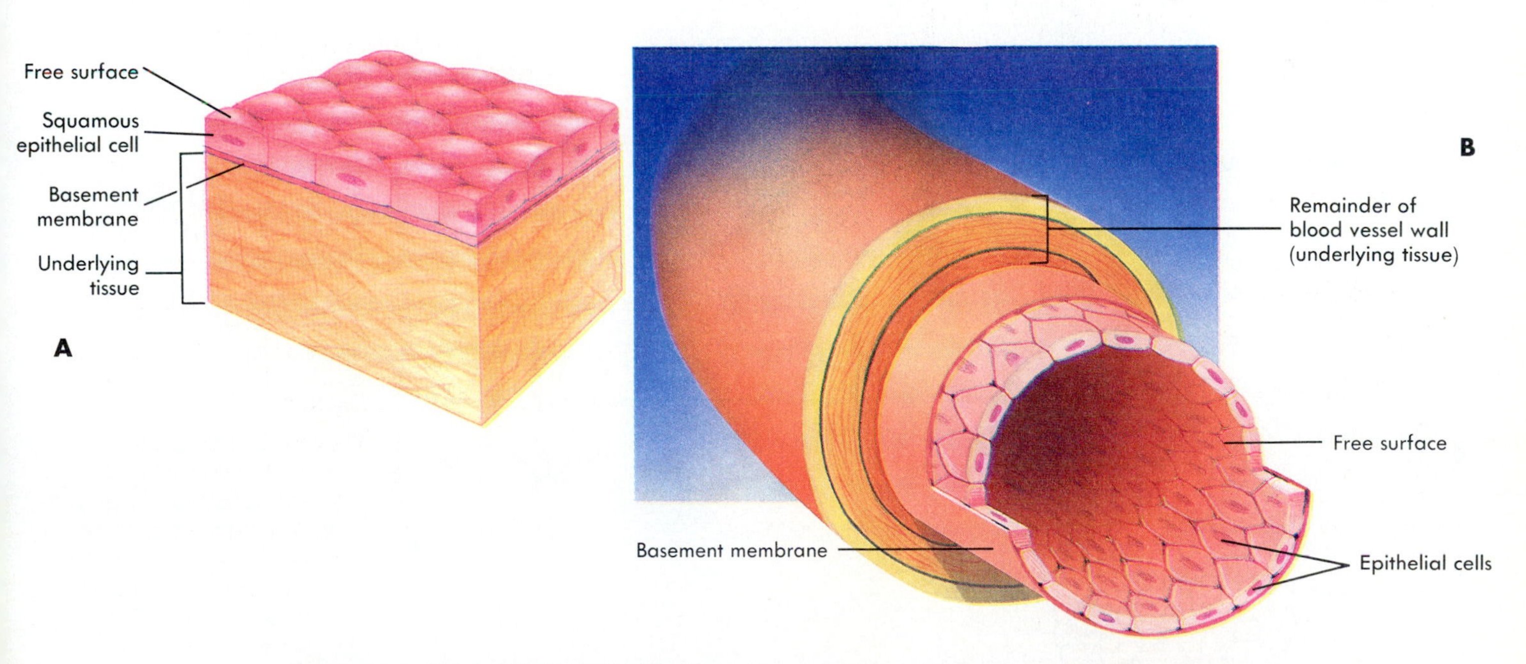

Figure 4-1 ● Characteristics of epithelium.
A, Epithelium has the following characteristics: little extracellular material between cells, a free surface, a basement membrane, and no blood vessels directly supplying the epithelium. **B,** Epithelial lining of a blood vessel.

Figure 4-2 • Epithelial tissue

	Tissue type and location		Structure/Function
A	**Simple squamous** Lining of blood vessels, heart, lymph vessels, and serous membranes, alveoli (air sacs) of lungs	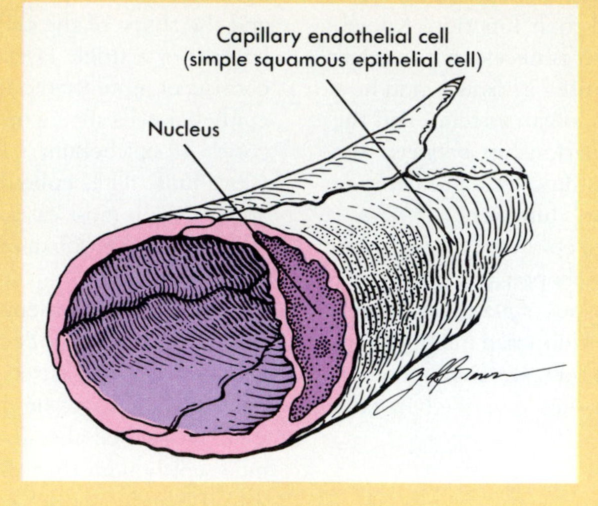 	**Structure** Single layer of thin, flat cells **Function** Diffusion, filtration, and protection against friction (secrete serous fluid)
B	**Simple cuboidal** Kidney tubules, many glands and their ducts , terminal bronchioles (air tubes) of lungs	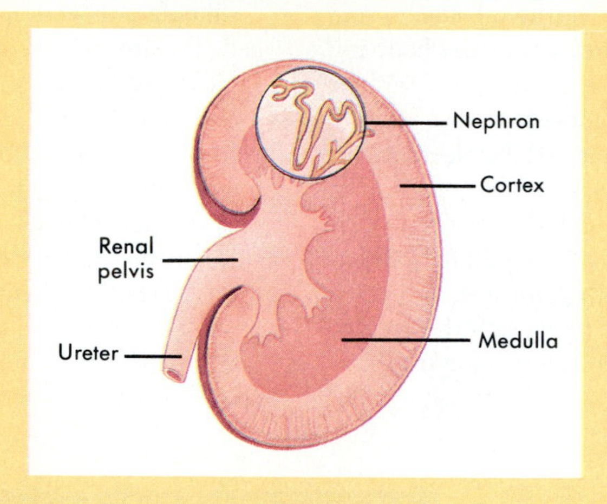 	**Structure** Single layer of cube-shaped cells; some cells have microvilli (kidney tubules) or cilia (terminal bronchioles of lungs) **Function** Secretion and absorption by cells of the kidney tubules; secretion by glands; movement of mucus-containing particles out of the terminal bronchioles by ciliated cells
C	**Simple columnar** Stomach, intestines, glands, some ducts of glands, bronchioles of lungs, uterus, and uterine tubes 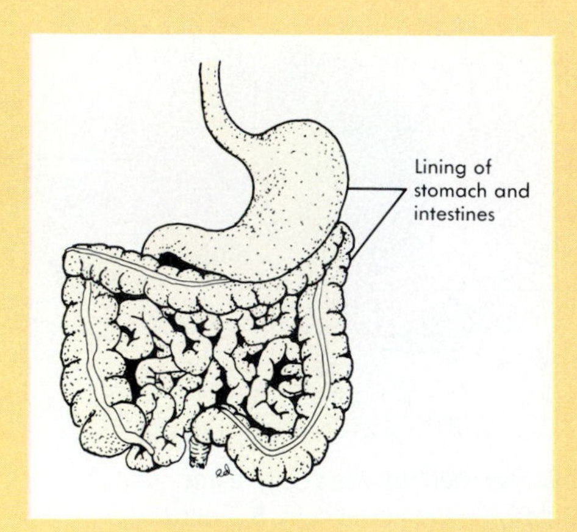 	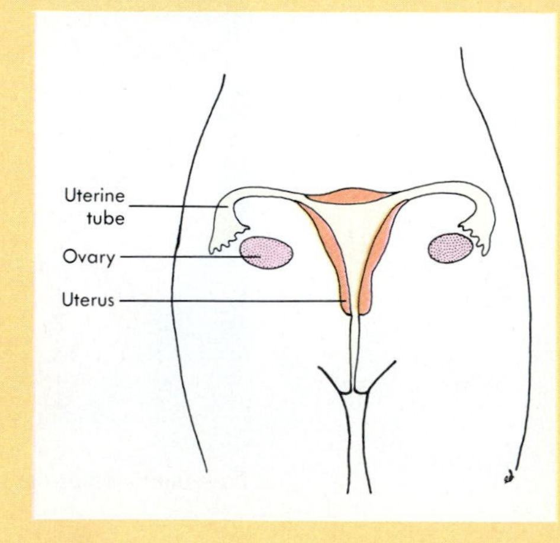 	**Structure** Single layer of tall, narrow cells; some have microvilli (stomach, intestines, and glands) or cilia (bronchioles of lungs, uterus, and uterine tubes) **Function** Secretion by cells of the stomach, intestines, and glands; absorption by cells of the intestine; movement of mucus by ciliated cells clears the lungs and is partially responsible for the movement of the ovum through the uterine tubes

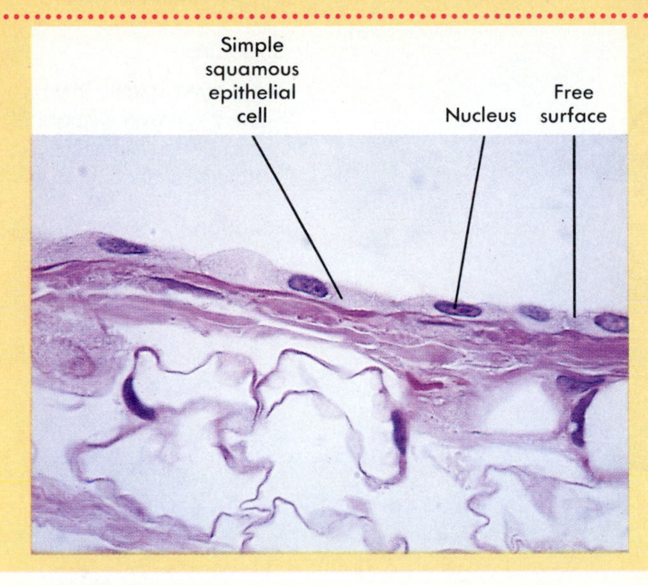
Simple squamous epithelial cell
Nucleus
Free surface

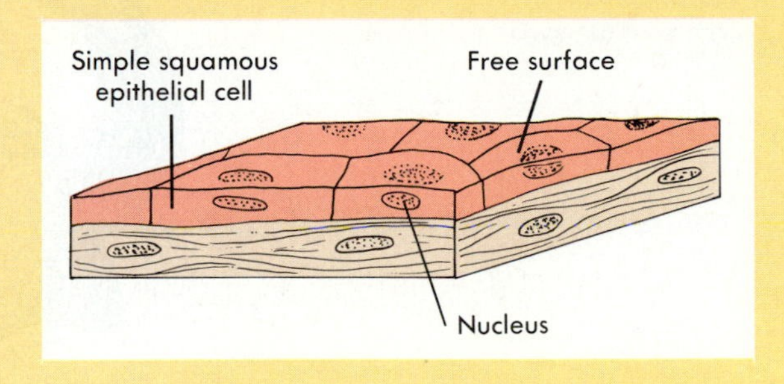
Simple squamous epithelial cell
Free surface
Nucleus

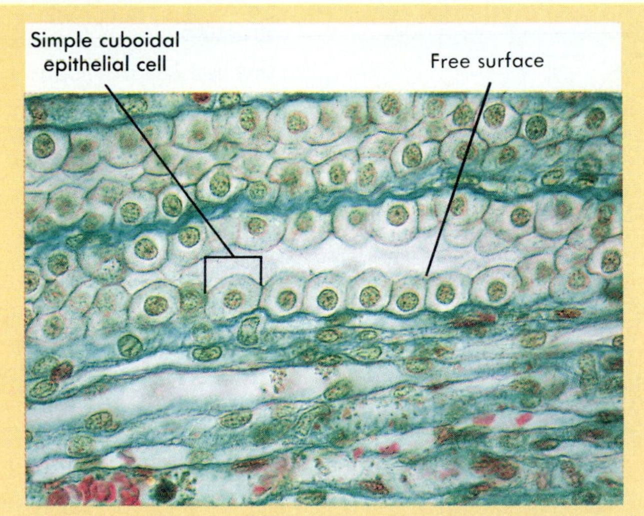
Simple cuboidal epithelial cell
Free surface

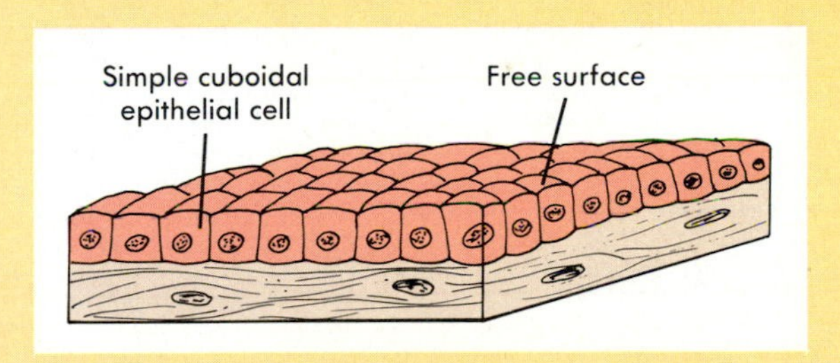
Simple cuboidal epithelial cell
Free surface

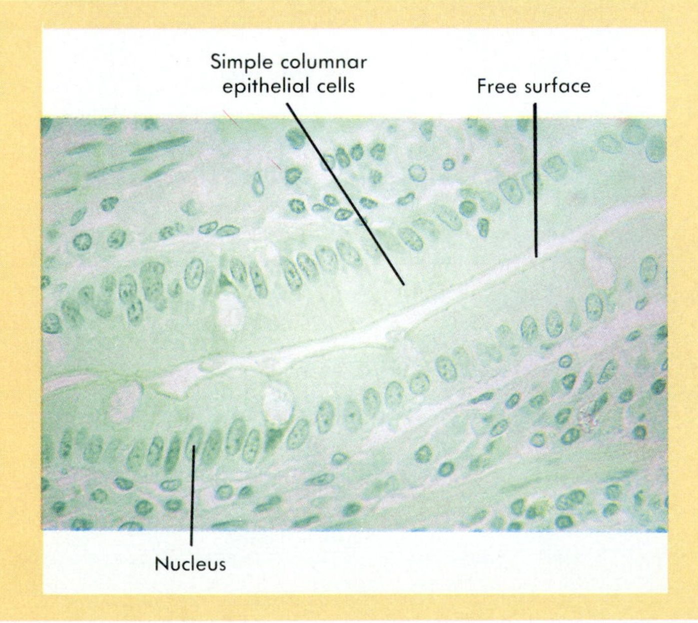
Simple columnar epithelial cells
Free surface
Nucleus

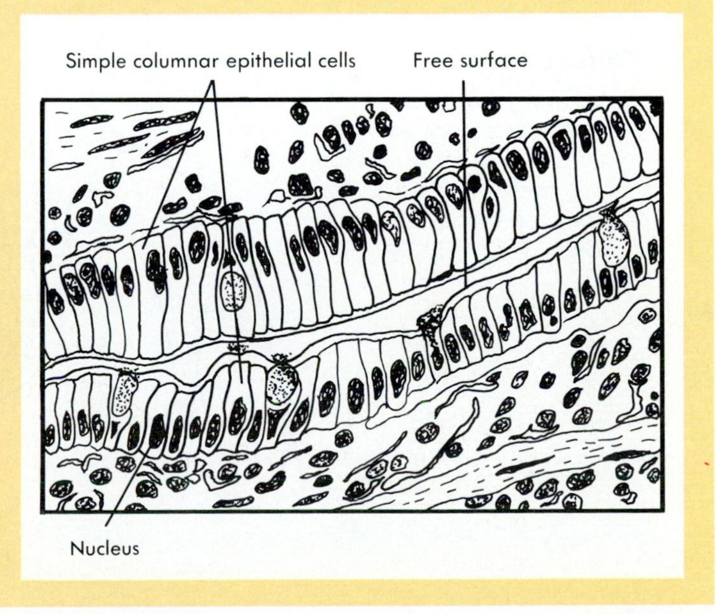
Simple columnar epithelial cells
Free surface
Nucleus

Continued.

Figure 4-2 ● Epithelial tissue—cont'd

	Tissue type and location		Structure/Function
D	**Stratified squamous** Skin, mouth, throat, esophagus, anus, vagina, and cornea of the eye		**Structure** Many layers of cells in which the basal layer is cuboidal and becomes flattened at the free surface **Function** Protect against abrasion and infection
E	**Transitional** Urinary bladder, ureters, and part of urethra	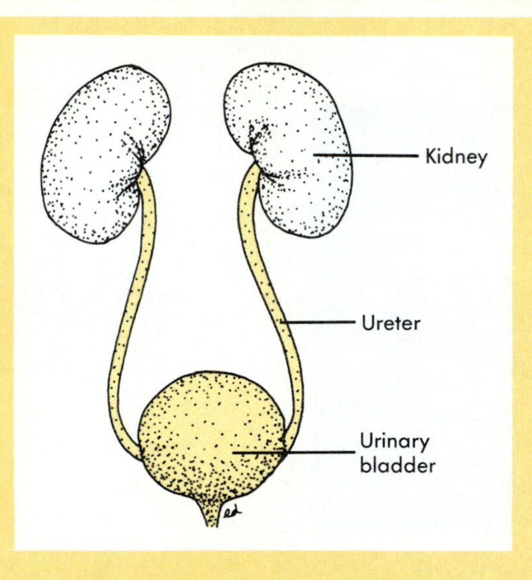	**Structure** Stratified cells that appear cubelike when the organ or tube is relaxed and appear squamous when the organ or tube is distended by fluid. **Function** Accommodates changes in the volume of an orgaan or tube; protection against the caustic effects of urine.
F	**Ciliated columnar** Nasal cavity, nasal sinuses, auditory tubes, pharynx (throat), larynx (voice box), trachea (wind pipe), and bronchi of lungs	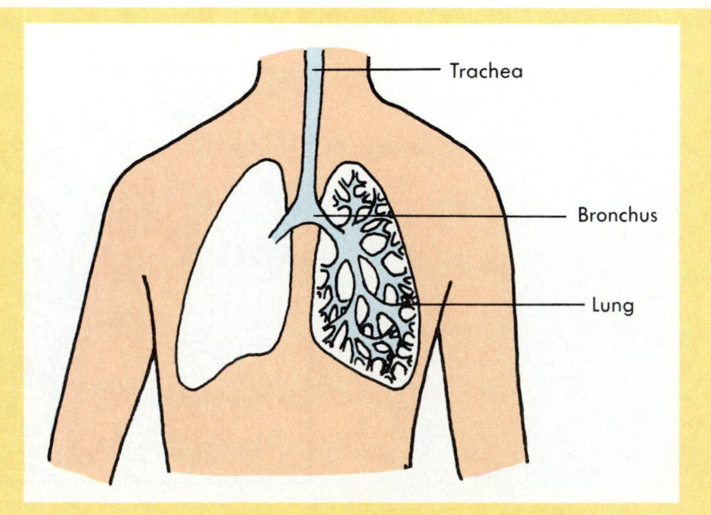	**Structure** Single layer of cells. Some cells are tall and thin and reach to the free surface, and others do not. The cells are almost always ciliated and are associated with goblet cells (single-celled glands) that produce mucus **Function** Movement of mucus (or fluid) that contains foreign particles

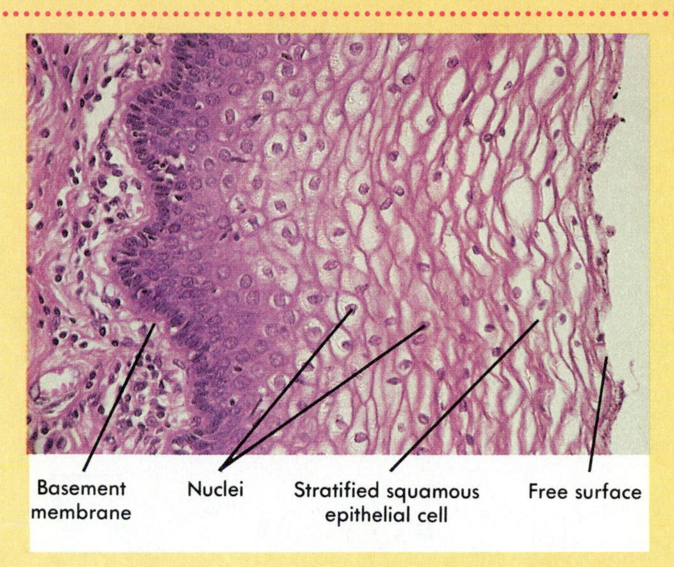
Basement membrane
Nuclei
Stratified squamous epithelial cell
Free surface

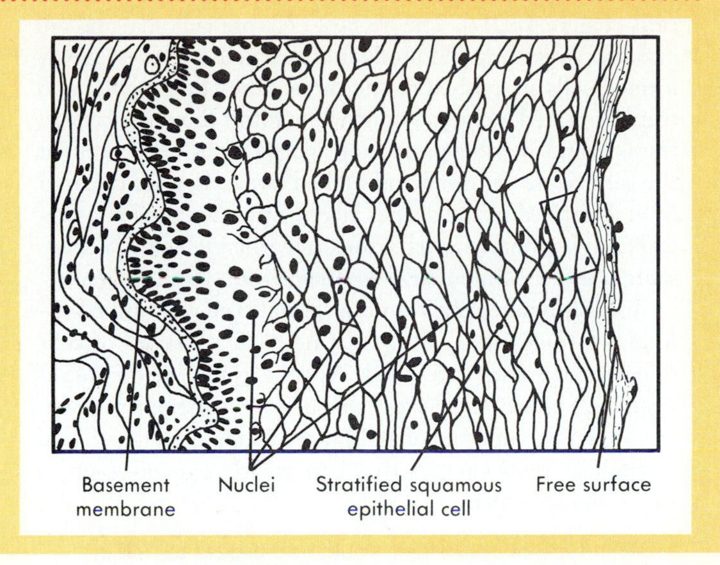
Basement membrane
Nuclei
Stratified squamous epithelial cell
Free surface

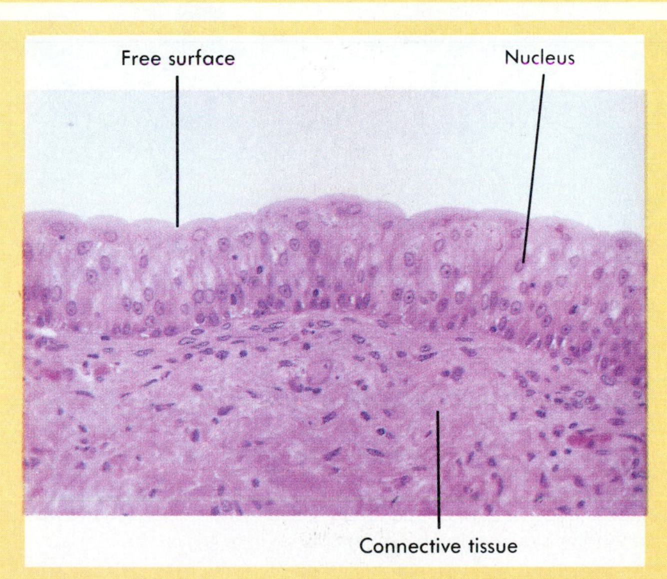
Free surface
Nucleus
Connective tissue

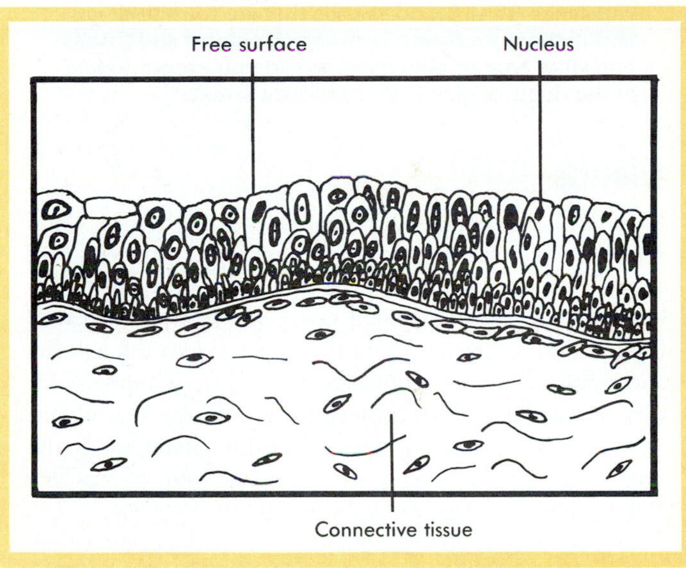
Free surface
Nucleus
Connective tissue

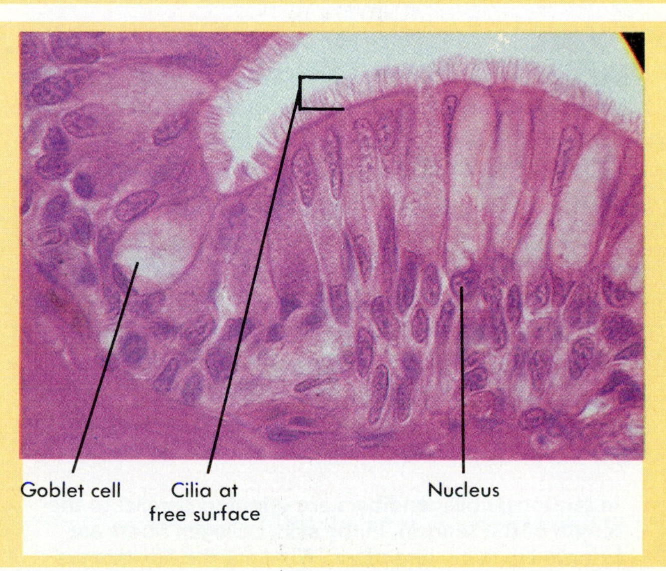
Goblet cell
Cilia at free surface
Nucleus

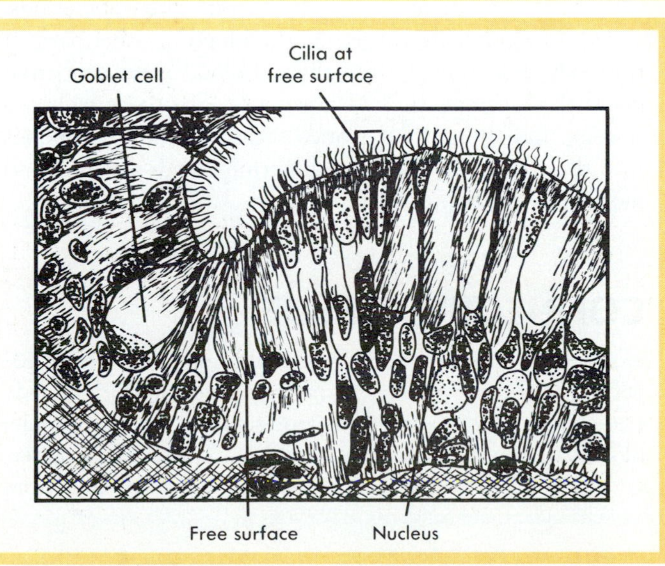
Goblet cell
Cilia at free surface
Free surface
Nucleus

Stratified squamous epithelium is thicker than simple squamous epithelium because it consists of many layers of cells (Figure 4-2, *D*). The deepest cells are cuboidal and are capable of dividing and producing new cells. As these newly formed cells are pushed to the surface they become flat and thin. Stratified squamous epithelium forms the outer layer of the skin and provides protection against abrasion. As cells at the surface are damaged and rubbed away, they are replaced by cells formed in the deeper layers. Stratified squamous epithelium is found in areas of the body where abrasion occurs, such as in the skin, mouth, esophagus, anus, and vagina.

Transitional epithelium is a special type of stratified epithelium (Figure 4-2, *E*) consisting of many layers of cells that can be greatly stretched. As transitional epithelium is stretched, the cells change shape from cuboidal to squamous, and the number of cell layers decreases. Transitional epithelium lines cavities that can greatly expand, such as the urinary bladder.

1 What type of epithelium would you expect to find in capillaries (small blood vessels that are the site of gas and nutrient exchange between the blood and tissues) and what type of epithelium would you expect to find in the ducts of glands that produce sweat?*

Glands

Most glands are composed primarily of epithelium. A **gland** is a single cell or a multicellular structure that secretes substances onto a surface, into a cavity, or into the blood. Glands with ducts are called **exocrine** (ek´so-krin) **glands** (Figure 4-3). Secretions from these glands pass through the ducts onto a surface or into an organ. For example, sweat from sweat glands flows through ducts onto the skin surface and secretions from the pancreas flow through ducts into the small intestine. See Figure 4-3 for some types of exocrine glands, their location, and secretions.

Endocrine (en´do-krin) **glands** have no ducts and empty their secretions directly into the blood. These secretions, called hormones, are carried by the blood to other parts of the body. For example, the thyroid gland is an endocrine gland that secretes thyroid hormones into the circulatory system, and the adrenal glands secrete adrenal hormones into the circulatory system. Endocrine glands are discussed more fully in Chapter 10.

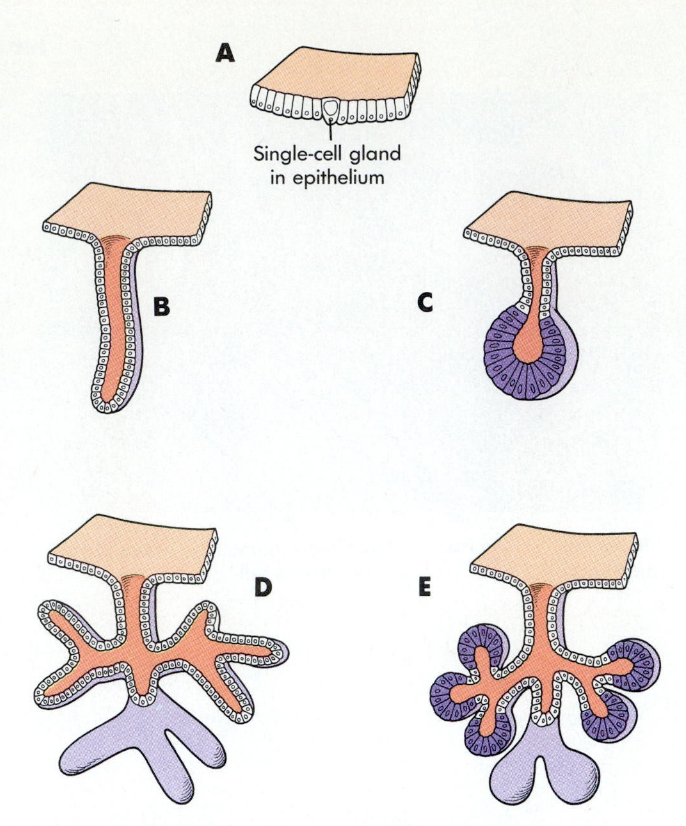

Figure 4-3 ● Types of exocrine glands.
A, Goblet cell. Single-cell gland that secretes mucus. Found in the epithelial lining of the lungs and intestine. **B,** Simple straight tubular glands. Found in the wall of the stomach and intestine. They secrete gastric acid in the stomach and digestive enzymes and mucus in the intestine. **C,** Simple acinar or alveolar (ending in a saclike structure). Found in the sebaceous glands of hair follicles. They secrete an oily substance that helps lubricate the skin. **D,** Compound tubular. Found in the wall of the intestine and secrete mucus. **E,** Compound acinar or alveolar. Found in the pancreas where they secrete enzymes, in the mammary glands where they secrete milk, and in the salivary glands where they secrete saliva.

CONNECTIVE TISSUE

Connective tissue functions to hold cells and tissues together. It also provides a supporting framework for the body (for example, bone) and transports substances (for example, blood). Connective tissue has extensive extracellular material often called **matrix** (ma´triks) that separates cells from each other. The structure of the extracellular matrix determines the functional characteristics of the connective tissue.

Dense connective tissue has an extracellular matrix consisting mostly of collagen fibers (Figure 4-4, A, on p. 50). **Collagen** (kol´lă-jen), which resembles microscopic ropes, is the most common protein in the body. It is flexible but resists stretching. The few cells found in dense connective tissue are **fibroblasts** (fi´bro-blastz), which are responsible for the production of the collagen fibers. Structures made up of dense connective tissue include tendons, which attach muscle to bones; ligaments, which attach bones to other bones; and the dermis, which is a layer of connective tissue that forms the inner layer of the skin.

2 In tendons, collagen fibers are oriented parallel to the length of the tendon. In the skin, collagen fibers are oriented in many directions. What are the advantages of the fiber arrangements in tendons and in the skin?

*Answers to predict questions appear in Appendix B at the back of the book.

Inflammation

The **inflammatory response,** or **inflammation,** occurs when tissues are damaged. It mobilizes the body's defenses, isolates and destroys microorganisms, and removes foreign materials and damaged cells so that tissue repair can proceed. Inflammation produces five major symptoms: redness, heat, swelling, pain, and disturbance of function. Although unpleasant, the processes producing the symptoms are usually beneficial.

Following an injury, chemical substances called **mediators of inflammation** are released from injured tissues and adjacent blood vessels. Mediators of inflammation include histamine (his´tă-mēn) and prostaglandins (pros´ta-glan´dinz). Some mediators cause dilation of blood vessels. This causes redness and heat, as occur when a person blushes. Dilation is beneficial because it increases the speed with which blood cells and other substances important for fighting infection and repairing the injury are brought to the injury site.

Mediators of inflammation also increase the permeability of blood vessels, allowing water, proteins, and blood cells to move out of the vessels and into the tissue where they can deal directly with the injury. **Edema** (e-de´mah), or swelling, of the tissues is a result. One type of blood cell that enters the tissues in large numbers is the **neutrophil** (nu´tro-fil). These cells take up and destroy bacteria and damaged tissue. Neutrophils are killed in this process and can accumulate as a mixture of dead cells and fluid, called **pus.**

Pain is produced in several ways. Nerve cell endings are stimulated by direct damage, by some of the chemical mediators of inflammation, and by increased pressure caused by edema and the accumulation of pus. Limitation of movement results from edema, pain, and tissue destruction, which all contribute to the disturbance of function. This disturbance of function can be adaptive because it warns the person to protect the injured area from further damage.

Sometimes the inflammatory response lasts longer or is more intense than is desirable, and drugs are used to suppress the symptoms by inhibiting the mediators of inflammation. For example, the effects of histamines released in hay fever are suppressed by antihistamines. Aspirin is an effective antiinflammatory agent and relieves pain by preventing the synthesis of prostaglandins.

Repair is not part of inflammation, but it immediately follows the inflammatory process and may begin while some symptoms of inflammation persist. **Tissue repair** is the substitution of viable cells for dead cells, and it can occur by regeneration or replacement. In **tissue regeneration,** the new cells are the same type as those that were destroyed, and normal function is usually restored. In **tissue replacement,** a new type of tissue develops that eventually causes scar production and the loss of some tissue function.

In contrast to dense connective tissue, the protein fibers in **loose** or **areolar** (ah-re´o-lar) **connective tissue** are widely separated from each other (Figure 4-4, *B*). Loose connective tissue is the "loose packing" material of the body, which fills the spaces between organs and holds them in place. It is found around glands, muscles, and nerves, and attaches the skin to underlying tissue.

Although **adipose** (ad´ĭ-pōs; fat) **tissue** has a matrix with protein fibers, it is not a typical connective tissue. There is very little matrix, and the adipose cells are large and closely packed together (Figure 4-4, *C*). Adipose cells are filled with lipids to store energy. Adipose tissue also pads and protects parts of the body and acts as a thermal insulator.

Cartilage (kar´tĭ-lij) is composed of cartilage cells, or **chondrocytes** (kon´dro-sītz), located in spaces called **lacunae** (la-ku´ne) within an extensive matrix (Figure 4-4, *D*). Collagen in the matrix gives cartilage strength, and water trapped in the matrix enables the cartilage to spring back after being compressed. Cartilage is relatively rigid and provides support, but if it is bent or slightly compressed it will resume its original shape. Cartilage heals slowly after an injury because blood vessels do not penetrate the cartilage. Thus cells and nutrients necessary for tissue repair do not easily reach the damaged area.

Bone is a hard connective tissue that consists of living cells embedded in a mineralized matrix (Figure 4-4, *E*). Bone cells, or **osteocytes** (os´te-o-sītz), are located within spaces in the matrix called lacunae. The strength and rigidity of the mineralized matrix enable bones to support and protect other tissues and organs of the body. Bone is considered in greater detail in Chapter 6.

Blood is unique because the matrix is liquid, enabling blood cells to move about freely (Figure 4-4, *F*). Some blood cells even leave the blood and wander into other tissues. The liquid matrix enables blood to flow rapidly through the body carrying food, oxygen, waste products, and other materials. Blood is discussed more fully in Chapter 11.

MUSCLE TISSUE

The main characteristic of **muscle tissue** is its ability to contract or shorten, making movements possible. Muscle contraction is due to contractile proteins inside muscle cells (see Chapter 7). The length of muscle cells is greater than their diameter. Because they often resemble tiny threads, muscle cells are called **muscle fibers.**

The three types of muscle are skeletal, cardiac, and smooth muscle. **Skeletal muscle** is what normally is thought of as "muscle" (Figure 4-5, *A*). It is the meat of animals and comprises about 40% of a person's body weight. Skeletal muscle, as the name implies, attaches to the skeleton and causes body movements. Skeletal muscle is normally under voluntary (conscious) control. Skeletal muscle cells tend to be long, cylindrical cells with several nuclei per cell. Some skeletal muscle cells extend the length of an entire muscle.

Text continued on p. 54.

Figure 4-4 • Connective tissue

	Tissue type and location		Structure/Function
A	**Dense connective** Tendons (attach muscle to bone), ligaments (attatch bone to bones), dermis of the skin, and capsules around organs	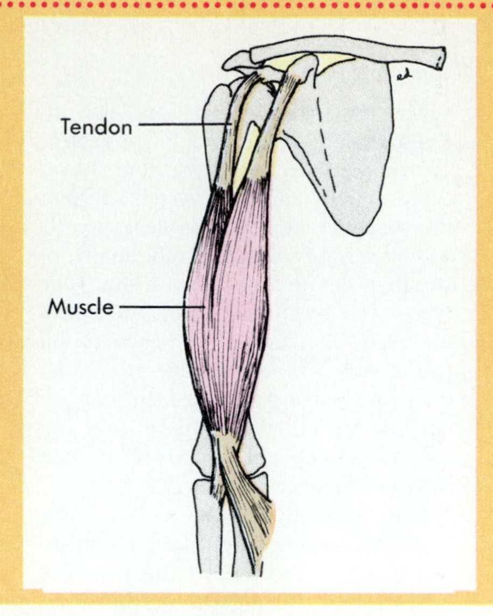	**Structure** Matrix consists almost entirely of collagen fibers produced by fibroblasts. The fibers can all be oriented in the same direction (tendons and ligaments), or in many different directions (dermis and capsules of organs) **Function** Able to withstand great pulling forces in the direction of fiber orientation
B	**Loose, or areolar** Widely distributed throughout the body, it is the substance on which most epithelial tissue rests. It is the packing between glands, muscles, and nerves and attaches the skin (dermis) to underlying tissues	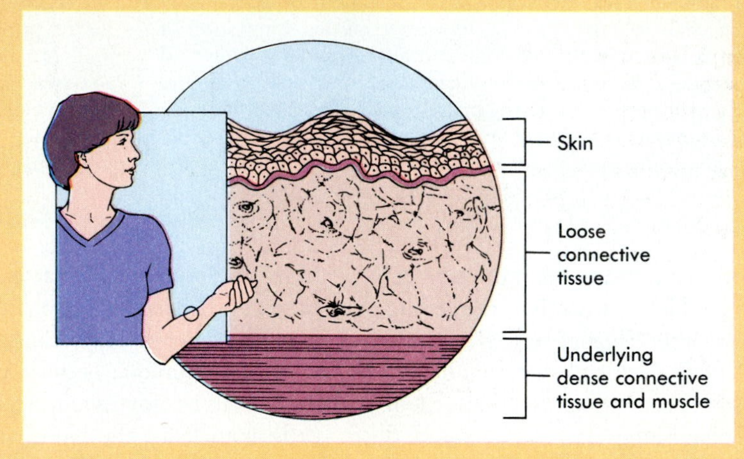	**Structure** Cells (fibroblasts, macrophages, and lymphocytes) within a fine network of mostly collagen fibers; the cells and fibers are separated from each other by fluid-filled spaces **Function** Loose packing, support, and nourishment for the structures with which it is associated
C	**Adipose tissue** Under skin, around organs such as the heart and kidneys, in the breasts, and in bones	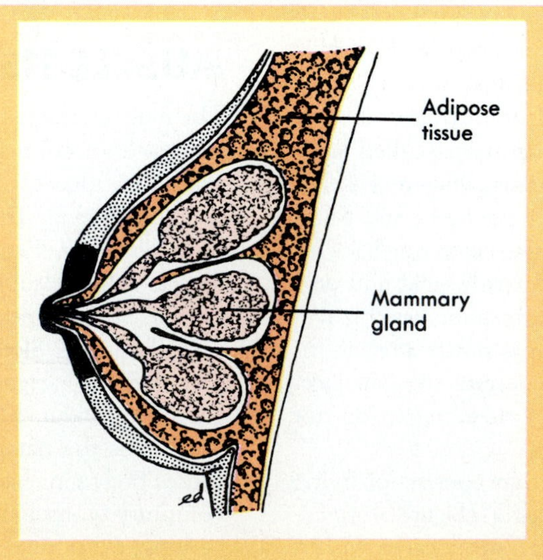	**Structure** Little extracellular material between adipose cells; the cells are so full of lipids that the cytoplasm is pushed to the periphery of the cell **Function** Energy storage, packing material that provides protection, and thermal insulator

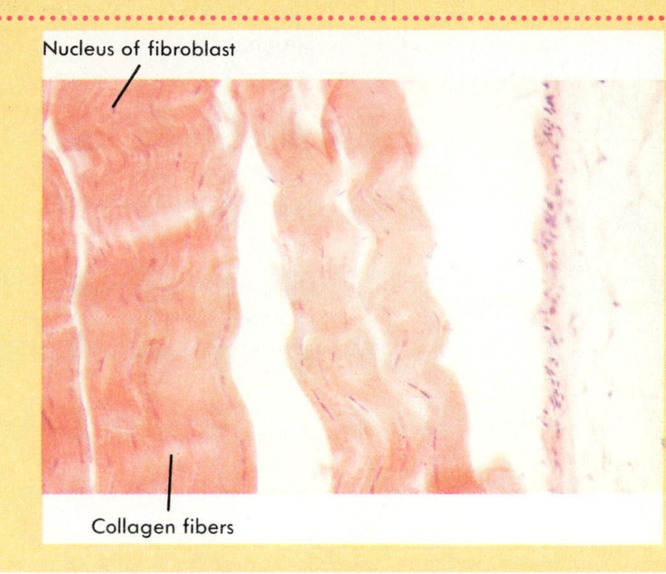
Nucleus of fibroblast
Collagen fibers

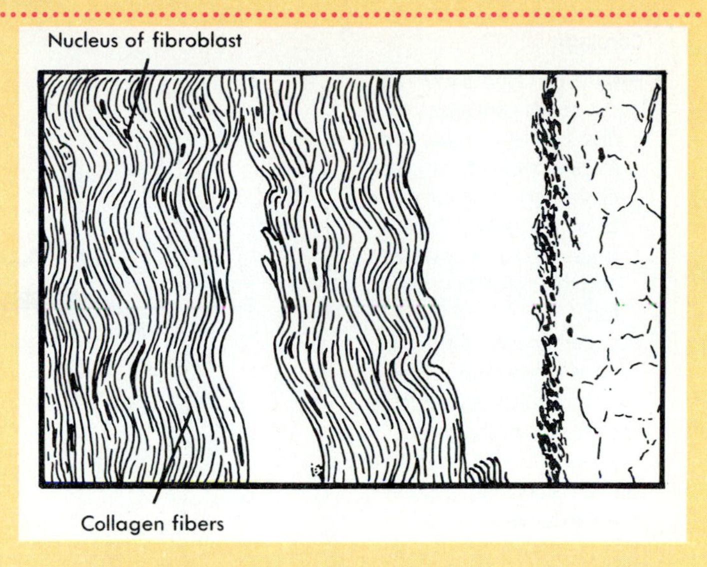
Nucleus of fibroblast
Collagen fibers

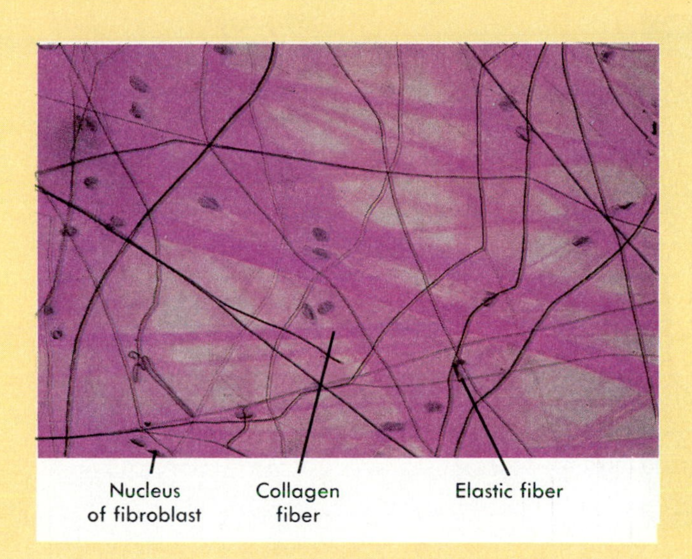
Nucleus of fibroblast
Collagen fiber
Elastic fiber

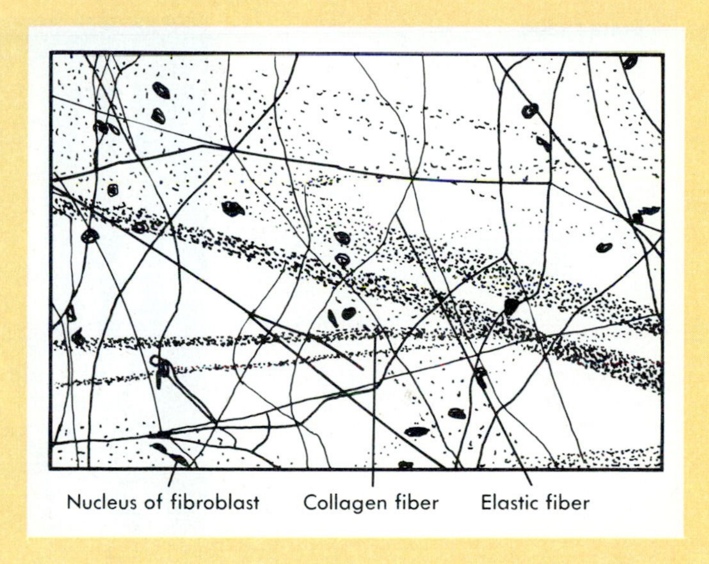
Nucleus of fibroblast
Collagen fiber
Elastic fiber

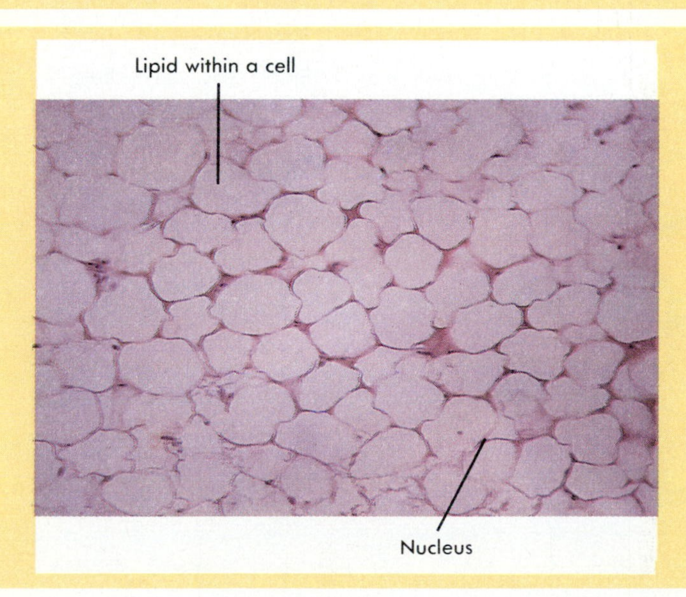
Lipid within a cell
Nucleus

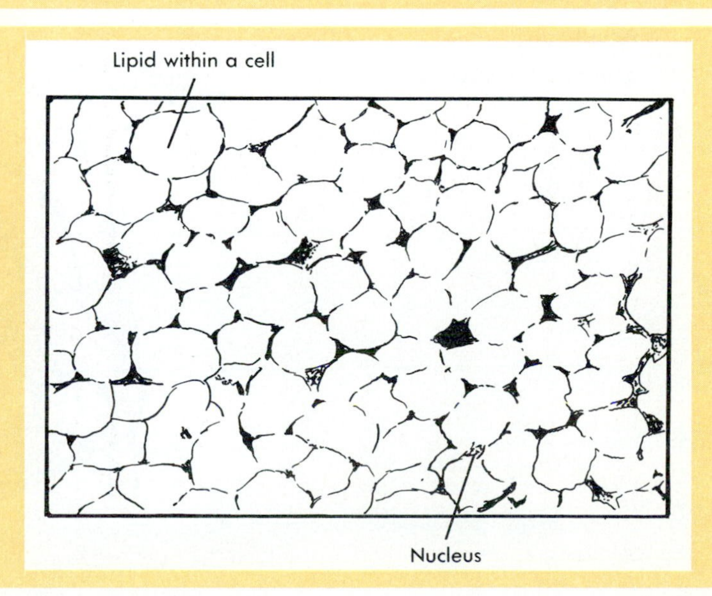
Lipid within a cell
Nucleus

Continued.

Figure 4-4 ● **Connective tissue—cont'd**

	Tissue type and location		Structure/Function
D	**Cartilage** Hyaline cartilage is found in the costal cartilages of ribs, cartilage rings of the respiratory tract, nasal cartilages, covering the ends of bones, growth (epiphyseal) plates of bones, and the embryonic skeleton. Fibrocartilage is found in intervertebral disks, symphysis pubis, and articular cartilages of the knee joints Elastic cartilage is found in the external ear	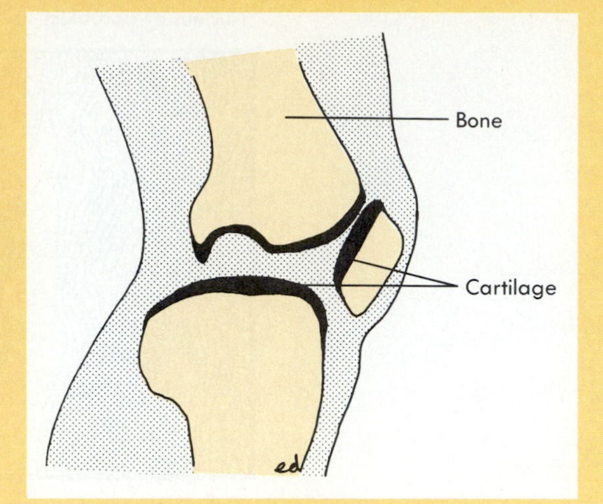	**Structure** Solid matrix with fibers dispersed throughout the ground substance. Cartilage cells are found within spaces called lacunae **Function** Hyaline cartilage forms a smooth surface in joints, is a site of bone growth, and forms most of the embryonic skeleton. Fibrocartilage can withstand great pressure, and elastic cartilage returns to its original shape when bent.
E	**Bone tissue** Bones	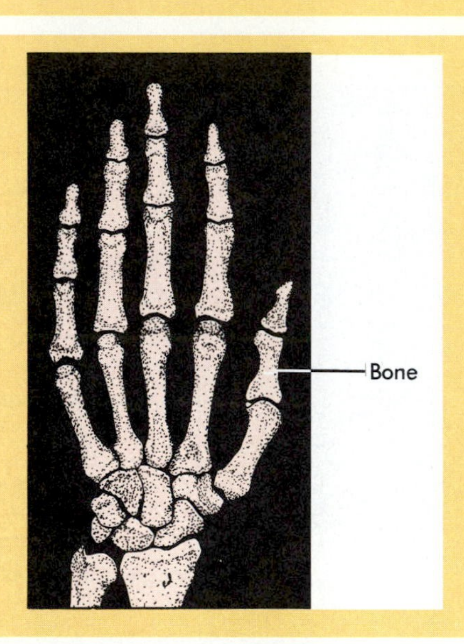	**Structure** Hard, mineralized matrix with osteocytes located within spaces called lacunae **Function** Provides great strength and support and protects internal organs such as the brain
F	**Blood** Within the blood vessels and heart	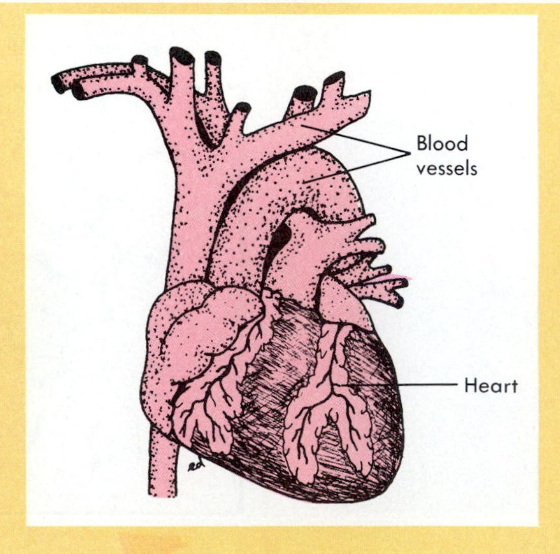	**Structure** Blood cells within a fluid matrix called plasma **Function** Transports oxygen, carbon dioxide, hormones, nutrients, waste products, and other substances; protects the body from infection and is involved in temperature regulation

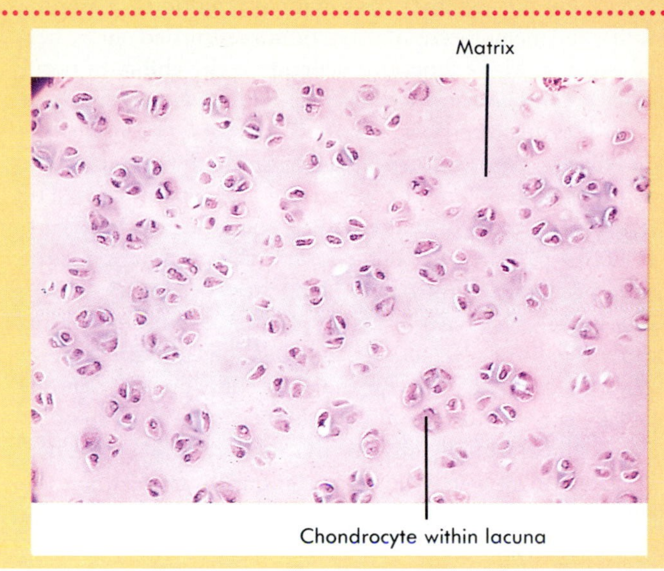
Matrix
Chondrocyte within lacuna

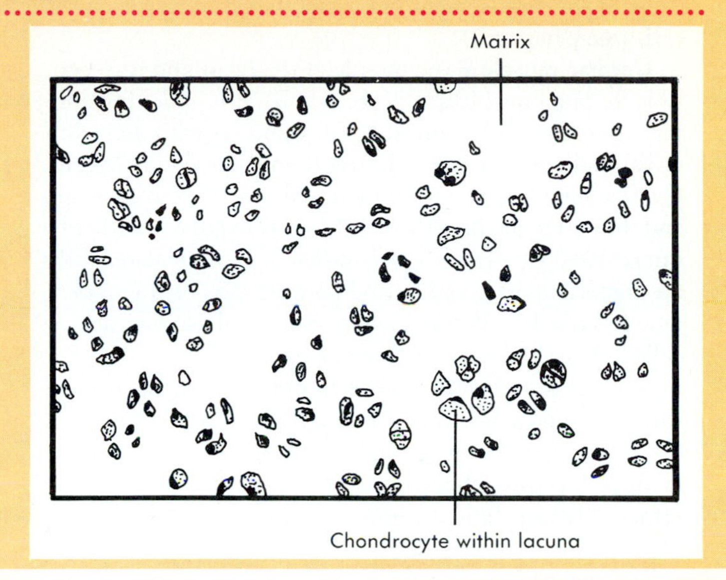
Matrix
Chondrocyte within lacuna

Osteocyte in a lacuna
Mineralized matrix

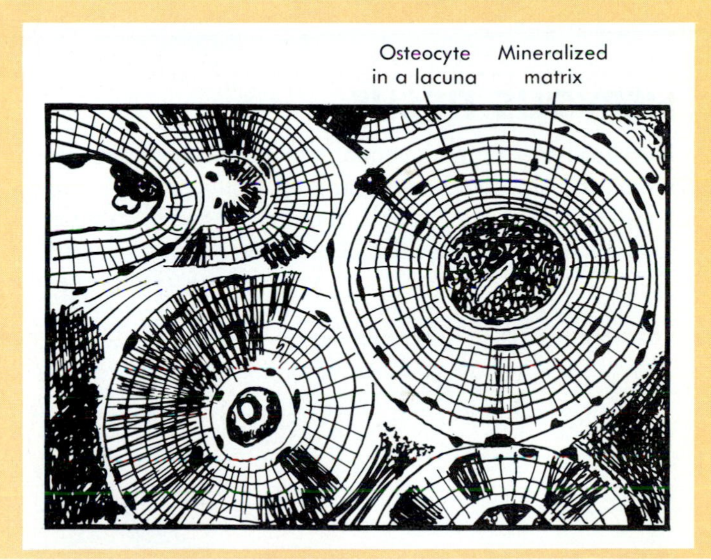
Osteocyte in a lacuna
Mineralized matrix

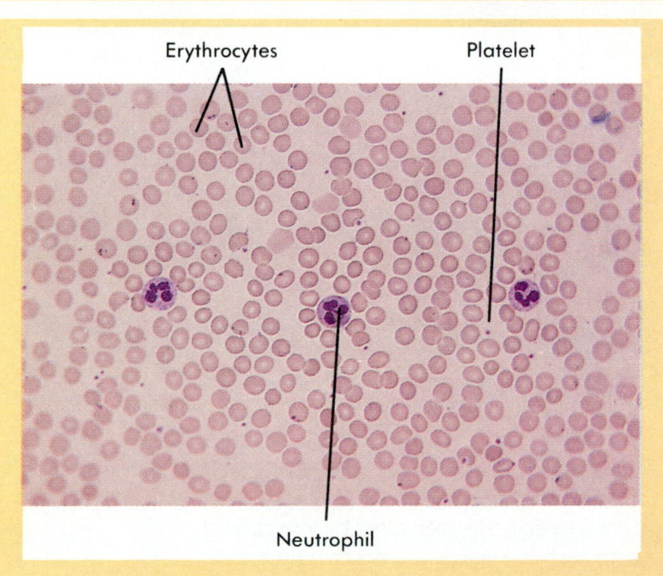
Erythrocytes
Platelet
Neutrophil

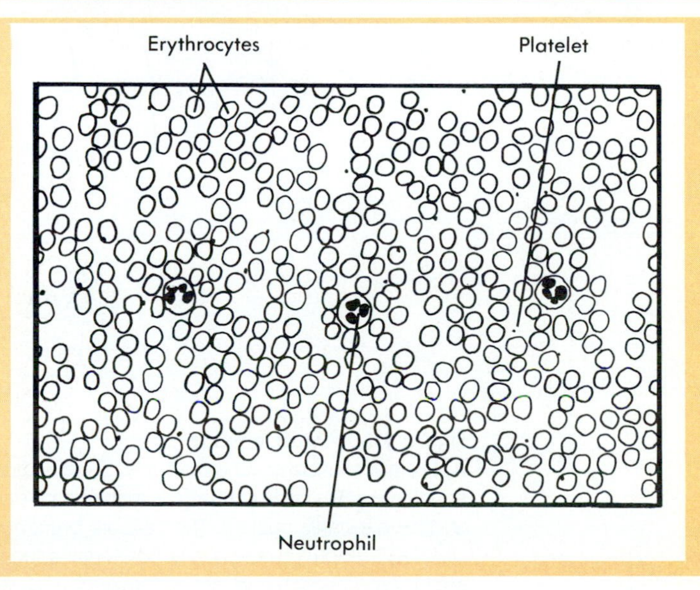
Erythrocytes
Platelet
Neutrophil

Skeletal muscle cells are **striated** (stri-āt-ed) or banded because of the arrangement of contractile proteins within the cells (see Chapter 7).

Cardiac muscle is the muscle of the heart and is responsible for pumping blood (Figure 4-5, *B*). It is under involuntary (unconscious) control. Cardiac muscle cells are cylindrical in shape, but much shorter than skeletal muscle cells. Cardiac muscle cells are striated with one nucleus per cell, and they often are branched and connected to each other by **intercalated** (in-ter´kă-la-ted) **disks.** The intercalated disks are important in coordinating contractions of the cardiac muscle cells by allowing many individual cardiac muscle cells to function as a unit (see Chapter 12).

Smooth muscle forms the walls of hollow organs (except the heart) and is found in the skin and the eyes (Figure 4-5, C). It is responsible for a number of functions such as movement of food through the digestive tract and emptying of the urinary bladder. Smooth muscle is under involuntary control. Smooth muscle cells are tapered at each end and have a single nucleus. They are not striated.

NERVOUS TISSUE

Nervous tissue forms the brain, spinal cord, and nerves. It is responsible for coordinating and controlling many of the body's activities. For example, the nervous system is responsible for the conscious control of skeletal muscles and the unconscious regulation of cardiac muscle. Awareness of ourselves and our external environment, emotions, reasoning skills, and memories result from processes carried out by nervous tissue. Many functions depend on the ability of nerve cells to communicate with each other and with other tissue types by using electrical signals called action potentials.

Nervous tissue consists of neurons, which are responsible for the action potential conduction, and support cells. The **neuron** (nu´ron), or **nerve cell,** is composed of three parts (Figure 4-6). The **cell body** contains the nucleus and is the site of general cell functions. **Dendrites** (den´drītz) and **axons** (ak´sonz) are nerve cell processes (extensions). Dendrites usually receive action potentials and conduct them toward the cell body, whereas the axon (only one per neuron) usually conducts action potentials away from the cell body. **Neuroglia** (nu-rog´le-ah) are the support cells of the nervous system, and they function to nourish, protect, and insulate the neurons. Nervous tissue is considered in greater detail in Chapter 8.

MEMBRANES

A **membrane** is a thin sheet or layer of tissue that covers a structure or lines a cavity. Most membranes consist of epithelium and the connective tissue on which the epithelium rests. The two major categories of membranes are the mucous membranes and the serous membranes.

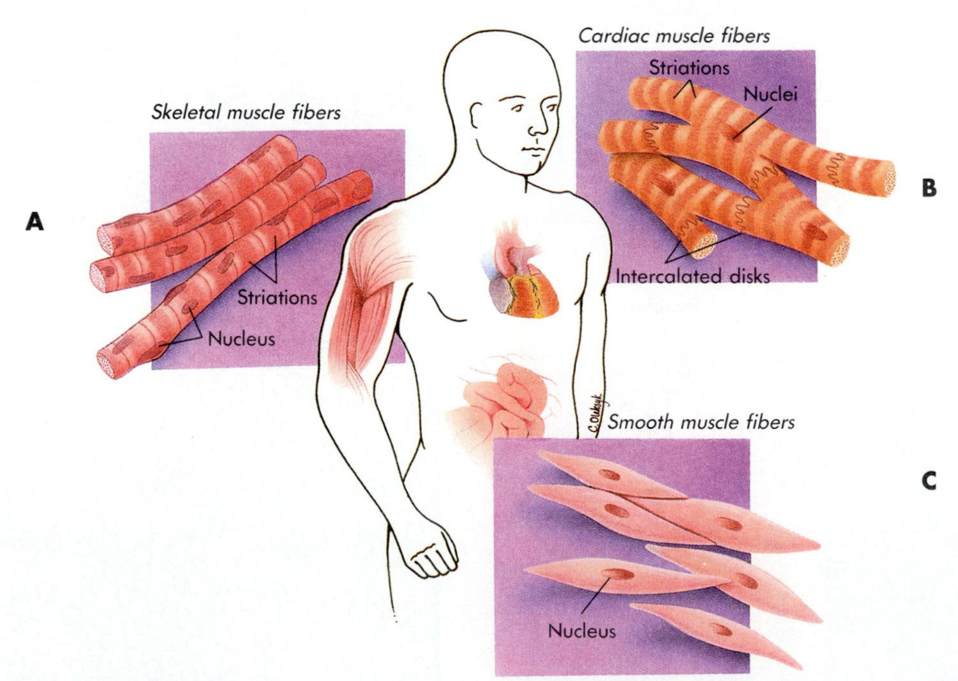

Figure 4-5 ● Muscle tissue.
A, Skeletal muscle attaches to bone. Skeletal muscle cells are cylindrical in shape, striated, and multinucleated. **B,** Cardiac muscle is in the heart. Cardiac muscle cells are cylindrical in shape, striated, and have a single nucleus. The cells are branched and are connected to each other by intercalated disks. **C,** Smooth muscle is in hollow organs such as the stomach and intestine. Smooth-muscle cells are tapered at each end, are not striated, and have a single nucleus.

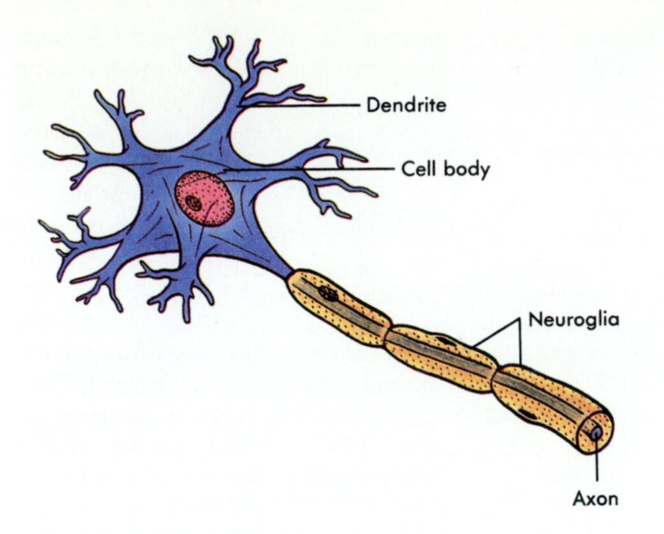

Figure 4-6 ● **Nervous tissue.**
The neuron consists of dendrites, a cell body, and a long axon. Neuroglia, or support cells, surround the axon. Nervous tissue is found in the brain, spinal cord, ganglia, and nerves.

Mucous membranes

Mucous (mu´kus) **membranes** consist of various kinds of epithelium resting on a thick layer of loose connective tissue. They line cavities that open to the outside of the body such as the digestive, respiratory, excretory, and reproductive tracts (Figure 4-7, *A*). Many, but not all, mucous membranes have mucous glands, which secrete mucus. The functions of mucous membranes vary depending on their location, and they include protection, absorption, and secretion. For example, the stratified squamous epithelium of the oral cavity (mouth) protects underlying tissues. The simple columnar epithelium of the intestine functions to absorb nutrients and to secrete digestive enzymes and mucus.

Serous membranes

Serous (sēr´us) **membranes** consist of simple squamous epithelium resting on a delicate layer of loose connective tissue. In contrast to the mucous membranes, serous membranes line trunk cavities that do not open to the exterior (Figure 4-7, *B* and see Chapter 1).

There are three main serous membranes named according to their location: the **pleural** (ploor´al) **membranes** line the cavities surrounding the lungs and cover the outer sur-

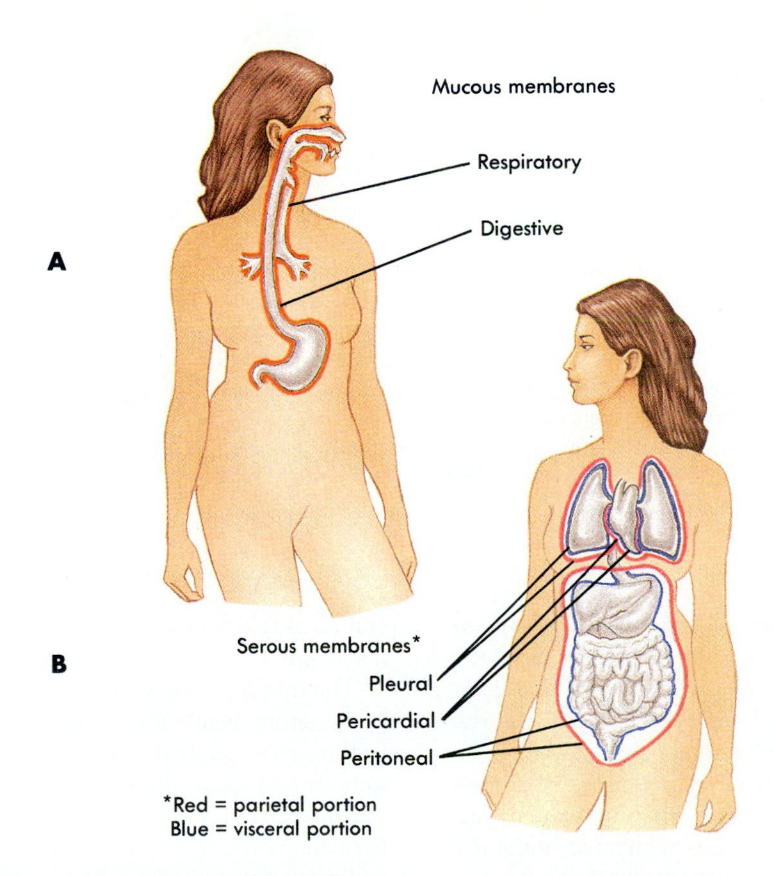

Figure 4-7 ● **Epithelial membranes.**
A, Mucous membranes line body cavities that open to the outside of the body, such as the digestive and respiratory tracts. **B,** Serous membranes line body cavities that do not open to the outside, such as the pleural, pericardial, and abdominopelvic cavities.

Understanding cancer

A **neoplasm** (ne´o-plazm), or **tumor,** is any tissue growth resulting from abnormal cell divisions that continue after normal cell divisions are complete. Tumors are either benign or malignant.

Benign (be-nīn´) tumors enlarge but do not spread to other sites in the body, and the cells are similar in structure to those in normal tissue. A benign tumor remains enclosed within a capsule of thick connective tissue and does not spread beyond its original site. When benign tumors are located, they usually can be removed surgically. If they are not removed they may compress surrounding tissues such as nerves and blood vessels as they enlarge. For example, a benign tumor originating from adipose tissue beneath the skin usually can be removed without difficulty, but a benign tumor in the brain can be difficult to remove and may compress surrounding brain tissue and cause it to malfunction.

Cancer, or **malignant** (mă-lig´nant) tumors, enlarge and spread to other sites in the body, are not surrounded by a thick connective tissue capsule, and consist of cells that can be very different than cells of normal tissue. Abnormal control of growth in malignant tumors is a result of an alteration in the genetic machinery of the cell and can be induced by viruses, environmental toxins, and other causes.

Malignant tumors can spread by local growth and expansion or by **metastasis** (mĕ-tas´tă-sis; moving to another place). Cells of malignant tumors divide continuously causing the tumors to enlarge rapidly. Since cells are not surrounded by a thick connective tissue capsule, they readily invade and destroy surrounding tissues. Metastasis occurs when tumor cells separate from the main mass and are carried by the lymphatic or circulatory system to new sites where additional malignant tumors form. The illness associated with cancer usually occurs as the tumor spreads by metastasis, destroying healthy tissues in several areas of the body and eliminating their functions. For example, cancer that originates in the colon can invade organs such as the lung, liver, and brain and destroy their functions.

Names of benign tumors frequently end with **"-oma,"** and names of malignant tumors usually are based on the tissue from which they originated. Names of malignant tumors originating from epithelial tissue end with **"carcinoma,"** and names of malignant tumors originating from connective tissue end with **"sarcoma."** The beginning of the names of tumors provide descriptive information also. For example, a benign tumor originating from fibrous connective tissue is a **fibroma** (fi-bro´mah), whereas a malignant tumor originating from fibrous connective tissue is called a **fibrosarcoma** (fi-bro-sar´ko-mah). A benign tumor originating from glandular epithelium is an **adenoma** (ad´ĕ-no´mah; adeno refers to glands), whereas a malignant tumor originating from glandular epithelium is an **adenocarcinoma** (ad´ĕ-no´kar-sĭ-no´-mah). Exceptions to this naming system do exist. For example, **leukemia** (lu´ke´me-ah) is a malignant tumor of blood-producing tissues. Names of tumors originating in several tissue types are listed in Table 4-1.

Anatomical imaging frequently is used to visualize suspected tumors. Techniques commonly used include x-rays, ultrasound, computerized axial tomography (CAT) scans, and magnetic resonance imaging (MRI). Once a tumor is detected, a small sample of tissue often is removed in a process called a **biopsy** (bi´op-se). The tissue can be microscopically examined to determine whether it is malignant or benign.

Cancer therapy is directed at confining and then killing malignant cells. This is accomplished by killing the tissue with radiation or lasers; by removing the tumor surgically; by **chemotherapy,** which consists of treating the patient with drugs that selectively kill rapidly dividing cells; or by stimulating the patient's immune system to destroy the tumor. **Oncology** (ong-kol´o-je) is the study of cancer and its associated problems.

3 A large malignant tumor was found in a woman's breast. Her breast was surgically removed along with many of the abundant lymphatic structures found in and around the breast. Six months later, several malignant tumors were identified in places such as her lungs and brain. Are the newly discovered tumors likely to be related to the original breast cancer or did they develop as new tumors? Explain.

face of the lungs; the **pericardial** (pĕr-ĭ-kar´de-al) **membrane** lines the cavity surrounding the heart and covers the outer surface of the heart; and the **peritoneal** (pĕr´ĭ-to-ne´al) **membrane** lines the abdominopelvic cavity and the surface of many organs of the abdominopelvic cavity.

Serous membranes are arranged so that one portion lines the body cavity and the other covers the surface of organs found in the body cavity. The serous membrane lining the surface of a body cavity is called the **parietal** (pă-ri´ĕ-tal) portion of a serous membrane. The serous membrane covering the surface of organs in a body cavity is the **visceral** (vis´er-al) portion of the serous membrane. The word parietal refers to a wall whereas the word viscera refers to internal organs.

The thin space between the parietal and visceral portions of the serous membranes is filled with **serous fluid** secreted by the serous membrane. The surface of the epithelial cells is smooth, and serous fluid acts as a lubricant so movement of organs in the body cavities can occur with little damage from abrasion.

Pericarditis (pĕr-ĭ-kar-di´tis) is inflammation of the pericardial membranes, **pleurisy** (ploor´ĭ-se) is inflammation of the pleural membranes, and **peritonitis** (pĕr´ĭ-to-ni´tis) is inflammation of the peritoneal membranes.

Table 4-1 • Terms used to describe some common tumors

Tissue of origin	Benign tumor	Malignant tumor
Epithelium		
Stratified squamous	Squamous cell papilloma	Squamous cell carcinoma
Glandular (e.g., mammary gland)	Adenoma	Adenocarcinoma
Melanocytes (pigment cells of the skin)	Nevus (mole)	Malignant melanoma
Connective tissue		
Fibrous	Fibroma	Fibrosarcoma
Adipose	Lipoma	Liposarcoma
Cartilage	Chondroma	Chondrosarcoma
Bone	Osteoma	Osteosarcoma
Nerve tissue		
Nerve sheath	Neurofibroma	Neurofibrosarcoma
Lymphoid tissue (lymphatic)	Lymphoma	Lymphocytic leukemia

Other membranes

In addition to mucous membranes and serous membranes, there are several other membranes in the body. The skin or **cutaneous membrane** is stratified squamous epithelium and dense connective tissue (see Chapter 5). Other membranes are made up of only connective tissue. **Synovial** (sĭ-no´ve-al) **membranes** line the inside of joint cavities (the space where bones come together with a movable joint), and the **periosteum** (pĕr´e-os´te-um) surrounds bone. These connective tissue membranes are discussed in Chapter 6.

SUMMARY

INTRODUCTION

A tissue is a group of cells with similar structure and function plus the extracellular substances located between the cells.

EPITHELIAL TISSUE

Epithelial tissue covers free surfaces, usually has a basement membrane, has little extracellular material, and has no blood vessels.

Classification

Simple epithelium has one layer of cells, whereas stratified epithelium has more than one layer of flat cells.

Categories of epithelium based on cell shape are squamous (flat), cuboidal (cubelike), and columnar (tall and thin).

Epithelium generally has two names: simple squamous epithelium (one layer of flat cells), simple cuboidal epithelium (one layer of cubelike cells), simple columnar epithelium (one layer of tall, thin cells), stratified squamous epithelium (several layers of flat cells), and transitional epithelium (a special type of epithelium consisting of many layers and well-adapted for stretching).

Glands

A gland is a single cell or a multicellular structure that secretes. Exocrine glands have ducts and endocrine glands do not.

CONNECTIVE TISSUE

Connective tissue holds cells and tissues together.

Connective tissue has a large amount of extracellular material that determines the functional characteristics of each type of connective tissue.

Dense connective tissue has large amounts of collagen fibers in the extracellular matrix and is found in tendons, ligaments, and the dermis of the skin.

Loose or areolar connective tissue has fewer collagen fibers and fills spaces between organs and holds them in place.

Adipose tissue consists of cells filled with lipids and it stores fat and protects parts of the body.

Cartilage has a rigid, collagen-rich matrix that provides support.

Bone has a hard mineralized matrix that enables bones to support and protect.

Blood has a liquid matrix and carries food, oxygen waste products, and other materials throughout the body.

MUSCLE TISSUE

Muscle tissue is specialized to shorten or contract. The three types of muscle tissue are skeletal, cardiac, and smooth muscle.

NERVOUS TISSUE

Nervous tissue is specialized to conduct action potentials (electrical signals).

Neurons conduct action potentials and neuroglia support the neurons.

MEMBRANES

Mucous membranes

Mucous membranes line cavities that open to the outside of the body (digestive, respiratory, excretory, and reproductive tracts). They contain glands and secrete mucus.

Serous membranes

Serous membranes line trunk cavities that do not open to the outside of the body (pleural, pericardial, and peritoneal cavities). The serous membranes secrete serous fluid.

Other membranes

Other membranes include the cutaneous membrane (skin), synovial membranes (line joint cavities), and periosteum (around bone).

CONTENT REVIEW

1. Define tissue.
2. Where is epithelium located? List four major characteristics of epithelium.
3. Explain how epithelium is classified according to number of cell layers and cell shapes. What is transitional epithelium?
4. What functions would a single layer of epithelium be expected to perform? A stratified layer?
5. Contrast the functions of squamous cells with cuboidal or columnar cells. Give an example of each.
6. Define gland. Distinguish between an exocrine and an endocrine gland.
7. What are the major functions of connective tissue? How does connective tissue differ from other types of tissue?
8. Describe dense connective tissue and give three examples of where it is found.
9. Describe areolar connective tissue and list its functions.
10. How is adipose tissue different from other connective tissues? What are the functions of adipose tissue?
11. Describe the structure of cartilage. What are the major functions of cartilage?
12. Describe the structure of bone tissue and state the functions of bone.
13. What is a unique structural feature of blood and what are its major functions?
14. What is the major functional characteristic of muscle?
15. List the three major types of muscle and their major characteristics.
16. What are the major functions of nervous tissue? What are dendrites and axons? What are the functions of neurons and neuroglia?
17. Compare mucous and serous membranes according to the type of cavity they line and their secretions. Name the serous membranes associated with the lungs, heart, and abdominopelvic cavity.
18. Describe where the following membranes are found: cutaneous membrane, synovial membranes, and periosteum.

CONCEPT QUESTIONS

1. The trachea or "wind pipe" is a tube that conducts air between the lungs and throat. Normally the trachea is lined with simple columnar epithelium that has cilia. Heavy smoking, which irritates the trachea, can result in the replacement of the normal epithelium with another type of epithelium. What type of epithelium would be expected as a replacement tissue?
2. Explain why adipose tissue is an exceptional connective tissue type and explain why it is still considered to be a connective tissue.
3. Part of the pancreas is an exocrine gland that produces digestive enzymes that are secreted into the small intestine. It has small sacs lined with epithelial cells that secrete the digestive enzymes. How many cell layers and what cell shape would be expected in the epithelium responsible for producing the digestive enzymes?
4. Some dense connective tissue has elastic fibers in addition to collagen fibers. This enables a structure to stretch and then recoil to its original shape. Examples are certain ligaments that hold together the vertebrae (bones of the back). When the back is bent (flexed) the ligaments are stretched. How does the elastic nature of these ligaments help the back function? How would the fibers be arranged in the ligaments?

CHAPTER TEST

Answers can be found in Appendix C

MATCHING For each statement in column A select the correct answer in column B (an answer may be used once, more than once, or not at all).

A	B
1. The layer most epithelial cells rest on.	adipose tissue
2. The surface of epithelium that is not in contact with other tissues.	basement membrane
3. Single layer of thin, flat epithelial cells.	bone
4. Single layer of tall, thin epithelial cells.	cardiac muscle
5. Single layer of epithelial cells that are nearly square in shape.	cartilage
6. Several layers of thin, flat epithelial cells.	dense connective tissue
7. Type of connective tissue found in tendons and ligaments.	free surface
8. Type of connective tissue with little extracellular material.	mucus membrane
9. Muscle type that is voluntary.	serous membrane
10. Lining of cavity that has no opening to the outside of the body.	simple columnar epithelium
	simple cuboidal epithelium
	simple squamous epithelium
	skeletal muscle
	stratified squamous epithelium
	transitional epithelium

FILL-IN-THE BLANK Complete each statement by providing the missing word or words.

1. Epithelia that are specialized to synthesize and secrete proteins include ___________ and ___________ epithelium.
2. ___________ is specialized to allow diffusion of material across it.
3. ___________ is found in areas of the body where abrasion occurs, such as in the mouth, esophagus, anus, and vagina.
4. ___________ is a special type of stratified epithelium consisting of many layers of cells that can be greatly stretched.
5. Glands with ducts are ___________ glands.
6. ___________, which resembles microscopic ropes, is the most common protein in the body.
7. ___________ is the "loose packing" material of the body, which fills the spaces between organs and holds them in place.
8. ___________ is relatively rigid and provides support, but if it is bent or slightly compressed it will resume its original shape.
9. ___________ is responsible for functions such as movement of food through the digestive tracts and emptying of the urinary bladder.

CHAPTER

5

The Integumentary System

KEY TERMS

dermis
Dense connective tissue that forms the deep layer of the skin; responsible for the structural strength of the skin.

epidermis
Outer portion of the skin formed of epithelial tissue that rests on the dermis; resists abrasion and forms a permeability barrier.

hair
A threadlike outgrowth of the skin consisting of columns of dead epithelial cells filled with keratin.

hypodermis
Loose connective tissue under the dermis; attaches the skin to muscle and bone.

keratin
Fibrous protein substance found in superficial skin cells, hair, and nails; provides structural strength.

melanin
Brown to black pigment responsible for skin and hair color.

nail
A thin, horny plate at the ends of the fingers and toes, consisting of several layers of dead epithelial cells containing a hard keratin.

sebaceous gland
Gland of the skin that produces sebum; usually associated with a hair follicle.

sweat gland
Secretory organ that usually produces a watery secretion called sweat that is released onto the surface of the skin; some sweat glands, however, produce an organic secretion.

vitamin D
Fat-soluble vitamin produced from precursor molecule in skin exposed to ultraviolet light.

OBJECTIVES

After reading this chapter you should be able to:

1. Describe the structure and function of the dermis, epidermis, and hypodermis.
2. Explain how melanin and blood affect skin color.
3. Describe the structure of a hair and discuss the phases of hair growth.
4. Name the glands of the skin and describe the secretions they produce.
5. Describe the parts of a nail and explain how nails grow.
6. Discuss the functions of skin, hair, glands, and nails.
7. List the changes the integumentary system undergoes with age.

FEATURES

THE INTEGUMENTARY SYSTEM consists of the skin and accessory structures such as hair, nails, and glands. Integument means covering and the integumentary system is familiar to most people because it covers the outside of the body. The integumentary system has many functions, including protecting internal structures, preventing the entry of infectious agents, regulating body temperature, and producing vitamin D.

SKIN

The skin is made up of two major tissue layers. The deep layer, the **dermis** (der´mis), is dense connective tissue. The superficial layer of the skin is the **epidermis** (ep´ĭ-der´mis), which is epithelial tissue that rests on the dermis (Figure 5-1). The skin varies in thickness from 0.5 mm in the eyelids to 5 mm in the back and shoulders.

The **hypodermis** (hi´po-der´mis), or **subcutaneous** (sub´ku-ta´ne-us) **tissue,** is not part of the integumentary system, but usually is considered with it because the hypodermis attaches the skin to underlying bone or muscle. The hypodermis is loose connective tissue that contains approximately half the body's stored fat, although the amount and location vary with age, sex, and diet. Fat in the hypodermis functions as padding and insulation, and it is responsible for some of the structural differences between men and women.

Dermis

The dermis is dense connective tissue (Figure 5-2). Collagen and elastic fibers in the dermis are responsible for most of the structural strength of the skin. The fibers are oriented in many different directions to resist stretch in many directions. If the skin is overstretched, the dermis can be damaged, leaving lines called **striae** (stri´e), or **stretch marks,** which are visible through the epidermis. For example, striae can develop on the abdomen and breasts of a woman during pregnancy.

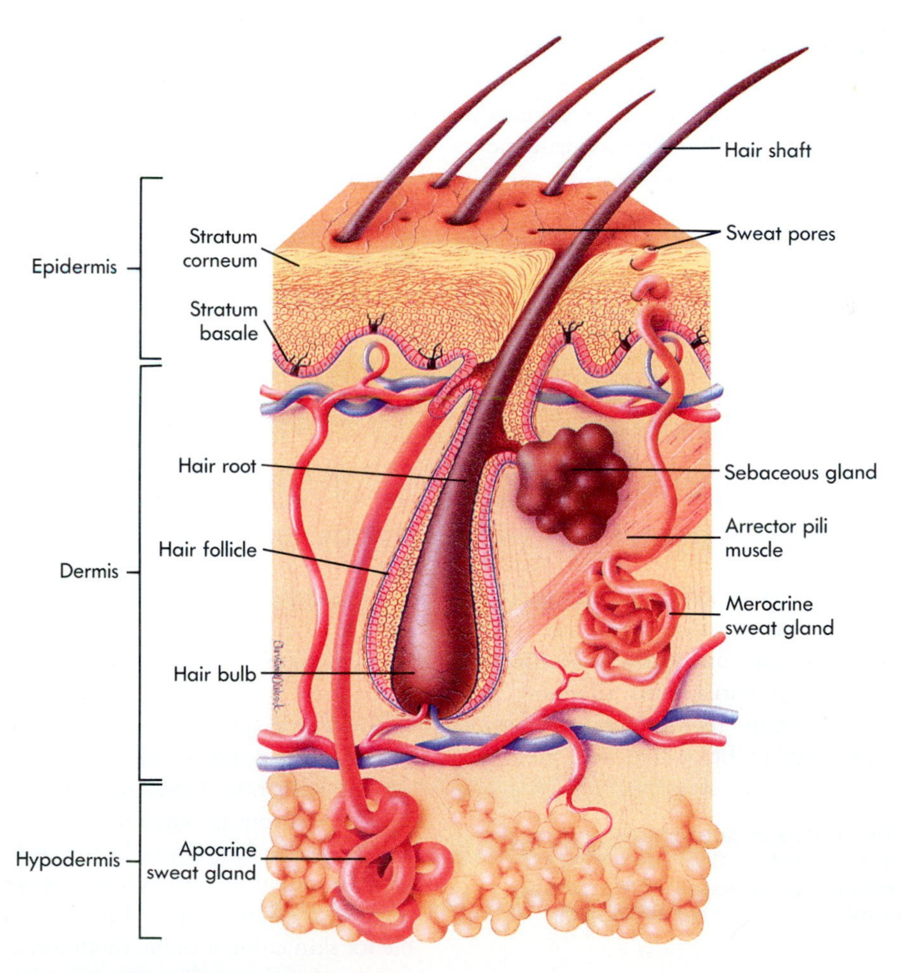

Figure 5-1 ● Skin and hypodermis.
The skin, consisting of the epidermis and the dermis, is connected by the hypodermis to underlying structures. Note the accessory structures (hairs, glands, and arrector pili) that project into the hypodermis, and the large amount of fat in the hypodermis.

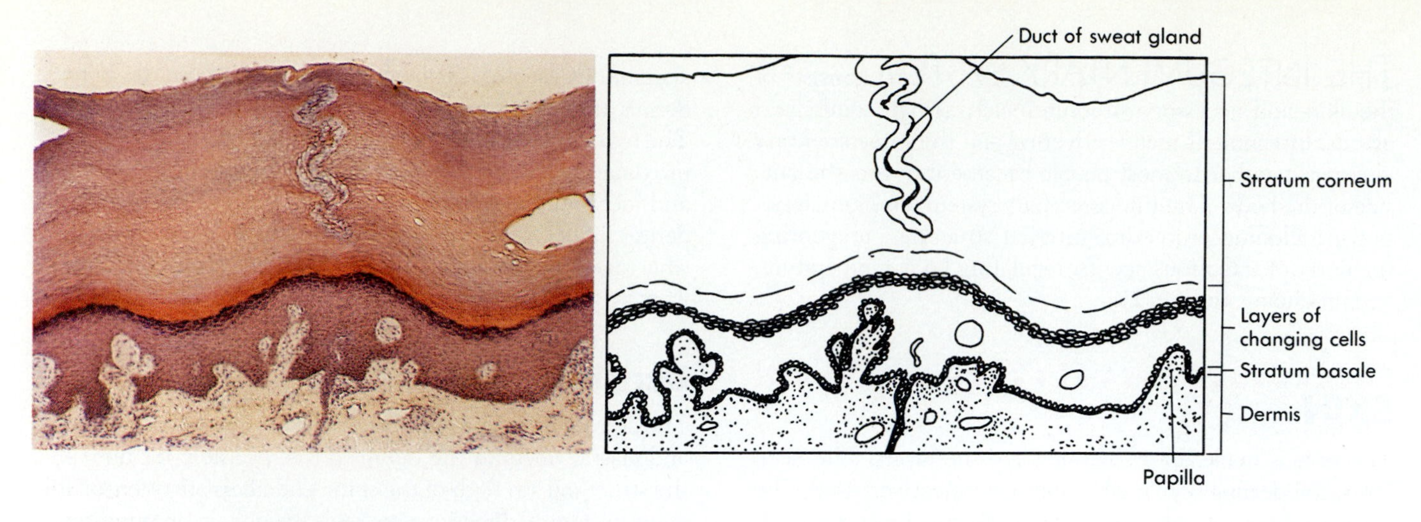

Figure 5-2 ● **Dermis and epidermis.**
The dermis is dense connective tissue. The epidermis is stratified squamous epithelium. Cells are produced in the stratum basale and move toward the surface of the skin to become stratum corneum cells.

The upper part of the dermis has projections called **papillae** (pă-pil´e), which are covered by the epidermis (see Figure 5-2). The papillae contain many blood vessels that supply the epidermis with nutrients, remove waste products, and aid in regulating body temperature. The papillae of the hands, soles of the feet, and tips of the digits are arranged to form parallel, curving ridges that shape the overlying epidermis into fingerprints and footprints. The ridges increase friction and improve the grip of the hands and feet.

Epidermis

The epidermis is stratified squamous epithelium which resists abrasion and forms a permeability barrier. The deepest part of the epidermis, the **stratum basale** (ba-să´le), consists of columnar cells that divide by mitosis (see Figure 5-2). Following cell division, one daughter cell becomes a new stratum basale cell and can divide again. The other daughter cell is pushed toward the surface. During their movement toward the surface the cells change shape and chemical composition and die. Groups of changed cells with similar characteristics form layers of cells within the epidermis. The outer layer of these cells, the **stratum corneum** (kor´ne-um), consists of dead, squamous cornified cells. Cornified cells are filled with the hard protein **keratin** (ker´ah-tin), which gives the stratum corneum its structural strength. These cells also are surrounded by lipids that help prevent fluid loss through the skin.

1 Some drugs are administered by applying the drug to the skin, for example, in a skin patch. The drug diffuses through the epidermis to blood vessels in the dermis. Are these drugs soluble in water or lipids? Explain.*

*Answers to predict questions appear in Appendix B at the back of the book.

The stratum corneum is composed of as many as 25 or more layers of dead squamous cells. Eventually the cells at the surface are sloughed from the skin. Dandruff is an example of stratum corneum cells sloughed from the surface of the scalp. In skin subjected to friction, the number of layers in the stratum corneum greatly increases, producing a thickened area called a **callus** (kal´us). Over a bony prominence, the stratum corneum can thicken to form a cone-shaped structure called a **corn,** which can be quite painful.

Skin color

Skin color is determined primarily by pigments in the skin and by blood circulating through the skin. **Melanin** (mel´ah-nin), a brown-to-black pigment, is responsible for most skin color. Melanin functions as a natural sunscreen that protects epidermal cells and underlying tissues from the harmful effects of ultraviolet light. **Melanocytes** (mel´ă-no-sīts) are specialized cells in the stratum basale that produce melanin. The melanocytes transfer melanin to other cells, so eventually all the cells in the deeper layers of the epidermis become pigmented. Moles and age spots are associated with a localized increase in the number of melanocytes, whereas freckles result from a localized increase in melanin production.

Melanin production is determined primarily by genetic factors and exposure to light. Genetic factors are responsible for the amounts of melanin produced in different races. Racial variations in skin color, such as black, brown, yellow, and white, are determined by the amount and distribution of melanin produced. All races have approximately the same number of melanocytes. Although many genes are responsible for skin color, a single mutation can prevent the manufacture of melanin. For example, **albinism** (al´bĭ-nizm) is a recessive genetic trait (see Chapter 20) that causes a deficiency or absence of melanin. Albinos have fair skin, white hair, and unpigmented irises in the eyes.

Exposure to ultraviolet light, for example in sunlight, stimulates melanocytes to increase melanin production. The result is a suntan, skin with additional melanin that absorbs the ultraviolet light before it damages cells. Melanin production in response to sunlight is an example of maintaining homeostasis. The amount of melanin needed to provide protection (homeostasis) is matched to the amount of exposure to ultraviolet light.

Blood flowing through the skin imparts a reddish hue, and when blood flow increases (for example, during blushing, anger, and the inflammatory response), the red color intensifies. A decrease in blood flow, such as occurs in shock, can make the skin appear pale. A decrease in the blood oxygen content produces a bluish color called **cyanosis** (si-ă-no´sis). **Birthmarks** are congenital (present at birth) disorders of the blood vessels (capillaries) in the dermis.

2 List the possibilities that would explain the difference in skin color between (A) the anterior and posterior surfaces of the forearm and (B) the palms of the hand and the lips.

ACCESSORY SKIN STRUCTURES

Hair

A **hair** is a threadlike structure consisting of columns of dead epithelial cells filled with keratin. The **shaft** of the hair protrudes above the surface of the skin, whereas the **root** and **hair bulb** are below the surface. The **hair follicle,** which holds the hair in place, is an extension of the epidermis deep into the dermis (see Figure 5-1). The hair follicle can play an important role in tissue repair. If the surface epidermis is damaged, the epithelial cells within the hair follicle can divide and serve as a source of new epithelial cells (see the section on understanding burns in this chapter).

The hair is produced in cycles in the hair bulb. During the **growth stage,** cells within the hair bulb divide. Similar to the epidermis, these cells become filled with keratin and die. The hair grows longer as cells are added to the base of the hair root. Thus the hair root and shaft consist of columns of dead epithelial cells filled with keratin. The average rate of hair growth is 0.3 mm per day. During the **resting stage,** growth stops, and the hair is held in the hair follicle. When the next growth stage begins, a new hair is formed, and the old hair falls out.

The loss of hair normally means that the hair is being replaced because the old hair falls out of the hair follicle when the new hair begins to grow. In some men, however, a permanent loss of hair results in "male pattern baldness." Although many of the hair follicles are lost, some remain and produce a very short, transparent hair, which for practical purposes is invisible. This conversion occurs when male sex hormones act on the hair follicles of men who have the genetic predisposition of "male pattern baldness."

Hair color, such as blond, brown, and black, is determined by varying amounts of melanin. Red hair color is produced by melanin containing iron. Similar to the epidermis, melanocytes within the hair bulb produce and distribute melanin to other cells. With age the amount of melanin in hair can decrease, causing the hair to become faded or white.

3 Marie Antoinette's hair supposedly turned white overnight after she heard she was to be sent to the guillotine. Is it reasonable to believe this story?

Muscles

Associated with each hair follicle are smooth muscle cells, the **arrector pili** (ah-rek´tor pi´li) (see Figure 5-1). Contraction of the arrector pili causes the hair to "stand on end" and also produces a raised area of skin called "goose bumps" or "goose flesh." In animals with fur, contraction of the arrector pili is beneficial because it functionally increases the thickness of the fur by raising the hairs. In the cold, the thicker layer of fur traps air and becomes a better insulator. The thickened fur can also make the animal appear larger and more ferocious, which might deter an attacker. It is unlikely that humans, with their sparse amount of hair, derive any important benefit from contraction of their arrector pili.

Glands

The major glands of the skin are the **sebaceous** (se-ba´shus) **glands** and the **sweat glands** (see Figure 5-1). Most sebaceous glands are connected by a duct to the upper part of a hair follicle. They produce **sebum,** an oily, white substance rich in lipids. The sebum lubricates the hair and the surface of the skin, which prevents drying and protects against some bacteria.

There are two kinds of sweat glands. The most common type are **merocrine** (mĕr´o-krin) sweat glands. They produce a secretion that is mostly water with a few salts. Merocrine sweat glands are located in almost every part of the skin, and they have ducts that open onto the surface of the skin through sweat pores (see Figure 5-1). When the body temperature starts to rise above normal levels, the sweat glands produce sweat, which evaporates and cools the body.

Merocrine sweat glands also release sweat in the palms, soles, axillae (armpits), and other places due to emotional stress. Emotional sweating is used in lie detector (polygraph) tests because sweat gland activity usually increases when a person tells a lie. The sweat produced, even small amounts, can be detected because the salt solution conducts electricity and lowers the electrical resistance of the skin.

The other type of sweat glands are **apocrine** (ăp´o-krin) sweat glands, which produce a thick secretion rich in organic substances. They open into hair follicles, but only in the axillae and genitalia (see Figure 5-1). These sweat glands become active at puberty because of the influence of sex hormones. Their secretion, which is essentially odorless when released, is quickly broken down by bacteria to form molecules responsible for body odor.

Understanding burns

Burns are classified according to the depth of the burn (Figure 5-A). In **partial-thickness burns** some portion of the stratum basale remains viable, and regeneration of the epidermis occurs from within the burn area as well as from the edges of the burn. Partial-thickness burns are divided into **first-** and **second-degree burns.**

First-degree burns involve only the epidermis. They are red and painful, and slight edema (swelling of tissues with fluid) can be present. First-degree burns can be caused by sunburn or brief exposure to hot or cold objects. Healing occurs without scarring in approximately 1 week and can be accompanied by flaking or peeling of the skin.

Second-degree burns damage the epidermis and the dermis. If there is minimal dermal damage, symptoms include redness, pain, edema, and blisters. Healing takes approximately 2 weeks, and there is no scarring. If the burn goes deep into the dermis, however, the wound appears red, tan, or white, can take several months to heal, and might scar. In all second-degree burns the epidermis regenerates from epithelial tissue in hair follicles and sweat glands, as well as from the edges of the wound.

In **full-thickness,** or **third-degree burns,** the epidermis and the dermis are completely destroyed, and recovery occurs from the edges. Third-degree burns often are surrounded by areas of first- and second-degree burns. Although the first- and second-degree burn areas are painful, the region of a third-degree burn is usually painless because sensory receptors in the epidermis and dermis have been destroyed. Third-degree burns appear white, tan, brown, black, or deep cherry red.

The skin normally prevents loss of water from the body and prevents the entry of microorganisms into the body. Deep partial- and full-thickness burns over large areas of the body result in excessive fluid loss. Proper fluid replacement is necessary to prevent dehydration, shock, and possibly death. Because microorganisms can infect the body more easily through these wounds, patients are also maintained under aseptic conditions and are given antibiotics. Third-degree burns also take a long time to heal, and they form disfiguring and debilitating scar tissue. To prevent these complications and to speed healing, skin grafts are performed. For example, in a split skin graft the epidermis and part of the dermis are removed from another part of the body and placed over the burn. Fluid from the burned area nourishes the graft until blood vessels grow into the graft and supply it with nourishment. Meanwhile, the donor tissue produces new epidermis from epithelial tissue in the hair follicles and sweat glands, such as occurs in superficial second-degree burns. Other types of skin grafts are also possible.

In cases in which a suitable donor site is not available, artificial skin, grafts from human cadavers, or grafts from pigs are used. These methods are often unsatisfactory because the body's immune system recognizes the graft as a foreign substance and rejects it. A solution to this problem is laboratory-grown skin. A piece of healthy skin from the burn victim is removed and placed in a flask with nutrients and hormones that stimulate rapid growth. The skin that is produced consists only of epidermis and does not contain glands or hair.

4 Dandy Chef has been burned on the arm. The doctor, using a forceps, pulls on a hair within the burned area. The hair easily pulls out. What degree of burn does the patient have and how do you know?

Nails

The **nail** is a thin plate, consisting of layers of dead stratum corneum cells that contain a very hard type of the protein keratin. The visible part of the nail is the **nail body,** and the part of the nail covered by skin is the **nail root** (Figure 5-3). The **cuticle** (ku´tĭ-kl) is stratum corneum that grows onto the nail body. The nail grows from the **nail matrix,** located under the proximal end of the nail. A small part of the nail matrix, the **lunula** (lu´nu-lah), can be seen through the nail body as a whitish, crescent-shaped area at the base of the nail. As it grows, the nail is pushed toward the tip of the finger at an average rate of 0.8 mm per day. Unlike hair, nails grow continuously and do not have a resting stage.

FUNCTIONS OF THE INTEGUMENTARY SYSTEM

Protection

The integumentary system performs many protective functions. The intact skin plays an important role in preventing water loss, and it prevents the entry of microorganisms and other foreign substances into the body. Secretions from skin glands also produce an environment unsuitable for some microorganisms. The stratified squamous epithelium of the skin protects underlying structures against abrasion. Melanin absorbs ultraviolet light and thus protects underlying structures from the damaging effects of the ultraviolet light. Hair provides protection in several ways: on the head it acts as a heat insulator, eyebrows keep sweat out of the eyes, eyelashes protect the eyes from foreign objects, and hairs in the nose and ears reduce the entry of dust and other materials. The nails protect the ends of the digits from damage and can be used in defense.

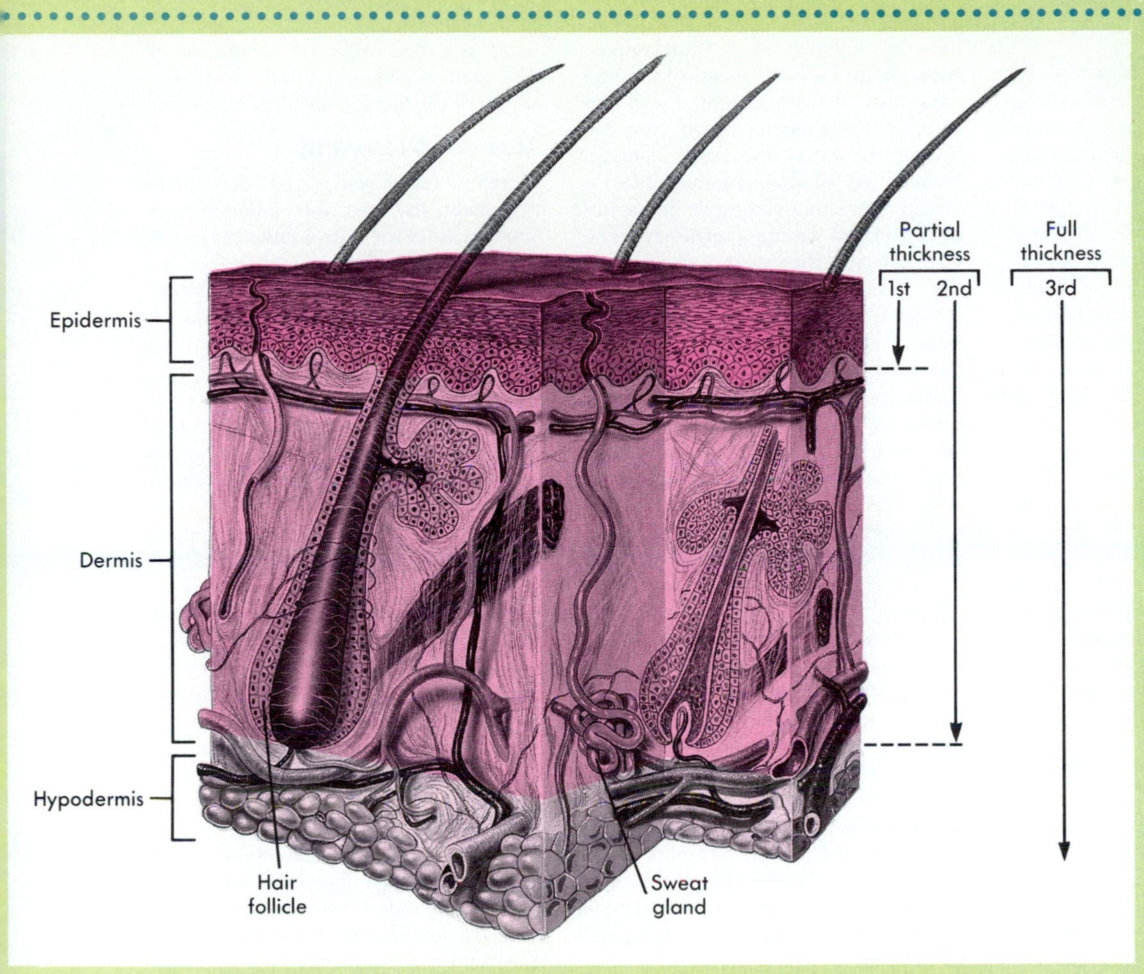

Figure 5-A Burns.
Parts of the skin damaged by different types of burns. Partial-thickness burns are subdivided into first-degree burns (damage to only the epidermis) and second-degree burns (damage to the epidermis and part of the dermis). Full-thickness burns or third-degree burns destroy the epidermis, dermis, and sometimes deeper tissues.

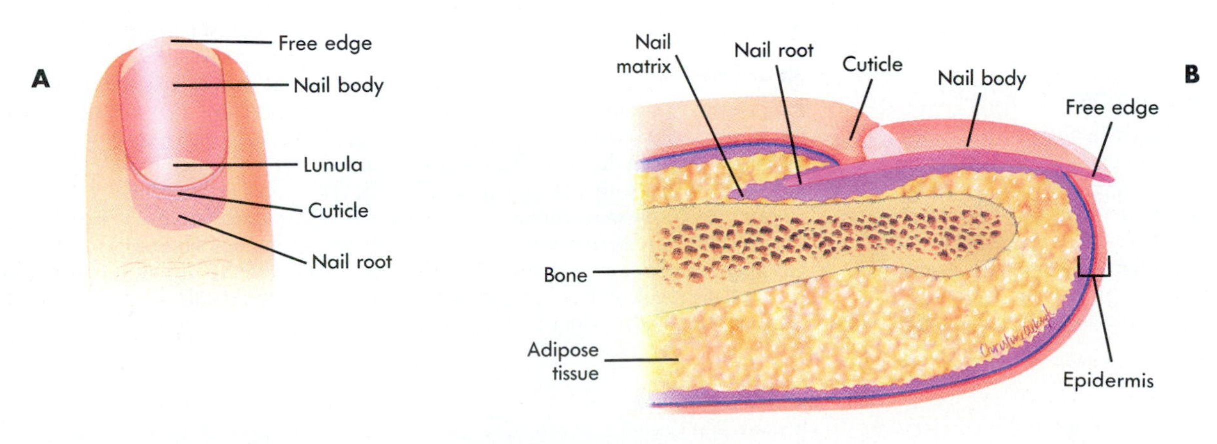

Figure 5-3 ● Nail.
A, Dorsal view. **B,** Lateral view.

Diseases of the skin

BACTERIAL INFECTIONS

Acne is a disorder of the hair follicles and sebaceous glands. Epidermal cells in the hair follicle stick to each other to form a mass of cells mixed with sebum that blocks the hair follicle. During puberty, hormones, especially testosterone, stimulate the sebaceous glands and sebum production increases. An accumulation of sebum behind the blockage produces a whitehead. A blackhead develops when the accumulating mass of cells and sebum pushes through the opening of the hair follicle. Although there is general agreement that dirt is not responsible for the black color of blackheads, the exact cause of the black color is disputed. If the wall of the hair follicle ruptures, bacteria enter the surrounding tissue and stimulate an inflammatory response that results in the formation of a red pimple filled with pus (see Chapter 4). If tissue damage is extensive, scarring occurs.

Impetigo (im-pĕ-ti´go) is a skin disease caused by *Staphylococcus aureus*. It usually affects children, producing small blisters containing pus that easily rupture to form a thick, yellowish crust. The bacteria are transmitted by direct contact (touching) and enter the skin through abrasions or small breaks in the skin.

Decubitus (de-ku´bĭ-tus) **ulcers,** also known as bedsores or pressure sores, can develop in people who are bedridden or confined to a wheelchair. The weight of the body, especially in areas over bony projections such as the hip bones and heels, compresses tissue and reduces circulation. The lack of blood flow results in the destruction of the hypodermis and the skin. After the skin dies, bacteria gain entry to produce an infected ulcer.

VIRAL INFECTIONS

Interestingly, many of the viruses that cause skin diseases do not enter the body through the skin. Instead the viruses enter through the respiratory system where they reside and multiply for approximately 2 weeks. Then they are carried by the blood to the skin where they cause lesions. Examples are rubeola, rubella, and chickenpox. **Rubeola** (ru-be´o-lah; measles) can be a dangerous disease because it can develop into pneumonia, or the virus can invade the brain and cause damage. **Rubella** (ru-bel´ah; German measles) is a mild disease but can prove dangerous if contracted during pregnancy. The virus can cross the placenta and damage the fetus, resulting in deafness, cataracts, heart defects, mental retardation, or death. **Chickenpox** is a mild disease if contracted as a child. **Herpes zoster** (zos´ter) or **shingles** is a disease caused by the chickenpox virus that occurs after the childhood infection. The virus remains dormant within nerve cells. Trauma, stress, or another illness somehow activates the virus, which moves through the nerve to the skin where it causes lesions along the nerve's pathway.

Cold sores or **fever blisters** are caused by a virus (herpes simplex I) related to the chickenpox virus. The initial infection usually does not produce symptoms. Dormant viruses can become active, however, and produce lesions in the skin around the mouth and in the mucous membrane of the mouth. The virus is transmitted by oral or respiratory routes. A related virus (herpes simplex II) is transmitted by sexual contact and produces genital lesions **(genital herpes)**.

Warts are uncontrolled growths caused by the human papilloma virus. Usually the growths are benign and disappear spontaneously, or they can be removed by a variety of techniques. The viruses are transmitted to the skin by direct contact with contaminated objects or an infected person. They also can be spread by scratching.

RINGWORM

Ringworm is a fungal infection that produces patchy scaling and an inflammatory response in the skin. The lesions are often circular with a raised edge and in ancient times were thought to be caused by worms. Several species of fungus cause ringworm in humans, and they usually are described by their location on the body. Ringworm in the scalp is called ringworm, ringworm of the groin is called jock itch, and ringworm of the feet is called athlete's foot.

ECZEMA AND DERMATITIS

Eczema (ek´zĕ-mah) and **dermatitis** (der´mă-ti´tis) describe inflammatory conditions of the skin. Causes of the inflammation can be allergy, infection, poor circulation, or exposure to physical factors such as chemicals, heat, cold, or sunlight.

PSORIASIS

The cause of **psoriasis** (so-ri´ă-sis) is unknown, but there may be a genetic component. In psoriasis there is increased cell division in the stratum basale, abnormal keratin production, and elongation of the dermal papillae toward the skin surface. The result is a thicker-than-normal stratum corneum that sloughs to produce large, silvery scales. If the scales are scraped away, bleeding occurs from the blood vessels at the top of the dermal papillae. Psoriasis is a chronic disease that can be controlled but as yet has no cure.

SKIN CANCER

Skin cancer is the most common type of cancer. Although chemicals and radiation (x-rays) are known to induce cancer, the development of skin cancer most often is associated with exposure to ultraviolet light from the sun. Consequently, most skin cancers develop on the face, neck, or hands.

The most common types of **skin cancer** involve the deep cells of the epidermis, producing open ulcers or nodular tumors (Figure 5-B, *A*). Surgical removal or radiation therapy cures these types of cancer. **Malignant melanoma** is a less common form of skin cancer that arises from melanocytes, usually in a preexisting mole. The **melanoma** can appear as a large, flat spreading lesion or as a deeply pigmented nodule (Figure 5-B, *B*). Spread or **metastasis** (mĕ-tas´tă-sis) of the cancer to other parts of the body is common, and unless diagnosed and treated early in development, this cancer is often fatal.

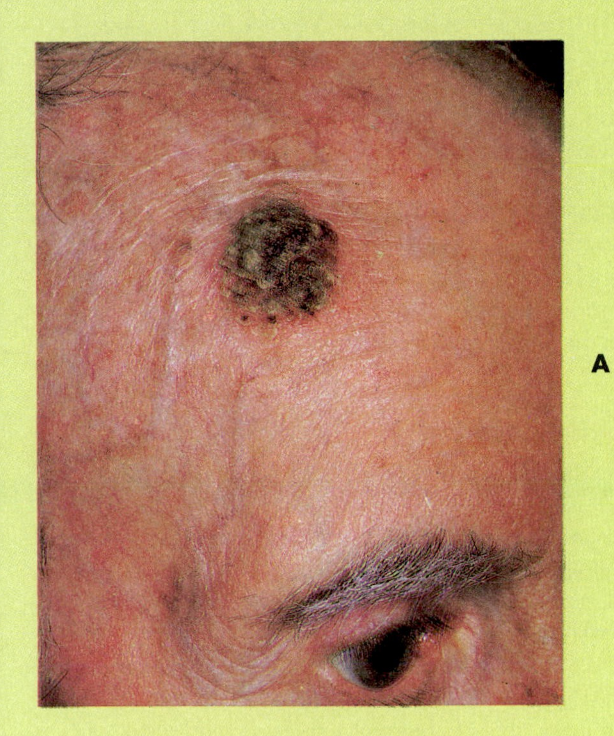

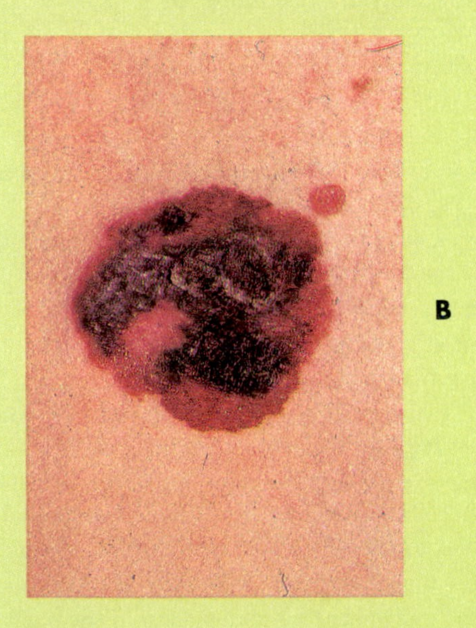

Figure 5-B.
A, Basal cell carcinoma begins with the cells of the stratum basale. **B,** Malignant melanoma begins with melanocytes.

Temperature regulation

The maintenance of a constant body temperature is an example of homeostasis. The integumentary system helps regulate body temperature by controlling heat loss or gain by the body. Exercise, fever, or an increase in environmental temperature tend to raise body temperature. The loss of excess heat can prevent a harmful increase in body temperature. Blood vessels in the dermis dilate and enable more blood to flow through the skin, thus transferring heat from deeper tissues to the skin, where the heat is lost. Sweat that spreads over the surface of the skin and evaporates also carries away heat and reduces body temperature.

If body temperature begins to drop below normal, heat can be conserved by constriction of dermal blood vessels, which reduces blood flow to the skin. Consequently less heat is transferred from deeper structures to the skin, and heat loss is reduced.

Vitamin D production

When the skin is exposed to ultraviolet light, an inactive form of **vitamin D** is produced in the skin. The inactive vitamin D is carried by the blood to the liver, where it is modified, and then to the kidneys where it is modified further to form active vitamin D. Adequate levels of vitamin D are necessary because vitamin D stimulates calcium and phosphate uptake in the small intestines. These substances are necessary for normal bone metabolism (see Chapter 6) and normal muscle function (see Chapter 7).

Sensation

The skin has receptors in the epidermis and dermis that can detect pain, heat, cold, and pressure (see Chapter 8). Although hair does not have a nerve supply, movement of the hair can be detected by sensory receptors around the hair follicle.

EFFECTS OF AGING ON THE INTEGUMENTARY SYSTEM

As the body ages, the blood flow to the skin is reduced, and the skin becomes thinner and appears more transparent. Because of decreased amounts of collagen in the dermis, skin is more easily damaged and repairs more slowly. Loss of elastic fibers in the skin and loss of fat from the hypodermis cause the skin to sag and wrinkle. A decrease in the activity of sebaceous and sweat glands results in dry skin and poor ability to regulate body temperature. The number of melanocytes generally decreases, but in some areas, the number of melanocytes increases to produce age spots. Gray or white hair also results because of a decrease in melanin production or a lack of melanin production. Skin that is exposed to sunlight ages more rapidly than nonexposed skin, so avoiding overexposure to sunlight and using sun blockers is advisable.

SUMMARY

The integumentary system consists of the skin, hair, glands, and nails.

SKIN

The hypodermis is loose connective tissue that attaches the skin to underlying tissues. Approximately half of the body's fat is stored in the hypodermis.

Dermis

The dermis is dense connective tissue.

Collagen and elastic fibers provide structural strength, and the blood vessels of the papillae supply the epidermis with nutrients.

Epidermis

The epidermis is stratified squamous epithelium divided into strata.

Cells are produced in the stratum basale and move to the stratum corneum, which is many layers of dead, cornified, squamous cells. The most superficial cells are sloughed.

Structural strength results from keratin inside the cells. Permeability characteristics result from lipids surrounding the cells.

Skin color

Melanocytes produce melanin that is responsible for different racial skin colors. Melanin production is determined genetically but can be modified by exposure to ultraviolet light (tanning).

Increased blood flow produces a red skin color, whereas a decreased blood flow causes a pale skin color. Decreased blood oxygen results in the blue color of cyanosis.

ACCESSORY SKIN STRUCTURES

Hair

Hairs are columns of dead, keratinized epithelial cells. Each hair consists of a shaft (above the skin), root (below the skin), and hair bulb (site of hair cell formation).

Hairs have a growth phase and a resting phase.

Muscles

Contraction of the arrector pili, which are smooth muscles, causes hair to "stand on end" and produces "goose bumps."

Glands

Sebaceous glands produce sebum, which oils the hair and the surface of the skin.

Merocrine sweat glands produce sweat that cools the body.

Apocrine sweat glands produce an organic secretion that can be broken down by bacteria to cause body odor.

Nails

The nail consists of the nail body and nail root.

The nail matrix produces the nail, which is stratum corneum containing hard keratin.

FUNCTIONS OF THE INTEGUMENTARY SYSTEM

Protection

The skin prevents the entry of microorganisms, acts as a permeability barrier, and provides protection against abrasion and ultraviolet light.

Temperature regulation

Through dilation and constriction of blood vessels, the skin controls heat loss from the body.

Evaporation of sweat cools the body.

Vitamin D production

Ultraviolet light stimulates the production of inactive vitamin D in the skin that is modified by the liver and kidneys into active vitamin D.

Vitamin D increases calcium uptake in the small intestines.

Sensation

The skin contains sensory receptors for pain, heat, cold, and pressure.

EFFECTS OF AGING ON THE INTEGUMENTARY SYSTEM

Blood flow to the skin is reduced, the skin becomes thinner, and elasticity is lost.

Sweat and sebaceous glands are less active, and the number of melanocytes decreases.

CONTENT REVIEW

1. Name the components of the integumentary system.
2. What type of tissue is the hypodermis, and what are its functions?
3. What type of tissue is the dermis? What is responsible for the structural strength of the dermis? What are papillae and what function do they perform?
4. What kind of tissue is the epidermis? Where are new epidermal cells formed and where are they sloughed?
5. What makes the skin resistant to abrasion and water loss?
6. Describe two factors that determine the amount of melanin produced in the skin.
7. How do melanin and blood affect skin color?
8. Define the root, shaft, and hair bulb of a hair. What kind of cells are found in a hair?
9. What is a hair follicle? Why is it important in the repair of skin?
10. Does hair grow from the tip of the hair shaft? What are the stages of hair growth?
11. What happens when the arrector pili of the skin contract?
12. What secretion is produced by the sebaceous glands? What is the function of the secretion?
13. Describe the two types of secretions produced by sweat glands. What happens to these secretions?
14. Name the parts of a nail. Where is the nail produced, and what kind of cells make up a nail? What is the lunula? How does nail growth differ from hair growth?
15. How does the integumentary system provide protection?
16. How does the integumentary system assist in the regulation of body temperature?
17. Describe the production of vitamin D by the body. What is the function of vitamin D?
18. List the types of sensations detected by receptors in the skin.
19. What changes occur in the skin as a result of age?

CONCEPT REVIEW

1. A woman has stretch marks on her abdomen, yet she states that she has never been pregnant. Is this possible?
2. Harry Fastfeet, a Caucasian, jogs on a cold day. What color would you expect his skin to be (1) after going outside and just before starting to run, and (2) during the run?
3. Consider the following statement: Dark-skinned children are more susceptible to rickets (insufficient calcium in the bones) than fair-skinned children. Defend or refute this statement.
4. Pulling on hair can be quite painful, yet cutting hair is not painful. Explain.

CHAPTER TEST

Answers can be found in Appendix C

MATCHING For each statement in column A select the correct answer in column B (an answer may be used once, more than once, or not at all.

A	B
1. Result from damage to the dermis.	albinism
2. Dermal projections covered by the epidermis.	cyanosis
3. Where epidermal cells are produced.	hair bulb
4. Protein that gives the stratum corneum its structural strength.	hair follicle
5. Skin pigment.	hair root
6. Inability to produce melanin.	keratin
7. Bluish color resulting from decreased oxygen.	melanin
8. Where cells of a hair are formed.	nail body
9. Produces sebum.	nail matrix
10. Where cells of the nail are formed.	nail root
	papillae
	sebaceous gland
	stratum basale
	stratum corneum
	striae

FILL-IN-THE-BLANK Complete each statement by providing the missing word or words.

1. The ________ is a layer of dense connective tissue that is responsible for most of the structural strength of the skin.
2. The ________ is a layer of loose connective tissue that attaches the skin to underlying bone and muscle.
3. Fluid loss through the skin is prevented by ________ that surround the cells of the epidermis.
4. ________ produce melanin and distribute the melanin to other cells of the epidermis.
5. The ________ is an epidermal projection into the dermis that functions to hold a hair in place.
6. The ________ contract and cause hair to "stand on end" and also produce "goose bumps."
7. There are two types of sweat glands. One type produces ________ secretions that function to cool the body, whereas the other type produces ________ secretions that are broken down by bacteria to produce body odor.
8. An inactive form of ________ is produced in skin exposed to ultraviolet light; once activated, this substance promotes calcium uptake in the small intestine.
9. Heat can be lost from the skin when blood vessels ________.
10. Skin exposed to ultraviolet light ages ________ rapidly than nonexposed skin.

CHAPTER

6

The Skeletal System

OBJECTIVES

After reading this chapter you should be able to:

1. Describe the functions of the skeletal system.
2. Name the parts of a long bone and describe the differences between spongy and compact bone.
3. Describe bone formation, growth, remodeling, and repair.
4. State the main features of the skull.
5. List the number and type of vertebrae that form the vertebral column.
6. List the bones of the thoracic cage, including the three types of ribs.
7. Name and describe the bones of the pectoral girdle and upper limb.
8. Name and describe the bones of the pelvic girdle and lower limb.
9. List and describe the various types of joints.
10. Describe the major types of joint movement.

KEY TERMS

appendicular skeleton
The limbs and the bones that attach the limbs to the trunk.

articulation
A place where two bones come together; a joint.

axial skeleton
Skull, vertebral column, and thoracic cage.

compact bone
Bone that is more dense and has fewer spaces than spongy bone.

osteoblast
Bone-forming cell.

osteoclast
Large multinucleated cell that breaks down bone.

osteocyte
Mature bone cell surrounded by bone matrix.

spongy bone
Bone with a latticelike appearance, having spaces filled with marrow.

synovial joint
Joint in which the ends of the bones are separated by a fluid-filled joint cavity and held together by a joint capsule.

FEATURES

- Understanding the vertebral column 80
- Disorders of bones 88
- Joint disorders 93

THE TERM SKELETON is derived from a Greek word meaning dried, indicating that the skeletal system is the dried, hard parts left after the softer parts are removed. The skeletal system consists of bones, cartilage, and ligaments. This chapter considers the functions of the skeletal system, emphasizes the anatomy of bones, and describes the connections or joints between bones that allow different movements. In addition, the dynamic nature of the skeletal system is considered. It is capable of growth, adapts to stress, and repairs itself after injury.

FUNCTIONS OF THE SKELETAL SYSTEM

The skeletal system provides support and protection, allows body movements, stores minerals and fats, and is the site of blood cell production.

1. *Support*. Rigid, strong bone is well suited for bearing weight and is the major supporting element of the body. Cartilage provides a firm, yet flexible support within certain structures such as the nose, external ear, and rib cartilages.
2. *Protection*. Bone is hard and protects the organs it surrounds. For example, the brain is enclosed and protected by the skull, and the spinal cord is surrounded by the vertebrae. The heart, lungs, and other organs of the thorax are protected by the rib cage.
3. *Movement*. Skeletal muscles are attached to bones by **tendons,** which are bands of dense connective tissue. Contraction of the skeletal muscles causes the bones to move, producing body movements. **Joints,** which are formed where two or more bones come together, permit and control the movement between bones. Smooth cartilage covers the ends of bones within some joints, allowing the bones to move freely. At the same time, **ligaments,** which are dense connective tissue bands, connect bones to each other and prevent excessive movements.
4. *Storage*. Some minerals in the blood are taken into bone and stored. Should blood levels of these minerals decrease, they are released from bone into the blood. The principal minerals stored are calcium and phosphorus. Bone also stores fat (adipose tissue) within cavities of the bone. If needed, the fat molecules are released into the blood and used by other tissues as a source of energy.
5. *Blood cell production*. The cavities of many bones contain bone marrow, which gives rise to blood cells and platelets (see Chapter 11) that leave the cavities and enter blood vessels.

GENERAL FEATURES OF BONE

Bone shapes

Individual bones can be classified according to their shape as long, short, flat, or irregular. **Long bones** are longer than they are wide. Most of the bones of the upper and lower limbs are long bones. **Short bones** are approximately as broad as they are long, such as the bones of the wrist and ankle. **Flat bones** have a relatively thin, flattened shape. Examples of flat bones are certain skull bones, ribs, scapulae (shoulder blades), and the sternum (breastbone). **Irregular bones** include the vertebrae (backbones) and facial bones with shapes that do not fit readily into the other three categories.

Bone anatomy

A typical long bone consists of a shaft, called the **diaphysis** (di-af´ĭ-sis), and an **epiphysis** (e-pif´ĭ-sis; pl. epiphyses) at each end of the bone (Figure 6-1, *A*). The epiphyses within joints are covered by **articular cartilage** (Figure 6-1, *B*). The remaining exterior bone surfaces are covered by a connective tissue membrane called the **periosteum** (pĕr´e-os´te-um), which contains blood vessels and nerves. Bones contain cavities such as the large **medullary cavity** in the diaphysis, and a network of small cavities in the epiphyses of long bones and in the interior of other bones. These spaces are filled with either **yellow marrow,** which consists mostly of fat, or **red marrow,** which consists of blood-forming cells (see Chapter 11). In adults, the medullary cavity contains yellow marrow.

Bone histology

Bone tissue consists of cells surrounded by matrix (see Chapter 4). The cells, called **osteocytes** (os´te-o-sītz), are located within spaces of the matrix called **lacunae** (lă-ku´ne). Bone matrix contains collagen and minerals, including calcium and phosphate. The matrix resembles reinforced concrete. Collagen, like reinforcing steel bars, lends flexible strength to the matrix. The mineral component, like concrete, gives the matrix compressional or weight-bearing strength.

1 What would happen to a bone if all the mineral were removed? What would happen if all the collagen were removed?*

There are two major types of bone, based on their structure: spongy and compact bone. **Spongy bone** forms the epiphyses of long bones and the interior of other bones. It has a network of cavities formed by interconnecting plates of bone (Figure 6-1, C). Like scaffolding, the plates of bone provide structural strength. The spaces between the bony plates are filled with red or yellow marrow. Usually no blood vessels enter the bony plates. Because the plates of bone are

*Answers to predict questions appear in Appendix B at the back of the book.

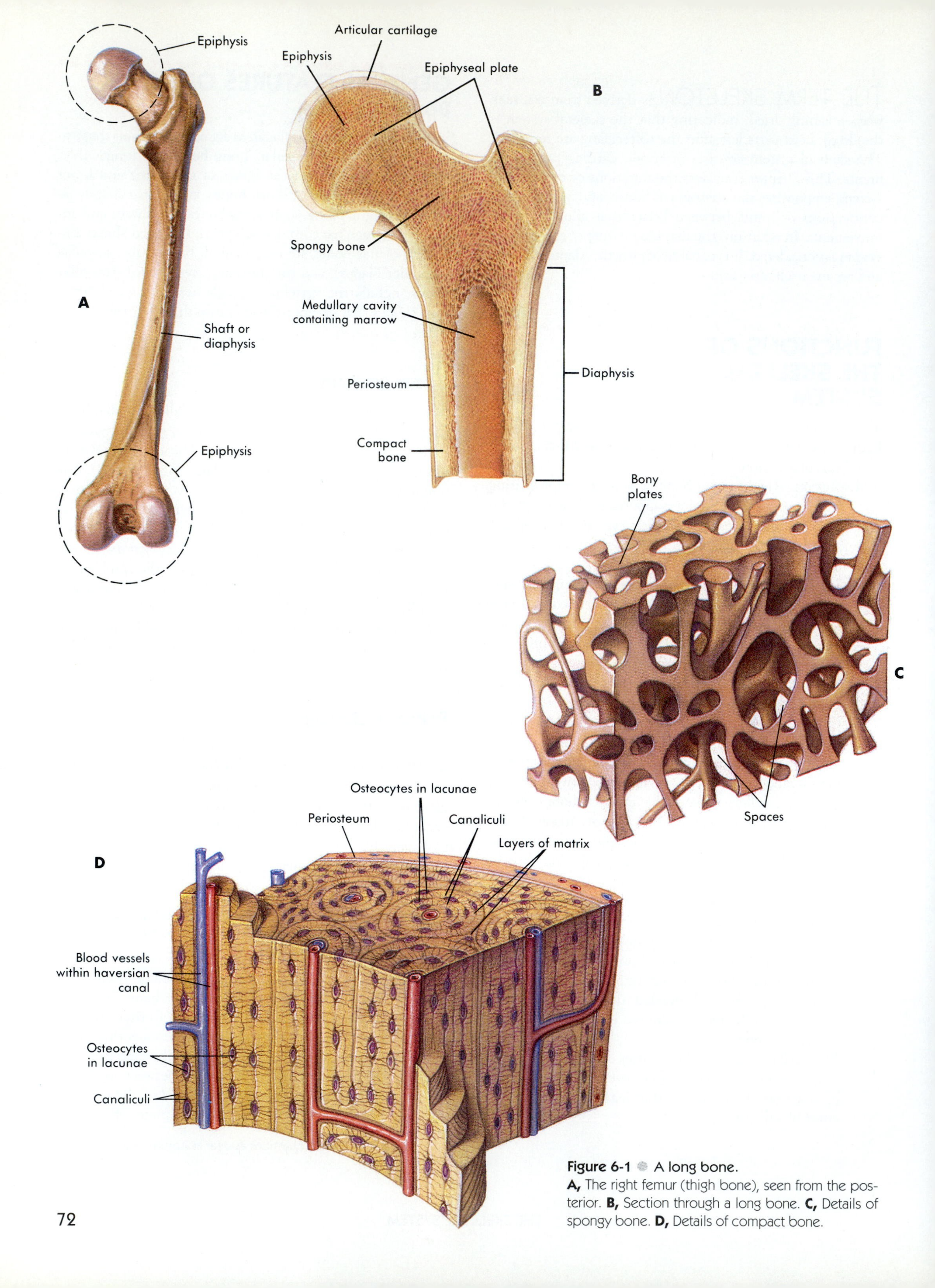

Figure 6-1 ● A long bone.
A, The right femur (thigh bone), seen from the posterior. **B,** Section through a long bone. **C,** Details of spongy bone. **D,** Details of compact bone.

thin, however, nutrients and wastes can diffuse between the osteocytes of the plates and blood vessels within the marrow.

Compact bone is mostly solid matrix and is very strong (Figure 6-1, *D*). It forms the diaphysis and the outer covering of bones. Compact bone consists of osteocytes located between thin sheets of matrix. The osteocytes are connected to each other by cell processes that extend through tiny canals called **canaliculi** (kan-ă-lik´u-li). Osteocytes receive nutrients from blood vessels that run parallel to the long axis of the bone within **haversian** (hă-ver´shan) **canals.** Nutrients leave the vessels of the haversian canals and diffuse to the osteocytes through the canaliculi. Waste products diffuse in the opposite direction.

Bone formation and growth

Before birth, the skeleton begins as either fibrous membranes or cartilage models. **Osteoblasts** (os´te-o-blastz), which are bone-forming cells, lay down bone matrix and replace the membranes or cartilage models with bone. After an osteoblast becomes completely surrounded by bone matrix, it is called an osteocyte.

The process of bone formation is not complete at birth. For example, the membranes that form around the brain during development are not completely ossified (converted to bone). The portions of the membrane that are not ossified at birth are called **fontanels** (fon´tă-nelz), or **soft spots** (Figure 6-2). They allow movement of skull bones during the birth process and allow growth of the head after birth. The

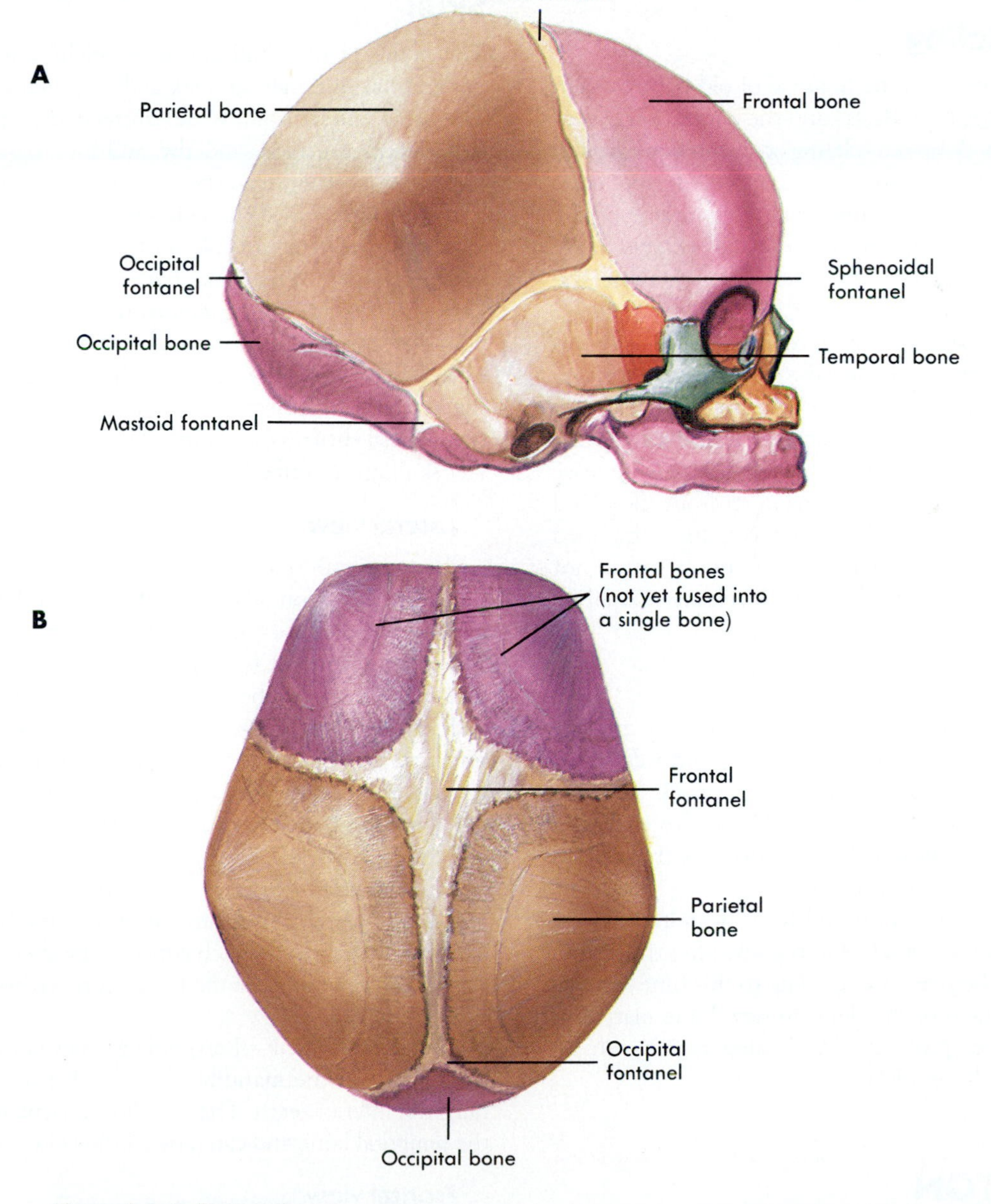

Figure 6-2 ● Fontanels.
Most of the fetal skull is bone. The membrane model that has not been converted to bone remains as the fontanels. **A,** Lateral view. **B,** Superior view.

fontanels are converted to bone by approximately 18 to 24 months of age.

After birth, bones continue to increase in diameter and length. Increased diameter results when osteoblasts add bone matrix to the outside surface of bone beneath the periosteum. Increased length occurs at the **epiphyseal** (e-pif´-ĭ-se-al) **plate,** a layer of cartilage between the epiphysis and diaphysis (see Figure 6-1, *B*). The cartilage grows, making the bone longer, and osteoblasts replace the cartilage with bone. Depending on the bone, between 12 and 25 years of age the cartilage of the epiphyseal plate stops growing and is converted to bone.

2 What would happen to a person if cartilage growth and bone formation at the epiphyseal plate did not occur?

Bone remodeling

Bone remodeling involves the removal of old bone by cells called **osteoclasts** (os´te-o-klastz) and the deposition of new bone by osteoblasts. Bone remodeling occurs in bone growth and calcium ion regulation. For example, during bone growth, osteoclasts remove bone from the surface of the medullary cavity. Consequently, as the bone becomes larger the size of the marrow cavity also increases; otherwise, the bone would be very heavy.

Bone is the major storage site for calcium in the body and bone remodeling is responsible for the movement of calcium into and out of bone. When blood calcium levels decrease below normal, osteoclasts break down bone and release calcium into the blood. When blood calcium levels increase above normal, osteoblasts take up calcium from the blood and deposit it in bone. Blood calcium levels are maintained within very narrow limits. Muscle and nerve tissue do not function properly unless blood calcium levels are normal.

Bone repair

When a bone is broken, blood vessels in the bone are also damaged. The vessels bleed, and a clot forms in the damaged area. Two to 3 days after the injury, blood vessels and cells from surrounding tissues begin to invade the clot to form a **callus** (kal´us). The callus is a fibrous network with cartilage that holds the bone fragments together.

Osteoblasts enter the callus and begin forming bone, a process that usually is complete 4 to 6 weeks after the injury. Immobilization of the bone is critical up to this time because movement can refracture the delicate new bone matrix. If bone healing occurs properly, the healed region can be stronger than the adjacent bone.

THE SKELETON

It is traditional to list 206 bones in the average adult skeleton, although the actual number varies from person to person and decreases with age as some bones become fused. Classified according to location, the bones are part of the axial or appendicular skeleton (Figure 6-3). The **axial** (ak´se-al) **skeleton** includes the bones of the head and trunk and the **appendicular** (ap´pen-dik´u-lar) **skeleton** consists of the limb bones and their **girdles** (attachments to the trunk).

AXIAL SKELETON

The axial skeleton consists of 80 bones divided into the skull (28 bones), hyoid bone, vertebral column (26 bones), and thoracic cage (25 bones).

Skull

The bones of the skull consist of eight brain case bones, fourteen facial bones, and six auditory ossicles (ear bones). The **brain case** surrounds and protects the brain, the **facial bones** form the face, and the **auditory ossicles** (os´ĭ-klz) transmit sound waves to the inner ear (see Chapter 9).

The **hyoid** (hi´oyd) bone is a small U-shaped bone that is located in the neck just inferior to the lower jaw. The hyoid bone is not part of the skull, but is attached to the skull and larynx by muscles and ligaments. It serves as the attachment point for several important neck and tongue muscles.

It is useful to think of the skull as a single unit because the bones are connected to each other by fibrous connections called **sutures** (su´churz). The major features of the intact skull are described as follows.

Lateral view

The **parietal** (pă-ri´ĕ-tal) and **temporal** bones join to form a large portion of the side of the head (Figure 6-4). The parietal bone is also joined to the **frontal** (forehead) bone and to the **occipital** (ok-sip´ĭ-tal) bone. A prominent feature of the temporal bone is an opening called the **external auditory meatus** (me-a´tus), which is a canal that enables sound waves to reach the eardrum. The **mastoid** (mas´-toyd) **process** and the **styloid** (sti´loyd) **process** are projections of the temporal bone that serve as attachment sites for neck and tongue muscles.

The **sphenoid** (sfe´noyd) bone and the **zygomatic** (zi-go-mat´ik), or **cheek, bone** are anterior to the temporal bone. The **zygomatic arch,** which consists of joined processes from the temporal and zygomatic bones, forms a bridge across the side of the face.

The **maxilla** (mak-sil´ah) or upper jaw contains the superior teeth and the **mandible** (man´dĭ-bl) or lower jaw contains the inferior teeth. The mandible articulates (joins) with the temporal bone and can move during chewing and speech.

Frontal view

From this view, the most prominent openings into the skull are the **orbits** (eye sockets) and the **nasal cavity** (Fig-

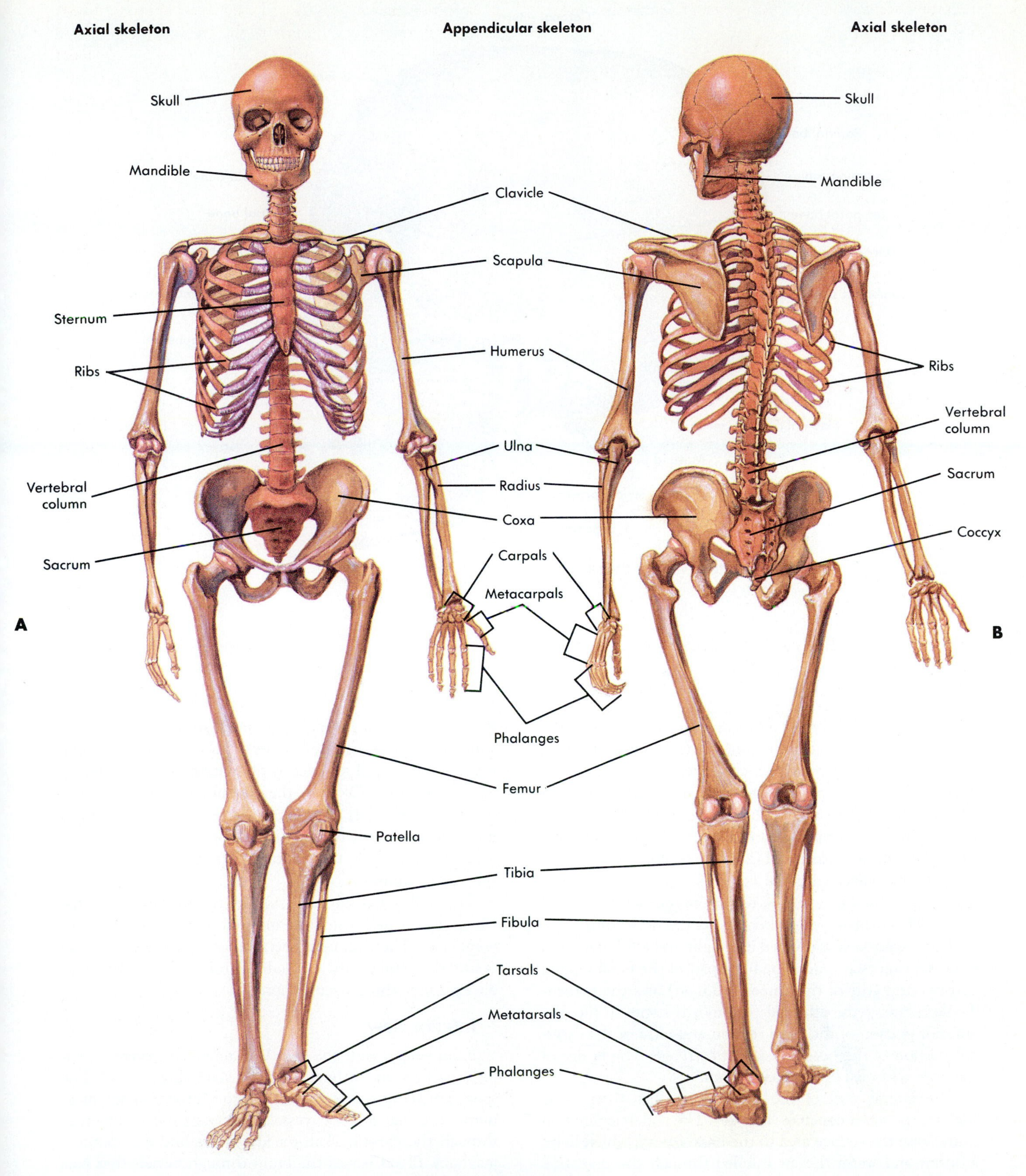

Figure 6-3 ● The skeleton.
A, Anterior view. **B,** Posterior view.

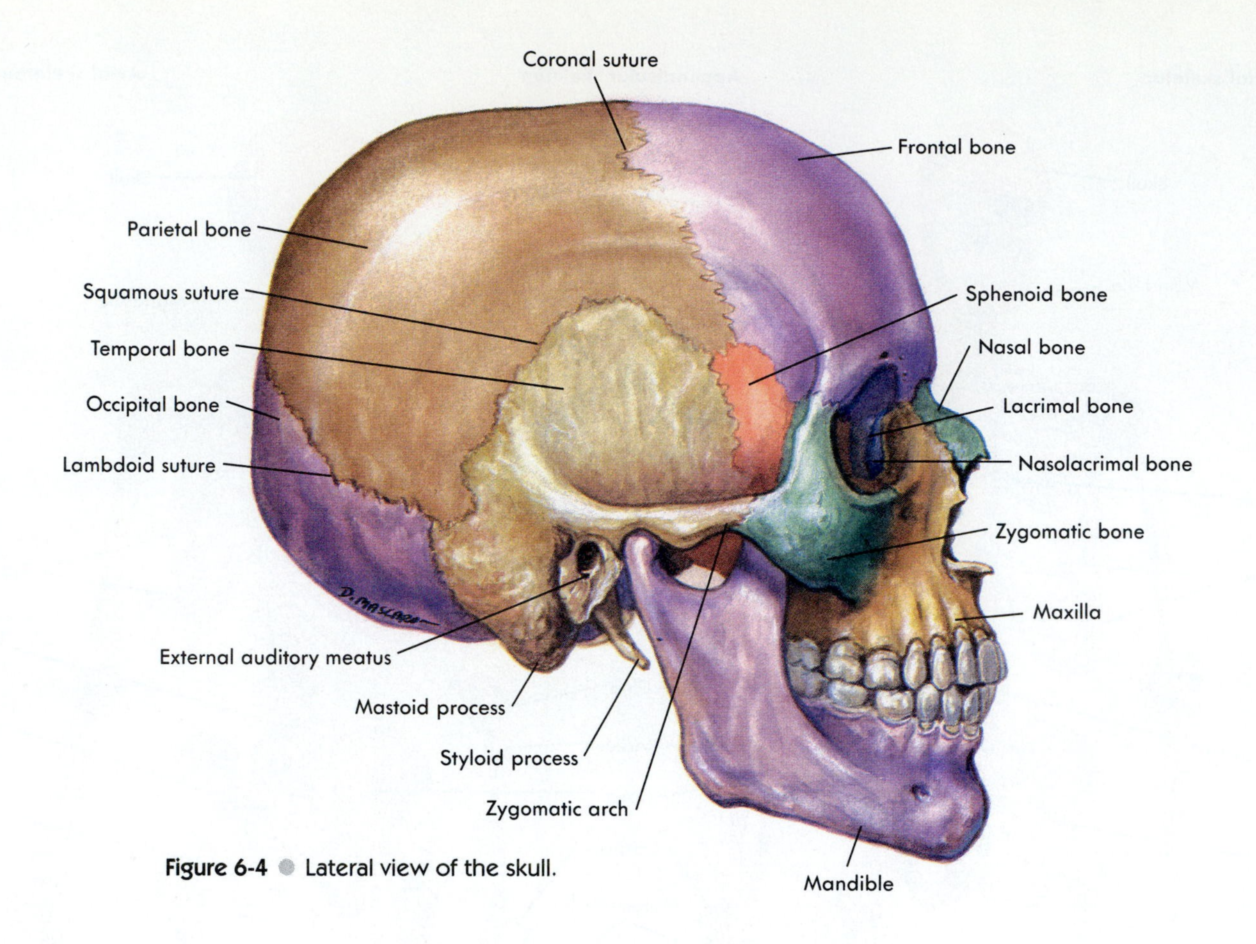

Figure 6-4 ● Lateral view of the skull.

ure 6-5). The bones of the orbits provide protection for the eyes, attachment points for the muscles that move the eyes, and have openings through which nerves, blood vessels, and ducts pass. For example, a **foramen** (fo-ra´men) is a hole in a bone. The optic nerve, for the sense of vision, passes from the eye through the **optic foramen** and enters the brain case. Just below the small **lacrimal** (lak-rĭ-mal) **bone** is a passageway, the **nasolacrimal canal** (see Figure 6-4), which contains a duct that carries excess tears from the eye to the nasal cavity. This explains why when we cry the nose "runs."

The nasal cavity is divided into right and left halves by a **nasal septum** (sep´tum). The bony part of the nasal septum consists primarily of the **vomer** (vo´mer;) and the perpendicular plate of the **ethmoid** (eth´moyd) bone. In life, the anterior portion of the nasal septum and most of the external portion of the nose consists of cartilage. The bridge of the nose is formed by the **nasal** bones.

The lateral wall of the nasal cavity has three bony shelves, the **nasal conchae** (kon´ke). The conchae function to increase the surface area in the nasal cavity, which helps moisten and warm the air inhaled through the nose (see Chapter 15).

Several of the bones associated with the nasal cavity have large cavities called the **paranasal sinuses** (Figure 6-6), which open into the nasal cavity. The sinuses decrease the weight of the skull and act as resonating chambers during voice production. Compare the normal voice with the voice of a person who has a cold and whose sinuses are "stopped up." The sinuses are named for the bones in which they are located and include the frontal, maxillary, ethmoidal, and sphenoidal sinuses.

The skull has additional sinuses, the **mastoid air cells,** which are located inside the mastoid processes of the temporal bone. These air cells open into the middle ear instead of into the nasal cavity. An auditory tube connects the middle ear to the throat (see Chapter 9).

Inferior view

From below, with the mandible removed, several openings can be seen in the base of the skull (Figure 6-7). The spinal cord connects to the brain through the **foramen magnum.** The major blood vessels supplying the brain enter through the **carotid** (kah-rot´id) **canals** and the foramen magnum. Blood leaves the brain through vessels that pass through the **jugular foramina** (jug´u-lar fo-ram´ĭ-nah).

Occipital condyles (kon´dilz), the smooth points of ar-

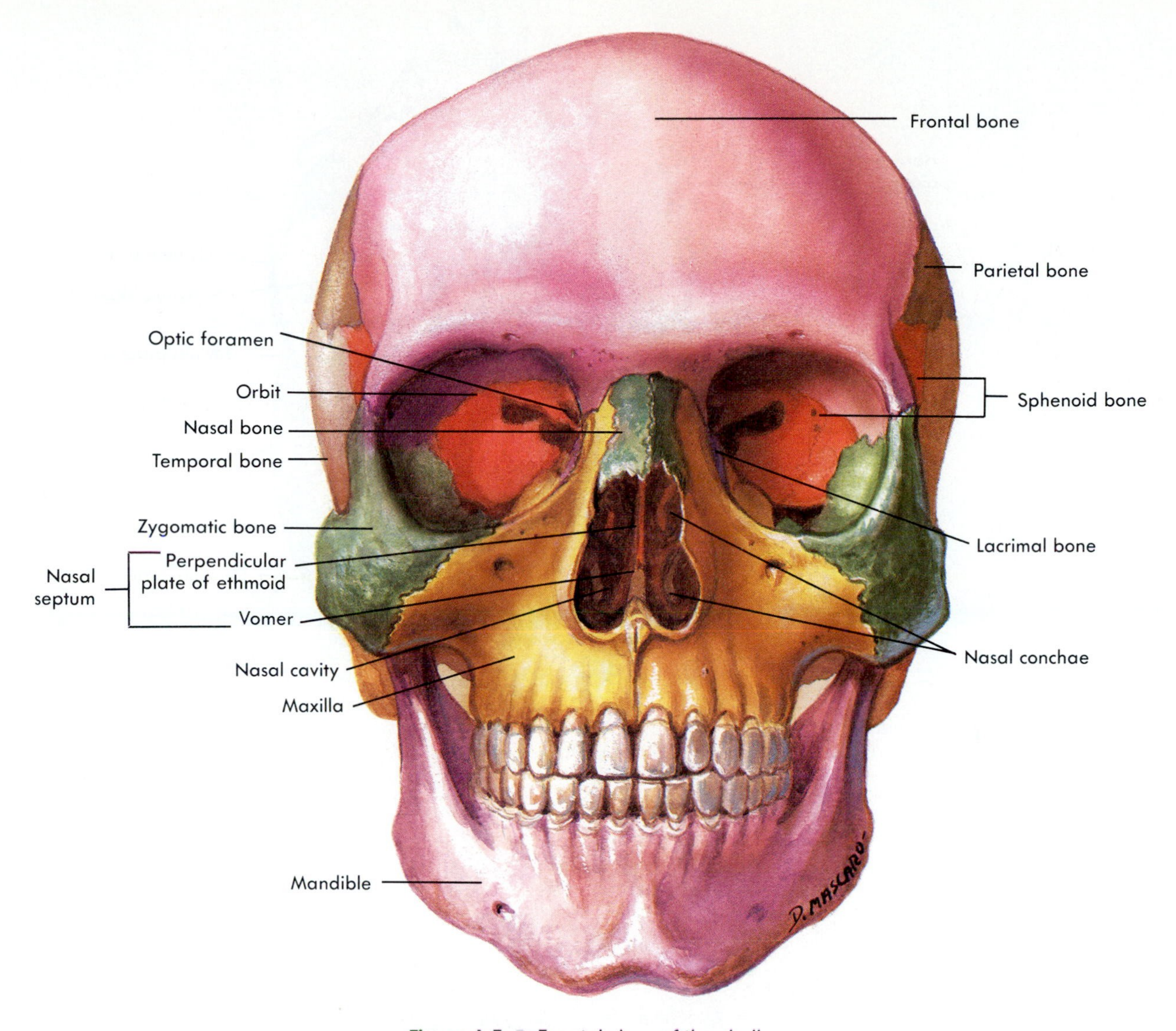

Figure 6-5 ● Frontal view of the skull.

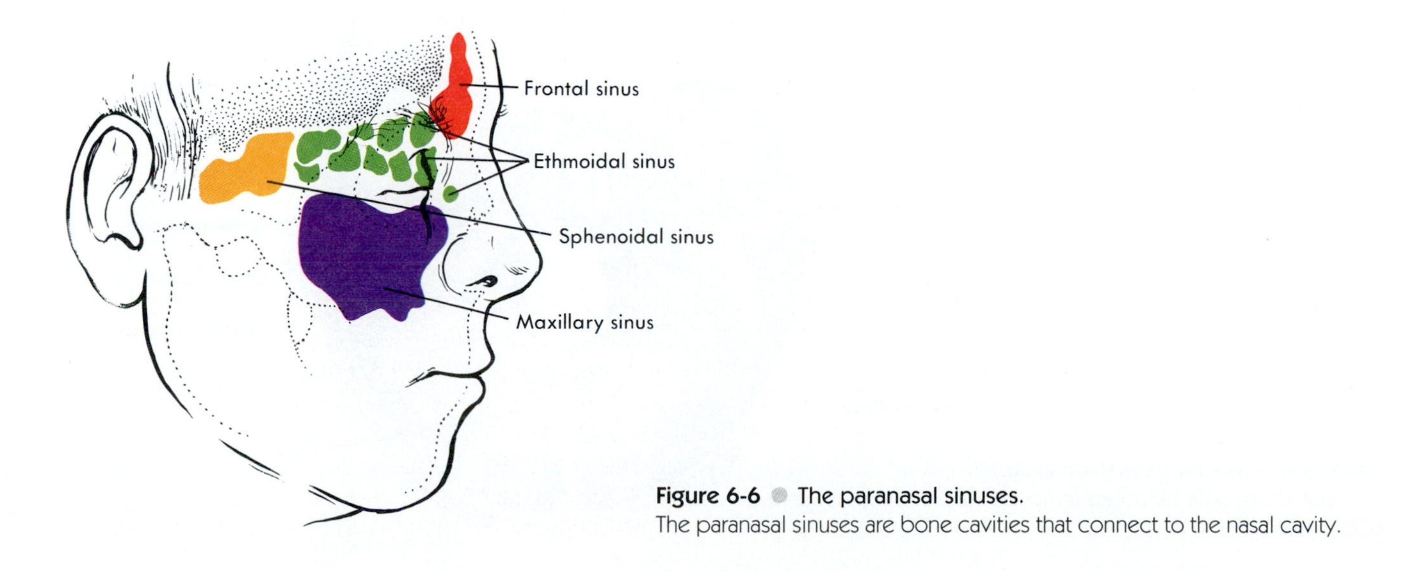

Figure 6-6 ● The paranasal sinuses.
The paranasal sinuses are bone cavities that connect to the nasal cavity.

Hard palate
Maxilla
Palatine bone
Vomer
Zygomatic arch
Sphenoid bone
Mandibular fossa
Styloid process
Temporal bone
Carotid canal
Jugular foramen
Mastoid process
Occipital condyle
Foramen magnum
Occipital bone
D. MASCARO

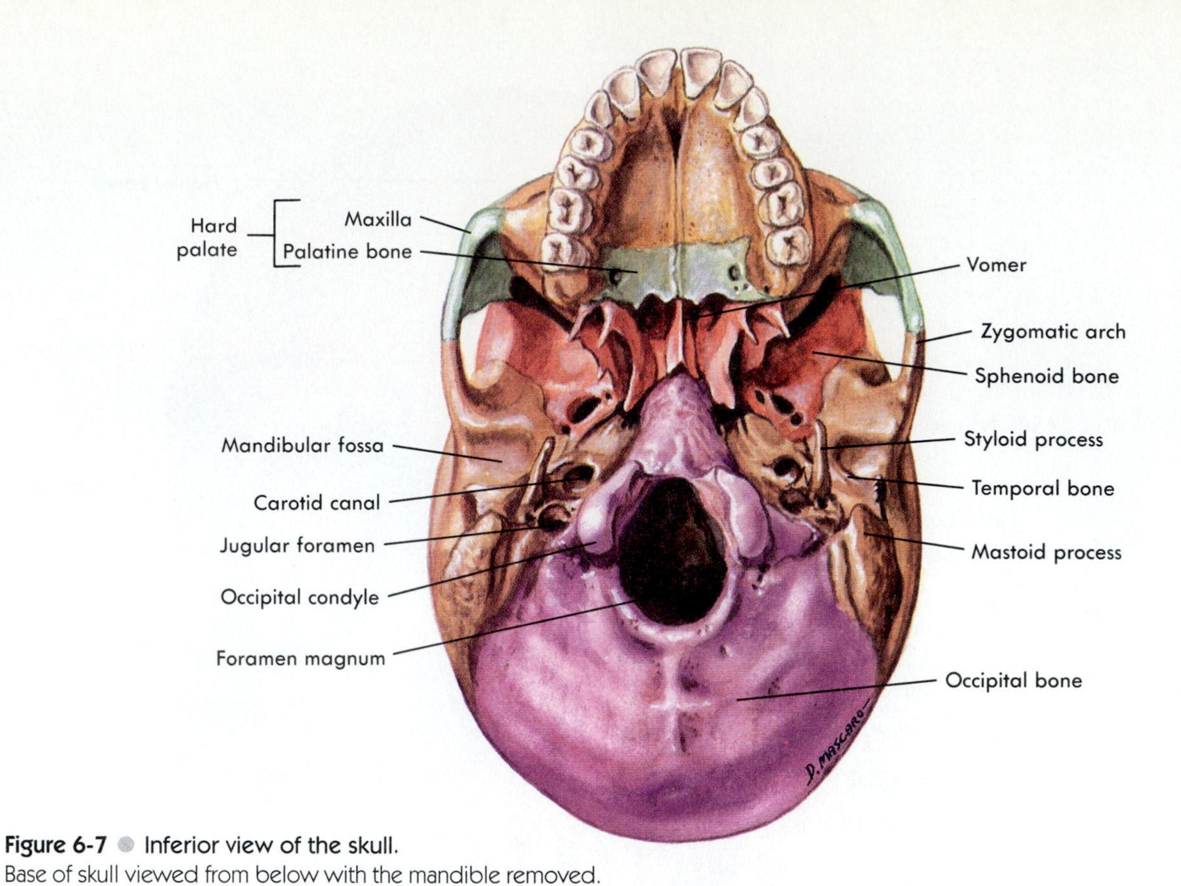

Figure 6-7 ● Inferior view of the skull.
Base of skull viewed from below with the mandible removed.

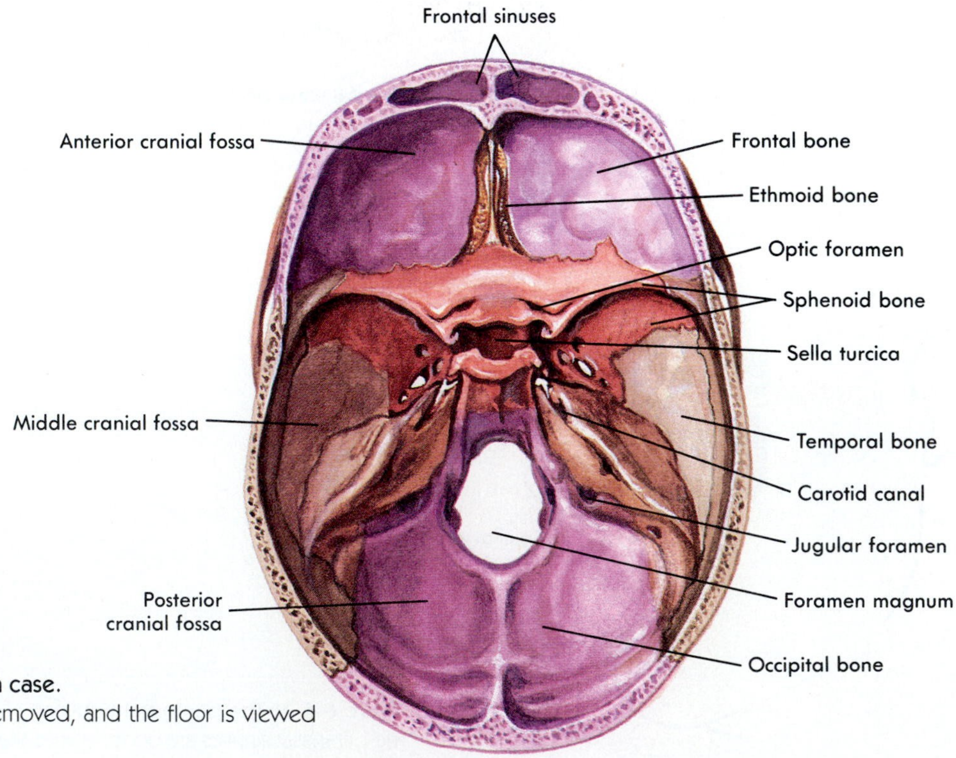

Frontal sinuses
Anterior cranial fossa
Frontal bone
Ethmoid bone
Optic foramen
Sphenoid bone
Sella turcica
Middle cranial fossa
Temporal bone
Carotid canal
Jugular foramen
Foramen magnum
Posterior cranial fossa
Occipital bone

Figure 6-8 ● Floor of the brain case.
The roof of the skull has been removed, and the floor is viewed from above.

ticulation between the skull and the vertebral column, are located on either side of the foramen magnum. The **mandibular fossa** (fos´ah), where the mandible articulates with the temporal bone, is anterior to the mastoid process.

The **hard palate** forms the floor of the nasal cavity. The anterior two thirds is formed by the maxillae (mak-sil´e) and the posterior one third by the **palatine** (pal´ă-tīn) **bones.** The hard palate functions to separate the nasal cavity from the mouth, enabling us to eat and breathe at the same time. A **cleft palate** is an opening between the oral cavity and the nasal cavity that results if the bones of the palate do not properly join together during development.

Interior of the skull

The floor of the brain case, when viewed from above with the roof cut away (Figure 6-8), consists of depressions called the **cranial fossae** (fos´e), which hold the brain. The **sphenoid** bone, which extends from one side of the skull to the other side, forms a structure resembling a saddle, the **sella turcica** (sel´ah tur´si-kah; Turkish saddle). The pituitary gland is located in the sella turcica (see Chapter 10).

Vertebral column

The **vertebral column** performs four major functions: (1) supports the weight of the head and trunk; (2) protects the spinal cord; (3) allows spinal nerves to exit the spinal cord; and (4) allows movement of the head and trunk (see Understanding the Vertebral Column).

The vertebral column consists of five regions, usually containing 26 bones (Figure 6-9): 7 **cervical vertebrae,** 12 **thoracic vertebrae,** 5 **lumbar vertebrae,** 1 **sacrum** (sa´krum), and 1 **coccyx** (kok´siks). Approximately 34 vertebrae form during development, but by adulthood five sacral vertebrae fuse to form the sacrum, and five coccygeal vertebrae fuse to form the coccyx.

Associated with the regions of the vertebral column, the adult vertebral column has four major curvatures: the cervical, thoracic, lumbar, and sacral/coccygeal curves (see Figure 6-9). The curves allow the vertebral column to flex and absorb shock. At birth, however, the vertebral column is curved in only one direction, which is posterior. This is called a **primary curvature.** The thoracic and sacral/coccygeal curves maintain the primary curvature throughout life, but the cervical and lumbar curves change as the baby matures. When the baby raises its head the cervical region becomes curved anteriorly. Later, when the infant walks, the lumbar region also becomes curved anteriorly. These changes in orientation are called **secondary curvatures.**

Abnormal vertebral curvatures are not uncommon. **Scoliosis** (sko-le-o´sis) is an abnormal lateral curvature of the **spine. Kyphosis** (ki-fo´sis), or hunchback, is an abnormal posterior curvature of the spine, and **lordosis** (lor-do´sis), or swayback, is an abnormal anterior curvature of the spine.

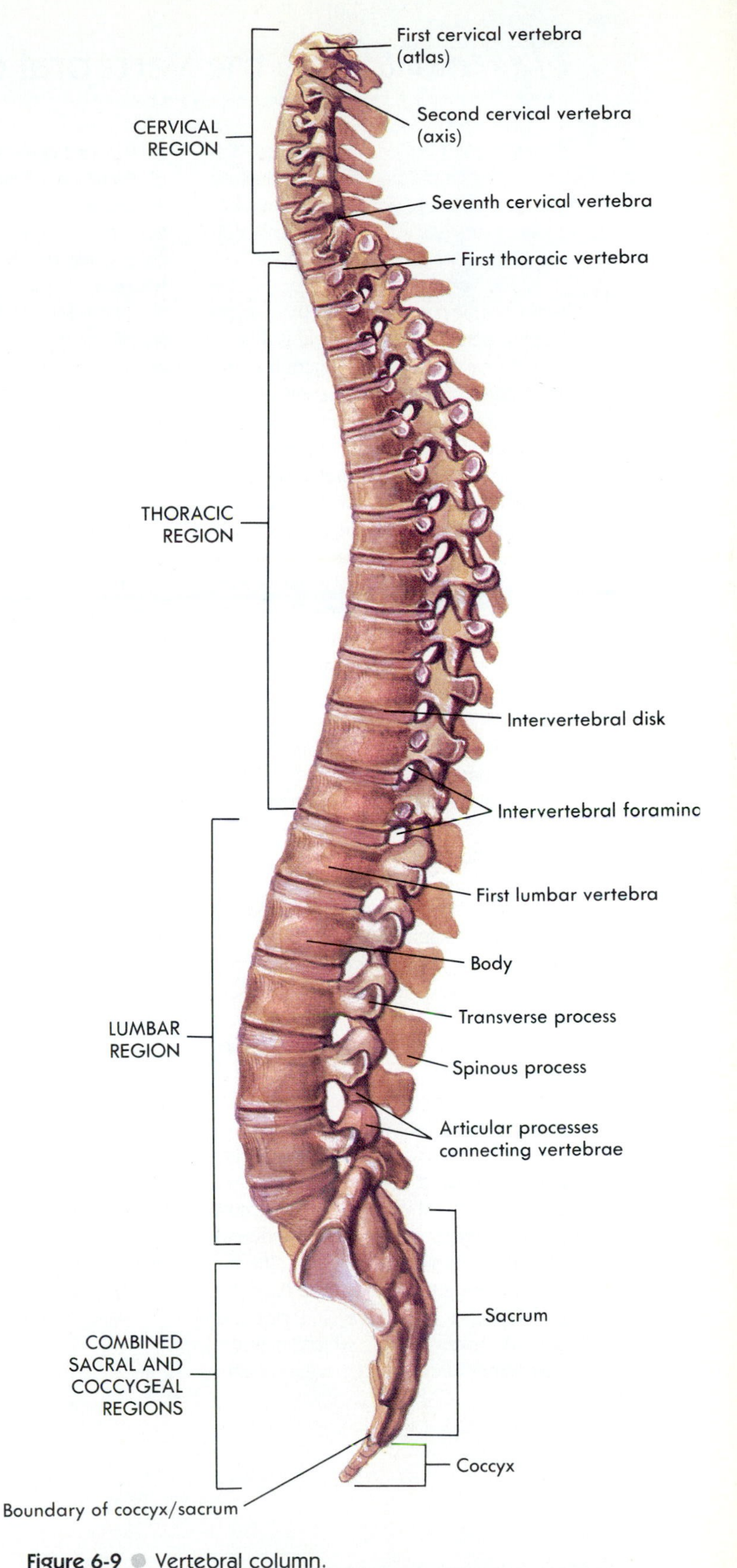

Figure 6-9 Vertebral column.
Viewed from the left side, the adult vertebral column has four curves.

Understanding the vertebral column

Structure and function are related to each other. This simple principle is beautifully illustrated by the vertebral column because its structure allows the vertebral column to perform four major functions:

1. *The vertebral column supports the weight of the head and trunk.* The weight-bearing portion of the vertebra is a bony cylinder called the **body** (Figure 6-A, *A*). The bodies of adjacent vertebrae are positioned over each other so that weight is transmitted from one body to the next.

During life, **intervertebral disks** of fibrocartilage, which are located between the bodies of the vertebrae, provide additional support and prevent the vertebral bodies from rubbing against each other. The disks consist of an external fibrous ring called the **anulus fibrosus** (an′u-lus fi-bro′sus), and an internal gelatinous material called the **nucleus pulposus** (pul-po′sus). The disk is like a jelly-filled donut with a tougher outer layer surrounding a softer, inner core. Because the disks are somewhat elastic, the body's weight daily compresses them. We are actually a half to three-quarters of an inch shorter at the end of the day than in the morning. During the night the disks are not under pressure and regain their shape. With increasing age, however, the disks become permanently compressed and height is permanently reduced.

A **herniated disk** results from the rupture of the anulus fibrosus with a partial or complete release of the nucleus pulposus (Figure 6-A, *B*). Just imagine squeezing a jelly-filled donut to understand what happens. A herniated disk is sometimes called a slipped disk, which is incorrect because the disk is tightly attached to the vertebrae and does not slip out of place. The nucleus pulposus that is squeezed out of the disk, however, can push against the spinal cord or spinal nerves, which interrupts their normal function and produces pain.

Herniated disks can be repaired in one of several ways. One procedure uses prolonged bed rest so that the herniated portion of the disk recedes and the anulus fibrosus repairs itself. In many cases, however, surgery is required, and the damaged disk is removed. To enhance the stability of the vertebral column, a piece of hipbone sometimes is inserted into the space previously occupied by the disk and the adjacent vertebrae are fused together.

2. *The vertebral column protects the spinal cord.* The vertebral column protects the spinal cord by surrounding it with bone. The **vertebral arch** and the dorsal portion of the vertebral body surround a large opening, the **vertebral foramen** (Figure 6-A, *C*). The vertebral foramen of adjacent vertebrae combine to form the **vertebral canal,** which contains the spinal cord.

Sometimes the vertebral arches can fail to form or can be incompletely formed during development, resulting in a condition called **spina bifida** (spi′nah bif′ĭ-dah). If the defect is severe, the coverings of the spinal cord and even the spinal cord itself can protrude through the opening to form a conspicuous bulge on the baby's back. Damage to the spinal cord can interfere with normal nervous system function below the point of the defect.

3. *The vertebral column provides an exit site for spinal nerves.* Because the spinal cord is surrounded by bone, there must be some way for spinal nerves to leave the spinal cord and extend to different parts of the body. An opening for this purpose, an **intervertebral foramen,** is formed by notches in the vertebral arches of adjacent vertebrae (see Figure 6-A, *A*).

4. *The vertebral column allows movement of the head and trunk.* Each vertebra has a **superior** and an **inferior articular process** (see Figure 6-A, *A*). The superior articular process of one vertebra joins the inferior articular process of the next superior vertebra. The processes can slide over one another, allowing a small amount of movement between adjacent vertebrae. In addition, when the vertebral column bends in one direction, the intervertebral disks are slightly compressed on one side and slightly expanded on the opposite side, allowing the bodies to move relative to each other. When the small amounts of movement between all the vertebrae are taken together, the vertebral column is capable of a considerable range of motion. At the same time, because the vertebrae are joined together, the stability of the vertebral column is increased, allowing the vertebral column to bear weight more effectively.

Movement of the vertebral column results from the contraction of skeletal muscles attached to the transverse and spinous processes (Figure 6-A, *D*). Each vertebra has a **transverse process** extending laterally from each side of the vertebral arch, and a single **spinous process** extending posteriorly from the vertebral arch. The spinous processes can be seen and felt as a series of lumps down the midline of the back.

3 If only the muscles attached to the spinous processes contracted, what movement of the vertebral column would result? If only the muscles attached to the transverse processes on one side of the vertebral column contracted, what movement of the vertebral column would result?

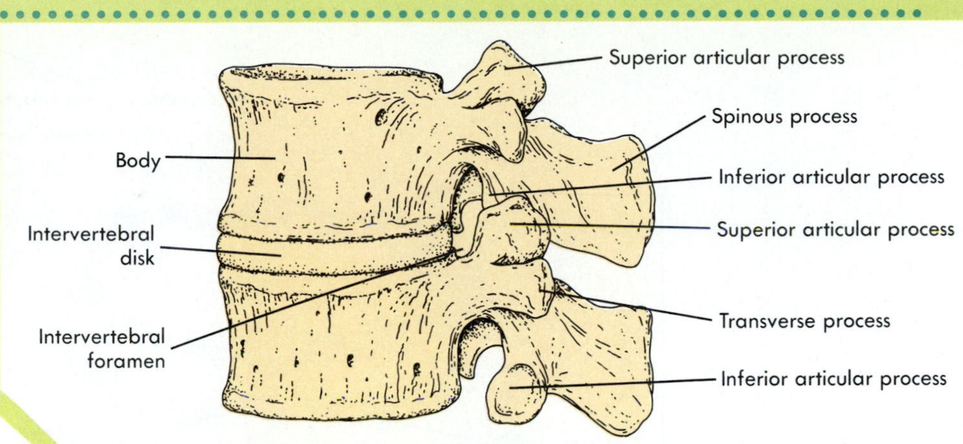

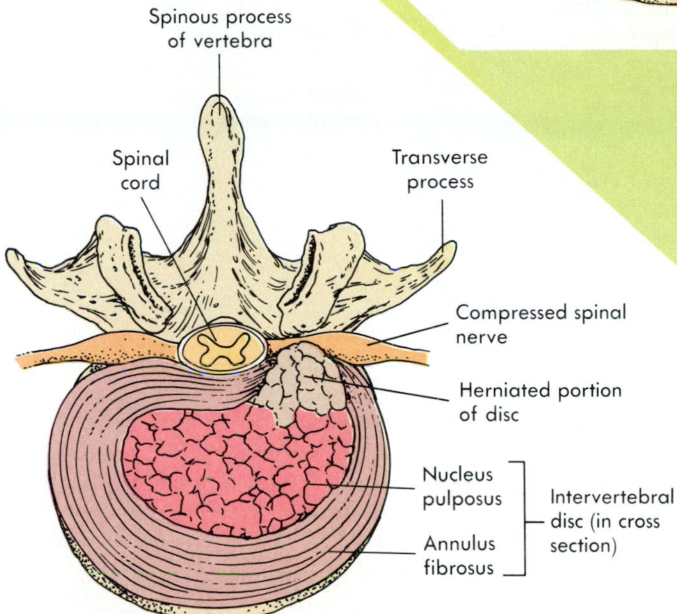

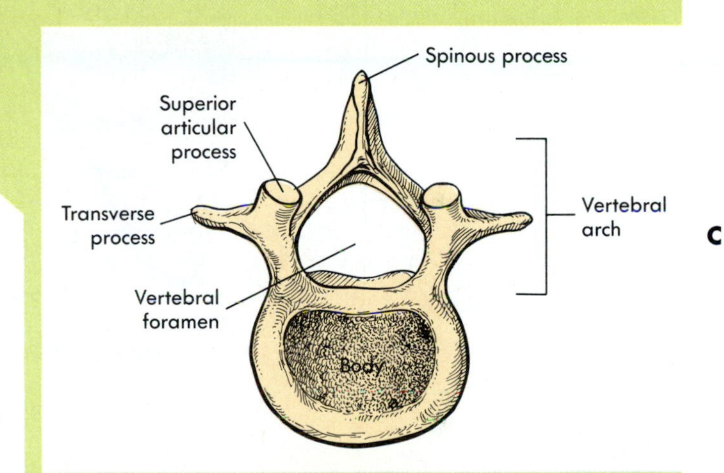

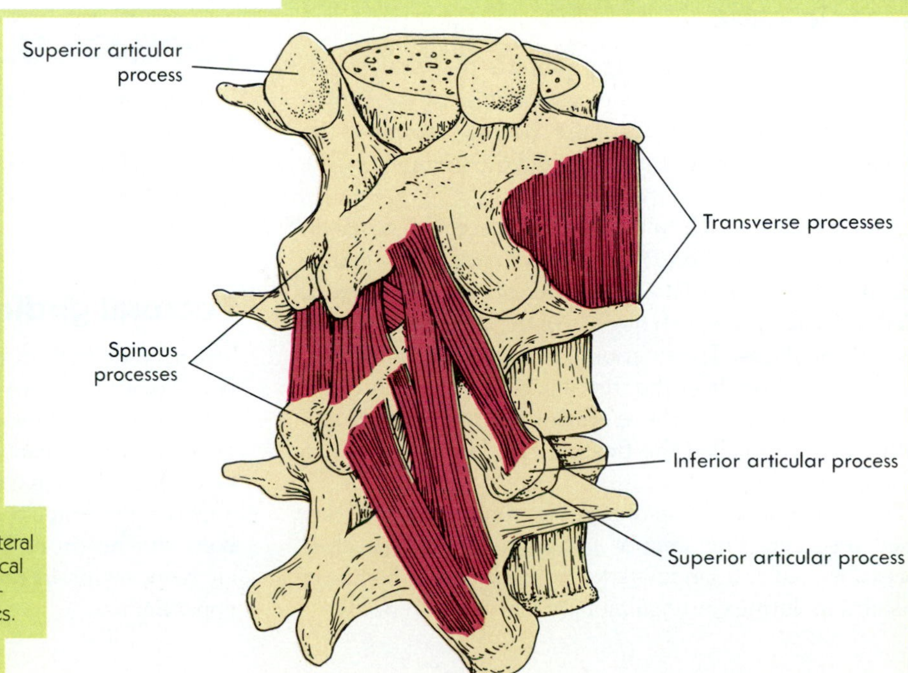

Figure 6-A Vertebrae.
A, Two vertebrae and an intervertebral disk, lateral view. **B,** Herniated disk, superior view. **C,** Typical vertebra, superior view. **D,** Skeletal muscles attached to the transverse and spinous processes.

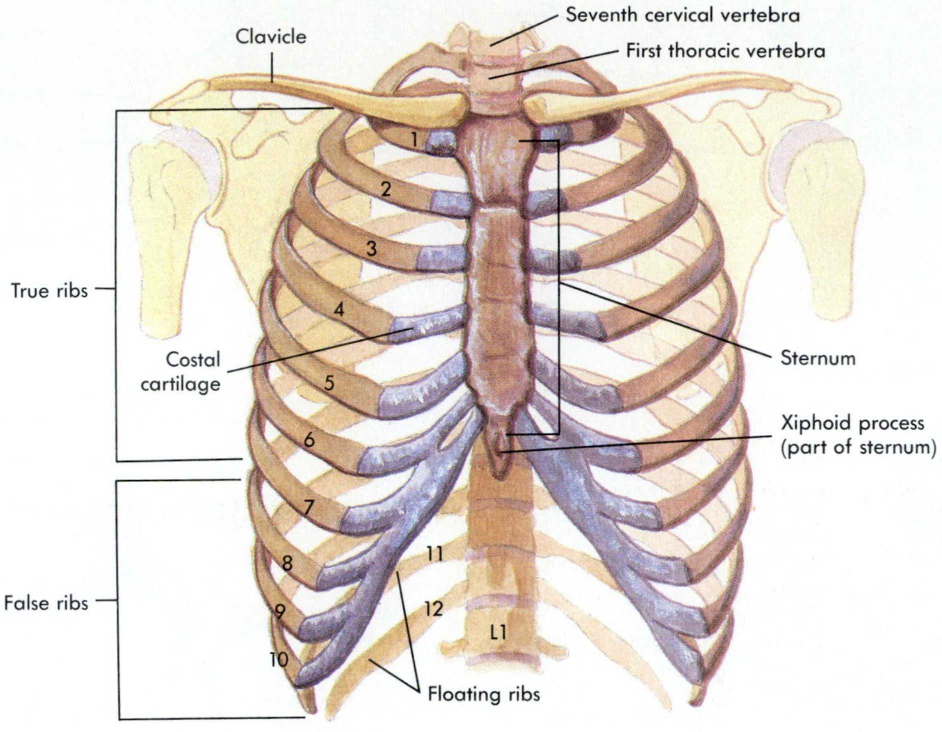

Figure 6-10 Thoracic cage. Anterior view.

Thoracic cage

The **thoracic cage,** or **rib cage,** protects the vital organs within the thorax and prevents the collapse of the thorax during respiration. It consists of the thoracic vertebrae, the ribs with their associated cartilages, and the sternum (Figure 6-10).

All of the ribs attach to the thoracic vertebrae. Based on their attachment to the sternum, the 12 pairs of ribs can be divided into true and false ribs. The superior seven pairs, called the **true ribs,** attach directly to the sternum by means of **costal cartilages.** The inferior five pairs, called **false ribs,** do not attach directly to the sternum. Ribs 8 through 10 attach to the cartilage of the seventh rib. The eleventh and twelfth ribs, also called the **floating ribs,** do not attach to the sternum.

The **sternum,** or **breastbone,** has three parts that are joined together. The inferior part, the **xiphoid** (zi´foyd) **process,** is used as a landmark for positioning the hands on the sternum during cardiopulmonary resuscitation (CPR).

APPENDICULAR SKELETON

The **appendicular** (ap´pen-dik´u-lar) **skeleton** consists of the bones of the pectoral girdle (4 bones), upper limbs (24 bones), pelvic girdle (2 bones), and lower limbs (26 bones). The girdles attach the limbs to the axial skeleton.

Pectoral girdle

The **pectoral** (pek´to-ral), or **shoulder, girdle** consists of two bones that attach the upper limb to the body: the **scapula** (skap´u-lah), or **shoulder blade,** and the **clavicle** (klav´ĭ-kl), or **collar bone** (Figure 6-11). The scapula functions as an attachment site for muscles that move the arm. The clavicle attaches the scapula to the sternum, providing the only bony attachment of the upper limb to the body. The clavicle helps maintain the normal position of the scapula and upper limb.

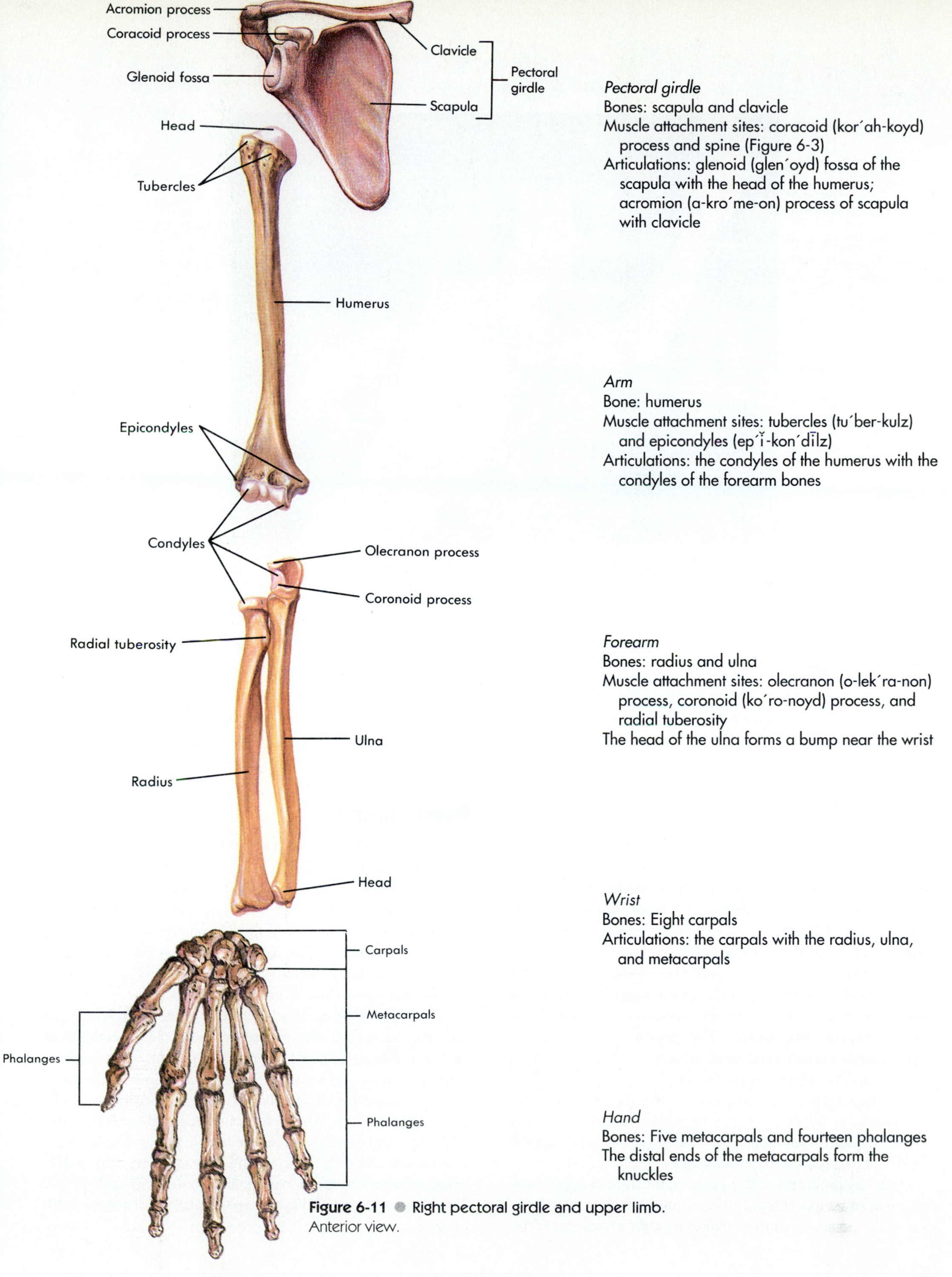

Pectoral girdle
Bones: scapula and clavicle
Muscle attachment sites: coracoid (kor´ah-koyd) process and spine (Figure 6-3)
Articulations: glenoid (glen´oyd) fossa of the scapula with the head of the humerus; acromion (a-kro´me-on) process of scapula with clavicle

Arm
Bone: humerus
Muscle attachment sites: tubercles (tu´ber-kulz) and epicondyles (ep´ĭ-kon´dīlz)
Articulations: the condyles of the humerus with the condyles of the forearm bones

Forearm
Bones: radius and ulna
Muscle attachment sites: olecranon (o-lek´ra-non) process, coronoid (ko´ro-noyd) process, and radial tuberosity
The head of the ulna forms a bump near the wrist

Wrist
Bones: Eight carpals
Articulations: the carpals with the radius, ulna, and metacarpals

Hand
Bones: Five metacarpals and fourteen phalanges
The distal ends of the metacarpals form the knuckles

Figure 6-11 ● Right pectoral girdle and upper limb. Anterior view.

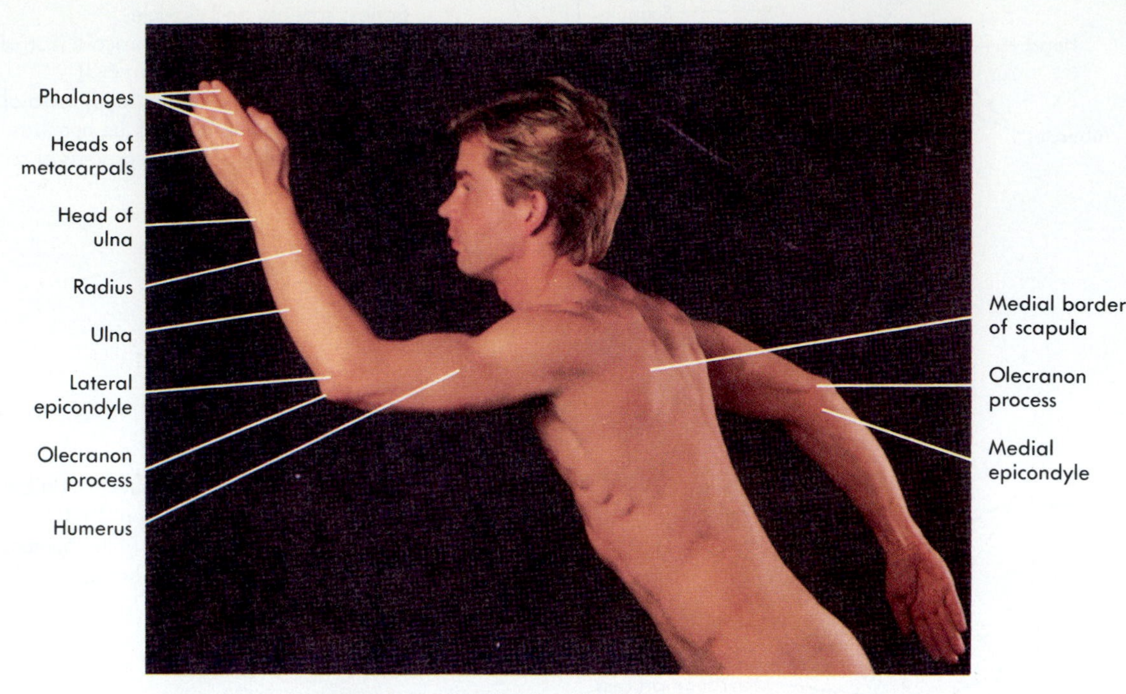

Figure 6-12 ● **Surface anatomy of the upper limb.** Bones and their projections can be seen or felt.

Upper limb

The upper limb consists of the bones of the arm, forearm, wrist, and hand (see Figure 6-11). The arm is the region between the shoulder and the elbow and contains the **humerus** (hu´mer-us). The humerus articulates with the scapula and the forearm bones.

The forearm is the region between the elbow and the wrist. It has two bones, the **ulna** (ul´nah) on the medial side of the forearm (the side with the little finger) and the **radius** on the lateral (thumb) side of the forearm.

The wrist is a relatively short region between the forearm and hand and is composed of eight **carpal** (kar´pul) bones. The bones and ligaments on the anterior side of the wrist form a carpal tunnel, which does not have much "give." Tendons and nerves pass through the carpal tunnel to the hand. If fluid accumulates in the carpal tunnel as a result of overuse or trauma, it may apply pressure to a major nerve passing through the tunnel. The pressure on this nerve causes **carpal tunnel syndrome,** which consists of tingling, burning, and numbness in the hand.

Five metacarpals are attached to the carpal bones and form the palm of the hand. The fingers or digits each consist of three small bones called **phalanges** (fă-lan´-jēz), except the thumb, which has two phalanges.

Many of the bones of the pectoral girdle and upper limb can be seen or felt (Figure 6-12). Knowing the exact location of the bones on yourself makes it easier to remember the bones. In addition many medical procedures or disorders are associated with specific parts of the skeletal system. For example, tennis elbow typically develops at the attachment site of muscles to the lateral **epicondyle** of the humerus.

Pelvic girdle

The **pelvic girdle** is a ring of bones formed by the sacrum and two **coxae** (kok´se) (Figure 6-13, *A*). Each coxa is a single bone formed by three bones fused to each other: the **ilium** (il´e-um), the **ischium** (ish´e-um), and the **pubis** (pu´bis). The **iliac crest** is the part of the hip one feels when placing the hand on the hip. The **anterior superior iliac spine** is an important hip landmark that is used as a reference point for finding the correct site to give an injection.

The coxae join each other at the **pubic symphysis** and join the sacrum at the **sacroiliac joint.** The **acetabulum** (as´ĕ-tab´u-lum) is the socket of the hip joint, and the **obturator** (ob´tu-ra´tor) **foramen** is the large hole in the coxa.

The female pelvis can be distinguished from the male pelvis (Figure 6-13, *B*). Even though the male pelvis is usually larger and more massive, the female pelvis is wider from side-to-side. The opening of the pelvic cavity, the **pelvic brim,** is larger and more oval in the female than in the male. The larger opening allows passage of a baby during the birth process.

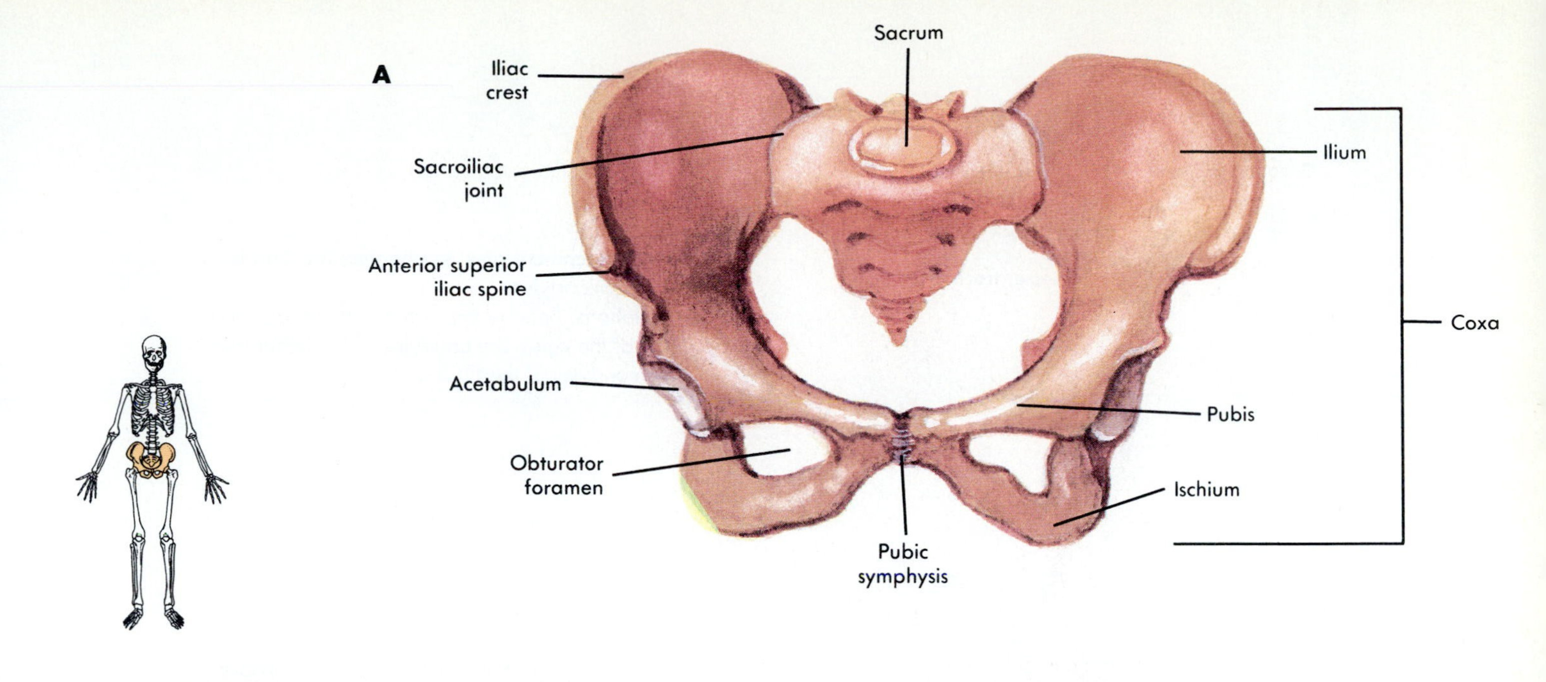

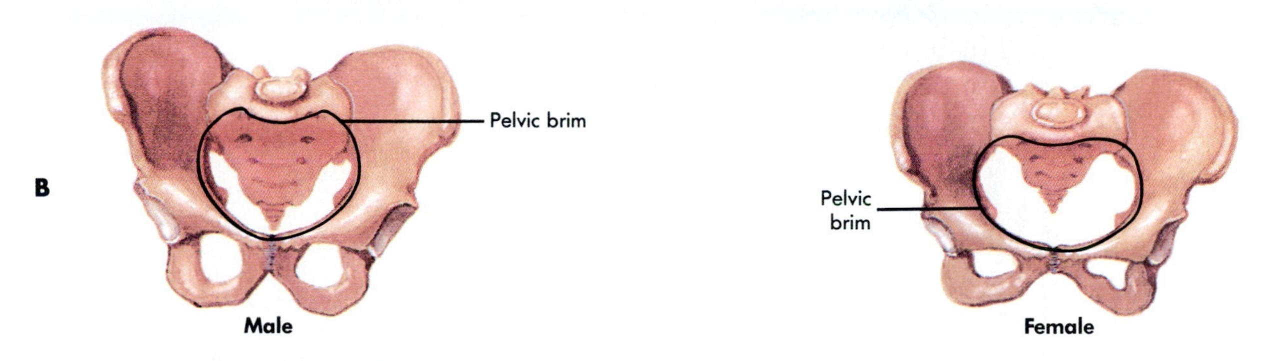

Figure 6-13 ● Pelvic girdle.
A, Anterior view. **B,** Comparison of female and male pelvic girdles. The pelvic brim is larger in the female.

Lower limb

The lower limb consists of the bones of the thigh, leg, ankle, and foot (Figure 6-14). The **thigh,** the region between the hip and the knee, contains a single bone, the **femur.** The **patella** (pa-tel´ah), or **kneecap,** is located within the major anterior tendon of the thigh muscles and enables the tendon to turn the corner over the knee.

The leg, the region between the knee and the ankle, contains two bones, the **tibia** (tib´e-ah), or **shin bone,** and the **fibula** (fib´u-lah). The tibia is the larger of the two bones and supports most of the weight of the leg.

The ankle consists of seven **tarsal** (tar´sal) bones.

The **metatarsals** and **phalanges** of the foot are arranged in a manner very similar to the metacarpals and phalanges of the hand. The distal ends of the metatarsals make up the ball of the foot.

There are two primary arches in the foot, formed by the positions of the tarsals and the metatarsals, and held in place by ligaments (Figure 6-15). A longitudinal arch extends from the heel to the ball of the foot, and a transverse arch extends across the foot. The arches function similar to the springs of a car, allowing the foot to give and spring back.

The arches of the foot normally form early in fetal life. Failure to form results in congenital (present at birth) **flat feet** or fallen arches in which the arches are depressed or collapsed. Flat feet can also occur when the muscles and ligaments supporting the arch tire and allow the arch to collapse.

Many of the bones of the lower limb can be seen and felt (Figure 6-16).

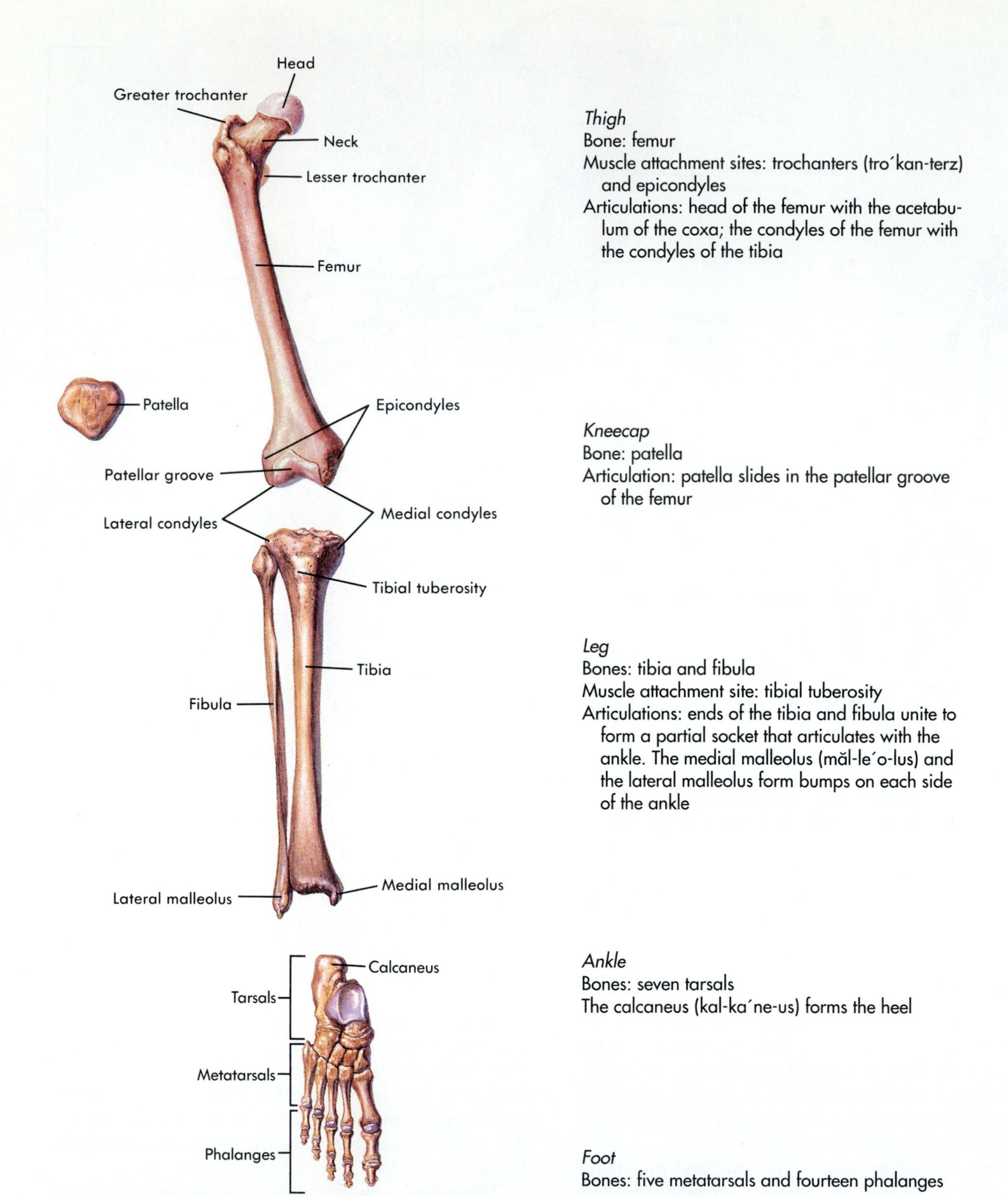

Thigh
Bone: femur
Muscle attachment sites: trochanters (tro´kan-terz) and epicondyles
Articulations: head of the femur with the acetabulum of the coxa; the condyles of the femur with the condyles of the tibia

Kneecap
Bone: patella
Articulation: patella slides in the patellar groove of the femur

Leg
Bones: tibia and fibula
Muscle attachment site: tibial tuberosity
Articulations: ends of the tibia and fibula unite to form a partial socket that articulates with the ankle. The medial malleolus (măl-le´o-lus) and the lateral malleolus form bumps on each side of the ankle

Ankle
Bones: seven tarsals
The calcaneus (kal-ka´ne-us) forms the heel

Foot
Bones: five metatarsals and fourteen phalanges

Figure 6-14 ● Right lower limb.
Anterior view.

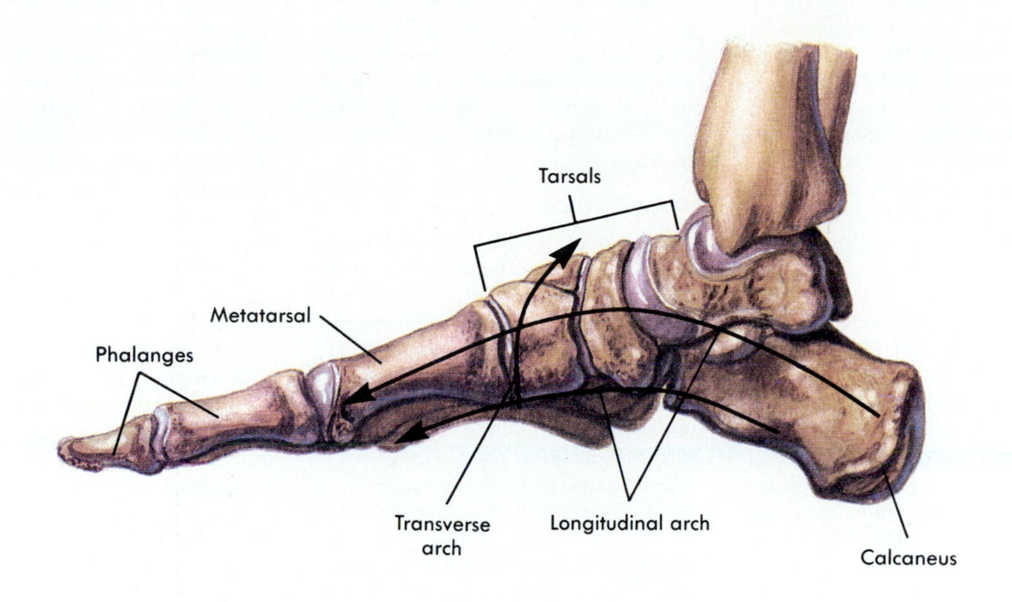

Figure 6-15 ● Arches of the foot.
Medial view of right foot.

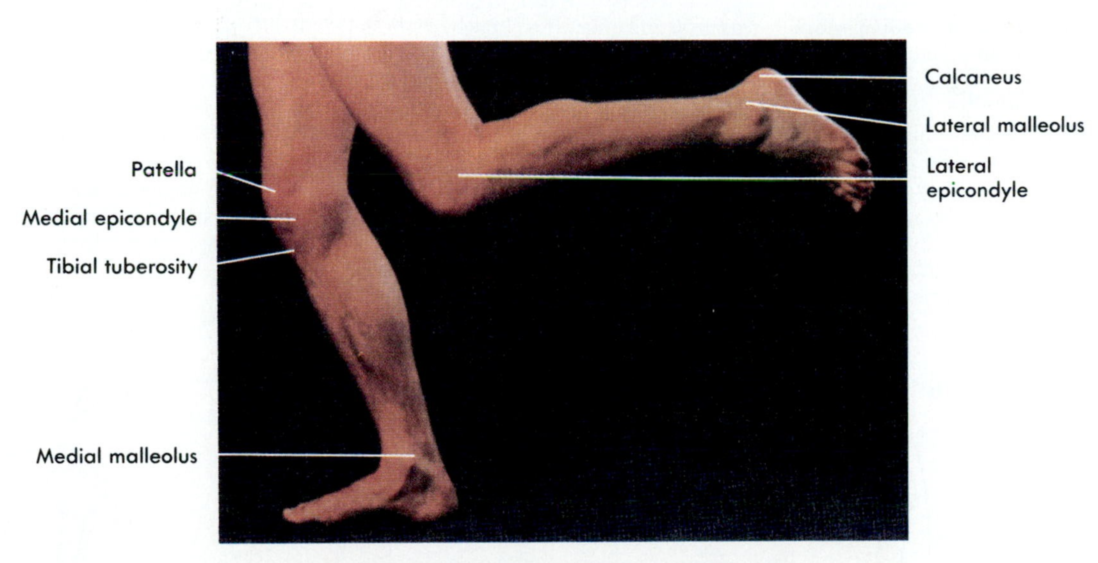

Figure 6-16 ● Surface anatomy of the lower limb.
Bones and their projections can be seen or felt.

Disorders of bones

GROWTH AND DEVELOPMENTAL DISORDERS

Giantism is a condition of abnormally increased height that involves excessive growth at the epiphyseal plate of long bones. It commonly results from overproduction of growth hormone by the pituitary gland (see Chapter 10). **Dwarfism,** the condition in which a person is abnormally short, is often the result of improper growth in the epiphyseal plates. It can be caused by genetic factors or insufficient production of growth hormone.

Osteogenesis imperfecta (os´te-o-jen´ĕ-sis im-per-fek´tah) is a group of genetic disorders producing bones that are easily fractured. The bones are brittle because insufficient collagen is formed to allow normal flexibility of the bones.

Rickets (rik´ets) involves growth retardation resulting from nutritional deficiencies of calcium or vitamin D. Calcium is necessary for normal bone formation and vitamin D is necessary for calcium absorption from the small intestine. Rickets produces bones that are soft, weak, and easily broken, because calcium is removed from bone to maintain normal blood calcium levels. It most often occurs in children whose diets are deficient in vitamin D.

BACTERIAL INFECTIONS

Osteomyelitis (os-te-o-mi-ĕ-li´tis) is bone inflammation that often results from bacterial infection, and it can lead to complete destruction of the bone. The introduction of bacteria into the body through wounds is the most common cause of osteomyelitis.

TUMORS

There are many types of bone tumors with a wide range of resultant bone defects. Tumors may be benign or malignant. Malignant bone tumors can metastasize (spread) to other parts of the body or they can spread to bone from metastasizing tumors elsewhere in the body.

DECALCIFICATION

Osteomalacia (os´te-o-mă-la´she-ah), or the softening of bones, results from calcium depletion from bones. If the body has an unusual need for calcium, for example, during pregnancy when fetal growth requires large amounts of calcium, it can be removed from the mother's bones, which consequently soften and weaken.

Osteoporosis (os´te-o-po-ro´sis), or porous bone, results from reduction in the overall quantity of bone tissue. In older people, bone can be broken down faster than it is produced, and the bones take on a mottled or porous appearance. Osteoporosis can be a severe problem in older people because it results in bones that are easily fractured. Postmenopausal osteoporosis can occur in women aged 50 and older. It is thought that the decrease in estrogen levels as a result of menopause is responsible for this form of osteoporosis. Osteoporosis is not limited to older people, however. It also can become a problem in young women athletes who "overtrain," because extreme exercise can result in decreased estrogen production.

Treatments for osteoporosis are designed to reduce bone loss and/or increase bone formation. Estrogen therapy appears to be effective in reducing bone loss in postmenopausal women. However, estrogen therapy may have disadvantages, such as increasing the risk of uterine and breast cancer. Increased dietary calcium and vitamin D can increase calcium uptake and reduce the amount of estrogen needed. Adequate calcium intake throughout life can also slow or prevent age-related osteoporosis. The greater the bone mass before the onset of osteoporosis, the greater is the tolerance for bone loss later in life. For this reason it is important for adults, especially women, in their twenties and thirties to ingest at least 1000 mg of calcium a day. Finally, exercise appears to be effective not only in reducing bone loss, but in increasing bone mass.

BONE FRACTURES

Bone fractures (Figure 6-B) can be classified as open if the bone protrudes through the skin, or closed if the skin is not perforated. If the fracture totally separates the two bone fragments, the fracture is called complete, and if not, it is called incomplete. A comminuted (kom´ĭ-nu-ted) fracture is one in which the bone breaks into more than two fragments. Fractures also can be classified according to the direction of the fracture line. Linear fractures are parallel to the long axis, transverse fractures are at right angles to the long axis, and oblique fractures are at an angle other than a right angle to the long axis.

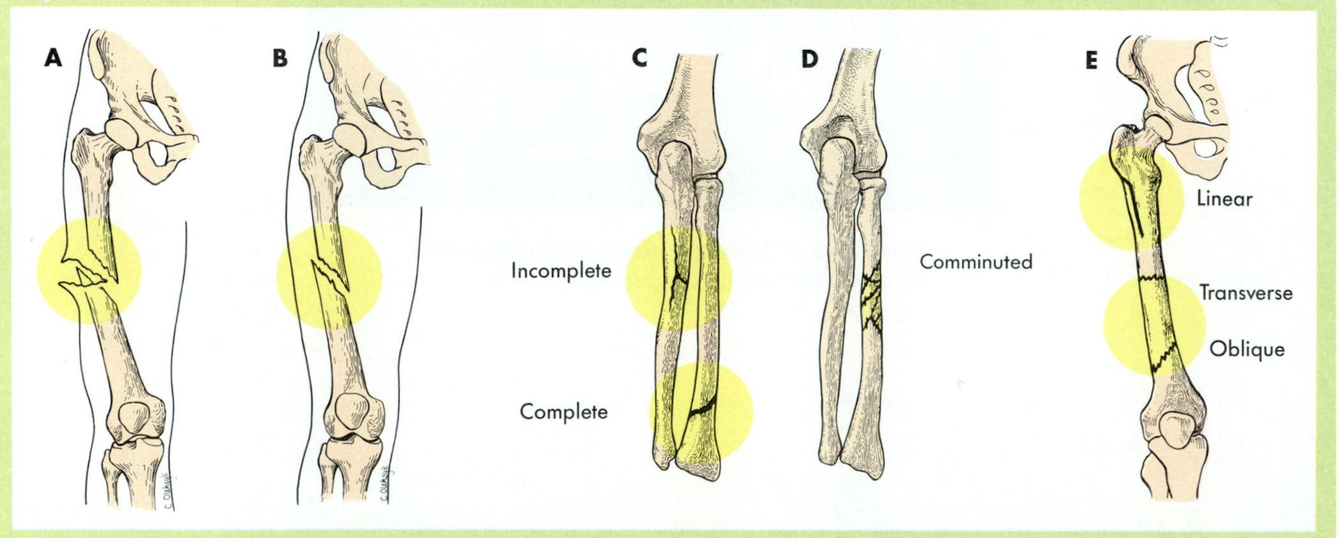

Figure 6-B Bone fractures.
A, Open. **B,** Closed. **C,** Complete and incomplete. **D,** Comminuted. **E,** Linear, transverse, and oblique.

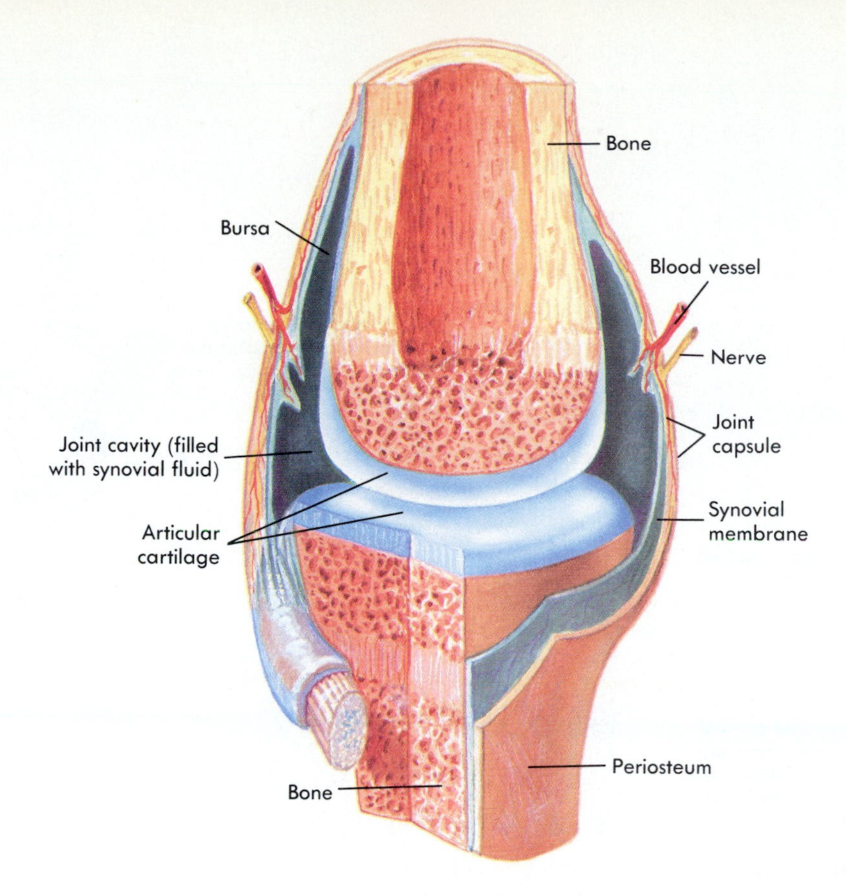

Figure 6-17 ● Structure of a synovial joint.

ARTICULATIONS

An **articulation,** or **joint,** is a place where two bones come together. Joints are classified according to the major connective tissue type that binds the bones together and according to whether there is a fluid-filled joint cavity. The three major classes of joints are fibrous, cartilaginous, and synovial.

An older method of classifying joints was based on the degree of movement of each joint: nonmovable, slightly movable, and freely movable. This classification scheme is no longer accepted in standard anatomy texts and is therefore not used in this text.

Fibrous joints

Fibrous joints consist of two bones that are united by fibrous tissue and that exhibit little or no movement. An example of a fibrous joint is a suture, a joint between bones of the skull (see Figure 6-4).

Cartilaginous joints

Cartilaginous joints unite two bones by means of cartilage. Only slight movement can occur at these joints. The intervertebral disks between vertebrae, the costal cartilages between the ribs and sternum, and the pubic symphysis are examples (see Figures 6-9, 6-10, and 6-13).

Synovial joints

Synovial (sĭ-no´ve-al) **joints** are freely movable joints. Most joints that unite the bones of the appendicular skeleton are synovial, whereas many of the joints that unite the bones of the axial skeleton are not. This pattern reflects the greater mobility of the appendicular skeleton compared with the axial skeleton.

Several features of synovial joints are important to their function (Figure 6-17). The articular surfaces of bones within synovial joints are covered with a thin layer of articular cartilage, which provides a smooth surface where the bones meet. The **joint cavity,** a small space between the bones, is enclosed by a **joint capsule,** which helps hold the bones together and allows for movement. Portions of the joint capsule may be thickened to form ligaments. In addition, ligaments and tendons outside the joint capsule contribute to the strength of the joint.

A **synovial membrane** lines the joint capsule everywhere except over the articular cartilage. The membrane produces synovial fluid, which fills the joint cavity and forms a lubricating film over the articular cartilages. In certain synovial joints, the synovial membrane may extend out of the joint cavity as a pocket or sac, called a **bursa** (bur´sah) (see Figure

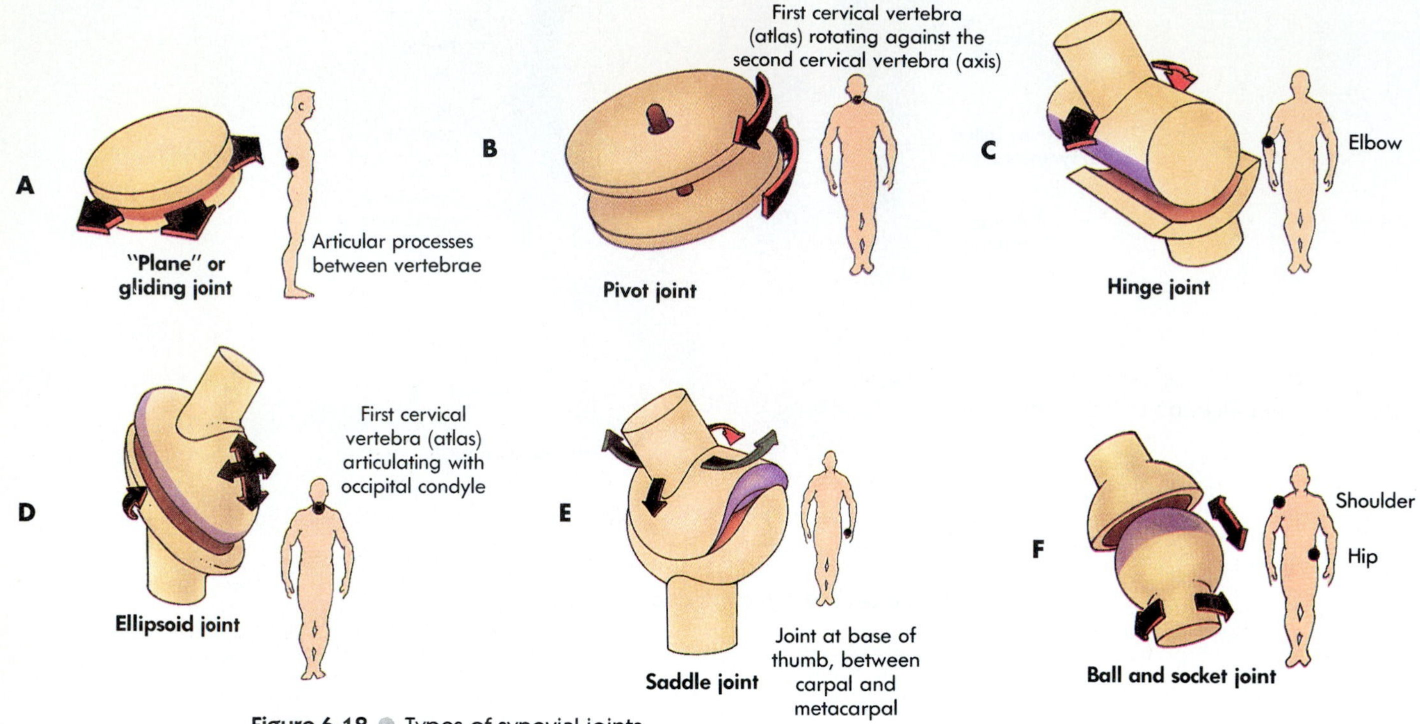

Figure 6-18 Types of synovial joints.
A, Plane, or gliding, joint. **B,** Pivot joint. **C,** Hinge joint. **D,** Ellipsoid joint. **E,** Saddle joint. **F,** Ball and socket joint.

6-17). Bursae function as fluid-filled "pillows" that reduce friction in areas of the body where structures would rub together. For example, when a tendon extends across the surface of a bone, a bursa can be located between the tendon and bone.

Types of synovial joints

Synovial joints are classified according to the shape of the adjoining articular surfaces (Figure 6-18). A **plane, or gliding joint,** consists of two opposed flat surfaces that glide over each other in many different directions. Examples are the articular processes between vertebrae. A **pivot joint** consists of a cylindrical bony process that rotates within a ring composed partly of bone and partly of ligament. Pivot joints restrict movement to rotation around a single axis. The rotation that occurs between the first and second cervical vertebrae when shaking the head "no" is an example. A **hinge joint** consists of a convex cylinder inserted into a concave surface of another bone. Hinge joints allow movement in one plane. Examples are the elbow and knee joints. An **ellipsoid,** or **condyloid, joint** is the football-shaped surface of one bone inserted into the corresponding concave surface of another bone. The shape of the joint limits its range of movement nearly to a hinge motion, but in two planes. The joint between the occipital condyles of the skull and the first cervical vertebra is an example. This joint allows the movements of shaking the head "yes" and tilting the head from side-to-side. A **saddle joint** consists of two saddle-shaped articulating surfaces oriented at right angles to one another. Movement in these joints can occur in two planes. The joint at the base of the thumb is a saddle joint. A **ball-and-socket joint** consists of a ball (head) at the end of one bone and a socket in an adjacent bone into which a portion of the ball fits. This type of joint allows a wide range of movement in almost any direction. Examples are the shoulder and hip joints.

Types of movement

The types of movement occurring at a given joint are related to the structure of that joint. Some joints are limited to only one type of movement, whereas others permit movement in several directions. Most movements are opposed by movements in the opposite direction and are therefore listed in pairs (Figure 6-19).

Flexion literally means to bend; extension means to straighten. As an example, flexion of the forearm is bending the elbow; **extension** of the forearm straightens the elbow. It may be helpful to think of the fetal position, because in such a position, nearly all the synovial joints are flexed.

Abduction (to take away) of most joints is movement away from the midline; **adduction** (to bring together) is movement toward the midline. Moving the legs away from the midline of the body, as in the outward movement of "jumping jacks," is abduction, and bringing the legs back together is adduction.

Pronation (pro-na´shun) is rotation of the forearm so that the palm faces posteriorly; **supination** (su´pĭ-na´shun) is rotation of the forearm so that the palm faces anteriorly.

Rotation is the turning of a structure around its long axis, as in shaking the head "no."

Circumduction occurs at freely movable joints such as the shoulder. In circumduction, the arm moves so that it describes a cone with the shoulder joint at the apex. The "windup" of a softball pitcher is an example of circumduction.

Most movements that occur in the course of normal activities are combinations of movements. A complex movement can be described by naming the individual movements involved.

4 What combination of movements is required at the shoulder and elbow joints for a person to move his right upper limb from the anatomical position and touch the side of his head with his fingers?

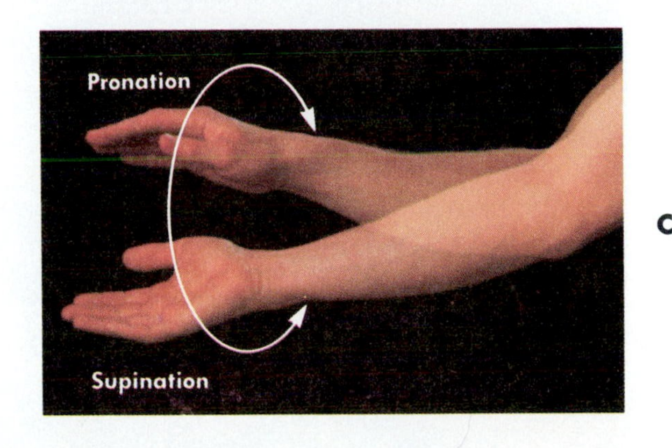

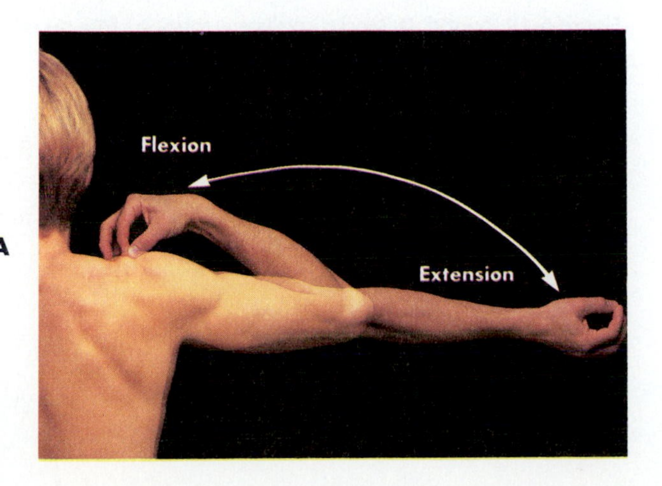

Figure 6-19 Movements.
A, Flexion and extension of the elbow. **B,** Abduction and adduction of the arm. **C,** Pronation and supination of the forearm and hand.

Continued.

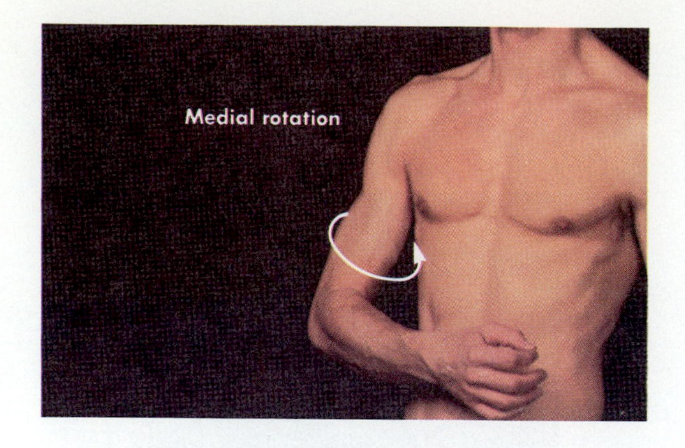

D

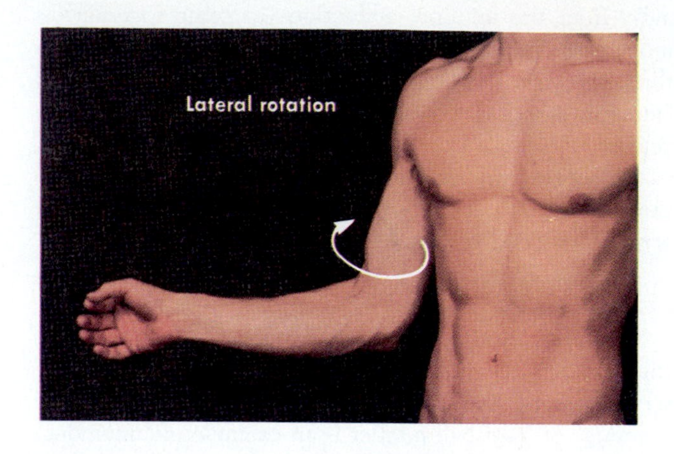

E

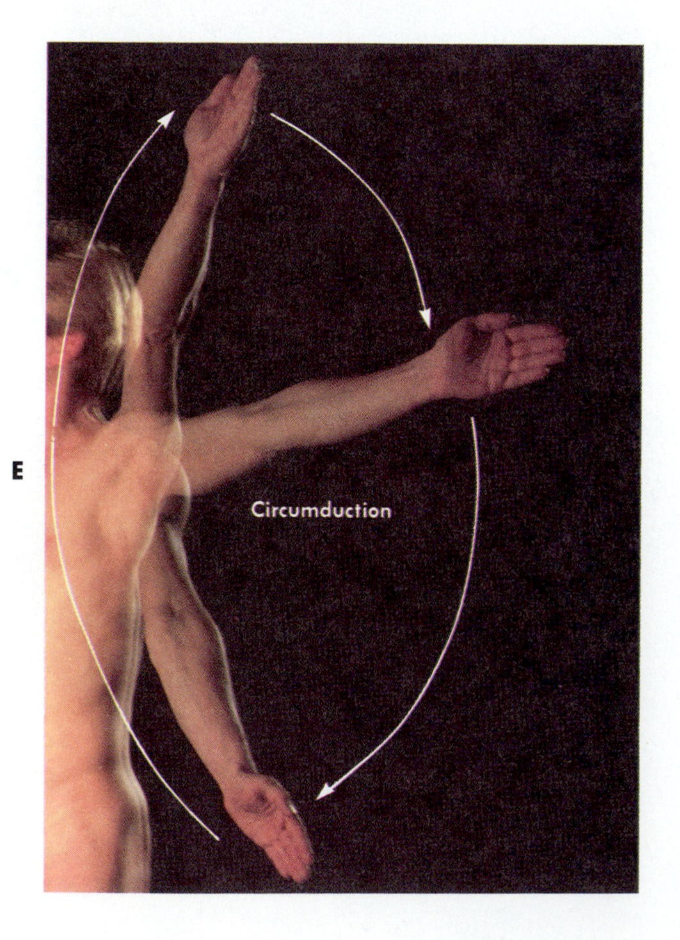

Figure 6-19, cont'd. D, Medial and lateral rotation of the humerus. **E,** Circumduction of the shoulder.

Joint disorders

ARTHRITIS

Arthritis (ar-thri´tis) is the inflammation of a joint. It is the most common and best known of the joint disorders, affecting 10% of the world's population. There are at least 20 different types of arthritis, which differ in cause and progress. Causes include infectious agents, metabolic disorders, trauma, and immune disorders.

Rheumatoid arthritis affects about 3% of all women and about 1% of all men in the United States. It is a chronic autoimmune disease in which one's own immune system destroys tissues. The cause is unknown, but may involve an abnormal response of the immune system to an infection. There may also be a genetic predisposition. Rheumatoid arthritis is a general connective tissue disorder that affects the skin, vessels, lungs, and other organs, but it is most pronounced in the joints (Figure 6-C). It is severely disabling and most commonly destroys small joints such as those in the hands and feet. The synovial membrane and associated connective tissue cells proliferate, forming a pannus (clothlike layer) in the joint capsule, which can grow into the articulating surfaces of the bones, destroying the articular cartilage. In advanced stages, the bones forming the joint can become fused.

DEGENERATIVE JOINT DISEASE

Degenerative joint disease (DJD), also called **osteoarthritis,** is the destruction and loss of articular cartilage in synovial joints. It results from the gradual "wear and tear" of a joint that occurs with advancing age. Slowed metabolic rates with increased age also seem to contribute to DJD. It is very common in older individuals and affects 85% of all people in the United States over the age of 70. It tends to occur in the weight-bearing joints such as the knees and is more common in overweight individuals. Mild exercise retards joint degeneration and enhances mobility.

GOUT

Gout is an inflammatory disease caused by an increase in uric acid in the body. Uric acid is a waste product of metabolism. It can be deposited as crystals in various tissues, including the kidneys and joint capsules, if its concentration in the blood becomes elevated. Gout is more common in males than in females. The most commonly affected joints (85% of the cases) are the base of the great toe and other foot and leg joints. Other joints may ultimately be involved, and damage to the kidney from crystal formation occurs in almost all advanced cases.

BURSITIS AND BUNIONS

Bursitis is the inflammation of a **bursa.** It usually results from injury to the bursa, but can be caused by infections. The bursae around the shoulders and elbows are common sites of bursitis. A **bunion** is a bursitis that develops over the joint at the base of the great toe. Bunions are frequently irritated by shoes that rub on them.

TRAUMA

A **sprain** results when the bones of a joint are forcefully pulled apart and the ligaments around the joint are pulled or torn. A **separation** exists when the bones remain apart after an injury to a joint. A **dislocation** is when the head of one bone is pulled out of the socket in a ball-and-socket or ellipsoid joint.

JOINT REPLACEMENT

As a result of recent advancements in biomedical technology, some joints of the body can now be replaced by artificial joints. Joint replacement, or arthroplasty, was developed in the late 1950s. It is used in patients with joint disorders to eliminate unbearable pain and to increase joint mobility. Degenerative joint disease is the leading disease requiring joint replacement, accounting for two thirds of the patients. Rheumatoid arthritis accounts for more than half the remaining cases.

Artificial joints usually are composed of metal in combination with modern plastics. Examples of metals used in artificial joints include stainless steel, titanium alloys, or cobalt-chrome alloys, and examples of plastics are high-density polyethylene, silicone rubber, or elastomer. The smooth metal surface rubbing against the smooth plastic surface provides a low-friction contact with a range of movement that depends on the design.

A
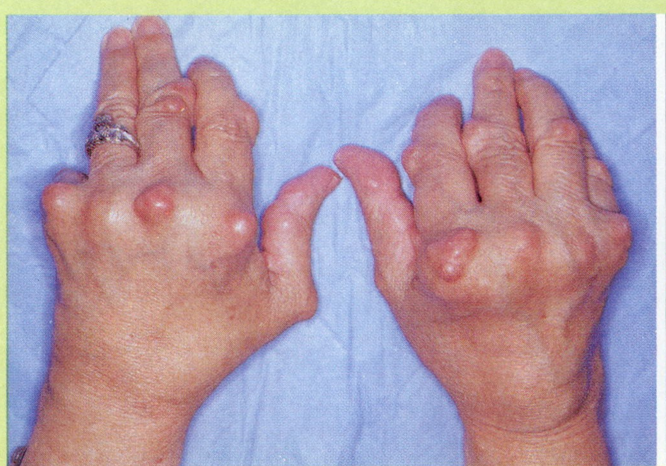
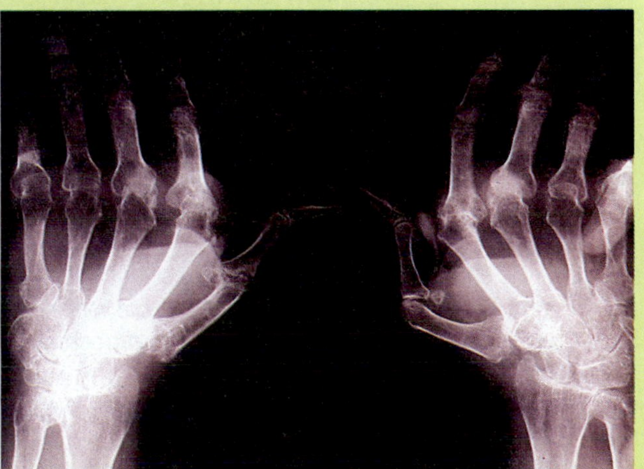
B

Figure 6-C Rheumatoid arthritis.
A, Photograph of hands with rheumatoid arthritis. **B,** X-ray of the same hands.

SUMMARY

The skeletal system consists of bone, cartilage, and ligaments.

FUNCTIONS OF THE SKELETAL SYSTEM

The skeletal system provides support and protection, allows body movements, stores minerals and fats, and is the site of blood cell production.

GENERAL FEATURES OF BONE

Bone shapes

Bones are classified as long, short, flat, or irregular.

Bone anatomy

Long bones consist of a diaphysis (shaft), epiphyses (ends), and cavities filled with yellow or red marrow.

Bone histology

Bone tissue is osteocytes in a matrix of collagen and minerals.

Spongy bone is interconnecting plates of bones with spaces containing marrow. Compact bone is mostly solid matrix.

Bone formation and growth

Bones are formed when osteoblasts lay down bone matrix in fibrous membranes or cartilage models.

Bones grow in diameter when bone is added to the outer surface of the bone; bones grow in length at the epiphyseal plate.

Bone remodeling

Bone remodeling consists of removal of old bone and deposition of new bone.

Bone repair

During bone repair, cells move in and form a callus, which is replaced by bone.

THE SKELETON

There are 206 bones.

AXIAL SKELETON

The axial skeleton includes the skull, hyoid bone, vertebral column, and thoracic cage.

Skull

The skull consists of 22 bones, divided between the brain case, face, and auditory ossicles.

From a lateral view, the parietal, temporal, and sphenoid bones can be seen.

From a frontal view, the orbits and nasal cavity can be seen, as well as associated bones and structures, such as the frontal bone, zygomatic bone, maxilla, and mandible.

The base of the skull has foramina for the passage of the spinal cord, blood vessels, and nerves; articulation sites for the vertebral column and mandible, and processes for muscle attachment.

The interior of the brain case has fossae that hold the brain.

Vertebral column

The vertebral column contains 7 cervical, 12 thoracic, and 5 lumbar vertebrae, plus 1 sacrum and 1 coccyx.

The vertebral column has cervical, thoracic, lumbar, and combined sacral and coccygeal curves.

Thoracic cage

The thoracic cage consists of thoracic vertebrae, ribs, and the sternum.

There are 12 ribs; 7 true and 5 false (two of the false ribs are also called floating ribs).

The sternum is an attachment site for the ribs

APPENDICULAR SKELETON

The appendicular skeleton consists of the bones of the upper and lower limbs and their girdles.

Pectoral girdle

The pectoral girdle includes the scapula and clavicle.

Upper limb

The upper limb consists of the arm (humerus), forearm (ulna and radius), wrist (eight carpal bones), and hand (five metacarpals, three phalanges in each finger, and two phalanges in the thumb).

Pelvic girdle

The pelvic girdle is made up of the sacrum and two coxae. Each coxa consists of an ilium, ischium, and pubis.

Lower limb

The lower limb includes the thigh (femur), leg (tibia and fibula), ankle (seven tarsals), and foot (metatarsals and phalanges, similar to the bones in the hand).

ARTICULATIONS

An articulation is a place where bones come together.

Fibrous joints

Fibrous joints consist of bones united by fibrous connective tissue. They allow little or no movement.

Cartilaginous joints

Cartilaginous joints consist of bones united by cartilage, and they exhibit slight movement.

Synovial joints

Synovial joints consist of articular cartilage over the uniting bones, a joint cavity lined by a synovial membrane and containing synovial fluid, and a joint capsule. They are highly movable joints.

Synovial joints can be classified as plane, pivot, hinge, ellipsoid, saddle, or ball-and-socket.

The major types of movement include flexion/extension, abduction/adduction, pronation/supination, rotation, and circumduction.

CONTENT REVIEW

1. What are the components of the skeletal system?
2. List the functions of the skeletal system and give an example for each function.
3. Define the terms diaphysis, epiphysis, medullary cavity, and periosteum.
4. Describe the structure of spongy and compact bone. How do nutrients reach the osteocytes in spongy and compact bone?
5. Describe the formation of bone before birth. What is a fontanel and what is the function of a fontanel?
6. How does bone grow in diameter? How do long bones grow in length?
7. What is bone remodeling? Give two examples.
8. Define the axial skeleton and the appendicular skeleton.
9. Name the three groups of bones that comprise the skull.
10. Give the locations of the paranasal sinuses. What are their functions?
11. Through what opening does the brain connect to the spinal cord?
12. What is the function of the hard palate?
13. Name and give the number of each type of vertebra. Name the four vertebral curves, state their function, and explain how they develop.
14. What is the function of the thoracic cage? Distinguish true, false, and floating ribs.
15. Name the bones that make up the pectoral girdle, arm, forearm, wrist, and hand. How many phalanges are in each finger and in the thumb?
16. Define the pelvis. What bones fuse to form each coxa? Where and with what bones do the coxae articulate?
17. Name the bones of the thigh, leg, ankle, and foot.
18. Define the term articulation or joint. Name the three major classes of joints.
19. Describe the structure of a synovial joint. How do the different parts of the joint function to permit joint movement?
20. On what basis are synovial joints classified? Describe the different types of synovial joints and give examples of each. What movements do each type of joint allow?
21. Describe and give examples of flexion/extension, abduction/adduction, and supination/pronation.

CONCEPT REVIEW

1. A 12-year-old boy fell while playing basketball. The physician explained that the head (epiphysis) of the femur was separated from the shaft (diaphysis). Although the bone was set properly, by the time the boy was 16 it was apparent that the injured lower limb was shorter than the normal one. Explain why this difference occurred.
2. A direct blow to the nose can result in a "broken nose." List at least three bones that could be broken.
3. Would you expect scoliosis, kyphosis, or lordosis to develop if the muscles on the left side of the back were weaker than the muscles on the right side of the back? Which of these conditions would you expect to develop as a result of pregnancy? Explain.
4. In what way could a complete fracture of the clavicle change the position of the upper limb?
5. Suppose you needed to compare the length of one lower limb to the other in an individual. Using bony landmarks, suggest an easy way to accomplish the measurement.
6. Justin Time leaped from his hotel room to avoid burning to death in a fire. If he landed on his heels, what bone was likely to fracture? Unfortunately for Justin, a 240-pound fireman, Hefty Stomper, ran by and stepped heavily on the distal part of Justin's foot (not the toes). What bones now could be broken?

CHAPTER TEST

Answers can be found in Appendix C

MATCHING 1 For each statement in column A select the correct answer in column B (an answer may be used once, more than once, or not at all).

A

1. Connects muscle to bone.
2. The shaft of a bone.
3. Site of blood cell production.
4. Cell that produces bone.
5. Cell that breaks down bone.
6. Bones of the head and trunk.
7. The coxae and sacrum.

B

appendicular skeleton
axial skeleton
diaphysis
epiphysis
ligament
osteoblast
osteoclast
osteocyte
pectoral girdle
pelvic girdle
red marrow
tendon
yellow marrow

MATCHING 2

A

1. Two flat surfaces that slide over each other.
2. A convex cylinder of bone inserted into a concave surface of another bone.
3. A football-shaped surface of bone inserted into a corresponding concave surface of another bone.
4. The joint between the first and second cervical vertebrae.
5. The joint between the occipital condyles and the first cervical vertebrae.
6. The hip joint.
7. Two joints, each of which restricts movement to two planes.
8. To move away from the midline.
9. Rotation of the forearm so that the palm faces posteriorly.
10. Turning of a structure around its long axis.

B

abduction
adduction
ball-and-socket joint
circumduction
ellipsoid joint
extension
flexion
hinge joint
pivot joint
plane joint
pronation
rotation
saddle joint
supination

FILL-IN-THE BLANK Complete each statement by providing the missing word or words.

1. Bone formed of interconnecting plates of bone with a network of small cavities containing marrow is called ___________.
2. In compact bone, nutrients diffuse to osteocytes from blood vessels in the bone. The nutrients leave the blood vessels, which are in ___________ , and diffuse through ___________ in the bone matrix.
3. The ___________ are the part of the membrane formed around the brain during development, which have not been converted to bone by the time of birth.
4. A layer of cartilage where long bones increase in length is the ___________.
5. A ___________ is a fibrous network of bone and cartilage that holds a broken bone together. It is later replaced with bone.
6. An example of a fibrous joint is a ___________.
7. An example of a cartilaginous joint is ___________.
8. The characteristics of a synovial joint include ___________ over the ends of the bones, a ___________ , which holds the bones together, and a joint cavity filled with ___________.

CHAPTER 7

The Muscular System

KEY TERMS

antagonist
A muscle that acts in opposition to another muscle.

insertion
The more movable attachment point of a muscle.

isometric contraction
Muscle contraction in which the length of the muscle does not change, but the amount of tension increases.

isotonic contraction
Muscle contraction in which the amount of tension is constant, and the muscle shortens.

origin
The less movable attachment point of a muscle.

oxygen debt
The amount of oxygen required to convert the lactic acid produced during anaerobic respiration to glucose.

prime mover
Muscle that plays a major role in accomplishing a movement.

sarcomere
The structural and functional unit of a muscle; the part of a muscle fiber formed of actin and myosin myofilaments.

sliding filament mechanism
The mechanism by which muscle contraction occurs wherein actin myofilaments slide past myosin myofilaments.

FEATURES

OBJECTIVES

After reading this chapter you should be able to:

1. Describe the microscopic structure of a muscle.
2. Describe the molecular events that result in muscle contraction and relaxation.
3. Define isotonic and isometric muscle contractions.
4. Distinguish between aerobic and anaerobic muscle contraction.
5. Define the following terms: origin, insertion, synergist, antagonist, and prime mover.
6. Explain how muscles work in opposition to produce controlled movement.
7. List the major muscle groups and describe the function of a major muscle from each group.

MUSCLE TISSUE has the ability to contract or shorten, making possible body movements. As described in Chapter 4, there are three types of muscle tissue: skeletal, cardiac, and smooth. This chapter deals with the structure and function of skeletal muscle. The structure of skeletal muscles and muscle cells is related to their ability to contract. Because contraction requires energy, how muscles obtain and use energy also is considered. In addition, the major muscles of the body are identified and the movements produced by contraction of these muscles are described.

FUNCTIONS OF SKELETAL MUSCLE

Skeletal muscle, with its associated connective tissue, comprises approximately 40% of the body's weight. It is responsible for body movements, posture, and heat production.

1. *Movement.* Skeletal muscles attach to bones by tendons, and contraction (shortening) of the muscles is responsible for body movements such as walking. Skeletal muscles also attach to the skin of the face, making possible different facial expressions.
2. *Posture.* Continual contraction of skeletal muscles maintains body position such as standing upright. If all the skeletal muscles relax, as during fainting, a person simply collapses.
3. *Heat production.* Contraction of skeletal muscles requires energy released by chemical reactions within muscle cells (see Chapter 2). A byproduct of these reactions is heat, which helps maintain body temperature. For example, shivering results from rapid muscle contractions that produce shaking rather than coordinated movements. The heat produced by the contracting muscles keeps the body warm when environmental temperatures become too cold.

MUSCLE STRUCTURE

A connective tissue sheath called **fascia** (fash´e-ah) surrounds and separates muscles (Figure 7-1). Connective tissue also extends into the muscle and divides it into numerous **muscle bundles.** The muscle bundles are composed of many elongated muscle cells called muscle fibers. Each **muscle fiber** is a cylindrical cell containing several nuclei located immediately beneath the cell membrane. The cytoplasm of each muscle fiber is filled with myofibrils. Each **myofibril** is a thread-like structure that extends from one end of the muscle fiber to the other. Myofibrils consist of two major kinds of protein fibers: **actin** (ak´tin), or **thin, myofilaments,** and **myosin** (mi´o-sin), or **thick, myofilaments.**

The actin and myosin myofilaments form highly ordered units called **sarcomeres** (sar´ko-merz), which are joined end-to-end to form the myofibrils (see Figure 7-1). The ends of a sarcomere are a network of protein fibers, which form the **Z lines** when the sarcomere is viewed from the side. The Z lines form an attachment site for actin myofilaments. The arrangement of the actin and myosin myofilaments in a sarcomere gives the myofibril a banded appearance because the myofibril appears darker where the actin and myosin myofilaments overlap. The alternating light and dark areas of the sarcomeres are responsible for the striations (banding pattern) seen in skeletal muscle cells observed through the microscope (see Figure 4-5, A).

MUSCLE CONTRACTION

Mechanism of skeletal muscle contraction and relaxation

Myosin myofilaments have small extensions that can join to actin myofilaments to form **cross bridges** (Figure 7-2). During muscle contraction the cross bridges move, pulling on the actin myofilaments. Just as a person can pull a rope "hand-over-hand," the cross bridges attach to actin, pull, release, reposition, pull, and so on. The effect of the cross bridge movements is to cause the actin myofilaments to slide past the myosin myofilaments. As the actin myofilaments at each end of the sarcomere are pulled toward each other, the Z lines to which the actin myofilaments are attached are pulled closer together, causing all of the sarcomeres in a muscle fiber to shorten. The movement of actin myofilaments past myosin during contraction is called the **sliding filament mechanism.**

1 The sarcomere is considered the functional unit of skeletal muscle. Explain how shortening of the sarcomeres results in contraction (shortening) of the muscle.*

When a contracted muscle relaxes, it can return to its original length if another contracting muscle or gravity pulls on the relaxed muscle. Just as a person can release a rope and let it slide through his hands, the cross bridges release the actin myofilaments and they slide past the myosin myofilaments, returning to their original position.

Stimulating skeletal muscle contractions

Muscle fibers are excitable, meaning they respond to stimuli by contracting. Normally the nervous system stimulates skeletal muscles to contract (see Chapter 8 for details). This stimulation allows voluntary control of skeletal muscle activity. It is also responsible for involuntary reflex contractions and muscle tone.

The force of contraction of a muscle can be varied by **recruitment,** which is a gradual increase in the number of muscle fibers stimulated. A weak muscle contraction results from only a few contracting muscle fibers. As the number of muscle fibers stimulated increases, the force of contraction increases. Maximal contraction of the muscle is achieved when all the muscle fibers in the muscle are stimulated to contract.

*Answers to predict questions appear in Appendix B at the back of the book.

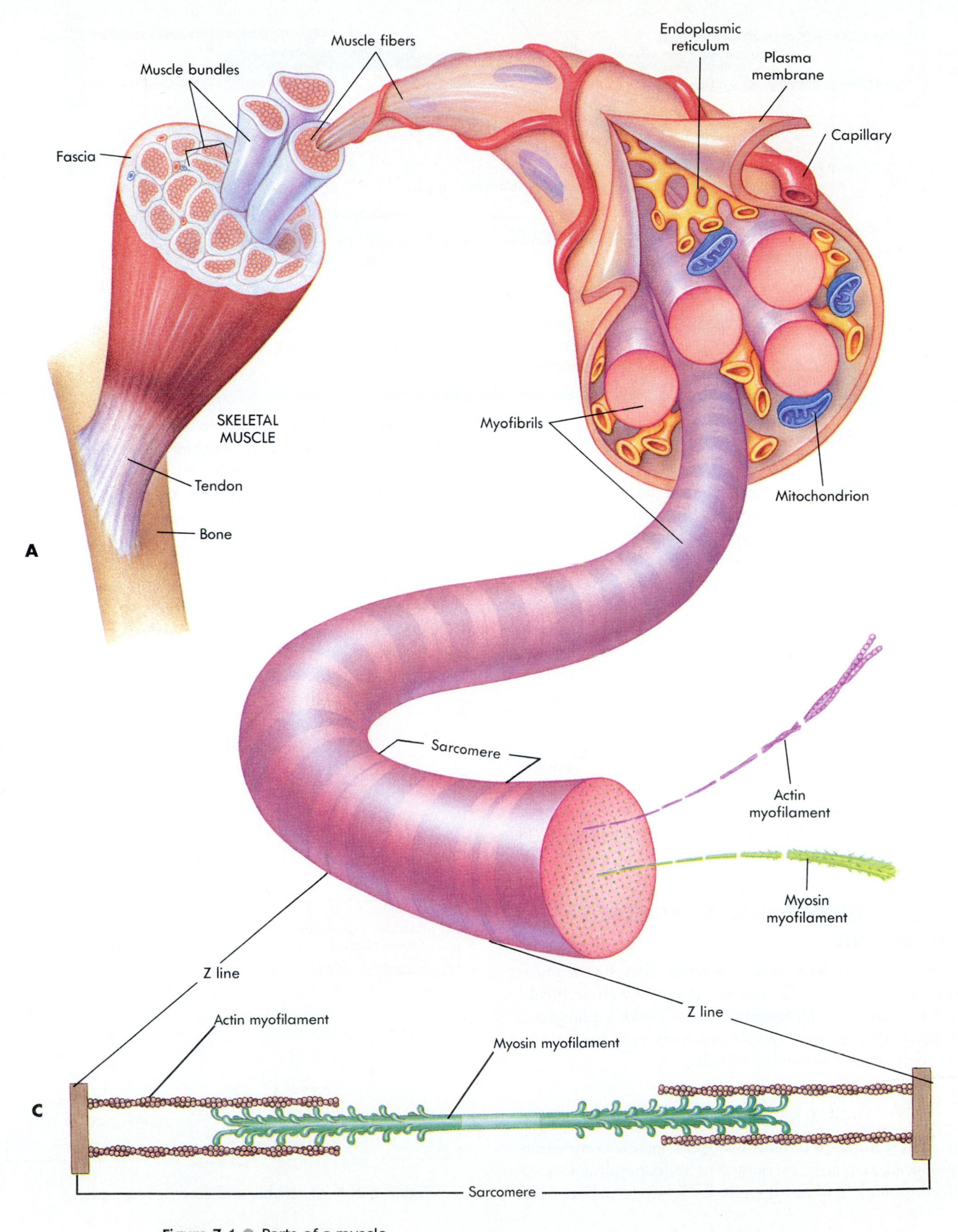

Figure 7-1 ● Parts of a muscle.
A, Muscle is subdivided into muscle bundles, which are composed of muscle fibers. **B,** Each muscle fiber contains myofibrils in which the banding patterns of the sarcomeres are seen. **C,** Sarcomeres consist of actin and myosin myofilaments.

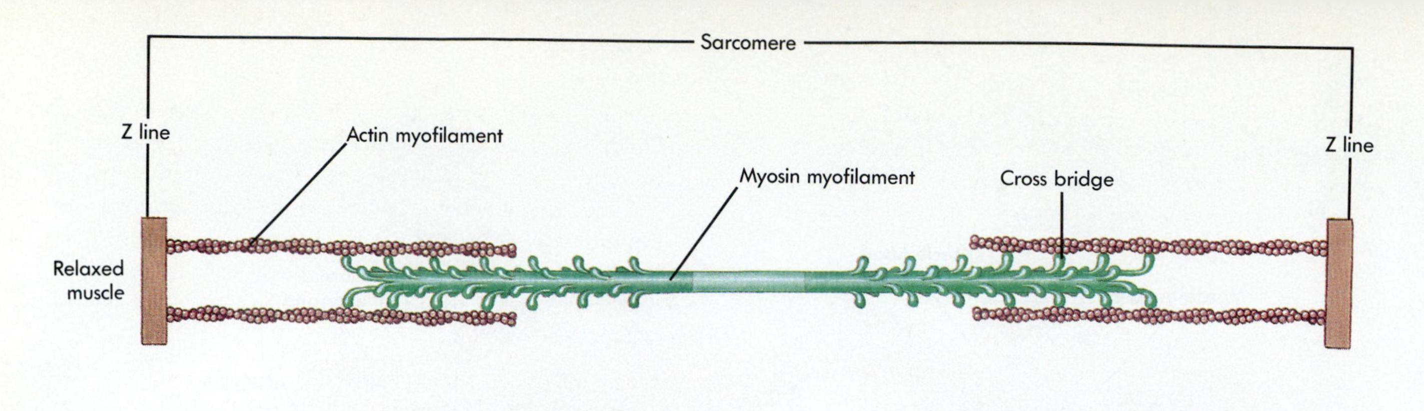

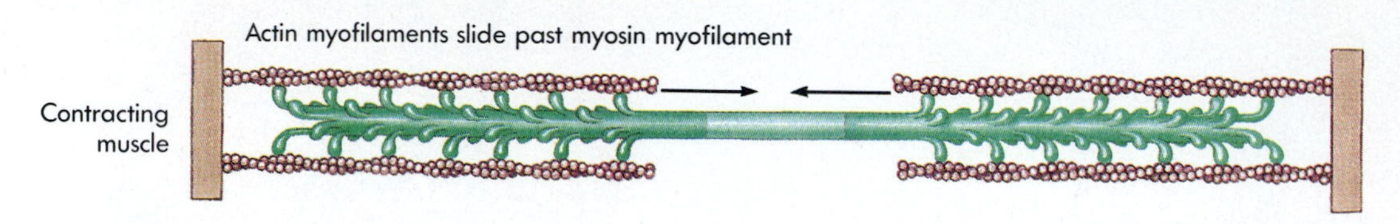

Figure 7-2 ● Sliding filament mechanism.
Movement of cross bridges causes actin myofilaments to slide past myosin myofilaments. Z lines are pulled closer together, causing the sarcomere to shorten.

Types of muscle contractions

Muscle contractions are classified as either isotonic or isometric. In **isotonic** (i´so-ton´ik; equal tension) **contractions,** the amount of tension produced by the muscle is constant during contraction, but the length of the muscle changes; for example, movement of the fingers to make a fist. In **isometric** (i´so-met´rik; equal distance) **contractions,** the length of the muscle does not change, but the amount of tension increases during the contraction process. Clenching the fist harder and harder is an example. Most movements are a combination of isometric and isotonic contractions. For example, when shaking hands, the muscles shorten some distance (isotonic contraction) and the degree of tension increases (isometric contraction).

Isometric contractions are also responsible for **muscle tone,** the constant tension produced by muscles of the body for long periods. Muscle tone is responsible for posture; for example, keeping the back and legs straight, the head held in an upright position, and the abdomen from bulging.

Energy requirements for muscle contraction

Contraction of skeletal muscle requires adenosine triphosphate (ATP). The ATP releases energy when it breaks down to adenosine diphosphate (ADP) and a phosphate (P). Some of the energy is used to move the cross bridges and some of the energy is released as heat.

ATP ⟶ ADP + P +
Energy (for cross bridge movement) + Heat

The ATP required to provide energy for muscle contraction is produced primarily in numerous mitochondria located within the muscle fibers. Because ATP is a very short-lived molecule and rapidly degenerates to the more stable ADP, it is necessary for muscle cells to constantly produce ATP.

ATP is produced by anaerobic or aerobic respiration. **Anaerobic** (an´ăr-o´bik) respiration, which occurs in the absence of oxygen, results in the breakdown of glucose to yield ATP and lactic acid. **Aerobic** (ăr-o´bik) respiration requires oxygen and breaks down glucose to produce ATP, carbon dioxide, and water (Figure 7-3). Compared with anaerobic respiration, aerobic respiration is much more efficient. The breakdown of a glucose molecule by aerobic respiration theoretically can produce 19 times as much ATP as

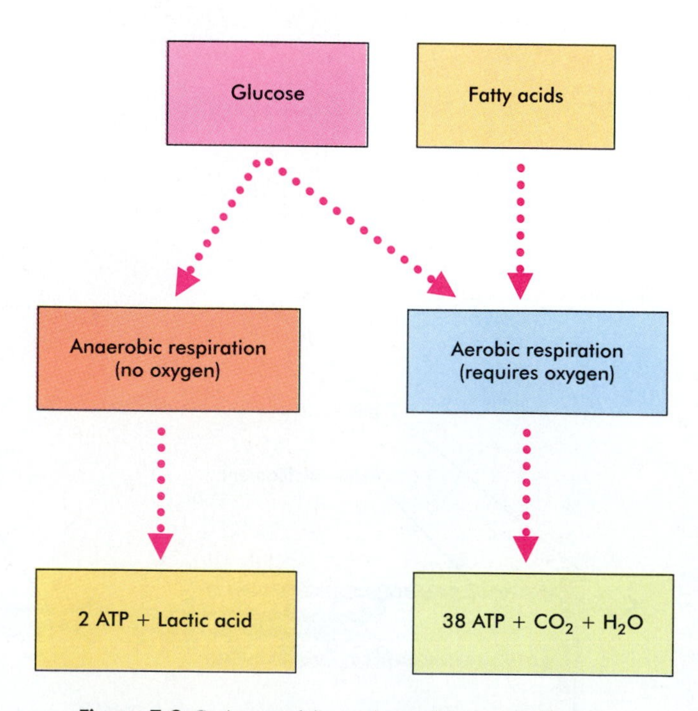

Figure 7-3 ● Anaerobic and aerobic respiration.

is produced by anaerobic respiration. In addition, aerobic respiration can utilize a greater variety of nutrient molecules to produce ATP than can anaerobic respiration. For example, aerobic respiration can use fatty acids to generate ATP. Although anaerobic respiration is less efficient than aerobic respiration, it can produce ATP when lack of oxygen limits aerobic respiration. By utilizing many glucose molecules, anaerobic respiration can rapidly produce much ATP, but only for a short period.

Resting muscles or muscles undergoing long-term exercise such as long-distance bicycling on level ground, depend primarily on aerobic respiration for ATP synthesis. Although some glucose is used as an energy source, fatty acids are a more important energy source during sustained exercise as well as during resting conditions. On the other hand, during intense exercise such as riding a bicycle up a steep hill, anaerobic respiration provides enough ATP to support intense muscle contractions for approximately 1 to 2 minutes. Anaerobic respiration is ultimately limited by depletion of glucose and a buildup of lactic acid within the muscle fiber. Lactic acid also can irritate muscle fibers, causing short-term muscle pain. Muscle pain that lasts for a couple of days following exercise, however, results from damage to connective tissue and muscle fibers within the muscle.

Muscle fatigue results when ATP is used during muscle contraction faster than it can be produced in the muscle cells, and lactic acid builds up faster than it can be removed. As a consequence, ATP levels are too low to sustain cross bridge movement and the contractions become weaker and weaker. For most of us, however, complete muscle fatigue is rarely the reason we stop exercising. Instead, we stop because of **psychological fatigue,** the feeling that the muscles have tired. A burst of activity in a tired athlete as a result of encouragement from spectators is an example of how psychological fatigue can be overcome.

After intense exercise, the respiration rate remains elevated for a period. Even though oxygen is not needed for the aerobic production of ATP molecules for contraction, oxygen is needed to convert the lactic acid produced by anaerobic respiration back to glucose. The increased amount of oxygen needed in chemical reactions to convert lactic acid to glucose is the **oxygen debt.** After the oxygen debt is paid, respiration rate returns to normal.

2 After a 1-mile run with a sprint at the end, a runner continues to breathe heavily for a period. Explain how the runner produced energy during the run and during the sprint, and explain the heavy breathing after the run.

MUSCLE ANATOMY

General principles

Most muscles extend from one bone to another and cross at least one movable joint. Muscle contraction causes most body movements by pulling one of the bones toward the other across the movable joint. Some muscles are not attached to bone at both ends. For example, some facial muscles attach to the skin, which moves as the muscles contract.

The points of attachment of each muscle are its origin and insertion (Figure 7-4). At these attachment points, the muscle is connected to the bone by a tendon. The **origin** is the most stationary end of the muscle and the **insertion** is the end of the muscle attached to the bone undergoing the

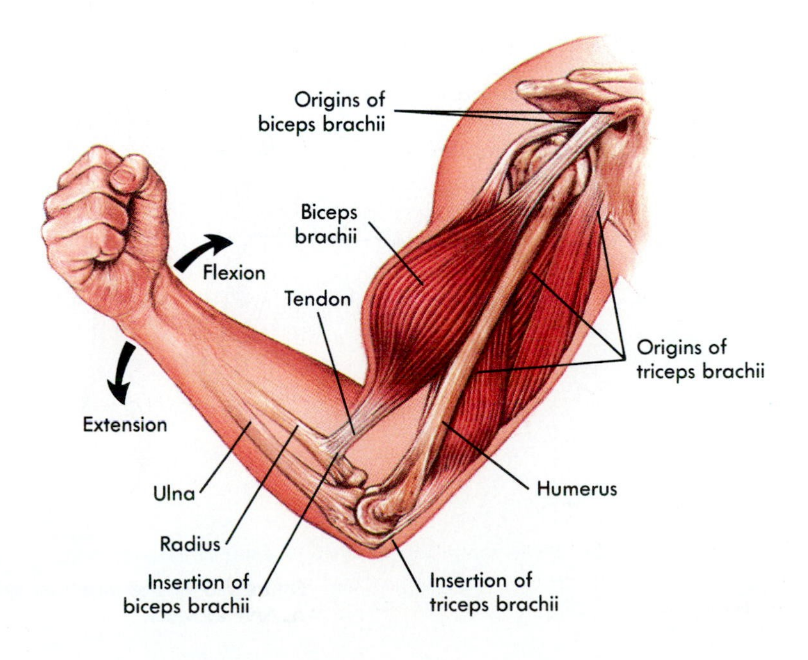

Figure 7-4 ● **Muscle attachment.** The origin is the less movable end of a muscle and the insertion is the more movable end. The biceps brachii causes flexion of the forearm and the triceps brachii causes extension of the forearm.

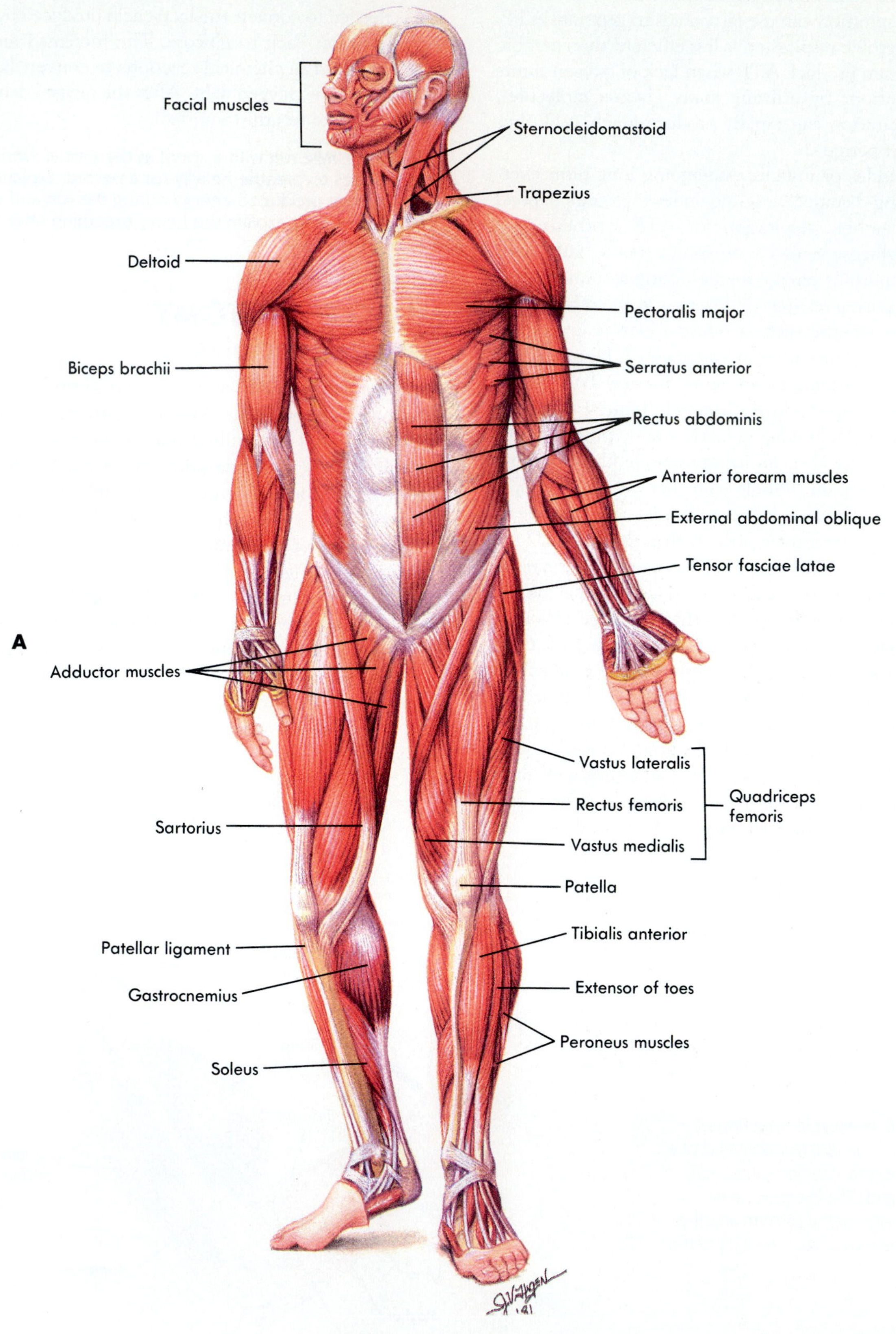

Figure 7-5 ● The muscular system.
A, Anterior view.

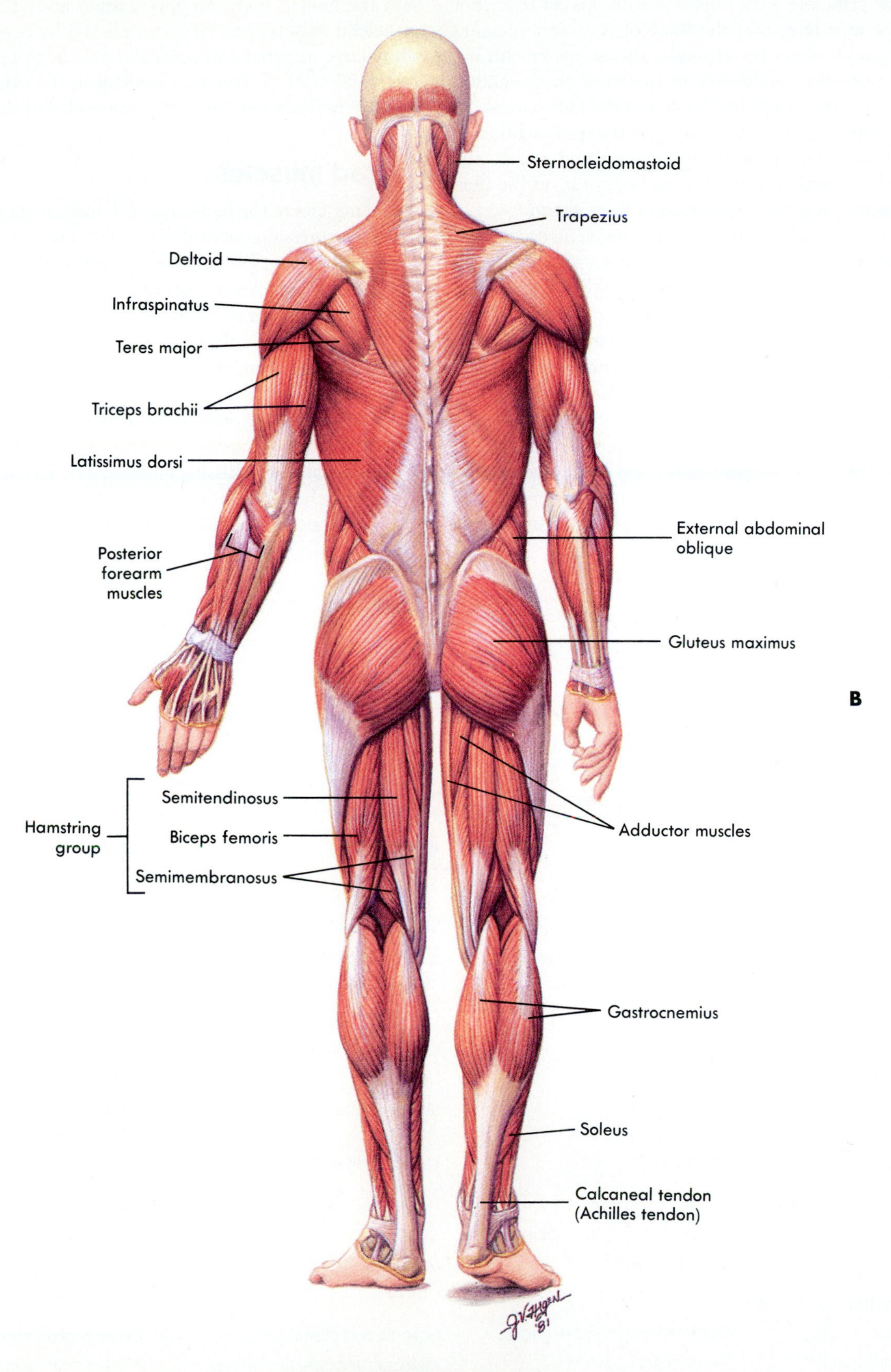

Figure 7-5, cont'd B, Posterior view.

greatest movement. Some muscles have more than one origin, but the principle is the same—the origins act to anchor or hold the muscle so that the force of contraction causes the insertion to move. For example, the biceps brachii has two origins on the scapula and an insertion on the radius. Contraction of the biceps brachii causes the radius to move, resulting in flexion of the forearm. The triceps brachii muscle has three origins; two on the humerus and one on the scapula. The insertion of the triceps brachii is on the ulna and contraction results in extension of the forearm.

Muscles that work together to cause movement are **synergists** (sin´er-jistz). For example, another arm muscle, the brachialis, flexes the forearm. The biceps brachii and the brachialis are synergists because they both cause the forearm to flex. If one of the synergists plays the major role in accomplishing a movement, it is the **prime mover.** The brachialis is the prime mover for flexing the forearm.

A muscle working in opposition to another muscle is called an **antagonist** (an-tag´o-nist). The triceps brachii is the antagonist of the biceps brachii and the brachialis because it extends the forearm. Because prime movers and antagonists produce opposite movements, the antagonist must relax when the prime mover contracts.

Although there are over 600 individual skeletal muscles in the human body, an appreciation and understanding of skeletal muscles can be accomplished by concentrating on the large superficial muscles and muscle groups (Figure 7-5, pp. 102-103). Tables that summarize the origin, insertion, and action of these muscles are provided in this chapter.

Head muscles

The muscles of the head and neck include those involved in facial expression, mastication (chewing), and moving the head and neck (Table 7-1 and Figure 7-6).

Several muscles act on the skin to produce the movements of facial expression. The **occipitofrontalis** (ok-sip´ĭ-to-fron-ta´lis) muscle raises the eyebrows. The **orbicularis oculi** (or-bik´u-lar´us ok´u-li) muscles close the eyelids and cause "crow's feet" wrinkles in the skin at the lateral corners of the eyes. The **orbicularis oris** (or´us) and **buccinator** (buk´sĭ-na´tor) muscles pucker the mouth and are frequently called the kissing muscles. The buccinator also flattens the cheeks as in whistling or blowing a trumpet and is sometimes called the trumpeter's muscle. Smiling is accom-

Table 7-1 ● Head and neck muscles

Muscle	Origin	Insertion	Action
● Muscles of facial expression			
Occipitofrontalis	Occipital bone	Skin of eyebrow	Elevates eyebrows
Orbicularis oculi	Maxilla and frontal bone	Skin around the eye	Closes eye
Orbicularis oris	Maxilla and mandible	Skin around the lips	Closes lips
Buccinator	Mandible and maxilla	Corner of mouth	Flattens cheeks
Zygomaticus muscles	Zygomatic bone	Corner of mouth	Elevates corner of mouth
Levator labii superioris	Maxilla	Upper lip	Elevates upper lip
Corrugator supercilii	Frontal bone	Skin of eyebrow	Lowers and draws together the eyebrows
Depressor anguli oris	Mandible	Lower lip near corner of mouth	Depresses corner of mouth
● Muscles of mastication			
Temporalis	Temporal region on side of skull	Mandible	Closes jaw
Masseter	Zygomatic arch	Mandible	Closes jaw
● Muscles that move the head			
Trapezius	Occipital bone and vertebrae	Scapula and clavicle	Extends head and neck
Sternocleidomastoid	Sternum and clavicle	Mastoid process of temporal	Rotates head and flexes neck

plished primarily by the **zygomaticus** (zi´go-mat´ĭ-kus) muscles. Sneering is accomplished by the **levator labii** (la´be-i) **superioris** muscles, and frowning by the **corrugator supercilii** (kor´ah-ga-ter su´per-si-le-i) and **depressor anguli** (ang´gu-li) **oris** muscles.

3 Harry Wolf, a notorious flirt, on seeing Sally Gorgeous, raises his eyebrows, winks, whistles, and smiles. Name the facial muscles he used to carry out this communication. Sally, thoroughly displeased with this exhibition, frowns and sneers in disgust. What muscles did she use?

The two pairs of muscles primarily responsible for chewing, or **mastication** (mas´tĭ-ka-shun), are among the strongest muscles of the body. The **temporalis** and **masseter** (mă-se-ter) muscles can be easily seen and felt on the side of the head during mastication.

Neck muscles

The neck muscles are responsible for moving the head and neck. The **trapezius** (tră-pe´ze-is) forms part of the posterior neck and functions to extend the head and neck. The **sternocleidomastoid** (ster´no-kli´do-mas´toyd) muscle is easily seen on the anterior and lateral sides of the neck. The neck is flexed when both sternocleidomastoid muscles contract. Rotation of the head is accomplished when one sternocleidomastoid muscle contracts and the other relaxes. For example, contraction of only the right sternocleidomastoid causes the head to rotate to the left.

4 Torticollis (tor´tĭ´kol´is), or wry neck, can result from injury to one of the sternocleidomastoid muscles. It is sometimes caused by damage to a baby's neck muscles during a difficult birth and usually can be corrected by exercising the muscle. Would damage to the right sternocleidomastoid muscle of a baby result in rotation of the head to the left or right? Explain. (Hint: feel the muscles of your neck as you rotate your head to the left and right.)

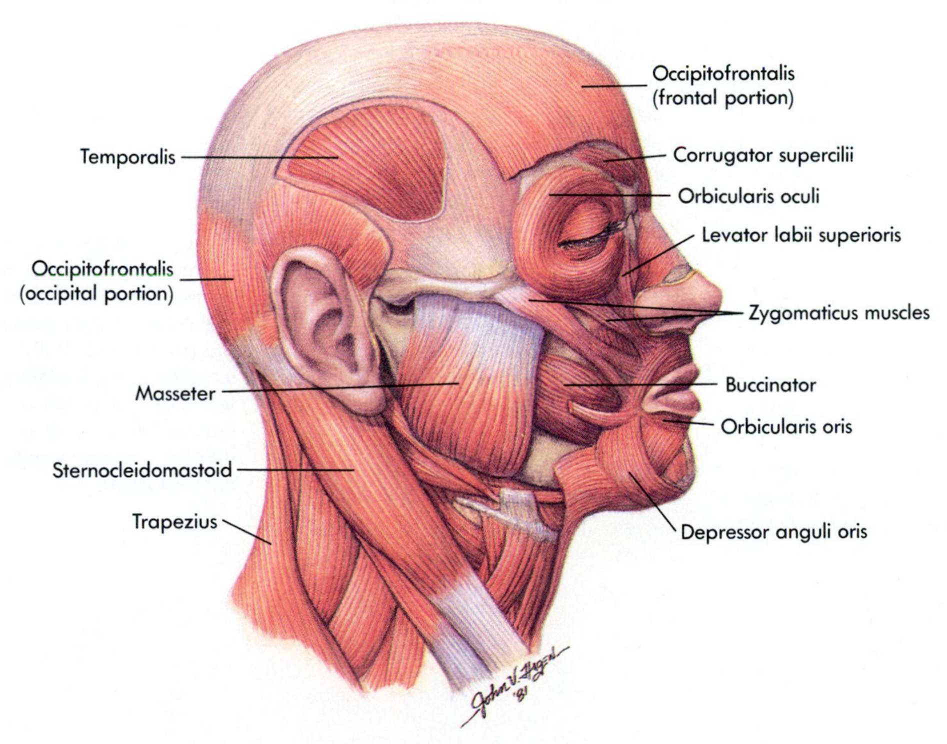

Figure 7-6 ● Muscles of facial expression and mastication. Lateral view.

Understanding muscle fibers

Muscles are sometimes classified as either fast-twitch or slow-twitch muscles. **Fast-twitch muscle fibers** contract quickly and fatigue quickly. They are well adapted to perform anaerobic respiration, which produces ATP rapidly for a short time. As ATP production decreases the fast-twitch muscle fibers fatigue. **Slow-twitch muscle fibers** contract more slowly and are more resistant to fatigue. They are better suited for aerobic respiration, which produces adequate ATP to sustain contractions for longer periods.

The white meat of a chicken's breast is comprised mainly of fast-twitch fibers. The muscles are adapted to contract rapidly for a short time but fatigue quickly. Chickens normally do not fly long distances. They spend most of their time walking. Ducks, on the other hand, fly for much longer periods and for greater distances. The red or dark meat of a chicken's leg or a duck's breast is composed of slow-twitch fibers. The darker appearance is due partly to the dark color of the enzyme system involved in aerobic respiration, partly to a richer blood supply, and partly to the presence of **myoglobin**. The richer blood supply ensures delivery of the oxygen necessary for aerobic respiration and the myoglobin functions to store oxygen temporarily. Myoglobin can continue to release oxygen in a muscle even when a sustained contraction has interrupted the flow of blood by compressing the blood vessels.

Humans exhibit no clear separation of slow-twitch and fast-twitch muscle fibers in individual muscles. Most muscles have both types of fibers, although the number of each type varies in a given muscle. The distribution of the fibers in a given muscle is constant for each individual and is established before birth. People who are good sprinters have a greater percentage of fast-twitch muscle fibers in their lower limbs, whereas good long-distance runners have a higher percentage of slow-twitch fibers. Athletes who are able to perform a variety of anaerobic and aerobic exercises tend to have a more balanced mixture of fast-twitch and slow-twitch muscle fibers.

5 Which muscles would have the greatest percentage of slow-twitch muscle fibers, the postural muscles of the back or the muscles of the arm? Explain.

The strength of a muscle is related to the size of the muscle. An increase in the size of a muscle, called **hypertrophy** (hi-per′tro-fe), results in a stronger muscle. For example, exercise causes both fast-twitch and slow-twitch fibers to hypertrophy and become stronger. The increase in muscle size and strength results from an increase in the number of myofibrils within muscle fibers, and is not due to an increase in the number of muscle fibers. Each muscle fiber becomes larger and stronger. Conversely, a muscle that is not used undergoes a decrease in size called **atrophy** (at′ro-fe). Atrophy usually involves a decrease in the number of myofibrils without a decrease in muscle fiber number. The muscular atrophy that occurs in a limb placed in a cast for several weeks is an example. Normally the number of muscle fibers remains relatively constant after birth. In old age, however, severe atrophy involves an irreversible decrease in the number of muscle fibers and can lead to paralysis.

Neither fast-twitch nor slow-twitch muscle fibers can normally be converted to muscle fibers of the other type. Although training can increase the capacity of both types of muscle fibers to perform more efficiently, the fast-twitch and slow-twitch muscle fibers respond differently to different types of training. Intense exercise resulting in anaerobic respiration has the greater effect on fast-twitch muscle fibers. The fast-twitch muscle fibers can greatly increase in size, resulting in large, strong muscles. Aerobic exercise, on the other hand, has the greatest effect on slow-twitch muscle fibers. The blood supply to the slow-twitch muscle fibers increases, resulting in increased endurance because the muscle is better able to aerobically produce ATP. However, there is only a moderate increase in size. Aerobic respiration also can convert fast-twitch muscle fibers that fatigue readily to fast-twitch muscle fibers that resist fatigue. This conversion is accomplished by increasing the number of mitochondria in the fast-twitch muscle fibers and by increasing the blood supply to the fast-twitch muscle fibers.

6 Suppose a man wished to develop very large, strong muscles and a woman wished to improve her muscular strength and endurance, but did not wish to "bulk up" and have large muscles. Explain to them, in terms of fast-twitch and slow-twitch muscle fibers, what kind of weight-training program would be best for each of them.

Table 7-2 ● Trunk muscles

Muscle	Origin	Insertion	Action
● Muscles that move the vertebral column			
Erector spinae	Ilium, sacrum, and vertebrae	Superior vertebrae and ribs	Extend, abduct, and rotate vertebral column
Deep back muscles	Vertebrae	Vertebrae	Extend, abduct, and rotate vertebral column
Rectus abdominis	Pubis	Xiphoid process of sternum and lower ribs	Flexes vertebral column; compresses abdomen
External abdominal oblique	Rib cage	Illiac crest and fascia of rectus abdominis	Flexes and rotates vertebral column; compresses abdomen
Internal abdominal oblique	Iliac crest and vertebrae	Lower ribs and fascia of rectus abdominis	Flexes and rotates vertebral column; compresses abdsomen
Transversus abdominis	Rib cage, vertebrae, and iliac crest	Xiphoid process of sternum, fascia of rectus abdominis, and pubis	Compresses abdomen

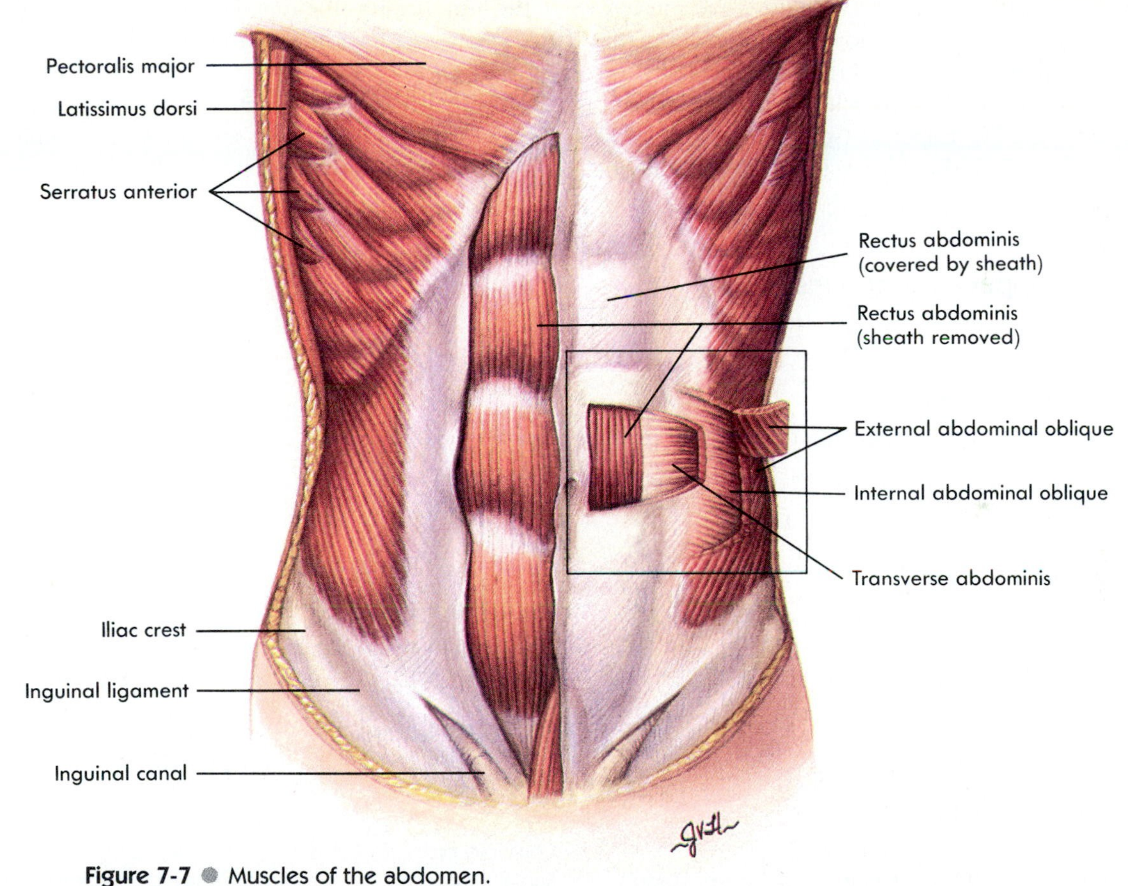

Figure 7-7 ● Muscles of the abdomen.
The rectus abdominis muscles are located to either side of the midline. Three layers of muscles comprise the lateral abdominal wall.

Trunk muscles

Trunk muscles include those of the vertebral column and the abdominal wall (Table 7-2). In humans, the back muscles are very strong to maintain erect posture. The **erector spinae** (e-rek´ter spi´ne) group of muscles are the muscles primarily responsible for keeping the back straight and the body erect. They also abduct and rotate the vertebral column. These muscles can be felt as a bulge just lateral to the vertebral column of the lower back. **Deep back muscles,** located between the spinous and transverse processes of adjacent vertebrae (see Figure 6-B, *D*), assist the erector spinae.

The muscles of the anterior abdominal wall (Figure 7-7) flex and rotate the vertebral column, compress the abdominal cavity, and hold in and protect the abdominal organs. On each side of the midline are the **rectus abdominis** (rek´tis ab-dom´i-nis) muscles. They are subdivided at three or more locations, causing the abdominal wall in a thin person with well developed abdominal muscles to appear segmented. Lateral to each rectus abdominis are three layers of muscle. From superficial to deep, these muscles are the **external abdominal oblique, internal abdominal oblique,** and **transversus abdominis muscles.** The muscle bundles of these three muscle layers are oriented in opposite directions to one another, producing a strong abdominal wall.

Upper limb muscles

The muscles of the upper limb include those that attach the limb and girdle to the body and those that are in the arm, forearm, and hand (Table 7-3).

Scapular movements

The connection of the upper limb to the body is accomplished primarily by muscles. These muscles act to hold the scapula firmly in position when the muscles of the arm contract. The scapular muscles also move the scapula into different positions, thereby increasing the range of movement of the upper limb. Two of these muscles are the trapezius and the serratus anterior. The **trapezius** (see Figure 7-5, *B*) forms the upper line from each shoulder to the neck, and the origin of the **serratus** (ser-a´tis) **anterior** from the first eight or nine ribs can be seen along the lateral thorax (see Figure 7-5, *A*).

Arm movements

The arm is attached to the thorax by the **pectoralis** (pek´tor-a-lis) **major** and **latissimus dorsi** (lah-tis´ĭ-mus dor´se) muscles (see Figure 7-5). The pectoralis major adducts and flexes (moves anteriorly) the arm. The latissimus dorsi adducts and powerfully extends (moves posteriorly) the arm. Because a swimmer uses these motions during the power stroke of the crawl, the latissimus dorsi is often called the swimmer's muscle. The **deltoid** (del´toyd) and other muscles attach the humerus to the scapula and clavicle (see Figure 7-5, *B*). The deltoid is the major abductor of the arm. It forms the rounded mass of the shoulder and is a common site for administering injections.

Forearm movements

The **triceps brachii** (tri´seps bra´ke-i) occupies the posterior part of the arm (Figure 7-8, *A*) and functions to extend

Table 7-3 Upper limb muscles

Muscle	Origin	Insertion	Action
Muscles that move the scapula			
Trapezius	Occipital bone and vertebrae	Scapula and clavicle	Holds scapula in place and rotates scapula
Serratus anterior	Ribs	Medial border of scapula	Rotates scapula and pulls scapula anteriorly
Muscles that move the arm			
Pectoralis major	Sternum, ribs, and clavicle	Tubercle of humerus	Adducts and flexes arm
Latissimus dorsi	Vertebrae	Tubercle of humerus	Adducts and extends arm
Deltoid	Scapula and clavicle	Shaft of humerus	Abducts, flexes, and extends arm
Teres major	Scapula	Tubercle of humerus	Adducts and extends arm
Infraspinatus	Scapula	Tubercle of humerus	Extends arm
Muscles that move the forearm			
Brachialis	Shaft of humerus	Coronoid process of ulna	Flexes forearm
Biceps brachii	Coracoid process of scapula and scapula superior to the glenoid fossa	Radial tuberosity	Flexes and supinates forearm
Triceps brachii	Shaft of humerus and lateral border of scapula	Olecranon process of ulna	Extends forearm
Muscles that move the wrist and fingers			
Anterior forearm muscles	Medial epicondyle	Carpals, metacarpals, and phalanges	Flex wrist, fingers and thumb; pronate forearm
Posterior forearm muscles	Lateral epicondyle	Carpals, metacarpals, and phalanges	Extend wrist, fingers, and thumb; supinate forearm
Intrinsic hand muscles	Carpals and metacarpals	Phalanges	Abduct, adduct, flex, and extend fingers and thumb

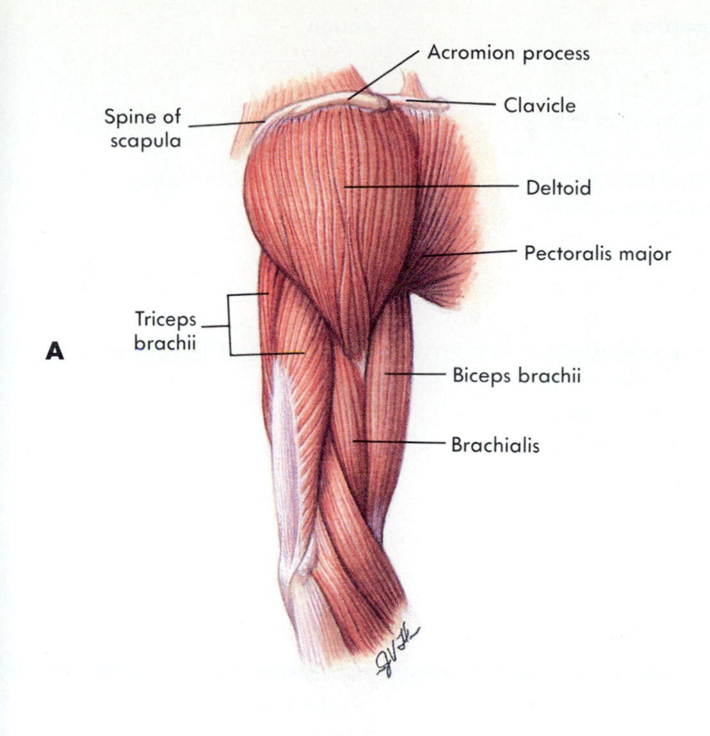

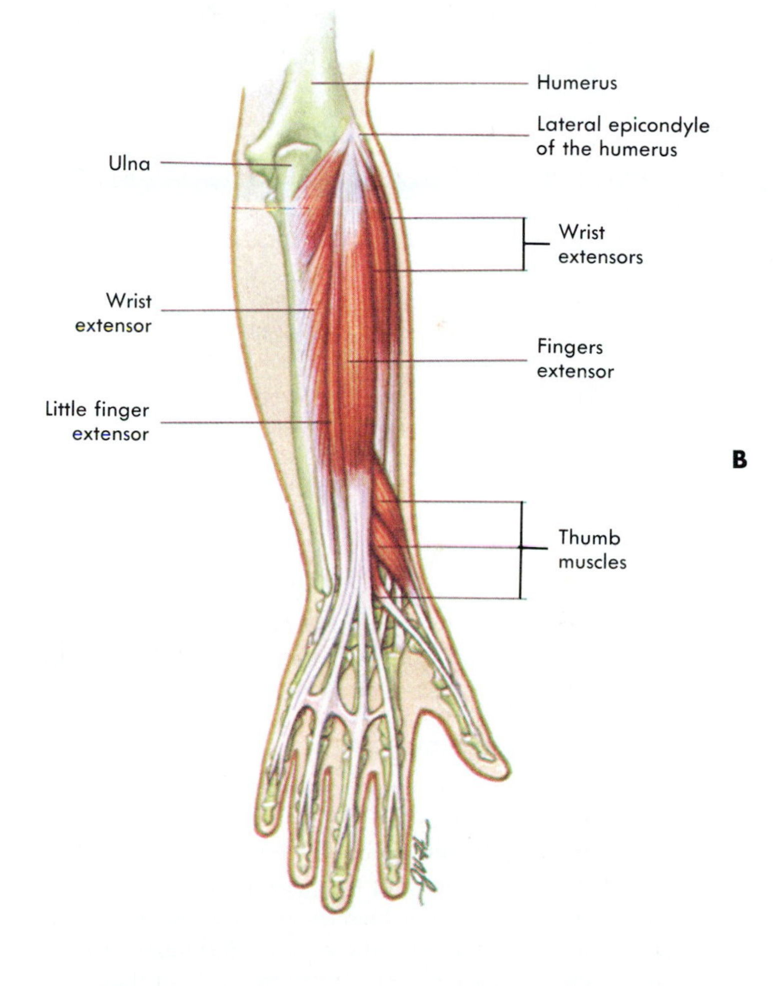

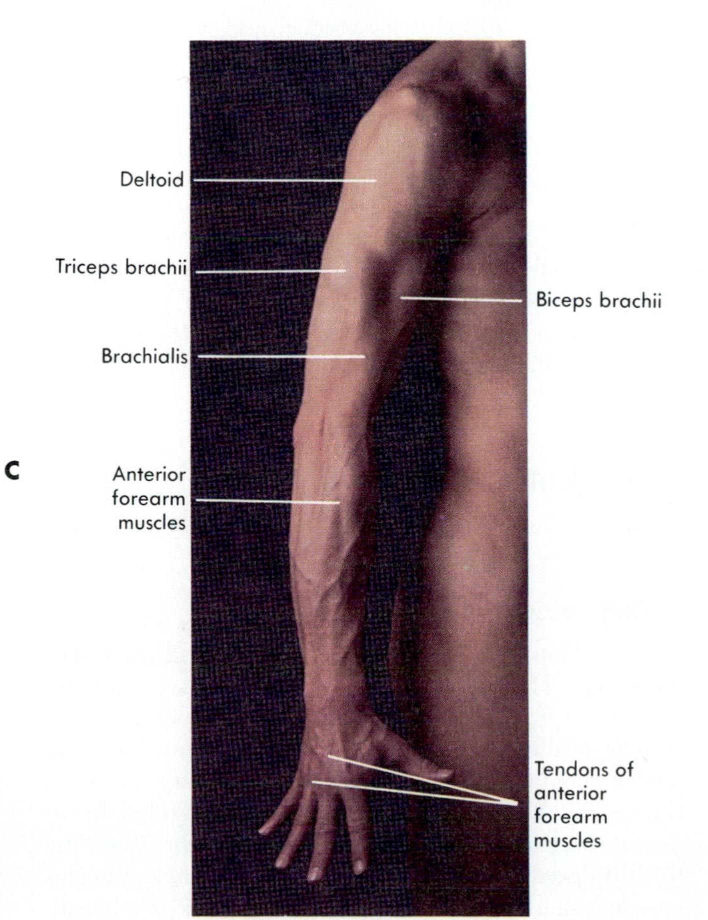

Figure 7-8 ● Muscles of the right upper limb.
A, Arm, lateral view. **B,** Forearm, posterior view. **C,** Surface anatomy of upper limb.

Table 7-4 • Lower limb muscles

Muscle	Origin	Insertion	Action
Muscles that move the thigh			
Iliopsoas	Ilium and vertebrae	Trochanter of femur	Flexes thigh
Tensor fascia latae	Anterior superior iliac spine	Lateral condyle of tibia	Abducts thigh
Gluteus maximus	Ilium, sacrum, and coccyx	Lateral side of femur	Extends and abducts thigh
Gluteus medius	Ilium	Trochanter of femur	Abducts thigh
Adductor muscles of thigh	Pubis	Femur	Adduct thigh
Muscles that move the leg			
Quadriceps femoris			
Rectus femoris	Anterior superior iliac spine	Tibial tuberosity	Extends leg and flexes thigh
Vastus lateralis	Femur	Tibial tuberosity	Extends leg
Vastus medialis	Femur	Tibial tuberosity	Extends leg
Vastus intermedius (Not shown in illustration)	Femur	Tibial tuberosity	Extends leg
Sartorius	Anterior superior iliac spine	Tibia	Flexes leg and thigh
Hamstring muscles			
Biceps femoris	Ischium and femur	Fibula	Flexes leg and extends thigh
Semimembranosus	Ischium	Tibia	Flexes leg and extends thigh
Semitendinosus	Ischium	Tibia	Flexes leg and extends thigh
Muscles that move the ankle and toes			
Tibialis anterior	Tibia	Tarsal and first matatarsal	Dorsiflexes foot
Deep anterior leg muscles	Tibia or fibula	Phalanges, metatarsals, or tarsals	Extend toes
Gastrocnemius	Medial and lateral epicondyles of femur	Calcaneus	Plantar flexes foot
Soleus	Tibia and fibula	Calcaneus	Plantar flexes foot
Deep posterior leg muscles	Tibia or fibula	Phalanges, metatarsals, or tarsals	Flex toes and invert foot
Peroneus muscles	Fibula and tibia	Tarsals and metatarsals	Evert foot
Intrinsic foot muscles	Tarsals or metatarsals	Phalanges	Abduct, adduct, flex, and extend toes

the forearm. The anterior part of the arm is occupied mostly by the **biceps brachii** (bi´seps bra´ke-i) and the **brachialis** (bra´ke-a-lis), which flex the forearm.

Wrist and finger movements

The 20 muscles of the forearm can be divided into anterior and posterior groups (see Figure 7-5). Most of the anterior forearm muscles are responsible for flexion of the wrist and fingers, whereas most of the posterior forearm muscles cause extension. Forearm muscles also are responsible for supination and pronation of the forearm. Although the forearm muscles are located in the forearm, many of their tendons extend to the wrist and fingers (Figure 7-8, *B*). The tendons extending the fingers, for example, are very visible on the back of the hand (Figure 7-8, *C*). The tendon for a wrist flexor on the radial side of the forearm is used as a landmark for taking the pulse.

Forceful extension of the wrist repeated over a period, such as occurs in the stroke of a tennis backhand, can result in inflammation and pain where the muscles attach to the lateral epicondyle of the humerus. This condition is sometimes referred to as "tennis elbow."

There are 19 muscles in the hand, which together are called **intrinsic hand muscles.** They are responsible for abduction and adduction of the fingers and for many other movements of the thumb and fingers. Some of these muscles account for the masses at the base of the thumb and little finger.

Lower limb muscles

The muscles of the lower limb include those located in the hip, thigh, leg, and foot (Table 7-4).

Thigh movements

Several hip muscles originate on the coxa and insert onto the femur. The **gluteus** (glu´te-us) **maximus** contributes most of the mass that can be seen as the buttocks, and the **gluteus medius,** a common site for injections, creates a smaller mass just superior and lateral to the gluteus maximus (Figure 7-9, *A*). The gluteal muscles abduct the thigh. In addition, the gluteus maximus extends the thigh. When the thigh is flexed at approximately a 45-degree angle, the gluteus maximus functions maximally to extend the thigh,

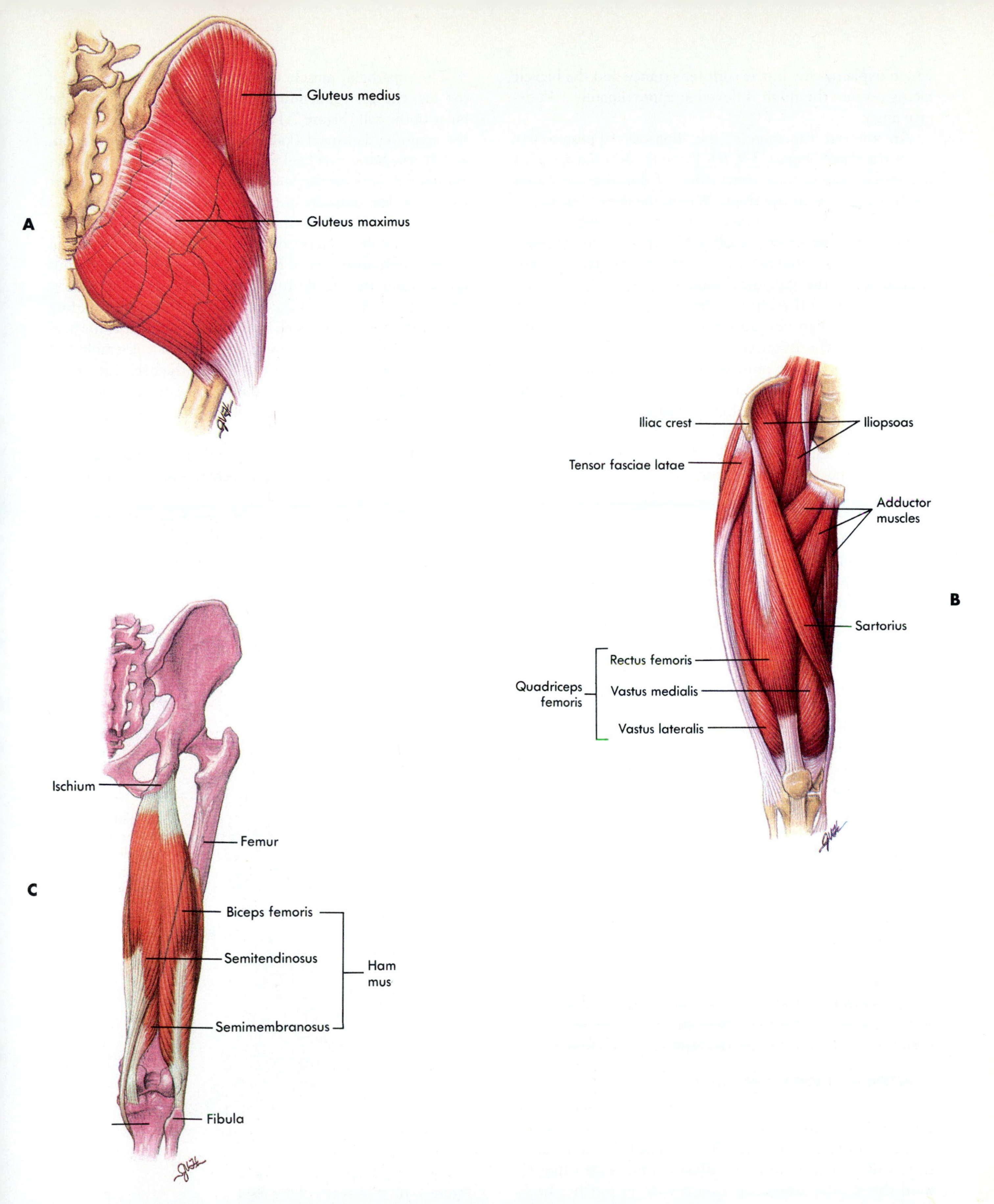

Figure 7-9 ● Muscles of the right hip and thigh.
A, Hip, posterior view. **B,** Hip and thigh, anterior view. **C,** Thigh, posterior view.

which explains why in the sprinter's stance and the bicycle racing posture the thigh is flexed at approximately a 45-degree angle.

An anterior hip muscle, the **iliopsoas** (il´e-op-so´us), flexes the thigh (Figure 7-9, *B*). If the thigh is fixed so that it does not move, then contraction of the iliopsoas causes the trunk to flex on the thigh. When one does a "sit-up," it is actually the powerful iliopsoas that is responsible for most of the movement. Only a small part of the movement is due to flexion of the vertebral column, which is mostly caused by contraction of the abdominal muscles.

In addition to the hip muscles, some of the muscles located in the thigh also attach to the coxa and can cause movement of the thigh (Figure 7-9, *B* and *C*). There are three groups of thigh muscles: the anterior thigh muscles can flex the thigh; the posterior thigh muscles can extend the thigh; and the medial thigh, or **adductor,** muscles, can adduct the thigh. A "pulled groin" consists of tearing one or more of the adductor muscles or their tendons, usually where the tendon attaches to the coxa.

Leg movements

The anterior thigh muscles are the **quadriceps femoris** (kwah´drĭ-seps feh-mor´is; four thigh muscles) and the **sartorius** (sar-to´re-us) (see Figure 7-9, *B*). The quadriceps femoris muscles are the primary extensors of the leg, but also function to flex the thigh. They have a common insertion, the patellar tendon, on and around the patella. The patellar ligament is an extension of the patellar tendon onto the tibial tuberosity. The patellar ligament is the point that is tapped with a rubber hammer when testing the knee-jerk reflex in a physical examination. One of the quadriceps femoris muscles (vastus lateralis) sometimes is used as an injection site, especially in infants who may not have well-developed deltoid or gluteal muscles. The sartorius flexes the leg and thigh.

The posterior thigh muscles are called **hamstring muscles,** and they are responsible for flexing the leg and extending the thigh (see Figure 7-9, *C*). Their tendons are easily felt and seen on the medial and lateral posterior aspect of a slightly bent knee. The hamstrings were so named because these tendons in pigs could be used to suspend hams during curing. Some animals such as wolves often bring down their prey by biting through the hamstrings. Therefore, "to hamstring" someone is to render them helpless. A "pulled hamstring" consists of tearing one or more of these muscles or their tendons, usually where the tendons attach to the coxa.

Ankle and toe movements

The 13 muscles in the leg, with tendons extending into the foot, can be divided into three groups: anterior, posterior, and lateral (Figure 7-10). The **tibialis** (tib´e-a´lis) **anterior** of the anterior group dorsiflexes or moves the foot toward the shin as when a person stands on just her heels. Other anterior leg muscles extend the toes. Their tendons can be seen on the top of the foot.

The superficial muscle of the posterior compartment of the leg, the **gastrocnemius** (gas-trok-ne´me-us) forms the bulge of the calf (Figure 7-11). It joins other muscles to form the common **calcaneal** (kal-ka´ne-al), or **Achilles tendon,** which attaches to the heel bone (calcaneus). These muscles plantar flex or point the foot as when a person stands on her toes. The deep muscles of the posterior group flex the toes and invert the foot (turn the medial side of the foot inward).

The Achilles tendon derives its name from a hero of Greek mythology. As a baby, Achilles was dipped into magic water that made him invulnerable to harm everywhere it touched his skin. His mother, however, holding him by the back of his heel, overlooked submerging his heel under the water. Consequently, his heel was vulnerable and proved to be his undoing; he was shot in the heel with an arrow at the battle of Troy and died. Thus, saying that someone has an "Achilles heel" means he has a weak spot that can be attacked.

The lateral muscles of the leg, called the **peroneus** (pĕr-o-ne´is) muscles (Figure 7-10), evert the foot (turn the lateral side of the foot outward).

The 20 muscles located within the foot, called the **intrinsic foot** muscles, flex, extend, abduct, and adduct the toes. They are arranged in a manner similar to the intrinsic muscles of the hand.

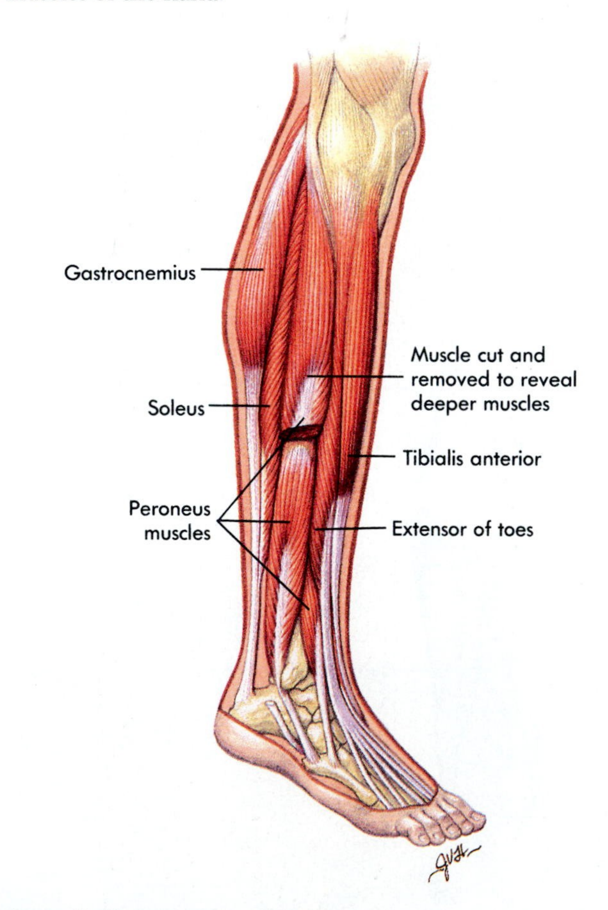

Figure 7-10 ● Muscles of the leg.
Lateral view. One of the peroneal muscles has been cut and removed to reveal a second peroneal muscle.

Figure 7-11 ● Surface anatomy of the right lower limb.
A, Anterior view. **B,** Posterior view.

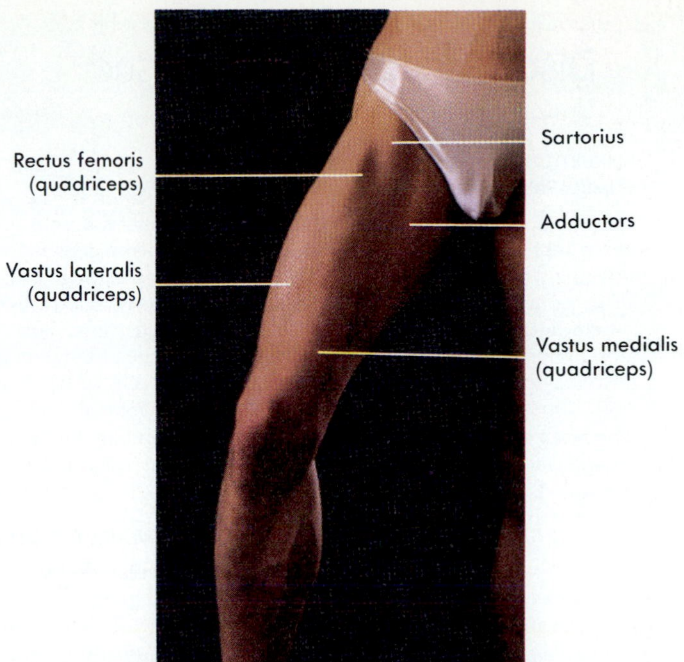

Gluteus medius
Gluteus maximus
Vastus lateralis (quadriceps)
Tendon of biceps femoris
Tendon of semitendinosus and semimembranosus
B
Gastrocnemius
Soleus
Calcaneal (Achilles) tendon

Disorders of muscle tissue

DISRUPTION OF NERVOUS SYSTEM STIMULATION

Denervation is an interruption of the nerve supply to a muscle, for example severing the nerve in an accident. Without stimulation from the nervous system, muscles lose their muscle tone and do not voluntarily contract, a condition called flaccid paralysis. Denervation also results in severe atrophy, which is a decrease in muscle size. If the muscle is reinnervated, muscle function is restored, and atrophy is stopped. Muscles that have been denervated sometimes are stimulated electrically to prevent severe atrophy. The strategy is to slow the process of atrophy while nerves grow toward the muscles and eventually reinnervate them.

Myasthenia gravis (mi´as-the´ne-ah grav´is) is characterized by muscular weakness and fatigue. Although the symptoms are expressed in the muscular system, the disorder results from a malfunction of the immune system. Proteins, called antibodies (see Chapter 17 for details), are produced by the immune system. Normally the antibodies attack and destroy bacteria and other disease-causing organisms. In myasthenia gravis, antibodies attack the connection between muscle fibers and nerve cells, gradually inhibiting the ability of the muscle fiber to respond to stimulation by the nervous system.

Polio is a viral disease that destroys the nerve cells supplying skeletal muscles, resulting in paralysis. **Multiple sclerosis** (skle-ro´sis) is a disorder of unknown cause, but viral infection and the immune system are suspected. It affects cells of the brain and spinal cord that control skeletal muscle activity. Depending on the part of the brain or spinal cord affected, the muscles are overstimulated or understimulated.

MUSCULAR DYSTROPHY

Muscular dystrophy (dis´tro-fe) refers to a group of muscle diseases that destroy skeletal muscle tissue. The diseases usually are inherited and are characterized by the progressive degeneration of muscle fibers leading to atrophy and eventual replacement by connective tissue. In the most common type of muscular dystrophy (Duchenne muscular dystrophy) there is an inability to produce a protein called dystrophin. According to one hypothesis, cells without dystrophin allow too much calcium to cross the cell membrane and enter the cell. The excess calcium causes the breakdown of muscle proteins. An experimental treatment injects the diseased muscle with immature muscle cells from a healthy person. The healthy muscle cells fuse with the diseased muscle cells and the nuclei of the healthy muscle cells cause dystrophin to be produced.

CRAMPS

Cramps are painful, spastic contractions of muscles that are usually the result of an irritation within a muscle. Local inflammation from buildup of lactic acid or connective tissue inflammation can cause contraction of muscle fibers surrounding the irritated region.

TENDONITIS

Tendonitis is an inflammation of a tendon and/or its attachment point. It usually occurs in athletes who overtax the muscle to which the tendon is attached.

FIBROMYALGIA

Fibromyalgia (fi´bro-mi-al´je-ah) is widespread, consistent aches and discomfort within muscles or where the muscles join their tendons. Fortunately the condition is not progressive, crippling, or life-threatening. The cause is unknown and there is no cure. Exercise, relaxation techniques, and correct posture help relieve symptoms.

SUMMARY

FUNCTIONS OF SKELETAL MUSCLE

Skeletal muscle is responsible for body movements, posture, and heat production.

MUSCLE STRUCTURE

Muscles are surrounded by fascia and subdivided into muscle bundles, which are composed of muscle fibers.

Muscle fibers contain myofibrils, which are composed of sarcomeres. A sarcomere is bounded by Z lines and contains actin and myosin myofilaments.

MUSCLE CONTRACTION

Mechanism of skeletal muscle contraction and relaxation

According to the sliding filament mechanism, movement of cross bridges causes actin myofilaments to slide past myosin myofilaments.

Relaxation occurs when cross bridges release actin myofilaments.

Stimulating skeletal muscle contractions

The nervous system stimulates skeletal muscle contraction.

Recruitment is a gradual increase in the number of muscle fibers stimulated.

Types of muscle contractions

In isotonic contractions tension is constant but muscle length shortens.

In isometric contraction muscle length is constant but tension increases.

Muscle tone consists of a small percentage of muscle fibers contracting isometrically and is responsible for posture.

Energy requirements for muscle contraction

Energy is produced by anaerobic (without oxygen) or aerobic (with oxygen) respiration.

After intense exercise, the rate of aerobic respiration remains elevated to repay the oxygen debt.

MUSCLE ANATOMY

General principles

Most muscles have an origin on one bone, an insertion onto another bone, and cross at least one joint.

Muscles working together are synergists, muscles working in opposition are antagonists.

A prime mover is the one muscle of a synergistic group that is primarily responsible for the movement.

Muscles of the head

Muscles of facial expression are associated primarily with the mouth and eyes.

Two pairs of muscles primarily are involved in mastication.

Neck muscles

Neck muscles flex, extend, abduct, and rotate the head.

Trunk muscles

Back muscles maintain an erect posture and extend, abduct, and rotate the vertebral column.

Muscles of the abdominal wall flex and rotate the vertebral column, compress the abdominal cavity, and hold in the abdominal organs.

Upper limb muscles

The upper limb is attached to the body primarily by muscles.

Scapular muscles hold the scapula in place and rotate the scapula.

Arm movements are accomplished by muscles attached to the trunk and scapula.

The forearm is flexed and extended by anterior and posterior arm muscles respectively.

Movements of the wrist and fingers are accomplished by most of the 20 forearm muscles and 19 intrinsic muscles in the hand.

Lower limb muscles

Hip muscles flex, extend, and abduct the thigh. Thigh muscles flex, extend, and adduct the thigh.

Thigh muscles flex and extend the leg.

Muscles of the leg and foot can be considered as being somewhat similar to those of the forearm and hand.

CONTENT REVIEW

1. List the three functions of skeletal muscles.
2. Define fascia, muscle bundle, muscle fiber, myofibril, myofilament, and sarcomere.
3. Describe the sliding filament mechanism of muscle contraction. How does a muscle relax?
4. How can the force of contraction of a muscle be varied?
5. Compare isometric and isotonic contraction. What is muscle tone?
6. Describe the two ways energy is produced in skeletal muscle.
7. What is the oxygen debt?
8. Define origin, insertion, synergist, antagonist, and prime mover.
9. Name the muscle that extends the head. What muscle can flex the neck and rotate the head?
10. Name the muscles primarily responsible for mastication.
11. Name the major muscles of the back and abdominal wall. What movements do they produce? In addition to movements, what functions do these muscle perform?
12. Name and give the function of the muscles that attach the scapula to the body.
13. Name the muscles that attach the arm to the scapula and to the trunk. For each muscle describe the action it produces.
14. Describe the muscles responsible for movements of the forearm, wrist, and fingers.
15. Describe the arrangement of hip and thigh muscles that cause movements of the thigh.
16. Name the major muscle groups that cause flexion and extension of the leg.
17. Describe the muscles responsible for movements of the foot and toes.

CONCEPT REVIEW

1. Muscles, especially in the limbs, often are arranged so that prime movers have antagonists. Explain why this arrangement is necessary.
2. A researcher was investigating the fast-twitch vs. slow-twitch composition of muscle tissue in the gastrocnemius muscle (in the calf of the leg) of athletes. Describe the general differences this researcher would see when comparing the muscles from athletes who were outstanding in the following events: 100-meter dash, weight lifting, the 10,000-meter run.
3. Name the muscle that acts as an antagonist for each of the following muscles: zygomaticus, sternocleidomastoid, erector spinae, deltoid, anterior forearm muscles, adductor muscles of thigh, quadriceps femoris, and tibialis anterior.
4. Describe an exercise routine that would build up each of the following groups of muscles: anterior arm, posterior arm, anterior forearm, anterior thigh, posterior leg, and abdomen.
5. Sherri Speedster started a 100-meter dash but fell to the ground in pain. Examination of her right lower limb revealed bulging of the muscles in the posterior thigh. In addition, Sherri is in pain and is unable to extend her thigh. Explain the nature of her injury. (Hint: an injured muscle is spasmodically contracting.)

CHAPTER TEST

Answers can be found in Appendix C

MATCHING 1 For each statement in column A select the correct answer in column B (an answer may be used once, more than once, or not at all).

A	B
1. Thread-like structure within muscle cells.	aerobic
2. The functional unit of skeletal muscle.	anaerobic
3. Movement of these structures within a sarcomere results in contraction.	cross bridges
4. Contraction in which tension remains constant.	insertion
5. Metabolism that requires oxygen.	isometric
6. Metabolism that produces lactic acid.	isotonic
7. The reason most people stop exercise.	muscle bundle
8. The most stationary end of a muscle.	muscle fatigue
	muscle fiber
	myofibril
	origin
	psychological fatigue
	sarcomere

MATCHING 2 For each muscle action in column A select the muscle that produces that action in column B (an answer may be used once, more than once, or not at all).

A	B
1. Flexes the neck.	anterior forearm muscles
2. Extends the vertebral column.	biceps brachii
3. Holds the scapula in place.	erector spinae
4. Trunk muscle that extends the arm.	gastrocnemius
5. Extends the forearm.	gluteus maximus
6. Flexes the wrist and fingers.	hamstring muscles
7. Abducts the thigh.	iliopsoas
8. Flexes the thigh.	latissimus dorsi
9. Extends the leg.	pectoralis major
10. Dorsiflexes the foot.	posterior forearm muscles
	quadriceps femoris
	rectus abdominis
	sternocleidomastoid
	tibialis anterior
	trapezius
	triceps brachii

FILL-IN-THE BLANK Complete each statement by providing the missing word or words.

1. The three major functions of skeletal muscles are ___________ , ___________ , and ___________.
2. The movement of actin myofilaments past myosin myofilaments during muscle contraction is called the ___________.
3. A gradual increase in the number of muscle fibers stimulated is called ___________.
4. ___________ respiration produces ATP molecules rapidly, but for a short time.
5. The amount of oxygen needed in chemical reactions to convert the lactic acid produced by anaerobic respiration to glucose is called the ___________.

CHAPTER 8

The Nervous System

KEY TERMS

autonomic nervous system
Composed of nerve fibers that send action potentials from the central nervous system to smooth muscle, cardiac muscle, and glands.

brainstem
Portion of the brain consisting of the medulla oblongata, pons, and midbrain; connects the superior part of the brain to the spinal cord and the cerebellum.

cerebellum
A part of the brain attached to the brainstem and important in maintaining muscle tone, balance, and coordination of movements.

cerebrum
The largest part of the brain, consisting of two hemispheres and including the cortex, nerve tracts, and basal ganglia.

ganglion
A group of nerve cell bodies in the peripheral nervous system.

meninges
Connective tissue membranes that surround and protect the brain and spinal cord; the dura mater, arachnoid, and pia mater.

nerve
Bundle of nerve fibers in the peripheral nervous system; there are 12 pairs of cranial nerves and 31 pairs of spinal nerves.

neuron
Cell specialized to receive and transmit action potentials; functional unit of the nervous system.

neurotransmitter
A chemical that a neuron releases into the synapse; serves to transmit information to another neuron or effector cell.

synapse
Junction between a neuron and some other cell.

FEATURES

OBJECTIVES

After reading this chapter you should be able to:

1. List and describe the divisions of the nervous system.
2. Describe the structure and function of neurons and neuroglia.
3. Define an action potential and explain how it moves along nerve fibers; explain how an action potential is transmitted across a synapse.
4. Describe the parts of the brain and give their functions.
5. Describe the spinal cord.
6. List the parts of a reflex arc and describe its function.
7. Describe the three meningeal layers surrounding the central nervous system.
8. List the twelve cranial nerves and give a brief description of their function.
9. Name the two divisions of the autonomic nervous system and describe the differences between them.

THE NERVOUS SYSTEM is one of the body's major regulatory and coordinating systems, and it is the seat of all mental activity, including consciousness, memory, and thinking. Homeostasis is maintained to a large degree by the nervous system's activities, which depend on the nervous system's ability to detect, interpret, and respond to changes in internal and external conditions.

This chapter considers the cells of the nervous system, the functions of the cells, and how the cells communicate through electrical signals called **action potentials** or nerve impulses. In addition, the parts of the nervous system and the functions performed by each part are described.

DIVISIONS OF THE NERVOUS SYSTEM

The nervous system can be divided into the central and the peripheral nervous systems (Figure 8-1). The central nervous system (CNS) consists of the brain and spinal cord. It functions to process information and initiate responses, and it is the site of mental activities. The **peripheral nervous system (PNS)** consists of receptors, nerves, and ganglia, which lie outside the CNS. It functions to detect stimuli and to transmit information to and from the CNS.

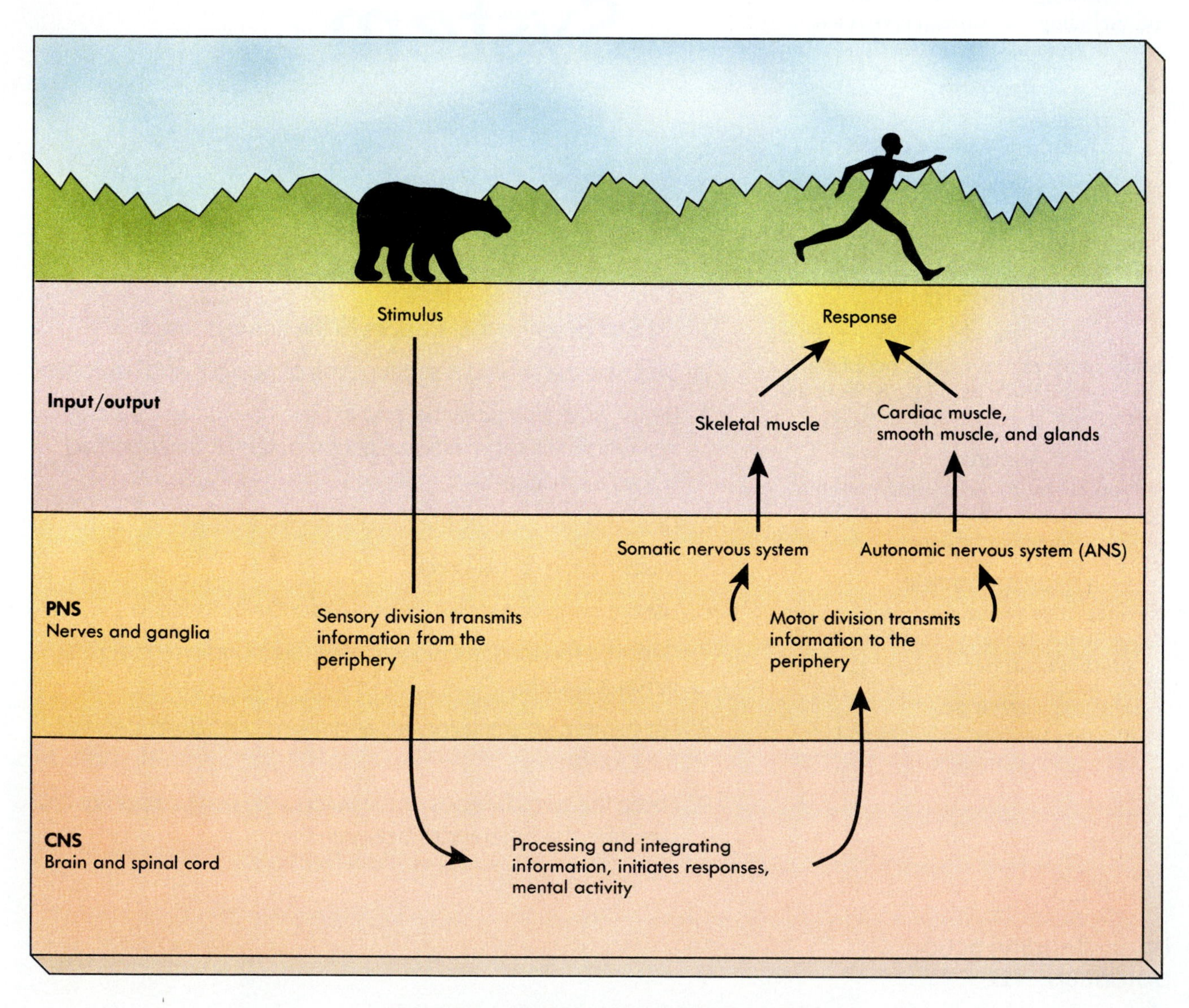

Figure 8-1 ● Divisions of the nervous system.

The PNS has two subdivisions: The **sensory division** transmits action potentials from sensory organs to the CNS. The **motor division** transmits action potentials from the CNS to **effector organs** such as muscles and glands. The motor division can be further subdivided into the **somatic** (so-mat´ik) **nervous system,** which transmits action potentials from the CNS to skeletal muscle, and the **autonomic nervous system** (ANS), which transmits action potentials from the CNS to smooth muscle, cardiac muscle, and glands.

CELLS OF THE NERVOUS SYSTEM

Neurons

Neurons (nu´ronz), or **nerve cells** (Figure 8-2), receive stimuli and transmit action potentials to other neurons or to effector organs. A neuron consists of **dendrites** (den´drītz), a **cell body,** and an **axon** (ak´son). Dendrites usually carry electric signals to the cell body and axons usually carry electric signals away from the cell body. Dendrites and axons are sometimes referred to as **nerve fibers.** The cell body has a single nucleus, ribosomes, and other organelles. The nucleus and ribosomes are necessary for protein synthesis. A nerve fiber dies if it is separated from the cell body because it no longer has a source of proteins.

Neurons are classified according to function. **Sensory neurons** conduct action potentials to the CNS and **motor neurons** conduct action potentials away from the CNS. **Association neurons,** which are located primarily within the CNS, conduct action potentials from one nerve cell to another nerve cell.

Neuroglia

Neuroglia (nu-rog´le-ah) cells support and insulate neurons. They are far more numerous than neurons and account for more than half of the brain's weight. In fact, the most common type of brain tumor, called a **glioma** (gli-o´mah), develops from neuroglia, not from neurons.

Neuroglia cells form **myelin** (mi´ĕ-lin) **sheaths** around axons (Figure 8-2). The myelin sheaths insulate the axons and increase the speed at which action potentials travel along the axon. Myelin is mainly lipoproteins formed from the plasma membrane of these cells, which wraps around the axon. The narrow gaps between the neuroglia cells are called the **nodes of Ranvier** (ron´ve-a).

Neuroglia cells also serve as the major supporting tissue in the CNS, surrounding and holding nerve cells in place. They also surround capillaries (tiny blood vessels) and influence their activities. These capillaries surrounded by neuroglia are responsible for the **blood-brain barrier,** which protects the brain and spinal cord from harmful substances. The capillaries are selectively permeable, preventing the passage of harmful substances from the blood to neurons, but allowing nutrient and waste product exchange between the blood and neurons. Sometimes, however, the blood-brain barrier can interfere with the treatment of CNS disorders because it prevents the entry of certain drugs into the CNS. For example, in Parkinson's disease there is a decrease of a chemical called dopamine in the brain. This decrease results in loss of muscle control and shaking movements. Administering dopamine is not effective because it cannot pass through the blood-brain barrier. Fortunately, another chemical called L-dopa can pass through the blood-brain barrier. Once in the brain, L-dopa is converted into dopamine, which temporarily reduces the symptoms of Parkinson's disease.

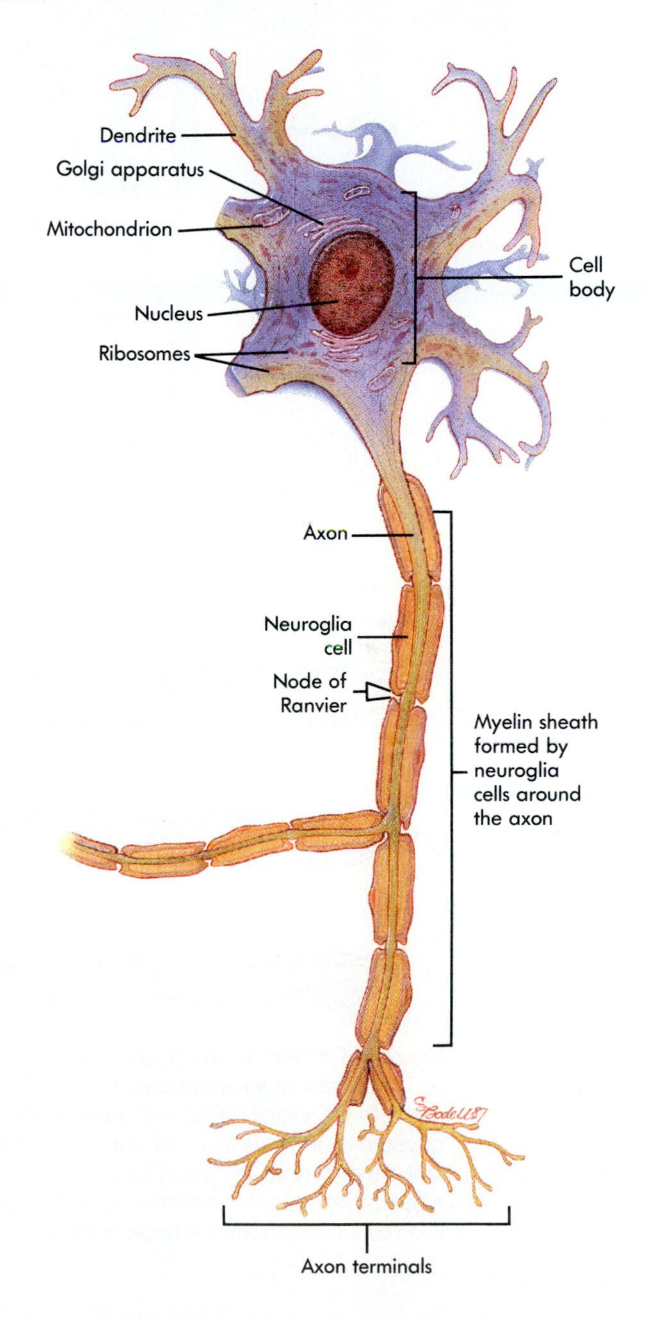

Figure 8-2 ● Neuron.
A neuron consists of dendrites, a cell body and an axon. Neuroglia cells form the myelin sheath, which surrounds the axon.

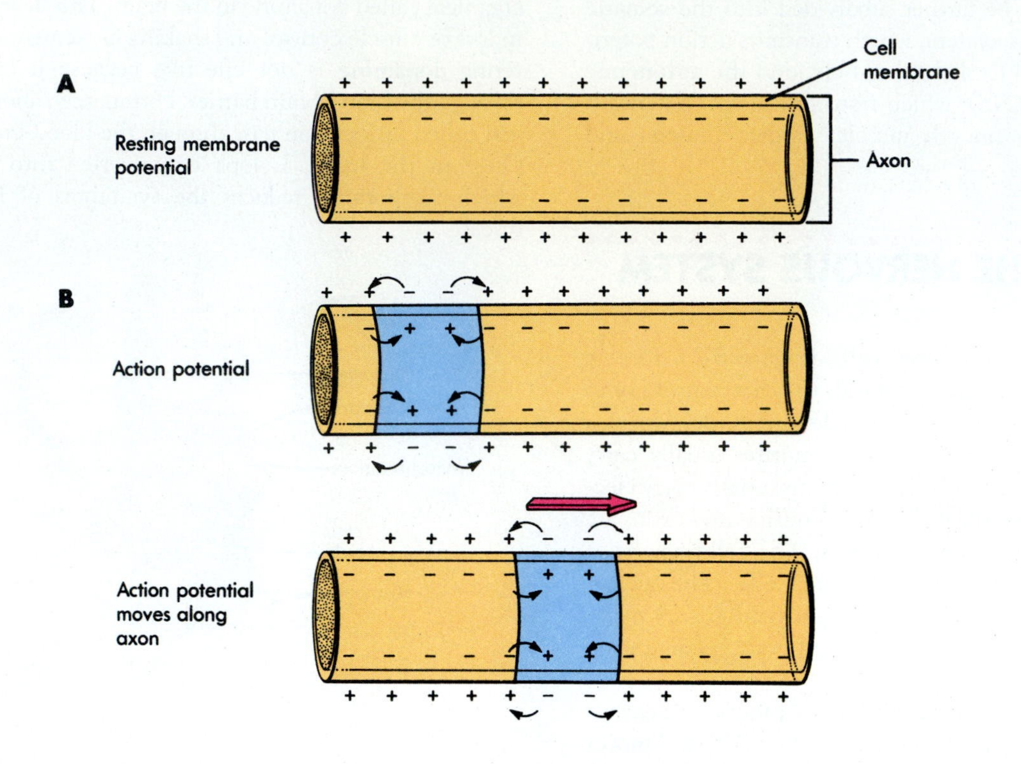

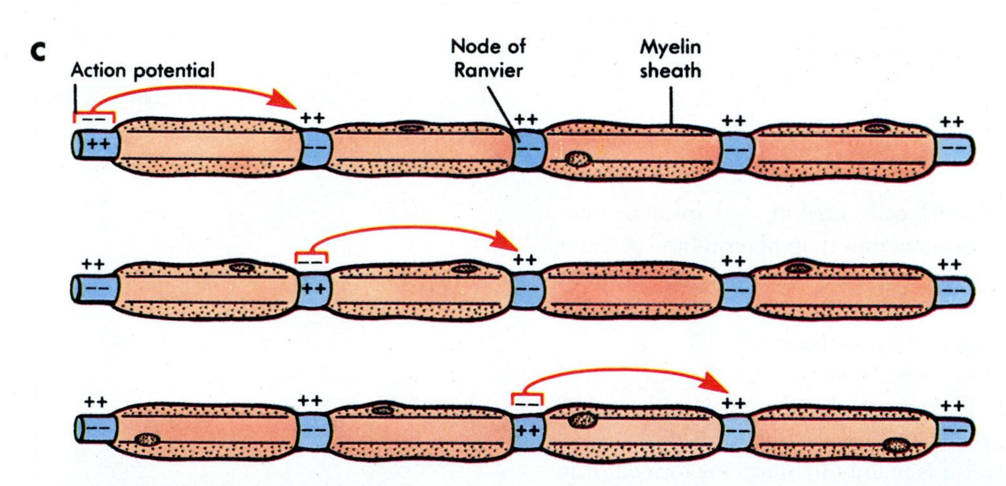

Figure 8-3 ● Membrane potentials.
A, The outside of an unstimulated resting cell membrane is positive compared with the inside, resulting in the resting membrane potential. An action potential is a brief reversal of the charges (represented by the blue area where a change in charge is shown across the membrane). **B,** Action potential moves relatively slowly down an unmyelinated axon by stimulating an action potential in the adjacent part of the membrane. **C,** Action potential moves relatively rapidly down a myelinated axon by jumping from one node of Ranvier to the next.

Organization of nervous tissue

Groups of nerve cell bodies and their dendrites form **gray matter.** Gray matter on the surface of the brain is called the **cortex,** and clusters of gray matter located deeper within the brain are called **nuclei.** In the PNS, a cluster of nerve cell bodies is called a **ganglion** (gan´gle-on; plural: ganglia). Bundles of nerve fibers with their myelin sheaths are whitish in color and are called **white matter.** White matter of the CNS forms conduction pathways or **nerve tracts,** which transmit action potentials from one area in the CNS to another. In the PNS, bundles of white matter and their connective tissue sheaths are called **nerves.**

ACTION POTENTIALS

Normally the outside of an unstimulated, resting cell membrane is positively charged compared with the inside. This condition is called the **resting membrane potential.** An **action potential** is a brief reversal of these charges in response to a stimulus such as chemicals, temperature, pressure, or light (Figure 8-3, *A*). The inside of the cell membrane temporarily becomes positive compared with the outside, then switches back to its original condition. The movement of the action potential along the membrane of a nerve fiber is called a **nerve impulse.**

The speed at which an action potential travels depends on the type of nerve fiber; conduction is more rapid in myelinated fibers than in unmyelinated fibers. **Myelinated fibers** have myelin sheaths formed by neuroglia cells, whereas **unmyelinated fibers** do not have myelin sheaths. In unmyelinated fibers, the action potential moves along the entire length of the nerve fiber (see Figure 8-3, *B*). In myelinated fibers, the myelin sheath insulates the segments of the nerve fiber it surrounds, preventing the production of action potentials. Action potentials are only produced at the uninsulated nodes of Ranvier. Consequently, an action potential at one node of Ranvier stimulates an adjacent node of Ranvier to produce an action potential. By this means action potentials jump from one node of Ranvier to the next and do not have to travel along every part of the nerve fiber's cell membrane (Figure 8-3, *C*). Therefore myelinated nerve fibers propagate action potentials much more rapidly than unmyelinated nerve fibers.

An analogy may make this concept clearer. Action potential conduction in a myelinated fiber is like a grasshopper jumping from one place to another, whereas action potential conduction in an unmyelinated axon is like a grasshopper walking. The grasshopper (action potential) moves more rapidly when jumping.

The importance of myelinated fibers is dramatically illustrated in diseases such as multiple sclerosis and diabetes mellitus in which the myelin sheath is gradually destroyed. Action potential transmission is slowed, resulting in impaired control of skeletal and smooth muscles. In severe cases, complete blockage of action potential transmission can occur.

THE SYNAPSE

A **synapse** (sin´aps) is a junction where action potentials can be transferred from one cell to another cell (Figure 8-4). The synapse between a neuron and a skeletal muscle fiber will be used to illustrate a typical synapse. The end of the axon from the neuron is separated from the muscle fiber by a small space called the **synaptic cleft.** Each axon ending contains many small vesicles filled with chemicals called **neurotransmitters.** When an action potential reaches the axon ending, some of the neurotransmitter molecules are released. They diffuse across the synaptic cleft and bind to receptors on the membrane of the muscle fiber, causing the production of an action potential in the muscle fiber membrane.

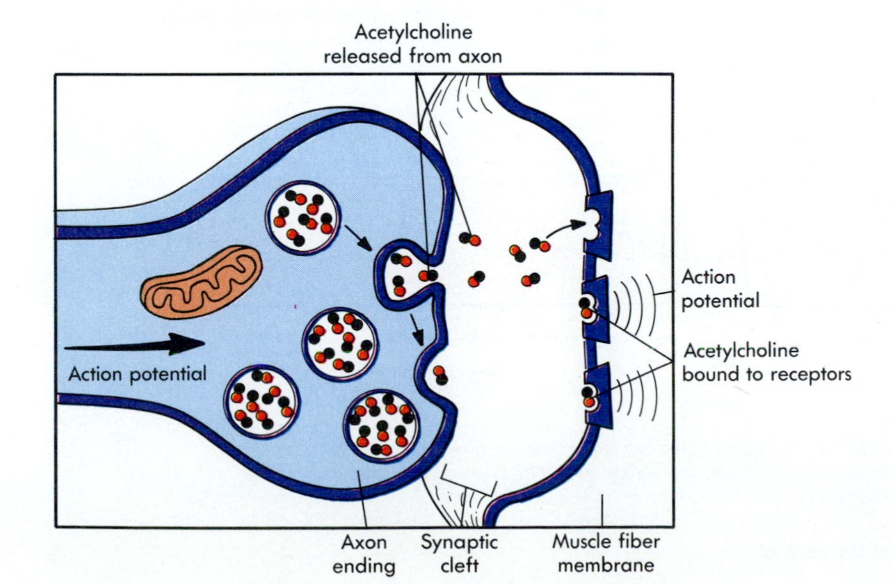

Figure 8-4 ● The synapse.
The end of an axon is on the left and the cell membrane of a muscle fiber is on the right. Acetylcholine is released from synaptic vesicles in the axon ending in response to an action potential. The acetylcholine diffuses across the synaptic cleft and binds to receptors on the muscle fiber membrane, causing an action potential in the muscle fiber.

Understanding action potentials

The production of action potentials is the first step leading to our awareness of stimuli. For example, you would not be aware of the characters on the page you are reading at this moment if it were not for the action potentials produced in nerve cells of your eyes in response to light. In addition, action potentials are stimuli that inform cells to begin an activity. For example, your eyes blink because action potentials stimulate the contraction of your eyelid muscles.

RESTING MEMBRANE POTENTIAL

The outside of most cell membranes is positively charged compared with the inside of the cell membrane, which is more negatively charged. This charge difference across the membrane of an unstimulated cell membrane is called the resting membrane potential (see Figure 8-3, *A*). The outside of the cell membrane can be thought of as the positive (plus) pole of a battery, and the inside as the negative (minus) pole. Thus a small voltage difference, or potential, can be measured across the resting cell membrane.

ACTION POTENTIAL

If the positive and negative poles of a battery are connected, electric current can flow between the poles. In a somewhat similar fashion, an electric current can flow across membranes. The current is generated by the diffusion of sodium (Na^+) and potassium (K^+) ions. There is a higher concentration of sodium ions outside the cell than inside, which means Na^+ ions diffuse from outside to inside the cell (see Figure 3-5). Conversely, there is a higher concentration of K^+ ions inside the cell than outside, so K^+ ions diffuse from inside to outside the cell.

In an unstimulated cell there is normally little movement of Na^+ or K^+ ions across the cell membrane because the membrane channels, through which the ions move, are mostly closed. This situation changes, however, when a cell is stimulated, because the membrane characteristics are changed for a brief time. Sodium channels open and positively charged sodium ions diffuse into the cell. Consequently, the inside of the cell membrane becomes positive compared with the outside, an event called **depolarization** (Figure 8-A, *A*). Closure of sodium channels and opening of potassium channels result in the diffusion of K^+ ions out of the cell. The movement of positively charged K^+ ions out of the cell makes the outside of the cell membrane positive compared with the inside. The return of the cell membrane to its resting membrane potential is called **repolarization** (Figure 8-A, *B*). Depolarization and repolarization together are called an **action potential.** The action potential results in a brief reversal of the charge across the cell membrane and a small electric current due to the movement of the ions.

During the action potential Na^+ ions move into and K^+ ions move out of the cell, slightly changing the concentrations of these ions inside and outside the cell. The normal concentrations of these ions on either side of the membrane are maintained by the **sodium-potassium exchange pump,** which continuously transports Na^+ ions out of the cell and K^+ ions into the cell.

1 Does the sodium-potassium exchange pump move Na^+ ions out of the cell and K^+ ions into the cell by diffusion or by active transport? Explain.*

Anything that affects the movement of Na^+ or K^+ ions through the cell membrane can affect the production of action potentials. For example, lidocaine is a local anesthetic that blocks sodium channels. Action potentials associated with painful stimuli are not transmitted to the CNS because no action potentials are produced.

Action potentials occur in an **all-or-none** fashion. If a stimulus is below a certain level, called **threshold,** there is no action potential (the none part of the response). A threshold stimulus produces an action potential (the all part of the response). Stimuli stronger than threshold produce a greater number of action potentials. For example, if someone touches you so lightly that you do not feel it, the stimulus was below threshold and no action potential was produced. The lightest touch that causes an action potential is a threshold stimulus, but it produces only one action potential. A heavy touch is an above threshold stimulus that causes the production of many action potentials. Thus weak versus strong stimulation of a neuron is differentiated by the number of action potentials produced.

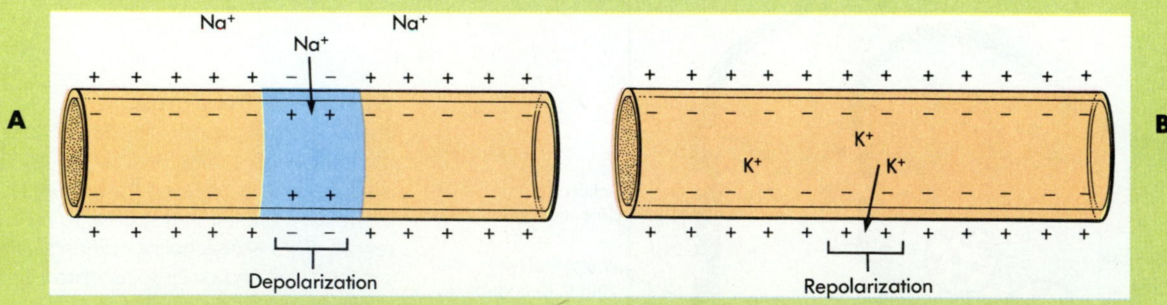

Figure 8-A Action potential.
A, Depolarization part of an action potential. The inside of the cell membrane becomes more positive than the outside because of the inward movement of Na^+ ions. **B,** Repolarization part of an action potential. The outside of the cell membrane once again becomes positive compared with the inside because of the outward movement of K^+ ions.

*Answers to predict questions appear in Appendix B at the back of the book.

Just as the fire from one torch can light another torch, an action potential can pass from one cell to another cell, allowing cells to communicate with each other. For example, if you touch a hot pan, action potentials produced in temperature and pain sensory neurons are sent to the CNS. Thus information in the form of action potentials is carried by nerve fibers from the finger toward the CNS. For the CNS to get this information, action potentials must pass from the sensory neurons to CNS neurons. Once the CNS has received the information, the CNS produces a response. One response is to remove the finger from the hot object by causing the appropriate skeletal muscles to contract. Action potentials pass from CNS neurons to motor neurons. The motor neurons then transmit action potentials to skeletal muscle fibers. The action potentials are transferred from the motor neurons to the skeletal muscle fibers, where they function as stimuli that cause the muscle fibers to contract.

Action potentials are transmitted only in one direction across a synapse, from the axon ending to the muscle fiber, because only the axon ending can release neurotransmitters. There are many known neurotransmitters and many other substances are suspected to be neurotransmitters. Two common neurotransmitters are **acetylcholine** (as-ĕ-til-ko´lēn) and **norepinephrine** (nor´ep-ĭ-nef´rin).

Once released, neurotransmitter substances are rapidly removed from the synaptic cleft so that they have very short-term effects. For example, acetylcholine is the neurotransmitter of synapses between neurons and skeletal muscle fibers. An enzyme called **acetylcholinesterase** (as´ĕ-til-ko-lin-es´-ter-ās) breaks down the acetylcholine soon after it is released. This ensures that one action potential in the neuron yields only one action potential in the muscle fiber, and only one contraction of the muscle fiber. The breakdown products of acetylcholine are then returned to the axon for reuse. The release and breakdown of neurotransmitters occurs so rapidly that a muscle fiber can be stimulated many times in 1 second.

Anything that affects the production, release, or degradation of acetylcholine or its ability to bind to receptors on the muscle fiber membrane will also affect the transmission of action potentials across the synapse. Some insecticides inhibit the activity of acetylcholinesterase. Consequently, acetylcholine is not broken down and accumulates in the space between the neuron and muscle fiber. The acetylcholine acts as a constant stimulus to muscle fibers and the insects die, partly because their muscles contract and cannot relax. Other poisons such as curare (ku-rah´re) bind to the acetylcholine receptors on the muscle fiber membrane preventing acetylcholine from binding to them. Therefore the muscle fiber cannot be stimulated by acetylcholine and does not contract.

CENTRAL NERVOUS SYSTEM

The **central nervous system (CNS)** consists of the brain and spinal cord. The brain is that part of the CNS housed within the brain case, and the spinal cord is surrounded by the vertebral column. The major regions of the brain are the brainstem, the diencephalon, the cerebrum, and the cerebellum (Figure 8-5).

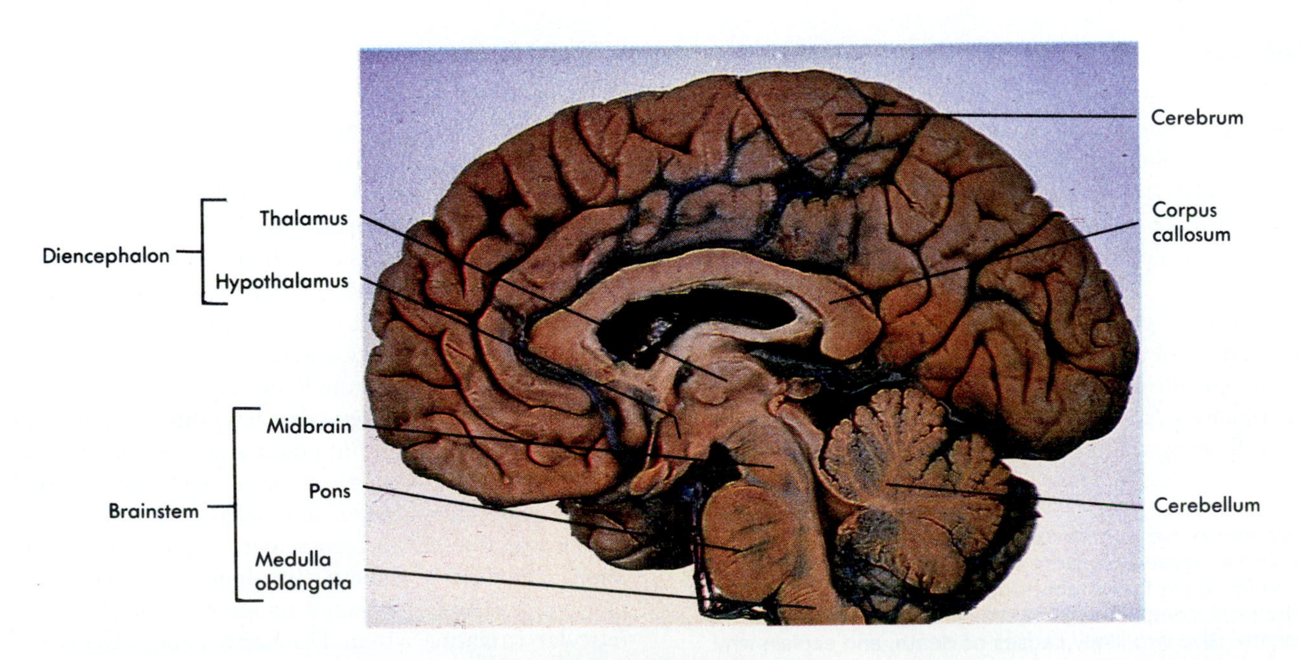

Figure 8-5 ● Regions of the brain.
Midsagittal section.

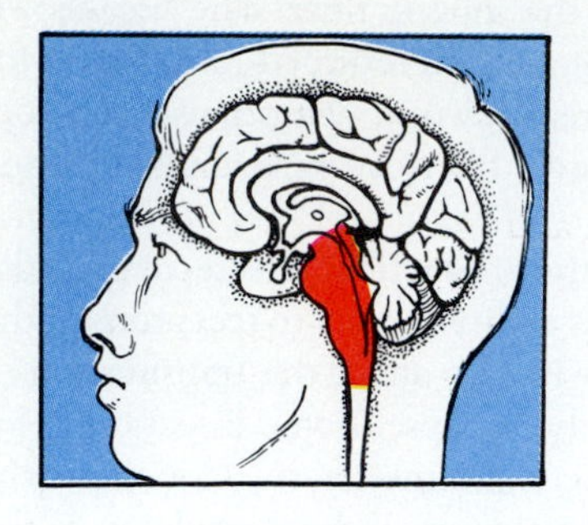

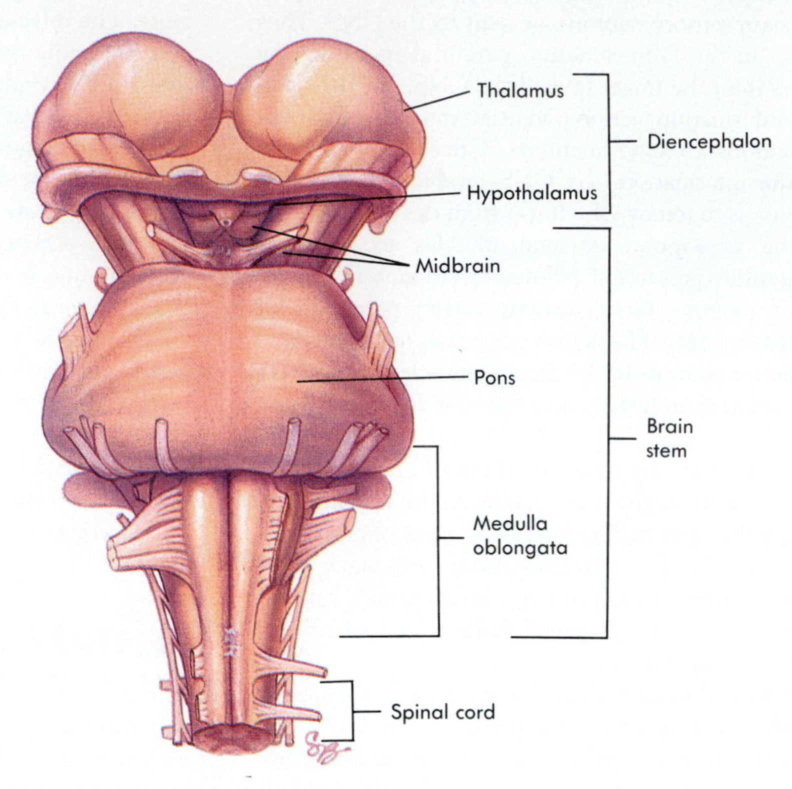

Figure 8-6 ● Brainstem and diencephalon. Anterior view.

Brainstem

The medulla oblongata, pons, and midbrain constitute the **brainstem** (Figure 8-6). The brainstem connects the spinal cord to the remainder of the brain and contains many ascending and descending nerve tracts. It also has nuclei that are responsible for many essential functions, as well as the nuclei from which all but two of the cranial nerves originate.

The **medulla oblongata** (ob´long-gah´tah) is the most inferior portion of the brainstem and is continuous inferiorly with the spinal cord. The medulla oblongata contains nuclei that regulate heart rate and blood vessel diameter, breathing, swallowing, vomiting, coughing, and sneezing.

2 A tumor or hematoma (bleeding into the tissues) can cause increased pressure within the skull. The pressure can force the brainstem through the foramen magnum of the skull, compressing the brainstem and leading to death. Give two likely causes of death, and explain why they would occur.

Immediately superior to the medulla oblongata is the **pons.** It contains ascending and descending nerve tracts as well as several nuclei and nerve tracts that relay information between the cerebrum and the cerebellum.

The **midbrain,** just superior to the pons, is the smallest region of the brainstem. It has four mounds of tissue that are involved with hearing and vision. For example, turning your head toward a loud sound, or the ability to track moving objects with the eyes, are coordinated here.

Scattered throughout the brainstem is a group of nuclei collectively called the **reticular activating system,** which plays an important role in arousing and maintaining consciousness. Stimuli such as an alarm clock ringing, sudden bright lights, ammonia (smelling salts), or cold water being splashed on the face can arouse consciousness. Conversely, removal of visual or auditory stimuli may lead to drowsiness or sleep. General anesthetics function by suppressing the reticular activating system. Damage to cells of the reticular activating system can result in coma.

Diencephalon

The **diencephalon** (di-en-sef´ă-lon) is the part of the brain between the brainstem and the cerebrum (Figure 8-7 and see Figure 8-6). Its main components are the thalamus and hypothalamus.

The **thalamus** (thal´ă-mus) is by far the largest portion of the diencephalon. It consists of a cluster of nuclei and is shaped somewhat like a yo-yo, with two large, lateral lobes connected in the center by a small intermediate mass. Most sensory nerve tracts that ascend through the cord and brainstem synapse in the thalamus. Thalamic neurons then send their axons to the cerebral cortex. The thalamus also has other functions such as influencing mood and registering unlocalized, uncomfortable perceptions of pain.

The **hypothalamus** is the most inferior portion of the diencephalon and contains several small nuclei. These nuclei are especially important in maintaining homeostasis. The hypothalamus plays a central role in the control of body temperature, hunger, and thirst. Sensations such as sexual pleasure, feeling relaxed and "good" after a meal, rage, fear, and responses to stress are also related to hypothalamic functions. Part of the hypothalamus is involved in memory and the emotional responses to odors, such as associating a particular fragrance with a special person. The hypothalamus is also connected to the pituitary gland and plays an important role in controlling the endocrine system by regulating hormone secretion from the pituitary gland (see Chapter 10).

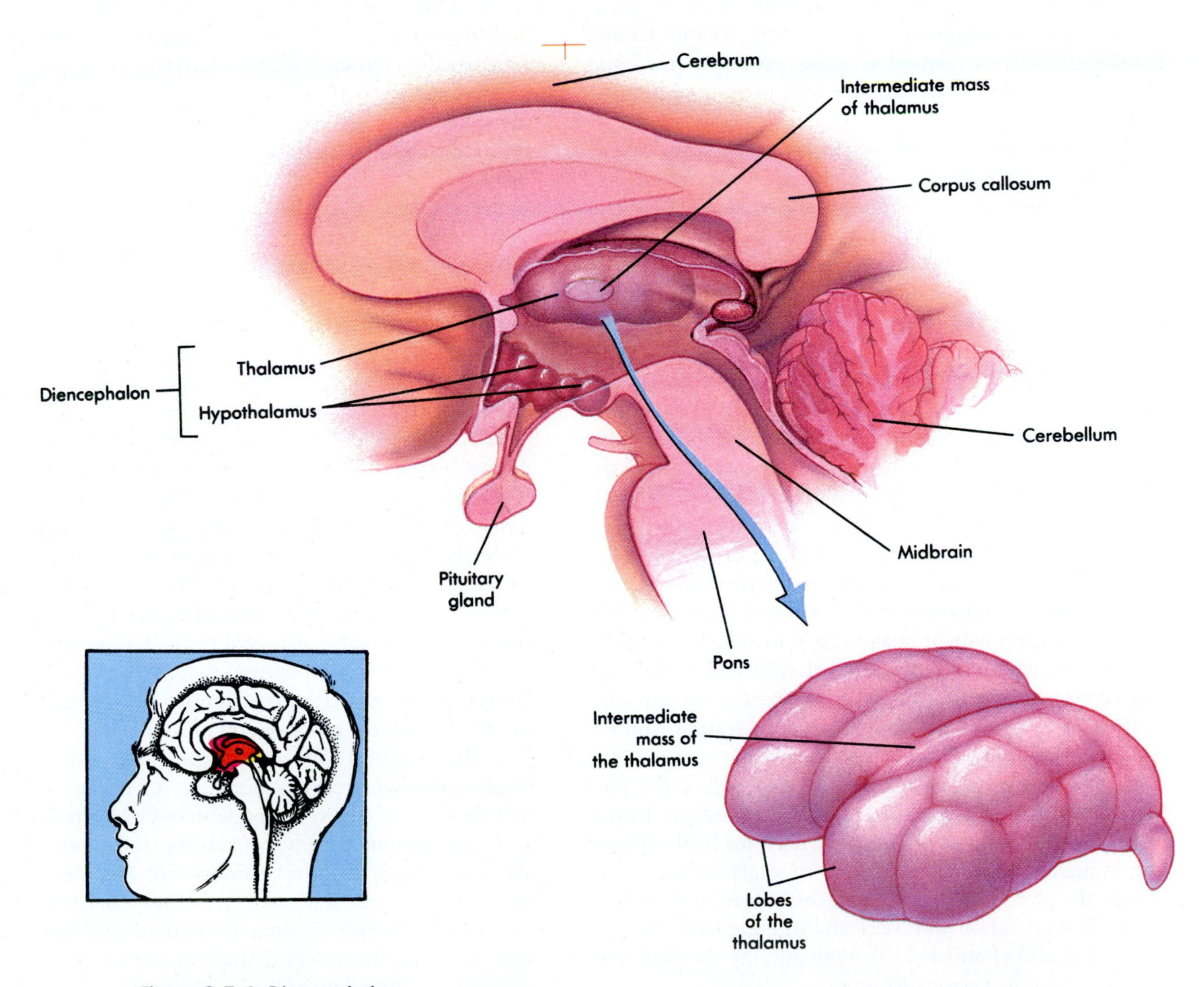

Figure 8-7 ● Diencephalon.
Sagittal view of the diencephalon in relation to the rest of the brain. The isolated thalamus is also shown.

Cerebrum

The **cerebrum** (sĕr´e-brum) is the largest portion of the brain (Figure 8-8). The gray matter of the cerebrum forms an outer cortex and internal clusters of nuclei. White matter nerve tracts connect the cortex and nuclei to each other, other parts of the brain, and the spinal cord.

The cerebrum is divided into left and right hemispheres. The most conspicuous features on the surface of each hemisphere are numerous folds called **gyri** (ji´ri; singular: gyrus), which greatly increase the surface area of the cortex, and intervening grooves called **sulci** (sul´si; singular: sulcus).

Each cerebral hemisphere is divided into lobes (see Figure 8-8, *A*), which are named for the skull bones overlying them. The **frontal lobe** is important in voluntary motor function, motivation, aggression, and mood. The **parietal lobe** is the principal center for the reception and evaluation of pain, temperature, pressure (touch), and taste. The **occipital lobe** functions in the reception and integration of visual input and is not distinctly separate from the other lobes. The **temporal lobe** evaluates olfactory and auditory (hearing) input and plays an important role in memory. Its anterior and inferior portions are referred to as the "psychic cortex," and they are associated with functions such as abstract thought and judgment.

Sensory nerve tracts project to specific regions of the cerebral cortex (see Figure 8-8, *B*) where sensations are perceived. The **general sensory areas** are located in the parietal lobe and receives general sensory input such as pain, temperature, and pressure. Other sensory areas include the **visual areas** in the occipital lobe, the **auditory areas** and the **olfactory areas** in the temporal lobe, and the **taste area** in the parietal lobe.

The **motor areas** in the frontal lobe initiate action potentials responsible for voluntary control of skeletal muscles. Part of the motor area, called **Broca's area,** is responsible for producing the muscle movements necessary for speech. Damage to Broca's area results in absent or defective speech called **aphasia** (ă-fa´ze-ah).

The motivation and the foresight to plan and initiate movements occur in the anterior portion of the frontal lobes, the **prefrontal area.** This area is well developed only in primates, especially in humans. It is involved in motivation and regulation of emotional behavior and mood. The large size of this area in humans may account for our relatively well-developed forethought and motivation, and for our emotional complexity.

In relation to its involvement in motivation, the prefrontal area is the functional center for aggression. In the past, one method used to eliminate uncontrollable aggression in mental hospital patients was to surgically remove or destroy the prefrontal regions of the brain (prefrontal lobotomy). This operation was successful in eliminating aggression, but it also eliminated the motivation to do much else and destroyed the personality.

The different areas of the cortex are constantly interacting with each other. For example, suppose you are about to ring a doorbell. The prefrontal area plans and initiates the process. Specific areas of the motor area involved with the movement of the upper limb are activated and action potentials travel down motor nerve tracts to the appropriate skeletal muscles. When your finger touches the doorbell, pressure receptors are activated and action potentials travel up sensory nerve tracts to the general sensory area, and you are aware of how much pressure is applied to the doorbell.

3 Describe the events that occur when you turn your head and look at someone sitting next to you.

Brain waves

Electrodes placed on a person's scalp and attached to a recording device can record the brain's electrical activity, producing an **electroencephalogram** (e-lek´tro-en-sef´ă-lo-gram) or **EEG.** These electrodes are not sensitive enough to detect individual action potentials but can detect the simultaneous action potentials in large numbers of neurons. As a result, the EEG displays wavelike patterns known as brain waves. This electrical activity is constant, but the intensity and frequency of electrical discharge differs from time to time based on the state of brain activity.

Distinct EEG patterns occur with specific brain disorders such as epileptic seizures. Doctors use these patterns to diagnose and determine the treatment for the disorders.

Memory

Memory is the mental ability to recall information. The temporal lobes are a major site for storing memories, although many other parts of the brain also are involved. **Short-term memory** is information retained for a few seconds to a few minutes. This memory is limited primarily by the number of bits of information that can be stored at any one time, which is usually about seven bits of information. When new information is presented, old information, previously stored in short-term memory, is eliminated.

Certain pieces of information are transferred from short-term memory to **long-term memory,** some of which may become permanent. Long-term memory may involve a physical change in neuron shape and the number of synapses between neurons. A whole series of neurons, called **memory engrams,** are probably involved in the long-term retention of a given piece of information, thought, or idea. Rehearsal of information assists in the transfer of information from short-term memory to long-term memory.

If the temporal lobe is damaged, the transition from short-term to long-term memory may not occur. The person will always live only in the present and in the more remote past, with memory already stored before the injury. This person is unable to add new memory. For example, a person with temporal lobe damage can be introduced to another person several times within a few minutes and believe each introduction is the first time they have met.

Right and left cerebral hemispheres

The right cerebral hemisphere controls muscular activity in and receives sensory input from the left half of the body.

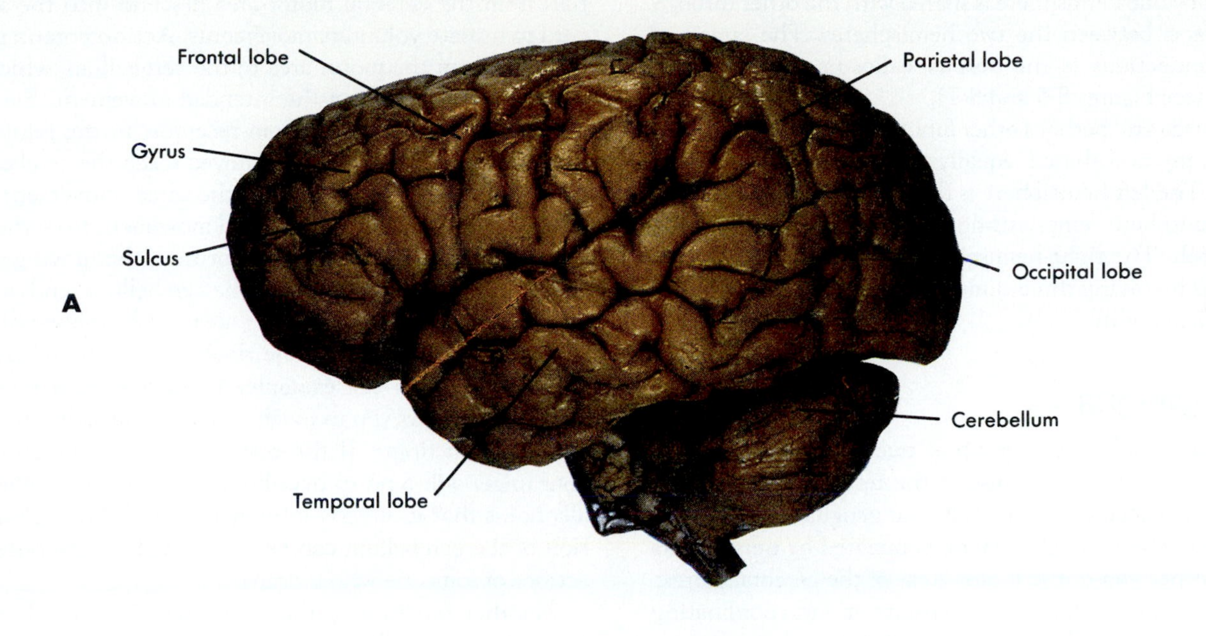

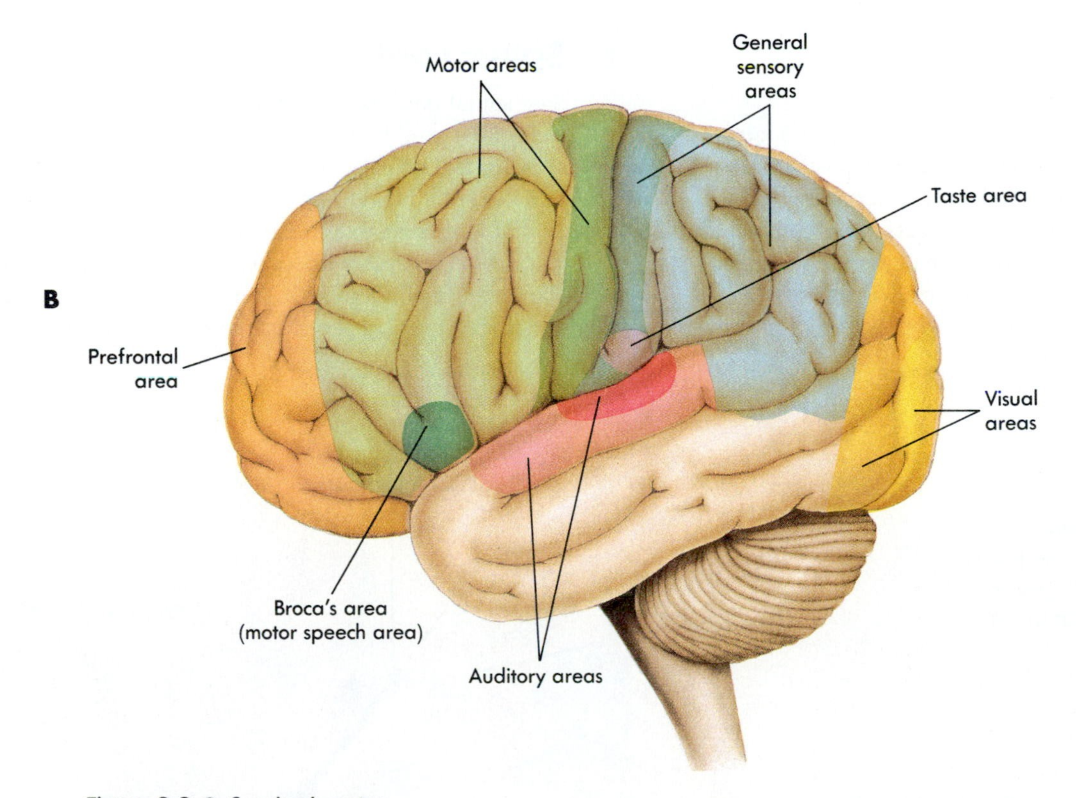

Figure 8-8 ● Cerebral cortex.
A, Lateral view of the cerebrum showing the gyri, sulci, and lobes. **B,** Cortical functional regions of the left cerebral hemisphere.

The left cerebral hemisphere controls muscles and receives input from the right half of the body. Sensory information received by one hemisphere is shared with the other through nerve tracts between the two hemispheres. The largest of these connections is the **corpus callosum** (kor´pus kah-lo´sum) (see Figures 8-5 and 8-7).

Language and perhaps other functions, such as artistic activities, are not shared equally between the two hemispheres. The left hemisphere is thought to be the most analytic hemisphere, emphasizing such skills as mathematics and speech. The right hemisphere is thought to emphasize functions involving three-dimensional or spatial perception and musical ability.

Basal ganglia

The **basal ganglia** are a group of nuclei located primarily within the cerebrum (the use of the term *ganglia* for these nuclei is an exception to the rule that ganglia are located in the PNS). The basal ganglia are connected by nerve tracts to each other and to the motor areas of the cerebral cortex. They play an important role in posture and in coordinating motor movements. The major effect of the basal ganglia is to decrease muscle tone and inhibit muscular activity. Disorders of the basal ganglia, such as Parkinson's disease, result in muscular rigidity; tremors; a slow, shuffling gait; and general lack of movement.

Cerebellum

Cerebellum (ser´ĕ-bel´um) (see Figures 8-5 and 8-8, *A*) means little brain. The cerebellar cortex is composed of gray matter, and it has gyri and sulci. Internally, the cerebellum consists of nuclei and nerve tracts. The cerebellum is involved in balance, maintenance of muscle tone, and coordination of fine motor movement. If the cerebellum is damaged, muscle tone decreases and fine motor movements become very clumsy.

A major function of the cerebellum is to compare intended movements with actual movements. Action potentials from the cerebral motor area descend into the spinal cord to initiate voluntary movements. Action potentials are also sent from the motor area to the cerebellum, which informs the cerebellum of the intended movement. Simultaneously, action potentials from receptors in the joints and tendons of the structure being moved reach the cerebellum, which informs the cerebellum of the actual movement. The cerebellum compares the intended movement from the motor area with the actual movement from the moving structure. If a difference is detected, the cerebellum sends action potentials to the cerebral motor area and to the spinal cord to correct the difference. The result is smooth and coordinated movements. For example, if you close your eyes, the cerebellum allows you to touch your nose smoothly and easily with your finger. If the cerebellum is not functioning, your finger will tend to overshoot the target. One effect of alcohol is that it directly inhibits the cerebellum. Dysfunction of the cerebellum can be understood by observing the actions of someone who is drunk.

Another function of the cerebellum involves learning complex motor skills such as playing the piano or riding a bicycle. Through repetition of the activity the cerebellum "learns" the skill. Then much of the activity is accomplished automatically by the cerebellum.

Spinal cord

The spinal cord extends from the foramen magnum at the base of the skull to the second lumbar vertebra. Seen in cross section, the spinal cord has a central gray portion shaped like the letter H and a peripheral white portion (Figure 8-9).

The gray matter consists of nerve cell bodies and dendrites that synapse with nerve fibers from spinal nerves or the brain. **Spinal nerves** consist of sensory nerve fibers, which carry action potentials to the spinal cord, and motor

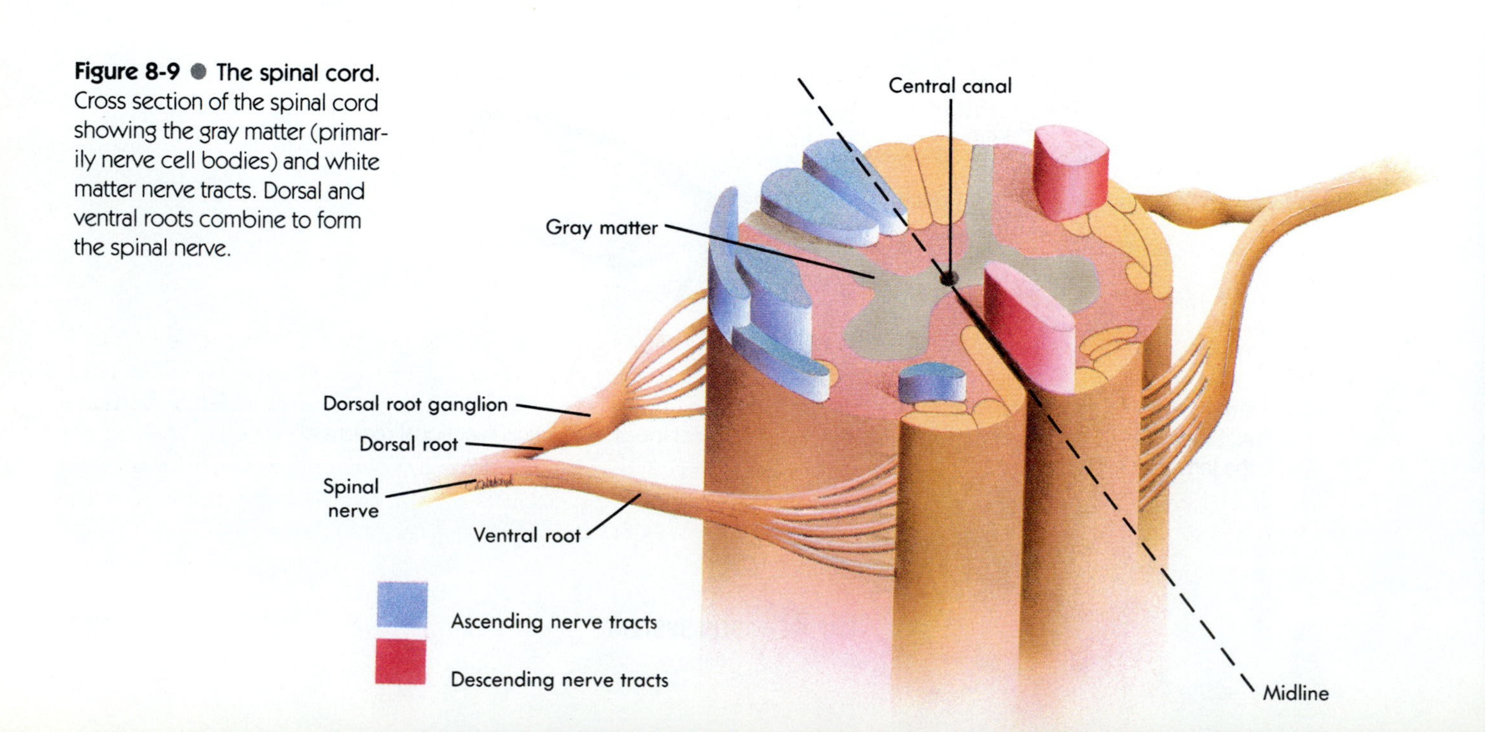

Figure 8-9 ● **The spinal cord.** Cross section of the spinal cord showing the gray matter (primarily nerve cell bodies) and white matter nerve tracts. Dorsal and ventral roots combine to form the spinal nerve.

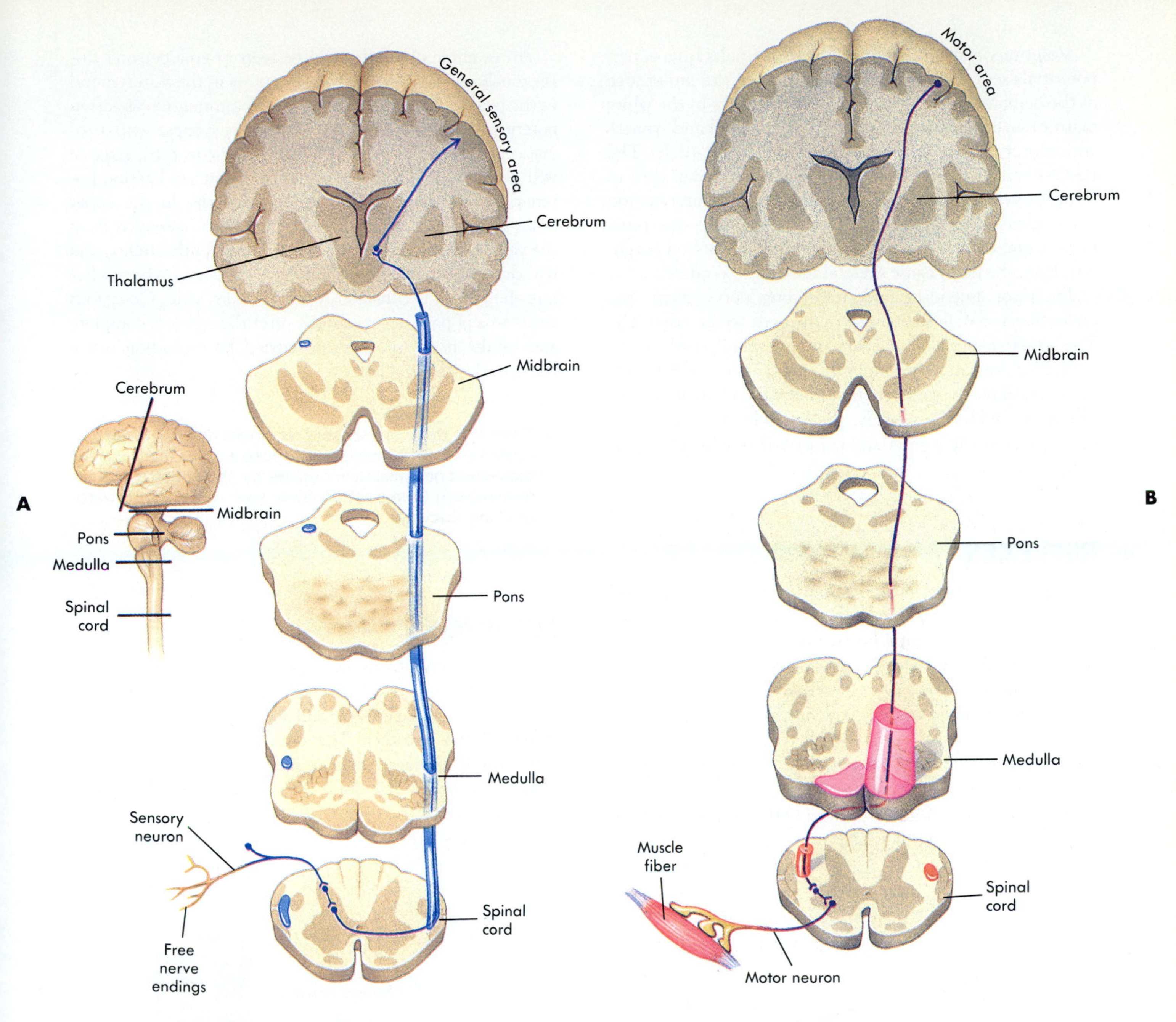

Figure 8-10 ● **Nerve tracts.**
A, The sensory nerve tract for pain and temperature. Nerve fiber from the skin enters the spinal cord and synapses with association neurons that cross over in the spinal cord and extend to the general sensory area. **B,** A motor nerve tract for controlling voluntary skeletal muscle activity. Nerve fiber travels from the motor area to the medulla where it crosses to the opposite side of the body and extends down a nerve tract to a motor neuron in the spinal cord. The motor neuron stimulates a skeletal muscle fiber.

nerve fibers, which carry action potentials away from the spinal cord. Close to the spinal cord, the sensory and motor nerve fibers separate into dorsal and ventral roots. A **dorsal root** consists of sensory nerve fibers and has a **dorsal root ganglion,** which contains the cell bodies of the sensory neurons. The nerve fibers of the sensory neurons project into the gray matter, where they synapse with association neurons (Figure 8-10, *A*). The association neurons form ascending nerve tracts in the white matter. These nerve fibers extend to the brain, especially the thalamus. From the thalamus nerve fibers project to the general sensory area of the cerebral cortex. Thus sensory stimuli received in the periphery of the body are transmitted to the brain in which awareness of the stimuli takes place.

Voluntary control of motor functions results from action potentials sent to the skeletal muscles from the motor area of the cerebral cortex. Descending nerve fibers in the white matter reach the gray matter of the spinal cord and synapse with motor neurons in the gray matter (Figure 8-10, *B*). The nerve fibers of the motor neurons leave the spinal cord to form the **ventral roots** of the spinal cord. The motor neuron nerve fibers pass through the ventral roots, into the spinal nerves, and extend to skeletal muscles. Thus action potentials from the brain cause skeletal muscles to contract.

For most ascending and descending nerve tracts, the nerve fibers cross from one side of the body to the other side. The cross over usually occurs in the spinal cord or the medulla oblongata. Because of this cross over, the right side of the brain receives sensory input from and controls the left side of the body. Conversely, the left side of the brain receives sensory input from and controls the right side of the body.

REFLEXES

A **reflex** is a response to a stimulus that does not involve conscious thought. Reflexes allow us to react to a stimulus more quickly than would be possible if conscious thought were involved. The **reflex arc** is the basic functional unit of the nervous system and is the smallest, simplest pathway capable of receiving a stimulus and yielding a response. A reflex arc has five basic components: (1) a sensory receptor, (2) a sensory neuron, (3) association neurons, (4) a motor neuron, and (5) an effector organ (Figure 8-11). Most reflexes involve the spinal cord or brainstem and don't involve higher brain centers.

An example of a reflex can be seen when a person's finger touches a hot pan. Sensory receptors in the skin respond to the heat. Sensory neurons carry this information as action potentials to the spinal cord, where they synapse with association neurons. The association neurons, in turn, synapse with motor neurons in the spinal cord that send action potentials along their axons to flexor muscles in the upper limb. These muscles contract and pull the finger away from the pan. No conscious thought is required for this reflex, and withdrawal of the finger from the painful stimulus begins before the person is consciously aware of any pain. Conscious awareness of pain occurs shortly after the reflex is complete and results from information carried by ascending nerve tracts to the cerebral cortex.

4 Given that there are sensory receptors that can detect stretch of skeletal muscles, propose a reflex mechanism that would help maintain posture by preventing the trunk from moving from a set position, such as leaning forward or tilting sideways.

MENINGES

Three connective tissue layers, the **meninges** (mĕ-nin´jez) (Figure 8-12), surround and protect the brain and spinal cord. The most superficial and thickest layer is the **dura mater** (du´rah ma´ter). The dura mater contains spaces called **dural sinuses,** which collect blood from the small veins of the brain. The dural sinuses empty into the jugular veins, which exit the skull.

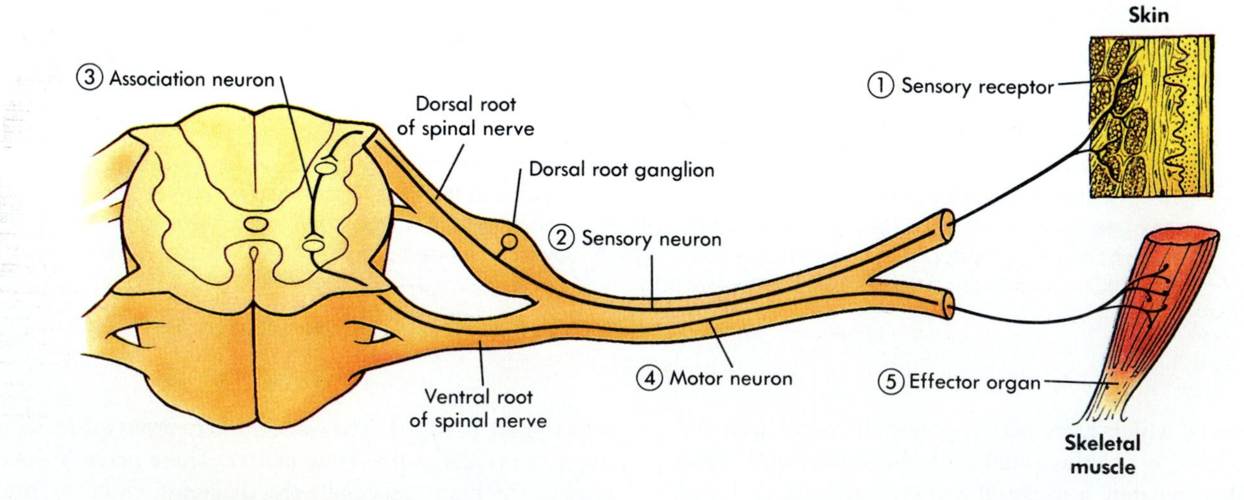

Figure 8-11 ● Reflex arc.
Basic diagram of a reflex arc, including the sensory receptor, sensory neuron, association neuron, motor neuron, and effector organ.

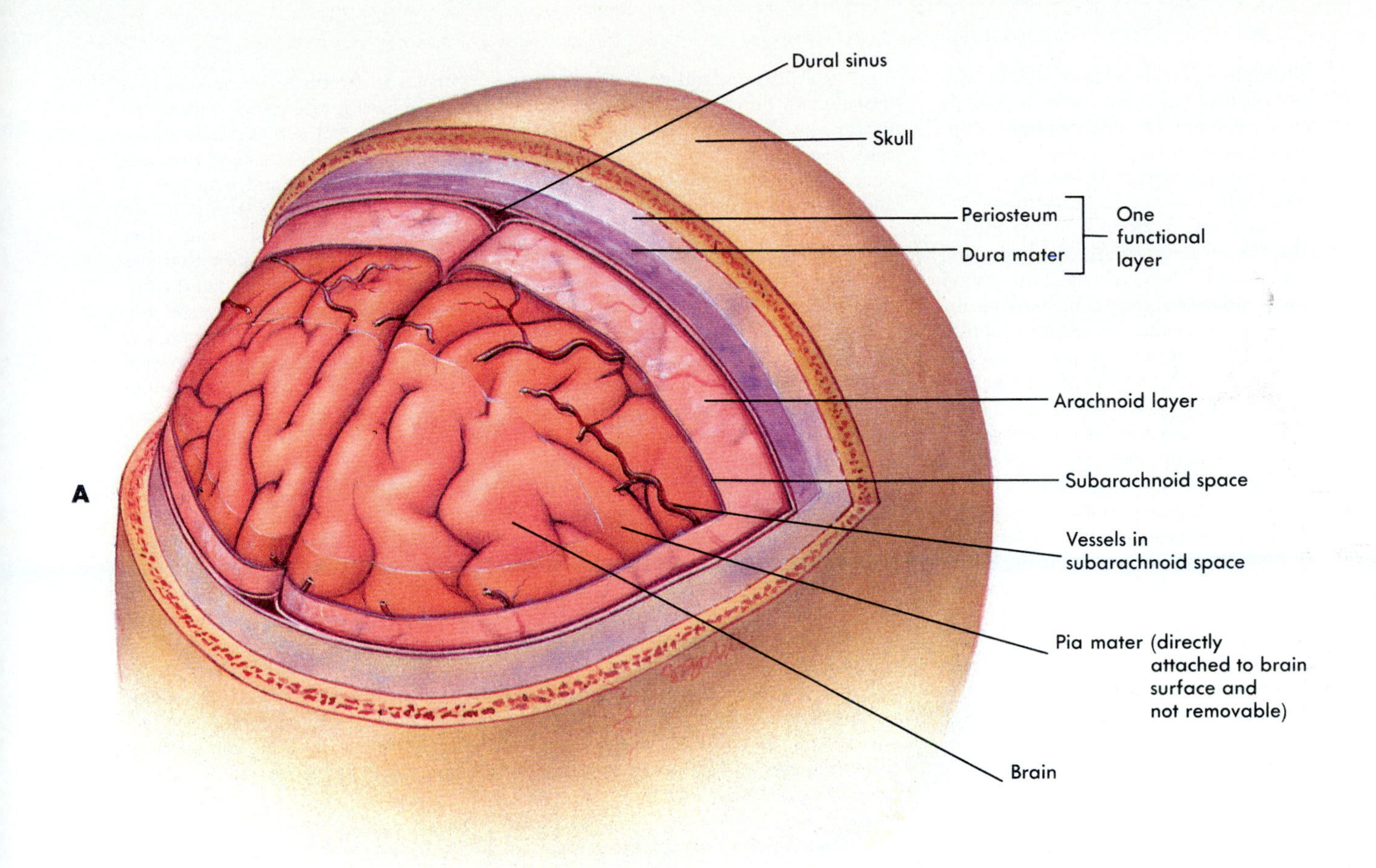

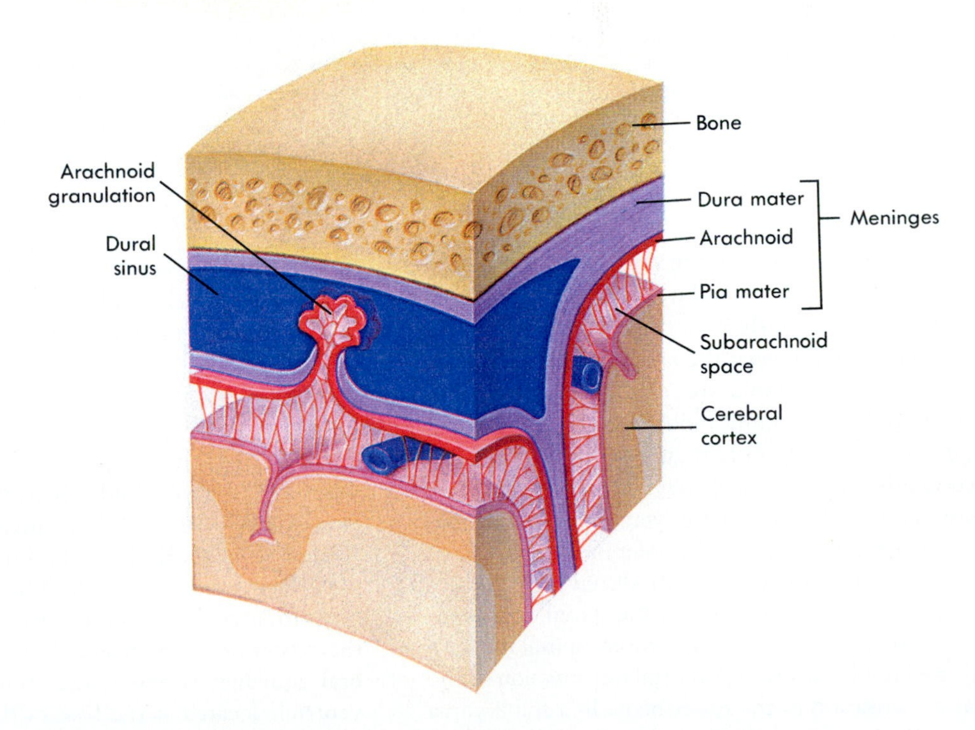

Figure 8-12 ● Meninges.
A, Meningeal coverings of the brain. **B,** More detailed view of the meninges.

Nervous system disorders

Multiple sclerosis (skle-ro´sis), or **MS,** although of unknown cause, is possibly viral in origin. The disease results in inflammation and an increased activity of the immune system (hyperimmunity). The inflammation and immune response result in damage to the myelin sheaths of neurons in the brain and spinal cord. The myelin sheaths around axons become sclerotic, or hard, resulting in poor conduction of action potentials. The symptomatic periods of MS are separated by periods of apparent remission. With each recurrence, however, many neurons are permanently damaged. Progressive symptoms of the disease include exaggerated reflexes, tremor, nystagmus (tremorous movement of the eyes), and speech defects.

Myasthenia gravis (mi´as-the´ne-ah grav-is) is an immune disorder in which the immune system attacks the synapse between motor neurons and skeletal muscles, resulting in extreme muscle weakness (see Chapter 7).

Cerebral palsy (pawl´ze) is a general term referring to defects in motor functions or coordination resulting from several types of brain damage, which can be caused by abnormal brain development or birth-related injury. Some symptoms of cerebral palsy are related to basal ganglia dysfunction. One of the features of cerebral palsy is the presence of slow, writhing, aimless movements. When the face, neck, and tongue muscles are involved, characteristics are grimacing, protrusion and writhing of the tongue, and difficulty in speaking and swallowing. Because cerebral palsy results from general brain damage, other neurological disorders, such as seizures and mental retardation, can be associated with it. It should be kept in mind, however, that persons with cerebral palsy can be of normal or above normal intelligence.

Parkinson's disease, characterized by muscular rigidity, tremor, a slow, shuffling gait, and general lack of movement, is caused by a lesion in the basal ganglia. A resting tremor called "pill-rolling" is characteristic of Parkinson's disease and consists of circular movement of the opposed thumb and index finger tip. The increased muscular rigidity in Parkinson's disease results from defective inhibition of muscle tone by some of the basal ganglia. In this disease, dopamine, an inhibitory neurotransmitter substance, is deficient. Parkinson's disease can be treated with L-dopa, a precursor to dopamine that crosses the blood-brain barrier from the capillaries of the brain into the brain tissue. Dopamine cannot cross the barrier.

Tumors destroy or compress brain tissue. Symptoms vary widely, depending on the location of the tumor, but may include headaches, neuralgia (pain along the distribution of a peripheral nerve), paralysis, seizures, coma, and death.

Stroke is a term meaning a sudden blow, suggesting the speed with which this type of defect can occur. It is also referred to clinically as a **cerebrovascular accident (CVA)** and is caused by hemorrhage, **thrombosis** (throm-bo´sis; a clot in a blood vessel), **embolism** (em´bo-lizm; a piece of clot that has broken loose and floats through the circulation until it reaches a vessel too small for it to pass through, which it blocks), or vasospasm of the cerebral blood vessels. These causes result in an **infarct** (in´farkt), a local area of cell death caused by a lack of blood supply. Symptoms depend on the location, but may include loss of sensation or paralysis on the side of the body opposite the cerebral infarct.

Spinal cord injury can disrupt or completely block the transmission of action potentials to and from the brain. Depending on the nerve tract involved there can be loss of sensations, loss of motor control, or loss of both sensations and motor control.

Senility once was thought to be a normal part of aging. The Latin term *senile* means old age. The severe symptoms of senility—general intellectual deficiency, mental deterioration (called dementia), memory loss, short attention span, moodiness, and irritability—result from several specific disease states. **Alzheimer's** (altz´hi-merz) **disease** is a severe type of senility, often affecting people under age 50. It results in

The second meningeal layer is the very thin, wispy **arachnoid** (ar-ak´noyd) layer. The third meningeal layer, the **pia** (pe´ah) **mater,** is very tightly bound to the surface of the brain and spinal cord. Between the arachnoid layer and the pia mater is the **subarachnoid space,** which is filled with cerebrospinal fluid and contains blood vessels.

The spinal cord extends only to approximately the level of the second lumbar vertebra. Spinal nerves surrounded by meninges extend to the end of the vertebral column. Because there is no spinal cord in the inferior portion of the vertebral column, a needle can be introduced into the subarachnoid space inferior to the end of the spinal cord without damaging the cord. In spinal anesthesia (spinal block) a drug is injected to block action potential transmission so the person has no sensation in the lower body. In a spinal tap a needle is used to take a sample of cerebrospinal fluid that is to be examined for infectious agents (meningitis) or for blood (hemorrhage). A radiopaque substance can be injected and a myelograph (x-ray of the spinal cord) can be taken to visualize spinal cord defects or damage.

Ventricles

The CNS contains fluid-filled cavities: **ventricles** (ven´trĭ-kulz) in the brain and the **central canal** of the spinal cord (Figure 8-13, *A*). Each cerebral hemisphere contains a relatively large **lateral ventricle.** The two lateral ventricles connect through small openings to a smaller **third ventricle** in the center of the diencephalon. A narrow canal, the **cerebral aqueduct** connects the third ventricle to the **fourth ventricle** located at the base of the cerebellum. The fourth ventricle is continuous with the **central canal of the spinal cord** and empties into the subarachnoid space.

general mental deterioration. The exact cause is unknown.

Epilepsy (ep´ĭ-lep´se) is actually a group of brain disorders that have seizure episodes in common. The seizure, a sudden massive neuronal discharge, can be either partial or complete, depending on the amount of brain involved and whether or not consciousness is impaired. The neuronal discharges may stimulate muscles innervated by the nerves involved, resulting in involuntary muscle contractions, that is, convulsions.

Headaches have a variety of causes that can be grouped into two basic classes: extracranial and intracranial. Extracranial headaches can be caused by inflammation of the paranasal sinuses, dental irritations, eye disorders, or tension in the muscles moving the head and neck. Intracranial headaches may result from inflammation of the brain or meninges, vascular problems, mechanical damage, or tumors.

Neuralgia (nur-al´je-ah) is a condition involving severe spasms of throbbing or stabbing pain along the pathway of a nerve. **Neuritis** (nu-ri´tis) is an inflammation of a nerve resulting from any one of a number of causes, including injury or infection. In sensory nerves, neuritis can result in the loss of sensation and/or pain. In motor nerves, neuritis can result in the loss of motor function. **Trigeminal neuralgia** involves the trigeminal nerve and consists of sharp bursts of pain in the face. **Facial palsy** involves the facial nerve and results in unilateral paralysis of the facial muscles. The affected side of the face droops because of the absence of muscle tone. **Sciatica** (si-at´ĭ-kah) is a neuralgia of the sciatic nerve, with pain radiating down the back of the thigh and leg. The most common cause of sciatica is a herniated lumbar disk putting pressure on the spinal nerves forming the lumbosacral plexus (see Chapter 6).

Anesthesia (an´es-the´ze-ah) is the loss of sensation. It can be a pathological condition, or it can be induced to facilitate surgery or some other medical treatment.

INFECTIONS

Encephalitis (en-sef´ă-li´tis) is an inflammation of the brain, most often caused by a virus and less often by bacteria or other agents. A large variety of symptoms result, including fever, coma, and convulsions. Encephalitis may also result in death. **Meningitis** (men-in-ji´tis) is an inflammation of the meninges. It may be caused by either a viral or a bacterial infection. Symptoms usually include stiffness in the neck, headache, and fever. In severe cases, meningitis also may cause paralysis, coma, or death.

Tetanus (tet´ă-nus) is an infection resulting from bacteria found in soil contaminated with animal wastes. The bacteria are often introduced through an open wound into the body in which they produce a potent neurotoxin. The toxin prevents muscle relaxation so that the body becomes rigid. The jaw muscles are affected early in the disease, and the jaw cannot be opened (lockjaw). Death results from spasms in the diaphragm and other respiratory muscles.

Herpes (her´pez) is a family of viral diseases characterized by skin lesions. The viruses apparently reside in the ganglia of sensory nerves and cause lesions along the course of the nerve. The herpes simplex I virus causes lesions of the lips and nose. The lesions are prone to occur during times of decreased resistance, such as during a cold. For this reason, they are called cold sores or fever blisters. The herpes simplex II virus (genital herpes) is responsible for a venereal (sexually transmitted) disease causing lesions on the external genitalia. The herpes zoster virus causes chicken pox in children and shingles in older adults. Shingles is an outbreak of skin lesions along the pathway of a nerve.

Poliomyelitis (po´le-o-mi´ĕ-li´tis) or **polio** is a viral infection that damages the motor neurons extending to skeletal muscles. Without stimulation from the CNS, the muscles are paralyzed and they atrophy, or waste away.

Cerebrospinal fluid

Cerebrospinal fluid (CSF) bathes the brain and spinal cord, providing a protective cushion around the CNS. It is produced by a specialized membrane called the **choroid** (ko´royd) **plexus** in each of the ventricles. The CSF flows from the lateral ventricles into the third ventricle and then through the cerebral aqueduct into the fourth ventricle (Figure 8-13, *B*). From the fourth ventricle CSF enters the subarachnoid space through small openings in the walls of the fourth ventricle or it enters the central canal of the spinal cord. Masses of arachnoid tissue, **arachnoid granulations,** extend from the subarachnoid space into a dural sinus (see Figure 8-12, *B*). CSF passes from the subarachnoid space into the blood through these granulations.

Blockage of the cerebral aqueduct or of the openings between the fourth ventricle and the subarachnoid space can result in accumulation of CSF in the ventricles, a condition known as **hydrocephalus** (hi´dro-sef´ă-lus). The accumulation of fluid inside the brain causes pressure that compresses the nervous tissue and dilates the ventricles. Compression of the nervous tissue usually results in irreversible brain damage. If the skull bones are not completely ossified when the hydrocephalus occurs, such as in a fetus or newborn, the pressure also may cause severe enlargement of the head. Hydrocephalus is treated by placing a drainage tube (shunt) into the brain ventricles in an attempt to eliminate the high internal pressures.

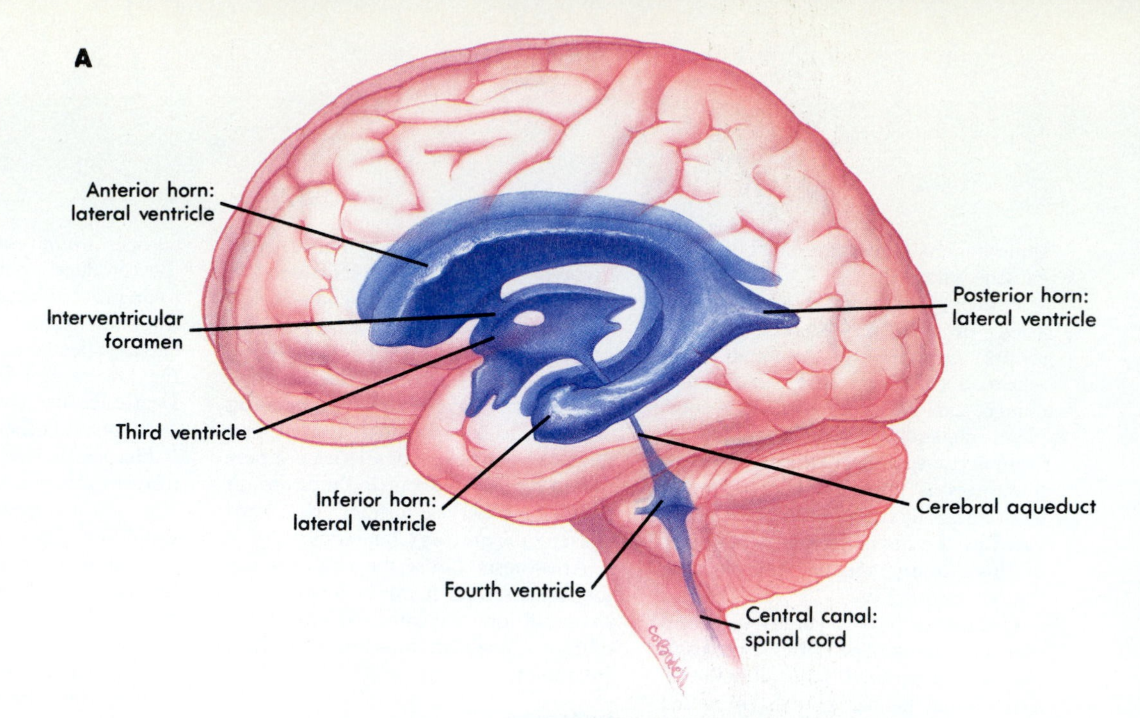

Figure 8-13 ● **Ventricles of the brain.**
A, Lateral view. **B,** Cerebrospinal fluid flows through the ventricles as shown. It exits the fourth ventricle and enters the subarachnoid space. Cerebrospinal fluid passes back into the blood through the arachnoid granulations, which penetrate the dural sinus.

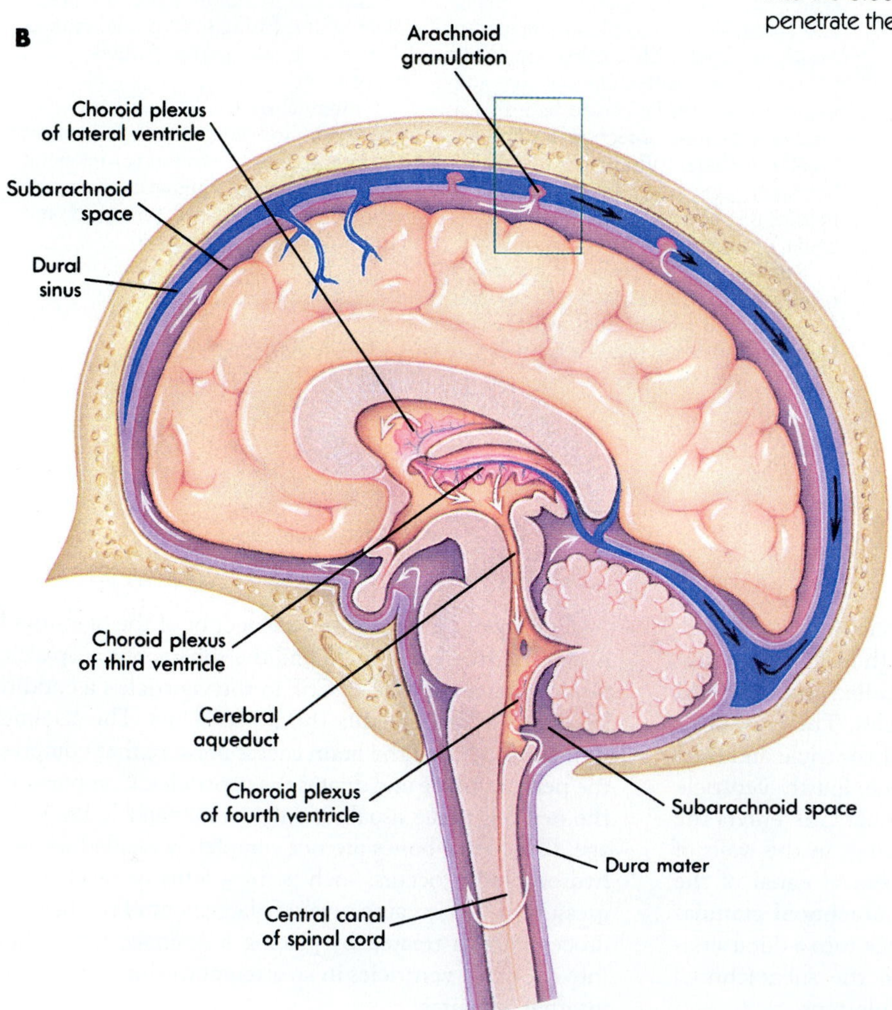

PERIPHERAL NERVOUS SYSTEM

The peripheral nervous system (PNS) consists of sensory and motor neurons. The sensory neurons collect information and convey it to the CNS, whereas the motor neurons convey information from the CNS to muscles and glands. The PNS can be divided structurally into two parts: a cranial part, consisting of 12 pairs of nerves, and a spinal part, consisting of 31 pairs of nerves.

Cranial nerves

The 12 pairs of **cranial nerves** arise from the brain (Figure 8-14). There are three general categories of cranial nerve function: sensory, somatic motor, and parasympathetic. **Sensory** functions include the special senses such as vision and the more general senses such as touch and pain in the face.

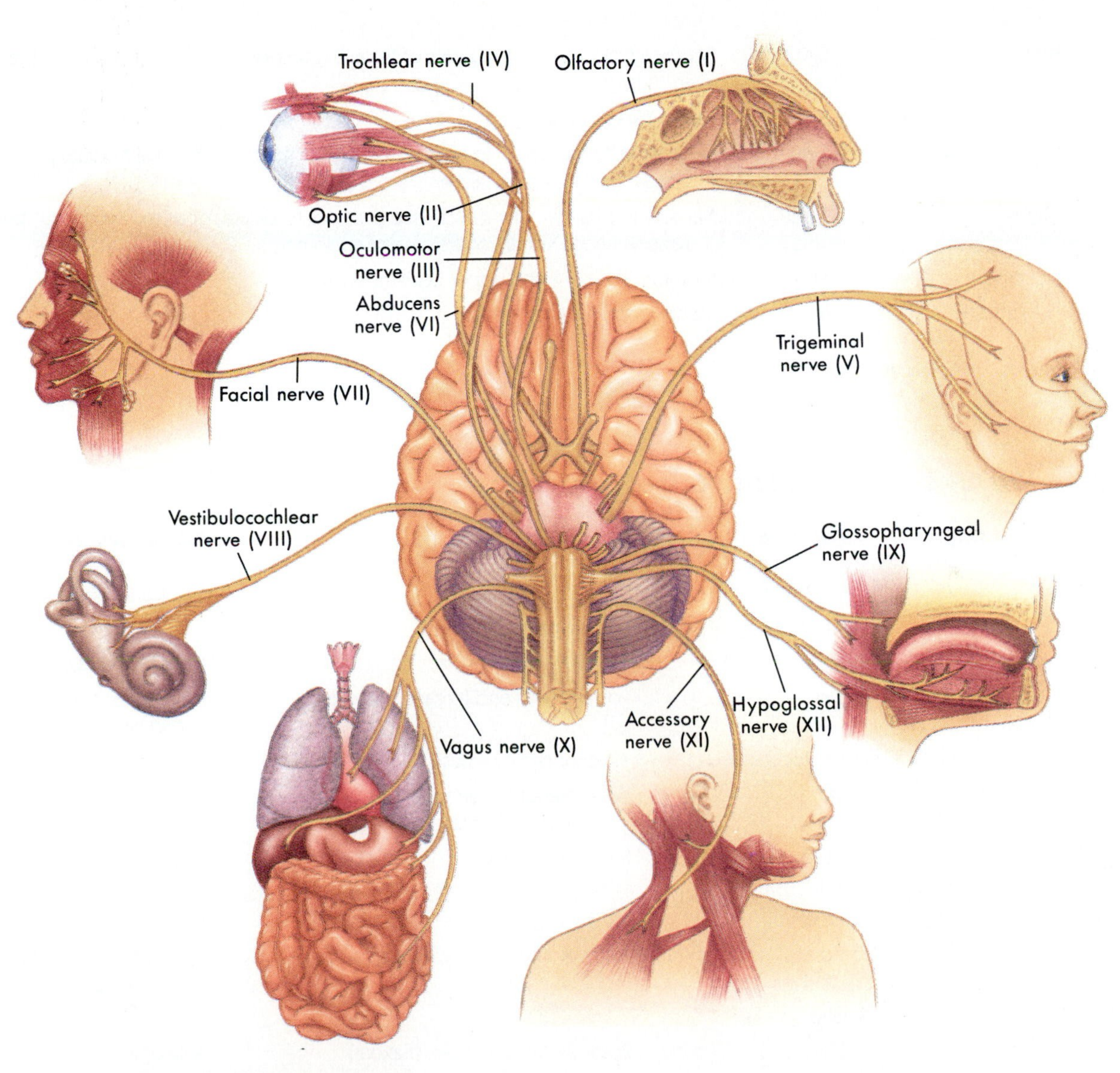

Figure 8-14 ● Cranial nerves.
Inferior surface of the brain showing the origins of the cranial nerves and the structures supplied by the cranial nerves.

Table 8-1 ● **Cranial nerves and their functions** (see Figure 8-14)

Number	Name	General function*	Specific function
I	Olfactory	S	Smell
II	Optic	S	Vision
III	Oculomotor	M,P	Motor to four of six eye muscles and upper eyelid; parasympathetic: constricts pupil, thickens lens
IV	Trochlear	M	Motor to one eye muscle
V	Trigeminal	S,M	Sensory to face and teeth; motor to muscles of mastication (chewing)
VI	Abducens	M	Motor to one eye muscle
VII	Facial	S,M,P	Sensory: taste; motor to muscles of facial expression; parasympathetic to salivary and tear glands
VIII	Vestibulocochlear	S	Hearing and balance
IX	Glossopharyngeal	S,M,P	Sensory: taste and touch to back of tongue; motor to pharyngeal muscle; parasympathetic to salivary glands
X	Vagus	S,M,P	Sensory to pharynx, larynx and viscera; motor to palate, pharynx, and larynx; parasympathetic to viscera of thorax and abdomen
XI	Accessory	M	Motor to two neck and upper back muscles
XII	Hypoglossal	M	Motor to tongue muscles

*S = sensory; M = motor; P = parasympathetic

Table 8-2 ● **Plexuses of the spinal nerves** (see Figure 8-15)

Plexus	Origin	Major nerves	Muscles innervated	Skin innervated
Cervical	C1–C4	Cervical branches	Several neck muscles	Neck and posterior head
		Phrenic	Diaphragm	
Brachial	C5–T1	Axillary	Two shoulder muscles	Part of shoulder
		Radial	Posterior arm and forearm muscles	Posterior arm, forearm, and hand
		Musculocutaneous	Anterior arm muscles	Radial surface of forearm
		Ulnar	Two anterior forearm muscles, most of the intrinsic hand muscles	Ulnar side of hand
		Median	Most anterior forearm muscles, some intrinsic hand muscles	Radial side of hand
Lumbosacral	L1–S4	Obturator	Medial thigh muscles	Medial thigh
		Femoral	Anterior thigh muscles (extensors)	Anterior thigh, medial leg and foot
		Sciatic		
		Tibial	Posterior thigh muscles (flexors), anterior and posterior leg muscles, most foot muscles	Sole of foot
		Common peroneal	Lateral thigh and leg, some foot muscles	Anterior and lateral leg, and dorsal foot

Somatic motor functions refer to the control of skeletal muscles in the head and neck. **Parasympathetic** (part of the autonomic nervous system) function involves the regulation of glands, smooth muscle, and cardiac muscle. Some cranial nerves have only one of the three functions, whereas others have more than one. Nerves that have both sensory and motor functions are called **mixed nerves.** Table 8-1 lists the names and functions of the cranial nerves.

Spinal nerves

The **spinal nerves** arise along the spinal cord from the union of the dorsal and ventral roots. All spinal nerves are mixed nerves, containing sensory and somatic motor nerve fibers. Most spinal nerves also have nerve fibers from the autonomic nervous system.

Each spinal nerve is named for the region of the vertebral column from which it emerges—cervical (C), thoracic (T), lumbar (L), sacral (S), and coccygeal (Cx). The spinal nerves are also numbered (starting superiorly) according to their order within each region. The 31 pairs of spinal nerves are, therefore, C1 to C8, T1 to T12, L1 to L5, S1 to S5, and Cx (Figure 8-15).

Most of the spinal nerves are organized into **plexuses** (plek´sus-ez), in which the nerves come together and then separate. The three major plexuses are the **cervical plexus,** the **brachial plexus,** and the **lumbosacral plexus** (Table 8-2 and see Figure 8-15). The major nerves of the neck and limbs are branches of these plexuses. Most of the thoracic spinal nerves do not join together to form plexuses. Instead they extend as individual nerves around the thorax between the ribs.

One of the most important branches of the cervical plexus is the **phrenic nerve,** which innervates the diaphragm. Contraction of the diaphragm is largely responsible for the ability to breathe (see Chapter 15).

5 Describe the effect on breathing of completely severing the spinal cord in the thoracic region (below the exit point of the phrenic nerve) versus the upper cervical region (above the exit point of the phrenic nerve).

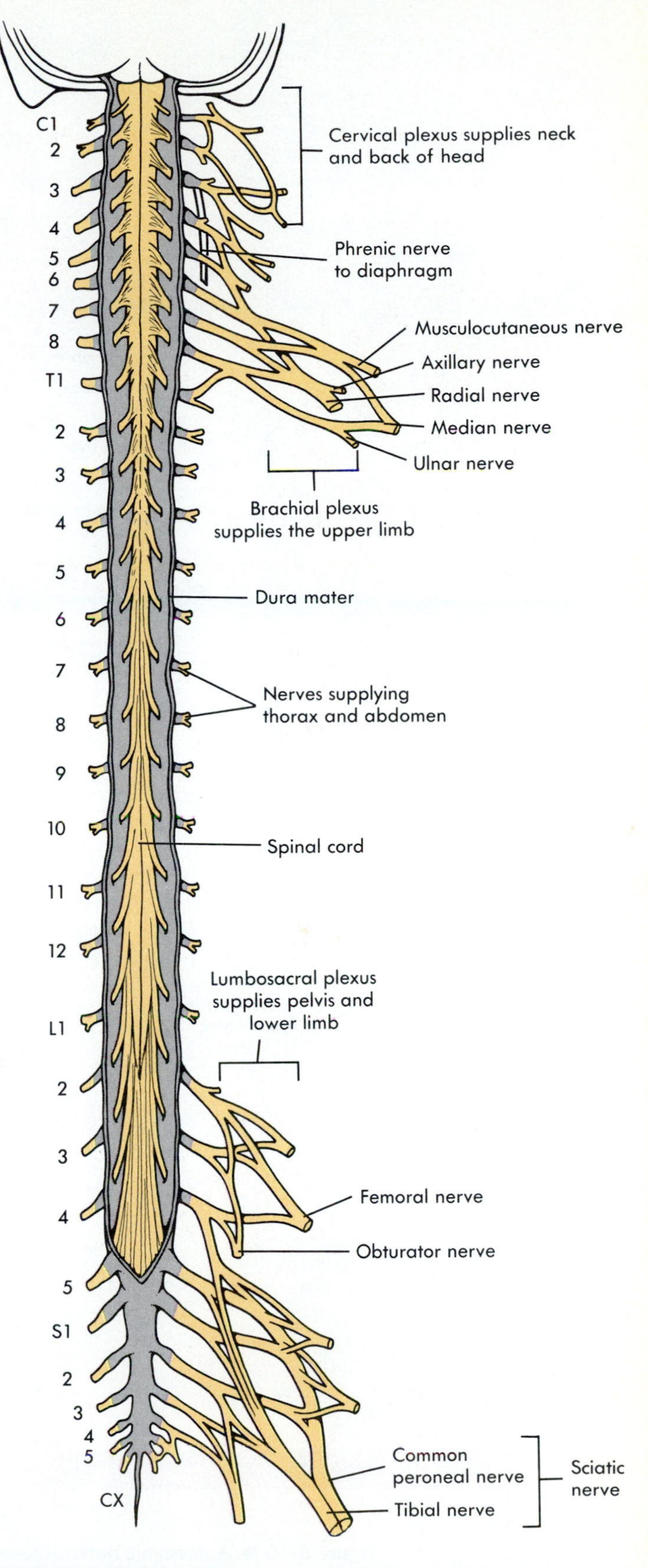

Figure 8-15 ● Spinal nerves.
Spinal cord, the spinal nerves, their plexuses, and the nerves that arise from the plexuses.

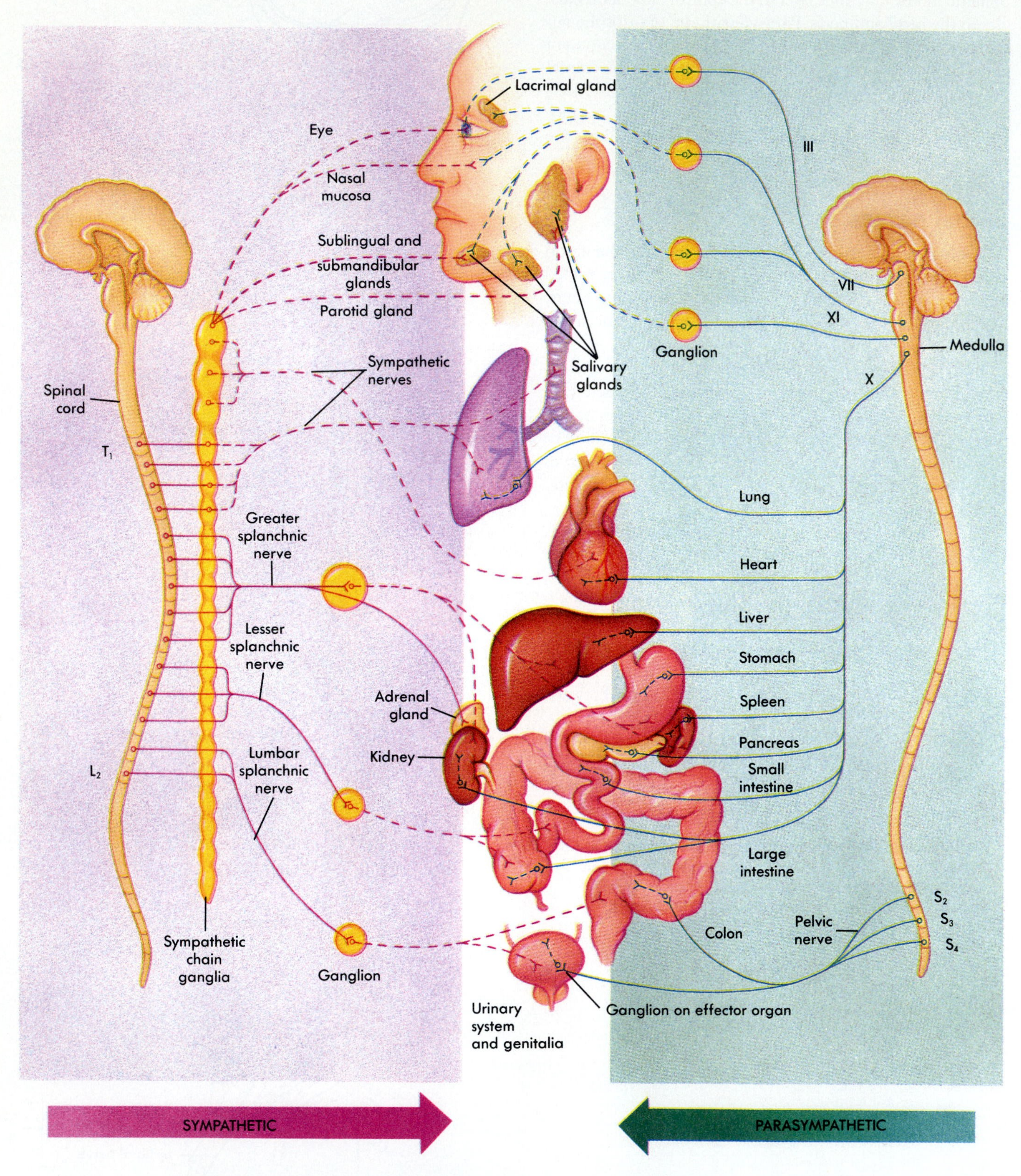

Figure 8-16 ● Autonomic nervous system.
The sympathetic neurons are red and the parasympathetic neurons are green. There are two ANS neurons in series. The first neuron is indicated by a solid line and the second by a broken line. The two neurons synapse in a ganglion.

AUTONOMIC NERVOUS SYSTEM

Motor neurons can be divided into two systems, the somatic nervous system and the autonomic nervous system (ANS). In somatic nervous system pathways one motor neuron extends from the central nervous system to skeletal muscles. Autonomic pathways, on the other hand, have two neurons in series. The first neuron extends from the CNS to a ganglion in which it synapses with the second neuron. The second neuron extends from the ganglion to smooth muscle, cardiac muscle, or glands. Autonomic functions are usually unconsciously controlled.

The autonomic nervous system is composed of sympathetic and parasympathetic divisions (Figure 8-16). Most organs that receive autonomic neurons are innervated by both divisions. Sweat glands and blood vessels, however, are innervated by sympathetic neurons almost exclusively. In most cases, the two autonomic divisions have opposite effects on structures that receive dual innervation. For example, the sympathetic division increases heart rate, whereas the parasympathetic division decreases heart rate.

Sympathetic division

The **sympathetic division** prepares a person for physical activity by increasing heart rate and respiration rate, and by stimulating sweating. The sympathetic division also stimulates the release of glucose from the liver for energy, and it stimulates mental activities. At the same time, it inhibits digestive activities. The sympathetic division is sometimes referred to as the "fight-or-flight" system because it prepares a person to either face a threat, or run from a threat.

The cell bodies of sympathetic neurons are in the spinal cord gray matter between the first thoracic (T1) and the second lumbar (L2) segments (see Figure 8-16). The nerve fibers of the neurons extend to **sympathetic chain ganglia,** which are interconnecting ganglia forming a chain along both sides of the spinal cord, or to ganglia located nearer target organs. Neurons from the ganglia project to target tissues such as glands, smooth muscles, or cardiac muscle.

Parasympathetic division

The **parasympathetic division** stimulates vegetative activities such as digestion, defecation, and urination, as well as slows the heart rate and respiration rate. It also causes the pupil of the eye to constrict and the lens to thicken, both of which help in close vision, such as in reading a book.

The cell bodies of the parasympathetic division are located within brainstem nuclei of certain cranial nerves (see Table 8-1) or within the spinal cord gray matter in the S2 to S4 regions of the spinal cord (see Figure 8-16). Axons of the parasympathetic neurons extend through cranial nerves or spinal nerves to ganglia located either near their target organs in the head or embedded in the walls of target organs in the rest of the body. The axons of the neurons in the ganglia extend a relatively short distance to the target organs.

6 List some of the responses stimulated by the autonomic nervous system in (A) a person who is extremely angry and (B) a person who has just finished eating and is now relaxing.

Autonomic neurotransmitter substances

The sympathetic and parasympathetic nerve endings secrete one of two neurotransmitters onto their target organs. All parasympathetic neurons secrete acetylcholine and most sympathetic neurons secrete norepinephrine (some secrete acetylcholine). Many body functions can be stimulated or inhibited by drugs that either mimic these neurotransmitters or prevent the neurotransmitters from activating their target organs. For example, drugs called beta blockers are used to treat an irregular, rapid heart rate because they prevent norepinephrine from stimulating the heart.

Biofeedback and meditation

Biofeedback takes advantage of electronic instruments or other techniques to monitor and change subconscious activities, many of which are regulated by the autonomic nervous system. Skin temperature, heart rate, and brain waves are monitored electronically. By watching the monitor and using biofeedback techniques, a person can learn to consciously control certain autonomic functions. Meditation is another technique that influences autonomic functions. Some people use biofeedback or meditation methods to relax by learning to reduce the heart rate or change the pattern of brain waves. The severity of stomach ulcers, migraine headaches, high blood pressure, anxiety, or depression also can be reduced.

SUMMARY

DIVISIONS OF THE NERVOUS SYSTEM

The central nervous system (CNS) consists of the brain and spinal cord, whereas the peripheral nervous system (PNS) consists of nerves and ganglia.

The sensory division of the PNS transmits action potentials to the CNS; the motor division carries action potentials away from the CNS.

The somatic nervous system innervates skeletal muscle and is mostly under voluntary control. The autonomic nervous system innervates cardiac muscle, smooth muscle, and glands, and it is mostly under involuntary control.

The autonomic nervous system has sympathetic and parasympathetic divisions.

CELLS OF THE NERVOUS SYSTEM

Neurons

Neurons receive stimuli and transmit action potentials.

Neurons consist of a cell body, dendrites, and an axon.

Neurons are sensory, motor, or association.

Neuroglia

Neuroglia are the support cells of the nervous system. They form myelin sheaths and are involved with the blood-brain barrier.

Organization of nervous tissue

Gray matter forms nuclei in the CNS and ganglia in the PNS.

White matter forms nerve tracts in the CNS and nerves in the PNS.

ACTION POTENTIALS

An action potential is a reversal of the normal charge difference across a cell membrane.

A nerve impulse is an action potential that moves along the membrane of a nerve fiber.

Action potentials move more rapidly down myelinated than unmyelinated fibers.

THE SYNAPSE

The synapse is a junction for transmitting action potentials between two cells.

An action potential arriving at the synapse causes the release of a neurotransmitter, which diffuses across the synaptic cleft and binds to the receptors of the cell membrane, causing the production of an action potential.

CENTRAL NERVOUS SYSTEM

The central nervous system consists of the brain and spinal cord.

Brainstem

The brainstem contains several nuclei as well as ascending and descending tracts.

The medulla oblongata contains nuclei that control such activities as heart rate, breathing, and swallowing.

The pons contains relay nuclei between the cerebrum and cerebellum.

The midbrain is involved in hearing and in visual reflexes.

The reticular formation is scattered throughout the brainstem and is involved in maintaining consciousness.

Diencephalon

The diencephalon consists of the thalamus (main sensory relay center) and the hypothalamus (the homeostatic control center of the body).

Cerebrum

The cerebrum has two hemispheres divided into lobes. The lobes are the frontal, parietal, occipital, and temporal.

Many CNS functions can be localized to specific areas of the cortex.

An EEG monitors brain waves, which are a summation of the brain's electrical activity.

Memory consists of short-term (lasting a few minutes) and long-term (permanent) memory.

Each hemisphere controls the opposite half of the body. The corpus callosum connects the two hemispheres. The left hemisphere is thought to be the dominant, analytical hemisphere, and the right hemisphere is thought to be dominant for spatial perception and musical ability.

Basal ganglia

The basal ganglia are cerebral nuclei that inhibit extraneous muscular activity, maintain posture, and coordinate motor movements.

Cerebellum

The cerebellum is involved in balance and muscle coordination. Its main function is to compare the intended action with what is occurring, and modifying to eliminate differences.

The cerebellum "learns" and then automatically controls complex, rapid movements.

Spinal cord

The spinal cord has a central gray portion and a peripheral white portion forming nerve tracts.

Reflexes

A reflex is a response to a stimulus that does not involve conscious thought.

A reflex arc consists of a sensory receptor, afferent neuron, association neurons, efferent neuron, and effector organ.

Meninges

Three connective tissue meninges cover the CNS: the dura mater, arachnoid layer, and pia mater.

Ventricles

The brain and spinal cord contain fluid-filled cavities: the lateral ventricles in the cerebral hemispheres, a third ventricle in the diencephalon, a cerebral aqueduct in the midbrain, a fourth ventricle at the base of the cerebellum, and a central canal in the spinal cord.

Cerebrospinal fluid

Cerebrospinal fluid is formed in the choroid plexuses in the ventricles, it exits through the fourth ventricle, and it reenters the blood through arachnoid granulations in the dural sinus.

PERIPHERAL NERVOUS SYSTEM

Cranial nerves

There are 12 pairs of cranial nerves that arise from the brain.

Some cranial nerves have sensory functions, some have somatic motor functions, and mixed nerves have sensory and somatic motor functions. Some cranial nerves also have parasympathetic functions.

Spinal nerves

The spinal nerves exit from the cervical, thoracic, lumbar, and sacral regions of the spinal cord.

The spinal nerves are grouped into plexuses.

Spinal nerves are mixed nerves. Spinal nerves also have autonomic nerve fibers.

AUTONOMIC NERVOUS SYSTEM

The autonomic nervous system consists of two neurons in series. The first neuron is located in the CNS and the second is in a ganglion.

The autonomic nervous system supplies smooth muscle, cardiac muscle, and glands.

Sympathetic division

The sympathetic division is involved in preparing the person for action, by increasing heart rate, blood pressure, and respiration rate.

Nerve cell bodies of the sympathetic division lie in the thoracic and upper lumbar regions of the spinal cord.

Parasympathetic division

The parasympathetic division is involved in vegetative activities such as the digestion of food, defecation, and urination.

Nerve cell bodies of the parasympathetic division are found in the nuclei of cranial nerves and the sacral region of the spinal cord.

Autonomic neurotransmitter substances

All parasympathetic neurons secrete acetylcholine onto their target organs.

Most sympathetic neurons secrete norepinephrine onto their target organs.

Biofeedback and meditation

It is possible to voluntarily control some autonomic nervous system functions.

CONTENT REVIEW

1. Define the sensory division, motor division, somatic nervous system and the autonomic nervous system.
2. What are the functions of neurons? Name the three parts of a neuron and describe their functions. List the three types of neurons based on their functions.
3. Define neuroglia. What is the function of myelin sheaths and the blood-brain barrier?
4. For nerve tracts, nerves, nuclei, and ganglia, name the cells or parts of cells found in each, state if they are white matter or gray matter, and name the part (CNS or PNS) of the nervous system in which they are found.
5. Explain how action potentials move more rapidly down myelinated axons than down unmyelinated axons.
6. Describe the function and operation of the synapse.
7. Name the three parts of the brainstem and describe their functions. What is the reticular activating system?
8. Name the two main components of the diencephalon and describe their functions.
9. Name four lobes of the cerebrum, describing their locations and functions.
10. Name the two types of memory and describe the processes that result in long-term memory.
11. Describe the function of the basal ganglia.
12. Describe the comparator activities of the cerebellum. How is the cerebellum involved with performing complex motor skills?
13. What are the functions of the spinal cord gray matter and white matter?
14. Differentiate dorsal root, ventral root, and spinal nerve. Which contain sensory fibers, and which contain motor fibers?
15. Name the five components of a reflex arc and explain the operation of a reflex arc.
16. Name the three meninges. Describe the production and circulation of the cerebrospinal fluid. Where does the cerebrospinal fluid return to the blood?
17. What are the three principal functional categories of the cranial nerves?
18. List the spinal nerves by name. Name the main plexuses derived from the spinal nerves.
19. Contrast the somatic and the autonomic nervous systems in terms of the number of motor neurons, conscious versus unconscious control, and types of effector organs.
20. Contrast the functions of the sympathetic and parasympathetic nervous systems.

CONCEPT REVIEW

1. Bob Canner improperly canned some home-grown vegetables. Consequently, he contracted botulism food poisoning after eating the vegetables. Botulism results from a toxin produced by bacteria. Symptoms included difficulty in swallowing and breathing. Eventually he died of respiratory failure (his respiratory muscles relaxed and would not contract). Botulism toxin affects the synapse between motor neurons and muscle fibers. There are several ways that a toxin could alter the normal operation of the synapse. Propose as many ways as you can that the botulism toxin could cause the observed symptoms. Please note that only one explanation is correct. The purpose of this question is to hypothesize as many logically correct explanations as possible that are consistent with what you know about synapses.
2. Louis Ville was accidentally struck in the head with a baseball bat. He fell to the ground unconscious. Later, when he had regained consciousness, he was unable to remember any of the events that happened during the 10 minutes before the accident. Explain.
3. A patient suffered brain damage in an automobile accident. It was suspected that the cerebellum was the part of the brain affected. Based on what you know about cerebellar function, how could you determine that the cerebellum was involved?
4. Suppose a patient suffers a spinal cord injury that completely severs all the nerve tracts on the left half of his spinal cord. What effect will this have on his ability to feel pain and temperature sensations and on his ability to voluntarily control skeletal muscles (Hint: see Figure 8-10).
5. Name the cranial nerve that, if damaged, would produce the following symptoms:
   A. The patient is unable to move the tongue.
   B. The patient is unable to see on one side.
   C. The patient is unable to feel one side of the face.
   D. The patient is unable to move the facial muscles on one side.
   E. The pupil of one eye is dilated and will not constrict.

CHAPTER TEST

Answers can be found in Appendix C

MATCHING 1 For each statement in column A select the correct answer in column B (an answer may be used once, more than once, or not at all).

A	B
1. Transmits action potentials to the CNS.	axon
2. Carries action potentials away from the neuron cell body.	dendrite
3. Supports CNS neurons, influences the blood-brain barrier.	ganglion
4. Forms myelin sheaths in the PNS.	motor division
5. Area of an axon not surrounded by myelin.	nerve
6. Space across which neurotransmitters diffuse.	neuroglia
7. Collection of cell bodies in the CNS.	nerve tract
8. Bundle of nerve fibers in the PNS.	node of Ranvier
	nucleus
	sensory division
	synaptic cleft

MATCHING 2 For each statement in column A select the correct answer in column B (an answer may be used once, more than once, or not at all).

A	B
1. Contains nuclei that regulate heart rate and respiration.	basal ganglia
2. Connects to the cerebellum.	Broca's area
3. Coordinates visual and auditory reflexes.	cerebellum
4. Most sensory nerve tracts synapse here.	hypothalamus
5. Regulates body temperature, thirst, and hunger.	medulla oblongata
6. Motor area necessary for speech.	midbrain
7. Inhibits muscular activity.	pons
8. Coordinates fine motor movements.	thalamus

MATCHING 3 For each statement in column A select the correct answer in column B (an answer may be used once, more than once, or not at all).

A	B
1. Contains the dural sinus.	arachnoid granulation
2. Innermost meningeal covering.	arachnoid layer
3. Produces cerebrospinal fluid.	choroid plexus
4. Returns cerebrospinal fluid to blood.	cranial nerves
5. Nerves that arise from the brain.	dura mater
6. Prepares the body for physical activity.	parasympathetic division
7. Has norepinephrine as a neurotransmitter.	pia mater
	spinal nerves
	sympathetic division

FILL-IN-THE BLANK Complete each statement by providing the missing word or words.

1. The type of nerve fiber that conducts action potentials the most rapidly is ___________.
2. The three parts of the brainstem are the ___________, ___________, and ___________.
3. The ___________ lobe of the cerebrum functions in the reception and integration of visual input.
4. The ___________ lobe of the cerebrum evaluates olfactory and auditory input and plays an important role in memory.
5. The ___________ lobe of the cerebrum is important in voluntary motor activities, motivation, and aggression.
6. The ___________ roots contains sensory nerve fibers, the ___________ roots contains motor nerve fibers, and spinal nerves contain both sensory and motor nerve fibers.
7. The five basic components of a reflex arc are ___________.
8. The ___________, the ___________, and the central canal of the spinal cord contain cerebrospinal fluid.

CHAPTER

The General and Special Senses

OBJECTIVES

After reading this chapter you should be able to:

1. Define general and special senses.
2. Describe olfactory neurons and explain how smells are detected.
3. Outline the structure and function of a taste bud.
4. List the accessory structures of the eye and explain their functions.
5. Name the three layers of the eye, and give the functions of each layer.
6. Describe the chambers of the eye and the fluids they contain.
7. Explain how images are focused on the retina.
8. Describe the structures of the outer and middle ears and state the function of each.
9. Describe the cochlea and explain how sounds are detected.
10. Explain how the structures of the inner ear function in balance.

KEY TERMS

choroid
Middle layer of the eye that covers the sclera within the posterior compartment of the eye; black in color in order to absorb light.

cochlea
The portion of the inner ear involved in hearing; shaped like a snail shell.

cornea
Transparent, anterior portion of the eye; refracts light entering the eye.

iris
The "colored" part of the eye that can be seen through the cornea; consists of smooth muscles that regulate the amount of light entering the eye through the pupil.

lens
Biconcave structure in the eye capable of being flattened or thickened to adjust the focus of light entering the eye.

retina
The inner, light-sensitive layer of the eye; contains rods and cones.

sclera
The dense, white, opaque outer layer of the eye; white of the eye.

semicircular canal
One of three canals in each temporal bone involved in the detection of motion of the head.

taste bud
Sensory structure found on the tongue; contains taste cells with receptors that respond to dissolved chemicals.

vestibule
Middle region of the inner ear responsible for static equilibrium.

FEATURES

THE SENSES are the means by which the brain receives information about the outer world and the body. Stimuli act on sensory receptors to produce action potentials that are transmitted along sensory nerve fibers to the central nervous system (CNS) in which sensations are actually perceived. The senses can be characterized as general or special. Sensations such as touch, temperature, pain, and proprioception (sense of position) are referred to as **general senses** because their receptors are found throughout the body. Taste, smell (olfaction), sight (vision), hearing, and balance (equilibrium) are usually referred to as the **special senses.** These senses depend on highly localized organs with very specialized sensory cells.

GENERAL SENSES

Many of the receptors for the general senses are associated with the skin (Figure 9-1) or deeper structures such as tendons, ligaments, and muscles. There are several types of receptors that respond to a wide range of stimuli, including pain, temperature, touch, pressure, and position. The sensory neurons associated with these receptors relay action potentials to the thalamus and on to the CNS (see Chapter 8).

Action potentials that reach the cerebrum can be perceived as conscious sensation. However, the cerebrum screens a large proportion of the action potentials that reach it and we are not consciously aware of the information carried by these action potentials. In addition, some action potentials are transmitted to other areas of the brain, such as the cerebellum, where they are not consciously perceived.

SPECIAL SENSES

The sensory organs that make up the special senses are highly specialized. Sensory organs that respond to taste and smell are specialized to respond to chemical stimuli, the eyes are specialized to respond to light, and the ears detect sound and monitor balance by responding to mechanical stimuli.

OLFACTION

Olfaction (ol-fak´shun) is the sense of smell. It occurs in response to airborne molecules called **odors** that enter the nasal cavity (Figure 9-2). Olfactory neurons are sensory neurons within the epithelial lining of the superior region of the nasal cavity. Airborne molecules dissolve in the fluid on the surface of the epithelium, bind to receptor molecules of the olfactory neurons, and stimulate the production of action potentials. Axons from the olfactory neurons form the **olfactory nerves** (cranial nerve I), which enter the **olfactory bulbs.** The axons synapse with association neurons that relay action potentials to the **olfactory areas** in the temporal lobes of the brain.

The threshold for the detection of odors is very low, so very few molecules are required to initiate an action potential. It has been proposed that a wide variety of detectable odors are actually combinations of a smaller number (perhaps as few as seven) of primary odors.

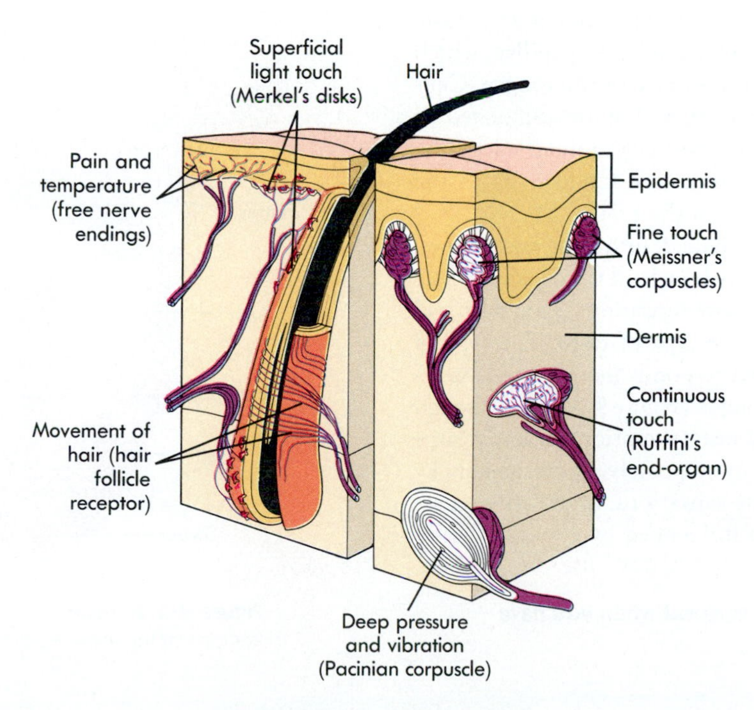

Figure 9-1 ● General senses.
Sensory nerve endings in the skin.

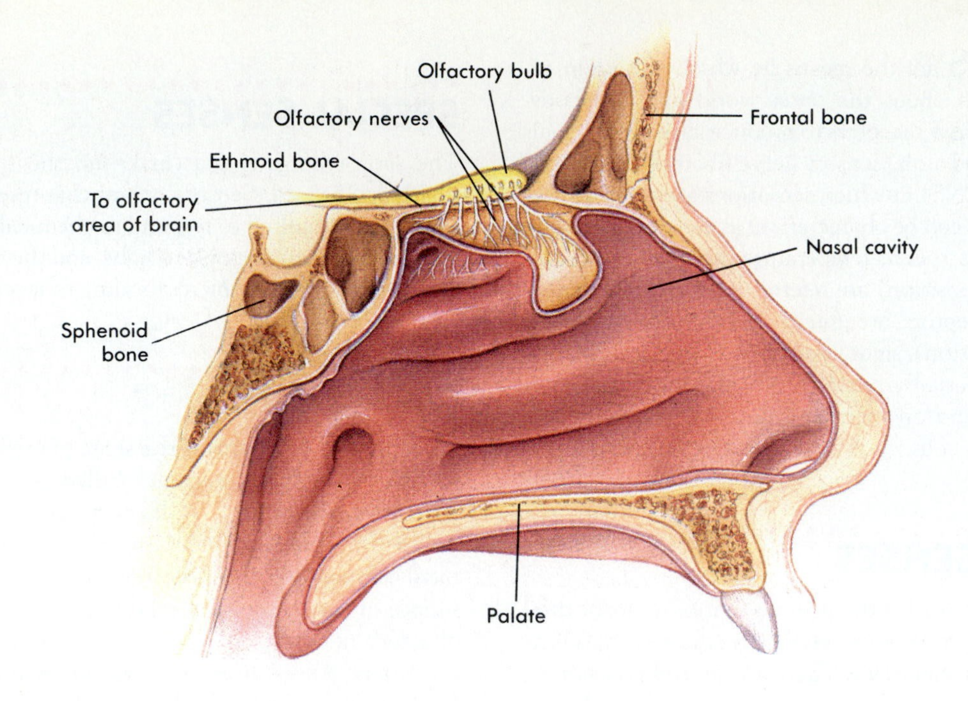

Figure 9-2 ● **Olfaction.**
Sagittal view. Lateral wall of the nasal cavity showing the olfactory nerves.

TASTE

The sensory structures that detect taste stimuli are the **taste buds.** Taste buds are usually associated with **papillae,** which are enlargements visible on the surface of the tongue. Cells within the taste buds have receptors that are stimulated by dissolved substances. Action potentials are carried by the neurons of several cranial nerves (see Chapter 8) to the **taste areas** in the parietal lobes of the brain.

Taste sensations can be divided into four basic types: sour, salty, bitter, and sweet. Although all taste buds are able to detect all four of the basic taste sensations, each taste bud is usually most sensitive to one type of taste. The type of taste to which each taste bud responds most strongly is related to its position on the tongue (Figure 9-3). Even though there are only four primary taste sensations, a fairly large number of different tastes can be perceived, presumably by combining the four basic taste sensations. Many taste sensations, however, are strongly influenced by olfactory sensations.

1 Why does food not taste as good when you have a cold?*

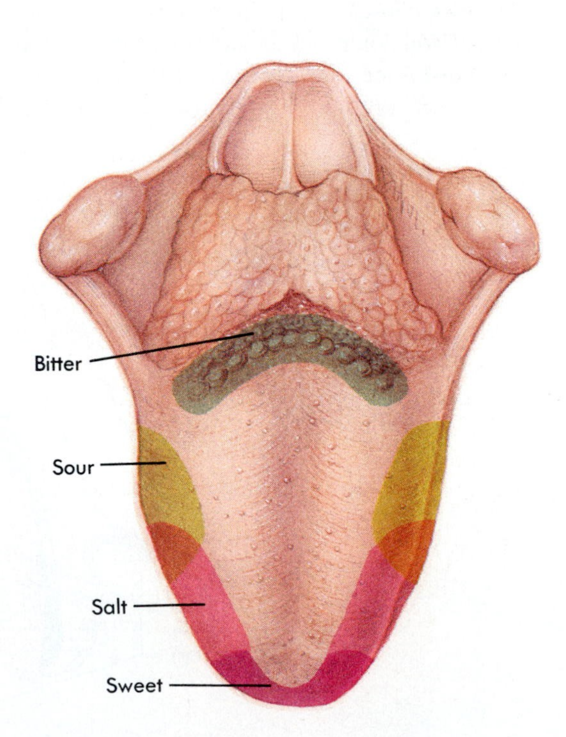

Figure 9-3 ● **Taste.**
Regions of the tongue sensitive to various tastes.

*Answers to predict questions appear in Appendix B at the back of the book.

VISION

The visual system includes the eyes, accessory structures, nerve pathways, and the visual areas of the cerebral cortex. Much of the information we obtain about the world around us is detected by the visual system. Visual input includes information about light and darkness, pattern and shape, color, and movement.

Accessory structures

Accessory structures protect, lubricate, clean, and move the eye. They include the eyebrows, eyelids, conjunctiva, lacrimal gland, and extrinsic eye muscles.

Eyebrows

The **eyebrows** protect the eyes by preventing perspiration from running down the forehead and into the eyes. Eyebrows also help shade the eyes from direct sunlight.

Eyelids

The **eyelids** with their associated lashes protect the eyes from foreign objects. If an object suddenly approaches the eye, the blink reflex causes the eyelids to rapidly close. Blinking, which normally occurs about 20 times per minute, also helps keep the eyes lubricated by spreading tears over the surface of the eye.

A **chalazion** (kal-a´ze-on) is a cyst caused by infection of the glands along the edge of the eyelid. A **sty** (stī) is an infection of an eyelash hair follicle.

Conjunctiva

The **conjunctiva** (kon-junk-ti´vah) is a thin mucous membrane that covers the inner surface of the eyelids and the anterior surface of the eye. **Conjunctivitis** is an inflammation of the conjunctiva, which often is caused by a bacterial infection.

Lacrimal gland

The **lacrimal** (lak´rĭ-mal) **gland** produces tears, which pass over the anterior surface of the eye (Figure 9-4). Most of the fluid produced by the lacrimal gland evaporates from the surface of the eye. Excess tears are collected in the medial corner of the eye and enter into the **nasolacrimal duct,** which opens into the nasal cavity. Tears lubricate the eye and cleanse it. In addition, tears contain an enzyme that helps combat eye infections.

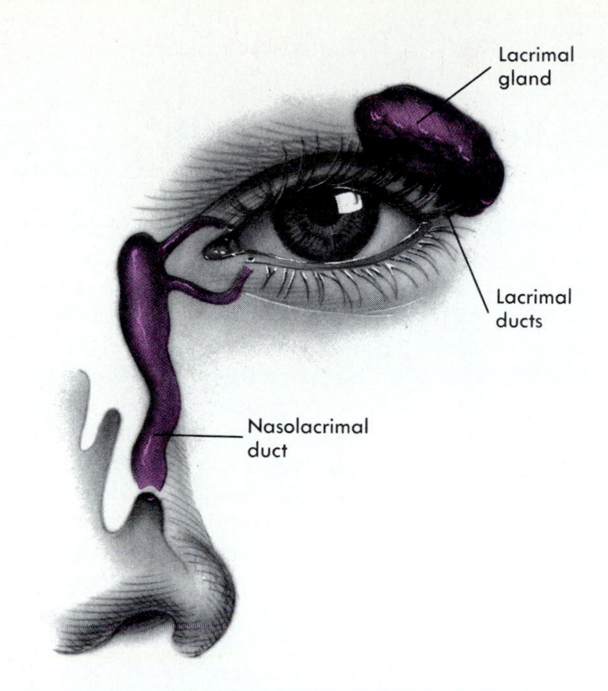

Figure 9-4 ● **The lacrimal gland.**
Tears produced in the lacrimal gland pass through lacrimal ducts to the surface of the eye. The tears evaporate or pass over the surface of the eye and enter the nasolacrimal duct to the nasal cavity.

2 Explain why it is often possible to "taste" a medication, such as eyedrops, that has been placed into the eyes. (Hint: how are taste and smell related?)

Extrinsic eye muscles

Movement of each eyeball is accomplished by six muscles, the **extrinsic eye muscles** (Figure 9-5). These muscles have their origins on the bones of the orbit and insert onto the eyeball.

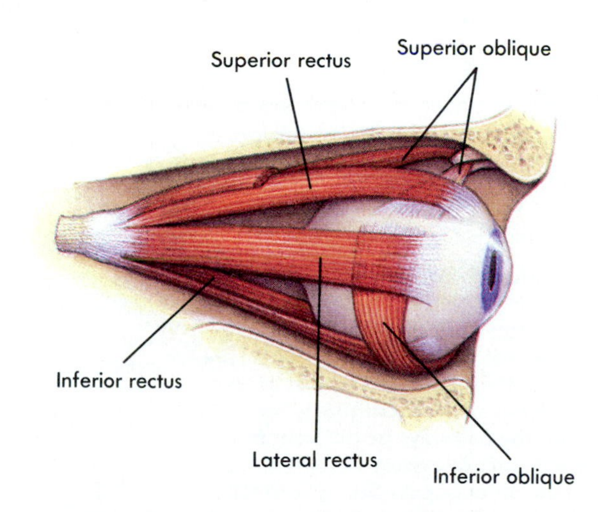

Figure 9-5 ● **Extrinsic eye muscles.**
Extrinsic muscles of the right eye as seen from a lateral view with the lateral wall of the orbit removed. One of the six extrinsic eye muscles, the medial rectus, cannot be seen from this view.

Strabismus (stră-biz´mus) is a condition in which one eye or both eyes are directed medially ("cross-eyed") or laterally ("cock-eyed"). The condition can result from over contraction or under contraction of extrinsic eye muscles.

Understanding pain

PAIN RECEPTORS

The simplest and most common sensory nerve endings are the **free nerve endings** (see Figure 9-1), which are distributed throughout almost all parts of the body. These nerve endings are responsible for a number of sensations, including pain, temperature, itch, and movement. Often, receptors for different stimuli interact to produce sensations. For example, damage to tissues usually activates pain receptors and touch receptors. The pain receptors send action potentials to the brain that are interpreted as pain, and the touch receptors send action potentials that help identify the location of the pain. The location of **superficial pain** is precisely identified as a result of the simultaneous stimulation of pain receptors and touch receptors in the skin. The location of **deep** or **visceral pain** is not precisely identified because of the absence of touch receptors in the deeper structures, and it is normally perceived as a diffuse, nonlocalized pain.

PAIN TRANSMISSION

Pain is an unpleasant sensory and emotional experience. There are two types of pain sensation: (1) sharp, well-localized, pricking, or cutting pain; and (2) diffuse, burning, or aching pain.

3 Based on your own experience, which type of pain sensation is perceived first following an injury? Given that these two types of pain are transmitted to the CNS by two different types of nerve pathways, how must the pathways be different from each other to explain the difference in the order of pain perception? (Hint: the order is determined by the speed of transmission of the sensation)

Anesthesia (an′es-the′ze-ah) is the loss of feeling or sensation, especially the loss of pain sensation. Local anesthesia is the suppression of pain transmission from part of the body by injecting drugs near nerves (see Understanding Action Potentials in Chapter 8). General anesthesia inhibits pain by causing a person to become unconscious. Drugs are used to suppress the reticular activating system, the part of the brain stem responsible for maintaining consciousness (see Chapter 8).

Administering drugs is not the only way to suppress pain transmission. There is a natural pain suppression system within the CNS. The brain and spinal cord produce **endorphins** (en′dor-fins) and related chemicals. Endorphins are so named because they are **end**ogenous (produced within the body) chemicals that are similar to **morphine.** For many years it was known that morphine suppressed pain, but it was not understood how this was accomplished. Then it was discovered that morphine is structurally similar to endorphins, which occur naturally in the body. Endorphins (and morphine) act at the synapse to prevent the transmission of action potentials from one neuron to another. When a person is subjected to painful stimuli, action potentials are transmitted to the CNS by sensory neurons. These sensory neurons synapse with association neurons that form the nerve tracts carrying painful sensations to the cerebral cortex and other parts of the brain (see Chapter 8). Within the CNS the endorphins interfere with action potential transmission by the nerve tracts associated with pain. Consequently, fewer action potentials reach the cerebral cortex and awareness of pain is reduced or eliminated. Decreased sensitivity to painful stimuli during exercise, pregnancy, and other types of stress, may result from increased production of endorphins and related chemicals. Acupuncture and electrical stimulation procedures also may act through this mechanism.

PAIN PERCEPTION

The stimulus strength that activates pain receptors is essentially the same for everyone. How painful a stimulus is perceived to be, however, can be very different from one person to the next. **Pain tolerance** is the ability to endure pain. Some individuals have a very high pain tolerance, whereas others are very sensitive to pain. Pain tolerance is also influenced by psychological factors. For example, under emergency conditions a person can tolerate more pain than under normal conditions.

Referred pain is a painful sensation in a region of the body that is not the source of the pain stimulus. Most commonly, referred pain is sensed in the skin or other superficial structures when internal organs are damaged or inflamed (Figure 9-A). This occurs because sensory neurons from the superficial area to which the pain is referred and the neurons from the deep, visceral area in which the actual damage occurs project to the same area of the cerebral cortex. The brain cannot distinguish between the two sources, and the painful sensation is referred to the most superficial structures innervated (such as the skin).

Referred pain is clinically useful in diagnosing the actual cause of the painful stimulus. For example, heart attack victims often feel cutaneous pain radiating from the left shoulder down the arm.

4 A man has constipation that causes distention and cramping of his colon (part of the large intestine). What kind of pain would he experience and where would the pain be located? Explain.

Phantom pain is the perception of pain in a structure that is not present, such as an amputated finger. Normally pain perception in the finger results from stimulation of pain sensory nerve endings in the finger. Action potentials are transmitted by the pain sensory neurons to the CNS, in which the pain is perceived as coming from the finger. If the finger is amputated, the nerve endings in the finger are removed. The rest of the pain sensory neurons, however, are still present. If they are stimulated at any point along their pathway, action potentials are transmitted to the CNS. Because the CNS cannot distinguish between action potentials generated at nerve endings or at other parts of the sensory neurons, the CNS interprets the action potentials as pain in the finger, even though the finger is gone. A similar phenomenon easily can be demonstrated by bumping the ulnar nerve as it crosses the elbow (the funny bone). A sensation of pain is felt in the fourth and fifth fingers, even though the neurons were stimulated at the elbow.

Although input from sensory neurons is a component of phantom pain, other factors are probably involved. It has recently been proposed that the brain has a neural network that is continuously aware of body parts. Even when that body part is removed, the brain creates an impression that the part is still there.

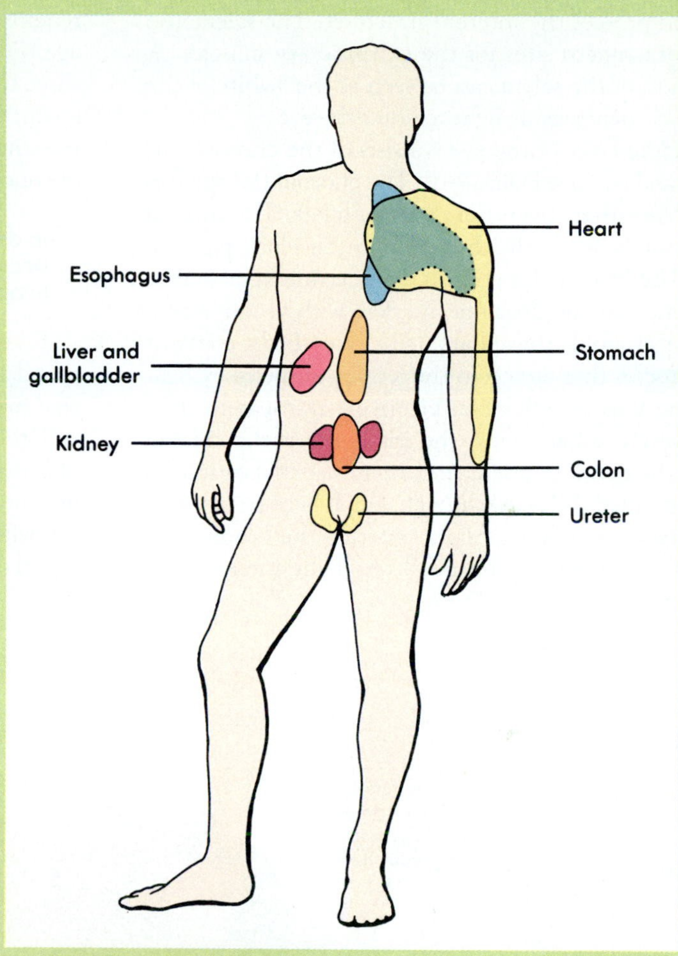

Figure 9-A Referred pain.
Pain from the indicated internal organs is referred to surface areas.

Anatomy of the eye

Layers of the eye

The adult eye is a fluid-filled sphere that is approximately 2.5 cm in diameter. The wall of the eye has three layers (Figure 9-6). The outer layer consists of the sclera and cornea. The **sclera** (skler´ah) is the firm, white, outer connective tissue layer of the posterior five sixths of the eye, and the **cornea** (kor´ne-ah) is the transparent anterior one sixth of the eye. The sclera and cornea help maintain the shape of the eye and protect the internal structures. The sclera also provides attachment sites for the extrinsic eye muscles. A small portion of the sclera can be seen as the "white of the eye." The cornea permits light to enter the eye.

The middle layer of the eye consists of the choroid, ciliary body, and iris (see Figure 9-6). The **choroid** (ko´royd) is a layer of tissue that covers the posterior sclera. The choroid has large numbers of cells that contain the black pigment melanin. The black color absorbs the entering light so that it is not reflected and does not interfere with vision. Anterior to the choroid, the **ciliary** (sil´e-ăr-e) **body** contains smooth muscles that attach to the lens by **suspensory ligaments.** The **lens** is a flexible, biconvex, transparent disc. Anterior to the ciliary body, the **iris** (i´ris) is the colored portion of the eye. It contains smooth muscle that surrounds an opening called the **pupil** through which light passes. The iris regulates the amount of light entering the eye by controlling the diameter of the pupil. As light intensity increases, the diameter of the pupil decreases, reducing the amount of light entering the eye. As light intensity decreases, the diameter of the pupil increases, allowing more light to enter the eye.

The **retina** (ret´ĭ-nah), the inner layer of the eye, rests on the choroid. The retina contains photoreceptor cells called rods and cones, and numerous association neurons (Figure 9-7). **Rods** are very sensitive to light and can function in very dim light, but do not provide a very sharp image. **Cones** require much more light and provide us with a very clear image. Cones are also responsible for color vision. There are three types of cones, each sensitive to a different color: blue, green, or red. The many colors that we can see result from the stimulation of various combinations of these three types of cones.

5 In dim light, colors seem to fade and objects seem to become colored by shades of gray. Explain how this happens.

Rods and cones have photosensitive molecules that respond to light. For example, rod cells contain a photosensitive molecule called rhodopsin. When light strikes a rod cell, rhodopsin changes shape, which causes the production of action potentials that are transmitted to the brain. Manufacture of rhodopsin in rod cells requires vitamin A. A person with a vitamin A deficiency may have difficulty seeing, especially in dim light—a condition called **night blindness.**

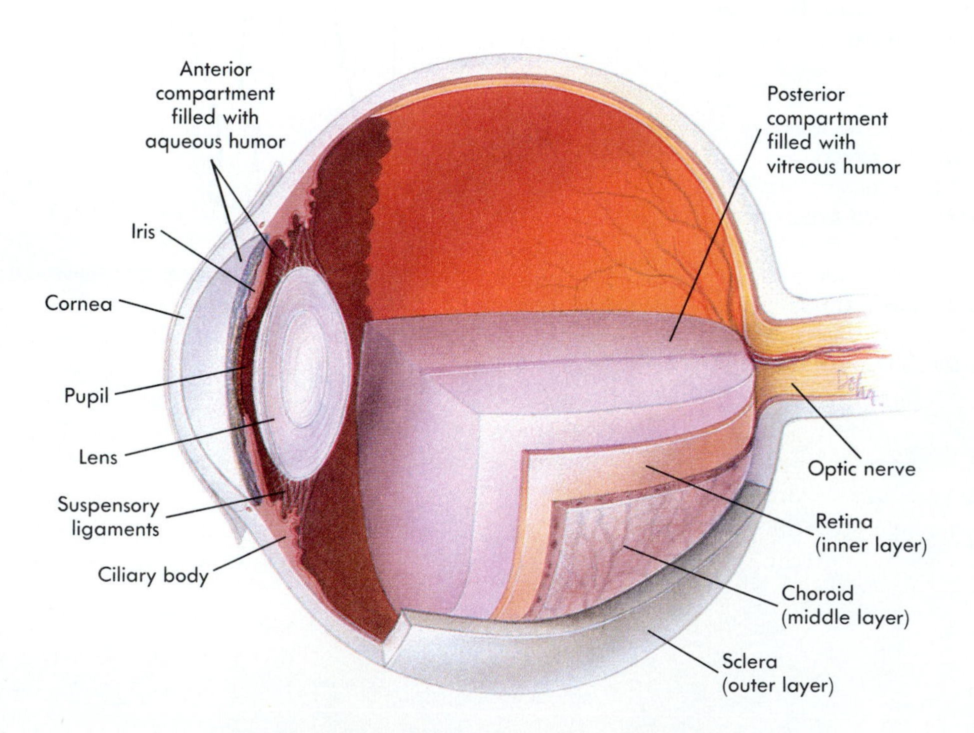

Figure 9-6 ● Layers and compartments of the eye. Sagittal section.

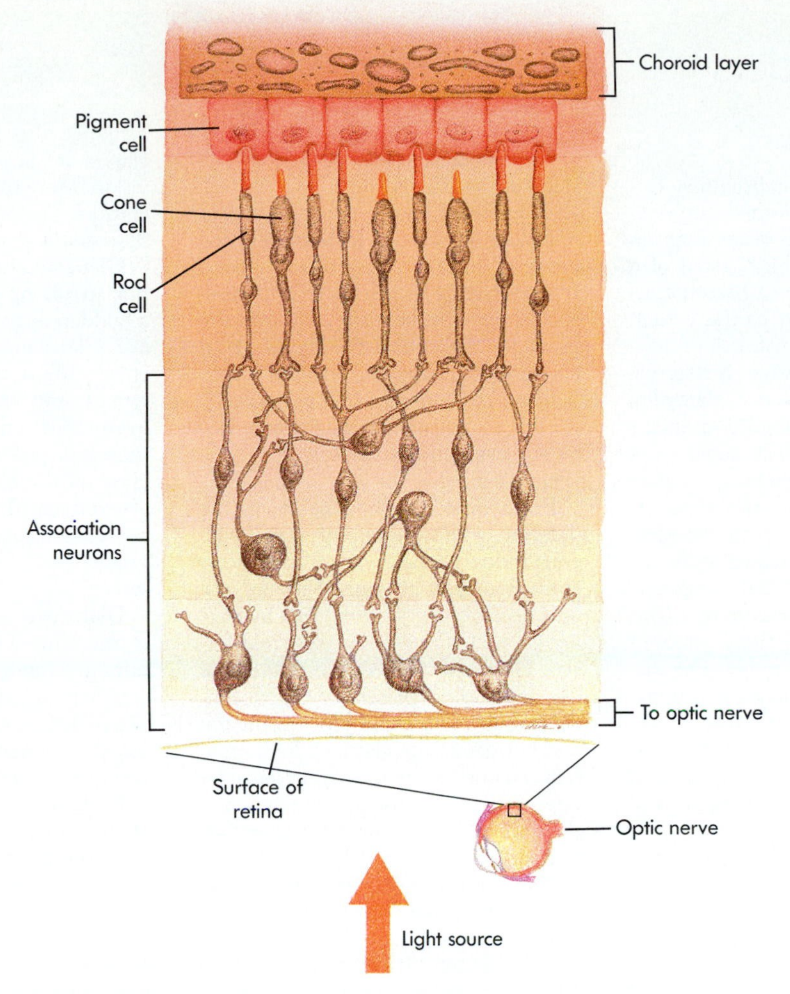

Figure 9-7 ● The retina.
Enlarged section of the retina showing its structure. The rods and cones are connected by association neurons to the optic nerve.

Compartments of the eye

The eyeball can be thought of as a sphere separated by the lens into two compartments (see Figure 9-6). The anterior compartment is filled with a watery fluid, the **aqueous humor,** which helps maintain pressure within the eye, refracts (bends) light, and provides nutrients to the inner surface of the eye. Aqueous humor is produced by the ciliary body from blood and is returned to the blood through a vein that encircles the cornea near the junction of the cornea and the sclera. Pressure within the eye resulting from the presence of aqueous humor keeps the eye inflated much like the air in a basketball. If circulation of the aqueous humor is blocked so that pressure in the eye increases, a defect called **glaucoma** (glaw-ko´mah) can result (see box, p. 152).

The posterior compartment of the eye is filled with a transparent jellylike substance, the **vitreous** (vit´re-us) **humor.** The vitreous humor helps maintain pressure within the eye, holds the lens and the retina in place, and functions to refract the light. Unlike aqueous humor, the vitreous humor does not circulate.

Neuronal pathways

Light striking rods and cones results in action potentials that are transmitted by association neurons in the retina toward the brain. The axons of the association neurons come together and exit the eye to form the **optic nerve** (cranial nerve I). The part of the retina where the optic nerve leaves the eye is called the **optic disc.** Because the optic disc consists of axons, but not rods or cones, light striking the optic disc does not generate action potentials. It is therefore called the **blind spot** of the eye. Even though there is a blind spot on the retina, there is no blank space in our vision because the brain fills in the missing space by using information from the retinal area immediately surrounding the blind spot.

The optic nerve leaves the eye and extends to the brain. Interpretation of the action potentials generated in the retina takes place in the visual areas in the occipital lobes of the brain.

Eye disorders

INFECTIONS

Neonatal gonorrheal ophthalmia (of-thal´me-ah) is a severe form of conjunctivitis that is contracted by an infant as it passes through the birth canal of a mother with gonorrhea (a bacterial infection). This infection carries a high risk of blindness. The treatment of newborn eyes with silver nitrate is effective in preventing the disease. **Chlamydial conjunctivitis** is contracted as an infant passes through the birth canal of a mother with a chlamydial (a type of bacteria) infection. This infection is not affected by silver nitrate, so in many places, newborns are treated with antibiotics for both chlamydial and gonorrheal infections. **Trachoma** (tra-ko´mah) is the greatest single cause of blindness in the world today. The disease, also caused by chlamydia, is transmitted by hand contact, flies, or objects such as towels. It is a conjunctivitis that leads to scarring of the cornea and blindness. It is most common in arid parts of Africa and Asia.

DEFECTS OF FOCUS

When a person's vision is tested, a chart is placed 20 feet from the eye, and the person is asked to read a line that has been standardized for normal vision. If the person can read the line, he has 20/20 vision, which means that he can see at 20 feet what people with normal vision see at 20 feet. If, on the other hand, the person only can read the letters at 20 feet that people with normal vision see at 40 feet, the person's eyesight is 20/40. Such a person would have myopia.

Myopia (mi-o´pe-ah), or **nearsightedness,** is the ability to see close objects but not distant ones. It is a defect of the eye in which the image is focused in front of the retina (Figure 9-B, *1*). Myopia can occur if the eyeball is too long or if the lens bends the light too much. It is corrected by a concave lens that spreads out the light rays approaching the eye so that when the light is focused by the eye, it is focused on the retina (Figure 9-B, *2*).

Hyperopia (hi´per-o´pe-ah), or **farsightedness,** is a disorder in which a person can clearly see far away objects, but not close-up objects. In hyperopia the image is focused "behind" the retina (Figure 9-B, *3*). It can occur if the eyeball is too short or if the lens does not bend the light enough. Hyperopia is corrected by a convex lens that causes light rays to converge as they approach the eye, so that when the light is focused by the eye, it is focused on the retina (Figure 9-B, *4*).

Presbyopia (prez-be-o´pe-ah) is the decrease in the ability of the eye to accommodate for near vision; this occurs as a normal part of aging. With age, the lens becomes less flexible. The average age of onset of presbyopia is the mid-forties. Presbyopia can be corrected by the use of "reading glasses" or by bifocals. Bifocals have different lenses in the top and bottom of the glasses. The bottom half is more convex to accommodate for near vision when the person reads, and the top half is less convex for distant vision.

Astigmatism (a-stig´mă-tizm) is a defect in which the cornea or lens is not uniformly curved, and the image is not sharply focused. Glasses may be made to adjust for the abnormal curvature as long as the curvature is not too irregular. If the curvature of the cornea or lens is too irregular, the condition is difficult to correct.

OTHER DISORDERS

Cataracts (kat´ă-raktz) are the most common cause of blindness in the United States. It is a condition in which clouding of the lens occurs as the result of advancing age, infection, or trauma. Excess exposure to ultraviolet radiation may be a factor in causing cataracts, so the wearing of sunglasses in bright sunshine is recommended. A certain amount of clouding occurs in 95% of people over age 65. Surgery to remove a cataract is actually the removal of the lens. Because the lens normally focuses light on the retina, after the lens is removed an artificial lens implant or glasses are necessary to restore normal vision. More than 400,000 cataracts operations are performed in the United States each year.

Glaucoma (glaw-ko´mah) is a condition involving excessive pressure due to a buildup of aqueous humor within the eye. Glaucoma results from an interference with normal reentry of aqueous humor into the blood stream, or an overproduction of aqueous humor. The increased pressure within the eye can close off the blood vessels entering the eye and may destroy the retina or optic nerve, resulting in blindness. Drugs or surgery can be used to reduce the pressure.

Diabetes is a major cause of blindness in the United States. Diabetes can result in optic nerve degeneration, cataracts, and detachment of the retina. These defects are often caused by blood vessel degeneration and hemorrhage, which are common in diabetic patients.

Retinal detachment is the separation of the photoreceptor layer of the retina from the layer of pigmented cells (see Figure 9-7). If a hole or tear occurs in the retina, fluid can accumulate within the retina. As a result, neurons can degenerate because they are separated from their normal blood supply. Permanent damage often can be prevented by reattaching the retina surgically or with lasers.

Damage to the cornea from injury or disease (for example, trachoma) can result in scarring of the cornea that inhibits light entry into the eye. A **corneal transplant** is the removal of the damaged cornea and its replacement with the cornea of a donor.

Color blindness is the absence of perception of one or more colors because the cone cells do not function properly. There may be a complete loss of color perception or only a decrease in perceptive ability. Most forms of color blindness occur more frequently in males, because they are inherited as sex-linked traits (see Chapter 20).

Functions of the complete eye

An important characteristic of light is that it can be refracted or bent by passing through a lens. A lens is **convex** if its surface bulges outward, and it is **concave** if its surface is depressed inward. As light passes through a convex lens the light rays bend inward or converge, whereas the light rays passing through a concave lens bend outward or diverge. When light rays converge they are focused. To see clearly, an image (light rays) must be correctly focused onto the retina.

The eye functions much like a camera. Light passes through the pupil into the eye, and the light is focused primarily by the cornea and lens onto the retina. The shape of the cornea and its distance from the retina are fixed, however, so that no adjustment in focus can be made by the cornea. Fine adjustments in focus are accomplished by changing the shape of the lens. When the smooth muscles of the ciliary body are relaxed, the elastic suspensory ligaments pull on the lens. Consequently, the lens is relatively flat, allowing for distant vision (Figure 9-8, *A*). When an object is brought closer than 20 feet to the eye, the smooth muscles of the ciliary body contract. This reduces the tension on the suspensory ligaments of the lens and allows the lens to assume a more spherical form because of its own internal elastic nature (Figure 9-8, *B*). The spherical lens then has a more convex surface, causing greater refraction of light. This process is called **accommodation,** and it enables the eye to focus objects closer than 20 feet onto the retina.

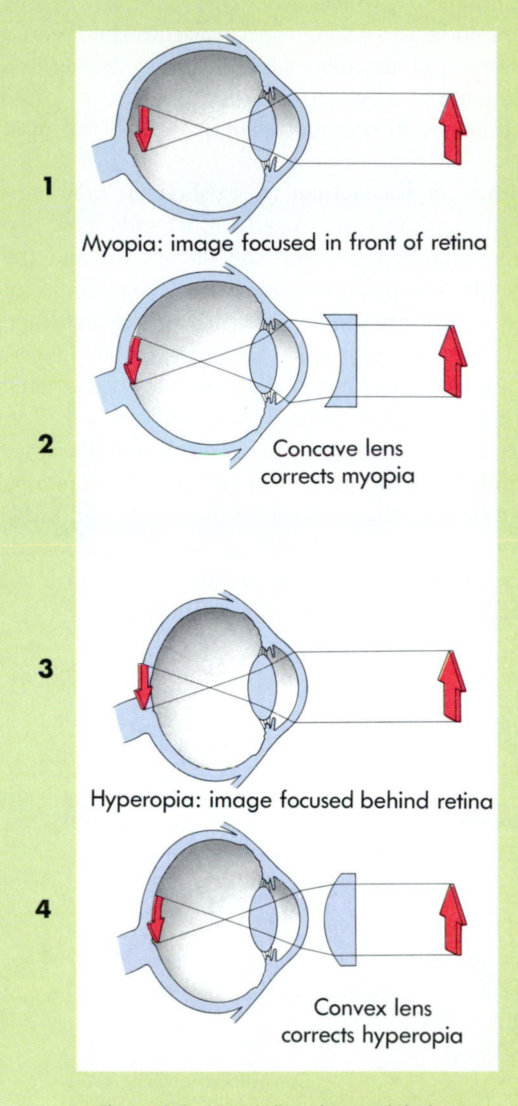

Figure 9-B Visual disorders and their correction by various lenses.
1, Myopia (nearsightedness). *2,* Correction of myopia with a concave lens. *3,* Hyperopia (farsightedness). *4,* Correction of hyperopia with a convex lens.

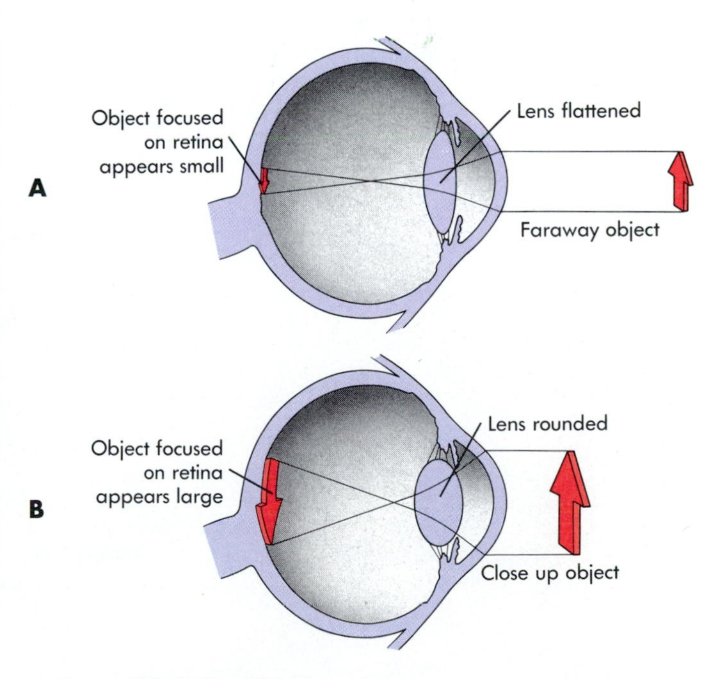

Figure 9-8 ● Focusing.
Sagittal section of the right eye. **A,** Distant image: The lens is flattened, and the image is focused on the retina. **B,** Near vision (accommodation): The lens is more rounded, and the image is focused on the retina.

HEARING AND BALANCE

The organs of hearing (auditory or acoustic organs) and balance or equilibrium can be divided into three portions: external ear, middle ear, and inner ear (Figure 9-9). The external and middle ears are involved in hearing only, whereas the inner ear functions in both hearing and balance.

Auditory structures and their functions

External ear

The **auricle** (aw´rĭ-kl) or **pinna** (pin´ah) is the fleshy part of the external ear on the outside of the head, which most people think of as the ear. The auricle opens into the **external auditory meatus** (me-a´tus), a passageway that leads to the eardrum. The meatus is lined with hairs and glands that produce **cerumen** (sĕ-roo´men) or **earwax.** The hairs and cerumen help prevent foreign objects from reaching the delicate eardrum. The **tympanic** (tim-pan´ik) **membrane** or **eardrum** is a thin membrane that separates the external ear from the middle ear. Sound waves collected by the auricle move down the external auditory meatus and cause the tympanic membrane to vibrate.

Middle ear

Medial to the tympanic membrane is the air-filled cavity of the middle ear. The **oval window** and the **round window** connect the middle ear to the inner ear. Another opening connects the middle ear to small spaces within the mastoid process—the mastoid air cells (see Chapter 6). Sometimes middle ear infections pass through this opening to cause **mastoiditis** (mas-toy-di´tis), an inflammation in the mastoid air cells.

The **auditory,** or **Eustachian** (u-sta´she-an), **tube** connects the middle ear to the pharynx (throat) and enables air pressure to be equalized between the outside air and the middle ear cavity. Unequal pressure between the middle ear and the outside environment can distort the eardrum, dampen its vibrations, and make hearing difficult. Distortion of the eardrum also stimulates pain fibers associated with that structure. That distortion is why, as a person changes altitude, sounds seem muffled and the eardrum may become painful. These symptoms can be relieved by opening the auditory tube, allowing air to enter the middle ear. Swallowing, yawning, chewing, and holding the nose and mouth shut while gently trying to force air out of the lungs are methods that can be used to open the auditory tube.

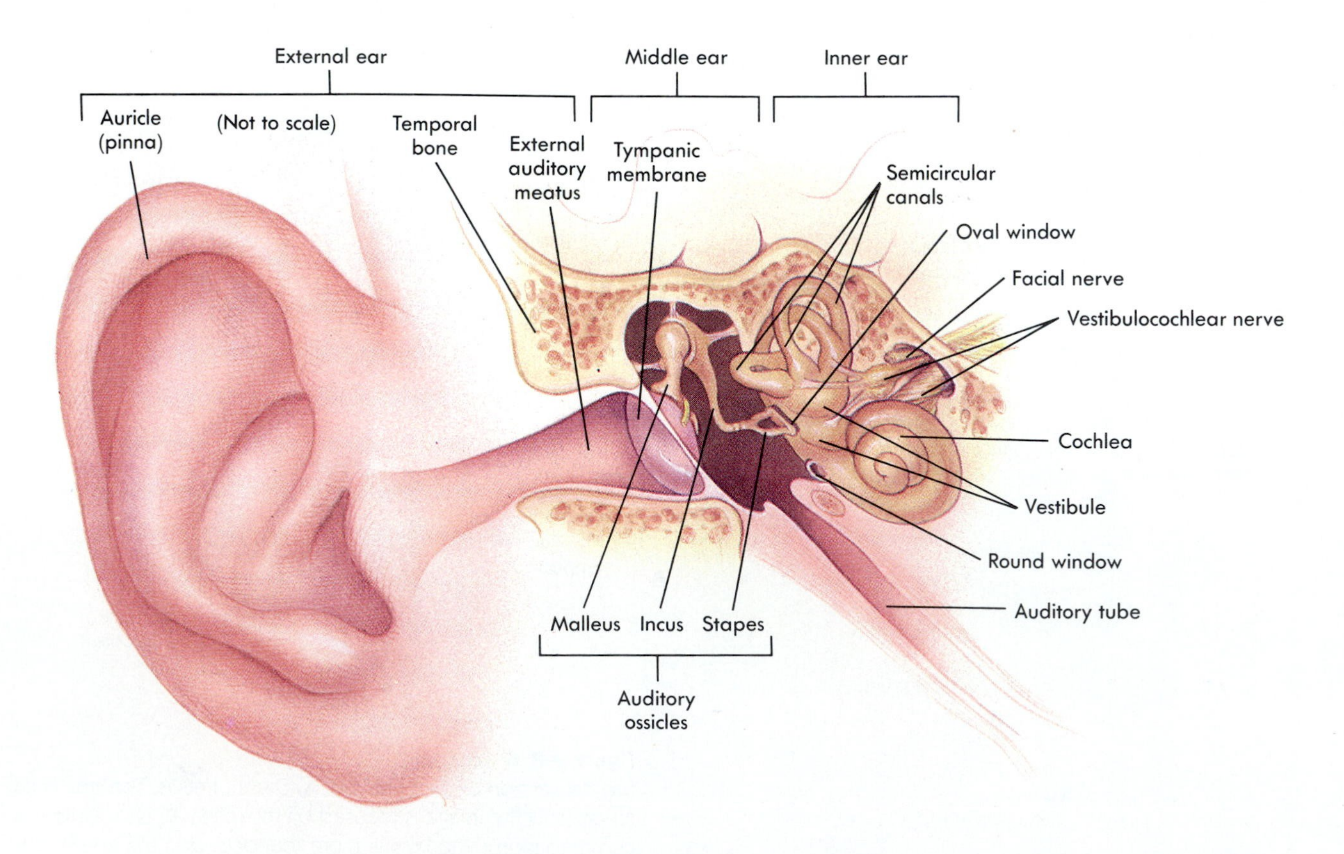

Figure 9-9 ● External, middle, and inner ears.

The middle ear contains three ear bones called **auditory ossicles** (os´ĭ-klz): the **malleus** (mal´e-us; hammer), **incus** (ing´kus; anvil), and **stapes** (sta´pez; stirrup). The auditory ossicles are joined together and transmit vibrations from the eardrum to the oval window.

Inner ear

The inner ear consists of interconnecting tunnels and chambers within the temporal bone. They can be divided into three regions: the cochlea, vestibule, and semicircular canals. The cochlea is involved in hearing, and the vestibule and semicircular canals are involved primarily in balance.

Hearing

The **cochlea** (kok´le-ah) is shaped like a snail shell (see Figure 9-9). Membranes divide the cochlea into three fluid-filled canals (Figure 9-10). Inside the middle canal is a specialized structure called the **organ of Corti** (kor´te). The organ of Corti contains specialized sensory cells called hair cells. Projections of the hair cells' membranes are called **hairs,** and the tips of the hairs are attached to a rigid gelatinous shelf.

Sound waves are collected by the auricle and are conducted through the external auditory meatus toward the tympanic membrane (see Figure 9-10). Sound waves strike

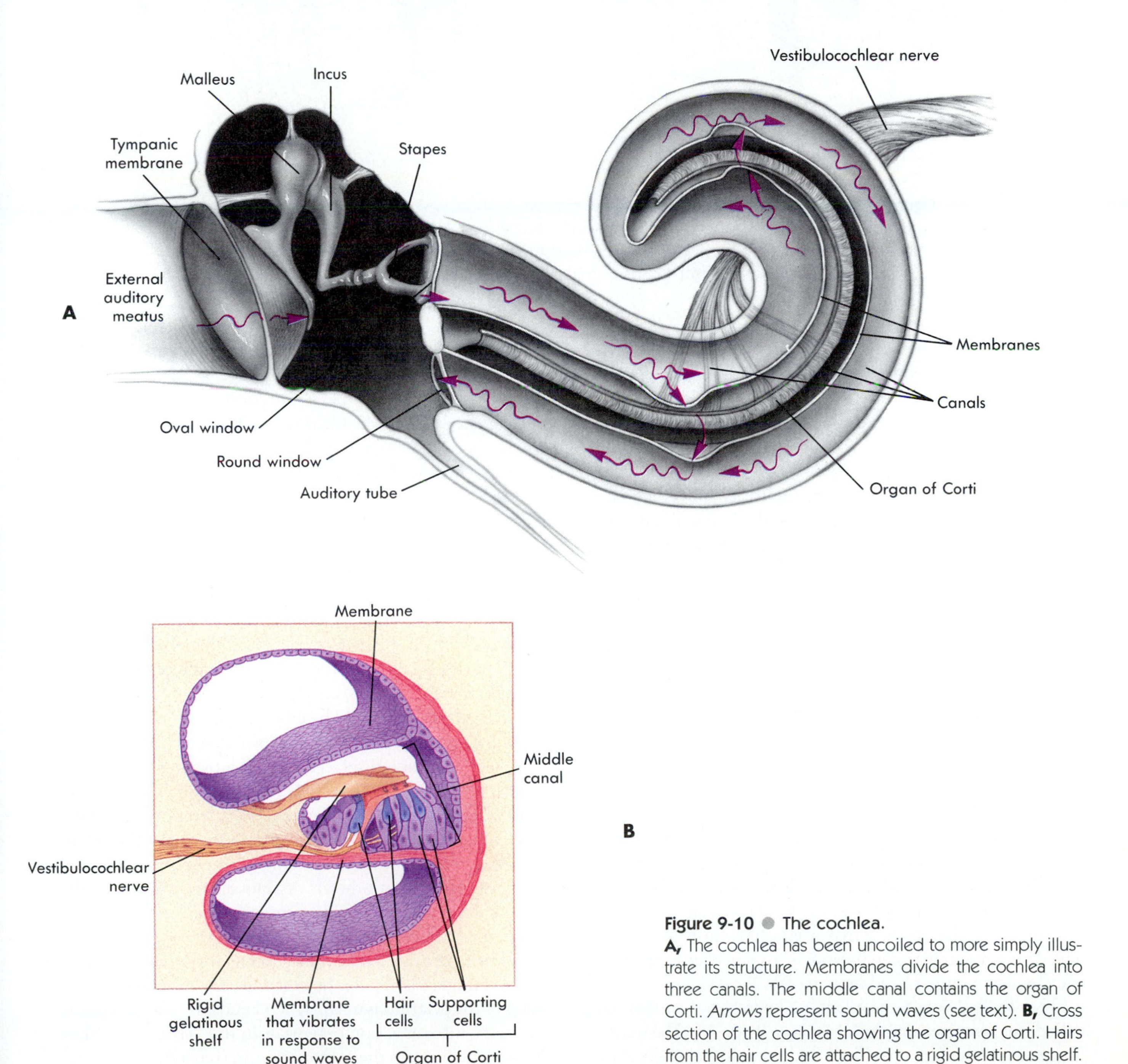

Figure 9-10 ● **The cochlea.**
A, The cochlea has been uncoiled to more simply illustrate its structure. Membranes divide the cochlea into three canals. The middle canal contains the organ of Corti. *Arrows* represent sound waves (see text). **B,** Cross section of the cochlea showing the organ of Corti. Hairs from the hair cells are attached to a rigid gelatinous shelf.

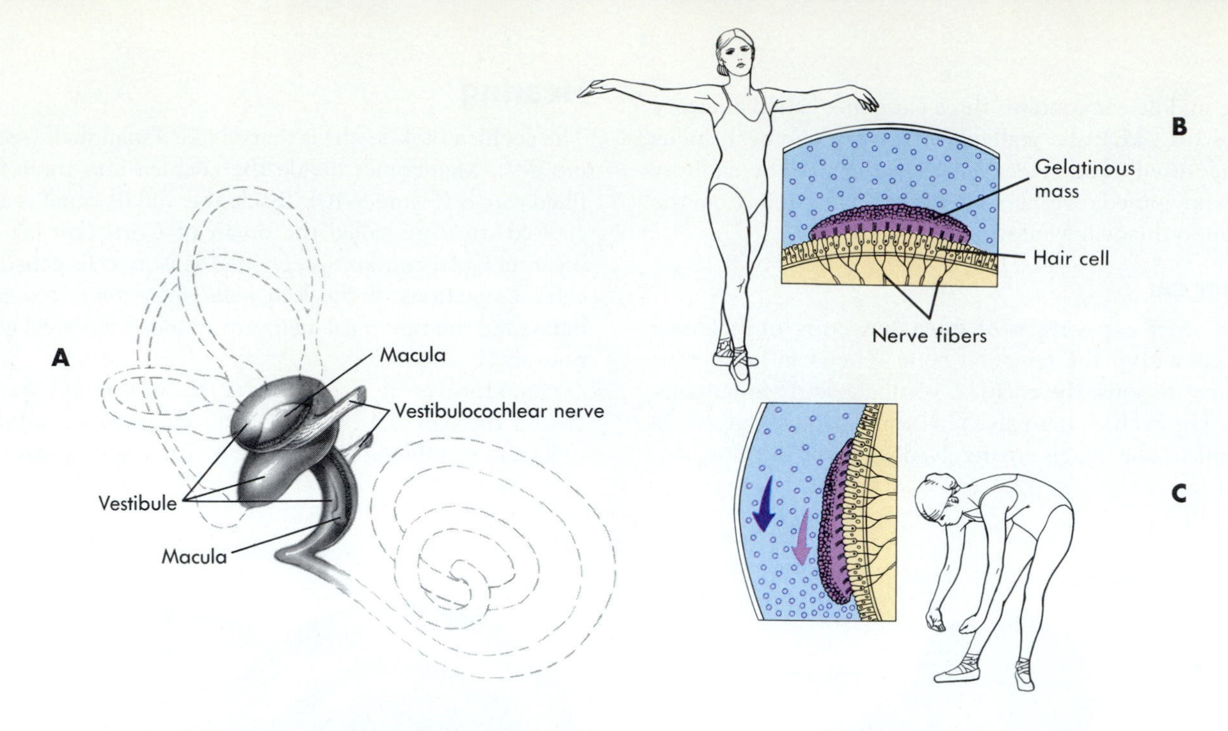

Figure 9-11 ● **Structure and function of the macula.**
A, The location of the maculae within the vestibule. **B,** The macula consists of hair cells with hairs embedded in a gelatinous mass. **C,** The macula responds to changes in position of the head relative to gravity, for example, as a person bends over from a vertical position the macula is displaced by the pull of gravity.

the tympanic membrane and cause it to vibrate. Vibrations of the tympanic membrane, in turn, cause the three ossicles of the middle ear to vibrate. By this mechanical linkage, vibration is transferred to the oval window. Movement of the stapes in the oval window produces vibrations in the fluid and membranes of the inner ear. The vibrations cause slight movement of the organ of Corti, resulting in bending and stretching of the hairs. Consequently, the hair cells produce action potentials that travel out the **vestibulocochlear** (ves-tib´u-lo-kok´le-ar) nerve to the brain. Auditory nerve tracts in the brain pass through the thalamus to the auditory area of the temporal lobe, in which the action potentials are interpreted as sounds (see Chapter 8).

The vibrations produced in the fluid of the cochlea eventually reach the round window, in which the vibrations are absorbed (see Figure 9-10). The vibrations cause the membrane of the round window to bulge outward into the middle ear.

Equilibrium

The sense of equilibrium, or balance, has two components: static equilibrium and kinetic equilibrium. **Static equilibrium** is associated with the **vestibule** (ves´tĭ-būl) and is involved in evaluating the position of the head relative to the surface of the earth. **Kinetic equilibrium** is associated with the semicircular canals and is involved in evaluating the change in rate of head movements.

The vestibule contains specialized patches of epithelium called the **maculae** (mak´u-le) (Figure 9-11, *A*). The maculae resemble the organ of Corti and contain hair cells. The hairs of these cells are embedded into a gelatinous mass that floats in a fluid and moves in response to gravity, bending the hair cells (Figure 9-11, *B*). Action potentials are initiated in the vestibulocochlear nerve and travel to the brain where they are interpreted. When a person bends over, the maculae are displaced by gravity and the resultant action potentials provide information to the brain concerning the position of the head relative to the surface of the earth.

There are three **semicircular canals,** which are involved in kinetic equilibrium and placed at nearly right angles to each other. The placement of the semicircular canals enables a person to detect movements of the head in almost any direction. At the base of each semicircular canal the epithelium is specialized to form a **crista ampullaris** (kris´tah am-pul-lar´is) (Figure 9-12, *A*), which has hair cells embedded in a gelatinous mass. The crista ampullaris is structurally and functionally similar to the maculae. The gelatinous mass is a float that is displaced by fluid movement within the semicircular canals (Figure 9-12, *B* to *D*). As the head begins to move in a given direction (acceleration), the fluid tends to remain stationary. This difference causes the gelatinous mass to be displaced in a direction opposite to that of the movement of the head. As movement continues the fluid "catches up." When movement of the head stops (deceleration), the fluid tends to continue to move. Movement of the head relative to the fluid (acceleration) or movement of the fluid relative to the head (deceleration) causes movement of the gelatinous mass, which bends the hairs. Bending the hairs initiates action potentials in the vestibulocochlear nerve, which relays the information to the brain.

Ear disorders

DEAFNESS

Deafness is a general term for loss of the ability to hear. There are two major categories of deafness: conduction and sensorineural deafness. **Conduction deafness** involves a mechanical deficiency in transmission of sound waves from the outer ear to the organ of Corti. Hearing aids may help people with such hearing deficiencies by boosting the sound volume reaching the ear. Some conduction deafness can be corrected surgically. For example, **otosclerosis** (o´to-skle-ro´sis) is an ear disorder in which bone grows over the oval window and immobilizes the stapes. This disorder can be surgically corrected by breaking away the bony growth and the stapes. The stapes is replaced by a small metal rod connected to the oval window at one end and to the incus at the other end.

Sensorineural deafness results from damage to the organ of Corti or nerve pathways. For example, exposure to sudden loud noises such as gunshots, or continual exposure to the noises of machinery, computers, or loud music, can result in neural damage. Sensorineural deafness is more difficult to correct than conduction deafness. Research is being conducted on ways to replace the hearing pathways with electrical circuits. One approach involves the direct stimulation of the vestibulocochlear nerve by electrical impulses. The mechanism consists of a microphone for picking up sound waves, a microelectronic processor for converting the sound into electrical signals, a transmission system for relaying the signals to the inner ear, and a long, slender electrode that is threaded into the cochlea. This electrode delivers electrical signals directly to the vestibulocochlear nerve.

EAR INFECTIONS

Otitis (o-ti´tis) **media,** infections of the middle ear, are quite common in young children. These infections usually result from the spread of infection from the mucous membrane of the pharynx (throat) through the auditory tube. The symptoms of low-grade fever, lethargy, and irritability often are not recognized by the parent as signs of middle ear infection. The infection also can cause a temporary decrease or loss of hearing because fluid buildup can dampen the eardrum. In extreme cases, the infection can damage or rupture the eardrum.

MOTION SICKNESS

Motion sickness consists of nausea and weakness caused by stimulation of the semicircular canals during motion (for example, in a boat, automobile, or airplane).

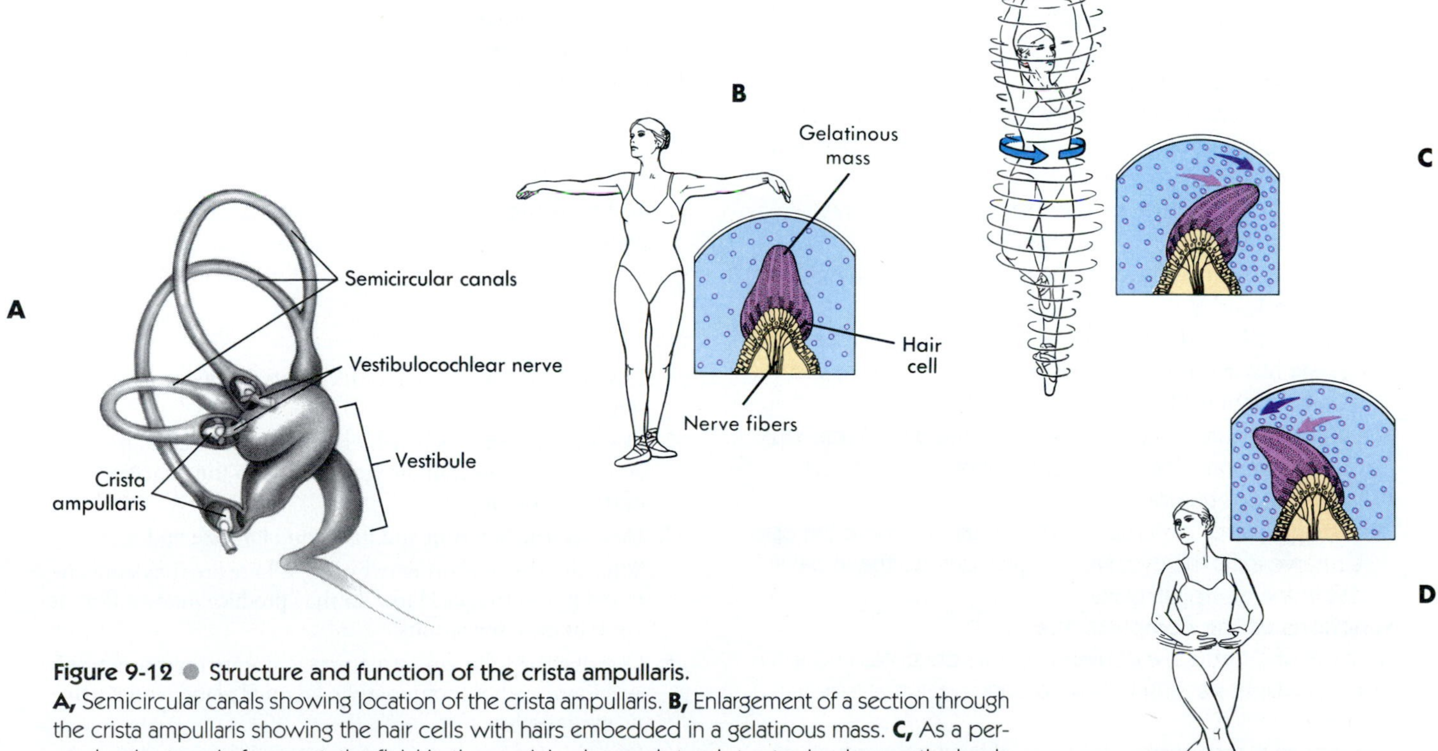

Figure 9-12 ● **Structure and function of the crista ampullaris.**
A, Semicircular canals showing location of the crista ampullaris. **B,** Enlargement of a section through the crista ampullaris showing the hair cells with hairs embedded in a gelatinous mass. **C,** As a person begins to spin from rest, the fluid in the semicircular canals tends to stay in place as the body and the crista ampullaris begin to move. As a result, the crista ampullaris is displaced in a direction opposite to the direction of spin. **D,** When the person stops spinning, the fluid continues to move and displaces the crista ampullaris in the direction of the spin.

SUMMARY

GENERAL SENSES

Receptors for general senses such as pain, temperature, touch, pressure, and position are scattered throughout the body.

SPECIAL SENSES

OLFACTION

Olfactory neurons have receptors that respond to dissolved substances.

Axons of the olfactory neurons enter the olfactory bulb. Association neurons carry action potentials to the olfactory areas of the brain.

The wide range of detectable odors results from combinations of a few primary odors.

TASTE

Taste buds contain cells with receptors that detect dissolved substances.

There are four basic types of taste: sour, salty, bitter, and sweet.

VISION

Accessory structures

The eyebrows prevent perspiration from entering the eyes.

The eyelids protect the eyes from foreign objects.

The conjunctiva covers the inner eyelid and the anterior surface of the eye.

Lacrimal glands produce tears that flow across the surface of the eye. Tears lubricate, clean, and protect the eye. Excess tears pass into the nasal cavity.

The extrinsic eye muscles move the eyeball.

Anatomy of the eye

The outer layer of the eye consists of the sclera and cornea.

The middle layer of the eye consists of the choroid, ciliary body, and iris.

The lens is held in place by the suspensory ligaments, which are attached to the ciliary body.

The retina is the inner layer of the eye and contains neurons sensitive to light.

Rods are responsible for vision in low light (night vision).

Cones are responsible for color vision, sharp vision, and vision in bright light.

The anterior compartment of the eye is filled with aqueous humor, whereas the posterior compartment is filled with vitreous humor.

The optic disc, or blind spot, is the location where the optic nerve exits the eye. Nerve fibers pass to the visual areas in the occipital lobes.

Functions of the complete eye

Light passing through a convex surface converges and is focused. Light passing through a concave surface diverges.

The cornea is responsible for most of the convergence of light in the eye, whereas the lens can adjust the focus by changing shape (accommodation).

HEARING AND BALANCE

Auditory structures and their functions

The external ear consists of the auricle and external auditory meatus, which ends at the eardrum.

The ossicles connect the eardrum to the oval window of the inner ear.

The auditory, or Eustachian, tube connects the middle ear to the pharynx and functions to equalize pressure. The middle ear is also connected to the mastoid air cells.

The inner ear has three parts: the semicircular canals, vestibule, and cochlea.

Hearing

The cochlea is a snail-shaped canal divided into three compartments by membranes.

The organ of Corti consists of hair cells.

Sound waves are funnelled by the auricle down the external auditory meatus, causing the eardrum to vibrate.

The eardrum vibrations are passed by the ossicles to the oval window, causing vibrations of the fluids and membranes of the inner ear.

Movement of the organ of Corti causes movement of the hair cells and generation of action potentials that travel along the vestibulocochlear nerve to the auditory areas of the brain.

Equilibrium

Static equilibrium evaluates the position of the head relative to gravity.

Maculae, located in the vestibule, consist of hair cells with the hairs embedded in a gelatinous mass. The gelatinous mass moves in response to gravity.

Kinetic equilibrium evaluates movements of the head.

There are three semicircular canals in the inner ear, arranged at right angles to each other.

The base of each semicircular canal contains a crista ampullaris, which has hair cells with hairs embedded in a gelatinous mass that responds to movements of the head.

CONTENT REVIEW

1. Describe the distribution of receptors for the general senses.
2. Describe the process by which airborne molecules initiate action potentials that are transmitted to the olfactory area of the cerebrum.
3. Describe the location and function of a taste bud.
4. What are the four primary tastes? Where are they concentrated in the tongue? How do they produce many different kinds of taste sensations?
5. Describe the following structures and state their functions: eyebrows, eyelids, conjunctiva, lacrimal gland, and extrinsic eye muscles.
6. Describe the structures composing each layer of the eye, and explain the functions of these structures.

7. Describe the lens of the eye, how the lens is held in place, and how the shape of the lens is changed.
8. What stimulus causes the pupils to constrict and dilate?
9. Name the two compartments of the eye, the substances that fill each compartment, and the functions of the substances.
10. What is the blind spot of the eye, and what causes it?
11. What kind of lens, convex or concave, causes light to be focused?
12. Define accommodation. What does accommodation accomplish?
13. Name the three regions of the ear, name the structures found in each region, and state the functions of the structures.
14. Describe how sound waves entering the auricle result in action potentials that are transmitted to the auditory area of the cerebrum.
15. Describe the location and function of the maculae.
16. What is the function of the semicircular canals? Describe the crista ampullaris and its mode of operation.

CONCEPT REVIEW

1. If the lateral rectus muscle of the right eye contracted (see Figure 9-5), in what direction would a person be looking?
2. Would a person with uncorrected myopia or uncorrected hyperopia be most likely to suffer from eyestrain (fatigue of the ciliary body) while reading a book? Explain.
3. An elderly man with normal vision developed cataracts. He was surgically treated by removing the lenses of his eyes. If he did not receive a lens implant, what kind of glasses would you recommend he wear to compensate for the removal of his lenses? Would his near or far vision be most affected by the removal of his lens?
4. Ima Diver takes a deep breath and dives toward the bottom of the ocean. The deeper she dives, the greater the water pressure around her becomes. It is possible that this pressure can lead to ear damage and loss of hearing. Describe the normal mechanisms that adjust for changes in pressure and explain how the increased pressure might cause loss of hearing.
5. If a vibrating tuning fork is placed against the mastoid process of the temporal bone, the vibrations will be perceived as sound, even if the external auditory meatus is plugged. Explain how this could happen.

CHAPTER TEST

Answers can be found in Appendix C

MATCHING 1 For each statement in column A select the correct answer in column B (an answer may be used once, more than once, or not at all).

A	B
1. Covers the inner eyelids and surface of the eye.	anterior compartment
2. Produces tears.	ciliary body
3. The firm outer layer of the eye.	cones
4. Transparent structure that allows light into the eye.	conjunctiva
5. Regulates the amount of light entering the eye.	cornea
6. Changes the shape of the lens.	iris
7. Responsible for color vision.	lacrimal gland
8. Contains aqueous humor.	posterior compartment
	rods
	sclera

MATCHING 2 For each statement in column A select the correct answer in column B (an answer may be used once, more than once, or not at all).

A	B
1. Located between the external ear and middle ear.	auditory ossicles
2. Transmits vibrations from the tympanic membrane to the oval window.	auditory tube
3. Equalizes pressure between the external ear and middle ear.	crista ampullaris
4. Structure that detects sound vibrations and produces action potentials.	macula
5. Absorbs vibrations in the inner ear.	organ of Corti
6. Structure that detects kinetic equilibrium.	oval window
	round window
	tympanic membrane

FILL-IN-THE BLANK Complete each statement by providing the missing word or words.

1. The __________ connect to the olfactory bulb. Axons from the olfactory bulb form the __________, which goes to the olfactory area of the temporal lobe.
2. The four basic types of taste are __________, __________, __________, and __________.
3. The __________ carry action potentials from the eye toward the brain. Interpretation of the action potentials takes place in the __________ of the occipital lobes of the brain.
4. The three auditory ossicles are the __________, __________, and the __________.
5. The structures of the vestibule involved with static equilibrium are the __________.

KEY TERMS

endocrine
Ductless gland that secretes internally, usually into the blood.

hormone
A substance secreted by endocrine tissues into the blood that acts on a target tissue to produce a specific response.

hypothalamic-pituitary portal system
A series of blood vessels that carry blood from the hypothalamus to the anterior pituitary.

hypothalamus
Important autonomic and endocrine control center of the brain located beneath the thalamus.

pituitary or hypophysis
An endocrine gland attached to the hypothalamus; secretes hormones that influence the function of several other endocrine glands and tissues.

receptor
A molecule in the cell of a target tissue to which a hormone binds. The binding of a hormone with its receptor initiates a response in the target tissue.

releasing hormone
Hormone that is released from neurons of the hypothalamus and flows through the hypothalamic-pituitary portal system to the anterior pituitary; regulates the secretion of hormones from the anterior pituitary.

target tissue
Tissue upon which a hormone acts.

FEATURES

CHAPTER 10

The Endocrine System

OBJECTIVES

After reading this chapter you should be able to:

1. Compare the means by which the nervous and endocrine systems regulate body functions.
2. Define the term hormone and describe how hormones interact with tissues to produce a response.
3. Describe how hormones maintain homeostasis.
4. State the location of each of the endocrine glands in the body.
5. List the hormones produced by each of the endocrine glands and describe their effects on the body.
6. Describe how the hypothalamus regulates hormone secretion from the pituitary.
7. Describe how the pituitary regulates the secretion of hormones from other endocrine glands.
8. Choose a hormone and use it to explain how negative feedback results in homeostasis.

THE ENDOCRINE SYSTEM and the nervous system are the two major regulatory systems in the body. Together they regulate and coordinate nearly all other body structures, yet they differ in several ways. The nervous system controls structures by sending action potentials along axons, whereas the endocrine system acts by sending hormones through the circulatory system. The nervous system usually acts more quickly and has short-term effects, whereas the endocrine system usually responds more slowly and has longer-lasting effects. In general, nervous stimuli control specific tissues or organs whereas the endocrine system has a more general effect on the body.

In this chapter the endocrine glands, their hormones, and the control of endocrine gland functions are described. To emphasize the importance of the endocrine system and to demonstrate its functions, some examples of endocrine system disorders are included.

THE ENDOCRINE SYSTEM

The **endocrine** (en´do-krin) **system** is made up of all the body's endocrine glands (Figure 10-1). **Endocrine glands** are glands without ducts that secrete chemicals into the blood. Examples of endocrine glands are the thyroid gland and the adrenal glands. In contrast, **exocrine** (ek´so-krin) **glands** secrete their products into ducts, which then carry the secretory products to an external or internal surface, such as the skin or digestive tract. Examples of exocrine glands are the sweat glands and the salivary glands.

Chemicals secreted by endocrine glands are called **hormones** (hor´monz), a term derived from the Greek word **hormon,** meaning to set into motion. Hormones are proteins and related substances, such as amino acids, and lipids such as steroids. Hormones act on tissues to either increase or decrease their activity. The hormones are distributed in the blood to all parts of the body, but only certain **target tissues** respond to each type of hormone. A target tissue for a hormone is made up of cells that have **receptors** for that hormone. Each hormone can bind only to its receptors and cannot influence cells that do not have receptors for it.

The secretion of hormones is controlled by negative-feedback mechanisms (see Chapter 1). Negative-feedback mechanisms keep the body functioning within a narrow range of values consistent with life. For example, insulin is a hormone that regulates the concentration of blood glucose, or blood sugar. When blood glucose levels increase after a meal, insulin is secreted. Insulin causes most body cells to take up glucose, resulting in a decrease in blood glucose levels. As the blood glucose levels decline, however, the rate of insulin secretion also decreases. As insulin levels decrease, the rate at which glucose is taken up by the tissues decreases, keeping the blood glucose levels from declining too much. This negative-feedback mechanism counteracts large increases and decreases in blood glucose and maintains blood

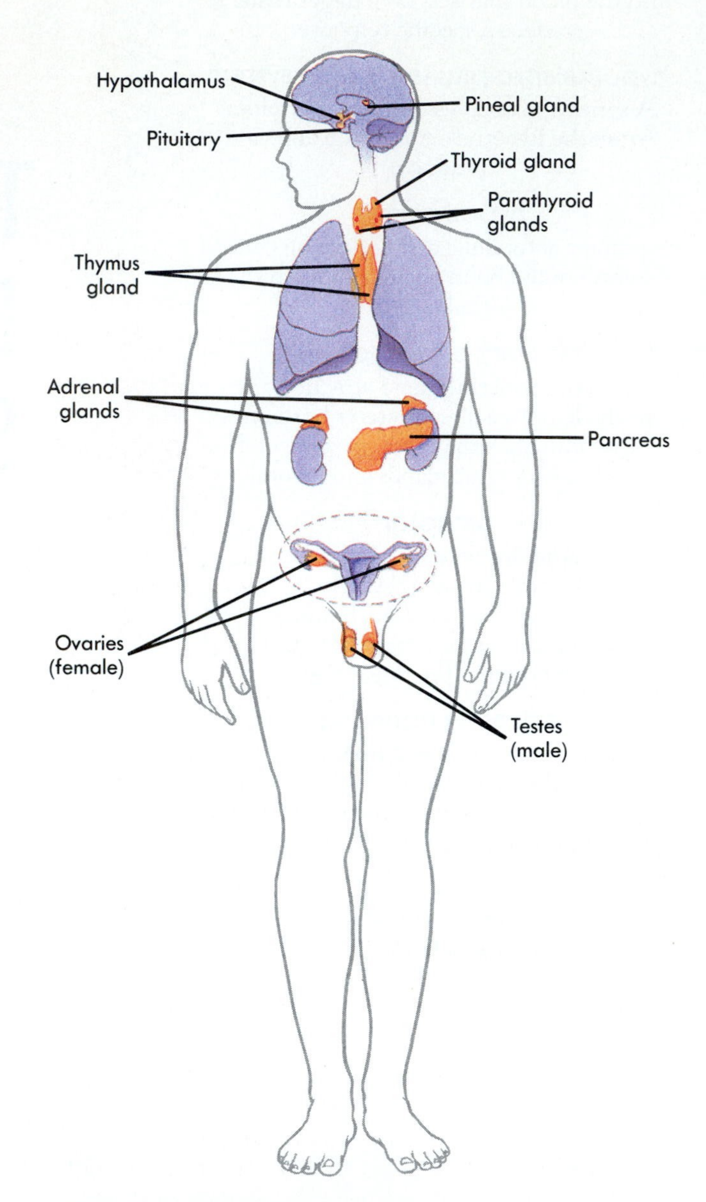

Figure 10-1 ● **The endocrine glands.**
The endocrine glands and their locations in the human body.

glucose levels within a narrow range of values consistent with homeostasis.

1 What would happen to blood glucose levels following a meal if insulin was not secreted?*

ENDOCRINE GLANDS AND THEIR HORMONES

The pituitary and hypothalamus

The **pituitary** (pit-u´ĭ-tĕr-e), or the **hypophysis** (hi-pof´ĭ-sis), is a small gland about the size of a pea (Figure 10-2). It rests in a depression of the sphenoid bone below the **hypothalamus** of the brain. The pituitary is connected to the hypothalamus by a stalk called the **infundibulum** (in-fun-dib´u-lum). The gland is divided into two parts: the **anterior pituitary** and the **posterior pituitary.** The hormones secreted from each part of the pituitary are listed in Table 10-1.

Hormones from the pituitary gland control the function of many other glands in the body, such as the ovaries, testes, thyroid gland, and adrenal glands. The pituitary gland also secretes hormones that directly influence growth, kidney function, delivery of infants, and milk production and release by the breasts.

The pituitary gland is controlled in two ways by the hypothalamus of the brain:

1. The secretion of hormones from the anterior pituitary is controlled by **releasing hormones** produced by the hypothalamus (see Figure 10-2, A). The releasing hormones enter a capillary bed in the hypothalamus and are transported through veins to a second capillary bed in the anterior pituitary. There the releasing hormones leave the blood and regulate hormone secretion. The capillary beds and veins that transport the hormones are called the **hypothalamic-pituitary portal system.**
2. Secretion of hormones from the posterior pituitary is controlled by nervous system stimulation of nerve cells within the hypothalamus (see Figure 10-2, B). These nerve cells have their cell bodies in the hypothalamus and their axons extend to the posterior pituitary. When these nerve cells are stimulated, action potentials from the hypothalamus travel along the axons and cause the release of hormones from the axon nerve endings in the posterior pituitary.

Within the hypothalamus and pituitary, the nervous and endocrine systems are closely interrelated. Emotions such as joy and anger, as well as chronic stress, influence the endocrine system through the hypothalamus. Also, hormones of the endocrine system can influence the functions of the hypothalamus and other parts of the brain.

*Answers to predict questions appear in Appendix B at the back of the book.

Hormones of the anterior pituitary

Growth hormone (GH) stimulates the growth of bones, muscles, and other organs by increasing protein synthesis (see Figure 10-2, A and Table 10-1). It also resists protein breakdown and favors fat breakdown during periods of food deprivation, such as between the evening meal and breakfast or during longer periods of fasting. The rate of growth hormone secretion increases when blood glucose levels are low. Growth hormone causes cells to use less glucose and to increase their use of fat as an energy source.

Growth hormone levels in the blood do not become greatly elevated during periods of rapid growth, but children have somewhat higher blood levels of growth hormone than adults. Other factors such as genetics, nutrition, and sex hormones also influence growth.

A young person suffering from a deficiency of growth hormone will remain a small, though normally proportioned, person called a **pituitary dwarf.** This condition can be treated by administering growth hormone. If excess growth hormone is present before bones complete their growth in length, exaggerated bone growth occurs. The result is **giantism,** and the person becomes abnormally tall. If excess hormone is secreted after growth in bone length is complete, growth in bone diameter continues. As a result, the facial features and hands become abnormally large, a condition called **acromegaly** (ak´ro-meg´al-e).

Thyroid stimulating hormone (TSH) binds to receptors on cells of the thyroid gland and causes the cells to secrete thyroid hormones (see Figure 10-2, A and Table 10-1). When too much TSH is secreted it causes the thyroid gland to enlarge and secrete too much thyroid hormone. When too little TSH is secreted, the thyroid gland decreases in size and too little thyroid hormone is secreted. The functions of the thyroid hormones are covered later in this chapter.

Adrenocorticotropic (ă-dre´no-kor´tĭ-ko-tro´pik) **hormone** (ACTH) increases the secretion of a hormone from the adrenal glands called **cortisol** (kor´tĭ-sol) (see Figure 10-2, A and Table 10-1). ACTH is also required to keep the adrenal gland from degenerating. ACTH binds to melanocytes in the skin and increases skin pigmentation. One symptom of too much ACTH secretion is a darkening of the skin.

Gonadotropins (gon´ă-do-tro´pinz) are hormones that bind to the gonads (ovaries and testes) (see Figure 10-2, A and Table 10-1). They regulate the growth, development, and functions of the gonads. **Luteinizing** (lu´te-ĭ-nīz-ing) **hormone** (LH) causes ovulation and sex hormone secretion from the ovary in females and sex hormone secretion from the testis in males. **Follicle stimulating hormone** (FSH) stimulates the development of oocytes in the ovary and sperm cells in the testis. Without LH and FSH, the ovary and testis decrease in size, no longer produce oocytes or sperm cells, and no longer secrete hormones.

Prolactin (pro-lak´tin) (PRL) promotes development of the breast during pregnancy and stimulates production of milk in the breast after pregnancy (see Figure 10-2, A and Table 10-1).

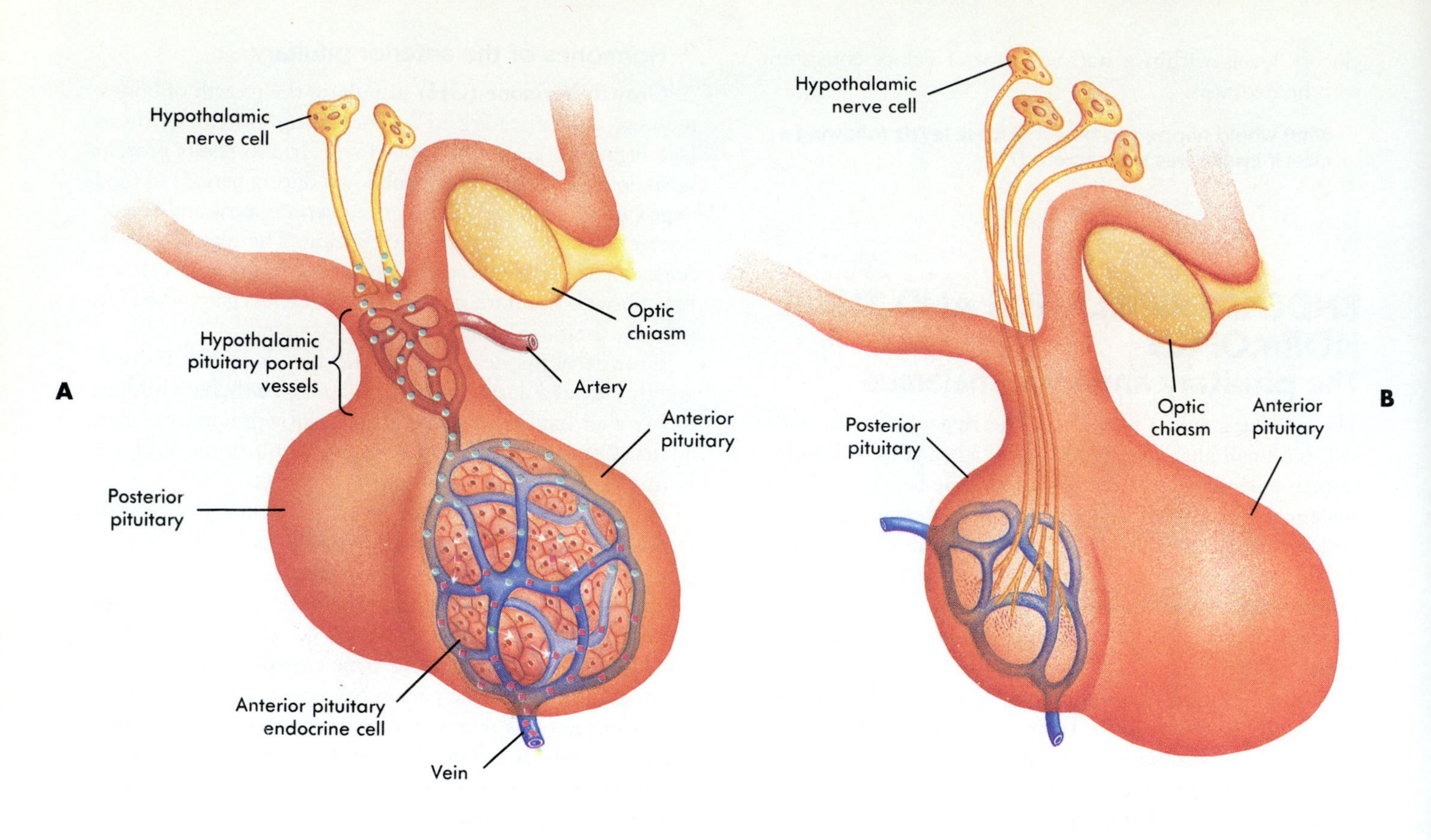
A
Hypothalamic
nerve cell
Optic
chiasm
Hypothalamic
pituitary portal
vessels
Artery
Anterior
pituitary
Posterior
pituitary
Anterior pituitary
endocrine cell
Vein
B
Hypothalamic
nerve cell
Optic
chiasm
Anterior
pituitary
Posterior
pituitary

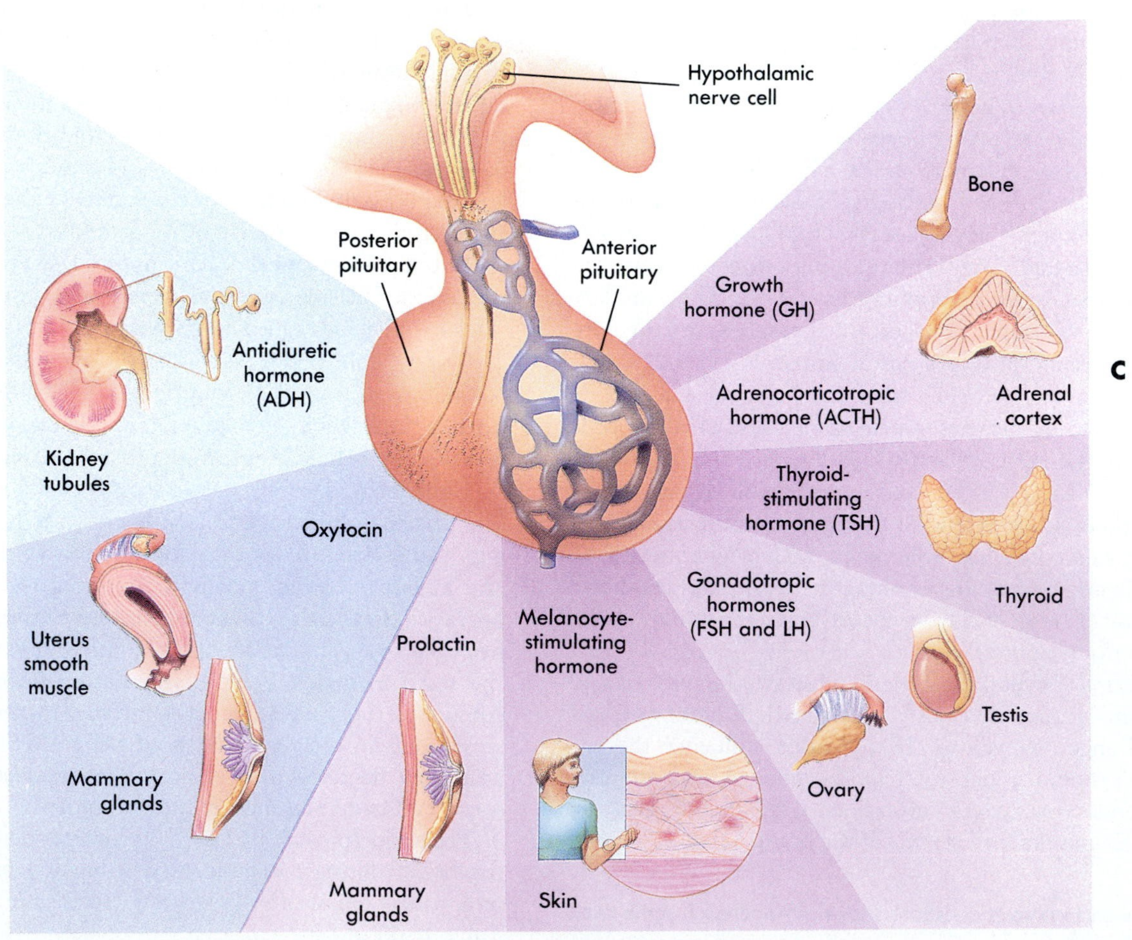
C
Hypothalamic
nerve cell
Posterior
pituitary
Anterior
pituitary
Bone
Growth
hormone (GH)
Antidiuretic
hormone
(ADH)
Kidney
tubules
Adrenocorticotropic
hormone (ACTH)
Adrenal
cortex
Thyroid-
stimulating
hormone (TSH)
Thyroid
Oxytocin
Gonadotropic
hormones
(FSH and LH)
Uterus
smooth
muscle
Prolactin
Melanocyte-
stimulating
hormone
Testis
Ovary
Mammary
glands
Mammary
glands
Skin

Table 10-1 • Pituitary hormones and their target tissues

Gland	Hormone	Target tissue	Response
• Pituitary gland			
Anterior	Growth hormone (GH)	Most tissues	Increases protein synthesis and breakdown of lipids
	Thyroid-stimulating hormone (TSH)	Thyroid gland	Increases thyroid hormone secretion
	Adrenocorticotropic hormone (ACTH)	Adrenal cortex	Increases secretion of glucocorticoid hormones such as cortisol; increases skin pigmentation at high concentrations
	Luteinizing hormone (LH)	Ovary in females; testis in males	Ovulation and progesterone production in the ovary; testosterone synthesis and support for sperm cell production in the testis
	Follicle-stimulating hormone (FSH)	Follicles in ovary in females; seminiferous tubules in males	Follicle maturation and estrogen secretion in ovary; spermatogenesis in the testis
	Prolactin (PRL)	Mammary gland in females	Milk production in women
	Melanocyte-stimulating hormone (MSH)	Melanocytes in skin	Increases melanin production in melanocytes to make the skin darker in color
Posterior	Antidiuretic hormone (ADH)	Kidney	Increases water reabsorption (less water is lost as urine); causes blood vessels to constrict and increases blood pressure
	Oxytocin	Uterus and mammary gland	Increases uterine contractions and increases milk "let down" from mammary glands

Figure 10-2 • The pituitary gland.
The structure of the pituitary gland and the relationship between the pituitary gland and the hypothalamus of the brain. **A,** Substances called releasing hormones are secreted from neurons of the hypothalamus as a result of certain stimuli acting on the brain. The releasing factors pass through the hypothalamic-pituitary portal vessels to the anterior pituitary. Within the anterior pituitary, the releasing hormones influence the secretion of anterior pituitary hormones. Hormones secreted from the anterior pituitary pass through the blood and influence the activity of their target tissues. **B,** Some neurons of the hypothalamus have axons that extend to the posterior pituitary. These neurons secrete the hormones of the posterior pituitary gland. In response to stimulation of the neurons, hormones are released from the axon endings of the neurons of the posterior pituitary gland and pass through the blood to their target tissues. **C,** Chart shows the pituitary gland's hormones and their target organs.

Melanocyte-stimulating hormone (MSH) binds to receptors on melanocytes and causes them to synthesize melanin (see Figure 10-2, A and Table 10-1). Over secretion of MSH causes the skin to darken. The structure of MSH is similar to the structure of ACTH, and both hormones cause the skin to darken.

Hormones of the posterior pituitary

Antidiuretic (an´tĭ-di-u-ret´ik) **hormone** (ADH) increases water retention by the kidneys, with the result that less water is lost as urine. ADH can also cause blood vessels to constrict when secreted in large amounts resulting in an increase in blood pressure (See Figure 10-2, *B* and Table 10-1). Consequently, ADH is sometimes called **vasopressin** (va-zo-pres´in). When too little ADH is secreted, a large volume of dilute urine is produced by the kidney.

Oxytocin (ok-sĭ-to´sin) causes contraction of the smooth muscle of the uterus and milk ejection or milk "let-down" from the breasts in women who are nursing (see Figure 10-2, *B* and Table 10-1). Commercial preparations of oxytocin are given under certain conditions to assist in childbirth and to constrict uterine blood vessels after childbirth.

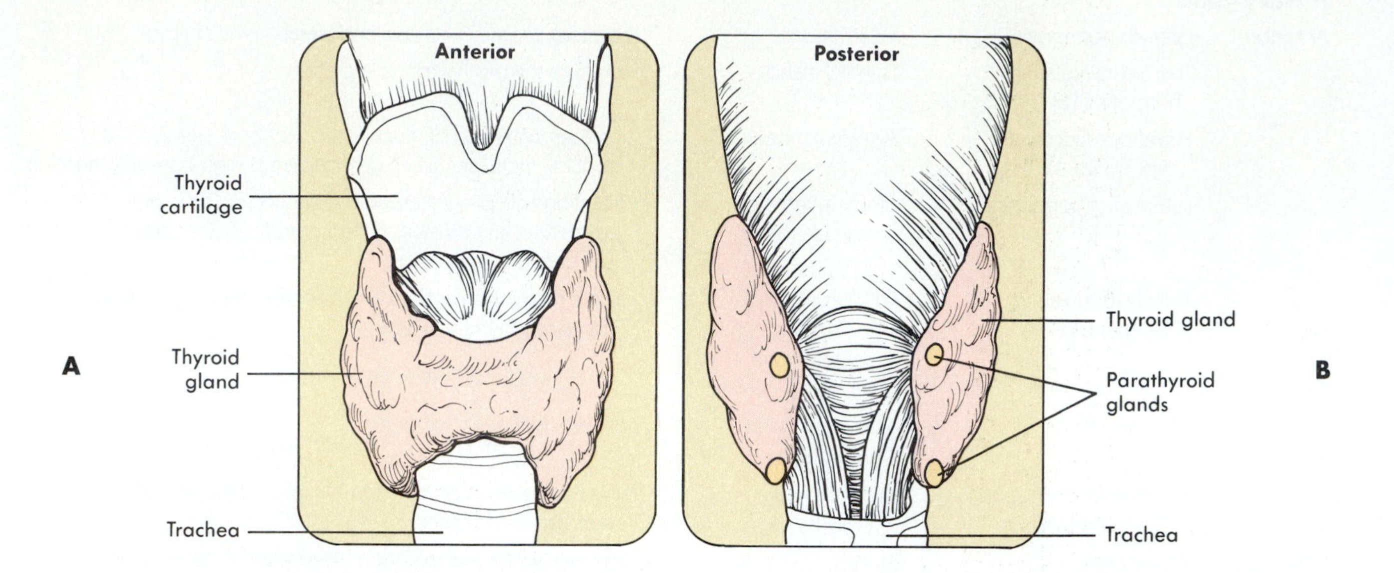

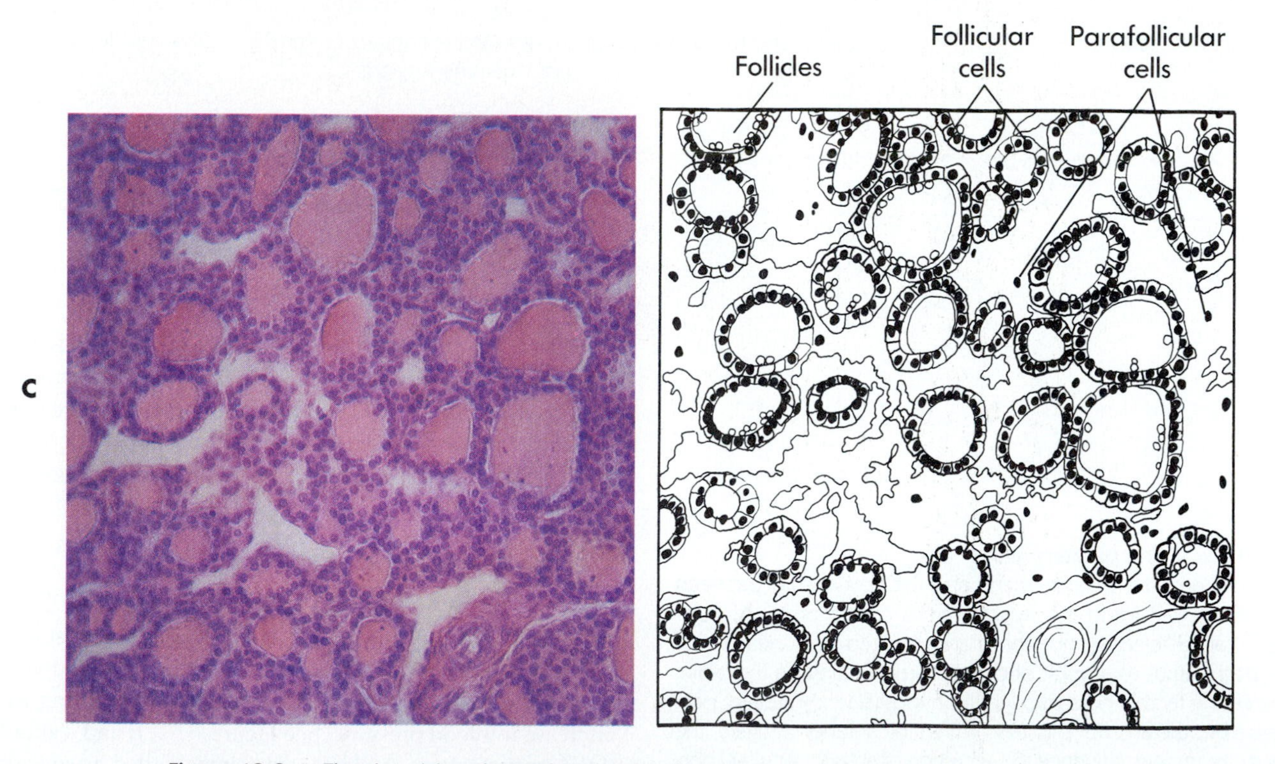

Figure 10-3 ● The thyroid and parathyroid glands.
A, Anterior view of the thyroid gland. **B,** Posterior view of the thyroid gland with the small parathyroid glands embedded in the thyroid tissue. **C,** Light micrograph and drawing of thyroid follicles and parafollicular cells.

Table 10-2	Thyroid gland hormones, parathyroid gland hormones, and their target tissues		
Gland	**Hormone**	**Target tissues**	**Response**
Thyroid gland	Thyroid hormones	Most cells of the body	Increase metabolic rate; essential for process of growth and maturation
	Calcitonin	Primarily bone	Decreases rate of breakdown of bone; prevents large increase in blood calcium levels
Parathyroid	Parathyroid hormone (PTH)	Bone, kidney	Increases rate of bone breakdown; increases vitamin D synthesis; essential for the maintenance of normal blood levels of calcium

The thyroid gland

The **thyroid** (thi´royd) **gland** is made up of two lobes of thyroid tissue connected by a narrow band. The lobes are located on either side of the trachea just below the larynx (Figure 10-3, A). The thyroid gland is one of the largest endocrine glands.

The thyroid gland contains numerous **thyroid follicles,** which are small spheres with walls that consist of single layers of cuboidal epithelial cells (Figure 10-3, C). The center of each thyroid follicle is filled with proteins to which thyroid hormones are attached. The cells of the thyroid follicles synthesize thyroid hormones, which are stored in the follicles. Between the follicles there are scattered **parafollicular** (păr´ah-fŏ-lik´u-lar) **cells.**

The main function of the thyroid gland is to secrete **thyroid hormones,** such as **thyroxine** (thi-rok´sin), which regulate the rate of metabolism in the body (Table 10-2). Without a normal rate of thyroid hormone secretion, growth and development cannot proceed normally. A lack of thyroid hormones is called **hypothyroidism.** In infants, hypothyroidism can result in **cretinism** (kre´tĭ-nizm), a condition in which the person is mentally retarded and has a short stature with abnormally formed skeletal structures. In adults, the lack of thyroid hormones results in a reduced rate of metabolism, sluggishness, and a reduced ability to perform routine tasks. Hypothyroidism also causes **myxedema** (mik-sĕ-de´mah), which is edema or swelling, resulting from a change in the structure of the connective tissue beneath the skin. An elevated rate of thyroid hormone secretion is known as **hyperthyroidism.** It results in an elevated rate of metabolism, extreme nervousness, increased heart rate, abnormal rhythms of the heart, diarrhea, weight loss, and chronic fatigue. Hyperthyroidism can also result in the accumulation of connective tissue behind the eyes which causes them to bulge forward, a condition called **exophthalmos** (eks-of-thal´mos).

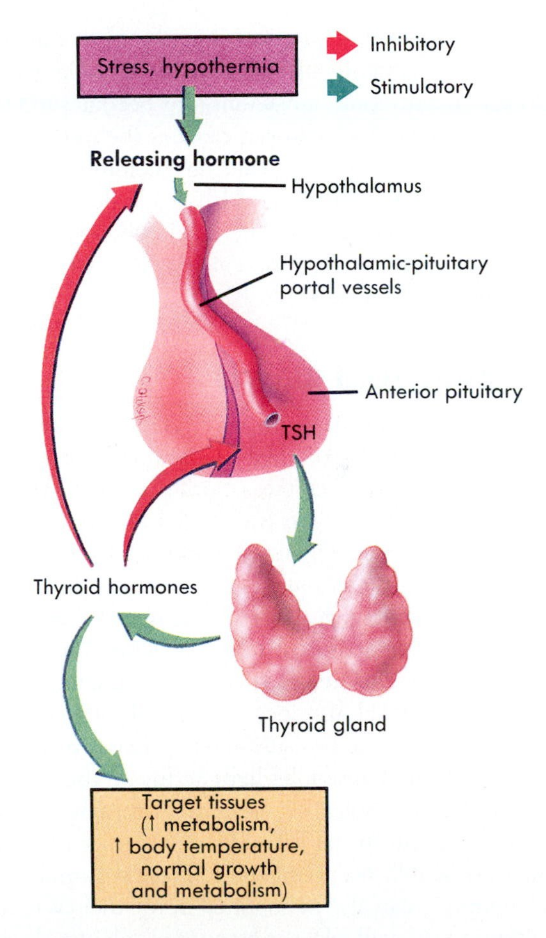

Figure 10-4 Regulation of thyroid hormone secretion. A releasing hormone passes from the hypothalamus through the hypothalamic-pituitary portal vessels to the anterior pituitary, where it causes secretion of thyroid-stimulating hormone (TSH). The thyroid-stimulating hormone passes through the blood to the thyroid gland where it stimulates the secretion of the thyroid hormones. The thyroid hormones inhibit the secretion of TSH by acting on both the hypothalamus and the anterior pituitary.

Thyroid hormone secretion is regulated by TSH from the anterior pituitary (Figure 10-4). Increasing blood levels of TSH increase the synthesis and secretion of thyroid hormones, and decreasing blood levels of TSH decrease the synthesis and secretion of thyroid hormones. The thyroid hormones, in turn, have a negative-feedback effect on the hypothalamus and pituitary. Because of the negative-feedback effect, the thyroid hormones fluctuate within a narrow concentration range in the blood.

The body requires iodine to synthesize the thyroid hormones. The iodine is taken up by the thyroid follicles in which hormone synthesis occurs. If the quantity of iodine present is not sufficient, the production and secretion of the thyroid hormones decrease. This decrease stimulates the pituitary to secrete TSH in large amounts. The excess TSH stimulates the thyroid gland and causes it to enlarge, a condition called a **goiter** (goy´ter).

2 What would happen to blood levels of thyroid hormones and to the structure of the thyroid gland if a person's immune system produced a large amount of a chemical that was so much like TSH it could bind to thyroid cells and act like TSH?

A hormone called **calcitonin** (kal-sĭ-to´nin) is secreted from the parafollicular cells of the thyroid gland. If the blood concentration of calcium ions becomes too high, calcitonin is secreted. Calcitonin reduces the rate at which calcium is released from bone, causing calcium ion levels to decrease to their normal range. Although calcitonin is important in reducing blood calcium levels when they get too high, calcitonin does not cause calcium levels to decrease below normal levels. (see Table 10–2).

The parathyroid glands

Four tiny **parathyroid** (păr-ă-thi´royd) **glands** are embedded in the posterior wall of the thyroid gland (Figure 10-3, *B*). The parathyroid glands secrete a hormone called **parathyroid hormone** (PTH), which is essential for the regulation of blood calcium levels (Figure 10-5 and see Table 10-2).

PTH acts on its target tissues to raise blood calcium levels to their normal range. PTH increases the resorption (breakdown) of bone tissue to release calcium into the circulatory system and decreases the rate at which calcium is lost in the urine. PTH also increases the absorption of calcium from the intestine by causing an increase in active vitamin D formation. Ultraviolet light acting on the skin is required for the first stage of vitamin D formation, other modifications occur in the liver. The final stage of vitamin D formation, which occurs in the kidney, is stimulated by PTH. Vitamin D can also be supplied in the diet. When too little vitamin D is available, the amount of calcium absorbed through the digestive system is very low.

PTH is more important than calcitonin in regulating blood levels of calcium. Decreased blood levels of calcium stimulate PTH secretion, and increased blood levels of calcium decrease PTH secretion (see Figure 10-5). For example, if too little calcium is consumed in the diet, blood levels of calcium decrease and PTH levels increase. Consequently, more calcium is absorbed from the small intestine, more calcium is released from bone, and less calcium is lost in the urine, all of which cause blood calcium levels to increase back to normal levels.

3 A chronic lack of vitamin D results in a condition called rickets (rik´ets). Rickets is most noticeable in young people in whom bones are so soft they become deformed. The weight-bearing bones of the thighs and legs are affected the most. Explain why rickets results in bones that are softer than normal. (Hint: how do low levels of vitamin D affect PTH levels?)

When blood levels of calcium decrease below their normal range of values, uncontrolled contraction of skeletal muscle, a condition called **tetany** (tet´ă-ne), and an increased activity of the nervous system result. The low blood levels of calcium cause cell membranes to increase their permeability to sodium ions. As a result spontaneous action potentials are produced in skeletal muscles and nervous tissue.

When blood levels of calcium increase above their normal range of values, a decrease in the activity of skeletal muscle, resulting in slow and weak movements, and a decrease in the activity of the nervous system result. The high blood levels of calcium cause cell membranes to decrease their permeability to sodium ions. As a result, fewer action potentials are produced in skeletal muscles and nervous tissue. Also, excess calcium may be deposited in the kidneys or other tissues resulting in inflammation of those tissues.

The adrenal glands

The **adrenal** (ă-dre´nal) **glands** are two small glands, each of which is located on top of a kidney. Each adrenal gland has an inner part, called the **adrenal medulla,** and an outer part, called the **adrenal cortex.** The adrenal medulla and the adrenal cortex function as separate endocrine glands.

The adrenal medulla

The main hormone released from the adrenal medulla is **epinephrine** (ep´ĭ-nef´rin) or **adrenalin** (ă-dren´a-lin). The adrenal medulla also secretes smaller amounts of **norepinephrine** (nor´ep-ĭ-nef´rin), or **noradrenalin.** Epinephrine and norepinephrine are secreted in response to stimulation by the sympathetic nervous system (Figure 10-6 and Table 10-3). Adrenal medulla hormones are referred to as the **fight-or-flight** hormones because they prepare the body for vigorous physical activity. Some of the major effects of epinephrine and norepinephrine are:

1. Constrict arteries supplying the internal organs and the skin, but not those supplying skeletal muscle. Blood flow decreases in internal organs and the skin but increases in skeletal muscles.
2. Increased heart rate, which increases the delivery of blood to tissues.
3. Increased metabolic rate of tissues, especially skeletal muscle, cardiac muscle, and nervous tissue.
4. Dilation of airway passages allowing air to move in and out of the lungs with greater ease.

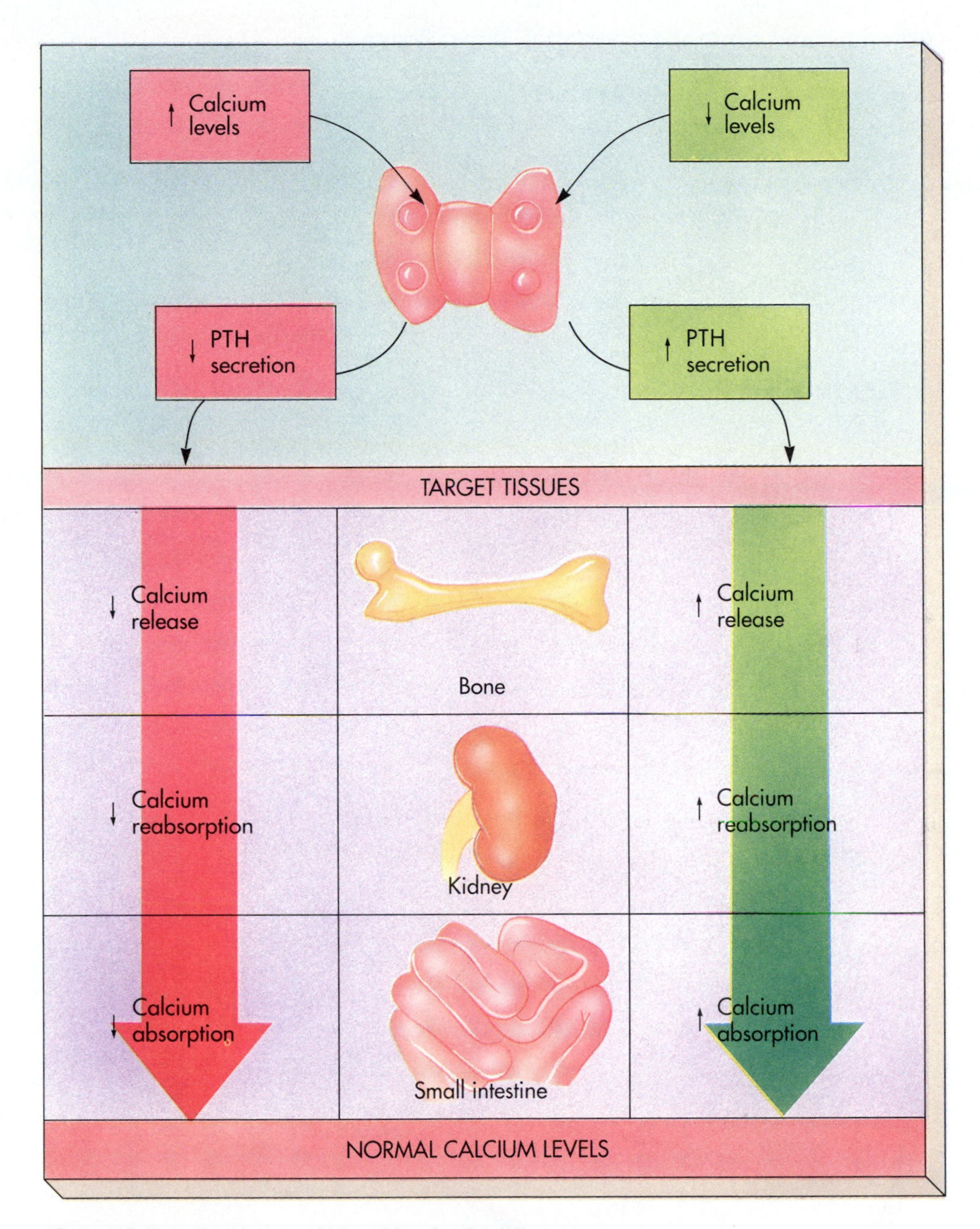

Figure 10-5 ● Regulation of blood levels of calcium.
When blood calcium levels increase, the rate of parathyroid hormone secretion decreases. The response of target tissues to these changes is to cause the blood calcium levels to decline toward their normal range. When blood calcium levels decrease, the rate of parathyroid hormone secretion increases. The response of target tissues to these changes is to increase blood calcium levels to their normal range.

Table 10-3 ● Adrenal gland hormones, and their target tissues

Gland	Hormone	Target tissue	Response
Adrenal glands			
Adrenal medulla	Epinephrine mostly, some norepinephrine	Heart, blood vessels, liver, fat cells	Increase cardiac output; increase blood flow to skeletal muscles and heart; increase release of glucose and fatty acids into blood; in general, prepare for physical activity
Adrenal cortex	Mineralocorticoids (aldosterone)	Kidneys; to lesser degree, intestine and sweat gland	Increases rate of sodium transport into body; increases rate of potassium excretion; favors water retention
	Glucocorticoid (cortisol)	Most tissues (e.g., liver, fat, skeletal muscle, immune tssues)	Increases fat and protein breakdown; increases glucose synthesis from amino acids; increases blood nutrient levels; inhibits inflammation and immune response
	Adrenal androgens	Most tissues	After puberty, increases sexual drive, pubic hair, and axillary hair growth in women; after puberty most androgens are produced by the testes in males

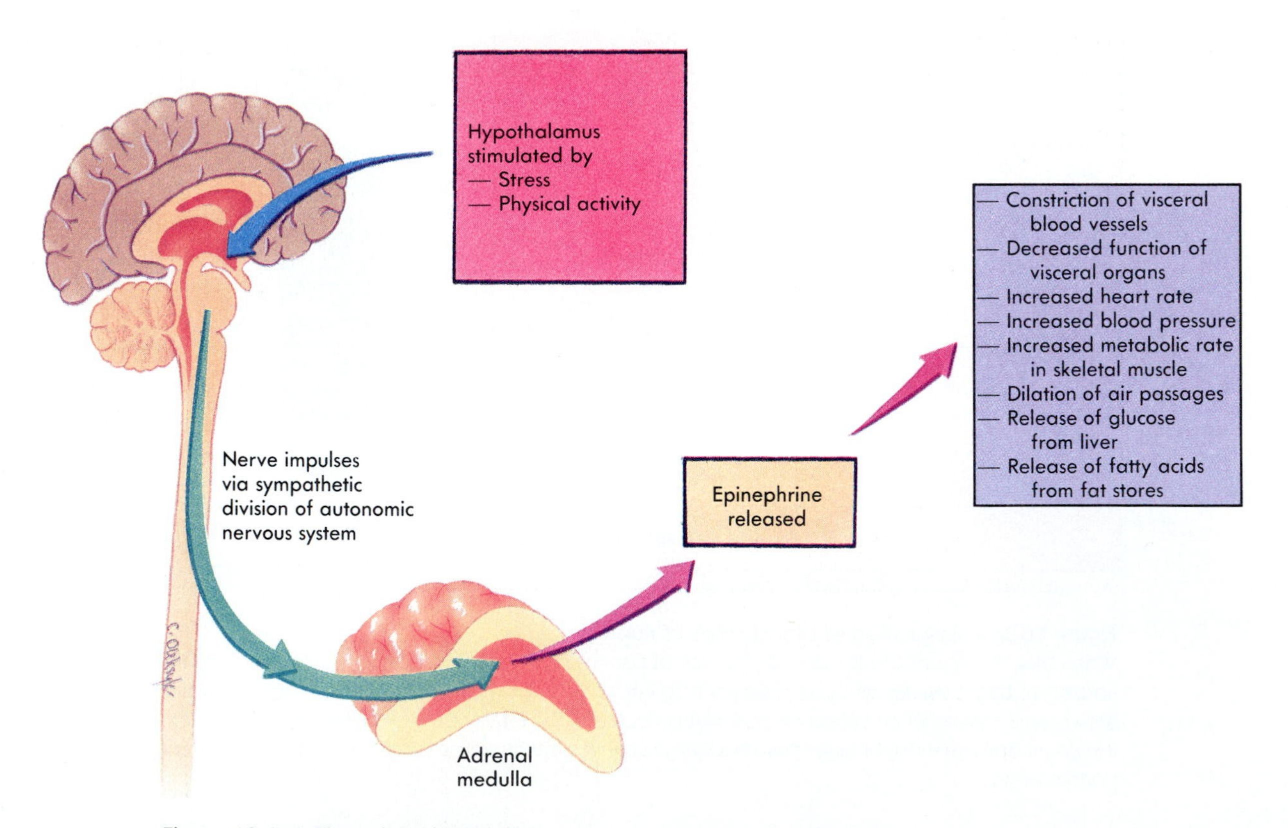

Figure 10-6 ● The adrenal medulla.
The adrenal medulla is a specialized portion of the sympathetic division of the autonomic nervous system. Conditions that cause increased activity of the sympathetic nervous system increase epinephrine release from the adrenal medulla. The responses to epinephrine prepare the body for physical activity.

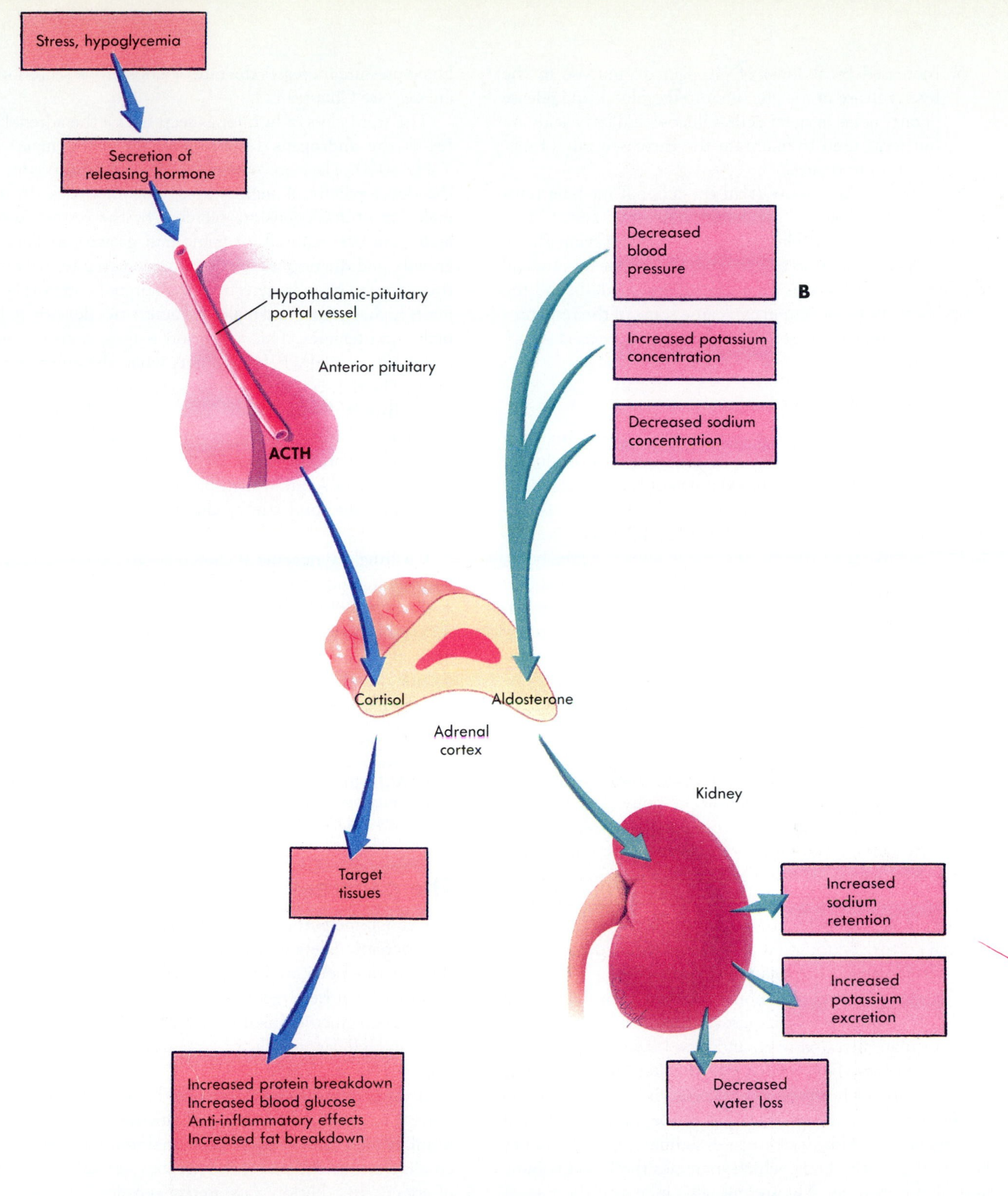

Figure 10-7 ● The adrenal cortex.

The adrenal cortex is the outer portion of the adrenal gland. **A,** Cortisol secretion is controlled by a releasing hormone from the hypothalamus that passes through the hypothalamic-pituitary portal system to the anterior pituitary, in which it binds to and stimulates cells that secrete ACTH. ACTH acts on the adrenal cortex and stimulates the secretion of cortisol. Cortisol acts on its target tissues to increase protein and fat metabolism, increase blood glucose levels, and reduce inflammation. **B,** Decreased blood pressure, increased blood potassium levels, and decreased blood sodium levels influence the adrenal cortex to increase the secretion of aldosterone. Aldosterone acts on the kidney to decrease water loss in the form of urine, increase potassium excretion, and reduce sodium excretion.

5. Increased breakdown of glycogen to glucose in the liver, release of the glucose into the blood, and release of fatty acids from fat cells. Glucose and fatty acids are nutrients used to maintain the increased rate of metabolism in tissues.

Responses to hormones from the adrenal medulla reinforce the effect of the sympathetic nervous system (see Chapter 8). Thus the adrenal medulla and the sympathetic nervous system function together to prepare one for physical activity. Also, hormones from the adrenal medulla and the sympathetic nervous system can cause some of the responses to stress, such as increased blood pressure, increased sweating, and nervousness.

The adrenal cortex

Three classes of steroid hormones are secreted from the adrenal cortex (Table 10-2). The **glucocorticoids** (glu´ko-kor´tĭ-koydz) help regulate blood nutrient levels in the body (Figure 10-7, *A*). The major glucocorticoid hormone is **cortisol** (kor´tĭ-sol), which increases the breakdown of proteins and fat and increases their use as energy sources by the body. For example, when blood glucose levels decline, cortisol secretion increases. Cortisol causes proteins to be broken down to amino acids and it causes the amino acids to be converted to glucose by the liver. Cortisol also acts on adipose tissue, causing fat to be broken down to fatty acids. The glucose and fatty acids are then metabolized and used as sources of energy by tissues.

In times of stress, cortisol is secreted in larger-than-normal amounts. In addition to increasing energy sources for tissues, it reduces the inflammatory response (see Chapter 4). **Cortisone** (kor´tĭ-sōn), a steroid closely related to cortisol, is often given as a medication to reduce inflammation such as occurs during certain allergic responses and injuries.

Adrenocorticotropic hormone (ACTH) from the anterior pituitary stimulates the secretion of cortisol from the adrenal cortex. Without ACTH, the adrenal cortex degenerates and loses most of its secretory capability.

The second class of hormones secreted from the adrenal cortex, the **mineralocorticoids** (min´er-al-o-kor´tĭ-koydz), helps regulate blood volume and blood levels of potassium and sodium ions (Figure 10-7, *B*). **Aldosterone** (al-dos´-ter-ōn) is the major hormone of this class. It acts primarily on the kidney, but it also affects the intestine, sweat glands, and salivary glands. Aldosterone causes sodium ions and water to be retained in the body, which increases the blood volume and blood pressure. Aldosterone also increases the rate at which potassium is eliminated from the body in the urine.

Blood levels of potassium and sodium ions act directly on the adrenal cortex to influence aldosterone secretion, although the adrenal cortex is much more sensitive to changes in blood potassium levels. The rate of aldosterone secretion increases when blood potassium levels increase and when blood sodium levels decrease. Changes in blood pressure also influence aldosterone secretion. When blood pressure decreases the rate of aldosterone secretion increases, and when blood pressure increases the rate of aldosterone secretion decreases (see Chapter 11).

The third class of hormones secreted by the adrenal cortex is the **androgens** (an´dro-jenz; Gr. *aner*, male) (see Table 10-3). They are named for their ability to stimulate the development of male sexual characteristics. In adult males, most androgens are secreted by the testes. Adrenal androgens are required for pubic hair growth, axillary hair growth, and the normal sex drive in women after puberty. If the secretion of androgens from the adrenal cortex is abnormally high, exaggerated male characteristics develop in both males and females. This condition is most apparent in females, and in males before puberty when the effects are not masked by the secretion of androgens by the testes.

Addison's disease results from abnormally low levels of aldosterone and cortisol. Symptoms include low blood levels of glucose, inability to use fat and protein for energy, low blood levels of sodium, high blood levels of potassium, low blood pressure, and the production of a large volume of urine.

Cushing's syndrome is caused by an excess secretion of cortisol and aldosterone. Symptoms include a higher-than-normal level of blood glucose, excessive destruction of protein resulting in destruction of muscle tissue, accumulation of fat tissue on the trunk of the body, high blood levels of sodium, low blood levels of potassium, high blood pressure, and the production of a small volume of concentrated urine.

4 ACTH precisely regulates the secretion of cortisol, which is only one of the adrenal hormones, but in the absence of ACTH the entire adrenal gland (cortex and medulla) degenerates. In the absence of ACTH, which adrenal hormones will be secreted in smaller-than-normal amounts?

The pancreas

The endocrine portion of the **pancreas** (pan´kre-us) consists of **pancreatic islets** (islets of Langerhans) dispersed among the exocrine portion of the pancreas. The islets secrete two hormones that help regulate blood levels of nutrients, especially blood glucose: **insulin** (in´su-lin) decreases blood glucose and **glucagon** (glu´ka-gon) increases blood glucose (Figure 10-8 and Table 10-4).

It is important to maintain blood glucose levels within a normal range of values. If blood glucose declines below its normal range, the nervous system malfunctions because glucose is the nutrient that is the nervous system's main source of energy. In addition, fats are broken down so rapidly by other tissues that acidic substances, released as byproducts of fat metabolism, cause the pH of the body fluids to decrease below normal, a condition called **acidosis.** If blood glucose levels are too high, the kidneys produce large volumes of urine containing substantial amounts of glucose. Because of the rapid loss of water in the form of urine, dehydration can result.

Insulin is released from the pancreatic islets in response to elevated blood glucose levels and increased blood levels of amino acids (Figure 10-8). The major target tissues for insulin are the liver, adipose tissue, muscles, and the area of the hypothalamus that regulates appetite, called the **satiety** (sa´ti-ĕ-te) **center.** Insulin increases the rate of glucose and amino acid uptake in these tissues. They are used as sources of energy or are converted to other molecules. Glucose is stored as glycogen or fat, and the amino acids are used to synthesize protein. The effects of insulin on target tissues are summarized in Table 10-4. A lack of insulin secretion results in a condition called **diabetes mellitus** (di-ă-be´tēz mel-li´tis) (see box, Understanding Diabetes Mellitus and Insulin Shock).

Glucagon is released from the pancreatic islets when blood glucose levels are low (see Figure 10-8). Glucagon acts primarily on the liver to cause the conversion of stored glycogen to glucose. The glucose is then released into the blood to increase blood glucose levels. After a meal, when blood glucose levels are elevated, glucagon secretion is reduced. Insulin and glucagon function together to regulate blood glucose levels (see Figure 10-8).

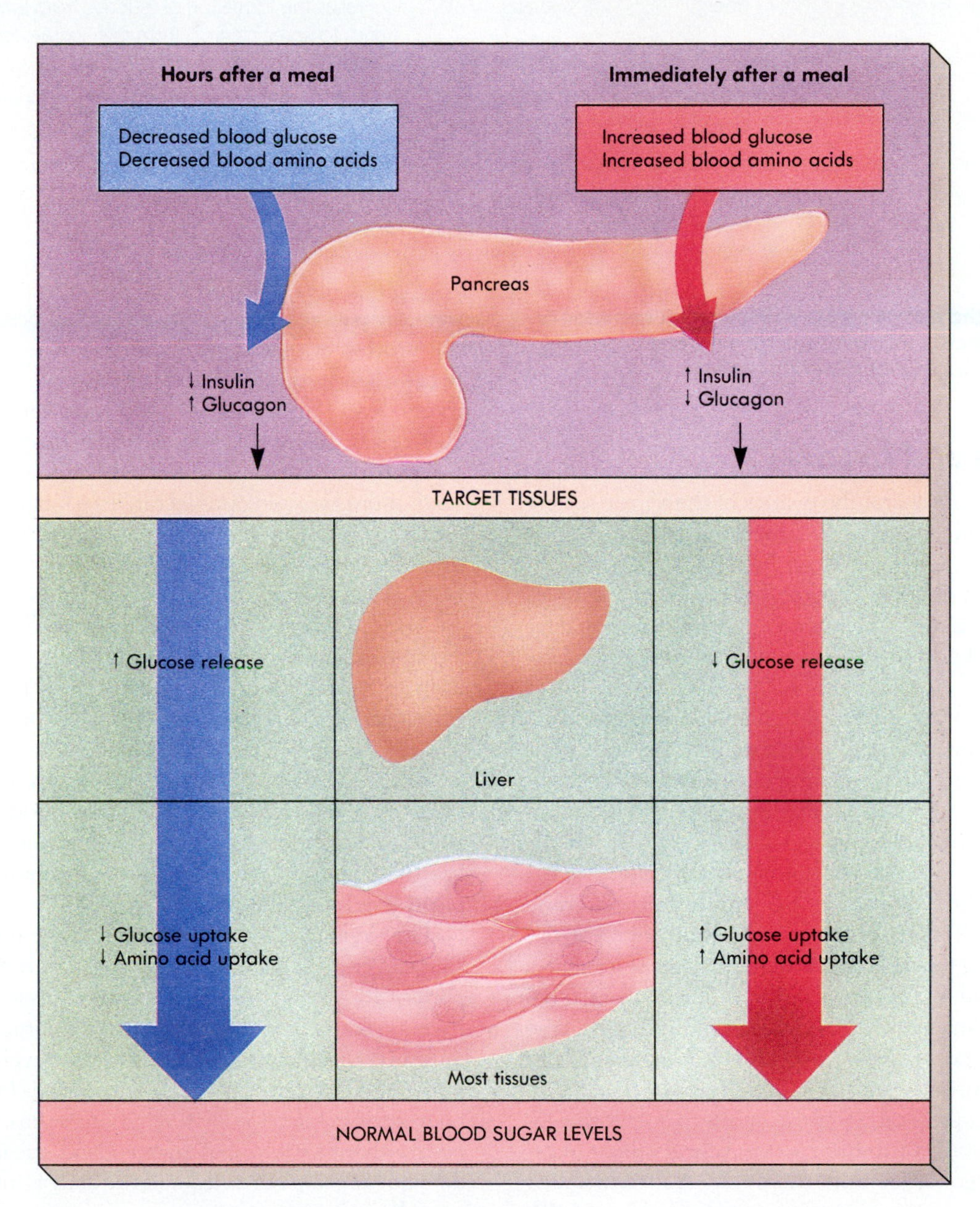

Figure 10-8 ● Regulation of insulin and glucagon secretion.
Decreasing concentrations of glucose increase the secretion of glucagon, which acts on liver cells to increase release of glucose from the liver into the blood. Release of glucose from the liver helps maintain blood glucose levels. Increasing blood glucose levels inhibit glucagon secretion. Increasing concentrations of blood glucose and amino acids stimulate insulin secretion. Insulin acts on most tissues to increase the uptake and use of glucose and amino acids. As the blood levels of glucose and amino acids decrease, the rate of insulin secretion also decreases.

Table 10-4 • Effects of insulin and glucagon on target tissues

Target tissue	Insulin responses	Glucagon responses
Skeletal muscle, cardiac muscle, cartilage, bone, fibroblasts, blood cells, and mammary glands	Increases glucose uptake and glycogen synthesis; increases uptake of amino acids	Little effect
Liver	Increases glycogen synthesis; increases use of glucose for energy	Causes rapid increase in the breakdown of glycogen to glucose and release of glucose into the blood; increases the formation of glucose from amino acids and, to some degree, from fats; increases metabolism of fatty acids
Adipose cells	Increases glucose uptake, glycogen synthesis, and fat synthesis	High concentrations cause breakdown of fats; probably unimportant under most conditions
Nervous system	Little effect except to increase glucose uptake in the satiety center	No effect

5 Predict how the rate of insulin and glucagon secretion would be affected following a large meal rich in carbohydrates, and after 12 hours without eating.

The testes and ovaries

The testes of the male and the ovaries of the female secrete sex hormones in addition to producing sperm cells or oocytes. The hormones produced by these organs play an important role in the development of sexual characteristics. Structural and functional differences between males and females and the ability to reproduce depend on the sex hormones.

The main hormone produced in the male is **testosterone** (tes-tos´tĕ-rōn), which is an androgen. It is responsible for the growth and development of the male reproductive structures, muscle enlargement, growth of body hair, voice changes, and increased male sexual drive. In the female, there are two main classes of hormones that affect sexual characteristics, the **estrogens** (es´tro-jenz) and **progesterone** (pro-jes´ter-ōn). Together these hormones contribute to the development and function of female reproductive structures and other female sexual characteristics. These characteristics include enlargement of the breasts and distribution of fat, which influences the shape of the hips, breasts, and legs. The female menstrual cycle, which results in the production of an oocyte and preparation of the uterus to accept the developing embryo, is controlled by the cyclical release of estrogens and progesterone from the ovary. A more detailed description of the sex hormones is presented in Chapter 19.

OTHER HORMONES

Cells in the lining of the stomach and small intestine secrete hormones that stimulate the production of digestive juices from the pancreas, stomach, and liver. These hormones aid the process of digestion by causing secretion of digestive juices when food is present in the digestive system but not at other times. Hormones secreted from the small intestine also help regulate the rate at which food passes from the stomach into the small intestine, so that food enters the small intestine at an optimal rate (see Chapter 16).

The **prostaglandins** (pros´tă-glan´dinz) are a group of lipid hormones that are widely distributed in tissues of the body. Some prostaglandins cause relaxation of smooth muscle, such as dilation of blood vessels, and others cause contraction. Prostaglandins produced during the delivery of a baby cause uterine smooth muscle to contract. Prostaglandins also play a role in inflammation. They are released by damaged tissues and cause blood vessel dilation, swelling, and pain. The ability of aspirin and related substances to reduce pain and inflammation, help prevent painful cramping of uterine smooth muscle, and treat headache may be a result of their inhibitory effect on prostaglandin synthesis.

The **thymus** (thi´mus) **gland** lies in the upper part of the thoracic cavity above the heart (see Figure 10-1). It is important in the function of the immune system. Part of the thymus gland's functions is to secrete a hormone called **thymosin** (thi´mo-sin), which helps in the development of certain white blood cells called T cells. T cells help protect the body against infection by foreign organisms. If an infant is born without a functional thymus gland, the immune system does not develop normally, and the body is incapable of successfully fighting infections.

The **pineal** (pin´e-al) **body** is a small pine-cone shaped structure located superior and posterior to the thalamus of the brain. The pineal body produces a hormone called **melatonin** (mel-ă-to´nin), which is thought to decrease the secretion of LH and FSH. Thus melatonin acts to inhibit the functions of the reproductive system. Melatonin may play

Understanding diabetes mellitus and insulin shock

Diabetes mellitus can result from any of the following: secretion of too little insulin from the pancreas, insufficient numbers of insulin receptors on target cells, or defective receptors that do not respond normally to insulin. In people who have diabetes mellitus, tissues cannot take up glucose effectively, causing blood glucose levels to become very high. The excess glucose is excreted in the urine and causes the urine volume to be very high because water follows the glucose by osmosis (see Chapter 3). The excessive loss of water in the urine results in an increased sensation of thirst. In addition, fats and proteins, instead of glucose, are broken down to provide an energy source for metabolism resulting in the wasting away of body tissues. Rapid fat breakdown produces acidic byproducts, which cause acidosis, a lowering of blood pH. People who have untreated diabetes mellitus have an increased appetite, because glucose cannot enter cells of the satiety center of the brain without insulin. Thus the satiety center responds by initiating the sensation of hunger as if there were very little blood glucose, even though blood glucose levels are very high. Major symptoms of diabetes, which are easy to observe, are high blood glucose levels, glucose in the urine, increased urine volume, increased thirst, wasting away of body tissues, acidosis, and an increased appetite. Lack of energy also results from the inability of tissues to take up and use glucose and from wasting away of the body tissues.

Type I diabetes mellitus (also called juvenile diabetes mellitus) develops in young people and is usually caused by a lack of insulin secretion. Viral infection of the pancreatic islets or an immune response may be responsible for the condition.

Type II diabetes mellitus (also called maturity onset diabetes mellitus) develops in older people and often does not result from a lack of insulin but from the reduced ability of the tissues to respond to insulin. In contrast to Type I diabetes, a high percentage of people who have Type II diabetes are obese. Fat cells continue to take up some glucose and convert it to fat in response to the high blood glucose levels. People who suffer from Type II diabetes mellitus are encouraged to lose weight and to restrict their intake of simple sugars. Many people who suffer from Type II diabetes mellitus do take insulin injections, but the treatment is not highly effective because their cells are less responsive to insulin. The age of onset for Type II diabetes may vary tremendously, and it appears to be, in part, hereditary.

Insulin shock can develop too much insulin is present, such as occurs when a diabetic is either injected with too much insulin or has not eaten after an insulin injection. Disorientation, convulsions, and loss of consciousness may result.

6 Joe suffers from Type I diabetes mellitus. One morning he was late for class. He rapidly got ready, took his normal insulin injection, and went to class without eating breakfast. During his second class he began to feel disoriented. Explain this symptom he exhibited and recommend a reasonable treatment. (Hint: what is the brain's primary source of energy?)

an important role in the onset of puberty in humans. Some tumors of the pineal body increase pineal secretions and others decrease pineal secretions. Those that increase pineal secretions generally result in inhibition of the reproductive system, and those that decrease pineal secretions stimulate the reproductive system.

SUMMARY

THE ENDOCRINE SYSTEM

Endocrine glands are glands without ducts that secrete chemicals into the blood, and the endocrine system is made up of all of the body's endocrine glands.

Hormones are chemicals secreted by the endocrine glands.

Hormones combine with molecules in cells called receptors, and the tissues containing cells with receptors for a hormone are called target tissues for that hormone.

The secretion of hormones by endocrine glands is controlled by negative-feedback mechanisms.

ENDOCRINE GLANDS AND THEIR HORMONES

The pituitary and hypothalamus

The pituitary is connected to the hypothalamus of the brain by the infundibulum. It is divided into the anterior and posterior pituitary.

Secretions from the anterior pituitary are controlled by releasing hormones that pass through the hypothalamic-pituitary portal system from the hypothalamus.

Hormones secreted from the posterior pituitary are controlled by action potentials that pass from the hypothalamus along axons to the posterior pituitary.

Hormones of the anterior pituitary include growth hormone (GH), thyroid-stimulating hormone (TSH), adrenocorticotropic hormone (ACTH), luteinizing hormone (LH) and follicle-stimulating hormone (FSH), prolactin, and melanocyte-stimulating hormone.

Hormones released from the posterior pituitary include antidiuretic hormone (ADH) and oxytocin.

The thyroid gland

The thyroid gland secretes thyroid hormones, which control the metabolic rate of tissues, and it secretes calcitonin, which helps regulate blood calcium levels.

The parathyroid glands

The parathyroid glands secrete parathyroid hormone, which helps regulate blood levels of calcium. Active vitamin D also helps regulate blood levels of calcium.

The adrenal glands

The adrenal medulla secretes epinephrine and smaller amounts of norepinephrine, which help prepare the body for physical activity.

The adrenal cortex secretes three classes of hormones.

A. Glucocorticoids (for example, cortisol) reduce inflammation and break down fat and proteins, making them available as energy sources to other tissues.

B. Mineralocorticoids (for example, aldosterone) help regulate sodium and potassium levels and water volume in the body. These hormones enhance sodium and water retention by the kidneys.

C. Adrenal androgens.

The pancreas

The pancreas secretes insulin in response to elevated levels of blood glucose and amino acids. Insulin increases the rate at which many tissues, including adipose tissue, liver, and skeletal muscles, take up glucose and amino acids.

The pancreas secretes glucagon in response to reduced blood glucose. Glucagon increases the rate at which the liver releases glucose into the blood.

The testes and ovaries

The testes secrete testosterone, and the ovaries secrete estrogens and progesterone. These hormones determine sexual characteristics and help control reproductive processes. LH and FSH from the pituitary gland control ovarian and testicular functions.

OTHER HORMONES

Hormones secreted by cells in the stomach and intestine help regulate stomach, pancreatic, and liver secretions.

The prostaglandins are hormones that produce numerous effects on the body and play a role in inflammation.

The thymus gland secretes thymosin, which helps in the development of T cells, and the thymus is important in the normal development of the immune system.

The pineal body secretes melatonin, which has an inhibitory effect on the reproductive system and may play a role in the development of puberty.

CONTENT REVIEW

1. Define endocrine gland and hormone.
2. What makes one tissue a target tissue and another not a target tissue for a hormone?
3. TSH from the anterior pituitary gland stimulates the thyroid gland to secrete thyroid hormones. When the blood levels of thyroid hormones increase, the thyroid hormones then inhibit the secretion of TSH. Explain why the control of thyroid hormones is called a negative-feedback system.
4. Describe how secretions of the anterior pituitary hormones are controlled.
5. Describe how secretions of the posterior pituitary hormones are controlled.
6. What are the functions of growth hormone? What happens when there is too little or too much growth hormone?
7. What is the function of thyroid-stimulating hormone?
8. What are the functions of adrenocorticotropic hormone? What happens when there is too little or too much adrenocorticotropic hormone?
9. What are the effects of gonadotropins on the ovary and testis?
10. What are the functions of prolactin?
11. What are the functions of melanocyte-stimulating hormone?
12. What are the functions of antidiuretic hormone?
13. What are the functions of oxytocin?
14. What are the functions of the thyroid hormones, and how is their secretion controlled? Describe the effects of hypersecretion and hyposecretion of the thyroid hormones.
15. Explain how calcitonin, parathyroid hormone, and vitamin D are involved in maintaining blood calcium levels.
16. List the hormones secreted from the adrenal gland, give their functions, and compare the means by which the secretion rate of each is controlled.
17. What are the major functions of insulin and glucagon? How is their secretion regulated?
18. List the two major types of diabetes mellitus and explain the cause of each.
19. List the effects of testosterone, progesterone, and estrogen.
20. List some functions regulated by prostaglandins.
21. Describe the function of the thymus gland.
22. Describe the function of the pineal body.

CONCEPT REVIEW

1. Predict the long-term effects of a prolonged diet that contains too little calcium on (a) the secretion rates of the hormones that regulate blood levels of calcium and (b) on body structures.
2. What symptoms would develop if the adrenal cortex degenerated and was no longer capable of secreting hormones?
3. In some people, the immune system of the body produces a protein in large amounts that acts like TSH. Predict the symptoms that would result from this condition.
4. Explain how you could tell, by examining a urine sample, if a person was suffering from diabetes mellitus or a lack of ADH secretion.
5. Occasionally the anterior pituitary gland of infants does not develop normally and, therefore, it does not secrete any anterior pituitary hormones. Please explain which hormones would have to be administered to such infants in order for them to grow to a normal height.

CHAPTER TEST

Answers can be found in Appendix C

MATCHING For each statement in column A select the correct answer in column B (an answer may be used once, more than once, or not at all).

A

1. Ductless glands.
2. Important autonomic and endocrine control center of the brain.
3. Secretes hormones in response to releasing hormones from the hypothalamus.
4. Tissue upon which a hormone acts.
5. Acts on the adrenal cortex and increases cortisol secretion.
6. Acts on the thyroid gland and increases thyroid hormone secretion.
7. Acts on the kidneys to retain water.
8. Cells that secrete calcitonin.
9. Secreted in response to decreasing blood levels of calcium.
10. A hormone called the fight-or-flight hormone.
11. Causes sodium ions and water to be retained in the body and increases the rate at which potassium is eliminated.
12. Causes target tissues to increase the rate they take up glucose.

B

adrenocorticotropic hormone (ACTH)
aldosterone
anterior pituitary
antidiuretic hormone (ADH)
calcitonin
cerebrum
endocrine glands
epinephrine
exocrine glands
follicle cells
glucagon
hypothalamus
insulin
luteinizing hormone (LH)
parafollicular cells
parathyroid hormone (PTH)
posterior pituitary
target tissue
thyroid-stimulating hormone (TSH)

FILL-IN-THE BLANK Complete each statement by providing the missing word or words.

1. The __________ is a series of blood vessels that carry blood from the hypothalamus to the anterior pituitary.
2. A __________ is a molecule in the cell of a target tissue to which a hormone binds.
3. __________ stimulates the growth of bones, muscles, and other organs by increasing protein synthesis.
4. __________ is required to keep the adrenal cortex from degenerating.
5. Prolactin is secreted from the __________ and helps promote development of the breast during pregnancy and stimulates the production of milk.
6. Lack of __________ in the diet or overproduction of thyroid-stimulating hormone results in the formation of a goiter.
7. __________ increases the absorption of calcium from the intestine by causing an increase in active vitamin D formation.
8. __________ increases the breakdown of protein and fat and increases their use as energy sources by the body.
9. In response to decreasing blood levels of glucose, __________ is released from the pancreas.
10. __________ is a type of hormone that reduces the intensity of inflammation.

CHAPTER 11

Blood

OBJECTIVES

After reading this chapter you should be able to:

1. Describe the functions of blood.
2. Name the components of plasma.
3. Describe the structure, function, and life history of erythrocytes.
4. Name and give the functions of the five different types of leukocytes.
5. Explain the formation and function of platelet plugs and clots.
6. Describe the regulation of clot formation and how clots are removed.
7. Describe the ABO and Rh blood groups and explain how transfusion reactions occur.
8. Describe diagnostic blood tests and the normal values for the tests, and give examples of disorders that produce abnormal test values.

KEY TERMS

anemia
Any condition that results in lower than normal hemoglobin in the blood or a lower than normal number of erythrocytes.

anticoagulant
Chemical that prevents coagulation or blood clotting.

erythrocyte
Red blood cell; contains hemoglobin, transports oxygen and carbon dioxide.

erythropoietin
Protein hormone that stimulates erythrocyte formation in red bone marrow.

fibrin
A threadlike protein fiber that aids in clot formation.

hematocrit
The percentage of total blood volume composed of erythrocytes.

hemoglobin
A substance in erythrocytes consisting of four globin proteins, each with an iron-containing red pigment heme; transports oxygen and carbon dioxide.

leukocyte
White blood cell; involved in immunity. The five types are neutrophils, eosinophils, basophils, lymphocytes, and monocytes.

plasma
Fluid portion of blood; blood minus the cells and cell fragments.

thrombocyte
Platelet; a cell fragment involved in platelet plug and clot formation.

FEATURES

CELLS require constant nutrition and waste removal because they are metabolically active. Most cells are located some distance from nutrient sources, such as the digestive tract, and sites of waste disposal, such as the kidneys. The cardiovascular system, which consists of the heart, blood vessels, and blood, provides the necessary connection between various tissues. The heart pumps blood through blood vessels that extend throughout the body. The blood is important in the maintenance of homeostasis in several ways: (1) transportation of oxygen, nutrients, enzymes, and hormones to tissues; (2) transportation of carbon dioxide and waste products away from tissues; (3) maintenance of body temperature by transporting heat between deep tissues and the body's surface; (4) maintenance of fluid, electrolyte, and pH balance; (5) protection of the body from disease-causing microorganisms, foreign substances, and tumors; and (6) prevention of blood loss when blood vessels are damaged.

Blood is connective tissue that consists of cells and cell fragments **(formed elements)** surrounded by a liquid matrix **(plasma)** (plaz´mah). The formed elements account for slightly less than half of the total blood volume, and plasma accounts for slightly more than half of the total blood volume. The total blood volume in the average adult is approximately 4 to 5 liters in females and 5 to 6 liters in males. Blood makes up approximately 8% of the body's total weight.

PLASMA

Plasma is a pale yellow fluid that consists of 92% water and 8% of substances such as proteins, ions, nutrients, gases, and waste products. Plasma proteins perform many important functions: albumin is important in osmosis and the normal movement of water between blood and tissues (see Chapter 3); globulins help protect against infections (see Chapter 14); and clotting proteins prevent blood loss. When the proteins that produce clots are removed from plasma, the remaining fluid is called **serum.**

FORMED ELEMENTS

About 95% of the volume of the formed elements consists of **red blood cells (RBCs),** or **erythrocytes** (ĕ-rith´ro-sītz) (Figure 11-1). The remaining 5% of the volume of the formed elements consists of **white blood cells (WBCs),** or **leukocytes** (lu´ko-sītz), and cell fragments called **platelets** (plāt´letz), or **thrombocytes** (throm´bo-sītz). RBCs are the most common of the formed elements in blood. They are 700 times more numerous than WBCs and 17 times more numerous than platelets. WBCs are the only formed elements possessing nuclei in healthy adults, whereas RBCs and platelets have few organelles and lack nuclei. After birth, the production of formed elements is confined primarily to red bone marrow, but some WBCs are produced in lymphatic tissue (see Chapter 14).

Red blood cells

Normal red blood cells are disk-shaped cells with edges that are thicker than the center area (see Figure 11-1). During their development, RBCs lose their nuclei and most of their organelles. Consequently, they are unable to divide. RBCs live for about 120 days in males and 110 days in females.

Function

The primary functions of RBCs are to transport oxygen from the lungs to the various tissues of the body and to assist in the transport of carbon dioxide from the tissues to the lungs. The transport of carbon dioxide is considered in Chapter 15. Oxygen transport is accomplished by **hemoglobin** (he´mo-glo´bin), which consists of red-pigment molecules called **hemes** and protein chains called **globins.** Each heme molecule contains iron. When hemoglobin is exposed to oxygen, oxygen molecules bind to the iron. Hemoglobin that is bound to oxygen is bright red in color, whereas hemoglobin without bound oxygen is a darker red color. Hemoglobin is responsible for 97% of the oxygen transport in

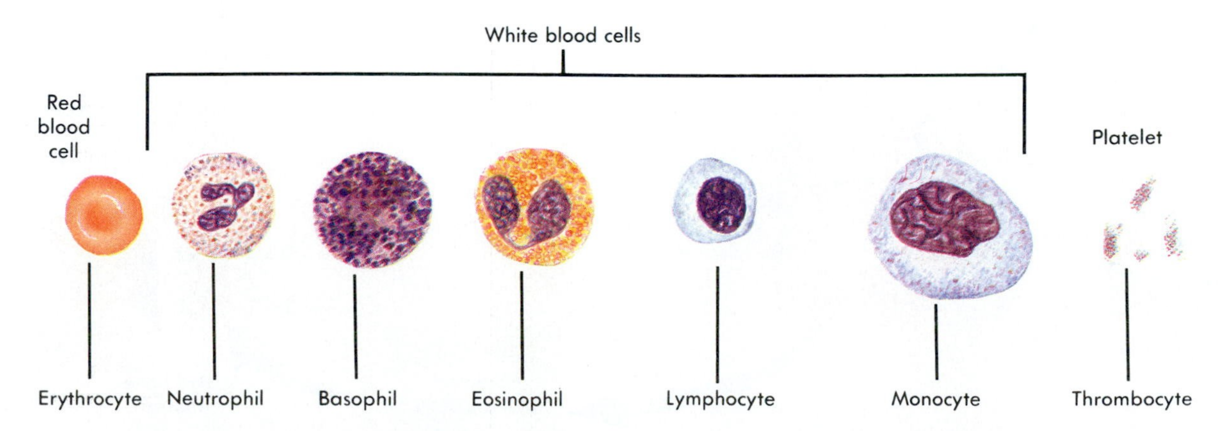

Figure 11-1 ● Formed elements.
Red blood cells (erythrocytes) are biconcave disk-shaped cells without a nucleus. There are five kinds of white blood cells (leukocytes): neutrophils, basophils, eosinophils, lymphocytes, and monocytes. Neutrophils, basophils, and eosinophils have a lobed nucleus and large granules that stain different colors. Lymphocytes and monocytes have a round nucleus and do not readily stain. Platelets (thrombocytes) are cell fragments.

blood. The remaining 3% is transported as a dissolved gas in the plasma.

Carbon monoxide is a gas produced by the incomplete combustion of gasoline. It binds to the iron in hemoglobin approximately 210 times as fast as oxygen and does not tend to unbind. As a result, the hemoglobin bound to carbon monoxide no longer transports oxygen. Nausea, headache, unconsciousness, and death are possible consequences of prolonged exposure to carbon monoxide.

Life history of red blood cells

Under normal conditions, about 2.5 million RBCs are destroyed every second. Fortunately, new RBCs are produced as rapidly as old RBCs are destroyed. RBCs are produced from a series of cell divisions that begin with cells in the red bone marrow. After each cell division the newly formed cells change and become more like a mature RBC. For example, after a certain cell division the newly formed cells manufacture large amounts of hemoglobin. After the final cell division the nucleus is lost from the cell and a completely mature RBC is formed.

The process of cell division requires the vitamins B_{12} and folic acid, which are necessary for the synthesis of DNA (see Chapter 3). Iron is required for the production of hemoglobin. Consequently, lack of vitamin B_{12}, folic acid, or iron can interfere with normal RBC production.

RBC production is stimulated by low blood oxygen levels. Typical causes of low blood oxygen are decreased numbers of RBCs, decreased or defective hemoglobin, diseases of the lungs, high altitude, inability of the cardiovascular system to deliver blood to tissues, and increased tissue demands for oxygen (for example, during endurance exercises).

Low blood oxygen levels increase RBC production by stimulating the release of the hormone **erythropoietin** (ĕ-rith´ro-poy´ĕ-tin) from the kidneys. Erythropoietin stimulates red bone marrow to produce more RBCs (Figure 11-2). Thus when oxygen levels in the blood decrease, the production of RBCs increases. Blood oxygen levels return to normal and homeostasis is maintained by increasing the delivery of oxygen to tissues. Conversely, if blood oxygen levels increase, less erythropoietin is released from the kidneys and RBC production decreases.

1 Cigarette smoke produces carbon monoxide. If a nonsmoker smoked a pack of cigarettes a day for a few weeks, what would happen to the number of RBCs in the person's blood? Explain. (Hint: what would happen to blood oxygen levels?)*

Old, damaged, or defective RBCs are removed from the blood by macrophages (large "eating" cells) located in the spleen and liver (Figure 11-3). Within the macrophage the globin part of the molecule is broken down into amino acids that are reused to produce other proteins. The iron is released and can be used to produce new hemoglobin in the red bone marrow. Only small amounts of iron are required in the daily diet because the iron is recycled. Women need more dietary iron than men because women lose more iron as a result of blood loss during menstruation.

The heme molecules are converted to **bilirubin** (bil-ĭ-

*Answers to predict questions appear in Appendix B at the back of the book.

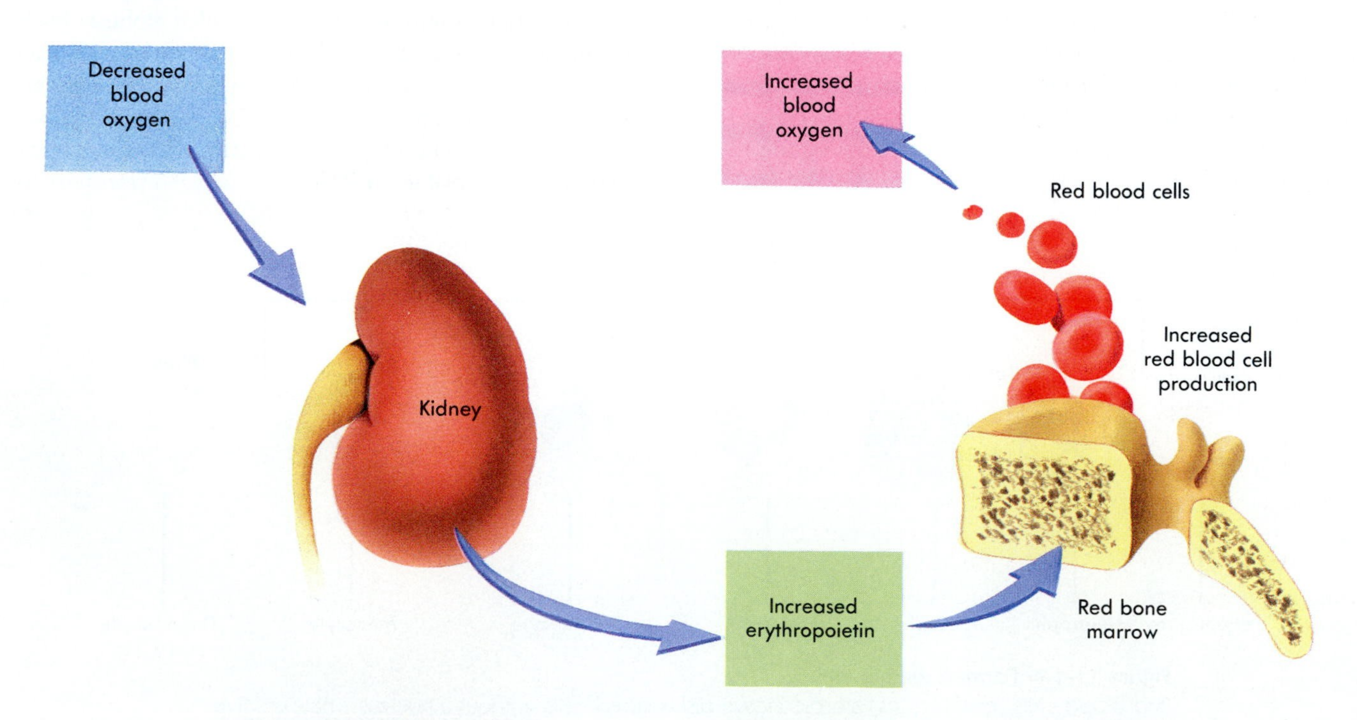

Figure 11-2 ● RBC production.
In response to decreased blood oxygen, the kidneys release erythropoietin, which stimulates RBC production in the red bone marrow.

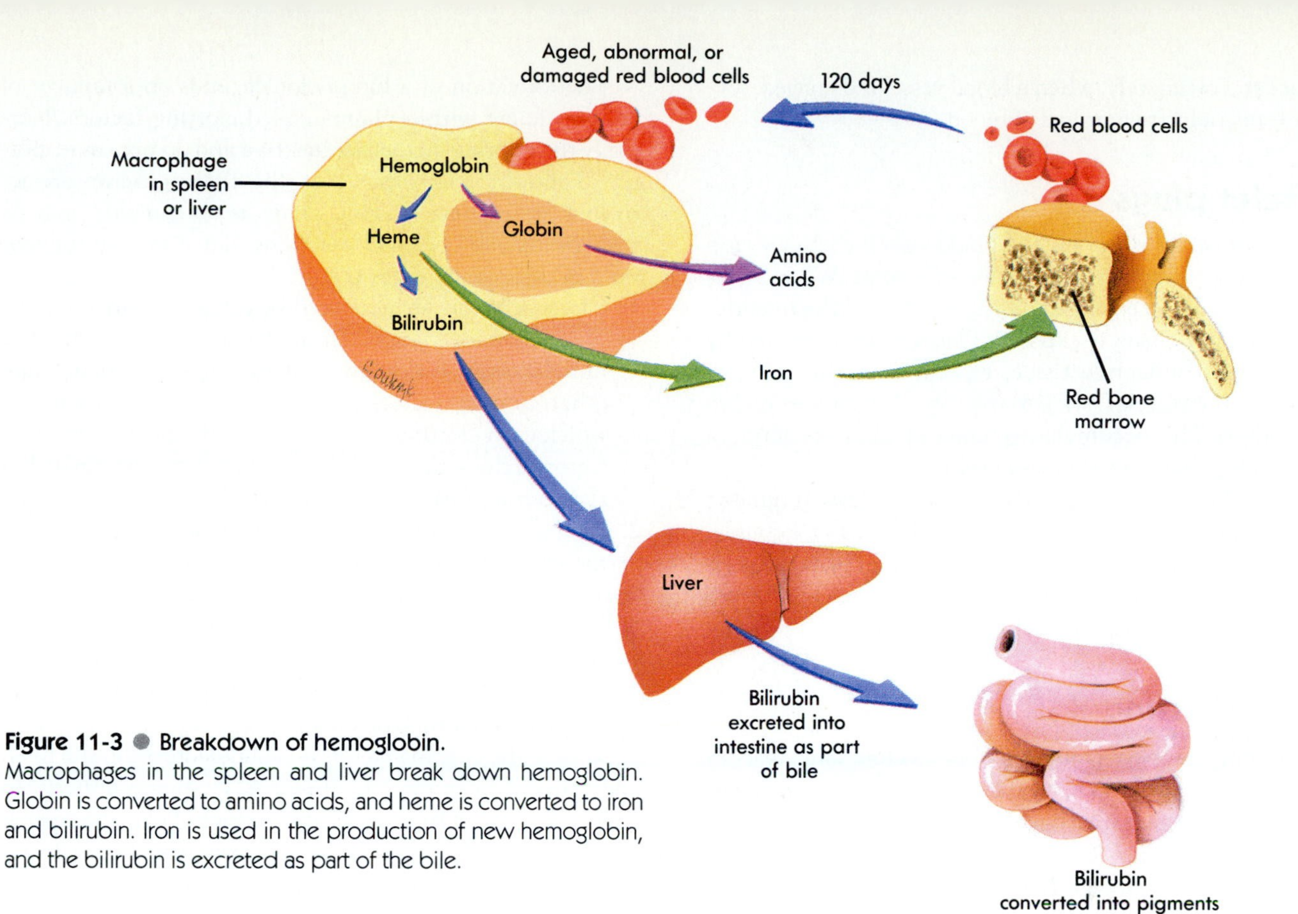

Figure 11-3 ● **Breakdown of hemoglobin.**
Macrophages in the spleen and liver break down hemoglobin. Globin is converted to amino acids, and heme is converted to iron and bilirubin. Iron is used in the production of new hemoglobin, and the bilirubin is excreted as part of the bile.

roo´bin), which is taken up by the liver. The bilirubin normally is excreted into the small intestine as part of the bile (see Chapter 16). If the liver is not functioning normally or if the flow of bile from the liver to the small intestine is hindered, bilirubin builds up in the circulation and produces **jaundice** (jawn´dis), a yellowish color of the skin. After it is in the intestine, bilirubin is converted by bacteria into other pigments that are responsible for the brown color of feces.

White blood cells

White blood cells (WBCs), or leukocytes, are spherical cells that are whitish in color because they lack hemoglobin. Although WBCs are part of blood, the blood serves primarily as a means to transport these cells to other tissues of the body. WBCs remove dead cells and debris from the tissues by phagocytosis (see Chapter 3), and they function to protect the body against invading microorganisms.

The important phagocytic WBCs are neutrophils and monocytes (see Figure 11-1). **Neutrophils** (nu´tro-filz) are the most common type of WBC. They usually remain in the blood for a short time (10 to 12 hours), move into other tissues, and phagocytize microorganisms and other foreign substances. Dead neutrophils, cell debris, and fluid can accumulate as **pus** at sites of infections. **Monocytes** (mon´o-sītz) are the largest of the WBCs. They leave the blood and transform to become **macrophages** (mak´ro-fāj´ez), which phagocytize bacteria, dead neutrophils, cell fragments, and any other debris within the tissues.

Lymphocytes (lim´fo-sītz) play an important role in the body's immune response (see Figure 11-1). They produce antibodies and other chemicals that destroy microorganisms, contribute to allergic reactions, reject grafts, control tumors, and regulate the immune system. **Basophils** (ba´so-filz) and **eosinophils** (e-o-sin´o-filz) release chemicals involved with inflammation. The activities of lymphocytes, basophils, and eosinophils are considered in greater detail in Chapter 14.

Platelets

Platelets, or thrombocytes, are minute fragments of cells consisting of a small amount of cytoplasm surrounded by a cell membrane. Platelets play an important role in preventing blood loss.

PREVENTING BLOOD LOSS

When a blood vessel is damaged, blood can leak into other tissues and interfere with normal tissue function, or blood can be lost from the body. Small amounts of blood loss from the body can be tolerated until new blood is produced to replace it. If large amounts of blood are lost, however, death can occur. Fortunately, when a blood vessel is damaged, several events help prevent loss of blood.

Platelet plugs

Small tears occur in the smaller blood vessels each day, and **platelet plug** formation quickly closes the tears. When a vessel is damaged, the epithelial lining is torn and the underlying connective tissue is exposed. Platelets are activated by the exposed connective tissue, especially by the collagen fibers, and the platelets stick to the connective tissue and to each other. The accumulating mass of platelets forms a platelet plug that seals the vessel shut.

When the platelets are activated, they release a number of chemicals that act to decrease blood loss. For example, prostaglandins function to activate additional platelets. Additionally, calcium and other chemicals released from platelets play a role in blood clotting.

Blood clotting

Platelet plugs alone are not sufficient to close large tears or cuts in blood vessels. When a blood vessel is severely damaged, blood clotting, or **coagulation** (ko-ag´u-la-shun), results in the formation of a clot. A **clot** is a network of thread-like protein fibers, called **fibrin** (fi´brin), that traps blood cells, platelets, and fluid.

The formation of a blood clot depends on a number of proteins found within plasma called **clotting factors.** Normally the clotting factors are inactive and do not cause clotting. Following injury, however, the clotting factors are activated to produce a clot. This is a complex process involving many chemical reactions, but it can be summarized in three main stages (Figure 11-4).

1. Clotting factors are activated by exposed connective tissue or chemicals released from damaged tissues. After the initial clotting factors are activated, they in turn activate other clotting factors. A series of reactions results in which each clotting factor activates the next clotting factor in the series until the clotting factor **prothrombin** (pro-throm´bin) **activator** is formed.
2. Prothrombin activator acts on an inactive clotting factor, called **prothrombin.** Prothrombin is converted to its active form, called **thrombin** (throm´bin).
3. Thrombin converts the inactive clotting factor **fibrinogen** (fi-brin´o-jen) into its active form, fibrin. Fibrin threads form a network, which traps blood cells and platelets and forms the clot.

Most of the clotting factors are manufactured in the liver, and many of them require vitamin K for their synthesis. In addition, many of the chemical reactions of clotting require calcium ions and chemicals released from platelets. Low levels of vitamin K or calcium, low numbers of platelets, or liver dysfunction can seriously impair the blood clotting process.

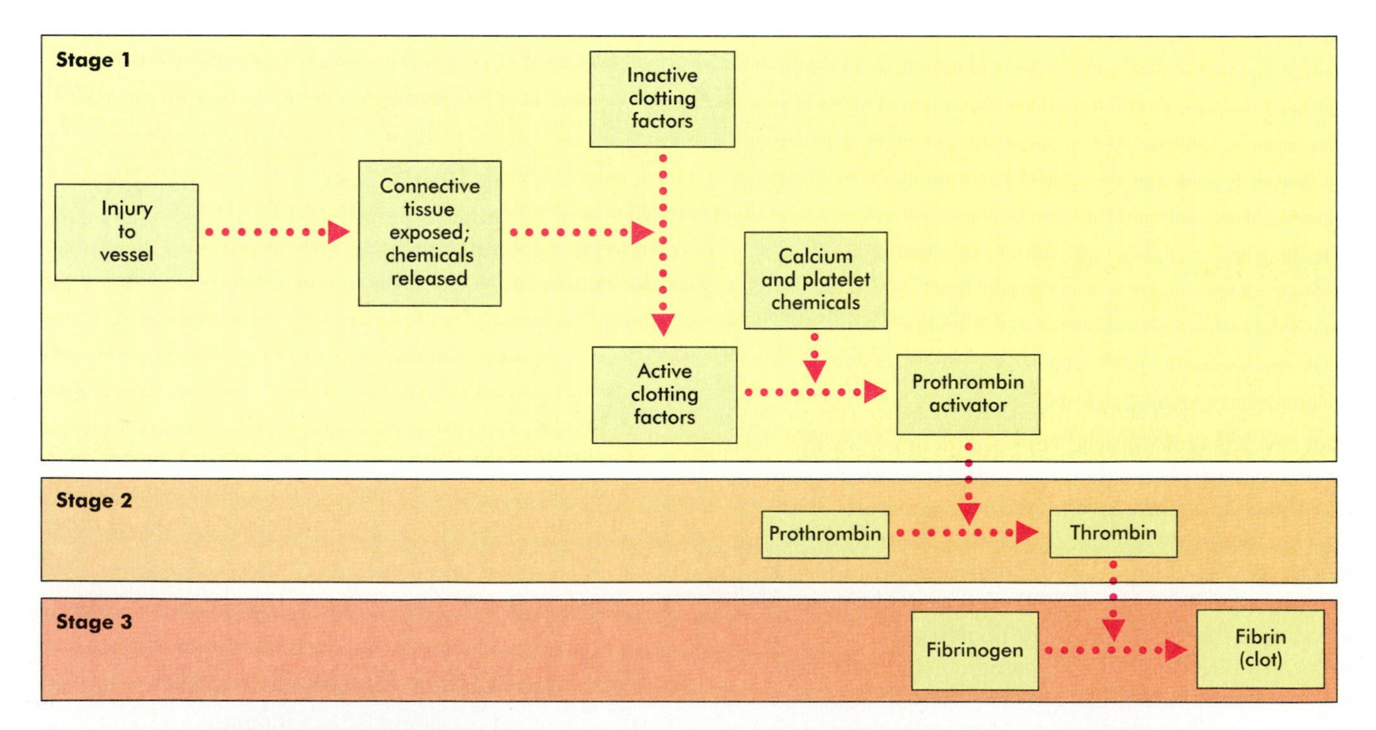

Figure 11-4 ● Clot formation.
In stage 1, inactive clotting factors are activated by exposure to connective tissue and by chemicals released from tissues. Through a series of reactions, the activated clotting factors form prothrombin activator. In stage 2, prothrombin is converted to thrombin by prothrombin activator. In stage 3, fibrinogen is converted to fibrin (the clot) by thrombin.

Control of clot formation

Without control, clotting would spread from the point of its initiation throughout the entire circulatory system. To prevent unwanted clotting, the blood contains several **anticoagulants** (an´tĭ-ko-ag´u-lantz), which prevent clotting factors from forming clots. Normally there are enough anticoagulants in the blood to prevent clot formation. At an injury site the stimulation for activating clotting factors is very strong, however, and so many clotting factors are activated that the anticoagulants no longer can prevent a clot from forming.

A clot that forms within a blood vessel is called a **thrombus** (throm´bus). An **embolus** (em´bo-lus) is a detached clot or substance that floats through the circulatory system and becomes lodged in a blood vessel. If a thrombus or embolus blocks blood flow to an essential organ such as the heart, death can result. One therapy for preventing clot formation is aspirin, which inhibits prostaglandin synthesis by platelets. This prevents platelet activation and reduces the likelihood of clotting because chemicals necessary for clotting are not released from the platelets.

Clot dissolution

After a clot is formed, it is slowly dissolved by a plasma protein called **plasmin** (plaz´min). Plasmin is formed from an inactive plasma protein called **plasminogen** (plaz-min´o-jen). Thrombin or **tissue plasminogen activator (t-PA)** released from surrounding tissues stimulate the conversion of plasminogen to plasmin. Over a period of a few days the plasmin slowly breaks down the fibrin.

A heart attack can result from blockage of blood vessels that supply blood to the heart. One treatment for a heart attack caused by a thrombus is to inject into the blood chemicals that activate plasmin. Streptokinase, a bacterial enzyme, and tissue plasminogen activator produced through genetic engineering have been successfully used to dissolve thrombi.

BLOOD GROUPING

If large quantities of blood are lost during surgery or in an accident, a person can die. A **transfusion** of blood is the transfer of blood from one person into another person. Early attempts to transfuse blood were often unsuccessful because they often resulted in **transfusion reactions,** which included clotting within blood vessels, kidney damage, and death. It is now known that transfusion reactions are caused by interactions between **antigens** (an´tĭ-jenz) and **antibodies** (see Chapter 14). In brief, the surfaces of RBCs have molecules called antigens, and in the plasma there are molecules called antibodies. Each antibody is very specific, meaning it can combine only with a certain type of antigen. When blood from two people is mixed, the antibodies in the plasma can bind to the antigens on the surface of the RBCs to form molecular bridges that connect the RBCs. As a result, **agglutination** (ă-glu´tĭ-na´shun), or clumping, of the cells occurs. The combination of the antibodies with the antigens also can initiate reactions that cause **hemolysis** (he-mol´ĭ-sis), or rupture of the RBCs.

The antigens on the surface of RBCs have been categorized into **blood groups.** Although many blood groups are recognized, the ABO and Rh blood groups are among the most important.

ABO blood group

The ABO antigens appear on the surface of RBCs. Type A blood has type A antigens, type B blood has type B antigens, type AB blood has both types of antigens, and type O blood has neither A nor B antigens (Figure 11-5). In addition, plasma from type A blood contains antibodies against type B antigens, and plasma from type B blood contains antibodies against type A antigens. Type O blood has both A and B antibodies, and type AB blood has neither.

The ABO blood types are not found in equal numbers. In Caucasians in the United States the distribution is type O, 47%; type A, 41%; type B, 9%; and type AB, 3%. Among blacks in the United States, the distribution is type O, 46%; type A, 27%; type B, 20%; and type AB, 7%.

A **donor** is a person who gives blood, and a **recipient** is a person who receives blood. Usually a donor can give blood to a recipient if they both have the same blood type. For example, a person with type A blood can donate to another person with type A blood. There would be no ABO transfusion reaction because the recipient would have no antibodies against the type A antigen. On the other hand, if type A blood were donated to a person with type B blood, there would be a transfusion reaction. This would occur because the person with type B blood would have antibodies against the type A antigen, and agglutination would result (Figure 11-6).

Historically, people with type O blood have been called universal donors because they usually can give blood to the other ABO blood types without causing an ABO transfusion reaction. Their RBCs have no ABO surface antigens and therefore do not combine with the recipient's A or B antibodies. For example, if type O blood is given to a person with type A blood, the type O RBCs do not react with the type B antibodies in the recipient's blood.

2 Historically people with type AB blood were called universal recipients. What is the rationale for this term?

It should be noted that the terms *universal donor* and *universal recipient* are misleading. For example, the transfusion of type O blood can produce a transfusion reaction for two reasons. First, there are other blood groups that can cause a transfusion reaction. To reduce the likelihood of a transfusion reaction, all the blood groups must be correctly matched. Second, antibodies in the blood of the donor can react with antigens in the blood of the recipient. For exam-

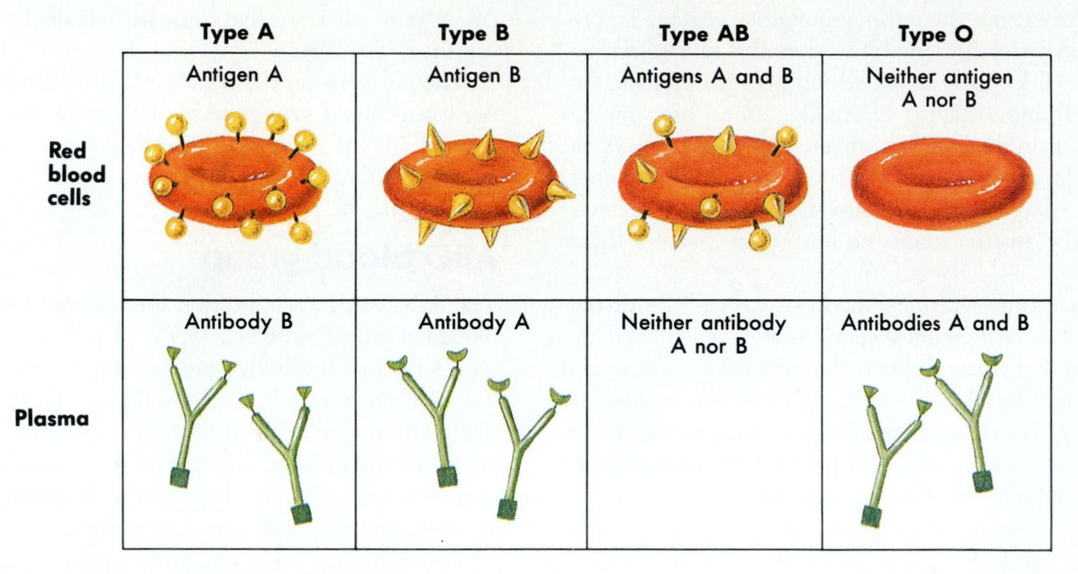

Figure 11-5 ● ABO blood group.
The antigens found on the surface of the RBCs of each blood type, and the antibodies found in each blood type are shown. The A antigens are depicted as round knobs and the B antigens as cones. The A antibodies have round-shaped endings and the B antibodies have cone-shaped endings.

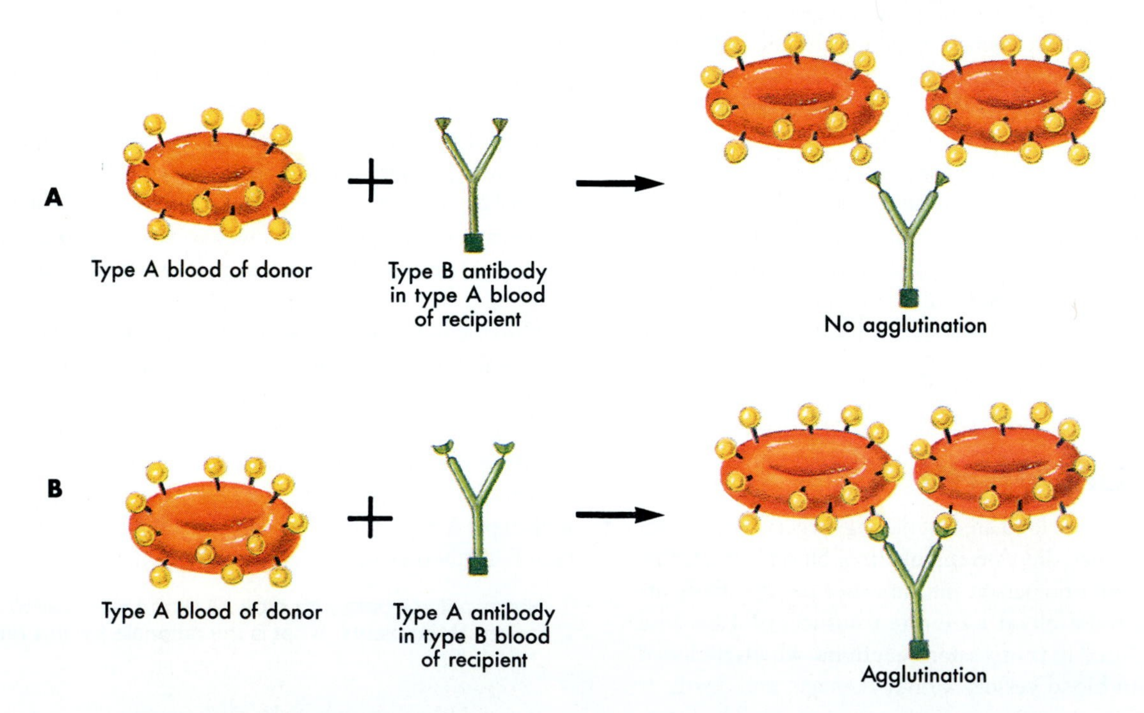

Figure 11-6 ● Agglutination reaction.
A, Type A blood donated to a type A recipient does not cause an agglutination reaction because the type B antibodies (with cone-shaped endings) in the recipient do not combine with the type A antigens (round knobs) in the donated blood. **B,** Type A blood donated to a type B recipient causes an agglutination reaction because the type A antibodies (with round-shaped endings) in the recipient combine with the type A antigens (round knobs) in the donated blood.

Understanding the Rh Blood Group

Antibodies against the Rh antigen do not develop unless an Rh-negative person is exposed to Rh-positive blood. This can occur through a transfusion or by transfer of blood between a mother and her fetus across the placenta. When an Rh-negative person receives a transfusion of Rh-positive blood, the Rh-negative recipient becomes sensitized to the Rh antigen and produces Rh antibodies. If the Rh-negative person is unfortunate enough to receive a second transfusion of Rh-positive blood after becoming sensitized, a transfusion reaction results.

Rh incompatibility can pose a major problem in some pregnancies when the mother is Rh-negative, and the fetus is Rh-positive (Figure 11-A). If fetal blood leaks through the placenta and mixes with the mother's blood, the mother becomes sensitized to the Rh antigen. The mother produces Rh antibodies that cross the placenta and cause agglutination and hemolysis of fetal RBCs. This disorder is called **hemolytic disease of the newborn (HDN)**, or **erythroblastosis fetalis** (ĕ-rith′ro-blas-to′sis fe-tă′lis), and it can be fatal to the fetus. In the first pregnancy there is often no problem. The leakage of fetal blood is usually the result of a tear in the placenta that takes place either late in the pregnancy or during delivery. Thus there is not enough time for the mother to produce enough Rh antibodies to harm the fetus. In later pregnancies there can be a problem because the mother has been sensitized to the Rh antigen. Consequently, if the fetus is Rh-positive and there is any leakage of fetal blood into the mother's blood, she rapidly produces large amounts of Rh antibodies and HDN develops.

Prevention of HDN is often possible if the Rh-negative woman is given an injection of a specific type of antibody preparation called anti-Rho(D) immune globulin (RhoGAM) immediately after each delivery. The injection contains antibodies against Rh antigens. The injected antibodies bind to the Rh antigens of any fetal RBCs that may have entered the mother's blood. This treatment inactivates the fetal Rh antigens and prevents sensitization of the mother.

If HDN develops, treatment consists of slowly removing the blood of the fetus or newborn and replacing it with blood that does not cause a transfusion reaction. Exposure of the newborn to fluorescent light is also used because it helps break down the large amounts of bilirubin formed as a result of RBC destruction. High levels of bilirubin are toxic to the nervous system and can cause destruction of brain tissue.

3 If a baby with HDN has type B positive blood, what type of blood should be used for replacement therapy? (Hint: what kind of antibodies from the mother have already been introduced into the baby?)

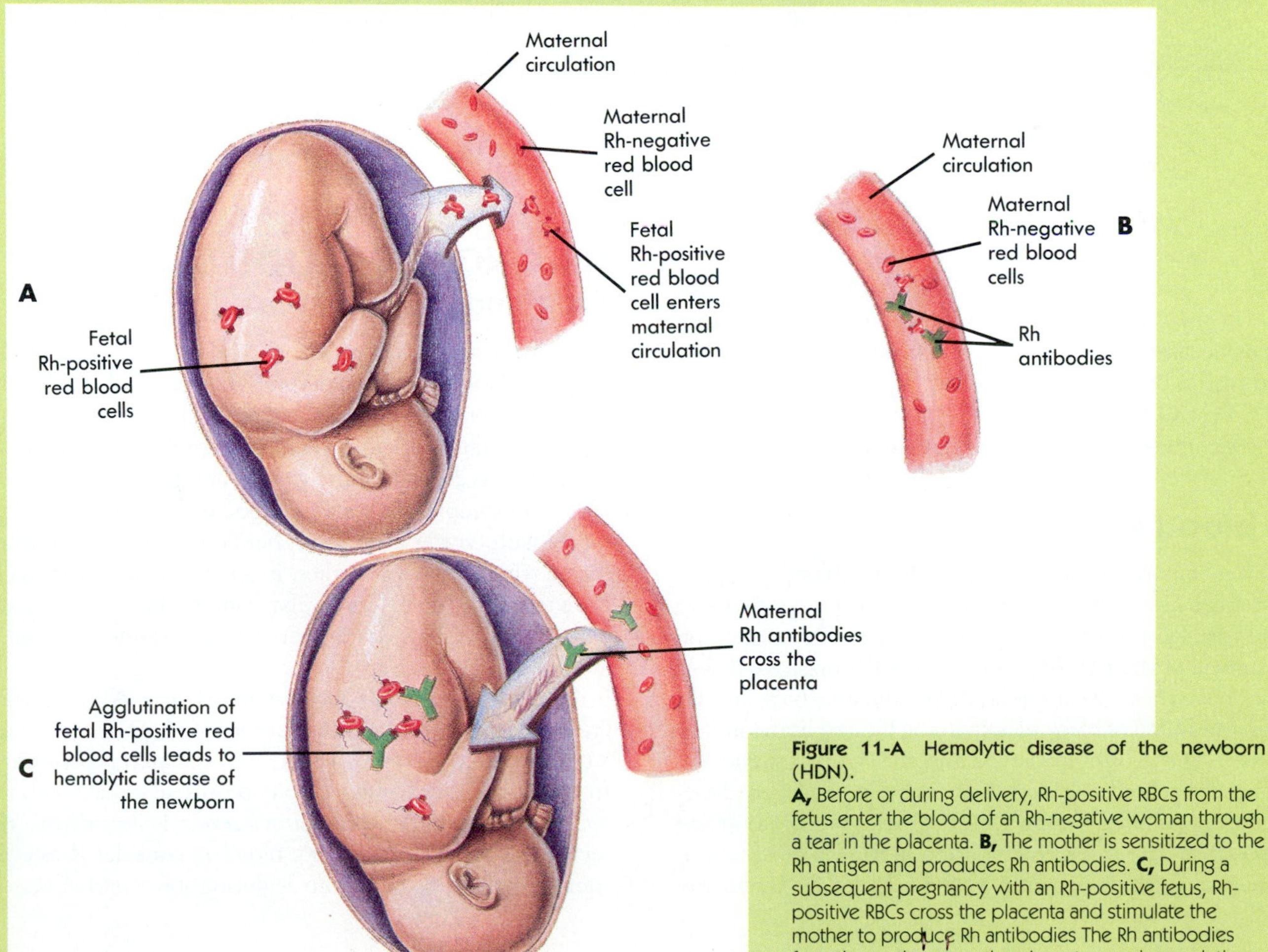

Figure 11-A Hemolytic disease of the newborn (HDN).
A, Before or during delivery, Rh-positive RBCs from the fetus enter the blood of an Rh-negative woman through a tear in the placenta. **B,** The mother is sensitized to the Rh antigen and produces Rh antibodies. **C,** During a subsequent pregnancy with an Rh-positive fetus, Rh-positive RBCs cross the placenta and stimulate the mother to produce Rh antibodies The Rh antibodies from the mother cross the placenta, causing agglutination and hemolysis of fetal RBCs, and HDN develops.

Some disorders of the blood

ANEMIA

Anemia (ă-ne´me-ah) is a deficiency of normal hemoglobin in the blood. The deficiency can be caused by an inadequate amount of normal hemoglobin. For example, a decreased number of RBCs, a decreased amount of hemoglobin in each RBC, or both. Or, the deficiency can be the result of abnormal hemoglobin production.

Anemia reduces the ability of the blood to transport oxygen. People with anemia suffer from a lack of energy and feel excessively tired and listless. They may appear pale and quickly become short of breath with only slight exertion.

One general cause of anemia is insufficient production of RBCs. **Aplastic anemia** is caused by an inability of the red bone marrow to produce RBCs. It is usually acquired as a result of damage to the red marrow by chemicals such as benzene, drugs such as certain antibiotics and sedatives, or radiation.

RBC production also can be lower than normal as a result of nutritional deficiencies. **Iron-deficiency anemia** results from a deficient intake or absorption of iron, or from excessive iron loss. Consequently not enough hemoglobin is produced, the number of RBCs decreases, and the RBCs that are manufactured are smaller than normal.

Another type of nutritional anemia is **pernicious** (per-nish´us) **anemia,** which is caused by inadequate vitamin B_{12}. Because vitamin B_{12} is necessary for the cell divisions that result in RBC formation, a shortage of vitamin B_{12} causes reduced RBC production. Although inadequate levels of vitamin B_{12} in the diet can cause pernicious anemia, the usual cause is insufficient absorption of the vitamin. Normally the stomach produces **intrinsic factor,** a protein that facilitates the absorption of vitamin B_{12}. Without adequate levels of intrinsic factor, insufficient vitamin B_{12} is absorbed, and pernicious anemia develops. Present evidence suggests that the inability to produce intrinsic factor is an autoimmune disease in which the body's immune system damages the cells in the stomach that produce intrinsic factor. **Folic acid deficiency** can also hinder cell divisions and cause anemia. A shortage of folic acid in the diet is the usual cause, with the disorder developing most often in the poor, in pregnant women, and in chronic alcoholics.

Another general cause of anemia is loss or destruction of RBCs. **Hemorrhagic** (hem-ŏ-raj´ik) **anemia** results from a loss of blood such as can result from trauma, ulcers, or excessive menstrual bleeding. Chronic blood loss in which small amounts of blood are lost over a period of time can result in iron-deficiency anemia. **Hemolytic** (he-mo-lit´ik) **anemia** is a disorder in which RBCs rupture or are destroyed at an excessive rate. It can be caused by inherited defects within the RBCs. For example, one kind of inherited hemolytic anemia results from a defect in the cell membrane that causes RBCs to rupture easily. Many kinds of hemolytic anemia result from unusual damage to the RBCs by drugs, snake venom, artificial heart valves, autoimmune disease, or hemolytic disease of the newborn.

Anemia can result from inadequate hemoglobin production. **Thalassemia** (thal-ă-se´me-uh) is a hereditary disease found in people of Mediterranean, Asian, and African ancestry. If hemoglobin production is severely depressed, death usually occurs before age 20. In less severe cases thalassemia produces a mild anemia.

Some anemias are caused by defective hemoglobin production. **Sickle cell anemia** is a hereditary disease found mostly in blacks that results in the formation of an abnormal hemoglobin. The RBCs assume a rigid, sickle shape

ple, type O blood has type A and B antibodies. If type O blood is transfused into a person with type A blood, the A antibodies (in the type O blood) react against the A antigens (in the type A blood). Usually such reactions are not serious because the antibodies in the donor's blood are diluted in the blood of the recipient, and few reactions take place. Type O blood is given to a person with another blood type only in life-or-death emergency situations.

Rh blood group

Another important blood group is the Rh blood group, so named because it was first studied in the rhesus monkey. People are Rh-positive if they have certain Rh antigens on the surface of their RBCs, and they are Rh-negative if they do not have these Rh antigens. Approximately 85% of Caucasians and 88% of black people in the United States are Rh positive. The ABO blood type and the Rh blood type are usually designated together. For example, a person designated as A positive is type A in the ABO blood group and Rh positive. The rarest combination in the United States is AB negative, which occurs in less than 1% of all Americans.

DIAGNOSTIC BLOOD TESTS

Type and cross match

To prevent transfusion reactions, the blood is typed, and a cross match is made. **Blood typing** determines the ABO and Rh blood groups of the blood sample. Typically, the cells are separated from the serum. The cells are tested with known antibodies to determine the type of antigen on the cell surface. For example, if a person's blood cells agglutinate when mixed with type A antibodies, but do not agglutinate when mixed with type B antibodies, it is concluded that the cells have type A antigen. In a similar fashion, the serum is mixed with known cell types (antigens) to determine the type of antibodies in the serum.

Normally, donor blood must match the ABO and Rh type of the recipient. Because other blood groups can cause a transfusion reaction, however, a cross match is performed. In a **cross match,** the donor's blood cells are mixed with the recipient's serum, and the donor's serum is mixed with the recipient's cells. The donor's blood is considered safe for transfusion only if there is no agglutination in either match.

and plug up small blood vessels. They are also more fragile than normal. In severe cases there is so much abnormal hemoglobin production that the disease is usually fatal before age 30. In many cases, however, the production of normal hemoglobin compensates for the abnormal hemoglobin and the person exhibits no symptoms.

HEMOPHILIA

Hemophilia (he´mo-fil´e-ah) is a genetic disorder in which clotting is abnormal or absent. It is most often found in people from northern Europe and their descendents. Hemophilia is a sex-linked trait, and it occurs almost exclusively in males. There are several types of hemophilia, each the result of a deficiency or dysfunction of a clotting factor. Treatment of hemophilia involves injection of the missing clotting factor taken from donated blood or produced by genetic engineering.

LEUKEMIA

Leukemia (lu-ke´me-ah) is a type of cancer in which abnormal production of one or more of the WBC types occurs. Because these cells are usually immature or abnormal and lack normal immunological functions, people with leukemia are very susceptible to infections. The excess production of WBCs in the red marrow may also interfere with RBC and platelet formation and thus lead to anemia and bleeding.

INFECTIOUS DISEASES OF THE BLOOD

After entering the body, many microorganisms are transported by the blood to the tissues they infect. For example, the polio virus enters through the small intestine and is carried to nervous tissue. After microorganisms are established at a site of infection, some of them can be picked up by the blood. These microorganisms can spread to other locations in the body, multiply within the blood, or be eliminated by the body's immune system.

Septicemia (sep´tĭ-se´me-ah), or blood poisoning, is the multiplication of microorganisms in the blood. Often septicemia results from the introduction of microorganisms by a medical procedure such as the insertion of an intravenous tube into a blood vessel. The release of toxins by bacteria can cause septic shock, which is a decrease in blood pressure that can result in death.

There are a few diseases in which microorganisms actually multiply within blood cells. **Malaria** (mă-la´re-ah) is caused by a protozoan that is introduced into the blood by the bite of the *Anopheles* mosquito. Part of the protozoan's development occurs inside RBCs. The symptoms of chills and fever are produced by toxins released when the protozoan causes the RBCs to rupture. **Infectious mononucleosis** (mon´o-nu´kle-o´sis) is caused by a virus that infects the salivary glands and lymphocytes. The lymphocytes are altered by the virus and the immune system attacks and destroys the lymphocytes. The immune system response is believed to produce the symptoms of fever, sore throat, and swollen lymph nodes. The **AIDS** virus also infects lymphocytes and causes immune system suppression (see Chapter 14).

The presence of microorganisms in blood is a concern when transfusions are made, because it is possible to infect the blood recipient. Blood is routinely tested in an effort to eliminate this risk, especially for AIDS and hepatitis. **Hepatitis** (hep´ă-ti´tis) is an infection of the liver caused by several different kinds of viruses. After recovering, hepatitis victims can become carriers. Although they show no signs of the disease, they release the virus into their blood or bile. To prevent infection of others, anyone who has had hepatitis is asked not to donate blood products.

Complete blood count

The **complete blood count (CBC)** is an analysis of the blood that provides much information. It consists of an RBC count, hemoglobin and hematocrit measurements, and a WBC count.

RBC count

A normal **red blood cell (RBC) count** for a male is 4.2 to 5.8 million RBCs per cubic millimeter of blood, and for a female it is 3.6 to 5.2 million RBCs per cubic millimeter of blood. **Polycythemia** (pol´e-si-the´me-ah) is an overabundance of RBCs. It can result from a decreased oxygen supply, which stimulates erythropoietin production, or from red bone marrow tumors. Polycythemia makes it harder for the blood to flow through blood vessels and increases the work load of the heart. It can reduce blood flow through tissues, and if severe, can result in plugging of small blood vessels (capillaries).

Hemoglobin measurement

The **hemoglobin measurement** determines the amount of hemoglobin in a given volume of blood, usually expressed as grams of hemoglobin per 100 milliliters of blood. The normal hemoglobin for a male is 14 to 18 grams per 100 milliliters of blood, and for a female it is 12 to 16 grams per 100 milliliters of blood. Abnormally low hemoglobin is an indication of anemia, which is a reduced number of RBCs, or a reduced amount of hemoglobin in each RBC.

Hematocrit measurement

The percentage of total blood volume composed of RBCs is the **hematocrit** (hem´ă-to-krit). One way to determine hematocrit is to place blood in a tube and spin the tube in a centrifuge. The formed elements are heavier than the plasma and are forced to one end of the tube (Figure 11-7). The RBCs account for 44% to 54% of the total blood volume in males and 38% to 48% in females.

White blood cell count

A **white blood cell (WBC) count** measures the total number of WBCs in the blood. There are normally 5,000 to 10,000 WBCs per cubic millimeter of blood. **Leukopenia** (lu-ko-pe´ne-ah) is a lower than normal WBC count and often indicates decreased production or destruction of the red

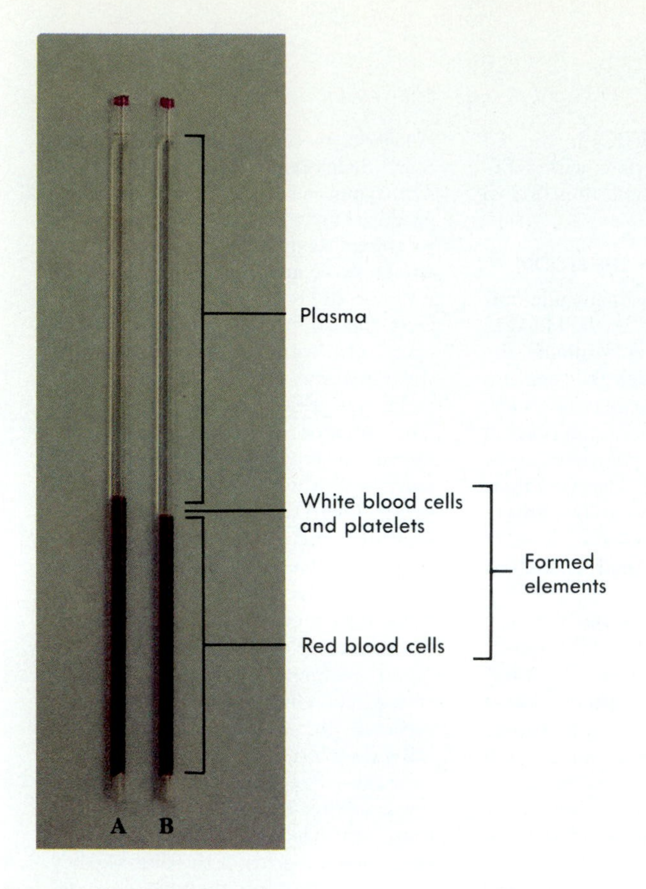

Figure 11-7 ● Hematocrit.
Blood is separated into plasma and formed elements. The relatively small amount of WBCs and platelets rest on the RBCs. **A,** Normal hematocrit of a male (44%). **B,** Normal hematocrit of a female (40%).

marrow. Radiation, drugs, tumors, viral infections, or a deficiency of vitamin B_{12} or folic acid can cause leukopenia. **Leukocytosis** (lu-ko-si-to´sis) is an abnormally high WBC count. **Leukemia** (lu-ke´me-ah), a tumor of the red marrow, and bacterial infections often cause leukocytosis.

White blood cell differential count

A **white blood cell differential count** determines the percentage of each of the five kinds of WBCs in the WBC count. Normally neutrophils account for 60% to 70%, lymphocytes 20% to 30%, monocytes 2% to 8%, eosinophils 1% to 4%, and basophils 0.5% to 1% of all the WBCs. Much insight about a person's condition can be obtained from a white blood cell differential count. For example, in bacterial infections, the neutrophil count is often greatly increased, whereas in allergic reactions the eosinophil and basophil counts are elevated.

Clotting

Two measurements that test the ability of the blood to clot are the platelet count and the prothrombin time.

Platelet count

A normal **platelet count** is 150,000 to 400,000 platelets per cubic millimeter of blood. **Thrombocytopenia** (throm-bo-si-to-pe´ne-ah) is a condition in which the platelet count is greatly reduced, resulting in chronic bleeding through small vessels and capillaries. It can be caused by decreased platelet production as a result of hereditary disorders, lack of vitamin B_{12} (pernicious anemia), drug therapy, or radiation therapy.

Prothrombin time measurement

Prothrombin time is a measure of how long it takes for the blood to start clotting, which is normally 9 to 12 seconds. Because many clotting factors have to be activated to form prothrombin, a deficiency of any one of them can cause an abnormal prothrombin time. Vitamin K deficiency, certain liver diseases, and drug therapy can cause an increased prothrombin time.

Blood chemistry

The composition of materials dissolved or suspended in the plasma can be used to assess the functioning of many of the body's systems. For example, high blood glucose levels can indicate that the pancreas is not producing enough insulin, high blood urea nitrogen (BUN) is a sign of reduced kidney function, increased bilirubin can indicate liver dysfunction, and high cholesterol levels can indicate an increased risk of developing cardiovascular disease. A number of blood chemistry tests are routinely done when a blood sample is taken, and additional tests are available.

4 When a person complains of acute pain in the abdomen, the physician suspects appendicitis, which is a bacterial infection of the appendix. What blood test can be done to help confirm the diagnosis?

SUMMARY

Blood transports gases, nutrients, waste products, and hormones.

Blood protects against disease and is involved in temperature, fluid, and electrolyte regulation.

PLASMA

Plasma is 92% water and 8% suspended or dissolved substances.

Plasma proteins maintain osmotic pressure, are involved in immunity, and prevent blood loss.

FORMED ELEMENTS

The formed elements are red blood cells (RBCs), white blood cells (WBCs) and platelets.

Red blood cells

RBCs are disk-shaped cells containing hemoglobin, which transport oxygen and carbon dioxide.

In response to low blood oxygen, the kidneys produce erythropoietin, which stimulates RBC production in red bone marrow.

Worn-out RBCs are phagocytized by macrophages in the spleen or liver. Hemoglobin is broken down, iron and amino acids are reused, and heme becomes bilirubin that is secreted in bile.

White blood cells

Neutrophils and macrophages (derived from monocytes) phagocytize microorganisms, dead cells, and debris.

Lymphocytes produce antibodies and are part of the immune system.

Basophils and eosinophils are involved in inflammation.

Platelets

Platelets are cell fragments involved with preventing blood loss.

PREVENTING BLOOD LOSS

Platelet plugs

Minor damage to blood vessels is repaired by platelet plugs.

Blood clotting

Blood clotting or coagulation is the formation of a clot (a network of protein fibers called fibrin).

There are three steps in the clotting process. (1) Activation of clotting factors by connective tissue and chemicals, resulting in the formation of prothrombin activator. (2) Conversion of prothrombin to thrombin by prothrombin activator. (3) Conversion of fibrinogen to fibrin by thrombin.

Control of clot formation

Clot formation is prevented by anticoagulants in the blood.

Clot dissolution

The breakdown of clots is accomplished by plasmin.

BLOOD GROUPING

Blood groups are determined by antigens on the surface of RBCs.

Antibodies can bind to RBC antigens, resulting in agglutination or hemolysis of RBCs.

ABO blood group

Type A blood has A antigens, type B blood has B antigens, type AB blood has A and B antigens, and type O blood does not have A or B antigens.

Type A blood has B antibodies, type B blood has A antibodies, type AB blood does not have A or B antibodies, and type O blood has A and B antibodies.

Mismatching the ABO blood group can result in transfusion reactions.

Rh blood group

Rh-positive blood has Rh antigens, whereas Rh-negative blood does not.

DIAGNOSTIC BLOOD TESTS

Type and cross match

Blood typing determines the ABO and Rh blood groups of a blood sample.

A cross match tests for agglutination reactions between donor and recipient blood.

Complete blood count

The complete blood count consists of the following: RBC count, hemoglobin measurement (grams of hemoglobin per 100 milliliters of blood), hematocrit measurement (percent volume of RBCs), and WBC count.

White blood cell differential count

The WBC differential count determines the percentage of each type of WBC.

Clotting

Platelet count and prothrombin time measure the ability of the blood to clot.

Blood chemistry

The composition of materials dissolved or suspended in plasma can be used to assess the functioning and status of the body's systems.

CONTENT REVIEW

1. List the major functions of blood.
2. Describe the composition of plasma. List the functions of plasma proteins.
3. Define the formed elements and name the different types of formed elements. Where are the formed elements produced?
4. Name the parts of a hemoglobin molecule. Which part binds to oxygen?
5. Why are vitamin B_{12}, folic acid, and iron important in RBC production?
6. Explain how low blood oxygen levels result in increased RBC production.
7. Where are RBCs broken down? What happens to the breakdown products?
8. Name the five types of WBCs and state a function for each type.
9. Describe platelets and how platelet plugs prevent bleeding.
10. What are clotting factors? Describe the three steps of their activation that result in the formation of a clot.
11. Explain the function of anticoagulants in the blood, and give an example of an anticoagulant.
12. How is a clot broken down?
13. What are blood groups, and how do they cause transfusion reactions?
14. List the four ABO blood types. Why is type O blood considered a universal donor and type AB blood a universal recipient?
15. What is meant by the term *Rh-positive?*
16. Define a thrombus and an embolus and explain why they are dangerous.
17. For each of the following tests, define the test and give an example of a disorder that would cause an abnormal test result.

a. Type and cross match
b. RBC count
c. Hemoglobin measurement
d. Hematocrit measurement
e. White blood cell count
f. White blood cell differential count
g. Platelet count
h. Prothrombin time
i. Blood chemistry tests

CONCEPT REVIEW

1. Red Packer, a physical education major, wanted to improve his performance in an upcoming marathon race. About 6 weeks before the race, 1 quart of blood was removed from his body, and the formed elements were separated from the plasma. The formed elements were frozen, and the plasma was reinfused into his body. Just before the race, the formed elements were thawed and injected into his body. Explain why this procedure, called blood doping, would help Red's performance.
2. Chemicals such as benzene can destroy red bone marrow. What symptoms would you expect to develop as a result of the lack of (1) RBCs, (2) platelets, and (3) WBCs?
3. E.Z. Goen habitually used barbiturates to depress feelings of anxiety. Because barbiturates suppress the respiratory centers in the brain, they cause hypoventilation (that is, slower than normal rate of breathing). What happens to the RBC count of a habitual user of barbiturates? Explain. (Hint: what would happen to E.Z.'s blood oxygen levels?)
4. According to the old saying, "Good food makes good blood." Name four substances in the diet that are essential for "good blood." What blood disorders develop if these substances are absent from the diet?
5. Why do people with anemia often have gray-colored feces (that is, not the normal brown color)?

CHAPTER TEST

Answers can be found in Appendix A

MATCHING For each statement in column A select the correct answer in column B (an answer may be used once, more than once, or not at all).

A	B
1. Liquid portion of blood.	bilirubin
2. The most numerous formed element.	erythropoietin
3. Cell fragment.	hemoglobin
4. Molecule that transports oxygen.	hemolysis
5. Stimulates RBC production.	iron
6. Breakdown product of heme.	jaundice
7. Yellowish color of the skin.	plasma
8. Element necessary for the production of hemoglobin.	platelet
9. Necessary for the cell divisions that form RBCs.	red blood cell
10. Necessary for the production of clotting factors.	vitamin B_{12}
	vitamin K
	white blood cell

FILL-IN-THE-BLANK Complete each statement by providing the missing word or words.

1. The two most important phagocytic WBCs are ___________ and ___________.
2. The WBCs that produce antibodies involved with the destruction of microorganisms are ___________.
3. Prothrombin activator converts ___________ to ___________.
4. Fibrinogen is converted to ___________ by ___________.
5. Excessive clot formation is prevented by ___________.
6. Plasminogen is converted to ___________, which dissolves clots.
7. Molecules called ___________ on the surface of RBCs react with molecules called ___________ in the plasma to cause transfusion reactions.
8. The clumping together of RBCs in a transfusion reaction is called ___________.
9. Type B blood has type ___________ antigens and type ___________antibodies.

CHAPTER 12

The Heart

KEY TERMS

atrium pl. atria
One of the two upper chambers of the heart; collects blood from the body or lungs during ventricular contraction and pumps blood into the ventricles during ventricular relaxation.

bicuspid valve
Valve consisting of two cusps of tissue; located between the left atrium and left ventricle of the heart.

coronary artery
Artery that carries blood to the muscle of the heart; arises from the base of the aorta.

diastole
Relaxation of the heart chambers during which they fill with blood; usually refers to ventricular relaxation.

electrocardiogram
Graphic record of the heart's electrical currents obtained with an electronic recording instrument.

semilunar valve
One of two valves in the heart composed of three semilunar-shaped cusps that prevent flow of blood back into the ventricles following ejection; located at the beginning of the aorta and pulmonary trunk.

sinoatrial (SA) node
Collection of specialized cardiac muscle fibers that acts as the "pacemaker" of the cardiac conduction system.

systole
Contraction of the heart chambers during which blood leaves the chambers; usually refers to ventricular contraction.

tricuspid valve
Valve consisting of three cusps of tissue; located between the right atrium and right ventricle of the heart.

ventricle
One of two lower chambers of the heart; pumps blood into arteries.

FEATURES

OBJECTIVES

After reading this chapter you should be able to:

1. Describe the size, shape, and location of the heart.
2. Describe the structure and name the components of the pericardial sac.
3. List the layers of the heart wall and describe the structure and function of each.
4. Describe the external anatomy of the heart.
5. Describe the flow of blood through the heart and name each of the chambers and structures through which the blood passes.
6. Explain the structure and function of the conduction system of the heart.
7. Define each wave of the electrocardiogram and relate each of them to contractions of the heart.
8. Name the heart sounds and explain their origin.
9. Describe the regulation of the heart.

THE HEART contracts forcefully to pump blood through the blood vessels of the body. In a healthy adult, at rest, the heart pumps about 5 liters of blood per minute, although this volume increases greatly during periods of exercise. For most people, the heart continues to pump for more than 75 years.

When a car's fuel pump stops, the motor also stops. Once the fuel pump is repaired the motor can be started again and it runs with no adverse effects. In contrast, if the heart loses its pumping ability for even a few minutes, the life of the individual is in danger.

Failure of the heart to function normally is a common health problem. Knowledge of the anatomy and physiology of the heart is required to understand the normal function of the heart and how it responds to disease. In this chapter the location and anatomy of the heart are described, followed by a discussion of its major functional characteristics and its regulation.

SIZE, FORM, AND LOCATION OF THE HEART

The adult heart has the shape of a blunt cone and is about the size of a closed fist. It is located in the thoracic cavity between the lungs. The blunt, rounded point of the cone is the **apex** (a´peks), and the larger, flat portion at the opposite end of the cone is the **base.** The apex is the most inferior and anterior part of the heart and it is directed to the left (Figure 12-1).

It is important to know the location of the heart. Placing a stethoscope to hear the heart sounds, placing electrodes on the chest to record an electrocardiogram (ECG), and performing cardiopulmonary resuscitation (CPR) depend on a knowledge of the heart's position.

FUNCTIONAL ANATOMY OF THE HEART

Pericardial sac

The **pericardial** (pĕr´ĭ-kar´de-al) **sac,** or **pericardium** (pĕr´ĭ-kar´de-um), surrounds the heart and anchors it within the thoracic cavity (Figure 12-2). A delicate layer of squamous epithelial cells, called the **serous pericardium,** lines the sac. The portion of the serous pericardium covering the heart surface is called the **visceral pericardium** or **epicardium** (ep´ĭ-kar´de-um). The portion of the serous pericardium lining the inner surface of the tough wall of the pericardial sac is called the **parietal pericardium.** The parietal pericardium and visceral pericardium are continuous with each other where the large arteries and veins enter or leave the heart. The tough outer wall of the pericardial sac is called the **fibrous pericardium.** The space between the visceral and parietal pericardium is the **pericardial cavity.** It contains a thin layer of **pericardial fluid** that helps reduce friction as the heart contracts and relaxes. Inflammation of the pericardium is called **pericarditis** (pĕr-ĭ-kar-di´tis).

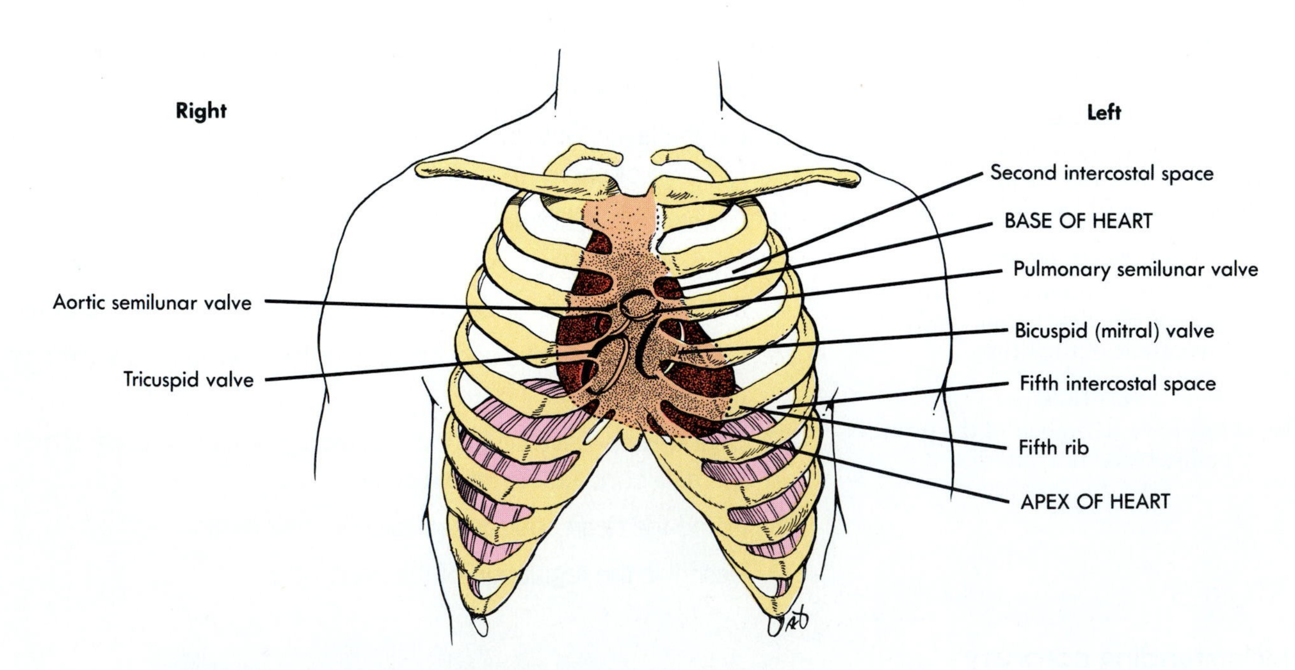

Figure 12-1 ● Location of the heart in the thorax.
The heart is deep to and slightly left of the sternum. The apex of the heart is in the fifth intercostal space about 9 cm from the midline. The base is in the second intercostal space. The valves of the heart are located near the base of the heart.

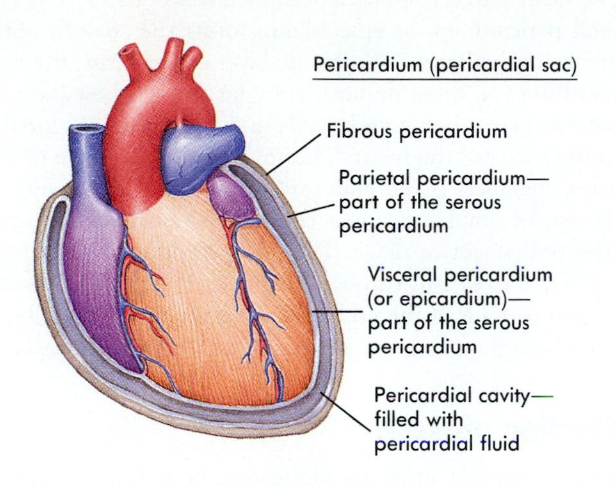

Figure 12-2 ● **Heart in the pericardial sac.**
The heart is located in the pericardial sac, which consists of an outer fibrous layer and an inner serous layer. The serous pericardium has two parts: the parietal pericardium lines the outer fibrous layer and the visceral pericardium (epicardium) lines the surface of the heart. The pericardial cavity, between the parietal and visceral pericardium, is filled with a small amount of pericardial fluid.

Heart chambers

The heart is a muscular pump consisting of four chambers: two **atria** (a´tre-ah; sing., **atrium,** a´tre-um) and two **ventricles** (ven´trĭ-kulz) (Figure 12-3). The thin-walled atria are located at the base of the heart, and the thick-walled ventricles extend from the base of the heart toward the apex. The two atria are separated by an **interatrial septum** and the two ventricles are separated by an **interventricular septum.**

Blood enters the right atrium from all parts of the body, except the lungs, through three major veins: the **superior vena cava** (ve´nah ca´vah), the **inferior vena cava,** and the smaller **coronary sinus.** Blood returning from the lungs enters the left atrium through the four **pulmonary veins.** Blood leaves the right ventricle through a large artery called the **pulmonary trunk** to flow to the lungs, and blood leaves the left ventricle through a large artery called the **aorta** to flow to all parts of the body (see Figure 12-3).

Heart valves

Valves located between the right atrium and right ventricle and between the left atrium and left ventricle allow blood to flow from the atria into the ventricles, but prevent blood from flowing back into the atria. Blood flowing from the atria into the ventricles pushes the valves open into the ventricles, but when the ventricles contract, blood pushes the valves back toward the atria and causes them to close.

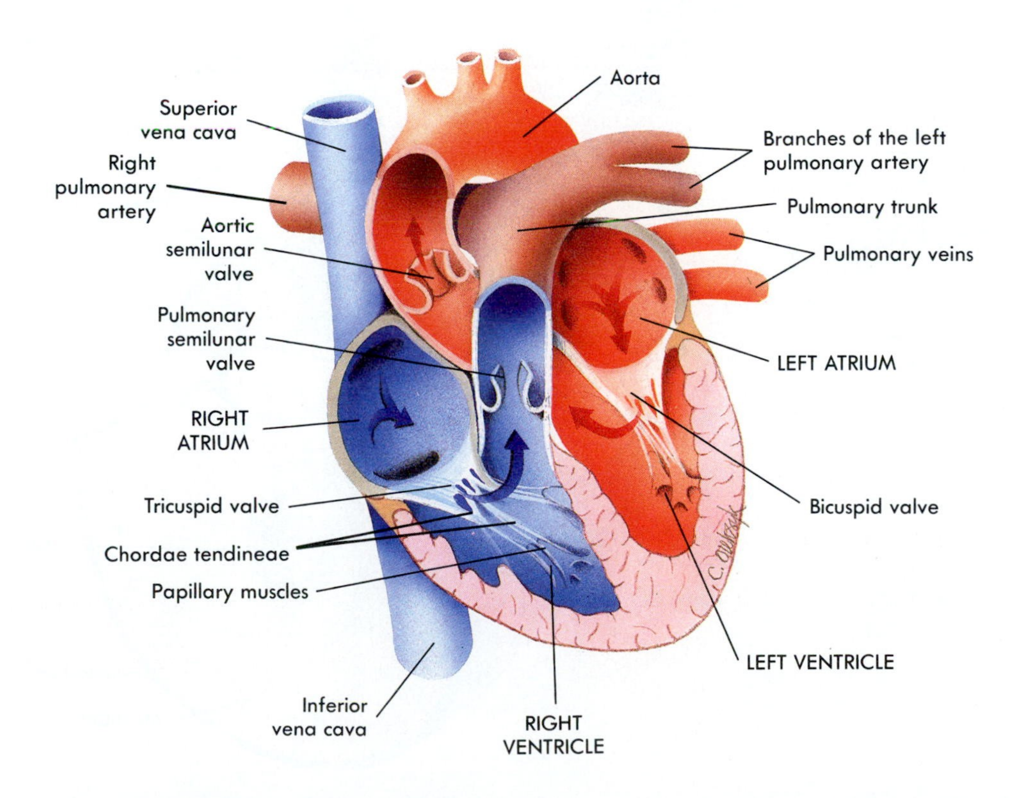

Figure 12-3 ● **Internal anatomy of the heart.**
Frontal section of the heart. The tricuspid and bicuspid valves and the semilunar valves are visible. *Arrows* show the direction of blood flow through the heart.

The valve between the right atrium and the right ventricle has three cusps and is called the **tricuspid** (tri-kus´ped) **valve.** The valve between the left atrium and left ventricle has two cusps and is called the **bicuspid** (bi-kus´ped), or **mitral** (mi´tral) (resembling a bishop's miter, a two-pointed hat), **valve** (see Figure 12-3).

The tricuspid and bicuspid valves are attached by thin, strong connective tissue strings, called **chordae tendineae** (kor´de ten´dĭ-ne-e), to cone-shaped muscular pillars called **papillary** (pap´ĭ-lĕr´e) **muscles** on the inner walls of the ventricles (see Figure 12-3). When the ventricles contract, blood pushes the cusps of the valves toward the atria, causing the valves to close. The chordae tendineae prevent the valves from inverting into the atria. Thus the valves close and stop blood from flowing back into the atria.

The aorta and pulmonary trunk have **aortic** and **pulmonary semilunar** (half moon-shaped) **valves,** respectively (see Figure 12-3). Each valve consists of three pocketlike semilunar cusps. Blood flowing out of the ventricles pushes against each valve, forcing it open. When blood flows back from the aorta or pulmonary trunk toward the ventricles, it enters the pockets of the cusps, causing them to meet in the center of the aorta or pulmonary trunk, thus closing the vessels and keeping blood from flowing back into the ventricles.

Heart wall

The heart wall is composed of three layers of tissue. The visceral pericardium, or epicardium, forms the smooth outer surface of the heart. The middle layer of the heart, the **myocardium** (mi´o-kar´de-um), is by far the thickest layer. It consists of cardiac muscle cells and is responsible for the contractions of the heart. The smooth inner surface of the heart chambers is the **endocardium** (en´do-kar´de-um). It consists of simple squamous epithelium that rests on a thin layer of connective tissue. The smooth endocardium allows blood to move easily through the heart. The heart valves are formed by folds of the endocardium, making a double layer of endocardium with connective tissue between the layers.

Cardiac muscle

Cardiac muscle cells are elongated, branching cells that contain one or occasionally two centrally located nuclei (Figure 12-4). Cardiac muscle cells appear striated, like in skeletal muscle, but the striations are less regularly arranged and less numerous than in skeletal muscle cells. The contractile structures in cardiac muscle are similar to those in skeletal muscle (see Chapter 7).

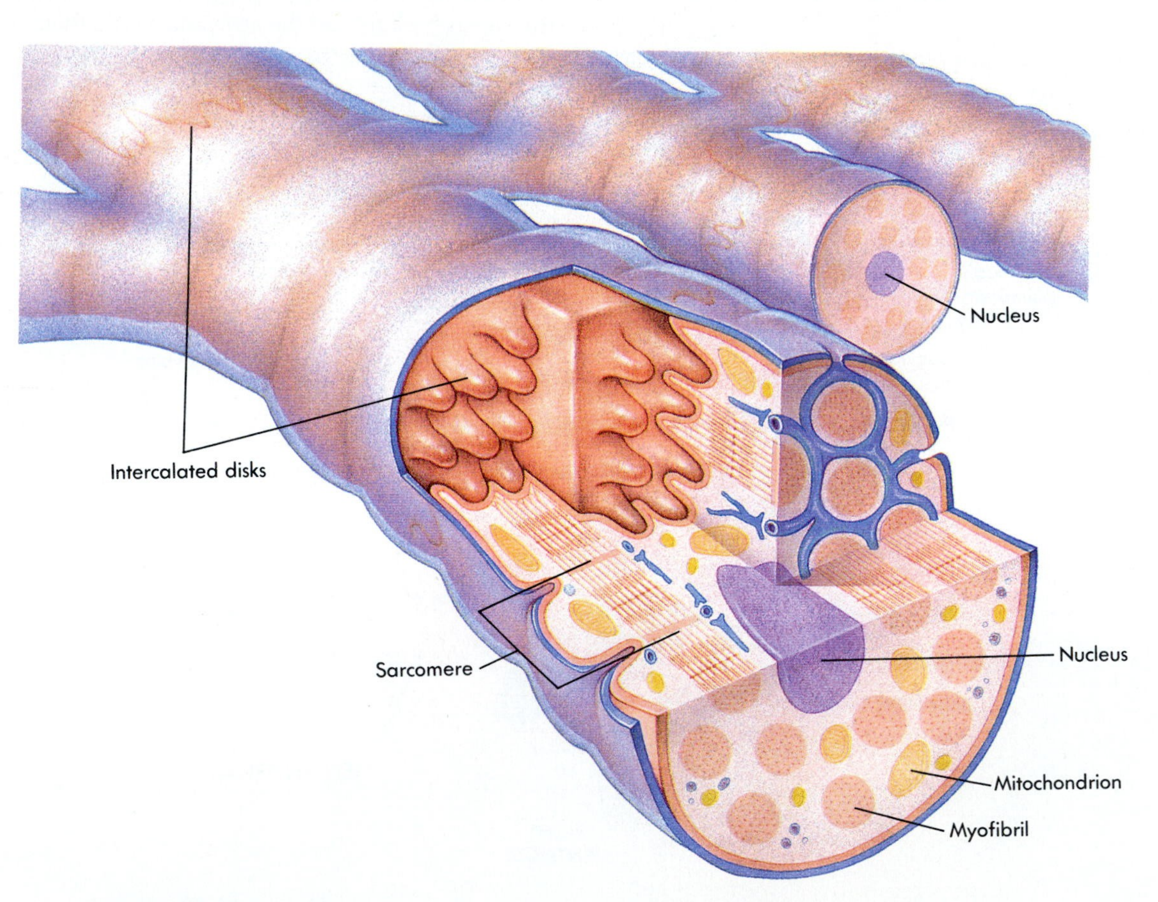

Figure 12-4 ● Cardiac muscle cells.
Cardiac muscle cells, like skeletal muscle cells, are striated. Intercalated disks, which bind cardiac muscle cells together, allow action potentials to pass from one cardiac muscle cell to the next.

Understanding coronary circulation

Cardiac muscle in the wall of the heart is thick and metabolically very active. The coronary arteries supply blood to the wall of the heart (Figure 12-A, *A* and *B*). Two **coronary arteries** originate from the base of the aorta just above the aortic semilunar valves. The left coronary artery originates on the left side of the aorta. Its branches supply much of the anterior wall of the heart and most of the left ventricle. The right coronary artery originates on the right side and supplies most of the wall of the right ventricle.

The **cardiac veins** drain blood from the cardiac muscle (see Figure 12-6, *B*). The cardiac veins are nearly parallel to the coronary arteries and they converge toward the posterior portion of the heart and empty into an enlarged vein called the **coronary sinus.** The coronary sinus, in turn, empties into the right atrium.

When a blood clot, or **thrombus** (throm´bus), suddenly blocks a coronary artery, a heart attack, or **coronary thrombosis,** occurs. The area cut off from its blood supply suffers from a lack of oxygen and nutrients, and dies if the blood supply is not quickly reestablished. The region of dead heart tissue is called an **infarct** (in´farkt). If the infarct is large enough, the heart may be unable to pump enough blood to keep the person alive.

In some cases it is possible to treat heart attacks with enzymes that break down blood clots. The enzymes are injected into the circulatory system of a heart attack patient, where they reduce or remove the blockage in the coronary artery. If the clot is broken down quickly, the blood supply to cardiac muscle is reestablished and the heart may suffer little permanent damage. There are a number of enzymes that can be used to break down clots, but it is not clear which one is most effective. In addition, the enzymes must be administered within a short time after the onset of the blockage to be effective.

Coronary arteries can become blocked gradually by **atherosclerotic** (ath´er-o-skle-rah´tik) **lesions.** Atherosclerotic lesions are thickenings in the walls of arteries containing deposits high in cholesterol and other lipids. The ability of cardiac muscle to function is reduced when it is deprived of an adequate blood supply. The person suffers from fatigue and often pain in the area of the chest and usually in the left arm with the slightest exertion. The pain is called **angina pectoris** (an-ji-nah pek´to-rus). Nitroglycerine is a drug that is often given to people suffering from angina pectoris. It dilates arteries and veins, which reduces the amount of work required of cardiac muscle.

If a coronary artery is partially blocked, **angioplasty** (an´je-o-plas´-te) is a procedure that is commonly used to open the blocked region of the coronary artery. A small balloon is threaded into the aorta and allowed to enter the coronary arteries. After the balloon has entered the coronary artery that is partially blocked, the balloon is inflated to open the blockage.

Another surgical procedure that relieves the effects of blocked coronary arteries is **coronary bypass surgery.** The technique involves taking healthy segments of blood vessels from other parts of the patient's body and using them to bypass obstructions in the coronary arteries. The technique is common for those who suffer from severe blockage of the coronary arteries.

1 Predict the effect of complete and sudden blockage of the left coronary artery on the heart's ability to pump blood.*

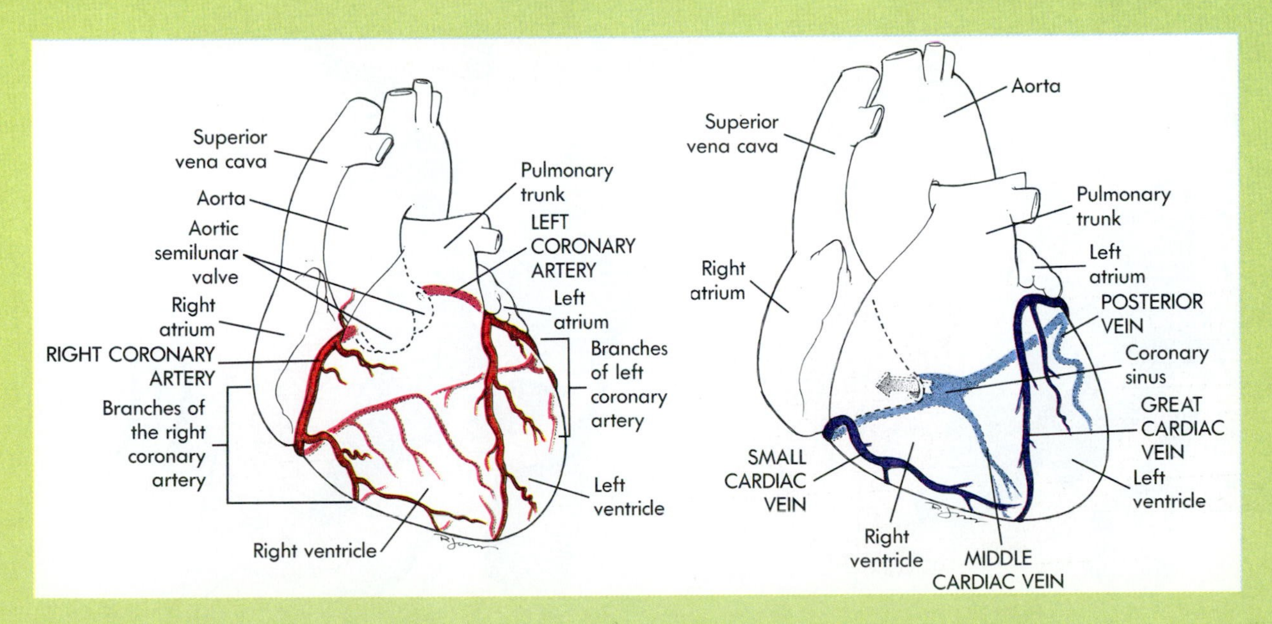

Figure 12-A Coronary vessels.
The vessels of the anterior surface are seen directly and have been given a darker color, whereas the vessels of the posterior surface are seen through the heart and have been given a lighter color. **A,** Coronary arteries supply blood to the wall of the heart. **B,** Cardiac veins carry blood from the wall of the heart back to the right atrium.

*Answers to predict questions appear in Appendix B at the back of the book.

Figure 12-5 ● Blood flow through the heart.
A, During diastole blood flows to the right atrium from systemic vessels (inferior and superior venae cavae) and cardiac veins into the right atrium, and from the pulmonary veins into the left atrium; the blood in the right atria pushes the tricuspid valve open and blood in the left atria pushes the bicuspid valve open. Blood flows directly from the right atrium into the right ventricle and from the left atrium into the left ventricle. The pulmonary and aortic semilunar valves remain closed during this part of the cardiac cycle. **B,** When the atria contract, blood is forced from the atria into the ventricles to complete ventricular filling. As long as the ventricles are relaxed, the semilunar valves (aortic and pulmonary) are closed. **C,** As the ventricles contract, the tricuspid and bicuspid valves close. After the pressure in the ventricles has increased until it exceeds the pressure in the pulmonary trunk and the aorta, the semilunar valves open, and blood is forced into the pulmonary trunk and aorta. When the ventricles begin to relax, the semilunar valves close, preventing blood from reentering the ventricles from the pulmonary trunk and aorta.

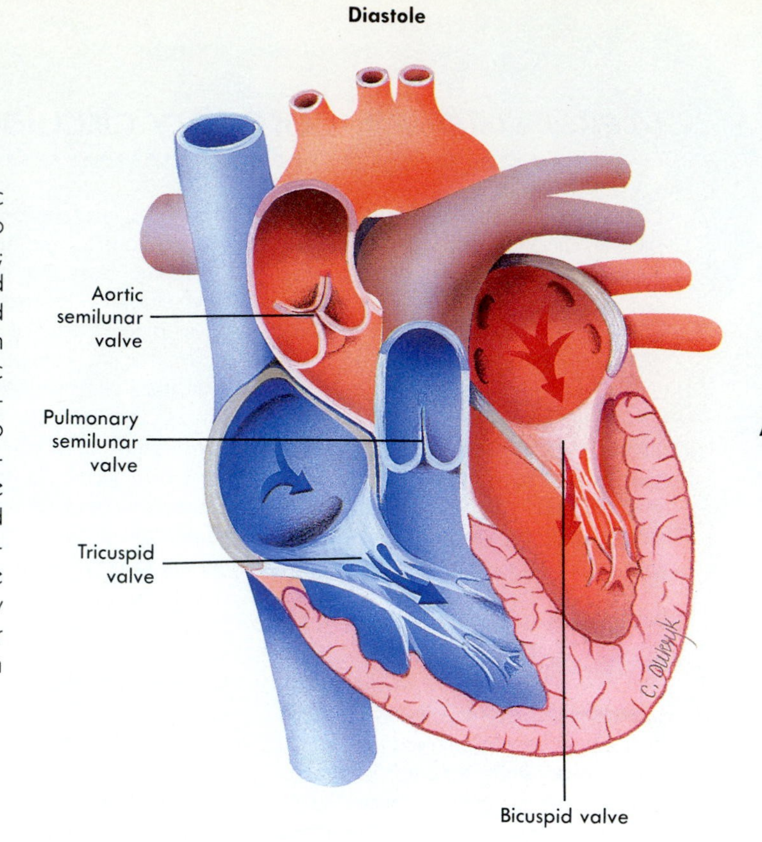

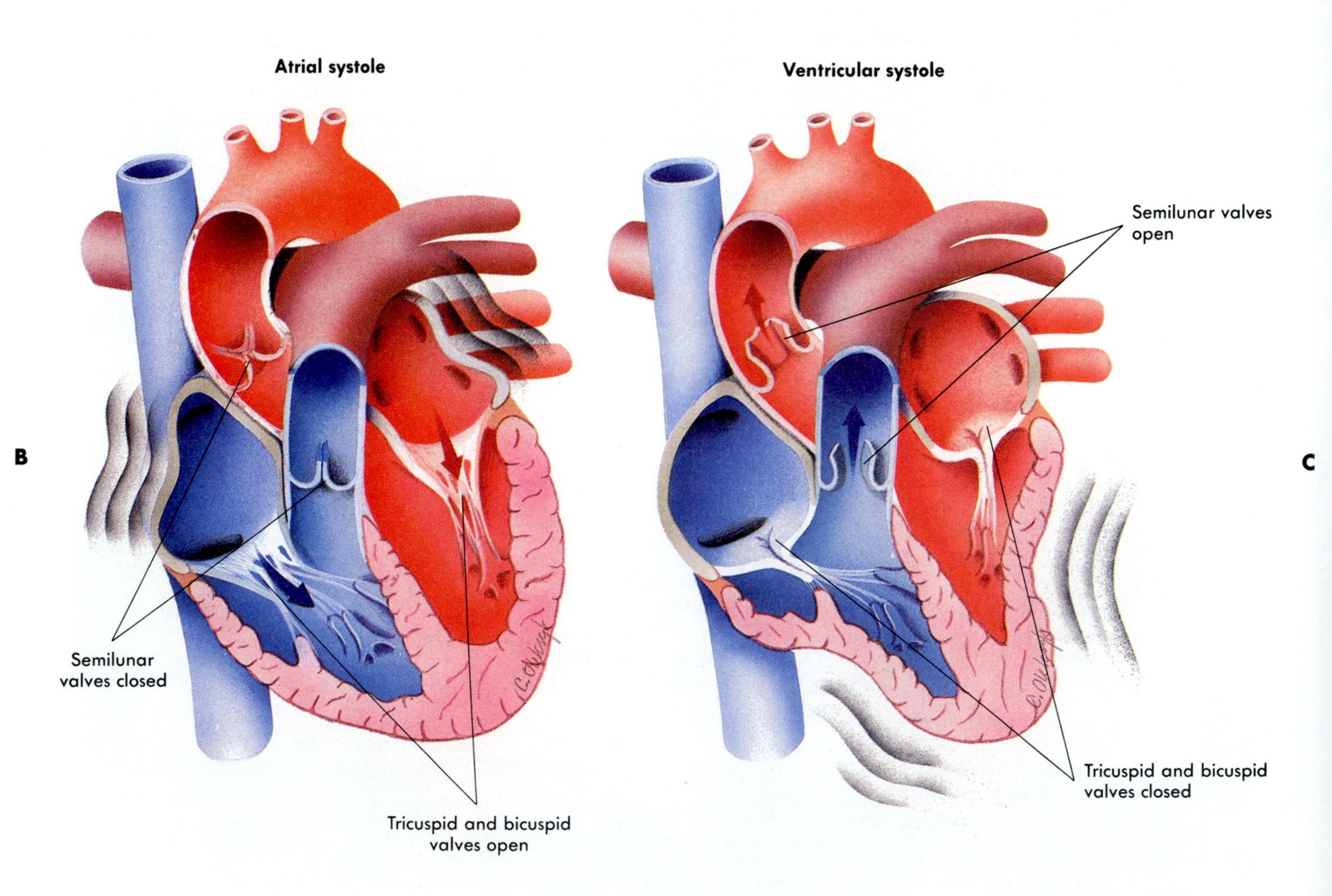

Cardiac muscle cells are joined end-to-end and laterally to adjacent cells by specialized cell-to-cell contacts called **intercalated** (in-ter´kă-la-ted) **disks** (see Figure 12-4). Action potentials cross from one cell to adjacent cells at the intercalated disks. As a result, cardiac muscle cells function as a unit rather than as individual cells. Once action potentials are produced in cardiac muscle, they pass from cell to cell and cause all of the cardiac muscle cells to contract.

Adenosine triphosphate (ATP) provides the energy for cardiac muscle contraction. Cardiac muscle cells have many mitochondria, which produce ATP at a rate rapid enough to sustain the normal energy requirements of cardiac muscle. An extensive capillary network provides an adequate oxygen supply to the cardiac muscle cells. Cardiac muscle requires a continuous supply of oxygen, and it will stop contracting after a very short time without oxygen. Unlike skeletal muscle, cardiac muscle cannot develop a significant oxygen debt (see Chapter 7).

BLOOD FLOW THROUGH THE HEART

The pumping action of the heart depends on the alternating contraction of the two atria and the two ventricles. Ventricular **diastole** (di-as´to-le) refers to relaxation of the two ventricles, and atrial diastole refers to relaxation of the two atria. Ventricular **systole** (sis´to-le) refers to the contraction of the two ventricles and atrial systole refers to contraction of the two atria. When the terms diastole and systole are used alone they refer to ventricular relaxation and contraction, because the ventricles contain more cardiac muscle and produce the pressure that forces the blood to circulate throughout the vessels of the body. Diastole and the next systole represent one **cardiac cycle.**

Blood flow through the heart is depicted in Figure 12-3. Blood enters the right atrium from all parts of the body except the lungs. Blood flows from the upper portion of the body through the superior vena cava, from the lower portion of the body through the inferior vena cava, and from the coronary circulation through the coronary sinus into the right atrium (see box, p. 195). Blood in the right atrium then flows into the right ventricle.

Contraction of the right ventricle pushes blood into the pulmonary trunk. The pulmonary trunk branches to form the two **pulmonary arteries,** which carry blood to the left and right lungs, where carbon dioxide is released and oxygen is picked up.

Blood returning from the lungs enters the left atrium through four pulmonary veins. Blood then flows into the left ventricle. Contraction of the left ventricle pushes blood into the aorta. Blood flowing through the aorta is distributed to all parts of the body except the lungs.

Even though it is convenient to describe blood flow through the heart one side at a time, it is important to understand that both atria contract at the same time followed by the simultaneous contraction of both ventricles. These coordinated contractions combined with the actions of the heart valves ensure effective movement of blood.

Blood enters the right atrium from the body and flows through the open tricuspid valve into the right ventricle at the same time that blood from the lungs enters the left atrium and flows through the open bicuspid valve into the left ventricle (Figure 12-5, *A*). Approximately 70% of ventricular filling occurs as a result. Then both atria contract simultaneously and the ventricles are filled with additional blood (Figure 12-6, *B*). After the atria contract, simultaneous contraction of the ventricles results in the movement of blood back toward the atria, which is prevented by closure of the tricuspid and bicuspid valves. As the ventricles continue to contract, the pulmonary semilunar valve opens and blood flows toward the lungs, and the aortic semilunar valve opens and blood flows to the body (Figure 12-5, C).

CONDUCTION SYSTEM OF THE HEART

Although the heart is innervated by both parasympathetic and sympathetic nerve fibers that can influence the heart's function, the heart is mainly regulated by specialized cardiac muscle cells in the wall of the heart that form the **conduction system of the heart** (Figure 12-6). The **sinoatrial (SA)** (si´no-a´tre-al) **node** is located in the upper wall of the right atrium (see Figure 12-6). It is called the **pacemaker** of the heart because it is the site where action potentials that initiate contraction of the heart originate. Because action potentials originate from the SA node automatically, without stimulation from the nervous or endocrine systems, the heart is said to be **autorhythmic.** After action potentials originate in the SA node, they spread over the right and left atria causing them to contract.

A second area of the heart, called the **atrioventricular (AV)** (a´tre-o-ven´trik´u-lar) **node,** is located in the lower portion of the right atrium. When action potentials reach the AV node, they spread slowly through the AV node and into a bundle of specialized cardiac muscle called the **bundle of His,** or **atrioventricular bundle.** The slower rate of action potential conduction in the AV node allows the atria to complete their contraction before action potentials are delivered to the ventricles.

After the action potentials pass through the AV node, they are conducted rapidly through the atrioventricular bundle, which projects into the ventricles. The atrioventricular bundle then forms two branches called the **left and right bundle branches,** which descend on either side of the interventricular septum. Near the apex of the heart the left and right bundle branches form many small bundles of

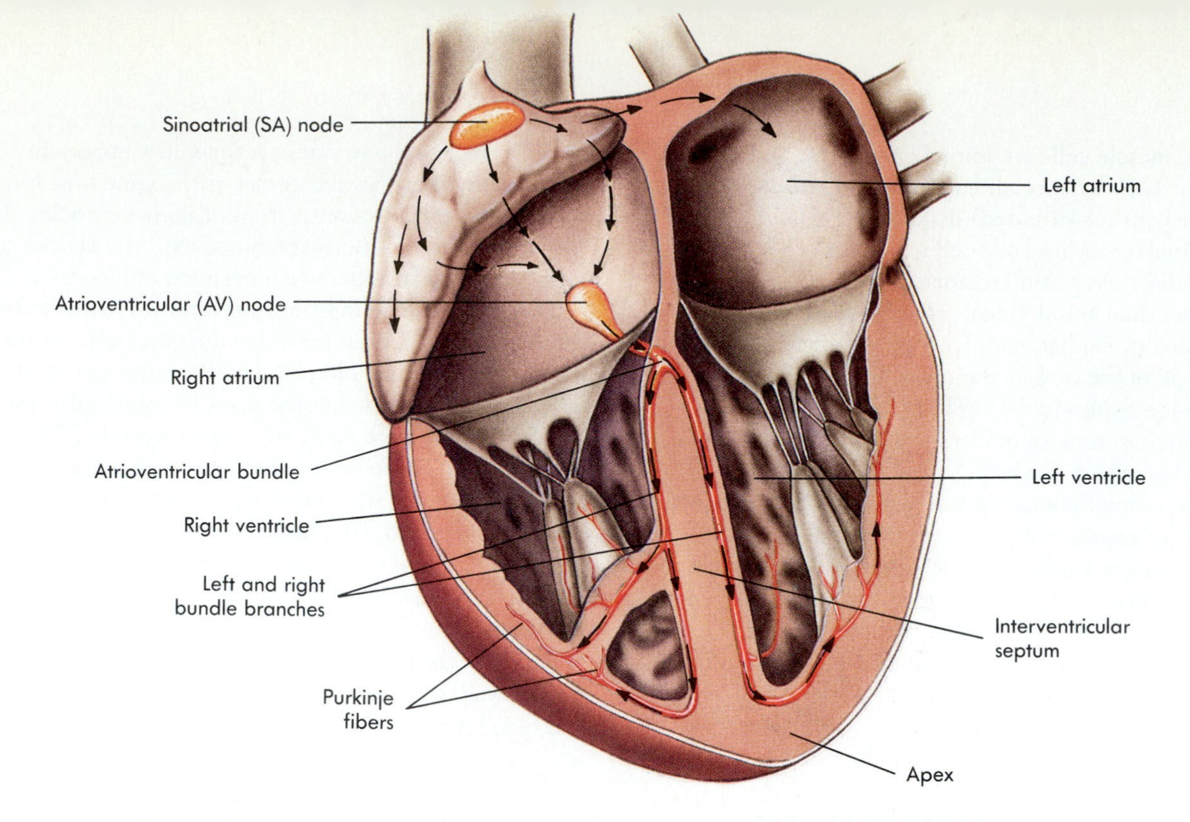

Figure 12-6 ● **Conducting system of the heart.**
Action potentials *(arrows)* travel across the wall of the right atrium from the SA node to the AV node. The atrioventricular bundle extends from the AV node to the interventricular septum, where it divides into right and left bundle branches. The bundle branches descend to the apex of each ventricle and then branch repeatedly to form Purkinje fibers, which are distributed throughout the ventricular walls.

Purkinje (per-kin´je) **fibers.** The Purkinje fibers pass around the apex of the ventricle and extend to the cardiac muscle of the ventricle walls. The atrioventricular bundle, the bundle branches, and the Purkinje fibers rapidly deliver action potentials to all the cardiac muscle of the ventricles. The coordinated contraction of the ventricles depends on the rapid conduction of action potentials by the conduction system within the ventricles.

2 Predict the effect on ventricular contractions if the blood supply is reduced to the area of the heart through which the left bundle branch passes.

Following their contraction, the ventricles begin to relax. After the ventricles have completely relaxed, another action potential originates in the SA node to begin the next cardiac cycle. The SA node is the pacemaker of the heart because action potentials originate spontaneously in the SA node faster (60 beats/minute or more) than in other areas of the heart. Other cardiac muscle cells, however, are capable of producing action potentials spontaneously. If the SA node is unable to function, another area of the heart, such as the AV node, becomes the pacemaker. The resulting heart rate is much slower than normal (approximately 40 beats/minute or less). When action potentials originate in an area of the heart other than the SA node, the result is an **ectopic** (ek-top´ik) **beat.**

Cardiac muscle can act as if there are thousands of pacemakers, each making a very small portion of the heart contract rapidly and independently of all other areas. This condition is called **fibrillation,** and it reduces the heart's output to only a few milliliters per minute when it occurs in the ventricles. Death results in a few minutes unless fibrillation of the ventricles is stopped. To stop the process of fibrillation, **defibrillation** is employed, in which a brief but strong electrical shock is applied to the chest region to depolarize all cardiac muscle cells at the same time. Cells of the SA node recover first and produce action potentials before any other area of the heart. Consequently, the normal pattern of action potential generation and the normal rhythm of contraction can be reestablished.

ELECTROCARDIOGRAM

Action potentials conducted through the heart during the cardiac cycle produce electrical currents that can be measured at the surface of the body. Electrodes placed on the surface of the body and attached to a recording device can detect the small electrical changes resulting from the action potentials in cardiac muscle. The record of these electrical events is an **electrocardiogram (ECG or EKG)** (Figure 12-7).

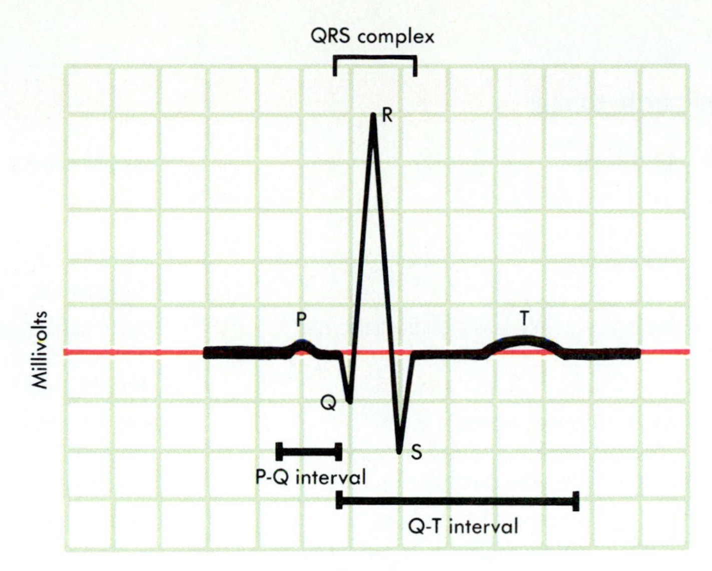

Figure 12-7 ● Electrocardiogram.
The major waves and intervals are labeled.

The ECG is not a direct measurement of cardiac muscle contractions and cannot be used to measure blood pressure. The action potentials that give rise to the ECG provide the stimulus that causes cardiac muscle to contract. Thus each deflection in the ECG record is an electrical event that provides information about the pattern of contractions in the atria and ventricles of the heart. Analysis of an ECG can be useful to identify abnormal heart rates or rhythms, abnormal conduction pathways, hypertrophy (enlargement) or atrophy (decreased size) of portions of the heart, and the approximate location of damaged cardiac muscle. The ECG is an extremely important diagnostic tool that is painless and easy to record, and does not require surgical procedures.

The normal ECG consists of a P wave, a QRS complex, and a T wave. The **P wave** results from depolarization of the atria and precedes atrial contractions. The **QRS complex** consists of three individual waves: the Q, R, and S waves. The QRS complex results from depolarization of the ventricles, and the beginning of the QRS complex precedes ventricular contraction. The **T wave** represents repolarization of the ventricles and precedes ventricular relaxation. A wave representing repolarization of the atria cannot be seen because it occurs during the QRS complex.

The time between the beginning of the P wave and the beginning of the QRS complex is the P-Q interval. The P-Q interval is commonly called the P-R interval because the Q wave is very small. During the P-Q interval the atria contract and begin to relax. The ventricles begin to depolarize at the end of the P-Q interval.

The Q-T interval extends from the beginning of the QRS complex to the end of the T wave and represents the length of time required for ventricular depolarization and repolarization. Table 12-1 describes several conditions associated with abnormal heart rhythms.

3 Predict the effect on the electrocardiogram if many ectopic action potentials occurred in the atria but not all of the ectopic action potentials stimulated QRS complexes in the ventricles.

HEART SOUNDS

A **stethoscope** (steth´o-skōp) is used to listen to heart sounds. There are two main heart sounds. The **first heart sound** can be represented by the syllable *lubb* and the **second heart sound** can be represented by *dupp*. The first heart sound has a lower pitch than the second heart sound. The first heart sound occurs at the beginning of ventricular systole and results from closure of the tricuspid and bicuspid valves. The second heart sound occurs at the beginning of ventricular diastole and results from closure of the semilunar valves. The valves usually do not make sounds when they open.

4 Compare the amount of blood flowing out of the ventricles between the first and second heart sounds of the same beat with the amount of blood flowing out of the ventricles between the second heart sound of one beat and the first heart sound of the next beat.

Abnormal heart sounds called **murmurs** are usually a result of faulty valves. A valve that fails to close tightly, allowing blood to leak through the valve when it is closed, is said to be an **incompetent valve.** The sound results from blood leaking through the incompetent valve when it is closed. A murmur caused by an incompetent valve is a swishing sound immediately after closure of the valve. For example, an incompetent bicuspid valve would result in a swishing sound immediately after the first heart sound, and

Table 12-1 • Major cardiac arrhythmias

Condition	Symptoms	Possible causes
• Abnormal heart rhythms		
Tachycardia	Resting heart rate in excess of 100 beats/min	Elevated body temperature, excessive sympathetic stimulation, toxic conditions
Bradycardia	Resting heart rate less than 60 beats/min	Elevated stroke volume in athletes, excessive vagus nerve stimulation, nonfunctional SA node (carotid sinus syndrome)
Sinus arrhythmia	Heart rate varies as much as 5% during respiratory cycle and up to 30% during deep respiration	Cause not always known; occasionally caused by ischemia, inflammation, or cardiac failure
Paroxysmal atrial tachycardia	Sudden increase in heart rate to 95-150 beats/min for a few seconds or even for several hours; P waves precede every QRS complex; P wave inverted and superimposed on T wave	Excessive sympathetic stimulation, abnormally elevated permeability of cardiac muscle to calcium ions
Atrial flutter	As many as 300 P waves/min and 125 QRS complexes/min; resulting in 2 or 3 P waves (atrial contractions) for every QRS complex (ventricular contraction)	Ectopic beats in the atria
Atrial fibrillation	No P waves, normal QRS and T waves, irregular timing, ventricles are constantly stimulated by atria, reduced ventricle filling; increased chance of ventricular fibrillation	Ectopic beats in the atria
Ventricular fibrillation	No QRS complexes or T waves, no rhythmic contraction of myocardium, many patches of asynchronously contracting ventricular muscle	Ectopic beats in the ventricles
• Heart blocks		
SA node block	No P waves, low heart rate resulting from AV node acting as pacemaker, normal QRS complexes and T waves	Ischemia, tissue damage resulting from infarction; cause sometimes not known
AV node blocks		
First degree	P-R interval greater than 0.2 sec	Inflammation of AV bundle
Second degree	P-R interval 0.25 to 0.45 sec; some P waves trigger QRS complexes and others do not; examples of 2:1, 3:1, and 3:2 P wave/QRS complex ratios	Excessive vagus nerve stimulation, AV node damage
Complete heart block	P wave dissociated from QRS complex, atrial rhythm about 100 beats/min, ventricular rhythm less than 40 beats/min	Ischemia of AV node or compression of AV bundle
• Premature contractions		
Premature ventricular contractions (PVCs)	Prolonged QRS complex, exaggerated voltage because only one ventricle may depolarize, possible inverted T wave, increased probability of fibrillation	Ectopic beat in ventricles, lack of sleep, too much coffee, irritability, occasionally occurs with coronary thrombosis

an incompetent aortic semilunar valve would result in a swishing sound immediately after the second heart sound.

A **stenosed** (sten´ōzd) **valve** is a valve in which the opening of the valve is narrowed. A murmur caused by a stenosed valve precedes closure of the stenosed valve. The sound is caused by blood being forced to flow through the narrowed opening of the valve. For example, when the bicuspid valve is stenosed, a swishing sound precedes the first heart sound, and when the aortic semilunar valve is stenosed, a swishing sound precedes the second heart sound.

5 If normal heart sounds are represented by *lubb-dupp, lubb-dupp,* what would a heart sound represented by *lubb-duppswish, lubb-duppswish* represent? What would *lubb-swishdupp, lubb-swishdupp* represent?

REGULATION OF HEART FUNCTION

Cardiac output is the volume of blood pumped by the ventricles of the heart each minute. The cardiac output can be calculated by multiplying the **stroke volume** (the volume of blood pumped per ventricle each time the heart contracts) by the **heart rate** (the number of times the heart contracts each minute).

$$\underset{(\text{ml/min})}{\text{Cardiac output}} = \underset{(\text{ml/beat})}{\text{Stroke volume}} \times \underset{(\text{beats/min})}{\text{Heart rate}}$$

For example, if the stroke volume is 70 ml/beat and the heart rate is 72 beats/minute, the cardiac output can be easily calculated:

$$\underset{(\text{Cardiac output})}{5{,}040 \text{ ml/min}} = \underset{(\text{Stroke volume})}{70 \text{ ml/beat}} \times \underset{(\text{Heart rate})}{72 \text{ beats/min}}$$

There are a number of control mechanisms that modify heart rate and stroke volume. Regulation of the cardiac output is necessary to maintain adequate blood flow to the tissues of the body under varying conditions. For example, blood flow to skeletal muscle increases dramatically during exercise to supply adequate oxygen and nutrients to the exercising skeletal muscles, but blood flow to skeletal muscles decreases during periods of rest.

Intrinsic regulation of the heart

Intrinsic regulation of the heart modifies the autorhythmic control of the heart and refers to those mechanisms contained within the heart itself. Intrinsic regulation is not due to either hormonal or nervous influences. If the amount of blood entering the heart increases, the heart fills to a greater volume, which stretches the cardiac muscle fibers. In response to stretch, cardiac muscle fibers contract with a greater force. The greater force of contraction causes an increased volume of blood to be pumped from the heart, resulting in an increased stroke volume. In addition to increasing the stroke volume, stretch of cardiac muscle causes a slight increase in the heart rate. Therefore, as the volume of blood entering the heart increases, cardiac output increases. Also, if the volume of blood entering the heart decreases, the cardiac output decreases. This relationship is called **Starling's law of the heart.**

Starling's law of the heart has a major influence on cardiac output. For example, muscular activity during exercise increases the volume of blood returning to the heart, which increases cardiac output. This is beneficial because an increased cardiac output is needed during exercise to supply oxygen to exercising skeletal muscles.

Extrinsic regulation of the heart

Extrinsic regulation of the heart modifies the autorhythmic control of the heart and refers to both nervous and hormonal regulation of the heart. Nervous influences are carried through the autonomic nervous system. Both sympathetic and parasympathetic nerve fibers innervate the SA node of the heart. Sympathetic stimulation causes the heart rate and the stroke volume to increase, whereas parasympathetic stimulation causes the heart rate to decrease.

The **baroreceptor** (bar´o-re-sep´tor) **reflex** plays an important role in regulating the function of the heart. **Baroreceptors** are stretch receptors that monitor blood pressure in the aorta and in the wall of the internal carotid arteries, which carry blood to the brain (Figure 12-8). As blood pressure increases, the walls of these arteries are stretched, in-

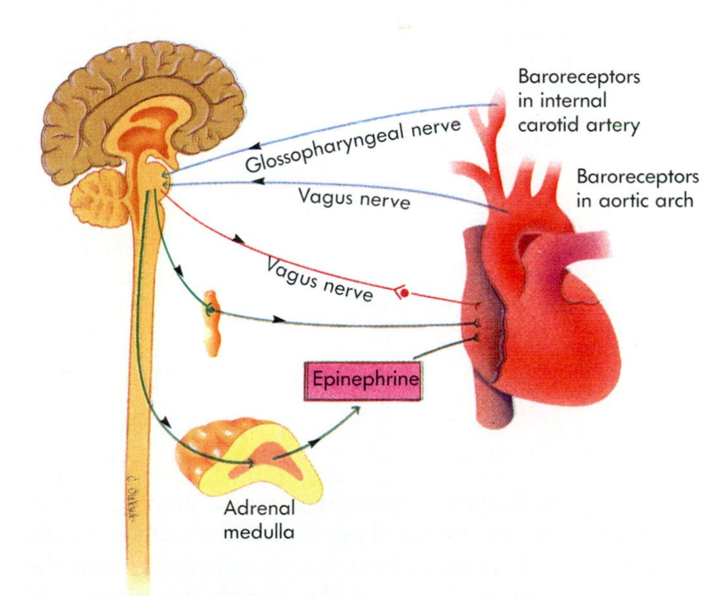

Figure 12-8 ● **Extrinsic regulation of the heart.** Sympathetic *(green)* and parasympathetic *(red)* nerves exit the spinal cord or medulla and extend to the heart to regulate its function. Hormonal influences such as epinephrine from the adrenal gland also help regulate the heart's action. Action potentials from baroreceptors in the aortic arch and carotid arteries are sent along nerve fibers to the medulla oblongata of the brain.

Conditions and diseases affecting the heart

HEART DISEASES

Endocarditis (en´do-kar-di´tis) is inflammation of the lining of the heart cavities. It affects the valves more frequently than other areas of the heart and may lead to deposition of scar tissue, which causes valves to become stenosed or incompetent. Endocarditis can result from bacterial infection or from the immune response to certain bacterial infections (see rheumatic heart disease).

Myocarditis (mi´o-kar-di´tis) is inflammation of the heart muscle and can lead to heart failure.

Pericarditis (pĕr-ĭ-kar-di´tis) is inflammation of the serous membrane of the pericardium due to either bacterial or viral infections and can be extremely painful.

Rheumatic (roo-mat´ik) **heart disease** may result from a **streptoccocal infection** in young people. A toxin produced by the bacteria causes an immune reaction about 2 to 4 weeks after the infection. The immune reaction causes inflammation of the endocardium. The inflamed valves, especially the bicuspid valve, can become stenosed or incompetent. The effective treatment of streptococcal infections with antibiotics has reduced the frequency of rheumatic heart disease.

Congenital heart disease is the result of abnormal development of the heart. The following conditions are common congenital defects:

1. **Septal defect** is a hole in one of the septums between the left and right sides of the heart. The hole can be between the atria or between the ventricles. These defects allow blood to flow from one side of the heart to the other and as a consequence, greatly reduce the pumping effectiveness of the heart.
2. **Patent ductus arteriosus** results when a blood vessel called the **ductus arteriosus,** which is present in the fetus, fails to close after birth. The ductus arteriosus extends between the pulmonary trunk and the aorta. It allows blood to pass from the pulmonary trunk to the aorta, thus bypassing the lungs. This is normal before birth because the fetal lungs are not functioning. If the ductus arteriosus fails to close after birth, blood flows in the opposite direction, from the aorta to the pulmonary trunk. Consequently blood flows through the lungs under high pressure and damages them. Also, the amount of work required of the left ventricle to maintain an adequate blood pressure is increased.
3. **Aortic** or **pulmonary stenosis** (stĕ-no´sis) results from a narrowed vessel or valve. In these cases the workload of the heart is increased because the ventricles must contract with a much greater force to pump blood from the ventricles. **Stenosis of the bicuspid valve** prevents the flow of blood into the left ventricle, causing blood to back up in the left atrium and in the veins carrying blood from the lungs to the left atrium. Filling of these veins causes the accumulation of fluid in the lungs (pulmonary edema). **Stenosis of the tricuspid valve** causes blood to back up in the right atrium and systemic veins, causing edema in the periphery. Filling of the veins with blood causes edema, the accumulation of fluid in tissues outside of the blood vessels.
4. **Cyanosis** (si-ă-no´sis), a bluish coloration of the skin or mucous membranes, is a symptom of inadequate heart function in babies suffering from certain congenital heart diseases. The blueness of the skin is caused by low oxygen levels in the blood in peripheral blood vessels. The term *blue baby* is sometimes used to refer to infants with cyanosis.

Heart failure results from a progressive weakening of the heart muscle. In heart failure, the heart is not capable of pumping all the blood that is returned to it. The ability of the Starling mechanism to increase cardiac output is exceeded so that further stretching of the cardiac muscle does not increase the stroke volume of the heart. Consequently, blood backs up in the veins. In right heart failure, the right ventricle pumps less blood than normal, blood backs up in the veins returning blood from the body to the heart, resulting in edema especially in the lower limbs. In left heart failure, the left ventricle pumps less blood than normal and blood backs up in the veins returning blood from the lungs to the heart, resulting in pulmonary edema. Hypertension (high blood pressure) can cause an enlargement of the heart and can finally result in heart failure. Malnutrition, chronic infections, toxins, severe anemias, or hyperthyroidism can also cause degeneration of the heart muscle.

Heart function in the elderly becomes less efficient, and by the age of

creasing the frequency of action potentials from the stretch receptors. When the blood pressure declines, there is a decrease in the frequency of action potential impulses from the stretch receptors. The action potentials are transmitted along nerve fibers from the stretch receptors to the medulla of the brain.

The baroreceptors provide information about rapid, short-term changes in blood pressure, such as those caused by changes in body position and in response to blood loss. In contrast, if blood pressure is increased for many hours, the baroreceptor reflex adapts to the new higher blood pressure. The baroreceptor reflex, therefore, regulates blood pressure in response to rapid, short-term changes in blood pressure. The baroreceptor reflex functions through negative-feedback mechanisms to maintain homeostasis.

Within the medulla of the brain there is a **cardioregulatory center,** which receives action potentials from the baroreceptors. As blood pressure decreases below normal, the lower frequency of action potentials sent to the medulla triggers a response in the cardioregulatory center that in-

70 there is often a decrease in cardiac output of about one third. Because of the decrease in reserve strength of the heart, many elderly people are often limited in their ability to respond to emergencies, infections, blood loss, or stress.

PREVENTION OF HEART DISEASE

Proper nutrition is important in reducing the risk of heart disease. A recommended diet is low in fats, especially saturated fats and cholesterol, and low in refined sugar. Diets should be high in fiber, whole grains, fruits, and vegetables. Total food intake should be limited to avoid obesity, and sodium chloride intake reduced (see Chapter 17). Diet may be particularly important in slowing the rate of development of **atherosclerosis** of the coronary arteries.

Tobacco and excessive use of alcohol should be avoided. Smoking increases the risk of heart disease by at least tenfold, and excessive use of alcohol also substantially increases the risk of heart disease.

Chronic stress, frequent emotional upsets, and a lack of physical exercise can increase the risk of cardiovascular disease. Remedies include relaxation techniques and aerobic exercise programs involving gradual increases in duration and difficulty in activities such as swimming, jogging, or aerobic dancing.

Hypertension is an abnormally high systemic blood pressure. It affects about one fifth of the population. Regular blood pressure measurements are important, because hypertension does not produce obvious symptoms. If hypertension cannot be controlled by diet, exercise, and the control of stress, it is important to treat the condition with prescribed drugs. The cause of hypertension in the majority of cases is unknown. Drugs commonly used to treat hypertension include the following either alone or in various combinations: diuretics that increase urine production and reduce blood volume, beta-blockers that block the stimulatory effect of epinephrine on the heart, drugs that reduce the rate of acetylcholine breakdown and inhibit the heart, and calcium-channel blockers that reduce the rate of calcium entry into cells and slow the heart.

HEART MEDICATIONS AND TREATMENTS

Digitalis(dig'ĭ-tal'is) slows and strengthens contractions of the heart muscle.

Nitroglycerin (ni'tro-glis'er-in) dilates arteries and veins, which reduces the amount of work required of cardiac muscle.

Beta blockers, or **beta-adrenergic blocking agents,** reduce the rate and strength of cardiac muscle contractions, thus reducing the heart's oxygen demand. Beta-adrenergic blocking agents are often used to treat people who suffer from rapid heart rates, certain types of arrhythmias and hypertension.

Calcium channel blockers reduce the rate at which calcium ions diffuse into cardiac muscle cells. Because the action potentials that produce cardiac muscle contractions depend, in part, on the flow of calcium ions into the cardiac muscle cells, the calcium channel blockers can be used to control the force of heart contractions and reduce arrhythmia, tachycardia (rapid heart rate), and hypertension.

Anticoagulants prevent clot formation in persons with damage to heart valves or blood vessels or in persons who have had a myocardial infarction. Aspirin functions as a weak anticoagulant. Some evidence suggests that one aspirin each day can benefit those who are likely to experience coronary thrombosis.

An **artificial pacemaker** is an instrument placed beneath the skin that has an electrode that extends to the heart. An artificial pacemaker provides an electrical stimulus to the heart. Artificial pacemakers are used in patients when the heart does not contract at a rate high enough to sustain normal physical activity.

A **heart lung machine** is a machine that serves as a temporary substitute for the patient's heart and lungs. It pumps blood throughout the body and oxygenates and removes carbon dioxide from the blood. It has made possible many surgeries on the heart and lungs.

Heart valve replacement or repair is a surgical procedure performed on those who have valves so deformed that they are severely incompetent or stenosed. Valves from pig hearts and valves made of synthetic materials such as plastic or Dacron are effective.

Heart transplants are possible when the immune characteristics of a donor and the recipient are closely matched. The healthy heart of a recently deceased donor is transplanted to the recipient, and the diseased heart of the recipient is removed. People who have received heart transplants must remain on drugs that suppress their immune responses for the rest of their lives to avoid rejection of the donor's heart by their immune system.

Artificial hearts have been used on an experimental basis. Currently artificial hearts do not function well enough to allow a high quality of life.

creases sympathetic stimulation and decreases parasympathetic stimulation of the heart. Consequently, the heart rate and stroke volume increase. These responses cause the blood pressure to increase toward its normal value.

When blood pressure increases above normal, the baroreceptors are stimulated. An increased frequency of action potentials, sent along the nerve fibers to the medulla of the brain, prompts the cardioregulatory center to increase parasympathetic stimulation of the heart and decrease sympathetic stimulation. As a result, the heart rate and stroke volume decrease, causing blood pressure to decrease toward its normal value.

6 In response to a severe hemorrhage, blood pressure is low, the heart rate is increased dramatically, and the stroke volume is low. If low blood pressure activates a baroreceptor reflex, which increases sympathetic stimulation of the heart, why is the stroke volume low?

Emotions integrated in the cerebrum of the brain can influence the heart. Excitement, anxiety, or anger can cause increased sympathetic stimulation of the heart and an increased cardiac output. Depression, on the other hand, can increase parasympathetic stimulation of the heart, causing a slight reduction in cardiac output.

Epinephrine and norepinephrine released from the adrenal medulla in response to emotional excitement or stress also influence the heart's function (see Figure 12-8). Epinephrine and norepinephrine cause an increased heart rate and stroke volume. Increased heart rate and blood pressure in response to emotional excitement result from increased sympathetic stimulation of the heart and epinephrine and norepinephrine secretion from the adrenal medulla.

The medulla oblongata of the brain contains **chemoreceptors** that are sensitive to changes in pH and carbon dioxide levels. A decrease in pH and an increase in carbon dioxide result in sympathetic stimulation of the heart. Decreases in pH and increases in carbon dioxide levels normally affect the heart under emergency conditions, such as when air passages to the lungs are blocked or the lungs are collapsed and cannot function.

Body temperature affects metabolism in the heart much as it affects other tissues. Elevated body temperature increases the heart rate, and reduced body temperature slows the heart rate. For example, during fever the heart rate is usually elevated, and during heart surgery the body temperature is sometimes intentionally reduced to slow the heart rate and metabolism of the body.

SUMMARY

The heart pumps blood through the blood vessels of the body.

SIZE, FORM, AND LOCATION OF THE HEART

The heart is about the size of a fist and is located between the lungs.

The apex is inferior to the base of the heart.

FUNCTIONAL ANATOMY OF THE HEART

Pericardial sac

The pericardial sac surrounds the heart.

The wall of the pericardial sac is lined by the parietal pericardium, and the outer surface of the heart is lined by the visceral pericardium (epicardium).

Between the visceral and parietal pericardium is the pericardial cavity, which is filled with pericardial fluid.

Heart chambers

The heart consists of four chambers: Two atria and two ventricles.

The atria are separated by the interatrial septum and the ventricles are separated by the interventricular septum.

The inferior and superior venae cavae and the coronary sinus enter the right atrium. The four pulmonary veins enter the left atrium.

The pulmonary trunk exits the right ventricle, and the aorta exits the left ventricle.

The left and right coronary arteries originate from the base of the aorta and supply blood to the wall of the heart.

Coronary veins are nearly parallel to the coronary arteries and carry blood from cardiac muscle to the coronary sinus.

Heart valves

The heart valves ensure one-way flow of blood.

The tricuspid valve (three cusps) separates the right atrium and right ventricle, and the bicuspid (two cusps), or mitral, valve separates the left atrium and left ventricle.

The papillary muscles attach by the chordae tendineae to the cusps of the tricuspid and bicuspid valves.

The aorta and pulmonary trunk are separated from the ventricles by the semilunar valves.

Heart wall

The heart wall consists of the outer epicardium, the middle myocardium, and the inner endocardium.

Cardiac muscle

Cardiac muscle is striated and depends on ATP for energy. It depends on aerobic metabolism and cannot develop an oxygen debt.

Cardiac muscle cells are joined by cell-to-cell contacts called intercalated disks that allow the cells to function as a single unit.

BLOOD FLOW THROUGH THE HEART

The cardiac cycle consists of diastole (relaxation) and systole (contraction) of the heart.

Blood flows into the atria while the atria are relaxed and much of the blood flows from the atria into the ventricles. During atrial contraction, filling of the right and left ventricle is completed.

During ventricular contraction, the tricuspid and bicuspid valves close, and blood forces open the semilunar valves; blood flows from the right ventricle into the pulmonary trunk and from the left ventricle into the aorta.

CONDUCTION SYSTEM OF THE HEART

The conduction system of the heart is made up of specialized cardiac muscle cells.

The SA node located in the upper wall of the right atrium is the normal pacemaker of the heart.

The AV node and atrioventricular bundle conduct action potentials to the ventricles.

The right and left bundle branches conduct action potentials from the atrioventricular bundle to Purkinje fibers, which supply the ventricular muscle.

An ectopic beat results from an action potential that originates in an area of the heart other than the SA node.

ELECTROCARDIOGRAM

The ECG is a record of electrical events within the heart.

The ECG can be used to detect abnormal heart rates or rhythms, conduction pathways, hypertrophy or atrophy of the heart, and the approximate location of damaged cardiac muscle.

The normal ECG consists of a P wave (atrial depolarization), the QRS complex (ventricular depolarization), and the T wave (ventricular repolarization).

Atrial contraction occurs during the P-Q interval, and the ventricles contract and relax during the Q-T interval.

HEART SOUNDS

The first heart sound results from closure of the tricuspid and bicuspid valves.

The second heart sound results from closure of the aortic and pulmonary semilunar valves.

Abnormal heart sounds are called murmurs. They can result from leaky (incompetent) valves or narrowed (stenosed) valves.

REGULATION OF HEART FUNCTION

Cardiac output (volume of blood pumped per ventricle per minute) is equal to the heart rate (beats per minute) times the stroke volume (volume of blood ejected per beat).

Intrinsic regulation of the heart

Intrinsic regulation refers to regulation that is contained in the heart.

As venous return to the heart increases, the heart wall is stretched, which increases stroke volume (Starling's law of the heart).

Extrinsic regulation of the heart

Extrinsic regulation refers to nervous and hormonal control of the heart.

Sympathetic stimulation increases stroke volume and heart rate; parasympathetic stimulation decreases heart rate.

Epinephrine and norepinephrine from the adrenal medulla increase the heart rate and stroke volume.

The baroreceptor reflex causes a decrease in heart rate and stroke volume in response to a sudden increase in blood pressure or an increase in heart rate and stroke volume in response to a sudden decrease in blood pressure.

Emotions influence heart function by increasing sympathetic stimulation of the heart in response to excitement, anxiety, or anger, and by increasing parasympathetic stimulation in response to depression.

Alterations in body fluid levels of carbon dioxide, pH, and body temperature influence heart function.

CONTENT REVIEW

1. Describe the size and location of the heart including its base and apex.
2. Describe the structure and function of the pericardial sac.
3. What chambers make up the left side and the right side of the heart?
4. Describe the structure and location of the tricuspid, bicuspid, and semilunar valves. What is the function of these valves?
5. Name the three layers of the heart wall and describe their function.
6. Describe the structure of cardiac muscle cells and the function of the intercalated disks.
7. Starting in the right atrium, describe the flow of blood through the heart.
8. What is the function of the conduction system of the heart? Starting with the SA node, describe the route taken by an action potential as it goes through the conduction system of the heart.
9. Describe the electrical events that generate each portion of the electrocardiogram. How do they relate to contraction events?
10. What portions of the heart contract during the P-Q interval and during the Q-T interval of the electrocardiogram?
11. What events cause the first and second heart sounds?
12. Describe how an incompetent and a stenosed valve can cause a heart murmur.
13. Define cardiac output, stroke volume, and heart rate.
14. Define Starling's law of the heart. What effect does an increase or a decrease in blood flow through veins to the heart have on cardiac output?
15. Describe the effect of parasympathetic and sympathetic stimulation on heart rate and stroke volume.
16. How does the nervous system detect and respond to the following: (1) a decrease in blood pressure and (2) an increase in blood pressure?
17. Explain how emotions affect heart function.
18. What is the effect of epinephrine and norepinephrine on the heart rate and stroke volume?
19. Describe the effect of a decreased blood pH and an increased blood carbon dioxide concentration on the heart rate.
20. How do changes in body temperature influence the heart rate?

CONCEPT REVIEW

1. A friend tells you that her son had an ECG, and it revealed that he had a slight heart murmur. Should you be convinced that he has a heart murmur? Explain.
2. A person who was required to stand without moving for a long period fainted because blood pooled in the veins in his legs. Explain why he fainted.
3. An experiment was performed on an animal in which the blood pressure in the aorta was monitored before and after blood flow to the baroreceptors was blocked. Predict the effect of blocking blood flow to the baroreceptors on the blood pressure.
4. A patient exhibited the following symptoms: chest pain, rapid and irregular pulse, and rapid and irregularly spaced P waves appear in the electrocardiogram. Explain each of the symptoms.
5. Explain why the walls of the ventricles are thicker than the walls of the atria.
6. Explain why transfusion is an effective way to increase blood pressure in a person who has experienced severe hemorrhage.

CHAPTER TEST

Answers can be found in Appendix C

MATCHING For each statement in column A select the correct answer in column B (an answer may be used once, more than once, or not at all).

A

1. The outer layer of the heart.
2. The layer of the heart consisting of cardiac muscle.
3. The valve between the left atrium and left ventricle.
4. The structure that allows action potentials to pass from one cardiac muscle cell to the next cardiac muscle cell.
5. The pacemaker of the normal heart.
6. The deflection of the electrocardiogram that represents depolarization of the atria.
7. A complete round of cardiac diastole and systole.
8. The artery into which the right ventricle pumps blood.
9. An instrument used to listen to the heart sounds.
10. An abnormal heart sound.
11. Regulation of the heart in which the amount of blood pumped out of the ventricles during each cardiac cycle increases as the amount of blood flowing into the atria increases.

B

aorta
aortic semilunar valve
AV node
bicuspid valve
cardiac cycle
endocardium
epicardium
extrinsic regulation
intercalated disk
murmur
myocardium
P wave
pulmonary semilunar valve
pulmonary trunk
QRS wave
SA node
Starling's law of the heart
stethoscope
T wave
tricuspid valve

FILL-IN-THE BLANK Complete each statement by providing the missing word or words.

1. Blood flows from the ___________ ___________ into the right ventricle.
2. The ___________ ___________ prevents blood from flowing from the right ventricle into the right atrium.
3. Cardiac muscle cells have many ___________, which produce ATP at a rate rapid enough to sustain cardiac muscle contractions.
4. When the left ventricle contracts, blood flows from the left ventricle through the ___________ ___________ valve into the ___________.
5. The two arteries that supply blood to cardiac muscle are the ___________ and ___________ ___________ arteries.
6. The region of dead heart tissue that is a result of coronary thrombosis is an ___________.
7. The stroke volume times the heart rate is the ___________ ___________.
8. The ___________ reflex regulates short-term changes in the blood pressure.

CHAPTER 13

Blood Vessels and Circulation

KEY TERMS

arteriole
The smallest diameter artery; blood flows from arterioles into capillaries.

artery
Blood vessel that carries blood away from the heart.

blood pressure
The force blood exerts against the blood vessel walls; expressed relative to atmospheric pressure and reported in the form of mm Hg pressure.

capillary
Tiny blood vessel consisting of only simple squamous epithelium and a basement membrane; major site for the exchange of substances between the blood and tissues.

hepatic portal circulation
Blood flow through the veins that begins in the small intestine, spleen, and stomach, and that carries blood to the liver.

peripheral circulation
Blood flow through the systemic circulation and the pulmonary circulation; includes all blood flow except that through the heart tissue itself.

pulmonary circulation
Blood flow from the right ventricle to the lungs and back to the left atrium.

systemic circulation
Blood flow from the left ventricle to the tissues of the body and back to the right atrium.

vein
Blood vessel that carries blood toward the heart.

venule
The smallest diameter vein; blood flows from capillaries into venules.

FEATURES

OBJECTIVES

After reading this chapter you should be able to:

1. Describe the structure and function of arteries, capillaries, and veins.
2. Describe the changes that occur in arteries as they age.
3. Describe the pulmonary portion of the circulatory system.
4. List the major arteries that supply each of the major body areas.
5. List the major veins that carry blood from each of the major body areas.
6. Describe how blood pressure can be measured.
7. Explain how blood pressure and resistance to flow change as blood flows through the blood vessels.
8. Describe the exchange of material across the capillary.
9. Explain how blood flow through tissues is regulated.
10. Describe the short-term and long-term regulation of arterial blood pressure.

THE PERIPHERAL CIRCULATION consists of blood vessels outside of the heart. Circulating blood delivers oxygen and other nutrients to cells and carries away carbon dioxide and waste products of metabolism. Without an adequate blood flow cells quickly die.

In this chapter, general features of blood vessels, the anatomy of major arteries and veins, and the physiology of circulation are discussed. Special emphasis is placed on regulation of local blood flow and the regulation of blood pressure.

GENERAL FEATURES OF BLOOD VESSEL STRUCTURE

Except for the capillaries and the venules, blood vessel walls consist of three relatively distinct layers. The thickness and the makeup of the layers vary with the diameter of the blood vessels and their type. From the inside to the outer wall of blood vessels, the layers, or **tunics,** are (1) the tunica intima, (2) the tunica media, and (3) the tunica externa (Figure 13-1).

The **tunica intima** consists of an **endothelium,** composed of simple squamous epithelial cells, and a small amount of connective tissue. The **tunica media,** or middle layer, consists mainly of smooth muscle cells arranged circularly around the blood vessel. It also contains variable amounts of elastic and collagen fibers, depending on the size and type of the vessel. The **tunica externa** is composed of connective tissue.

Arteries (ar´ter-ez) are blood vessels that carry blood away from the heart. The arteries form a continuum from the largest to the smallest branches. The elastic arteries branch repeatedly to form the smaller muscular arteries, which in turn branch and decrease in size to become arterioles.

Blood is pumped from the ventricles of the heart into **elastic arteries,** which are the largest-diameter arteries and have the thickest walls. The abundant elastic fibers in the tunica media allow the elastic arteries to stretch and then recoil. The elastic arteries are stretched when the ventricles of the heart pump blood into them. The recoil of the elastic arteries maintains blood pressure and blood flow in blood vessels while the ventricles of the heart are relaxed.

Muscular arteries include medium sized and small-diameter arteries. The abundant smooth muscle in the tunica media enables these vessels to control blood flow to different regions of the body by either constricting (vasoconstriction) or dilating (vasodilation).

Arterioles (ar-te´re-ōlz) transport blood from small arteries to capillaries and are the smallest arteries. Vasodilation and vasoconstriction in the arterioles help control the amount of blood flowing to specific tissues.

Capillaries (kap´ĭ-ler-ēz) are the smallest blood vessels and they branch to form networks (Figure 13-2). The capillary wall is endothelium consisting of a single layer of squamous epithelial cells (see Figure 13-1, C). The endothelium is surrounded by a delicate layer of loose connective tissue. Capillaries are so small in diameter that red blood cells (RBCs) flow through them in single file. As blood flows

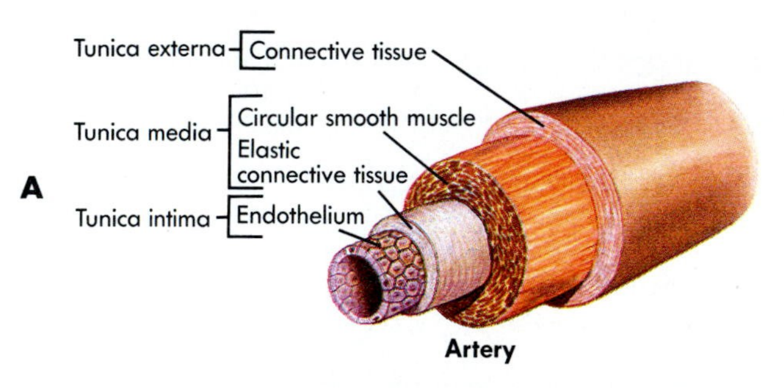

Figure 13-1 ● **Blood vessel structure.**
The tunica intima of arteries and veins includes the endothelium and connective tissue. The tunica media includes a smooth muscle layer and variable amounts of elastic and collagen fibers. The tunica externa is the connective tissue surrounding the vessel. **A,** An artery. The tunica media with its smooth muscle and varying amounts of connective tissue is the dominant layer. **B,** A vein. Veins have thinner walls than arteries, and the tunica media of a vein contains less smooth muscle than an artery of equal diameter. The dominant layer in the wall of a vein is the tunica externa. **C,** A capillary. Capillaries have a smaller diameter than arteries and veins, and their walls consist of a thin layer of endothelium surrounded by a delicate layer of connective tissue. The connective tissue is not shown in the illustration.

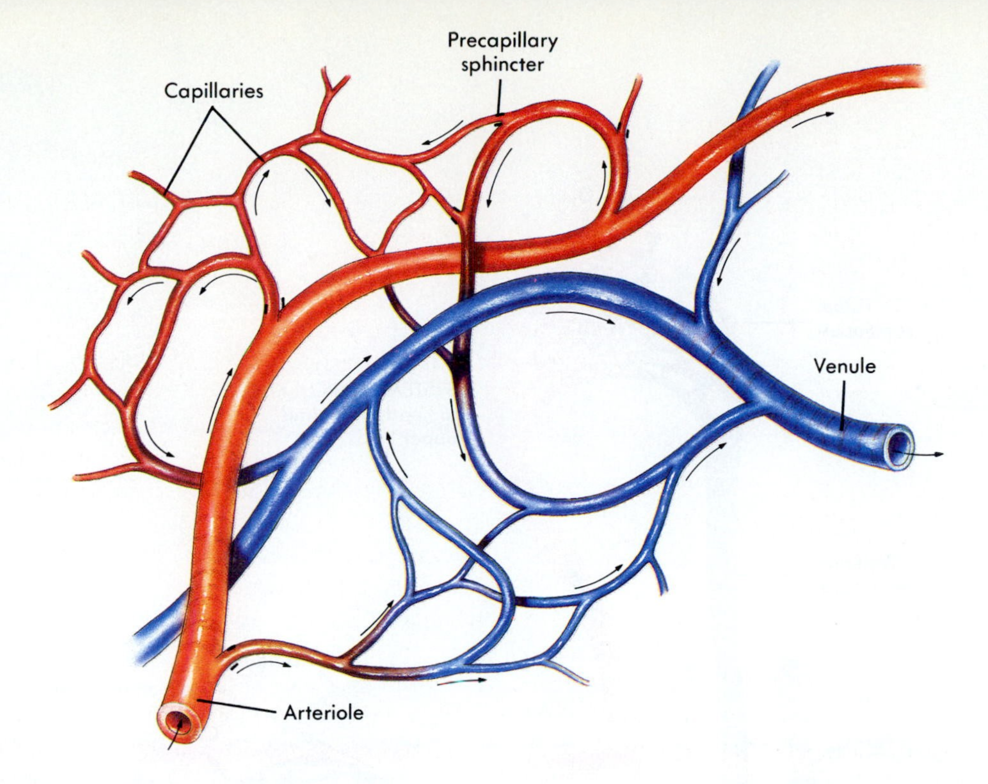

Figure 13-2 ● Capillary network.
An arteriole, giving rise to a capillary network. The capillary network forms numerous branches. Blood flow into the network is regulated by the precapillary sphincters. Blood flows from capillaries into venules.

through capillaries, blood gives up oxygen and nutrients to surrounding cells, and it takes up carbon dioxide and other byproducts of metabolism. Blood flow from the arterioles into the capillaries is regulated by smooth muscle cells called **precapillary sphincters.**

From the capillaries, blood flows into venules. **Venules** are similar in structure to capillaries except that they have a larger diameter, and blood from several capillaries enters a single venule.

From the venules, blood flows into veins. **Veins** are blood vessels that carry blood toward the heart. When compared with arteries of the same size, the walls of the veins are thinner and contain less elastic tissue and fewer smooth muscle cells (see Figure 13-1, *B*). Veins are generally larger in diameter than arteries, and blood pressure is much lower in veins than in arteries. The thin-walled veins readily collapse when they are compressed or when blood drains from them, and they easily inflate when blood flows into them.

As blood flows through veins toward the heart, the veins become larger in diameter and fewer in number. Veins having diameters greater than 2 mm contain valves. Each valve consists of folds in the tunica intima that form two flaps, which are shaped like and function like the semilunar valves of the heart. The valves allow blood to flow toward the heart but not in the opposite direction. When veins are compressed by the contraction of surrounding skeletal muscle tissue, blood can flow through them in only one direction.

There are more valves in veins of the lower limbs than in veins of the upper limbs. The valves prevent the flow of blood toward the feet in response to the pull of gravity.

PULMONARY CIRCULATION

The peripheral circulatory system consists of the pulmonary circulation and the systemic circulation. **Pulmonary circulation** is the flow of blood through vessels that carry blood from the right ventricle of the heart to the lungs and back to the left atrium. Blood from the right ventricle is pumped into the **pulmonary** (pul´mo-nĕr-e) **trunk** (Figure 13-3). This short vessel branches into the **right and left pulmonary arteries,** which extend to the right and left lungs, respectively. Poorly oxygenated blood is carried through pulmonary arteries to the pulmonary capillaries in which oxygen is taken up by the blood and carbon dioxide is released. Four **pulmonary veins** exit the lungs (two from each lung) and carry the oxygenated blood to the left atrium.

SYSTEMIC CIRCULATION: ARTERIES

Systemic circulation is the flow of blood through the vessels that carry blood from the left ventricle of the heart to the tissues of the body and back to the right atrium. Oxygenated blood is pumped from the left ventricle into the aorta. Blood is distributed from the aorta to all portions of the body. After passing through capillaries, blood returns to the right atrium of the heart through the veins of the systemic circulation (see Figure 13-3).

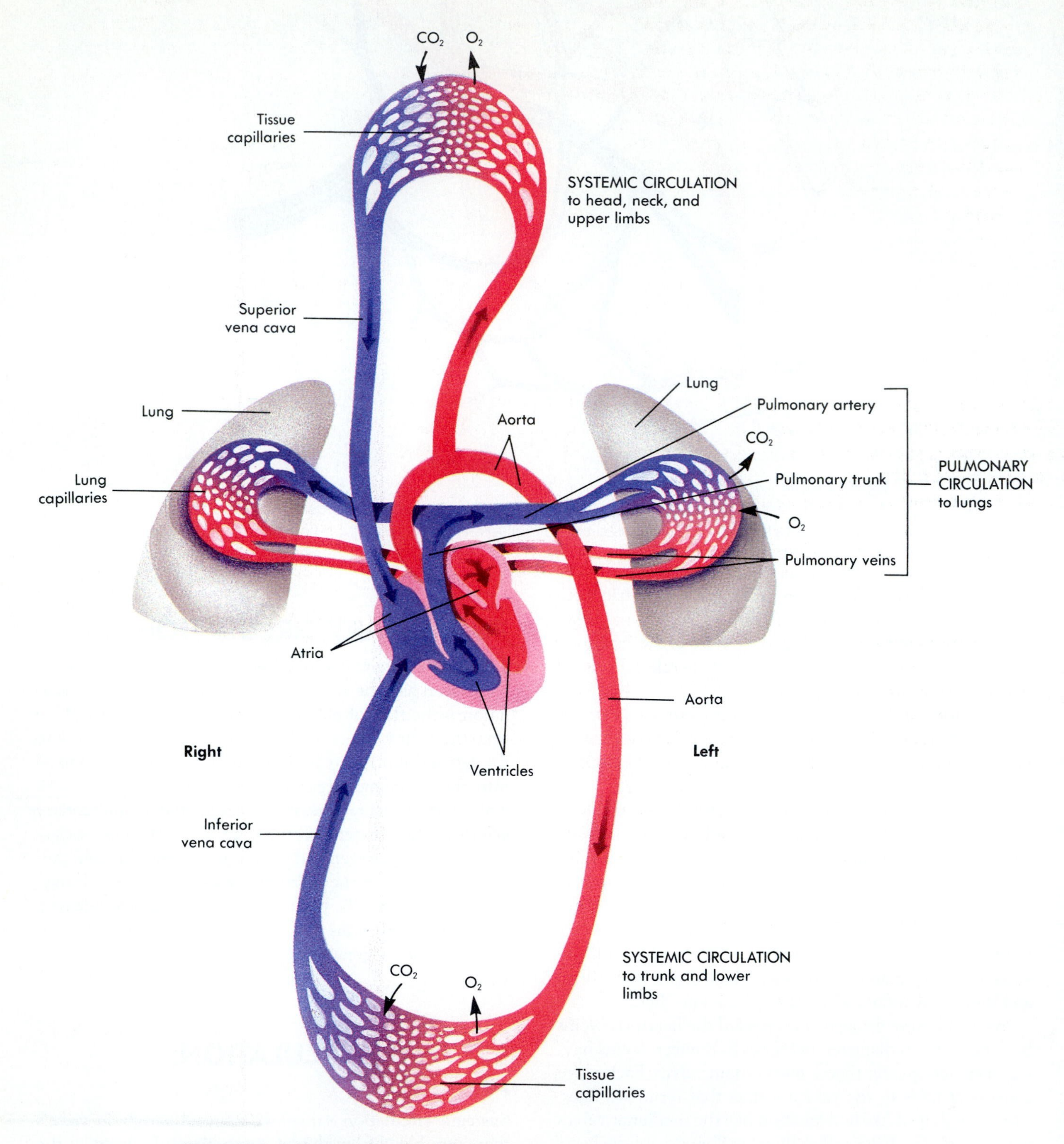

Figure 13-3 ● The peripheral circulation.

The peripheral circulation consists of the blood vessels outside of the heart. The peripheral circulation is divided into the pulmonary circulation and the systemic circulation. Pulmonary circulation is blood flow to and from the lungs. The systemic circulation is blood flow to and from the tissues of the body.

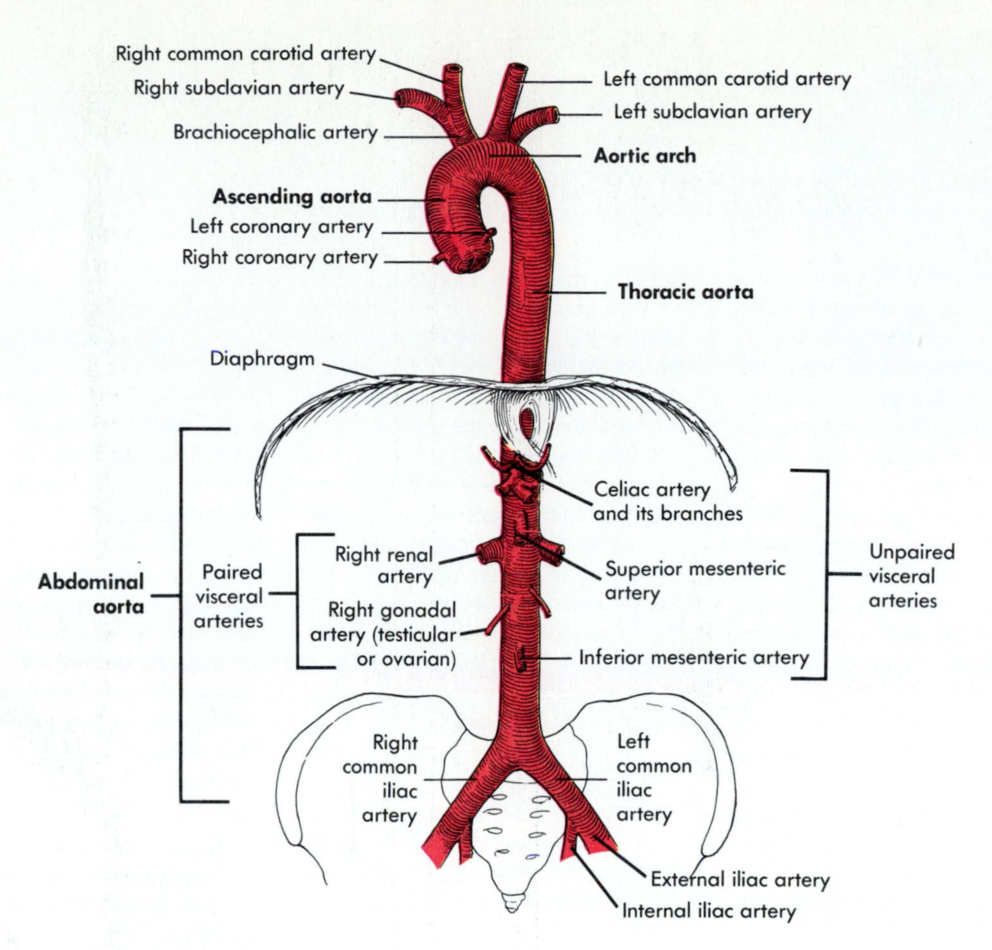

Figure 13-4 ● Branches of the aorta.
The aorta is considered in three portions: the ascending aorta, the aortic arch, and the descending aorta (thoracic and abdominal aorta).

Aorta

All arteries of the systemic circulation branch directly or indirectly from the **aorta** (a-or´tah) (Figure 13-4). The portion of the aorta that passes superiorly from the left ventricle is called the **ascending aorta.** The right and left **coronary arteries** arise from the base of the ascending aorta and supply blood to the cardiac muscle.

From the ascending aorta, the aorta arches posteriorly and to the left as the **aortic arch.** Three major arteries, which carry blood to the head and upper limbs, originate from the aortic arch. They are the **brachiocephalic** (bra´ke-o-sě-fal´ik) **artery,** the **left common carotid** (kă-rot´id) **artery,** and the **left subclavian** (sub-kla´ve-an) **artery** (see Figure 13-4).

The **descending aorta** is the longest portion of the aorta. It extends through the thorax and abdomen to the pelvis. The portion of the descending aorta that extends through the thorax to the diaphragm is called the **thoracic aorta.** The **abdominal aorta** extends from the diaphragm to the point where it divides into the two common iliac arteries.

Arteries of the head and neck

The first vessel to branch from the aortic arch is the brachiocephalic artery. It is a short artery that branches to form the **right common carotid artery,** which transports blood to the right side of the head and neck, and the **right subclavian**

Understanding blood vessels

AGING OF THE ARTERIES

The walls of all arteries undergo changes as they age. The most significant effects of aging occur in the large elastic arteries that carry blood to organs such as the brain, heart, and kidneys. The changes can result in increased blood pressure and reduced blood flow to tissues.

Changes in arteries that make them less elastic are referred to as **arteriosclerosis** (ar-te′re-o-sklĕ-ro′sis), or hardening of the arteries. These changes occur in nearly every individual and they become more severe with advancing age. **Atherosclerosis** (ath′er-o-sklĕ-ro′sis) is a type of arteriosclerosis in which fatty materials such as cholesterol are deposited in the walls of arteries to form plaques. In addition to fatty materials, dense connective tissue and calcium deposits form as plaques enlarge.

Atherosclerosis hampers blood flow because the plaques reduce the inside diameter of the arteries and the elasticity of blood vessel walls. Consequently, blood flow to tissues is reduced, and the work the heart must perform is greatly increased. The rough surface of the arteries caused by atherosclerosis also attracts platelets, which adhere to the walls of the arteries and increase the chance of clot formation (see Chapter 11).

Atherosclerosis develops at a relatively young age in some people. Lack of exercise, smoking, obesity, and a diet high in cholesterol and other fats increase the severity and the rate at which atherosclerosis develops. Severe atherosclerosis also appears to be more prevalent in some families than in others, which suggests a genetic influence.

VARICOSE VEINS

Varicose (var′ĭ-kōs) **veins** result when the veins in the lower limbs become so dilated that the cusps of the valves no longer overlap to prevent the backflow of blood. The varicose veins commonly can be seen as large veins through the skin. Gravity causes blood to pool in the veins of the lower limbs resulting in an increased pressure in the veins. The increased pressure causes fluid to leak from the veins into the interstitial spaces, thereby resulting in edema. In severe cases, blood flow in the veins can become so stagnant that the blood clots. As a consequence, **phlebitis** (flĕ-bi′tis), or inflammation of the veins, can result. If the condition becomes severe enough, lack of blood flow can lead to **gangrene,** (gang′-grēn), which is tissue death and infection of the tissue with bacteria. Conditions that elevate the pressure in veins can increase the severity of varicose veins. An example is pregnancy, in which compression of the pelvic veins by the enlarged uterus results in increased pressure in the veins that drain the lower limbs.

1 Explain why people who suffer from varicose veins are told by their physicians to elevate their lower limbs.*

*Answers to predict questions can be found in Appendix B at the back of the book.

artery, which transports blood to the right upper limb (see Figures 13-4 and 13-5).

There is no brachiocephalic artery on the left side of the body. Instead, the left common carotid artery and left subclavian artery branch directly off the aortic arch. The **left common carotid artery** transports blood to the left side of the head and neck, and the **left subclavian artery** transports blood to the left upper limb (see Figure 13-4).

The common carotid arteries extend superiorly along each side of the neck to the level of the mandible, where they branch into **internal** and **external carotid** arteries (see Figure 13-5). The base of each internal carotid artery is slightly dilated to form a **carotid sinus,** which contains structures important in monitoring blood pressure (baroreceptors). The external carotid arteries have several branches that supply the structures of the face, nose, and mouth. The internal carotid arteries pass through the carotid canals and supply most of the blood to the brain.

Some of the blood to the brain is supplied by the **vertebral arteries,** which branch from the subclavian arteries, pass through the transverse foramina of the cervical vertebrae, and enter the cranial vault through the foramen magnum (see Figure 13-5). Within the cranial vault, branches of the vertebral arteries and the larger right and left internal carotid arteries form a system of vessels called the **circle of Willis** at the base of the brain (Figure 13-6). The vessels that supply blood to most of the brain branch off of the circle of Willis. Even though blood can reach the brain through both the vertebral and internal carotid arteries, blood flow from the carotid arteries is critical. The vertebral arteries cannot supply enough blood to the brain to maintain life if the carotid arteries are blocked.

2 The term *carotid* means to put to sleep, implying that if the carotid arteries are blocked for several seconds, the patient can lose consciousness. Interruption of the blood supply for a few minutes can result in permanent brain damage. What is the physiological significance of atherosclerosis in the carotid arteries?

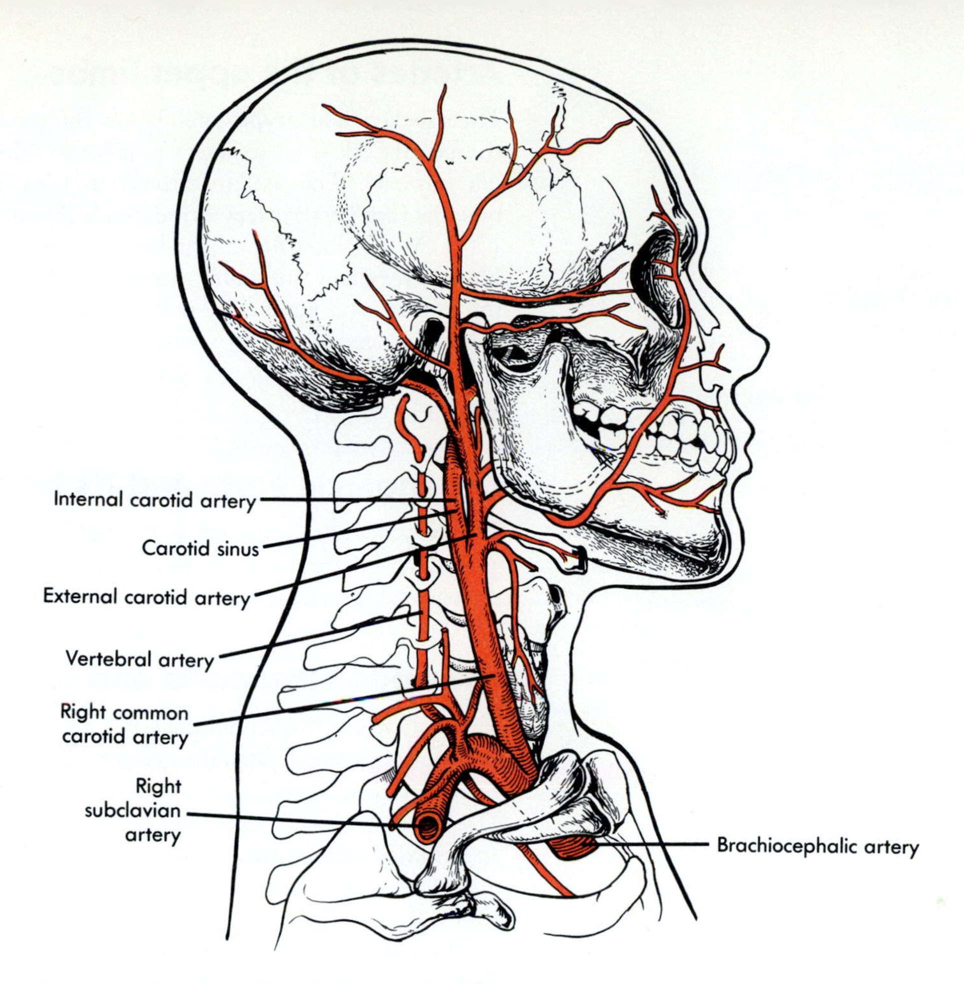

Figure 13-5 ● Arteries of the head and neck.
The brachiocephalic artery, the right common carotid artery, the right subclavian artery, and their branches. The major arteries to the head are the common carotid and vertebral arteries.

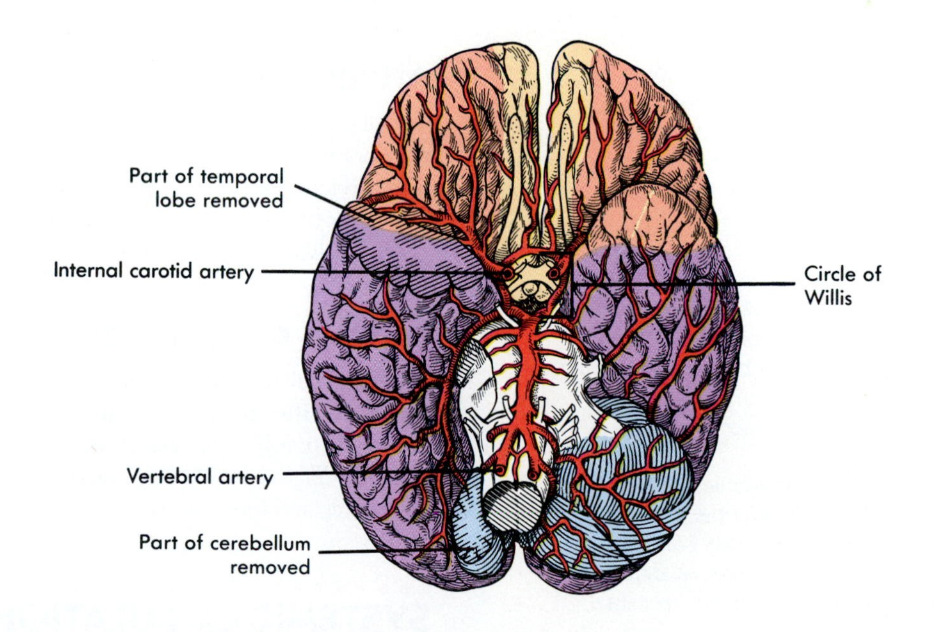

Figure 13-6 ● Arteries supplying the brain.
Inferior view of the brain showing the vertebral and internal carotid arteries and their relationship to the circle of Willis. Most of the blood supply to the brain is through arteries that branch off the circle of Willis.

Arteries of the upper limbs

The **subclavian artery,** located below the clavicle, becomes the **axillary** (ak´sĭ-ler´e) **artery** as it courses through the axilla (armpit). The axillary artery, in turn, becomes the **brachial** (bra´ke-al) **artery** as it extends into the arm (Figure 13-7). At the elbow the brachial artery branches to form the **ulnar artery** and the **radial artery,** which supply blood to the forearm and hand. The radial artery is the artery most commonly used for taking a pulse. The pulse can be detected conveniently on the thumb side of the anterior surface of the wrist.

Table 13-1 • Major branches of the aorta (see Figures 13-4 and 13-7)

Arteries	Tissues supplied
Ascending aorta	
Coronary arteries	Heart
Aortic arch	
Brachiocephalic	Right arm, right side of head, and right side of neck
Left common carotid	Left side of head and left side of neck
Left subclavian	Left arm
Descending aorta	
Thoracic aorta	
Visceral branches	Lung tissues Esophagus
Parietal branches	Thoracic wall Diaphragm
Abdominal aorta	
Visceral branches (unpaired)	
Celiac	Stomach, esophagus, duodenum, liver, spleen, and pancreas
Superior mesenteric	Pancreas, small intestine, and first part of colon
Inferior mesenteric	Last part of colon and the rectum
Visceral branches (paired)	
Suprarenal	Adrenal gland
Renal	Kidney
Gonadal	
Testicular (male)	Testis and ureter
Ovarian (female)	Ovary, uterer, and uterine tube
Parietal branches	Diaphragm Lumbar vertebrae and back muscles Inferior vertebrae
Common iliac	
External iliac	Lower limb
Internal iliac	Lower back, hip, pelvis, urinary bladder, vagina, uterus, rectum, and external genitalia

The thoracic aorta and its branches

The branches of the thoracic aorta can be divided into two groups: The **visceral arteries** supply the thoracic organs, and the **parietal arteries** supply the thoracic wall (Table 13-1).

The abdominal aorta and its branches

The branches of the abdominal aorta, like those of the thoracic aorta, can be divided into **visceral** and **parietal** groups. There are three major visceral branches that are unpaired: The **celiac** (se´le-ak), **superior mesenteric** (mes´en-ter´ik), and **inferior mesenteric arteries,** which supply the digestive tract, liver, spleen, and pancreas (see Table 13-1 and Figures 13-4 and 13-7).

The paired visceral branches of the abdominal aorta include the **renal arteries** to the kidneys, the **suprarenal arteries** to the adrenal glands, and the **testicular** or **ovarian arteries** to the gonads (testes or ovaries) (see Figure 13-4). The parietal arteries of the abdominal aorta supply the diaphragm and abdominal wall.

Arteries of the pelvis

The abdominal aorta divides at the level of the fifth lumbar vertebra into two **common iliac** (il´e-ak) **arteries.** They, in turn, divide to form the **external iliac arteries,** which enter the lower limbs, and the **internal iliac arteries,** which supply the pelvic area (see Figure 13-7).

Arteries of the lower limbs

The external iliac artery becomes the **femoral** (fem´o-ral) **artery** in the thigh. The femoral artery becomes the **popliteal** (pop´lĭ-te-al) **artery** in the posterior region of the knee (see Figure 13-7). Branches of the popliteal artery supply the leg and foot.

SYSTEMIC CIRCULATION: VEINS

The **superior vena cava** (ve´nah ka´vah) returns blood from the head, neck, thorax, and upper limbs to the right atrium of the heart; and the **inferior vena cava** returns blood from the abdomen, pelvis, and lower limbs to the right atrium.

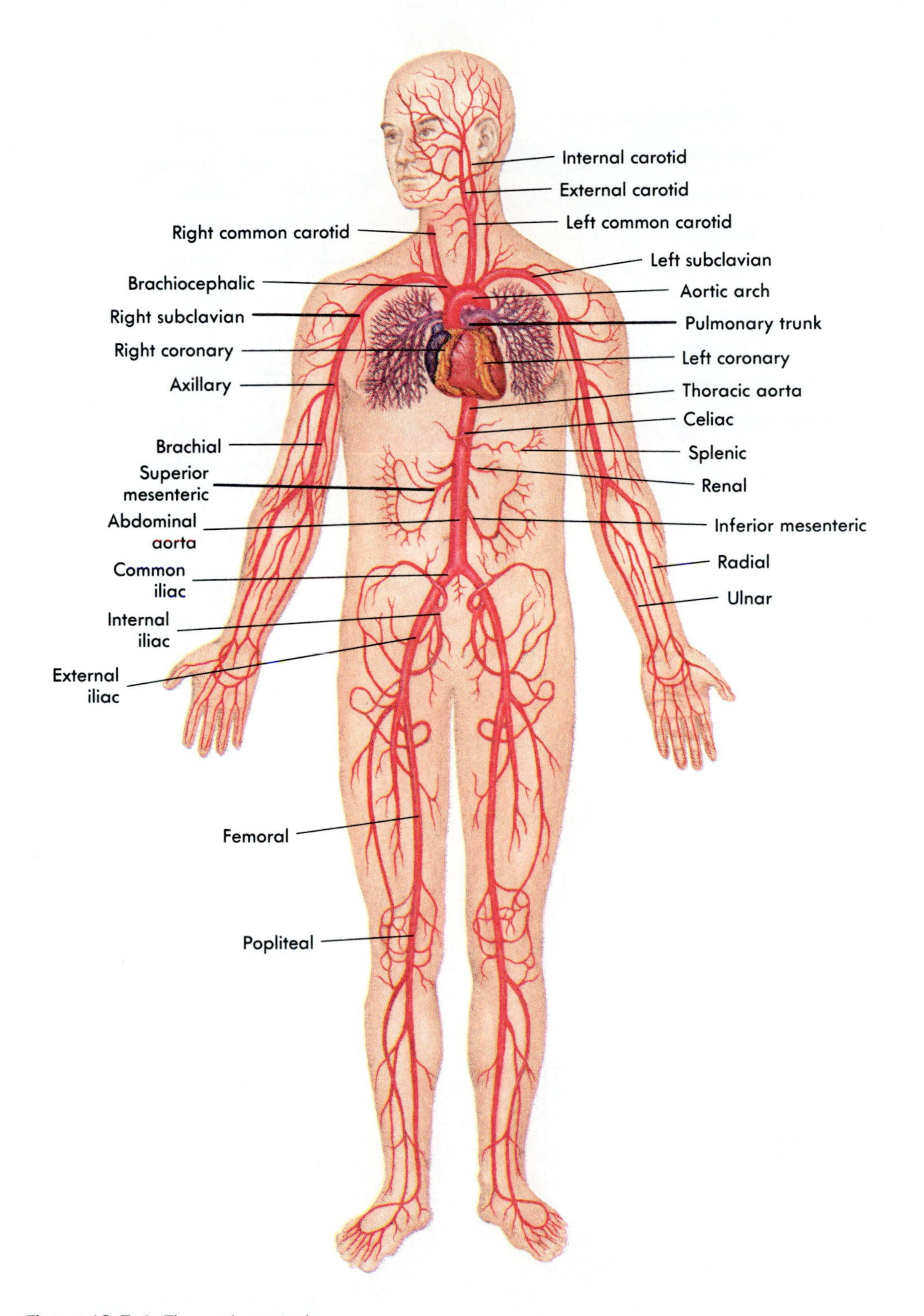

Figure 13-7 ● The major arteries.
The major arteries that carry blood from the left ventricle of the heart to the tissues of the body.

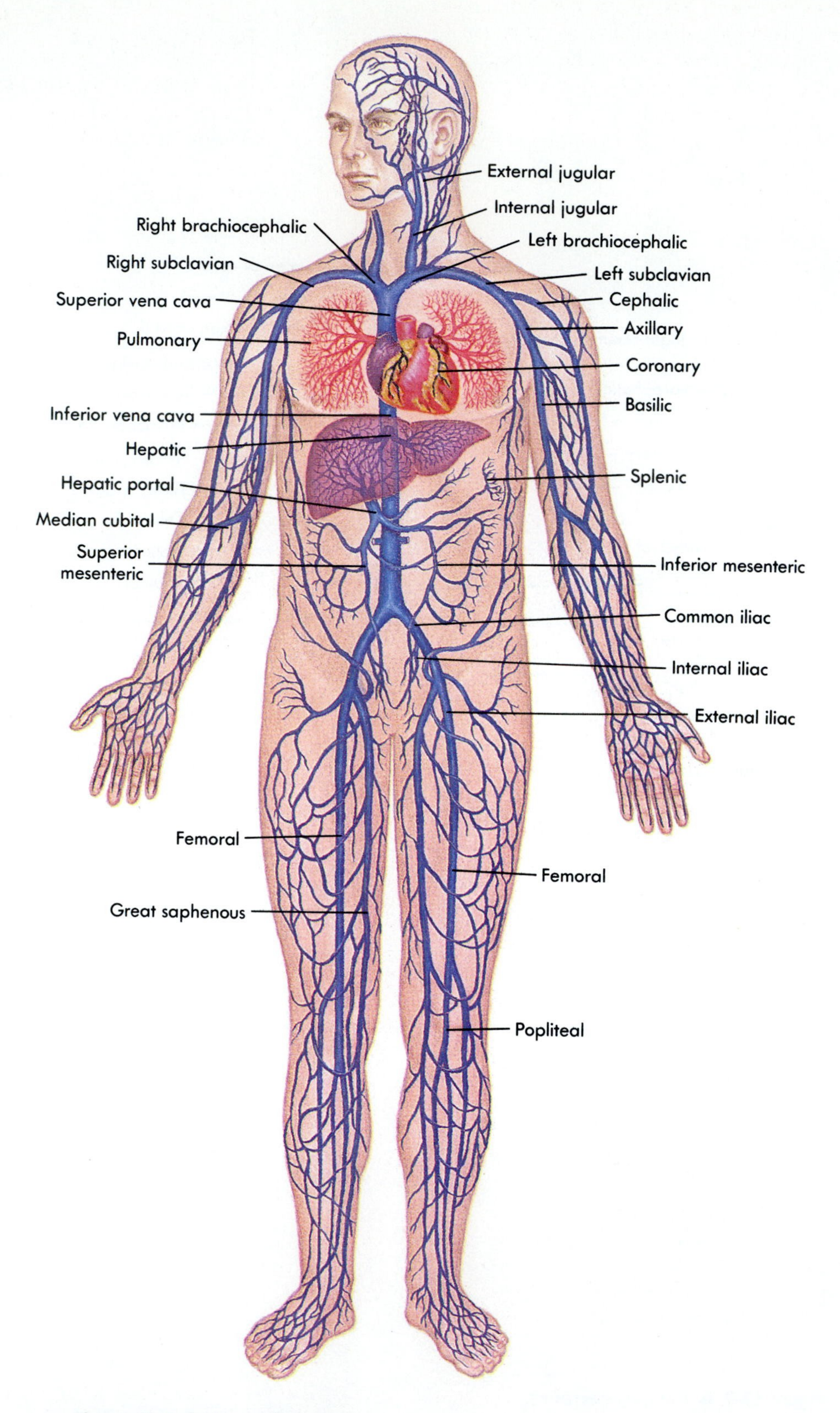

Figure 13-8 ● The major veins.
The major veins that carry blood from the tissues of the body to the right atrium.

Veins of the head and neck

The two pairs of major veins that drain blood from the head and neck are the **external** and **internal jugular** (jug´u-lar) **veins** (Figure 13-8). The external jugular veins are more superficial and drain blood from the posterior head and neck. The larger and deeper internal jugular veins drain blood from the brain and the anterior head, face, and neck. The internal jugular veins join the **subclavian veins** on each side of the body to form the **brachiocephalic veins,** which empty into the superior vena cava.

Veins of the upper limbs

The veins of the upper limbs (see Figure 13-8) can be divided into deep and superficial groups. The deep veins, which drain the deep structures of the upper limb, follow the same course as the arteries and are named for the arteries they accompany. The main deep veins are the **brachial veins,** which accompany the brachial artery and empty into the **axillary vein.**

The superficial veins drain the superficial structures of the upper limb and then empty into the deep veins. The **cephalic** (sĕ-fal´ik) **vein** and the **basilic** (bah-sil´ik) **vein** empty into the **axillary vein.** Many of their tributaries in the forearm and hand can be seen through the skin. For example, the **median cubital** (ku´bĭ-tal) **vein** is usually prominent on the anterior surface of the upper limb at the level of the elbow and is often used as a site for drawing blood.

Veins of the thorax

In the thorax three major veins return blood to the superior vena cava. The **right** and **left brachiocephalic veins** drain blood from the arms, head, neck, and upper thorax to the superior vena cava. The **azygos** (az´ĭ-gus) **vein** receives blood from the thoracic wall and also empties into the superior vena cava.

Veins of the abdomen and pelvis

Blood from the lower limbs, pelvis, and abdomen returns to the heart through the inferior vena cava. The **external iliac veins** drain the lower limbs and join the **internal iliac veins** from the pelvis to form the **common iliac veins.** The common iliac veins combine to form the inferior vena cava (Table 13-2; see Figure 13-8).

Blood from the abdomen returns by two routes. The gonads (testes or ovaries), kidneys, and adrenal glands drain directly into the inferior vena cava. Blood from the stomach, intestines, and spleen drains to the liver through a specialized system of blood vessels called the **hepatic** (hĕ-pat´ik) **portal system** (Figure 13-9 and Table 13-3).

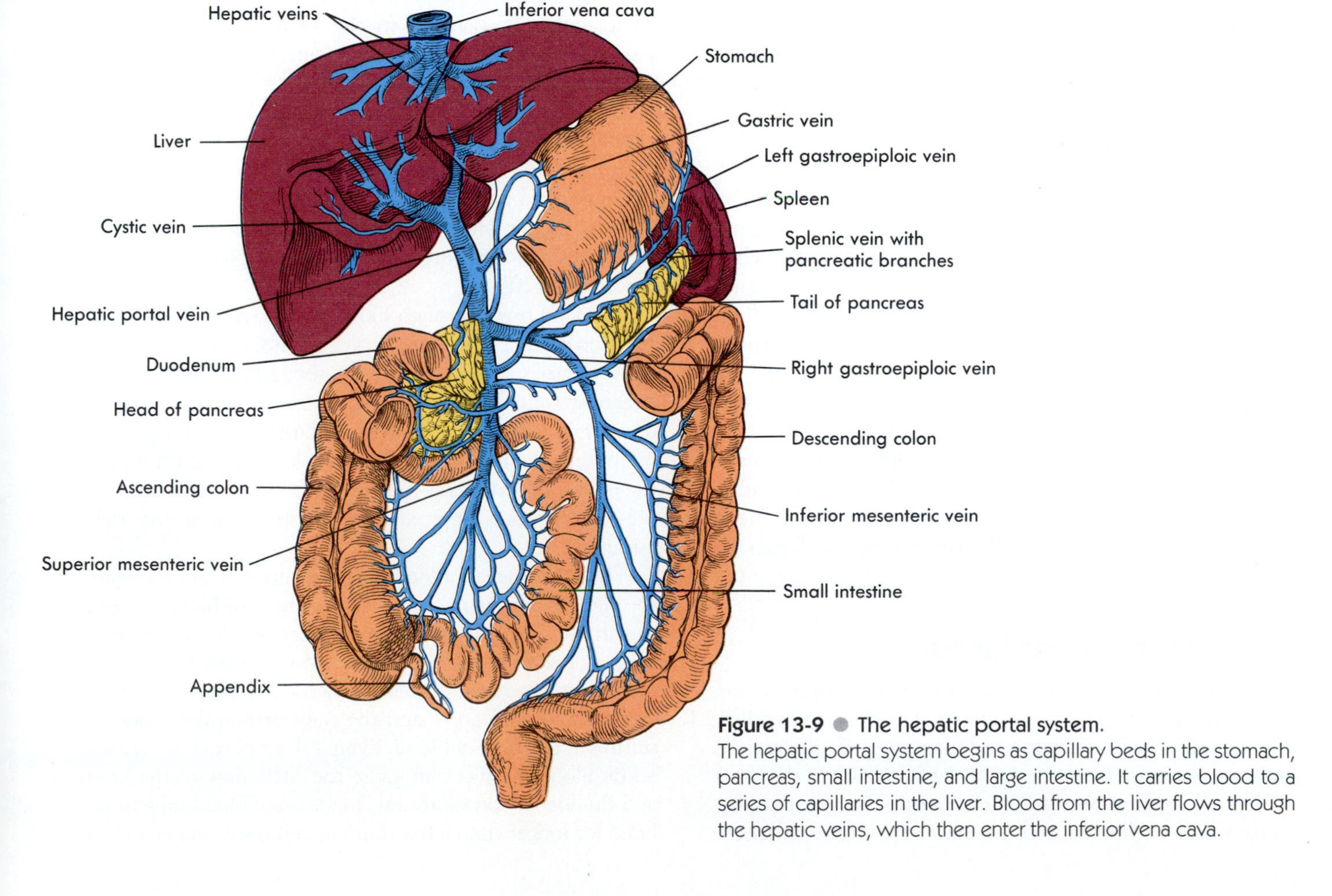

Figure 13-9 ● The hepatic portal system.
The hepatic portal system begins as capillary beds in the stomach, pancreas, small intestine, and large intestine. It carries blood to a series of capillaries in the liver. Blood from the liver flows through the hepatic veins, which then enter the inferior vena cava.

Table 13-2 Veins draining the abdomen, pelvis, and lower limbs (see Figure 13-8)

Veins	Tissues drained
Inferior vena cava	
Hepatic veins	Liver (see hepatic portal system)
Common iliac	
External iliac	Lower limb
Internal iliac	Pelvis and its viscera
Renal	Kidney
Suprarenal	Adrenal gland
Gonadal	
Testicular (male)	Testis
Ovarian (female)	Ovary
Phrenic	Diaphragm

Table 13-3 Hepatic portal system (see Figure 13-9)

Veins	Tissues drained
Hepatic portal	
Superior mesenteric	Small intestine and most of the colon
Splenic	Spleen
Inferior mesenteric	Descending colon and rectum
Pancreatic	Pancreas
Gastric	Stomach
Cystic	Gallbladder

A portal system is a vascular system that begins and ends with capillary beds and has no pumping mechanism such as the heart between them. The hepatic portal system begins with capillaries in the digestive tract and ends with capillaries in the liver. Blood entering the liver through the **hepatic portal vein** is rich with nutrients collected from the intestines, but it also may contain a number of toxic substances harmful to the tissues of the body. Within the liver, nutrients are taken up and stored or modified so they can be used by other cells of the body. Toxic substances are either metabolized by the liver or removed from the blood. Blood from the liver is collected into **hepatic veins,** which join the inferior vena cava.

Veins of the lower limbs

The veins of the lower limbs, like those of the upper limbs, consist of deep and superficial groups. The deep veins follow the same path as the arteries and are named for the arteries they accompany. The superficial veins consist of the great and small **saphenous** (să-fē´nus) **veins.** The **great saphenous vein,** the longest vein in the body, originates in the foot and ascends along the medial side of the leg and thigh to empty into the femoral vein (see Figure 13-8). The much smaller **small saphenous vein** begins in the foot and empties into the popliteal vein, which, in turn, empties into the femoral vein. The femoral vein empties into the external iliac vein.

It is often the great saphenous vein that is surgically removed and used in coronary bypass surgery. Portions of the great saphenous vein are grafted to coronary arteries or their branches to create a route of blood flow that bypasses blocked portions of those arteries. The circulation that is interrupted by the removal of the saphenous vein flows through other veins of the leg.

THE PHYSIOLOGY OF CIRCULATION

The function of the circulatory system is to maintain adequate blood flow to all tissues to provide nutrients and oxygen and to remove waste products of metabolism.

Blood pressure

Blood pressure is a measure of the force blood exerts against the blood vessel walls. Contractions of the ventricles of the heart are mainly responsible for blood pressure. As the ventricles contract, pressure in the ventricles increases and blood is forced out of them. As a result, blood pressure in the aorta and pulmonary trunk increases to a maximum value called the **systolic** (sis-tol´ik) **pressure.** When the ventricles relax, blood pressure falls to a minimum value called the **diastolic** (di-ă-stol´ik) **pressure.** A normal blood pressure in the aorta, and its larger branches, for a resting young adult male is 120 mm Hg for the systolic pressure and 80 mm Hg for the diastolic pressure. A normal blood pressure is usually reported as 120/80.

As blood flows through blood vessels, resistance to flow (friction) causes a drop in blood pressure. From the aorta to the capillaries, the pressure decreases and the difference between systolic and diastolic pressure is reduced. Consequently, blood pressure in the capillaries is a steady pressure of approximately 30 mm Hg. After leaving the capillaries, the pressure decreases as it passes through the veins until the pressure becomes approximately 0 mm Hg at the right atrium.

The difference in blood pressure between the aorta and the right atrium is necessary for the flow of blood through the arteries, capillaries, and veins, because blood flows from areas of higher pressure to areas of lower pressure. If blood pressure in the aorta decreases, then the difference in pressure between the aorta and the right atrium decreases, resulting in less flow of blood. Even a short period of very low aortic blood pressure can cause too little flow to the brain and the loss of consciousness. Inadequate blood flow to the brain for longer than a few minutes can result in permanent brain damage. On the other hand, increased blood flow through tissues can be achieved by increasing blood pressure. For example, during exercise increased heart rate and stroke volume cause blood pressure to increase. Consequently, the increased blood flow supplies exercising muscles with needed oxygen and nutrients and carries away carbon dioxide and waste products.

The **auscultatory** (aws-kul´tah-to´re) method of determining systolic and diastolic blood pressure is used under most clinical conditions (Figure 13-10). A blood pressure cuff is placed around the patient's arm and a stethoscope is placed over the brachial artery. The blood pressure cuff is then inflated until the brachial artery is completely collapsed. Because no blood flows through the constricted area, no sounds can be heard at this point. The pressure in the cuff is then gradually lowered. As soon as the pressure in the cuff declines below the systolic pressure, blood flows through the constricted area each time the left ventricle contracts. The blood flow is turbulent, and the turbulent blood flow produces vibrations in the blood and surrounding tissues that can be heard through the stethoscope. The pressure at which the first sound is heard represents the systolic pressure.

As the pressure in the blood pressure cuff is lowered still more, the sounds change in tone and loudness. When the pressure has dropped until continuous blood flow is reestablished, the sound disappears completely. The pressure at which the sounds disappear is the diastolic pressure.

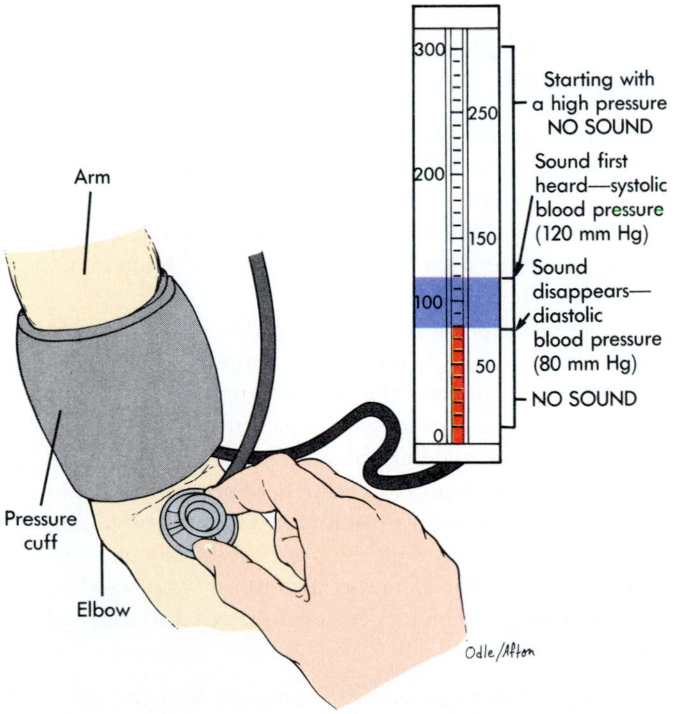

Figure 13-10 ● **Blood pressure measurement.**
Blood pressure measurement using a sphygmomanometer. The blood pressure cuff is inflated to a high pressure, and the pressure is then decreased slowly. The pressure at which turbulent blood flow is first heard is the systolic blood pressure. The pressure at which sounds disappear is the diastolic blood pressure.

The pulse

Ejection of blood from the left ventricle into the aorta produces a pressure wave, or **pulse,** which travels rapidly along the arteries. A pulse can be felt at locations where large arteries are close to the surface of the body. It is helpful to know the major locations where the pulse can be detected because monitoring the pulse is important clinically. For example, the heart rate and rhythm of the heart can be determined by feeling the pulse (Figure 13-11).

Capillary exchange

The main function of the circulatory system is to provide an adequate blood flow to the capillaries. There are about 10 billion capillaries in the body that carry blood to all tissues. Nutrients are delivered to cells through capillaries, and

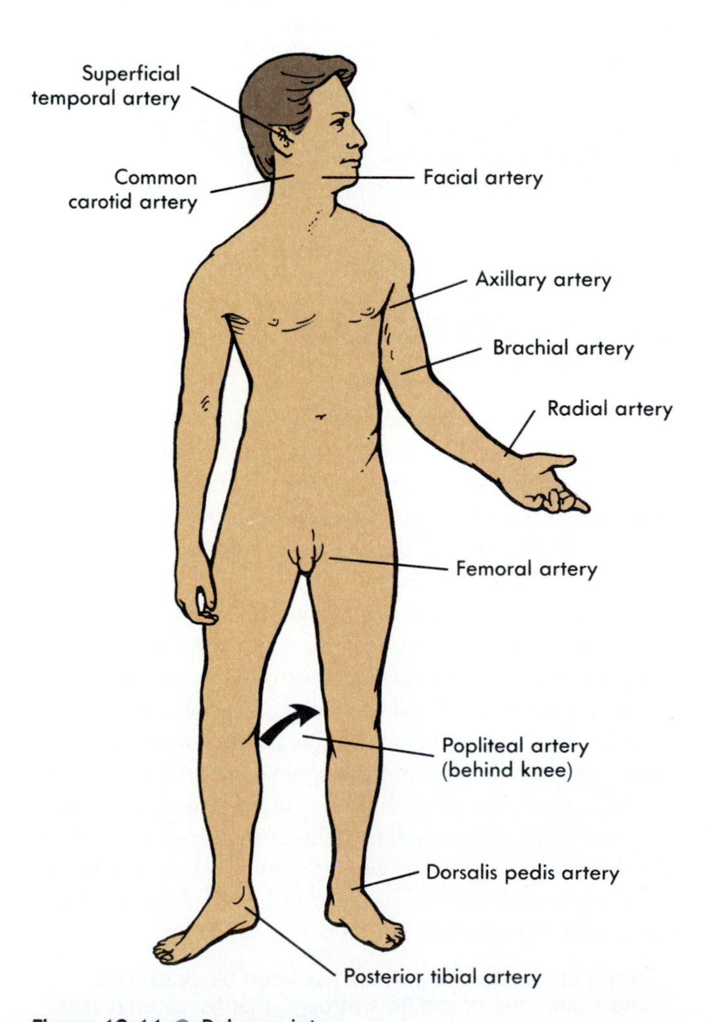

Figure 13-11 ● **Pulse points**
Major points where the pulse can be monitored. Each pulse point is named after the artery on which the pulse can be detected.

waste products are carried away. The nutrient and waste product exchange is essential for the survival of cells.

The circulatory system is controlled in several ways so that adequate blood flow to the capillaries is maintained. If the activity of a tissue increases, such as skeletal muscle during exercise, blood flow through the capillaries supplying the tissue increases to meet its increased metabolic needs. Also, when the activity of a tissue decreases, blood flow through the capillaries supplying the tissue decreases.

Nutrients and waste products mainly diffuse across the walls of capillaries. For example, movement of oxygen and other nutrients from capillaries to cells depends on a higher concentration of oxygen and nutrients in the capillaries than in the cells supplied by the capillaries. Waste products diffuse from cells to the capillaries because they are present in higher concentration in the cells than in the capillaries.

In addition to the diffusion of nutrients and waste products, a small amount of fluid is forced out of the capillaries at their arteriolar ends. Most of that fluid, but not all, reenters the capillaries at their venous ends. The remaining fluid enters lymphatic capillaries and is eventually returned to the general circulation (see Chapter 14).

Edema, or swelling, results from a disruption in the normal exchange of fluid across the capillary walls. For example, inflammation results in an increase in the permeability of capillaries. Plasma proteins leak out of the capillaries and fluid follows, resulting in edema.

LOCAL CONTROL OF BLOOD VESSELS

Blood flow through capillaries of a given tissue are precisely controlled so that the blood flow is closely matched to the metabolic needs of the tissue. Two mechanisms control blood flow through tissues. One mechanism involves local activities in the tissues and the other mechanism involves regulation by the nervous system.

Local control of blood flow through tissues is achieved by contraction and relaxation of the precapillary sphincters. The precapillary sphincters are controlled by the metabolic needs of the tissues. Blood flow increases when oxygen levels or levels of nutrients such as glucose and amino acids decrease. Blood flow also increases when by-products of metabolism, such as carbon dioxide, build up in tissue spaces. For example, the precapillary sphincters in resting skeletal muscle remain constricted when the muscle is at rest, but in exercising muscle the precapillary sphincters dilate resulting in a greatly increased blood flow.

3 When blood flow to a tissue has been blocked for a short time, the blood flow through that tissue increases to as much as five times its normal value after the removal of the blockage. Explain that response based on what you know about the local control of blood flow.

Nervous control regulation of blood flow through tissues is carried out primarily through the sympathetic division of the autonomic nervous system. Sympathetic vasoconstrictor fibers innervate the smooth muscle in the walls of most blood vessels of the body (Figure 13-12). Nervous control of blood vessels does not affect the capillaries and precapillary sphincters because they have no nerve supply.

An area of the lower pons and upper medulla oblongata, called the **vasomotor** (va´so-mo´tor) **center,** continually transmits a low frequency of nerve impulses to the sympathetic vasoconstrictor fibers. As a consequence, the blood vessels are continually in a partially constricted state, a condition called **vasomotor tone.**

Nervous control of blood vessels causes blood to be shunted from one area of the body to another. For example, during exercise blood is routed to organs necessary for exercise such as skeletal muscle, and away from organs not necessary for exercise such as the viscera. The different blood flows result from changes in vasomotor tone. In skeletal muscle, vasomotor tone decreases, blood vessels dilate, and blood flow increases. In the viscera, vasomotor tone increases, blood vessels further constrict, and blood flow decreases. The combination of local control and nervous control of blood flow through tissues assures an adequate blood supply to meet the needs of tissues under varying conditions.

REGULATION OF ARTERIAL PRESSURE

An adequate arterial blood pressure is required to maintain blood flow through the blood vessels of the body and, therefore, is essential for life. Without an adequate arterial blood pressure, neither dilation of the precapillary sphincters nor dilation of arteries that supply a tissue will result in increased blood flow to that tissue.

Arterial blood pressure is determined by the amount of blood pumped by the heart and by peripheral resistance. Blood pressure increases if the heart rate and stroke volume increase, and blood pressure decreases if the heart rate and stroke volume decrease. **Peripheral resistance** is the total resistance to blood flow and it is influenced primarily by vasoconstriction and vasodilation of small arteries and arterioles. Blood pressure increases if the peripheral resistance increases. Therefore, blood pressure increases if arteries and arterioles constrict, and blood pressure decreases if arteries and arterioles dilate.

4 In a long-distance runner, blood pressure does not decrease below resting values even though the flow of blood through arteries of the lower limbs increases greatly. Explain how that occurs.

Mechanisms that assure an adequate blood pressure do so by influencing the heart rate, stroke volume, and peripheral resistance. Three types of control mechanisms that regulate

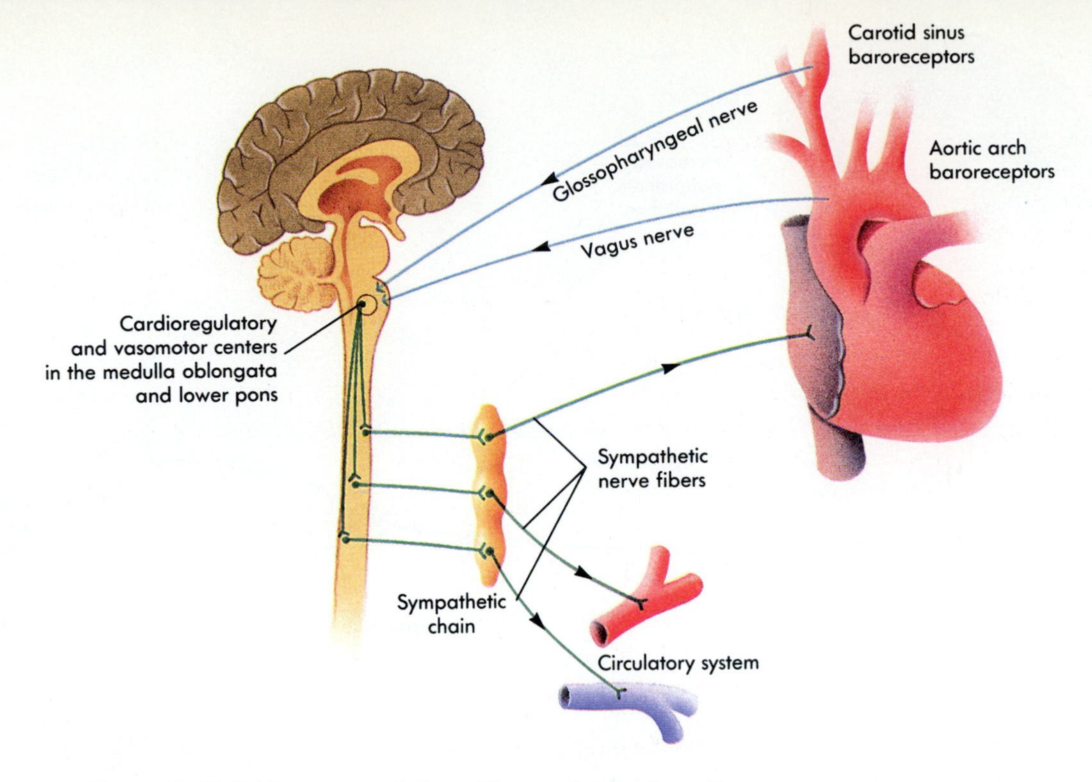

Figure 13-12 ● Nervous regulation of the cardiovascular system. Baroreceptors located in the carotid sinuses and aortic arch detect changes in blood pressure. Action potentials from the baroreceptors are conducted to the medulla oblongata. In response to a sudden decrease in blood pressure, sympathetic stimulation of the heart increases, and vasomotor tone increases. Both of these responses function to increase blood pressure back to its normal value. In response to a sudden increase in blood pressure sympathetic innervation of the heart and vasomotor tone decrease, and the blood pressure returns to its normal value.

blood pressure are the baroreceptor reflexes, chemoreceptor reflexes, and hormonal mechanisms.

Baroreceptor reflexes

Baroreceptors are pressure receptors that respond to stretch in arteries caused by increased blood pressure. Baroreceptors are located in the carotid sinus at the base of the internal carotid artery and in the walls of the aortic arch (see Figure 13-12). Action potentials are transmitted from the baroreceptors to the medulla oblongata of the brain.

A sudden decrease in blood pressure results in a decreased action potential frequency in the baroreceptors. The decrease in action potential frequency delivered to the medulla oblongata produces responses in the vasomotor center and cardioregulatory center, which function to raise blood pressure. There is an increase in vasomotor tone and an increase in sympathetic stimulation of the heart (see Figure 13-12). The result is an increase in peripheral resistance and an increased volume of blood pumped by the heart. The increased peripheral resistance and volume of blood pumped by the heart raises the blood pressure toward its normal value.

The **baroreceptor reflex** is important in regulating blood pressure on a moment-to-moment basis. For example, when a person rises rapidly from a sitting or lying position to a standing position, blood pressure in the neck and thoracic regions drops dramatically as a result of the pull of gravity on the blood. This reduction can be so great that blood flow to the brain is reduced enough to cause dizziness. The falling blood pressure activates the baroreceptor reflex, which reestablishes normal blood pressure within a few seconds. In a healthy person, a temporary sensation of dizziness is all that may be experienced.

5 Explain how the baroreceptor reflex responds when the blood pressure in the area of the baroreceptors increases, such as when a person does a headstand.

Chemoreceptor reflexes

Carotid bodies are small structures that lie near the carotid sinuses, and **aortic bodies** are structures near the aortic arch. These structures contain sensory receptors that respond to changes in oxygen concentration, carbon dioxide concentration, and blood pH. Because they are sensitive to chemi-

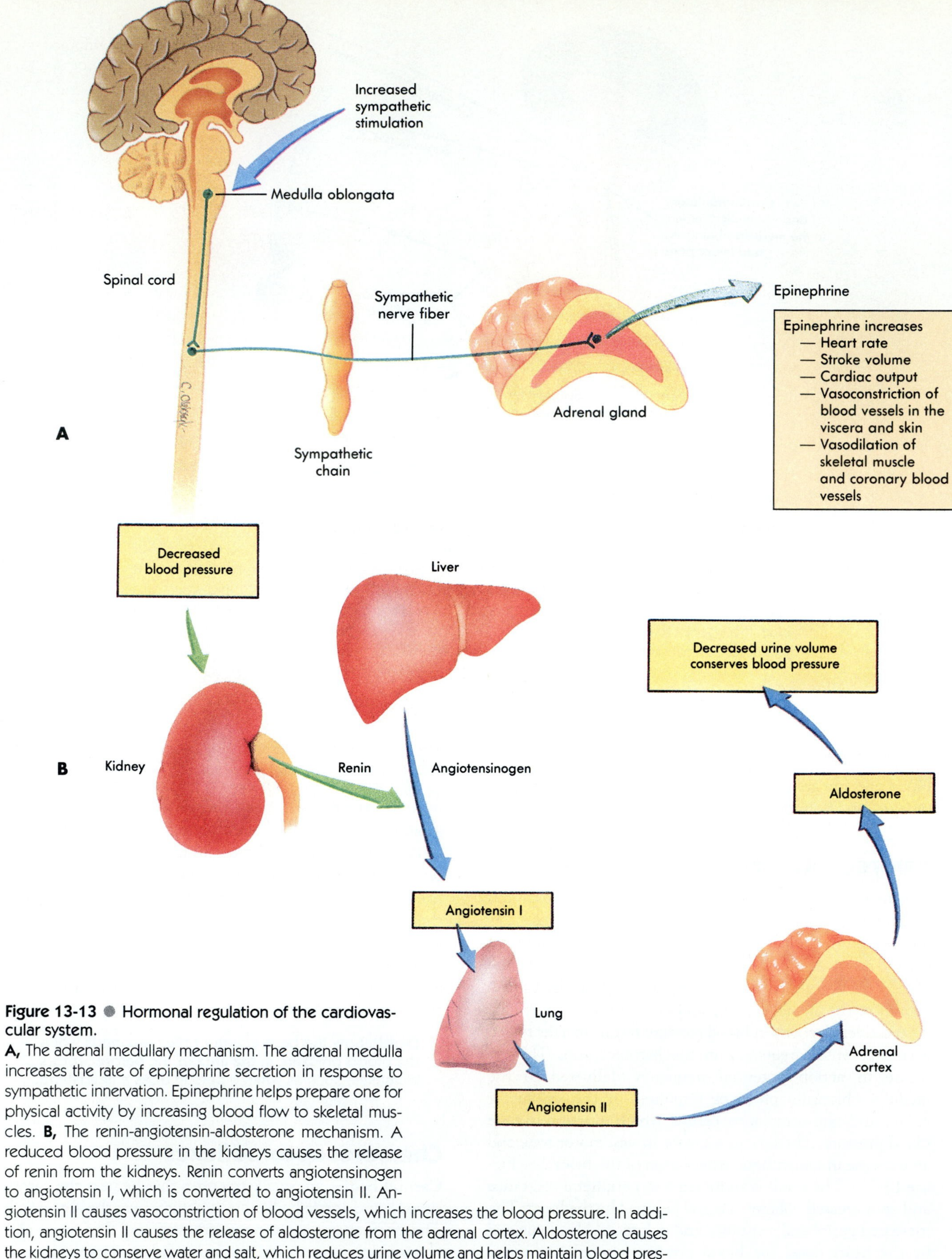

Figure 13-13 ● **Hormonal regulation of the cardiovascular system.**

A, The adrenal medullary mechanism. The adrenal medulla increases the rate of epinephrine secretion in response to sympathetic innervation. Epinephrine helps prepare one for physical activity by increasing blood flow to skeletal muscles. **B,** The renin-angiotensin-aldosterone mechanism. A reduced blood pressure in the kidneys causes the release of renin from the kidneys. Renin converts angiotensinogen to angiotensin I, which is converted to angiotensin II. Angiotensin II causes vasoconstriction of blood vessels, which increases the blood pressure. In addition, angiotensin II causes the release of aldosterone from the adrenal cortex. Aldosterone causes the kidneys to conserve water and salt, which reduces urine volume and helps maintain blood pressure by preventing the loss of water from the blood through the kidneys as urine.

cal changes in the blood, they are called **chemoreceptors.** They send action potentials along nerve fibers to the medulla oblongata. There are also chemoreceptors in the medulla oblongata (see Chapter 12).

When oxygen levels decrease, carbon dioxide levels increase, or when pH decreases, the chemoreceptors respond with an increased frequency of action potentials, called the **chemoreceptor reflex.** In response, there is an increase in vasomotor tone, resulting in an increased blood pressure. The increased blood pressure causes a greater rate of blood flow to the lungs, which helps increase blood oxygen levels and reduce blood carbon dioxide levels. The chemoreceptors function under emergency conditions. They respond strongly only when the oxygen levels in the blood fall to very low levels, or carbon dioxide levels become substantially elevated.

Hormonal mechanisms

In addition to the rapidly acting nervous mechanisms that regulate arterial pressure, there are important hormonal mechanisms that help control blood pressure.

Adrenal medullary mechanism

Stimuli that result in increased sympathetic stimulation of the heart and the blood vessels also increase the secretion of epinephrine and norepinephrine from the adrenal medulla. Epinephrine and norepinephrine increase heart rate and stroke volume and cause vasoconstriction of blood vessels in the skin and viscera. Epinephrine also causes vasodilation of blood vessels in skeletal muscle and cardiac muscle. Epinephrine and norepinephrine therefore increase the supply of blood flowing to the skeletal muscle and cardiac muscle, and this prepares one for physical activity (Figure 13-13, *A*).

Renin-angiotensin-aldosterone mechanism

In response to reduced blood pressure, the kidneys release an enzyme called **renin** (ren´in) into the circulatory system (Figure 13-13, *B*). Renin acts on the blood protein **angiotensinogen** (an´je-o-tin´sin-o-jen) to produce **angiotensin I.** Other enzymes in the circulatory system act on angiotensin I to convert it to **angiotensin II,** its most active form. Angiotensin II is a potent vasoconstrictor substance. Thus in response to a reduced blood pressure, the release of renin by the kidney acts to increase the blood pressure toward its normal value.

Angiotensin II also acts on the adrenal cortex to increase the secretion of **aldosterone** (al-dos-ter-ōn). Aldosterone causes the kidneys to conserve sodium ions and water. Thus the volume of water lost from the blood into the urine is reduced. For example, a large decrease in blood volume results in a decreased blood pressure. If blood pressure declines, conserving or increasing the blood volume will help bring the blood pressure back up to a normal range of values. Aldosterone helps prevent any additional decrease in blood pressure because it helps maintain normal blood volume.

Long-term and short-term regulation of blood pressure

The baroreceptor reflex and the adrenal medullary hormones are most important in controlling blood pressure on a short-term basis. They are sensitive to sudden changes in blood pressure, and they respond quickly. The renin-angiotensin-aldosterone system is more important in the maintenance of blood pressure on a long-term basis. It is influenced by small changes in blood pressure and it responds by gradually bringing the blood pressure back to its normal range.

FETAL CIRCULATION

There are some major differences between the circulation of blood before birth and after birth. Prior to birth, most of the blood bypasses the lungs, and vessels exist to provide a blood supply to the placenta.

Blood bypasses the lungs by flowing through an opening in the interatrial septum called the **foramen ovale** (for-ra´men o-val´e), which allows blood to flow directly from the right atrium into the left atrium without passing through the right ventricle or the lungs (Figure 13-14, *A*). Also, a short artery called the **ductus arteriosus** extends from the pulmonary trunk to the aorta. Blood flows through the ductus arteriosus from the pulmonary trunk into the aorta and bypasses the lungs.

The fetal blood supply to the placenta is through the **umbilical arteries,** which originate from the internal iliac arteries. As fetal blood flows through the placenta, substances are exchanged with the mother's blood. Oxygen and nutrients from the mother's blood move into fetal blood; carbon dioxide and waste products produced by the fetus move into the mother's blood. As oxygen- and nutrient-rich blood returns from the placenta to the fetus, it flows through the **umbilical vein,** bypasses the liver, and flows into the inferior vena cava (see Figure 13-14, *A*).

Following birth, the foramen ovale and the ductus arteriosus close to establish an adult pattern of circulation. In addition, the blood flow through the umbilical arteries and vein is cut off, and the umbilical blood vessels atrophy (Figure 13-14, *B*).

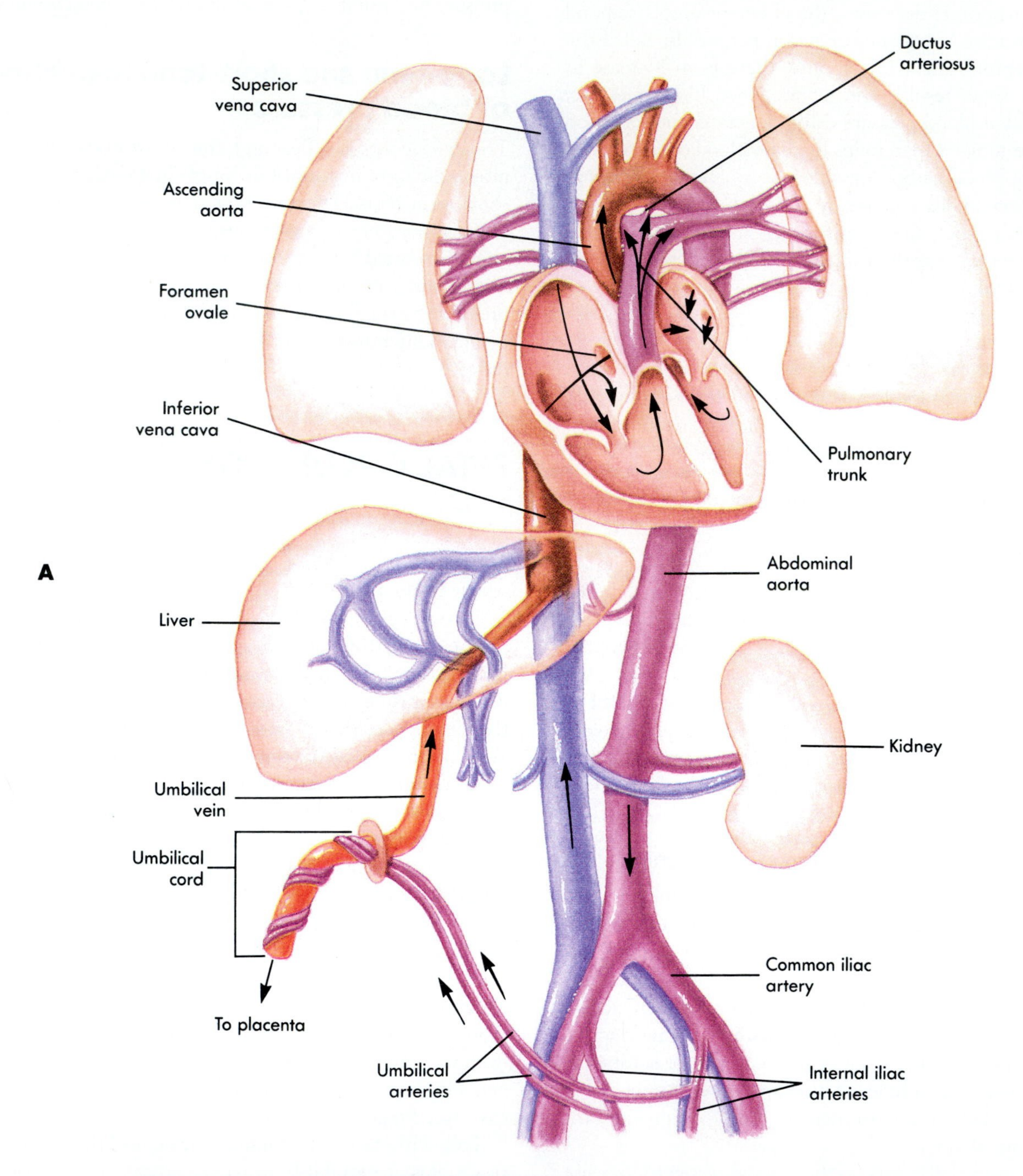

Figure 13-14 ● Circulatory changes at the time of birth.
A, Before birth the umbilical vein and arteries attach to the placenta, blood passes through the foramen ovale in the heart, and the ductus arteriosus shunts blood from the pulmonary trunk to the aorta.

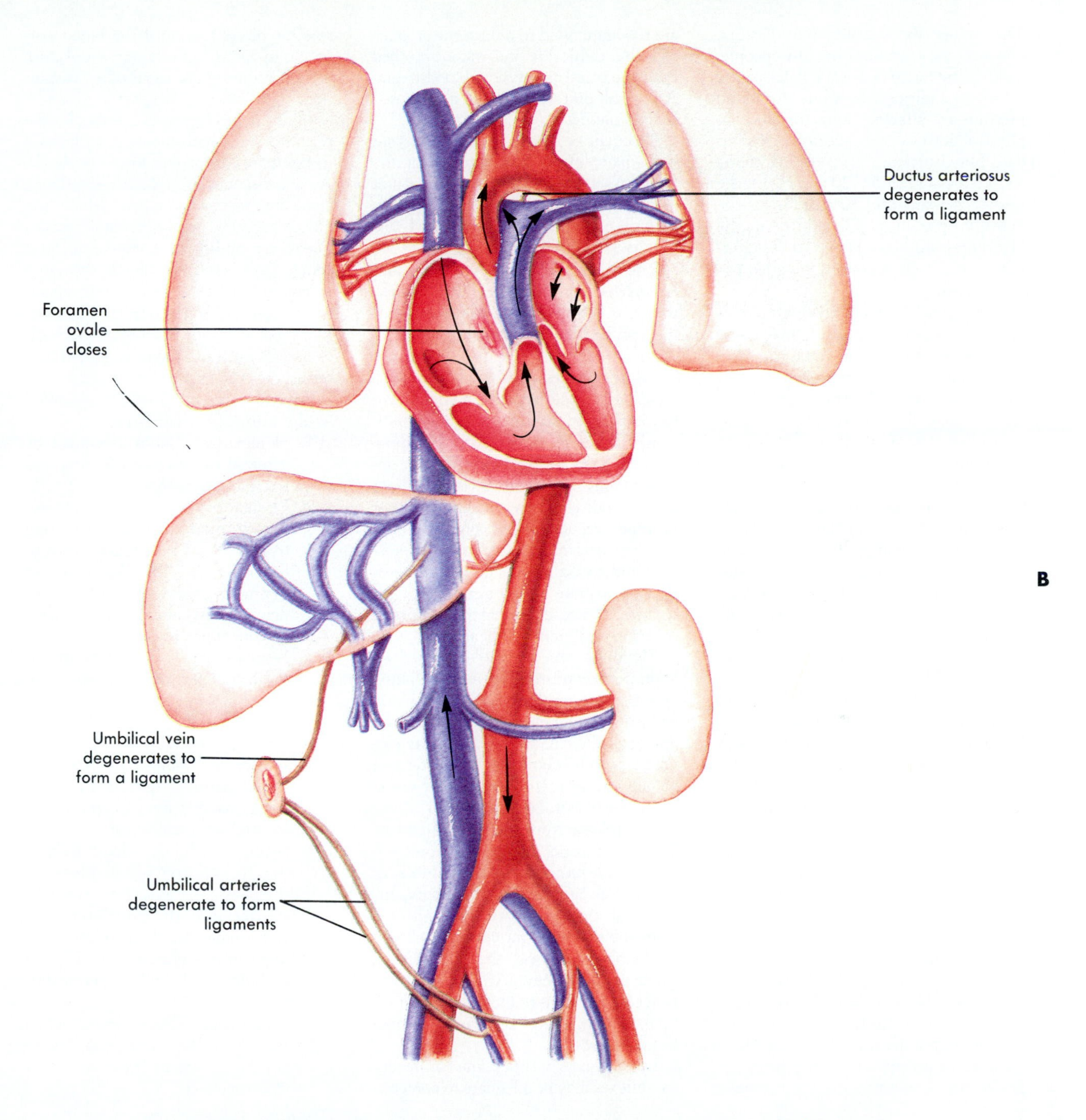

Figure 13-14, cont'd. B, After birth the placenta is gone, the umbilical vein and arteries degenerate, the foramen ovale closes, and the ductus arteriosus closes.

Hypertension and circulatory shock

The values for systolic and diastolic pressure vary among healthy people, making the range of normal values quite broad. In addition, the values for blood pressure are affected by factors such as physical activity and emotions. A standard blood pressure for a resting young adult male is 120/80 (120 mm Hg for the systolic pressure and 80 mm Hg for the diastolic pressure). Generally, if the blood pressure exceeds 140/90, it is considered too high, a condition called **hypertension.**

Blood pressure that is too high is dangerous for several reasons. It may cause blood vessels to rupture, and it increases the chance that either emboli or thrombi may form. Ruptured blood vessels or blocked blood vessels can result in strokes or heart attacks.

Hypertension affects about 20% of all people at some time in their lives. Chronic hypertension has an adverse effect on the heart and blood vessels because hypertension requires the heart to perform a greater-than-normal amount of work. The extra work leads to hypertrophy of the cardiac muscle, especially in the left ventricle, and can lead to heart failure. **Heart failure** is the progressive weakening of cardiac muscle so that it eventually becomes incapable of pumping enough blood to maintain life. Hypertension also increases the rate at which arteriosclerosis develops. Arteriosclerosis, in turn, increases the chance that blood clots will form and that blood vessels will rupture. Common conditions associated with hypertension are cerebral hemorrhage, coronary infarction, hemorrhage of renal blood vessels, and poor vision resulting from burst blood vessels in the retina.

Treatments that decrease peripheral resistance by dilating blood vessels, decrease total blood volume by increasing the rate of urine production, or decrease cardiac output are used to reduce blood pressure in people who suffer from hypertension. Low-salt diets also normally are recommended to reduce the amount of sodium chloride and water absorbed from the intestine into the bloodstream. A low-salt diet will result in a decreased blood volume.

Circulatory shock is defined as an inadequate blood flow throughout the body. As a consequence, tissues suffer damage resulting from a lack of oxygen. Severe shock can damage vital body tissues and lead to death.

There are several causes of shock, but **hemorrhagic** (hem-ŏ-raj´ik) **shock,** which is the result of extensive bleeding, can be used to illustrate the general characteristics of shock. Extensive bleeding reduces the blood volume, a condition called **hypovolemia,** and can result from injury or any other condition that results in extensive bleeding. If shock is not severe, blood pressure decreases only a moderate amount. Under these conditions, the mechanisms that normally regulate blood pressure function to reestablish normal blood pressure and blood flow. The baroreceptor reflexes initiate strong sympathetic responses that result in intense vasoconstriction and increased heart rate.

Blood pressure is also increased by several mechanisms that increase blood volume. As a result of the reduced blood pressure in the kidneys, increased amounts of renin are released. The elevated renin levels result in a greater rate of angiotensin II formation, causing vasoconstriction and increased aldosterone release from the adrenal cortex. The aldosterone, in turn, promotes water and salt retention by the kidneys. In response to reduced blood pressure, antidiuretic hormone (ADH) is released from the posterior pituitary gland, and ADH also enhances the retention of water by the kidneys. An intense sensation of thirst leads to increased water intake, which helps to restore the normal blood volume.

In mild cases of shock, the baroreceptor reflexes may be adequate to compensate for blood loss until the blood volume is restored, but in more severe cases all mechanisms are required to sustain life.

In more severe cases of shock, the regulatory mechanisms are not adequate to compensate for the effects of shock. As a consequence, a positive-feedback cycle begins to develop in which the blood pressure regulatory mechanisms lose their ability to compensate, and shock worsens. As shock becomes worse, the effectiveness of the regulatory mechanisms deteriorate. The positive-feedback cycle proceeds until death occurs, or until treatment is applied that terminates the cycle.

Several types of shock are classified by the cause of the condition:

1. Hemorrhagic shock is caused by internal or external bleeding and leads to hypovolemia.
2. Plasma loss shock is caused by reduced blood volume resulting from a loss of plasma and leads to hypovolemia. Examples are loss of plasma from severely burned areas, dehydration, and severe diarrhea or vomiting.
3. Neurogenic shock results in vasodilation in response to emotional upset or anesthesia.
4. Anaphylactic (an-a-fi-lak´tik) shock is caused by an allergic response that results in the release of inflammatory substances that cause vasodilation and an increase in capillary permeability.
5. Septic shock, or "blood poisoning," results from severe infections that cause the release of toxic substances into the circulatory system, depressing the activity of the heart and leading to vasodilation and increased capillary permeability.
6. Cardiogenic shock results when the heart stops pumping due to conditions such as heart attack or electrocution.

SUMMARY

The peripheral circulatory system consists of blood vessels outside of the heart.

GENERAL FEATURES OF BLOOD VESSEL STRUCTURE

Except for capillaries and venules, blood vessels have three layers:

(1) The inner tunica intima consists of simple squamous epithelium called endothelium and connective tissue; (2) The tunica media, the middle layer, contains circular smooth muscle, collagen, and elastic fibers; and (3) The outer tunica externa is connective tissue.

Arteries carry blood away from the heart:

Large elastic arteries have many elastic fibers and help maintain blood pressure; muscular arteries have much smooth muscle and regulate blood flow to different regions of the body; and arterioles are the smallest arteries and have smooth muscle cells and a few elastic fibers.

Capillaries consist of endothelium. Nutrient and waste product exchange is the principal function of capillaries. Precapillary sphincters regulate blood flow through capillary networks.

Veins carry blood toward the heart. Veins have three thin tunics. Valves prevent the backflow of blood in the veins.

PULMONARY CIRCULATION

The pulmonary circulation moves blood to and from the lungs.

Pulmonary arteries carry oxygen-poor blood from the right ventricle to the lungs, and pulmonary veins carry oxygen-rich blood from the lungs to the left atrium.

SYSTEMIC CIRCULATION: ARTERIES

The systemic circulation moves blood from the left ventricle to the tissues of the body and back to the right atrium.

Aorta

The aorta leaves the left ventricle to form the ascending aorta, aortic arch, and descending aorta, which consists of the thoracic and abdominal aorta.

Arteries of the head and neck

The brachiocephalic, left common carotid, and left subclavian arteries branch from the aortic arch to supply the head and the upper limbs.

The common carotid arteries and the vertebral arteries supply the head. The common carotid arteries divide to form the external carotids (which supply the face, nose, and mouth) and the internal carotids (which supply the brain).

Arteries of the upper limbs

The subclavian artery continues as the axillary artery and then as the brachial artery, which branches to form the radial and ulnar arteries.

The thoracic aorta and its branches

The thoracic aorta has visceral branches, which supply the abdominal organs, and parietal branches, which supply the thoracic wall.

The abdominal aorta and its branches

The abdominal aorta has visceral branches, which supply the abdominal organs, and parietal branches, which supply the abdominal wall.

Arteries of the pelvis

The abdominal aorta divides to form the common iliac arteries, which divide to form the external and internal iliac arteries. Branches of the internal iliac arteries supply the pelvis.

Arteries of the lower limbs

The external iliac artery continues as the femoral artery and then as the popliteal artery to the leg.

SYSTEMIC CIRCULATION: VEINS

The superior vena cava drains the head, neck, thorax, and upper limbs. The inferior vena cava drains the abdomen, pelvis, and lower limbs.

Veins of the head and neck

The external jugular veins drain the posterior head and posterior neck.

The internal jugular veins drain the brain, anterior head, and anterior neck.

Veins of the upper limbs

The main deep vein is the brachial vein; the superficial veins are the basilic, cephalic, and median cubital.

Veins of the thorax

The left and right brachiocephalic veins and the azygos veins return blood to the superior vena cava.

Veins of the abdomen and pelvis

Vessels from the kidneys, adrenal glands, and gonads directly enter the inferior vena cava.

Vessels from the stomach, intestines, spleen, and pancreas connect with the hepatic portal vein. The hepatic portal vein transports blood to the liver for processing. Hepatic veins from the liver join the inferior vena cava.

Veins of the lower limbs

The deep veins follow the same path as the deep arteries and are named after the arteries they accompany.

The main superficial vein is the great saphenous vein.

THE PHYSIOLOGY OF CIRCULATION

Blood pressure

Blood pressure is a measure of the force exerted by blood against the blood vessel wall.

Blood pressure fluctuates between the systolic (highest value) and the diastolic (lowest value) blood pressure.

Blood moves through the circulatory system from areas of higher to lower pressure.

Blood pressure in the aorta is measured by the auscultatory method under clinical conditions and fluctuates between approximately 120 (systolic) and 80 (diastolic) mm Hg. Blood pressure at the right atrium is approximately 0 mm Hg.

The pulse

A pulse can be detected when large arteries are near the surface of the body.

Capillary exchange

Most exchange across the wall of the capillary is by diffusion.

There is a movement of fluid from the blood into the tissues at the arterial end of the capillary and back into the capillary at its venous end. Some of the fluid gained by the tissues is removed by the lymphatic system.

LOCAL CONTROL OF BLOOD VESSELS

Blood flow through a tissue is usually proportional to the metabolic needs of the tissue and is controlled by the precapillary sphincters.

The vasomotor center (sympathetic nervous system) controls blood vessel diameter.

Vasomotor tone is a state of partial contraction of blood vessels.

The nervous system is responsible for routing the flow of blood from one region of the body to another.

REGULATION OF ARTERIAL PRESSURE

Arterial blood pressure is controlled by the amount of blood pumped by the heart and by the total peripheral resistance.

Baroreceptor reflexes

Baroreceptors are sensory receptors that are sensitive to stretch.

Baroreceptors are located in the carotid sinuses and the aortic arch.

The baroreceptor reflex changes peripheral resistance and the volume of blood pumped by the heart in response to changes in blood pressure.

Chemoreceptor reflexes

Chemoreceptors are sensory receptors sensitive to oxygen, carbon dioxide, and pH levels in the blood.

Chemoreceptors are located in the carotid bodies, the aortic bodies, and the medulla.

The chemoreceptor reflex increases peripheral resistance and is most important under emergency conditions.

Hormonal mechanisms

Epinephrine and norepinephrine released from the adrenal medulla as a result of sympathetic stimulation increases heart rate, stroke volume, and vasoconstriction.

Renin is released by the kidneys in response to low blood pressure. Renin promotes the production of angiotensin II, which causes vasoconstriction and an increase in aldosterone secretion. Aldosterone reduces urine output.

Long-term and short-term regulation of blood pressure

The baroreceptor reflex and the adrenal medullary hormones are most important in controlling blood pressure on a short-term basis.

The renin-angiotensin-aldosterone is more important in controlling blood pressure on a long term-basis.

FETAL CIRCULATION

Blood bypasses the lungs by flowing from the right to the left atrium through the foramen ovale and by flowing from the pulmonary trunk to the aorta through the ductus arteriosus.

Blood flows to the placenta from the infant through two umbilical arteries that arise from the internal iliac arteries and from the placenta through a single umbilical vein to the inferior vena cava.

CONTENT REVIEW

1. Name the three layers of a blood vessel. What kinds of tissue are in each layer?
2. Name, in order, all the types of blood vessels, starting at the heart, going to the tissues, and returning to the heart.
3. Describe the structure and function of elastic arteries, muscular arteries, and arterioles.
4. Describe the structure of capillaries and explain their major function. Name the structure that regulates blood flow through capillaries.
5. Describe the structure of veins.
6. What is the function of valves in blood vessels and in which blood vessels are valves found?
7. Name the blood vessels of the pulmonary circulation.
8. List the different parts of the aorta.
9. Name the arteries that supply the head, upper limbs, thorax, abdomen, pelvis, and lower limbs. Describe the specific areas each artery supplies.
10. Name the major vessels that return blood to the heart. What area of the body does each drain?
11. List the veins that drain blood from the head, upper limbs, thorax, abdomen, pelvis, and lower limbs. Describe the hepatic portal system.
12. Define blood pressure and describe how it is normally measured.
13. Describe the changes in blood pressure starting in the aorta, moving through the vascular system, and returning to the right atrium.
14. Define pulse and explain what information can be determined from monitoring the pulse.
15. Describe the exchange of nutrients and fluid across the wall of capillaries.
16. Explain what is meant by the local control of blood flow through tissues and describe what carries out local control.

17. Describe nervous control of blood vessels. Define the vasomotor center and vasomotor tone.
18. How is blood pressure related to heart rate, stroke volume, and peripheral resistance?
19. Where are baroreceptors located? Describe the baroreceptor reflex when blood pressure increases and when it decreases.
20. Where are the chemoreceptors located? Describe what happens when blood oxygen levels decrease, blood carbon dioxide levels increase, or blood pH decreases.
21. For epinephrine, norepinephrine, and renin, state where each is produced, what stimulus causes an increased production, and what effect each have on the circulatory system.
22. Name the mechanisms responsible for short-term regulation of blood pressure.
23. Name the mechanisms responsible for long-term regulation of blood pressure.
24. Describe the differences in blood flow between the fetus and in a person after birth.

CONCEPT REVIEW

1. For each of the following destinations, name all the arteries that a red blood cell would encounter if it started its journey in the left ventricle.
   a. The brain
   b. External portion of the skull
   c. The left arm
   d. The foot
   e. Kidney
   f. Small intestine
2. For each of the following starting places, name all the veins that a red blood cell would encounter on its way back to the right atrium.
   a. The brain
   b. External portion of the skull
   c. The left arm
   d. The foot
   e. Kidney
   f. Small intestine
3. A person suffering from severe hemorrhagic shock was treated by health care professionals and his blood pressure was restored to normal. However, it became clear that permanent kidney damage resulted. Explain how permanent kidney damage can occur as a result of shock.
4. Hugo Faster ran a race. During the race his stroke volume and heart rate increased. Vasoconstriction occurred in his viscera and his blood pressure increased, but not dramatically. Explain these changes in his circulatory system.
5. A common treatment for hypertension is to give a diuretic drug to the patient that increases urine volume. Explain how a drug that increases urine volume would help lower blood pressure.

CHAPTER TEST

MATCHING For each statement in column A select the correct answer in column B (an answer may be used once, more than once, or not at all).

A

1. A blood vessel that carries blood away from the heart.
2. The innermost layer of a blood vessel wall.
3. The blood vessel into which blood flows from capillaries.
4. The process resulting in the deposition of fatty materials containing cholesterol in the walls of arteries to form plaques.
5. The portion of the aorta from which the brachiocephalic artery, the left common carotid artery, and the left subclavian artery arise.
6. The blood vessels that supply most of the blood to the brain.
7. The artery that is most frequently used to monitor the pulse.
8. The major branches of the aorta that supply the lower limbs and the pelvic area.
9. Large vein that returns blood from the head, neck, thorax, and upper limbs to the right atrium of the heart.
10. In a healthy person at rest the maximum blood pressure.
11. The reflex that brings the blood pressure back to normal in response to a sudden decrease in blood pressure.
12. The mechanism that is most important in long-term regulation of blood pressure.

B

adrenal medullary mechanism
aortic arch
arteriosclerosis
artery
ascending aorta
atherosclerosis
baroreceptor reflex
chemoreceptor reflex
common iliac arteries
descending aorta
diastolic pressure
external carotid arteries
femoral arteries
inferior vena cava
internal carotid arteries
precapillary sphincter
radial artery
renin-angiotensin-aldosterone mechanism
superior vena cava
systolic pressure
tunica externa
tunica intima
tunica media
vein
venule
vertebral arteries

FILL-IN-THE BLANK Complete each statement by providing the missing word or words.

1. Blood flow through the __________ __________ __________ begins as capillary beds in the small intestine, spleen, and stomach, and carries blood to the liver, in which it ends in a capillary bed.
2. The system of blood vessels that carry blood from the right ventricle of the heart to the lungs and back from the lungs to the left atrium makes up the __________ circulation.
3. The smooth muscle in the tunica media of __________ arteries enables these vessels to control blood flow to different regions of the body by either constricting or dilating.
4. The brachiocephalic artery gives rise to the right __________ __________ artery, which transports blood to the right side of the head and neck, and the right __________ artery, which transports blood to the right upper limb.
5. The three major unpaired branches of the abdominal aorta that supply blood to the viscera are the __________ artery, the __________ __________ artery, and the __________ __________ artery.
6. Generally, if the blood pressure exceeds 150/90 mm Hg, it is considered too high, a condition called __________.
7. Control of blood flow through capillaries is achieved by contraction and relaxation of the __________ __________.

CHAPTER 14

The Lymphatic System and Immunity

KEY TERMS

antibody
Protein found in blood plasma that binds to a specific, usually foreign or toxic, substance.

antigen
Any substance that stimulates the specific immune system.

B cell
Lymphocyte responsible for antibody-mediated immunity.

complement
Group of plasma proteins that stimulates phagocytosis, inflammation, and lysis of cells.

immunity
Ability to resist damage from foreign substances such as microorganisms and harmful chemicals.

interferon
A protein that inhibits viral replication.

lymph
Clear or yellowish fluid derived from tissue fluid and found in lymph vessels.

lymph node
Encapsulated mass of lymphatic tissue found along lymph vessels; functions to filter lymph and produce lymphocytes.

nonspecific resistance
Immune system response that is the same upon each exposure to an antigen; there is no ability to remember a previous exposure to a specific antigen.

specific resistance
Immune system response in which there is an ability to recognize, remember, and destroy a specific antigen.

T cell
Lymphocyte responsible for cell-mediated immunity.

OBJECTIVES

After reading this chapter you should be able to:

1. Describe the functions of the lymphatic system.
2. Explain how lymph is formed and transported.
3. Describe the structure and function of tonsils, lymph nodes, the spleen, and the thymus.
4. Define nonspecific resistance and describe the cells and chemicals involved.
5. List the events that occur during an inflammatory response and explain their significance.
6. Define antibody-mediated immunity and cell-mediated immunity.
7. Discuss the primary and secondary response to an antigen, and explain the basis for long-lasting immunity.
8. Describe the functions of T cells.
9. Explain the four ways that specific resistance can be acquired.

FEATURES

THE LYMPHATIC SYSTEM includes lymph, lymphocytes, lymph vessels, lymph nodes, tonsils, the spleen, and the thymus gland (Figure 14-1). The lymphatic system performs three basic functions. First, it helps maintain fluid balance in the tissues. Approximately 30 liters of fluid pass from the blood capillaries into the tissue spaces each day, whereas only 27 liters pass from the tissue spaces back into the blood capillaries. If the extra 3 liters of tissue fluid were to remain in the tissue spaces, edema would result, causing tissue damage and eventual death. These 3 liters of fluid enter the lymphatic capillaries, where the fluid is called **lymph** (limf), and it passes through the lymph vessels to return to the blood.

Second, the lymphatic system absorbs fats and other substances from the digestive tract (see Chapter 16). Special lymph vessels called **lacteals** (lak´te-als) are located in the lining of the small intestine. Fats enter the lacteals and pass through the lymph vessels to the venous circulation. The lymph passing through these lymph vessels has a milky appearance because of its fat content, and it is called **chyle** (kīl).

Third, the lymphatic system is part of the body's defense system. Lymphatic organs such as lymph nodes filter lymph, and the spleen filters blood, removing microorganisms and other foreign substances. In addition, lymphatic organs contain lymphocytes and other cells that are capable of destroying microorganisms and foreign substances.

LYMPHATIC SYSTEM

Lymph vessels

The lymphatic (lim-fat´ik) system, unlike the circulatory system, only carries fluid away from the tissues. The lymphatic system begins in tissues as **lymph capillaries,** which are tiny, close-ended vessels consisting of simple squamous

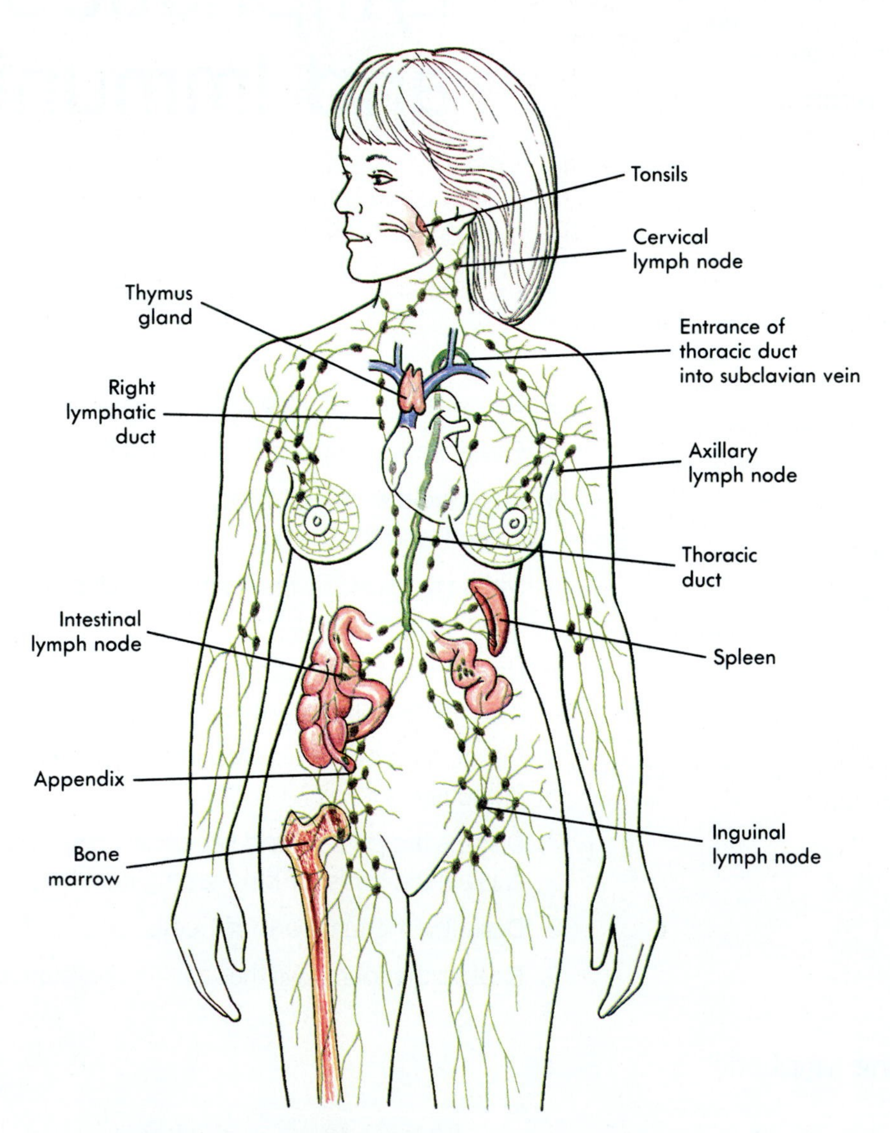

Figure 14-1 ● Lymphatic system.
The lymph vessels and major lymphatic organs are shown.

epithelium. Fluid tends to move out of the blood capillaries into tissue spaces and then out of the tissue spaces into lymph capillaries (Figure 14-2, *A*).

The lymph capillaries join to form larger **lymph vessels** that resemble small veins. Small lymph vessels have one-way valves similar to the valves of veins. When a lymph vessel is compressed, backward movement of lymph to the lymph capillaries is prevented by the valves. Consequently, lymph moves forward through the lymph vessels. Three factors assist in the transport of lymph through the lymph vessels: (1) contraction of surrounding skeletal muscle during activity, (2) contraction of smooth muscle in the lymph vessel wall, and (3) pressure changes in the thorax during respiration.

The lymph vessels converge and eventually empty into the blood at two locations in the body. Lymph vessels from the upper right limb and the right half of the head, neck, and chest form the **right lymphatic duct,** which empties into the right subclavian vein. Lymph vessels from the rest of the body enter the **thoracic duct,** which empties into the left subclavian vein (see Figure 14-1 and Figure 14-2, *B*).

Lymphatic organs

Lymphatic organs contain **lymphatic tissue,** which consists of many lymphocytes, macrophages, and other cells. The **lymphocytes** originate from red bone marrow and are carried by the blood to lymph organs. When the body is exposed to microorganisms or foreign substances, the lymphocytes divide and increase in number. The lymphocytes are part of the immune system response that causes the destruction of microorganisms and foreign substances.

Tonsils

There are three groups of **tonsils.** The **palatine tonsils** usually are referred to as "the tonsils," and they are located on each side of the posterior opening of the oral cavity. The **pharyngeal tonsil** is located near the internal opening of the nasal cavity. An enlarged pharyngeal tonsil is called an **adenoid** (ad´ĕ-noid), and it can interfere with normal breathing. The **lingual tonsil** is near the posterior margin of the tongue.

The tonsils provide protection against pathogens and other potentially harmful material entering the nose and mouth. Sometimes the tonsils or adenoids become chronically infected and must be removed. In adults the tonsils decrease in size and may eventually disappear.

Lymph nodes

Lymph nodes are small, round structures distributed along the various lymph vessels (see Figure 14-1). Most lymph passes through at least one lymph node before entering the blood. Although lymph nodes are found throughout the body, there are three superficial aggregations of lymph nodes on each side of the body: the inguinal nodes in the groin, the axillary nodes in the axillary (armpit) region, and the cervical nodes of the neck.

Lymph nodes are divided into compartments that contain lymphatic tissue (Figure 14-3). Lymph enters the lymph node through afferent vessels, passes through the lymphatic

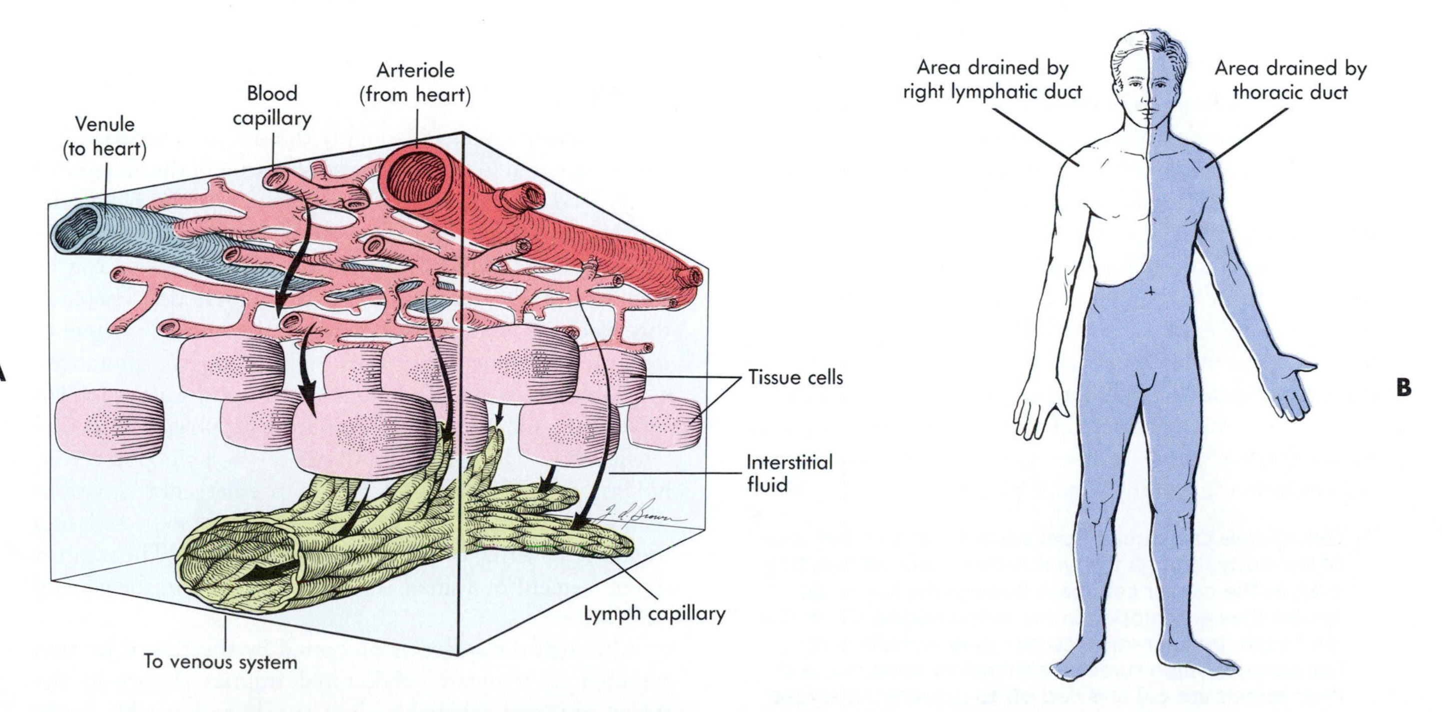

Figure 14-2 ● Lymph formation and drainage.
A, Movement of fluid from blood capillaries into tissue spaces and from tissue spaces into lymph capillaries to form lymph. **B,** Overall lymph drainage. Lymph from the blue area drains through the thoracic duct. Lymph from the white area drains through the right lymphatic duct.

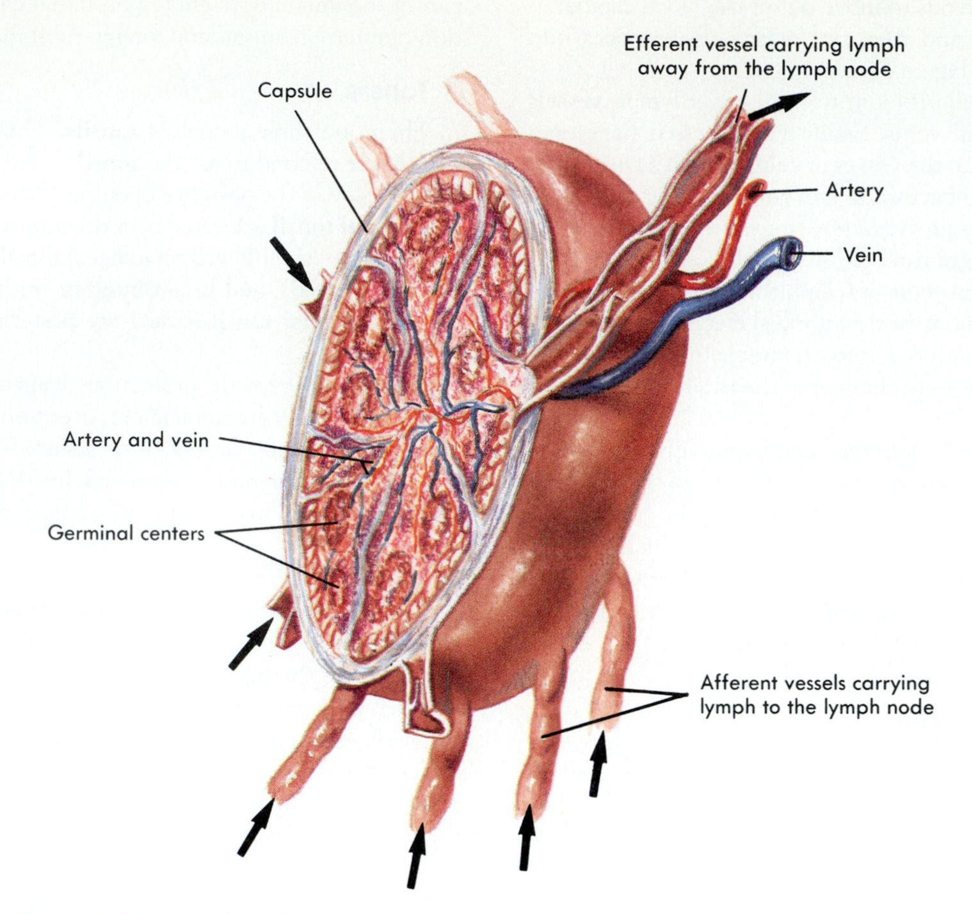

Figure 14-3 ● Lymph node.
Arrows indicate the direction of lymph flow. The germinal centers are sites of lymphocyte production. As lymph moves through the lymph sinuses macrophages remove foreign substances.

tissue, and exits through efferent vessels. As lymph moves through the lymph nodes, two functions are performed. One function is activation of the immune system. Microorganisms or other foreign substances in the lymph can stimulate lymphocytes in the lymphatic tissue to start dividing. These areas of rapidly dividing lymphocytes are called **germinal centers.** The newly produced lymphocytes are released into the lymph and eventually reach the blood where they circulate and enter other lymphatic tissues. The lymphocytes are capable of destroying microorganisms and other foreign substances. Another function of the lymph nodes is the removal (phagocytosis) of microorganisms and foreign substances from the lymph by macrophages.

1 Cancer cells can spread from a tumor site to other areas of the body through the lymphatic system. At first, however, as the cancer cells pass through the lymphatic system they are trapped in the lymph nodes, which filter the lymph. During radical cancer surgery, malignant (cancerous) lymph nodes are therefore removed, and their vessels are cut and tied off to prevent the spread of the cancer. Predict the consequences of tying off the lymph vessels.*

*Answers to predict questions appear in Appendix B at the back of the book.

Spleen

The **spleen** (splēn) is roughly the size of a clenched fist, and it is located in the left, superior corner of the abdominal cavity (see Figure 14-1). The spleen filters the blood, destroying microorganisms, foreign substances, and worn out red blood cells (RBCs). Lymphocytes in the spleen can be stimulated in the same manner as in lymph nodes. Therefore the spleen also functions to activate the immune system in response to microorganisms and foreign substances. Macrophages in the spleen remove microorganisms, foreign substances, and worn out RBCs through phagocytosis (see Chapter 11). The spleen also functions as a blood reservoir, holding a small volume of blood. In emergency situations such as hemorrhage, smooth muscle in the blood vessels and the outer covering of the spleen can contract. The result is the movement of a small amount of blood into the general circulation.

Although the spleen is protected by the ribs, it is often ruptured in traumatic abdominal injuries. Injury to the spleen can cause severe bleeding, shock, and possible death. A splenectomy, removal of the spleen, is performed to stop the bleeding. Other lymphatic tissue and the liver compensate for the loss of the spleen's functions.

Disorders of the lymphatic system

Because the lymphatic system is involved with the production of lymphocytes that fight infectious diseases, and because the lymphatic system filters blood and lymph to remove microorganisms, it is not surprising that many infectious diseases produce symptoms associated with the lymphatic system. **Lymphadenitis** (lim-fad-ĕ-ni´tis) is an inflammation of the lymph nodes, causing them to become enlarged and tender. It is an indication that microorganisms are being trapped and destroyed within the lymph nodes. **Lymphangitis** (lim-fan-ji´tis) is an inflammation of the lymph vessels. This often results in visible red streaks in the skin that extend away from the site of infection. If the microorganisms pass through the lymph vessels and nodes to reach the blood, septicemia, or blood poisoning, can result (see Chapter 11). A **lymphoma** (lim-fo´mah) is a neoplasm (tumor) of lymphatic tissue that is almost always malignant. Hodgkin's disease is an example. Typically lymphomas begin as an enlarged, painless mass of lymph nodes. Enlargement of the lymph nodes, however, can compress surrounding structures and produce complications. The immune system is depressed, and the patient has an increased susceptibility to infections. Fortunately, treatment with drugs and radiation is effective for many people who suffer from lymphoma.

Bubonic (bu-bon´ik) **plague** and **elephantiasis** (el-ĕ-fan-ti´ă-sis) are diseases of the lymphatic system. In the sixth, fourteenth, and nineteenth centuries the bubonic plague killed large numbers of people. Fortunately, there are relatively few cases today. Bubonic plague is caused by bacteria that are transferred to humans from rats by the bite of the rat flea. The bacteria localize in the lymph nodes, causing the lymph nodes to enlarge. The term *bubonic* is derived from a Greek word referring to the groin, because the disease often causes the inguinal lymph nodes of the groin to swell. Without treatment, septicemia followed by death rapidly occurs in 70% to 90% of those infected. Elephantiasis is caused by long, slender roundworms. The adult worms lodge in the lymph vessels and can cause such a blockage of lymph flow that a limb can become permanently swollen and enlarged. The resemblance of the affected limb to that of an elephant's leg is the basis for the name of the disease. The offspring of the adult worms pass through the lymphatic system into the blood. They can be transferred from an infected person to other humans by mosquitoes.

Thymus

The **thymus** (thi´mus) is a bilobed gland roughly triangular in shape (see Figure 14-1). It is located in the superior mediastinum, the partition dividing the thoracic cavity into left and right parts. The size of the thymus differs markedly depending on the age of the individual. In a newborn the thymus may extend halfway down the length of the thorax. The thymus continues to grow until puberty, although not as rapidly as other structures in the body. After puberty, the thymus decreases in size, and in older adults the thymus may be so small that it is difficult to find.

Large numbers of lymphocytes are produced in the thymus, but for unknown reasons most degenerate. The thymus functions as a site for the processing and maturation of lymphocytes. While in the thymus, lymphocytes do not respond to foreign substances. However, after thymic lymphocytes have matured, they enter the blood and travel to other lymphatic tissues, where they help protect against microorganisms and other foreign substances.

IMMUNITY

Immunity is the ability to resist damage from foreign substances such as microorganisms and harmful chemicals such as toxins released by microorganisms. Immunity is categorized as **nonspecific resistance** or **specific resistance**. The distinction between nonspecific resistance and specific resistance involves the concepts of specificity and memory. In nonspecific resistance, each time the body is exposed to a substance, the response is the same. For example, each time a bacterial cell is introduced into the body, it is phagocytized with the same speed and efficiency. In specific resistance, the response during the second exposure is faster and more effective than the response to the first exposure. For example, upon first exposure to a virus, the body may take many days to destroy the virus. During this time the virus damages tissues, producing the symptoms of disease. After the second exposure to the same virus, however, the response is very rapid and effective because the immune system can now recognize and destroy this particular virus. The virus is eliminated before any symptoms develop, thus the person is immune. Specific resistance is possible because of specificity and memory, the ability of the system to recognize and remember a particular substance.

NONSPECIFIC RESISTANCE

Nonspecific resistance includes mechanical mechanisms, chemical mediators, cells, and the inflammatory response.

Mechanical mechanisms

Mechanical mechanisms prevent the entry of microorganisms into the body. Microorganisms cannot cause a disease if they cannot get into the body. The skin and mucous membranes form barriers that prevent the entry of microorganisms; and tears, saliva, and urine wash away microorganisms. In the respiratory tract, mucous membranes are ciliated (see Chapter 4), and microorganisms trapped in the mucus are swept to the back of the throat and swallowed. Coughing and sneezing also remove microorganisms from the respiratory tract.

Chemical mediators

Chemical mediators are substances that bring about immune system responses. Some chemical mediators that are found on the surface of cells kill microorganisms. Lysozyme in tears and saliva is an example. Other chemical mediators, for example histamine, promote inflammation by causing vasodilation and increasing vascular permeability.

Complement is a group of at least 11 proteins found in plasma. The operation of complement proteins is similar to that of clotting proteins (see Chapter 11). Normally, complement proteins circulate in the blood in an inactive form. Certain complement proteins can be activated by combining with foreign substances (for example, parts of a bacterial cell) or by combining with antibodies (see the discussion of the specific immune system). Once activation begins, a series of reactions results, in which each complement protein activates the next complement protein. The activated complement promotes inflammation and phagocytosis, and can directly lyse (rupture) bacterial cells.

Interferons (in´ter-fēr´onz) are proteins that protect the body against viral infections. When a virus infects a cell, the cell produces viral nucleic acids and proteins, which are assembled into new viruses. The new viruses are released from the infected cell to infect other cells. Because infected cells usually stop their normal functions or die during viral replication, viral infections are clearly harmful to the body. Fortunately, viruses often stimulate infected cells to produce interferons. Interferons do not protect the cell that produces them. Instead, interferons bind to the surface of neighboring cells where they stimulate those cells to produce antiviral proteins. These antiviral proteins stop viral reproduction in the neighboring cells by preventing the production of new viral nucleic acids and proteins.

Cells

White blood cells (WBCs) and the cells derived from WBCs (see Chapter 11) are the most important cellular components of the immune system. WBCs are produced in red bone marrow and lymphatic tissue and are released into the blood. Chemicals released from microorganisms or damaged tissues attract the WBCs, and they leave the blood and enter affected tissues. Important chemical mediators known to attract WBCs include complement and histamine.

Neutrophils are small phagocytic cells that usually are the first cells to enter infected tissues from the blood. However, neutrophils often die after phagocytizing a single microorganism. **Pus** is primarily an accumulation of dead neutrophils at a site of infection.

Macrophages are monocytes that leave the blood, enter tissues, and enlarge about fivefold. Macrophages can ingest more and larger items than can neutrophils. Macrophages usually appear in infected tissues after neutrophils and are responsible for most of the phagocytic activity in the late stages of an infection, including the cleanup of dead neutrophils.

Basophils release chemical mediators such as histamine that promote inflammation. In addition, there are cells found in connective tissue, called **mast cells,** that release inflammatory chemicals. On the other hand, **eosinophils** release chemicals that break down inflammatory chemicals. At the same time that inflammation is initiated, eosinophils act to reduce and contain the inflammatory response. Inflammation is beneficial in the fight against microorganisms, but too much inflammation can be harmful, resulting in the unnecessary destruction of healthy tissues as well as the destruction of the microorganisms.

Inflammatory response

The **inflammatory response** is a complex sequence of events involving many of the chemical mediators and cells previously discussed. Although they may vary in some details depending on the events producing them, most inflammatory responses are similar.

A bacterial infection is used here to illustrate an inflammatory response (Figure 14-4). The bacteria, or damage to tissues, cause the release or activation of chemical mediators such as histamine, prostaglandins, complement, and others. The chemical mediators produce several effects: (1) vasodilation, which increases blood flow and brings phagocytes and other WBCs to the area; (2) attraction of phagocytes, which leave the blood and enter the tissue; and (3) increased vascular permeability, which allows fibrin and complement to enter the tissue from the blood. Fibrin prevents the spread of infection by walling off the infected area. Complement further enhances the inflammatory response and attracts additional phagocytes. This process of releasing chemical mediators and attracting phagocytes and other WBCs continues until the bacteria are destroyed. Phagocytes (mainly macrophages) remove microorganisms and dead tissue, and the damaged tissues are repaired.

Typically, inflammation produces the symptoms of redness, heat, swelling, pain, and disturbance of function. Vasodilation and increased vascular permeability are responsible for redness, heat, and swelling. Inflammatory chemical mediators and pressure created by swelling are mainly responsible for pain, which results in disturbance of function (see Chapter 4 for details).

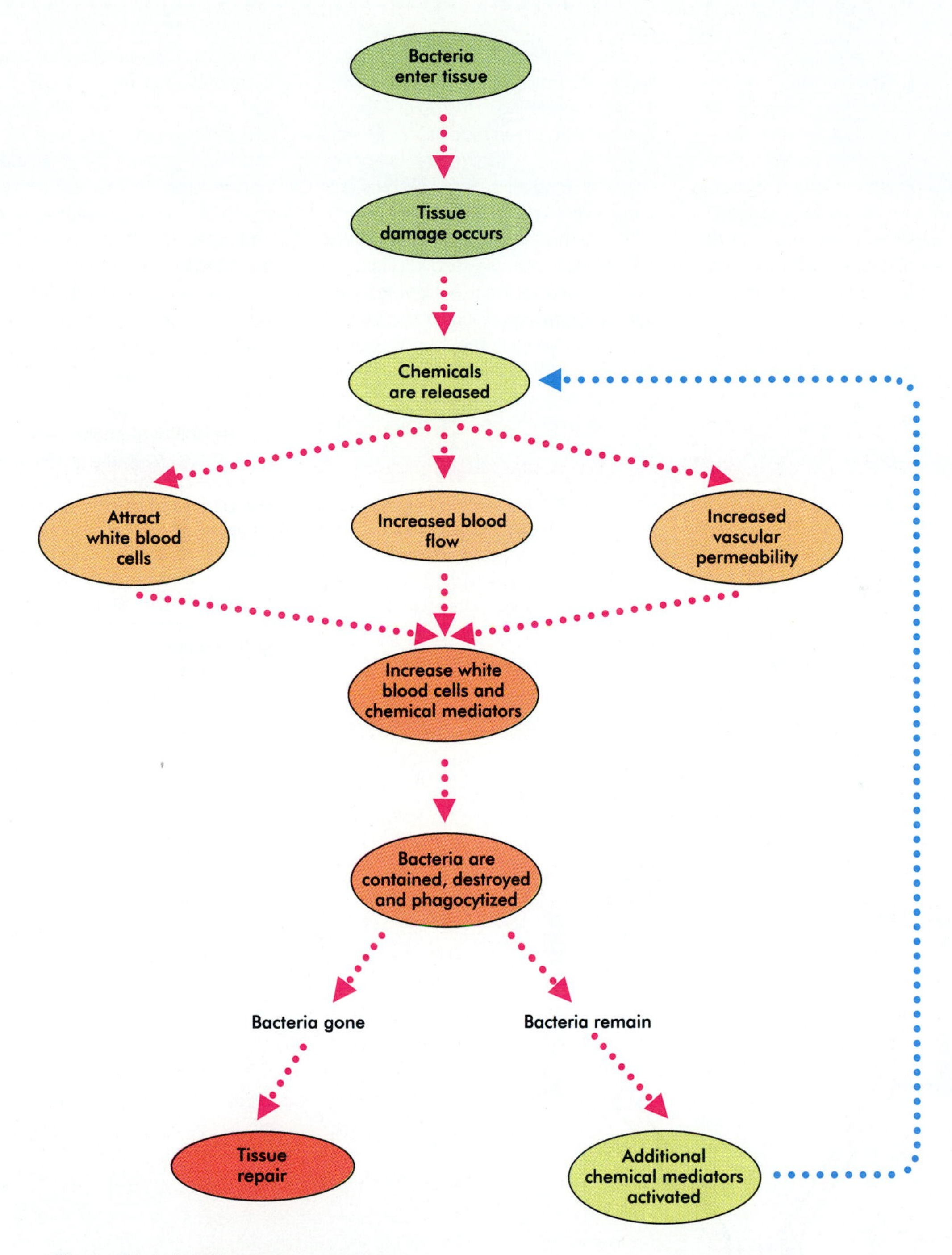

Figure 14-4 ● Inflammatory response.
Bacteria cause tissue damage and the release of chemical mediators that initiate inflammation and phagocytosis, resulting in the destruction of the bacteria.

Understanding antibodies

Large amounts of antibodies are in plasma, although plasma also contains other proteins. The plasma proteins can be separated into albumin and alpha, beta, and gamma globulin portions. Antibodies are sometimes called **gamma globulins** because they are found mostly in the gamma globulin portion of plasma. Antibodies are also called **immunoglobulins** (Ig) because they are globulin proteins involved in immunity.

Antibodies are proteins produced in response to an antigen. They are Y-shaped molecules (Figure 14-A). The end of each "arm" of the antibody is the part of the antibody that combines with the antigen. This part of a particular antibody can only join with a particular antigen. This is similar to the lock and key model of enzymes (see Chapter 2). Thus a particular antibody is specific for, and can only combine with, a particular antigen.

Antibodies normally are produced as part of an immune system response that eliminates antigens. The binding of antibodies to antigens is a signal to begin this process. In effect, the antigens are identified as something to be eliminated when the antibodies bind to them. For example, the binding of antibodies to antigens stimulates an inflammatory response. Chemical mediators and cells are activated, eventually destroying the antigens.

The ability of antibodies to bind to specific antigens is being used clinically. For example, **monoclonal antibodies** are a pure antibody preparation that is specific for only one antigen. The antigen, injected into a laboratory animal, activates a group of identical B cells called a clone. The clone only produces antibodies against the antigen. The antibodies are called monoclonal, meaning there is one (mono) kind of antibody produced from a clone of cells. The B cells are removed from the animal and fused with tumor cells, which typically are rapidly dividing cells. The resulting clone-tumor cells have two ideal characteristics: they divide to form large numbers of cells and they produce only one kind of antibody.

Monoclonal antibodies have many applications. They are used for determining pregnancy and for diagnosing diseases such as gonorrhea, syphilis, hepatitis, rabies, and cancer. These tests are specific and rapid because the monoclonal antibodies bind only to the antigen being tested. Monoclonal antibodies also can be used to treat cancer. Anti-cancer drugs are attached to monoclonal antibodies that bind to cancer cells. The drug then kills the cancer cell. This approach has the advantage of selectively destroying cancer cells while sparing normal, healthy cells.

2 The ability of antibodies to join antigens together is the basis for many clinical tests. The antigens are joined together when one "arm" of the antibody binds to one antigen and the other "arm" of the antibody binds to a second antigen. Using blood typing as an example, explain how the ability to join antigens together makes some clinical tests possible. (Hint: individual antigens are too small to be seen.)

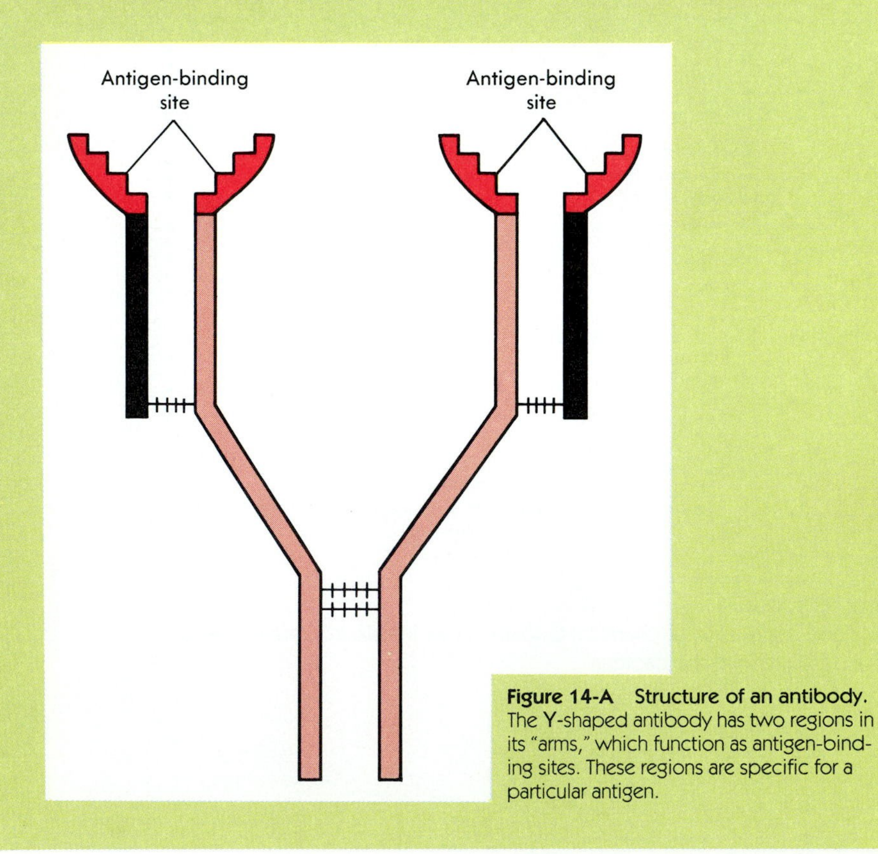

Figure 14-A Structure of an antibody. The Y-shaped antibody has two regions in its "arms," which function as antigen-binding sites. These regions are specific for a particular antigen.

SPECIFIC RESISTANCE

Specific resistance involves the ability to recognize, respond to, and remember a particular substance. Molecules that stimulate specific resistance responses are called **antigens** (an´ti-jenz). Antigens can be divided into two groups: foreign antigens and self antigens. **Foreign antigens** are introduced from outside the body. Components of bacteria, viruses, and other microorganisms are examples of foreign antigens that cause disease. Pollen, animal hairs, foods, and drugs are also foreign antigens and can cause an overreaction of the immune system, producing allergies. Transplanted tissues and organs that contain foreign antigens result in the rejection of the transplant. **Self antigens** are molecules produced by the body that stimulate an immune system response. The response to self antigens can be beneficial. For example, the recognition of tumor antigens can result in destruction of the tumor. The response to self antigens can also be harmful. **Autoimmune disease** results when self antigens stimulate unwanted destruction of normal tissue. An example is rheumatoid arthritis, which results in the destruction of tissue within joints.

Specific resistance results from the activities of lymphocytes. **B cells** are lymphocytes that give rise to cells responsible for the production of proteins called antibodies. The antibodies are found in the plasma. Because antibodies are involved, this type of specific resistance is called **antibody-mediated immunity. T cells** are lymphocytes responsible for **cell-mediated immunity.** In addition, some T cells are involved with regulating both antibody-mediated and cell-mediated immunity. For example, **helper T cells** stimulate the activities of B cells and other T cells.

Although the immune system is divided into different parts, it should be noted that this is an artificial division used to emphasize particular aspects of immunity. Actually, immune system responses often involve more than one part of the immune system. For example, it is possible for both the antibody-mediated and cell-mediated immune system to recognize and respond to an antigen. Once recognition has occurred, many of the events that lead to destruction of the antigen are nonspecific resistance activities such as inflammation and phagocytosis. It is the coordinated activities of all parts of the immune system that protect the body from disease.

Antibody-mediated immunity

Exposure of the body to an antigen can stimulate the production of antibodies. The antibodies bind to the antigens and the antigens can be destroyed (see box, p. 238). Because antibodies are in body fluids, antibody-mediated immunity is effective against extracellular antigens such as bacteria, viruses (when they are outside cells), and toxins. Antibody-mediated immunity also is involved with some allergic reactions.

The production of antibodies after the first exposure to an antigen is different from that following a second or subsequent exposure (Figure 14-5). The **primary response** re-

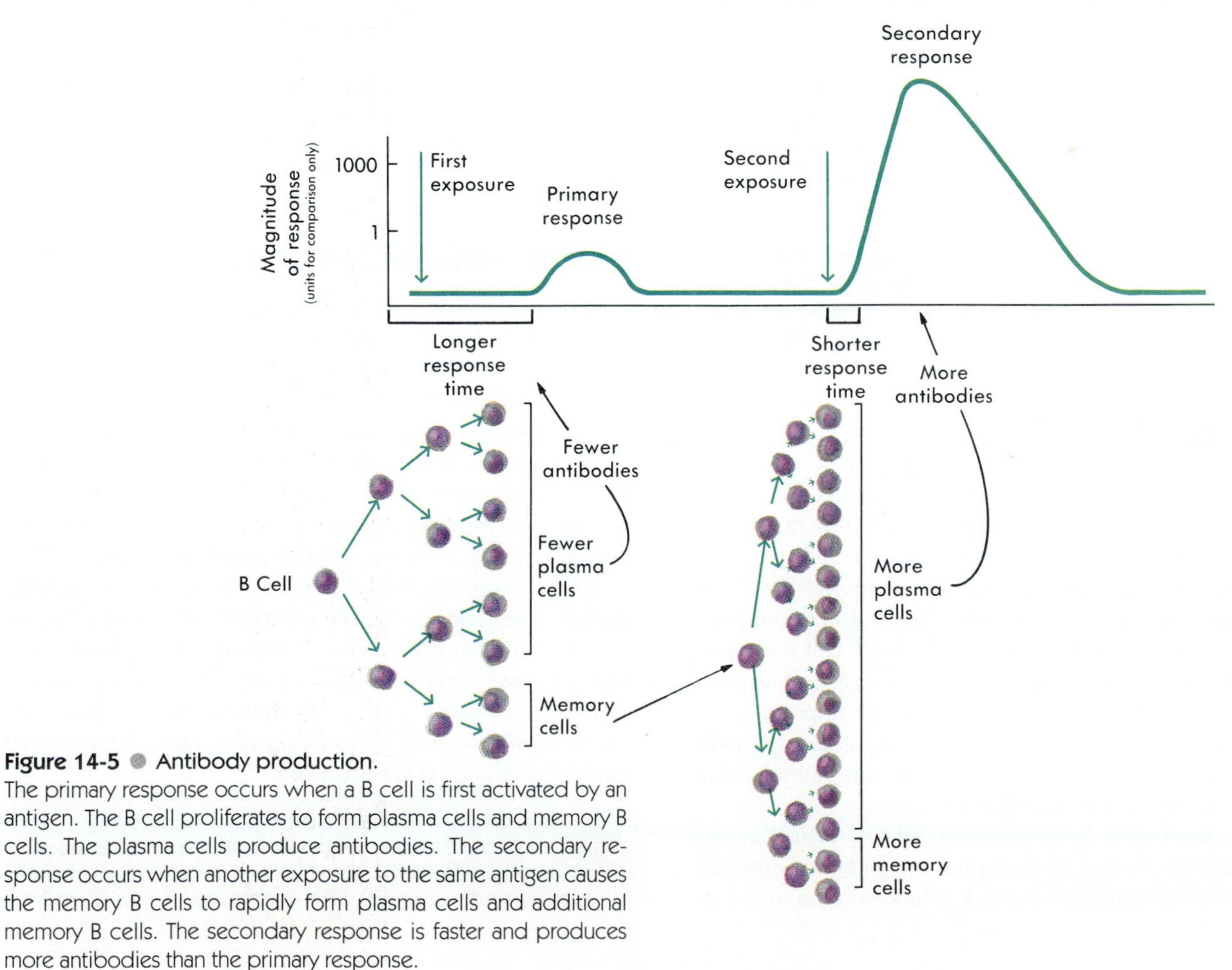

Figure 14-5 ● **Antibody production.**
The primary response occurs when a B cell is first activated by an antigen. The B cell proliferates to form plasma cells and memory B cells. The plasma cells produce antibodies. The secondary response occurs when another exposure to the same antigen causes the memory B cells to rapidly form plasma cells and additional memory B cells. The secondary response is faster and produces more antibodies than the primary response.

sults from the first exposure of a B cell to an antigen. When the antigen binds to the antigen-binding receptor on the B cell, the B cell undergoes several divisions to form plasma cells and memory B cells. **Plasma cells** produce antibodies. The primary response normally takes 3 to 14 days to produce enough antibodies to be effective against the antigen. In the meantime, the individual usually develops disease symptoms because the antigen has had time to cause tissue damage.

Memory B cells are responsible for the **secondary,** or **memory, response,** which occurs when the immune system is exposed to an antigen against which it already has produced a primary response. When exposed to the antigen again, the memory B cells rapidly divide to form plasma cells, which rapidly produce antibodies. The secondary response provides better protection than the primary response for two reasons. First, the time required to start producing antibodies is less (from a few hours to a few days), and second, more antibodies are produced. As a consequence, the antigen is quickly destroyed, no disease symptoms develop, and the person is immune.

The memory response also includes the formation of new memory cells, which provide protection against additional exposures to a specific antigen. Memory cells are the basis of specific resitance. After destruction of the antigen, plasma cells die, the antibodies they released are degraded, and antibody levels decline to the point where they can no longer provide adequate protection. However, memory cells persist for many years and probably for life in some cases. If memory-cell production is not stimulated, or if the memory cells produced are short-lived, it is possible to have repeated infections of the same disease. For example, the same cold virus can cause the common cold more than once in the same person.

3 One theory for long-lasting immunity assumes that humans are continually exposed to the same disease-causing agent. Explain how this exposure could produce lifelong immunity. (Hint: what happens each time a person is exposed to the antigen?)

Cell-mediated immunity

Cell-mediated immunity is a function of T cells and is most effective against microorganisms that live inside the cells of the body. Viruses and some bacteria are examples of intracellular microorganisms. Cell-mediated immunity also is involved with some allergic reactions and control of tumors.

After an antigen activates a T cell, the T cell undergoes a series of divisions to produce cytotoxic T cells and memory T cells (Figure 14-6). **Cytotoxic T cells** are responsible for the cell-mediated immunity response. The **memory T cells** provide a secondary response and long-lasting immunity in the same fashion as memory B cells.

Cytotoxic T cells have two main effects. First, they release chemical mediators that promote inflammation and phagocytosis. Second, cytotoxic T cells can come into contact with other cells and cause them to lyse. Virus-infected cells have viral antigens, tumor cells have tumor antigens, and tissue transplants have foreign antigens that can stimulate cytotoxic T cell activity. The cytotoxic T cell binds to these antigens on a cell and causes the cell to lyse (rupture).

ACQUIRED IMMUNITY

There are four ways to acquire specific resistance: active natural, active artificial, passive natural, and passive artificial. Natural and artificial refer to the method of exposure. Natural exposure implies that contact with the antigen occurred as part of everyday living and was not deliberate. Artificial exposure is a deliberate introduction of an antigen or antibody into the body.

Active immunity results when an individual is exposed to an antigen (either naturally or artificially) and the response of the individual's own immune system is the cause of the immunity. Passive immunity occurs when another person or animal develops immunity and the immunity is transferred to a nonimmune individual.

Active natural immunity

Active natural immunity results from natural exposure to an antigen such as a disease-causing microorganism that causes an individual's immune system to respond against the antigen. Because the individual is not immune during the first exposure, he usually develops the symptoms of the disease.

Active artificial immunity

In **active artificial immunity,** an antigen is deliberately introduced into an individual to stimulate his immune system. This process is called **vaccination,** and the introduced antigen is a **vaccine.** Injection of the vaccine is the usual mode of administration. Examples of injected vaccinations are the DTP injection against diphtheria, tetanus, and pertussis (whooping cough); and the MMR injection against mumps, measles, and rubella (German measles). Sometimes the vaccine is ingested, as in the oral poliomyelitis vaccine (OPV).

The vaccine usually consists of some part of a microorganism, a dead microorganism, or a live, altered microorganism. The antigen has been changed so that it will stimulate the immune system but will not cause the symptoms of disease. Because active artificial immunity produces long-lasting immunity without disease symptoms, it is the preferred method of acquiring specific resistance.

4 In some cases, a booster shot is used as part of a vaccination procedure. A booster shot is another dose of the original vaccine given some time after the original dose was administered. Why are booster shots given?

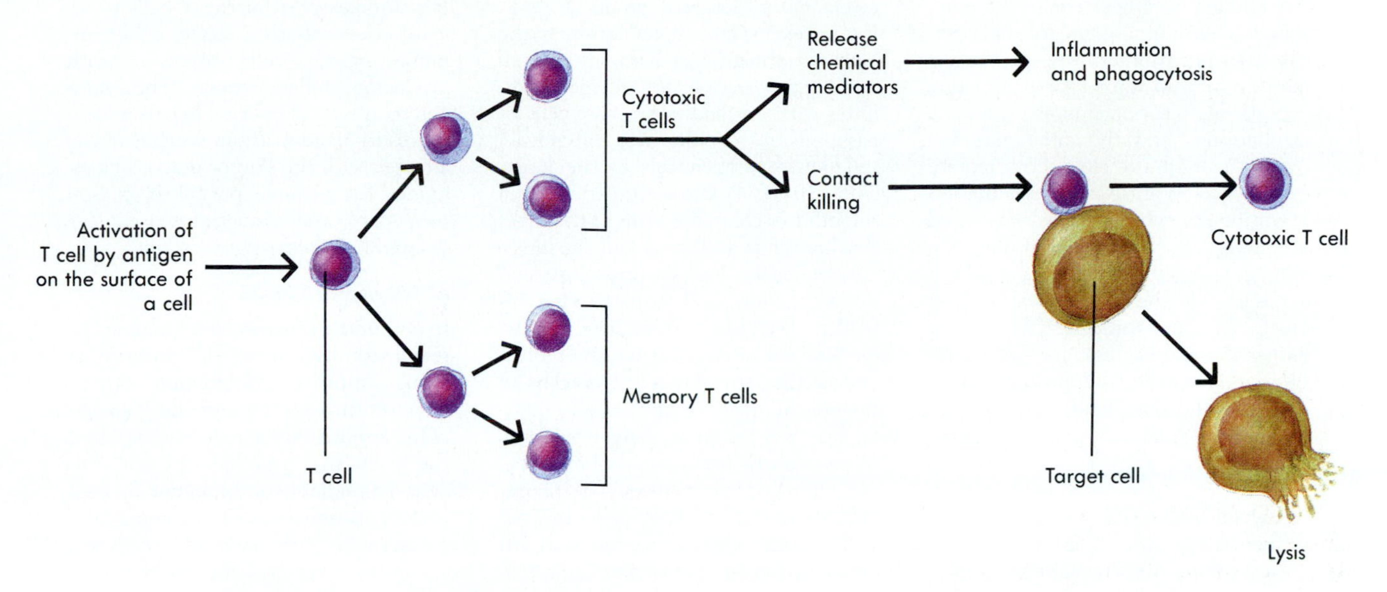

Figure 14-6 ● Stimulation of and effects of T cells.
When activated, T cells form cytotoxic T cells and memory T cells. The cytotoxic T cells release chemical mediators that promote inflammation and phagocytosis, processes that can result in the destruction of the antigen. Cytotoxic T cells can also lysis target cells. The memory T cells are responsible for the secondary response.

Passive natural immunity

Passive natural immunity results from the transfer of antibodies from a mother to her fetus across the placenta before birth. During her life, the mother has been exposed to many antigens, either naturally or artificially, and she has antibodies against many of these antigens. These antibodies protect the mother and the developing fetus against disease. Some antibodies can cross the placenta and enter the fetal blood. Following birth, these antibodies provide protection for the first few months of the baby's life. Eventually the antibodies are broken down, and the baby must rely on its own immune system. If the mother nurses her baby, antibodies in the mother's milk may also provide some protection for the baby.

Passive artificial immunity

Achieving **passive artificial immunity** begins with vaccinating an animal such as a horse. After the animal's immune system responds to the antigen, antibodies are removed from the animal and are injected into the individual requiring immunity. Alternatively, a human who has developed immunity through natural exposure or vaccination is used as a source of antibodies. Passive artificial immunity provides immediate protection because the antibodies are immediately available. Remember that the primary response to an antigen normally takes days to develop enough antibodies to destroy the antigen. Therefore passive artificial immunity is the preferred treatment when there may not be enough time for the individual to develop his own active immunity. However, the technique provides only temporary immunity because the antibodies are used or eliminated by the recipient. There is no permanent immunity because the antigen is eliminated before the immune system can produce memory cells.

Antiserum is the general term used for antibodies that provide passive artificial immunity, because the antibodies are found in serum, which is plasma minus the clotting factors. Antisera are available against microorganisms that cause disease such as rabies, hepatitis, and measles; bacterial toxins such as tetanus, diphtheria, and botulism; and venoms from poisonous snakes and black widow spiders.

Immune system problems of clinical significance

ALLERGY

An **allergy,** or **hypersensitivity reaction,** is a harmful response to an antigen that does not stimulate the specific immune system of most people. Immune and allergic reactions involve the same mechanisms, and the differences between them are not clear. Both require exposure to an antigen and stimulation of antibody-mediated or cell-mediated immunity. In allergic reactions, the antigen is called the **allergen,** and later exposure to the allergen stimulates much the same processes that occur during the normal immune system response. However, the processes that eliminate the allergen also produce undesirable side effects such as a strong inflammatory reaction. This immune system response can be more harmful than beneficial and can produce many unpleasant symptoms.

Immediate hypersensitivities produce symptoms within a few minutes of exposure to the allergen and are caused by antibodies. The reaction takes place rapidly because the antibodies are already present as a result of prior exposure to the allergen. For example, in persons with hay fever, the allergens, usually plant pollens, are inhaled and absorbed through the respiratory mucous membrane. The resulting localized inflammatory response produces swelling and excess mucus production. In the kind of **asthma** (az´mah) resulting from an allergic reaction, the allergen combines with antibodies on mast cells or basophils. As a result, these cells release inflammatory chemicals in the lungs. The chemicals cause constriction of smooth muscle in the walls of the tubes that transport air throughout the lungs. Consequently, less air flows into and out of the lungs, and the patient has difficulty breathing. **Urticaria** (ur´ti-kăr´e-ah) or hives is a skin rash or localized swelling that is usually caused by an ingested allergen. **Anaphylaxis** (an-ă-fi-lak´sis) is a systemic allergic reaction, often resulting from drugs (for example, penicillin) or insect stings. The chemicals released from mast cells and basophils cause systemic vasodilation, increased vascular permeability, a drop in blood pressure, and possibly death. Transfusion reactions and hemolytyic disease of the newborn (see Chapter 11) are also examples of immediate hypersensitivity reactions.

Delayed hypersensitivities take hours to days to develop and are caused by T cells. It takes some time for this reaction to develop because it takes time for the T cells to move to the allergen. It also takes time for the T cells to release chemicals that attract other immune system cells involved with producing inflammation. The most common type of delayed hypersensitivity reactions result from contact of the allergen with the skin or mucous membranes. For example, poison ivy, poison oak, soaps, and cosmetics can cause a delayed hypersensitivity reaction.

AUTOIMMUNE DISEASE

In autoimmune disease the immune system incorrectly treats self antigens as foreign antigens. Autoimmune disease operates through the same mechanisms as hypersensitivity reactions except that the reaction is stimulated by self antigens. Examples of autoimmune diseases include thrombocytopenia, lupus erythematosus, rheumatoid arthritis, rheumatic fever, and myasthenia gravis.

IMMUNODEFICIENCY

Immunodeficiency is a failure of some part of the immune system to function properly. Congenital (present at birth) immunodeficiencies usually involve failure to form adequate numbers of B cells, T cells, or both. **Severe combined immunodeficiency (SCID),**

SUMMARY

The lymphatic system consists of lymph, lymph vessels, lymphocytes, lymph nodes, tonsils, the spleen, and the thymus gland.

The lymphatic system maintains fluid balance in tissues, absorbs fats from the small intestine, and defends against foreign substances.

LYMPHATIC SYSTEM

Lymph vessels

Lymph vessels carry lymph away from tissues. Valves in the vessels ensure the one-way flow of lymph.

Skeletal muscle contraction, contraction of lymph vessel smooth muscle, and thoracic pressure changes move the lymph through the vessels.

The thoracic duct and right lymphatic duct empty lymph into the blood.

Lymphatic organs

Lymphatic tissue produces lymphocytes when exposed to foreign substances, and it filters lymph and blood.

The tonsils protect the openings between the nasal and oral cavities and the pharynx.

Lymph nodes, located along lymph vessels, filter lymph.

Lymphocytes in the spleen respond to foreign substances in the blood and macrophages phagocytize foreign substances and worn out RBCs. The spleen also functions as a reservoir for blood.

The thymus processes lymphocytes that move to other lymphatic tissue to respond to foreign substances.

IMMUNITY

Immunity is the ability to resist the harmful effects of microorganisms and other foreign substances.

NONSPECIFIC RESISTANCE

Mechanical mechanisms

The skin and mucous membranes are barriers that prevent the entry of microorganisms into the body.

Tears, saliva, and urine act to wash away microorganisms.

Chemical mediators

Chemicals kill microorganisms and increase inflammation.

Interferons prevent the replication of viruses.

in which both B cells and T cells fail to form, is probably the best known. Unless the person suffering from SCID is kept in a sterile environment or is provided with a compatible bone marrow transplant, death from infection results.

Acquired immune deficiency syndrome (AIDS) is caused by the human immunodeficiency virus (HIV). HIV infects primarily T cells, eventually resulting in the death of the T cells. Without adequate numbers of T cells, the immune system is depressed, and there is an inability to deal with intracellular microorganisms and cancer. Patients infected with HIV usually die of secondary infections (caused by microorganisms other than HIV) or Kaposi's sarcoma (cancerous growths in the skin and lymph nodes).

HIV is transmitted primarily by intimate sexual contact, blood (for example, shared drug needles), or from a mother to her fetus or nursing infant, or blood transfusions. There are usually no symptoms at first. It can be 6 weeks to 1 year before the antibody test for HIV is positive. The HIV antibody test detects HIV antibodies in the blood, but does not directly detect HIV. Commonly, the first sign of AIDS is swollen lymph nodes caused by overstimulation of B cells and antibody production. As the disease progresses, T cell numbers decrease, and susceptibility to infections increases. At present there is no cure and all AIDS patients eventually die as a result of the disease. Present therapies interfere with HIV replication or attempt to prevent or treat secondary infections. For example, zidovidine, also known as azidothymidine (AZT), prevents the virus from replicating, and pentamidine is used to treat pneumonia caused by an intracellular protozoan. Several strategies are being explored to produce an AIDS vaccine.

TUMOR CONTROL

Tumor cells have tumor antigens that distinguish them from normal cells. According to the concept of **immune surveillance,** the immune system detects tumor cells and destroys them before a tumor can form. Failure of the immune system to destroy tumors as they form can result in cancer.

TRANSPLANTATION

The surface of cells in the human body contains antigens called **human lymphocyte antigens (HLAs).** The immune system can distinguish between self and foreign cells because self cells have self HLAs, whereas foreign cells have foreign HLAs. Rejection of a graft is caused by a normal immune system response to foreign HLAs.

Graft rejection can occur in two different directions. In **host vs. graft rejection,** the recipient's immune system recognizes the donor's tissue as foreign and rejects the transplant. In a **graft vs. host** rejection, the donor tissue (for example, bone marrow) recognizes the recipient's tissue as foreign, and the transplant rejects the recipient, causing destruction of the recipient's tissue and death.

To reduce graft rejection, a tissue match is performed. Only tissue with HLAs similar to the recipient's have a chance of being accepted. An exact match is possible only for a graft from one part to another part of the same person, or between identical twins. For all other graft situations, drugs that suppress the immune system must be administered throughout the patient's life to prevent graft rejection.

Cells

Neutrophils are the first phagocytic cells to respond to microorganisms.

Macrophages are large phagocytic cells that are active in the latter part of an infection.

Basophils and mast cells promote inflammation, whereas eosinophils inhibit inflammation.

Inflammatory response

Chemical mediators cause vasodilation and increase vascular permeability, allowing the entry of chemicals into damaged tissues. Chemicals also attract phagocytes.

The amount of chemical mediators and the number of phagocytes increase until the cause of the inflammation is destroyed. Then the tissues undergo repair.

SPECIFIC RESISTANCE

Antigens are molecules that stimulate specific resistance.

B cells are responsible for antibody-mediated immunity. T cells are involved with cell-mediated immunity.

Antibody-mediated immunity

The primary response results from the first exposure to an antigen. B cells form plasma cells, which produce antibodies, and memory cells.

The secondary (memory) response results from exposure to an antigen after a primary response. Memory cells quickly form plasma cells and memory cells.

Cell-mediated immunity

Exposure to an antigen activates cytotoxic T cells and produces memory cells.

Cytotoxic T cells lyse virus-infected cells, tumor cells, and tissue transplants. Cytotoxic T cells release chemical mediators that promote inflammation and phagocytosis.

ACQUIRED IMMUNITY

Active natural immunity results from everyday exposure to an antigen against which the person's own immune system mounts a response.

Active artificial immunity results from deliberate exposure to an antigen (vaccine) to which the person's own immune system responds.

Passive natural immunity is the transfer of antibodies from a mother to her fetus or baby.

Passive artificial immunity is the transfer of antibodies (antiserum) from an animal or another person to a person requiring immunity.

CONTENT REVIEW

1. List the parts of the lymphatic system, and describe the three main functions of the lymphatic system.
2. What is the function of valves in lymph vessels? What causes lymph to move through lymph vessels?
3. Which parts of the body are drained by the right lymphatic duct and which by the thoracic duct?
4. Name the three groups of tonsils. What is their function?
5. Where are lymph nodes found? What is the function of the germinal centers within lymph nodes?
6. Where is the spleen located? What are the functions of the spleen?
7. Where is the thymus gland located and what function does it perform?
8. What is the difference between nonspecific resistance and specific resistance?
9. How do mechanical mechanisms and chemical mediators provide protection against microorganisms? Describe the effects of complement and interferons.
10. Describe functions of the two major phagocytic cell types of the body.
11. Name the cells involved in promoting and inhibiting inflammation.
12. Describe the effects that take place during an inflammatory response.
13. Define antigen. What is the difference between a self antigen and a foreign antigen?
14. What are the functions of plasma cells and memory B cells?
15. Define the primary and memory (secondary) response. How do they differ from each other in regard to speed of response and amount of antibody produced?
16. What are the functions of cytotoxic T cells and memory T cells?
17. Define active natural, active artificial, passive natural, and passive artificial immunity. Give an example of each.

CONCEPT REVIEW

1. A patient is suffering from edema in the lower-right limb because of inadequate lymph return. Explain why massage would help remove the excess fluid. (Hint: consider the factors that move lymph through the lymph vessels.)
2. Durability is a measure of how long immunity lasts. Compare the durability of active immunity and passive immunity. Explain the difference between the two types of immunity. In what situations would one type be preferred over the other type?
3. Tetanus is caused by bacteria that enter the body through wounds in the skin. The bacteria produce a toxin that causes spastic muscle contractions. Death often results from failure of the respiration muscles. A patient comes to the emergency room after stepping on a nail. If the patient has been vaccinated against tetanus, the patient is given a tetanus booster shot, which consists of the toxin altered so that it is harmless. If the patient has never been vaccinated against tetanus, the patient is given an antiserum shot against tetanus. Explain the rationale for this treatment strategy. Sometimes both a booster and an antiserum shot are given, but at different locations of the body. Explain why this is done, and why the shots are given in different locations of the body.
4. A patient has an allergic reaction (see the box on p. 242). As part of the treatment scheme, it was decided to try to identify the allergens that stimulated the allergic reaction. A series of solutions, each containing an allergen that commonly causes a reaction, was composed. Each solution was then injected into the skin at different locations on the patient's back. The following results were obtained: (1) at one location, within a few minutes the injection site became red and swollen, (2) at another injection site, swelling and redness did not appear until 2 days later, and (3) no redness or swelling developed at the other sites. Explain what happened for each observation and what caused the redness and swelling.

CHAPTER TEST

Answers can be found in Appendix C

MATCHING 1 For each statement in column A select the correct answer in column B (an answer may be used once, more than once, or not at all).

A	B
1. Fluid derived from tissue spaces.	chyle
2. Lymph with a high fat content.	germinal center
3. Has one-way valves.	lymph
4. Filters lymph.	lymph capillary
5. Area of rapidly dividing lymphocytes.	lymph node
6. Filters blood.	lymph vessel
7. Site of lymphocyte processing and maturation.	spleen
	thymus

MATCHING 2 For each statement in column A select the correct answer in column B (an answer may be used once, more than once, or not at all).

A	B
1. Group of proteins that promote inflammation, phagocytosis, and lysis.	active artificial immunity
2. Protein that protects the body against viral infections.	active natural immunity
3. Molecule that stimulates specific resistance.	antigen
4. Divides to form plasma cells and memory cells.	B cell
5. Produces antibodies.	complement
6. Lyses virus-infected cells, tumor cells, and transplanted cells.	cytotoxic T cell
7. Vaccines produce this kind of immunity.	interferon
8. Antisera produce this kind of immunity.	passive artificial immunity
	passive natural immunity
	plasma cell

FILL-IN-THE BLANK Complete each statement by providing the missing word or words.

1. The lymphatic system maintains ___________ balance in tissues, absorbs ___________ from the digestive tract, and is part of the body's ___________ system against microorganisms.
2. The ___________ drains the upper right limb and the right half of the head, neck, and chest.
3. The three factors that result in lymph movement through lymph vessels are ___________, ___________, and ___________.
4. The type of immunity in which the response to a second exposure to an antigen is faster and more effective than the response to the first exposure is called ___________.
5. The two major phagocytic WBCs are ___________ and ___________.
6. The two cell types that promote inflammation are ___________ and ___________; whereas ___________ reduce inflammation.
7. The ___________ causes vasodilation, attracts phagocytes, and increases vascular permeability.

CHAPTER

15

The Respiratory System

KEY TERMS

alveolus
The terminal sac-like endings of the respiratory system, where gas exchange occurs.

bronchiole
One of the finer subdivisions of the bronchial tubes.

bronchus
Any one of the tubes conducting air from the trachea to the bronchioles.

diaphragm
Muscular partition between the abdominal and thoracic cavities; its movement produces most of the changes in thoracic volume responsible for respiration.

larynx
Organ of voice production located between the pharynx and the trachea; consists of a framework of cartilages and elastic ligaments containing the vocal cords and the muscles that control the position and tension of the vocal cords.

lung
One of a pair of organs in the thoracic cavity; the principal organs of respiration.

pharynx
Upper expanded portion of the digestive tube located between the esophagus and the oral and nasal cavities.

pleura
Serous membrane covering the lungs and lining the thoracic wall, diaphragm, and mediastinum.

respiratory center
Neurons in the medulla oblongata and pons of the brain that control inspiration and expiration.

trachea
Air tube extending from the larynx into the thorax, where it divides to form the two primary bronchi.

OBJECTIVES

After reading this chapter you should be able to:

1. Describe the anatomy of the respiratory passages beginning at the nose and ending with the alveoli.
2. Describe the lungs, the membranes that cover the lungs, and the cavities that surround the lungs.
3. Explain how contraction of the muscles of respiration causes air to flow into and out of the lungs during respiration.
4. Define the pulmonary volumes and vital capacities.
5. Name the components of the respiratory membrane and explain how gases move across it.
6. Describe how oxygen and carbon dioxide are transported in the blood.
7. Name the neural mechanisms that control respiration and describe how they work.
8. Explain how alterations in blood carbon dioxide levels, blood pH, and blood oxygen levels affect respiration.

FEATURES

ALL LIVING CELLS OF THE BODY require oxygen and produce carbon dioxide. The respiratory system and the cardiovascular system take oxygen from the air and transport it to the cells of the body, transport carbon dioxide from cells and release it into the air, and help regulate the pH of the body fluids. The term *respiration* refers to the following processes: (1) ventilation, the movement of air into and out of the lungs; (2) gas exchange between the air in the lungs and the blood; (3) transport of oxygen and carbon dioxide in the blood; and (4) gas exchange between the blood and the tissues.

ANATOMY OF THE RESPIRATORY SYSTEM

The respiratory system is divided into two parts. The **upper respiratory tract** consists of the nose, nasal cavity, and pharynx. The **lower respiratory tract** includes the larynx, trachea, bronchi, and lungs (Figure 15-1).

Nose and nasal cavity

The **nasal cavity** extends from the nostrils (openings) of the **nose** to the openings into the pharynx (see Figure 15-1). The nasal septum divides the nasal cavity into right and left sides (see Figure 6-5). The hard palate forms the floor of the nasal cavity and the lateral walls of the nasal cavity are modified by the presence of bony ridges called **conchae** (kon´ke) (see Figure 6-5). The conchae increase surface area within the nasal cavity.

After air enters the nasal cavity through the nostrils, the air comes into contact with the surfaces of the nasal cavity. Epithelial cells of the mucous membrane lining the nasal cavity produce mucus that traps debris in the air. The cilia on the surface of the cells sweep the mucus to the pharynx, where it is swallowed (see Chapter 3). The air also is humidified and warmed by the mucous membrane before the air passes into the pharynx.

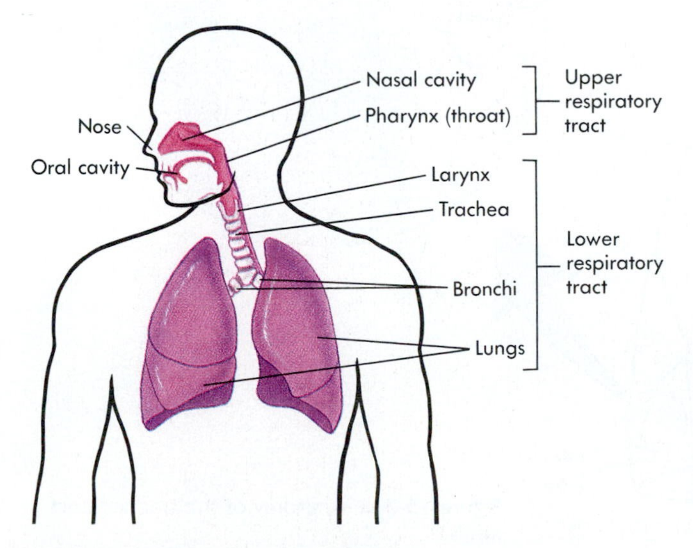

Figure 15-1 ● The respiratory system.
The upper respiratory tract consists of the nose, nasal cavity, and pharynx (throat). The lower respiratory tract consists of the larynx, trachea, bronchi, and lungs.

1 Explain what happens to your throat when you sleep with your mouth open, especially when your nasal passages are plugged as a result of having a cold.*

Paranasal sinuses are air-filled spaces within the bones of the skull (see Figure 6-6). The paranasal sinuses are lined with a mucous membrane and open into the nasal cavity. When the mucous membranes become swollen because of infections or allergies, these openings can become blocked. The mucus then accumulates within the sinuses, and the increasing pressure can produce a painful sinus headache.

Pharynx

The **pharynx** (făr´ingks), or throat, is the common passageway of both the digestive and respiratory systems (see Figure 15-1). It receives air from the nasal cavity and air, food, and water from the mouth. Inferiorly, the pharynx allows air to pass into the larynx and allows food and water to enter the esophagus (see Chapter 16).

Larynx

The **larynx** (lăr´ingks), or voice box, consists of a hollow casing of nine cartilages that are connected to each other by muscles and ligaments (see Figure 15-1). The largest and most superior of the cartilages, the **thyroid cartilage,** forms a protrusion called the Adam's apple.

The **vocal cords** are a pair of ligaments that extend across the larynx (Figure 15-2). Air moving past the vocal cords causes them to vibrate, producing sound. The force of air moving past the vocal cords controls the loudness, and the tension of the vocal cords controls the pitch of the voice. An inflammation of the mucous epithelium of the vocal cords is called laryngitis. Swelling of the vocal cords during laryngitis inhibits voice production.

Superior to the vocal cords is a flaplike cartilage of the larynx called the **epiglottis** (ep-ĭ-glot´is) (see Figure 15-2). During swallowing, the epiglottis covers the opening of the larynx and prevents materials from entering it.

Trachea and bronchi

The **trachea** (tra´ke-ah), or windpipe, is a 12-cm ($4^3/4$ in) long membranous tube that extends from the larynx into the thoracic cavity. Columnar epithelium with numerous cilia

*Answers to predict questions appear in Appendix B at the back of the book.

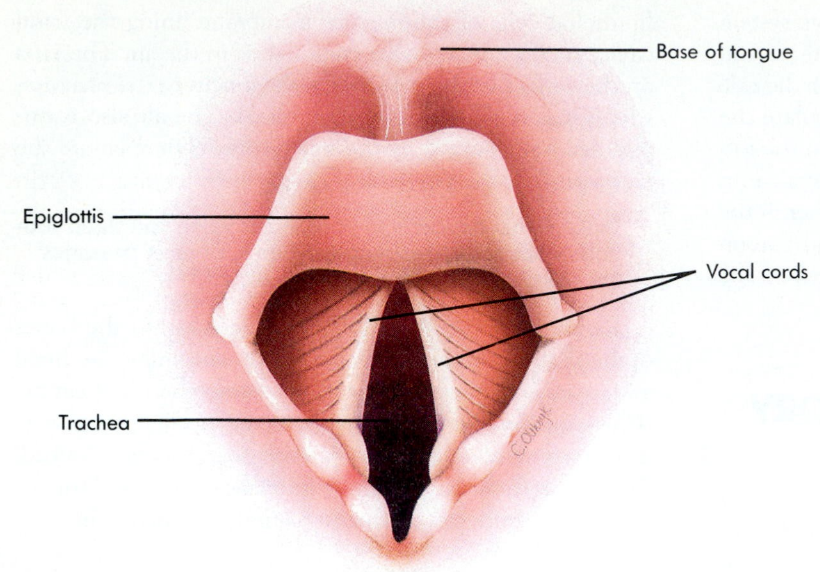

Figure 15-2 ● The vocal cords.
The vocal cords viewed from above, showing their relationship to the epiglottis and trachea.

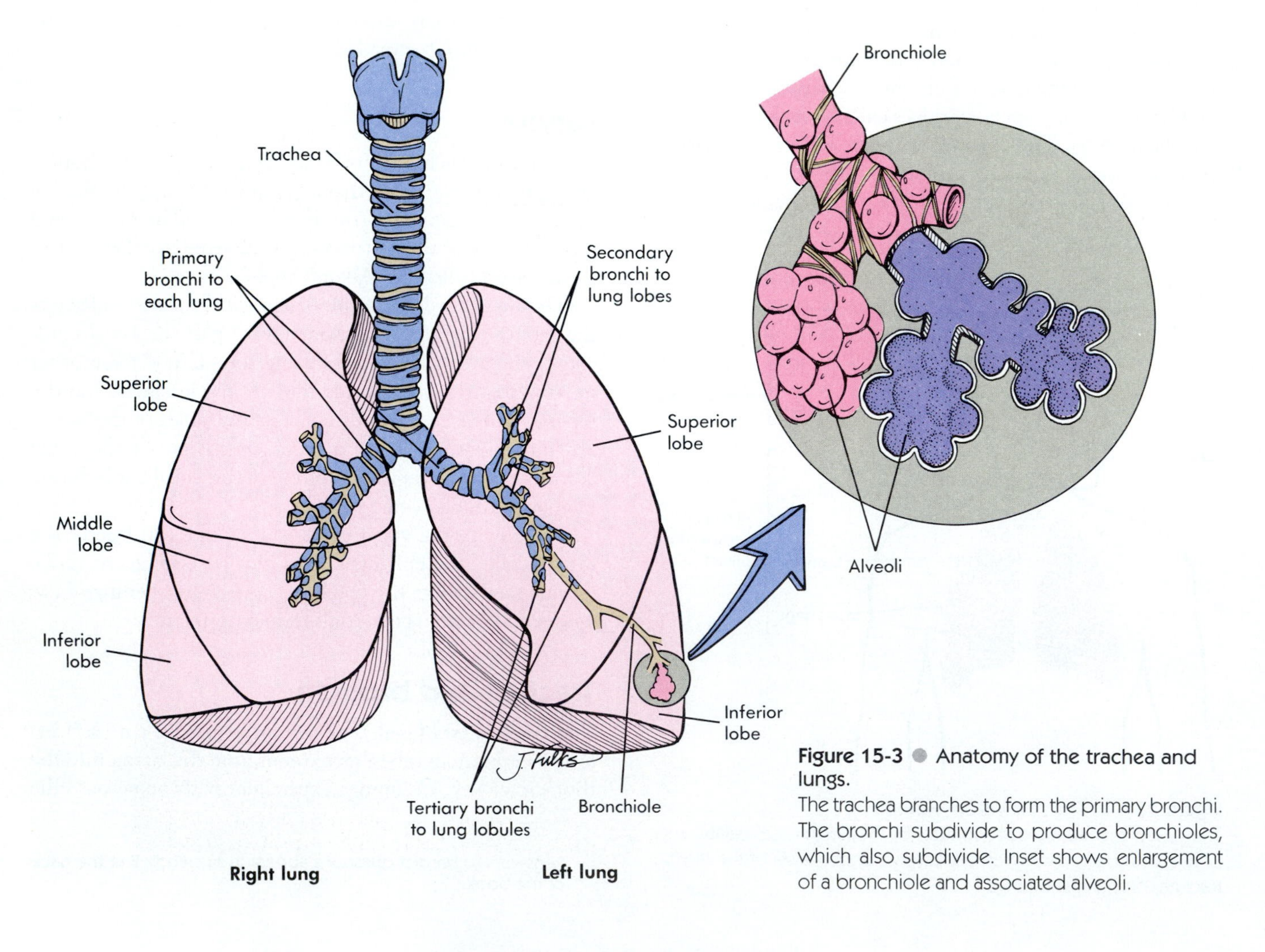

Figure 15-3 ● Anatomy of the trachea and lungs.
The trachea branches to form the primary bronchi. The bronchi subdivide to produce bronchioles, which also subdivide. Inset shows enlargement of a bronchiole and associated alveoli.

and mucous-producing goblet cells line the trachea. The cilia propel mucus and foreign particles toward the pharynx, where they enter the esophagus and are swallowed.

The trachea consists of connective tissue and smooth muscle reinforced with 15 to 20 C-shaped pieces of cartilage (Figure 15-3). The rigid cartilages reinforce the anterior and lateral sides of the trachea and maintain an open passageway for air. The esophagus lies immediately posterior to the cartilage-free posterior wall of the trachea. During swallowing, food within the esophagus causes the diameter of the esophagus to enlarge. The enlarged esophagus protrudes into the non-rigid cartilage-free part of the trachea as food passes through the esophagus.

The trachea divides into the left and right **primary bronchi** (brong´ki), which extend from the trachea to the lungs (see Figure 15-3). Like the trachea, the primary bronchi are lined with ciliated columnar epithelium and are supported by C-shaped cartilage rings.

Lungs

The **lungs** are the principal organs of respiration. The right lung has three lobes and the left lung has two lobes (Figure 15-4). The lobes are separated by deep, prominent fissures on the surface of the lung. Each lobe is divided into lobules that are separated from each other by connective tissue. Because major blood vessels and bronchi do not cross the connective tissue, individual diseased lobules can be surgically removed, leaving the rest of the lung relatively intact. There are 9 lobules in the left lung and 10 lobules in the right lung.

Within the lungs there is a system of air passageways formed by the branching of the bronchi (see Figures 15-3 and 15-4). There is a primary bronchus extending from the trachea into each lung. The primary bronchi divide into secondary bronchi, which conduct air to each lobe. The secondary bronchi, in turn, give rise to tertiary bronchi, which extend to the lobules of the lungs. The bronchi within the lobules continue to branch many times, finally giving rise to **bronchioles** (brong´ke-ōlz). The bronchioles also subdivide numerous times, eventually giving rise to clusters of air sacs called **alveoli** (al´ve-o´li). The walls of the alveoli are thin, simple squamous epithelium across which gas exchange with blood occurs.

In contrast to the bronchi, bronchioles have no cartilage in their walls. Abundant smooth muscle, however, regulates the diameter of the bronchioles. Relaxation of the smooth muscle, for example during exercise, allows more air to flow through the bronchioles. Conversely, in an asthma attack, contraction of the smooth muscle restricts air flow, making breathing difficult.

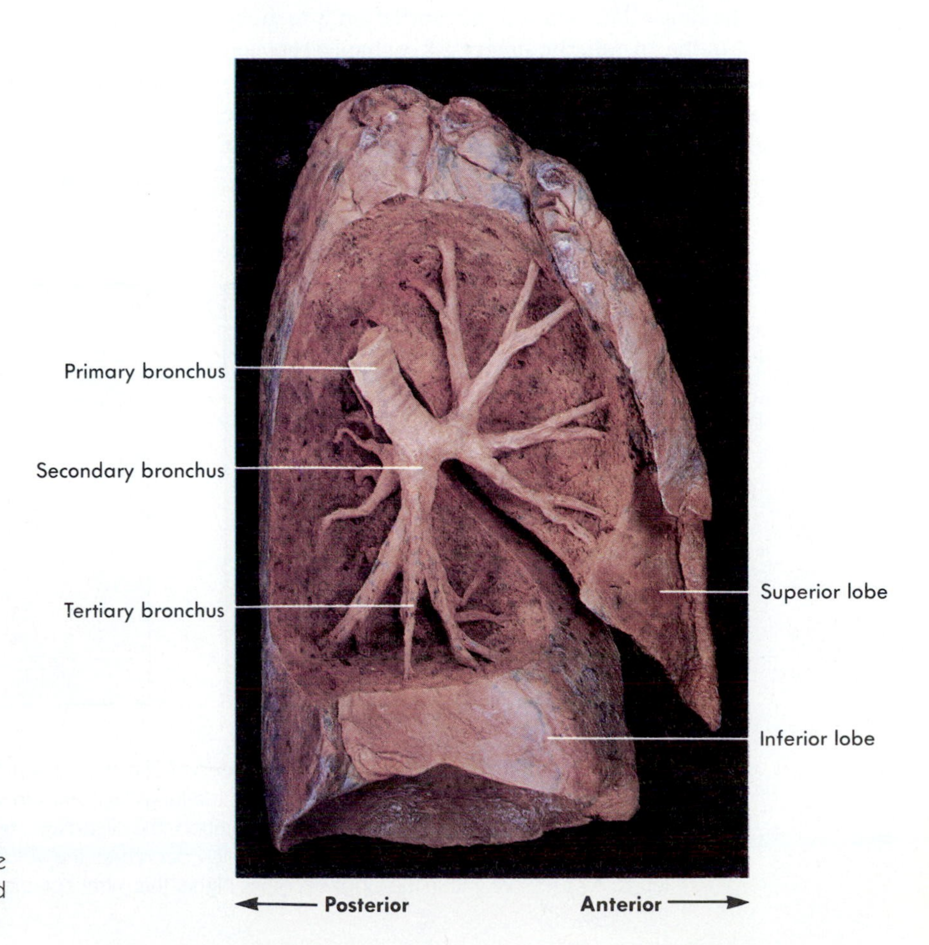

Figure 15-4 ● Air passageways.
Medial view of left lung showing branching of the bronchi. The division between the superior and inferior lobes is also visible.

Pleural cavities

The lungs are contained within the thoracic cavity. In addition, each lung is surrounded by a separate pleural (ploor´al) cavity. Each pleural cavity is lined with a serous membrane called the **pleura,** which has two parts. The **parietal pleura** lines the inner surface of the thoracic cavity and the mediastinum, whereas the **visceral pleura** covers the surface of the lung (see Figure 4-7).

The **pleural cavity,** between the parietal and visceral pleurae, is filled with a small volume of **pleural fluid** produced by the pleural membranes. The pleural fluid performs two functions: it acts as a lubricant, allowing the pleural membranes (the visceral and parietal pleurae) to slide past each other as the lungs and the thorax change shape during respiration, and it helps hold the pleural membranes together. The pleural fluid acts like a thin film of water between two sheets of glass (the visceral and parietal pleurae); the glass sheets can slide over each other easily, but it is difficult to separate them.

2 Pleurisy is an inflammation of the pleural membranes. Explain why this condition is so painful, especially when a person takes deep breaths.

VENTILATION

Ventilation, or breathing, is the process of moving air into and out of the lungs. The function of ventilation is to move air to and from the alveoli, the sites of gas exchange between the air and the blood. There are two phases of ventilation: **inspiration,** or inhalation, is the movement of air into the lungs; **expiration,** or exhalation, is the movement of air out of the lungs.

Two physical principles are involved with the movement of air into and out of the lungs. First, movement of air through a tube such as a bronchus is caused by a pressure difference. If the pressure is higher at one end of the tube than at the other end, air will flow from the area of higher pressure toward the area of lower pressure. Second, changes in the volume of a container such as the thoracic cavity cause changes in the air pressure within the container. As the volume of a container increases, the pressure within the container decreases, and as the volume of a container decreases, the pressure within the container increases.

During inspiration, muscles of respiration cause the volume of the thoracic cavity to increase. The **diaphragm** (di´ă-fram) is a large dome of skeletal muscle that separates the thoracic cavity from the abdominal cavity. When the diaphragm contracts, the dome is flattened, thus increasing the volume of the thoracic cavity (Figure 15-5, *A*). Other muscles of respiration also alter the volume of the thoracic cavity. For example, the external intercostal muscles increase thoracic volume during inspiration by lifting the anterior ends of the ribs and the sternum.

The increased volume of the thoracic cavity causes the lungs to expand because the visceral pleura adheres to the parietal pleura. As the lungs expand, the alveolar volume increases, resulting in a decrease in air pressure in the alveoli below atmospheric pressure. Because of this pressure difference, air flows into the lungs.

During expiration the muscles of inspiration relax. The diaphragm returns to its resting position and the thoracic wall passively returns to its original position, resulting in a decrease in thoracic volume (Figure 15-5, *B*). Consequently there is a decrease in alveolar volume, an increase in alveo-

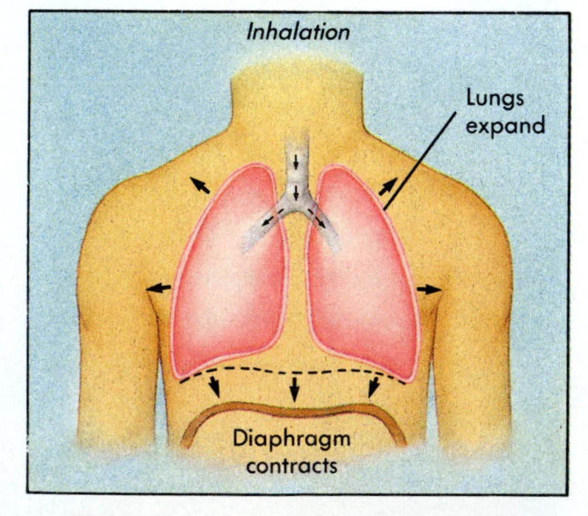

Figure 15-5 ● The respiratory movements.
A, For inspiration, the diaphragm contracts, becoming flattened, which increases the volume of the thoracic cavity and lungs. Air pressure within the lungs becomes lower than atmospheric air pressure and air moves into the lungs. **B,** For expiration, the diaphragm relaxes and the diaphragm and the thorax assume their resting positions, which decreases the volume of the thoracic cavity and lungs. Air pressure within the lungs becomes higher than atmospheric pressure and air moves out of the lungs.

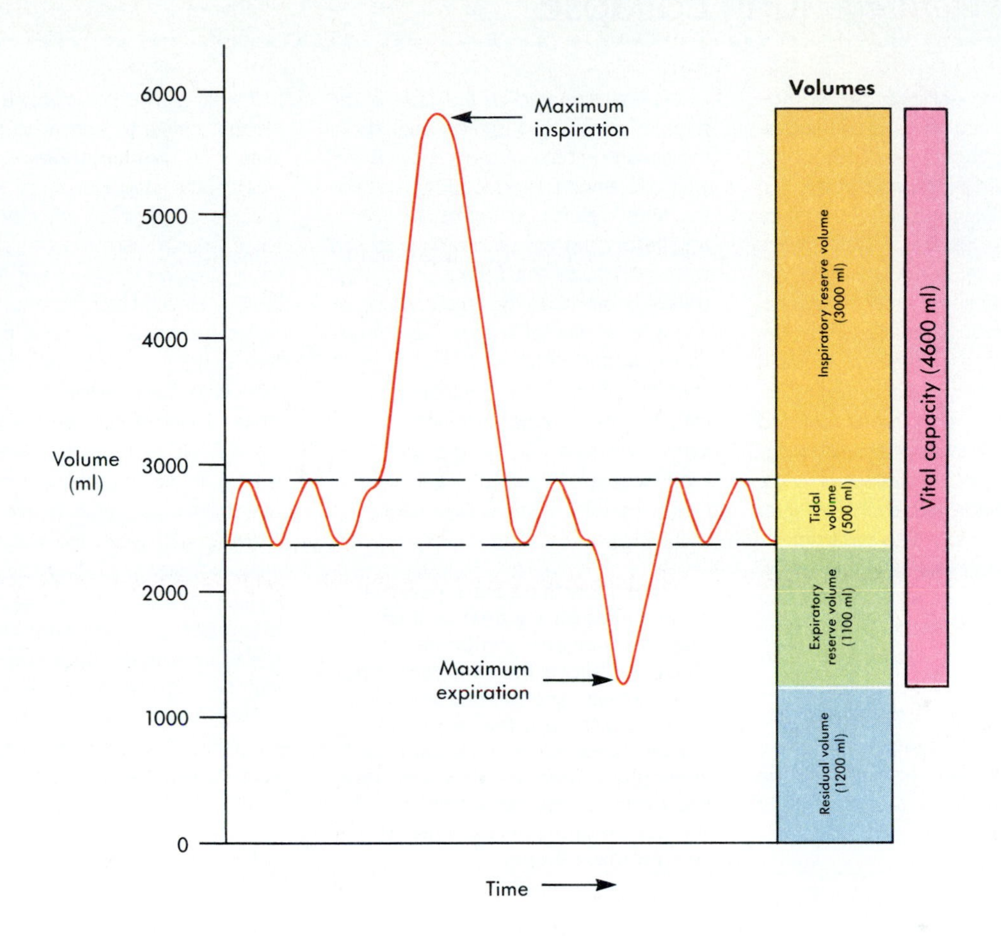

Figure 15-6 ● Pulmonary volumes and vital capacity.
The values for the pulmonary volumes. The values shown in the diagram are values during quiet breathing.

lar pressure above atmospheric pressure, and air flows out of the lungs. During labored breathing muscles of expiration also contract. For example, the internal intercostals depress the ribs and the sternum, decreasing the volume of the thorax to a greater extent, which results in the expiration of a greater volume of air.

PULMONARY VOLUMES

Spirometry (spi-rom´ĕ-tre) is the process of measuring volumes of air that move into and out of the respiratory system, and the **spirometer** (spi-rom´ĕ-ter) is the device that is used to measure these pulmonary volumes. Measurements of the respiratory volumes can provide information about the health of the lungs. The four pulmonary volumes and their normal values (Figure 15-6) for a young adult male are as follows:

1. **Tidal volume:** The volume of air inspired or expired in a single breath during regular breathing; usually measured at rest (approximately 500 ml) but can refer to air movement during exercise.
2. **Inspiratory reserve volume:** the amount of air that can be inspired forcefully after inspiration of the normal tidal volume (approximately 3000 ml).
3. **Expiratory reserve volume:** the amount of air that can be expired forcefully after expiration of the normal tidal volume (approximately 1100 ml).
4. **Residual volume:** the volume of air still remaining in the respiratory passages and lungs after a maximum expiration (approximately 1200 ml).

Vital capacity is the sum of the inspiratory reserve volume, the tidal volume, and the expiratory reserve volume. It is the maximum volume of air that a person can expel from his respiratory tract after a maximum inspiration (approximately 4600 ml).

Understanding lung collapse

The lungs tend to collapse for two reasons: (1) elastic recoil caused by the elastic fibers in the connective tissue of the lungs and (2) surface tension of the thin layer of water lining the alveoli. **Surface tension** is the attraction of water molecules to each other at the boundary between water and some other substance. For example, when water falls as rain, the water molecules are attracted to each other more than they are attracted to air. Consequently, the water forms rounded water droplets. Because water molecules lining the alveoli are also attracted to the surface of the alveoli, formation of a water droplet causes the alveoli to collapse.

Two factors keep the lungs from collapsing: (1) surfactant and (2) the tendency of the visceral pleura to adhere to the parietal pleura. **Surfactant** (sur-fak′tant) is a mixture of lipoprotein molecules secreted by cells of the alveolar epithelium. The surfactant molecules form a layer over the thin layer of water within the alveoli, reducing surface tension and the tendency of the lungs to collapse. Surfactant is not produced by the lungs of the fetus in adequate quantities until about the seventh month of pregnancy. Thereafter, the amount produced increases as the fetus matures. In premature infants, **respiratory distress syndrome** is caused by too little surfactant. Because of the high surface tension the lungs tend to collapse, and a great deal of energy must be exerted by the muscles of respiration to keep the lungs inflated. Without specialized treatment, babies with severe respiratory distress syndrome die soon after birth because fatigue of the respiratory muscles results in inadequate ventilation of the lungs.

3 Fortunately there are successful therapies for the treatment of respiratory distress syndrome. One possibility is the administration of surfactant. Another possibility is respiratory therapy that involves the administration of air at greater than atmospheric pressure. Based on your knowledge of the cause of respiratory distress syndrome, explain the rationale for these therapies.

The fluid in the pleural cavity causes the visceral pleura to adhere to the parietal pleura. A **pneumothorax** (nu-mo-tho′raks) is the introduction of air into the pleural cavity. Air can enter by an external route when a sharp object, such as a bullet or broken rib, penetrates the thoracic wall, or air can enter the pleural cavity by an internal route if alveoli at the lung surface rupture, such as may occur in a patient with emphysema. The introduction of air into the pleural space separates the visceral pleura from the parietal pleura, leaving an air-filled pleural cavity. Consequently, the lung does not adhere to the thoracic wall when thoracic volume increases. Instead, elastic fibers and water surface tension in the lung cause the lung to collapse, even when respiratory movements are exaggerated. To reinflate a lung that has collapsed, the hole must be closed. Then, if a tube is placed into the pleural cavity and suction is applied, the lung will reinflate.

GAS EXCHANGE

Ventilation supplies atmospheric air to the alveoli. The next step in the process of respiration is the diffusion of gases between the alveoli and the blood in the pulmonary capillaries. There are approximately 300 million alveoli in each lung. Surrounding each alveolus is a network of capillaries arranged so air in the alveolus is separated by a thin respiratory membrane from the blood contained within the capillaries (Figure 15-7, *A*). The **respiratory membrane** consists of (1) a thin layer of fluid containing surfactant that lines the alveolus; (2) the wall of the alveolus, which consists of simple squamous epithelium; (3) a thin interstitial space; and (4) the wall of the capillary, which consists of simple squamous epithelium.

Gases diffuse across the respiratory membrane from areas of higher to lower concentrations (Figure 15-7, *B*). Because the cells of the body use oxygen, there is a lower concentration of oxygen in the blood than in the alveoli, and oxygen diffuses from the alveoli into the blood. Conversely, the cells of the body produce carbon dioxide and there is a higher concentration of carbon dioxide in the blood than in the alveoli. Therefore carbon dioxide diffuses out of the blood into the alveoli.

Gases can readily diffuse across the respiratory membrane because the respiratory membrane is thin and because it has a large surface area. Changes in the respiratory membrane as a result of disease can markedly affect gas exchange. For example, in patients with pneumonia, fluid accumulates in the alveoli, and gases must diffuse through a thicker-than-normal layer of fluid. If the thickness of the respiratory membrane is doubled or tripled, the rate of gas exchange is markedly decreased.

In normal adults, the total surface area of the respiratory membrane is approximately 70 m^2, which is approximately the area of one half of a singles tennis court. The surface area of the respiratory membrane is decreased by several respiratory diseases, including emphysema and lung cancer. Even small decreases in this surface area adversely affect the respiratory exchange of gases during strenuous exercise. When the surface area of the respiratory membrane is decreased to one third or one fourth of normal, the exchange of gases is significantly restricted even under resting conditions.

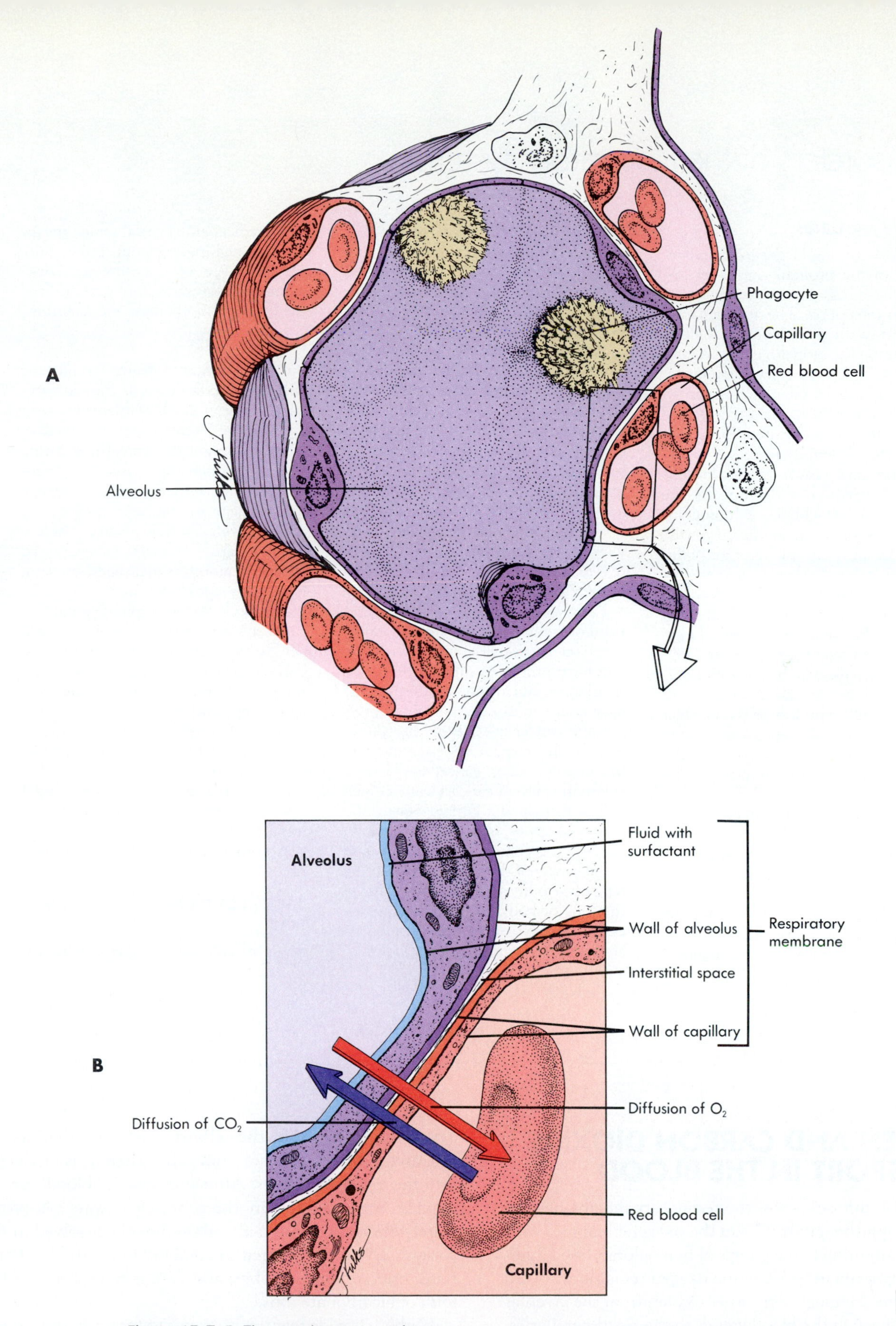

Figure 15-7 ● The respiratory membrane.
A, An alveolus surrounded by capillaries. **B,** A magnification of the respiratory membrane. Oxygen diffuses across the respiratory membrane from the alveolus into the capillary. Carbon dioxide diffuses across the respiratory membrane from the capillary into the alveolus.

Disorders of the respiratory system

BRONCHI AND LUNGS

Bronchitis (brong-ki´tis) is an inflammation of the **bronchi** caused by irritants, such as cigarette smoke, air pollution, or infections. The inflammation results in swelling of the mucous membrane lining the bronchi, increased production of mucus, and decreased movement of mucus by cilia. Consequently, the diameter of the bronchi is decreased and ventilation is impaired. Bronchitis can progress to emphysema.

Emphysema (em-fĭ-se´mah) results in the destruction of the walls of the alveoli and bronchioles. Chronic inflammation of the bronchioles, usually caused by cigarette smoke or air pollution, probably initiates emphysema. Narrowing or collapse of the bronchioles restricts air movement and air tends to be retained in the lungs. Coughing to remove accumulated mucus increases pressure in the alveoli, resulting in rupture and destruction of alveolar walls. The loss of the alveolar walls has two important consequences. There is decreased surface area of the respiratory membrane, resulting in decreased gas exchange, and there is loss of the elastic fibers in the alveolar walls, resulting in a decreased ability of the lungs to recoil and expel air. Symptoms of emphysema include shortness of breath and enlargement of the thoracic cavity. Treatment involves removing sources of irritants (for example, stopping smoking), promoting the removal of bronchial secretions, retraining people to breathe so that expiration of air is maximized, and administering antibiotics to prevent infections. The progress of emphysema can be slowed, but there is no cure.

Asthma (az´mah) is a disorder in which there are periodic episodes of contractions of bronchial smooth muscle, which restrict air movement. Many cases of asthma are allergic responses to pollen, dust, animal dander, or other substances (see Chapter 14). Treatment includes drugs that relax the smooth muscles and reduce inflammation. Sometimes injections are given to reduce the sensitivity of the immune system to the substances that stimulate an asthma attack.

Pulmonary fibrosis is the replacement of lung tissue with fibrous connective tissue, making the lungs less elastic and breathing more difficult. Exposure to asbestos, silica, and coal dust are the most common causes.

Lung cancer arises from the epithelium of the respiratory tract, often in the bronchi. Lung cancer is the most common cause of cancer death in men and women in the United States, and almost all cases are found in smokers. Because of the rich lymphatic and blood supply of the lungs, cancer in the lung can readily spread to other parts of the lung or body. In addition, the disease is often advanced before symptoms become severe enough for the victim to seek medical aid. Typical symptoms include coughing, sputum production, and blockage of the airways. Treatments include removal of part or all of the lung, chemotherapy, and radiation.

NERVOUS SYSTEM

Sudden infant death syndrome (SIDS), or crib death, is the most frequent cause of death of infants between 2 weeks and 1 year of age. Death results when the infant stops breathing during sleep. Although the cause of SIDS remains controversial, there is evidence that damage to the respiratory center during development is a factor. There is no treatment, but at-risk babies can be placed on monitors that sound an alarm if the baby stops breathing.

Paralysis of the respiratory muscles can result from transection of the spinal cord in the cervical or thoracic regions. The damage interrupts nerve tracts that transmit action potentials to the muscles of respiration. Transection of the spinal cord often results from trauma such as automobile accidents or diving into water that is too shallow. Another cause of paralysis is poliomyelitis, a viral infection that damages neurons of the respiratory center or motor neurons that stimulate the muscles of respiration.

DISEASES OF THE UPPER RESPIRATORY TRACT

Strep throat is caused by a streptococcal

OXYGEN AND CARBON DIOXIDE TRANSPORT IN THE BLOOD

After oxygen diffuses across the respiratory membrane into the blood, approximately 97% of the oxygen combines with the iron-containing heme groups of hemoglobin (see Chapter 11). Approximately 3% of the oxygen remains dissolved in the plasma. Hemoglobin carries oxygen from the alveolar capillaries through the blood vessels to the tissue capillaries. The combination of oxygen with hemoglobin is reversible, and in the tissues, oxygen diffuses away from hemoglobin molecules and into the tissue spaces. Oxygen then diffuses into cells and is used by the cells in aerobic metabolism.

Carbon dioxide diffuses from cells, where it is produced, into the tissue capillaries. After it enters the blood, carbon dioxide is transported in three principal ways: Approximately 8% is transported as carbon dioxide dissolved in the plasma; 20% is transported in combination with blood proteins, primarily hemoglobin; and 72% is transported in the form of bicarbonate ions.

Carbon dioxide reacts with water to form a hydrogen ion (H^+) and a bicarbonate ion (HCO_3^-)

$$CO_2 + H_2O \longleftrightarrow H^+ + HCO_3^-$$

bacteria (see Chapter 21) and is characterized by inflammation of the pharynx and fever. Frequently, inflammation of the tonsils and middle ear are involved. Without a throat analysis, the infection cannot be distinguished from viral causes of pharyngeal inflammation. Current techniques allow rapid diagnosis within minutes to hours, and antibiotics are effective in treating strep throat.

Diphtheria (dif-the´re-ah) was once a major cause of death among children. It is caused by a bacterium. A grayish membrane forms in the throat and can block the respiratory passages totally. A vaccine against diphtheria is part of the normal immunization program for children in the United States.

The **common cold** is the result of a viral infection. Symptoms include sneezing, excessive nasal secretions, and congestion. The infection easily can spread to sinus cavities, lower respiratory passages, and the middle ear. Laryngitis and middle ear infections are common complications. The common cold usually runs its course to recovery in about 1 week.

DISEASES OF THE LOWER RESPIRATORY TRACT

Laryngitis (lăr-in-ji´tis) is an inflammation of the larynx, especially the vocal cords, and **bronchitis** (brong-ki´tis) is an inflammation of the bronchi. Bacterial or viral infections can move from the upper respiratory tract to cause laryngitis or bronchitis. Bronchitis also is often caused by continually breathing air containing harmful chemicals, such as those found in cigarette smoke.

Whooping cough (pertussis) is a bacterial infection. The infection causes a loss of cilia of the respiratory epithelium. Mucus accumulates and the infected person attempts to cough up the mucus. The coughing can be severe. A vaccine for whooping cough is part of the normal vaccination procedure for children in the United States.

Tuberculosis (tu-ber´ku-lo´sis) is caused by a tuberculosis bacterium. In the lung, the tuberculosis bacteria form lesions called tubercles. The small lumps contain degenerating macrophages and tuberculosis bacteria. An immune reaction is directed against the tubercles, which causes the formation of larger lesions and inflammation. The tubercles can rupture, releasing bacteria that infect other parts of the lung or body. Recently, a strain of the tuberculosis bacteria has developed that is resistant to treatment, and there is concern that tuberculosis will again become a widespread infectious disease.

Pneumonia (nu-mo´ne-ah) refers to many infections of the lung. Most pneumonias are bacterial, but some are viral. Symptoms include fever, difficulty in breathing, and chest pain. Inflammation of the lungs results in the accumulation of fluid within alveoli (pulmonary edema) and poor inflation of the lungs with air. A protozoal infection that results in pneumocystosis pneumonia is rare except in persons who have a compromised immune system. This type of pneumonia has become one of the infections commonly suffered by persons who have acquired immune deficiency syndrome (AIDS).

Flu (influenza) is a viral infection of the respiratory system and does not affect the digestive system as is commonly assumed. Flu is characterized by chills, fever, headache, and muscular aches in addition to respiratory symptoms. There are several strains of flu viruses. Mortality is approximately 1%, and most of those deaths are among the very old and very young. During a flu epidemic the infection rate is so rapid and the disease is so widespread, the total number of deaths is substantial even though the percentage of deaths is relatively low. Flu vaccines can provide protection against the flu.

A number of fungal diseases affect the respiratory system. The fungal spores usually enter the respiratory system through dust particles. Spores in soil and feces of certain animals make the rate of infection higher in farm workers and gardeners in certain areas of the country (usually the south and southwest). The infections usually result in minor respiratory infections, but in some cases they can cause infections throughout the body.

An enzyme called **carbonic anhydrase** inside red blood cells (RBCs) increases the rate at which carbon dioxide reacts with water to form hydrogen ions and bicarbonate ions. In the tissue capillaries, carbon dioxide diffuses into RBCs and hydrogen and bicarbonate ions are formed (Figure 15-8, *A*). In capillaries of the lungs, the process is reversed and the bicarbonate and hydrogen ions form carbon dioxide and water. The carbon dioxide diffuses from the RBCs into the alveoli and is expired (Figure 15-8, *B*).

Carbon dioxide has an important effect on the pH of blood. As carbon dioxide levels increase, the blood pH decreases (becomes more acidic) because carbon dioxide reacts with water to form hydrogen ions (see Figure 15-8, *A*). An increase in hydrogen ions causes a decrease in pH (see Chapter 2). Conversely, as blood levels of carbon dioxide decline, the blood pH increases (becomes more basic). Hydrogen and bicarbonate ions react to form carbon dioxide, and the decrease in hydrogen ions causes pH to increase (see Figure 15-8, *B*).

4 What effect would a rapid rate of respiration have on blood pH? What effect would holding one's breath have on blood pH? Explain. (Hint: consider the changes in blood carbon dioxide levels.)

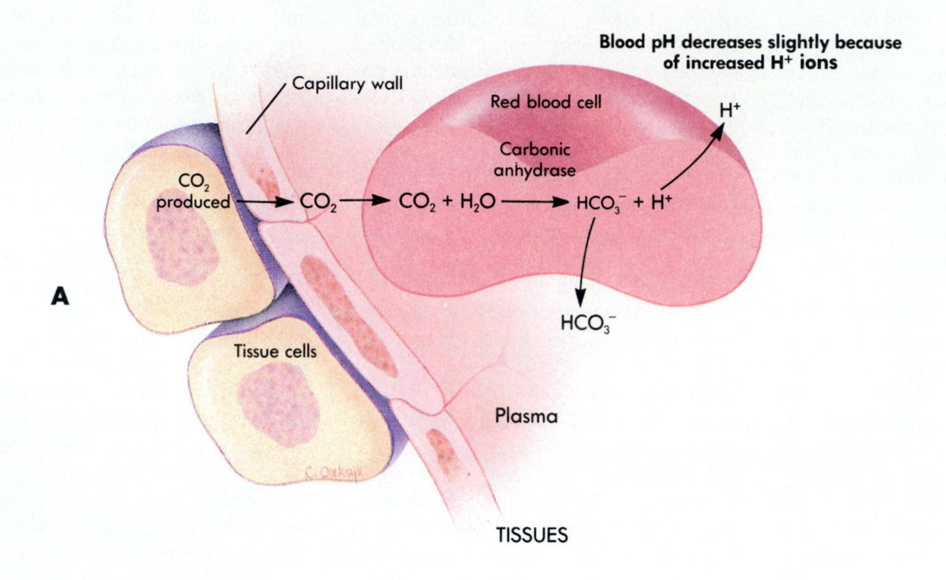

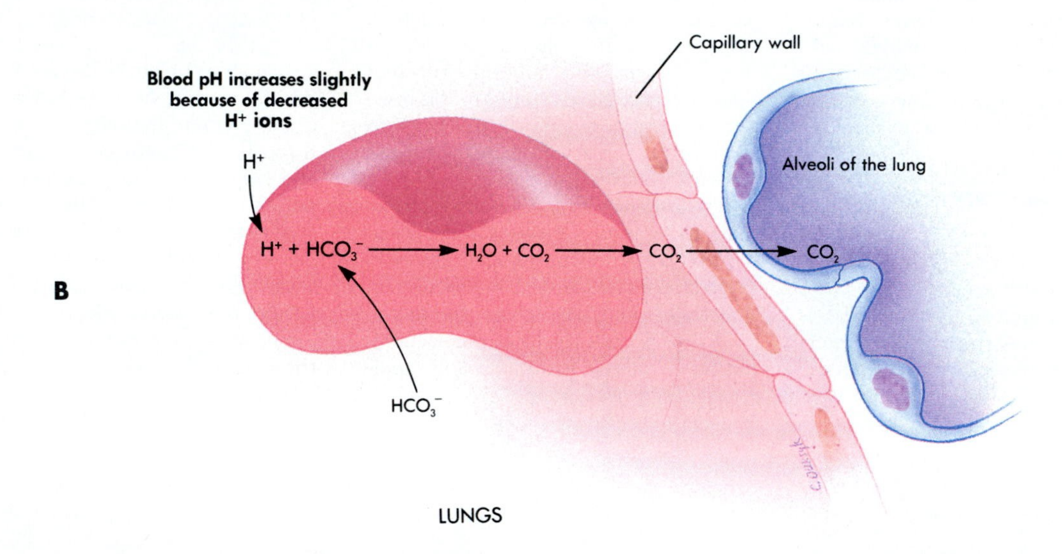

Figure 15-8 ● Carbon dioxide transport as bicarbonate.

A, In the tissue, carbon dioxide enters the RBCs and combines with water to form hydrogen ions and bicarbonate ions. The ions diffuse out ot the RBCs. The increase in hydrogen ions in the blood causes a decrease in blood pH. **B,** In the lung capillaries, hydrogren ions and bicarbonate ions diffuse into RBCs and combine to form water and carbon dioxide. The carbon dioxide diffuses into the alveoli of the lungs. Blood pH increases as hydrogen ions in the blood decrease.

CONTROL OF RESPIRATION

The normal rate of respiration in adults is between 12 and 20 respirations per minute. In children the rates are higher and may vary from 20 to 40 respirations per minute.

A collection of neurons in the medulla oblongata and in the lower portion of the pons together make up the **respiratory center,** which controls the rate and depth of respiration (Figure 15-9). If the respiratory center is damaged, respiratory movements stop. Action potentials pass from the respiratory center to the diaphragm along nerve fibers of the **phrenic** (fren´ik) **nerves.** The phrenic nerves arise from the cervical region of the spinal cord and pass to the diaphragm. Action potentials pass from the respiratory center to the other muscles of respiration along nerve fibers of the spinal nerves.

Stimuli that influence respiration, such as blood levels of carbon dioxide, blood pH, blood oxygen levels, and emotions, do so by altering the activity of the respiratory center (see Figure 15-9). Carbon dioxide levels in the blood and blood pH are the most important stimuli affecting the respiratory center. Chemoreceptors in the medulla oblongata are sensitive to small changes in blood carbon dioxide levels and pH. Increasing carbon dioxide levels in the blood and a decreasing pH strongly stimulate the respiratory center, resulting in an increased rate and depth of respiration. In contrast, decreasing carbon dioxide levels in the blood and an increasing pH result in a slower rate and depth of respiration.

5 Ima Gasper hyperventilates (breathes rapidly and deeply) for several minutes. Hyperventilation "blows off" or removes carbon dioxide from the body at a faster-than-normal rate. After stopping the hyperventilation, Ima experiences a short period in which respiration does not occur (apnea). Then normal breathing resumes. Explain.

Chemoreceptors in the carotid and aortic bodies (see Chapter 12) also provide input to the respiratory center (see Figure 15-9). These chemoreceptors are most sensitive to changes in blood oxygen levels. Unlike blood carbon dioxide levels, small changes in blood levels of oxygen do not normally act as a stimulus to the respiratory center. When blood oxygen levels decline to low levels, however, the chemoreceptors of the carotid and aortic bodies are strongly stimulated. They send action potentials to the respiratory center and produce an increase in the rate and depth of respiration. Conditions that activate the carotid and aortic body chemoreceptors include exposure to high altitudes, diseases such as emphysema, or emergency conditions such as shock.

There is some conscious control over respiration. It is possible to breathe voluntarily or to stop respiratory movements voluntarily. Some people can hold their breath until they lose consciousness as a result of the lack of oxygen in the brain. Some children use this strategy to encourage parents to give them what they want. As soon as conscious control of respiration is lost, however, the automatic control of respiration resumes, and the person starts to breathe again.

Emotions also can have an important influence on respiratory movements. Some people respond to stressful situations with uncontrolled and exaggerated respiratory movements (hyperventilation). In response to hyperventilation, blood carbon dioxide levels become very low, and blood pH increases. These changes can cause the person to faint. The increased pH is detected by the vasomotor center, which causes blood vessels in the periphery to dilate, and blood pools in the abdomen and legs instead of returning to the heart. As a consequence, blood pressure falls dramatically, the brain suffers from a lack of blood flow, and the brain malfunctions as a result of the lack of oxygen, causing the person to faint. Depression, in contrast to stress, can cause a slight reduction in the respiratory rate.

During exercise, respiration rate and depth greatly increase. Only during very heavy exercise, however, do blood carbon dioxide, oxygen, and pH levels change very much from their normal values. Therefore the increased respiration of exercise results from other stimuli. When the motor cortex actively stimulates skeletal muscles to contract during exercise, it also stimulates the respiratory center. In addition, the movement of the limbs stimulates sensory receptors (proprioceptors) in joints that detect movements. The movement of the limbs has a strong stimulatory effect on respiration. Touch, thermal, and pain receptors in the skin also stimulate the respiratory center, which explains the gasp in response to being splashed with cold water or being pinched (see Figure 15-9).

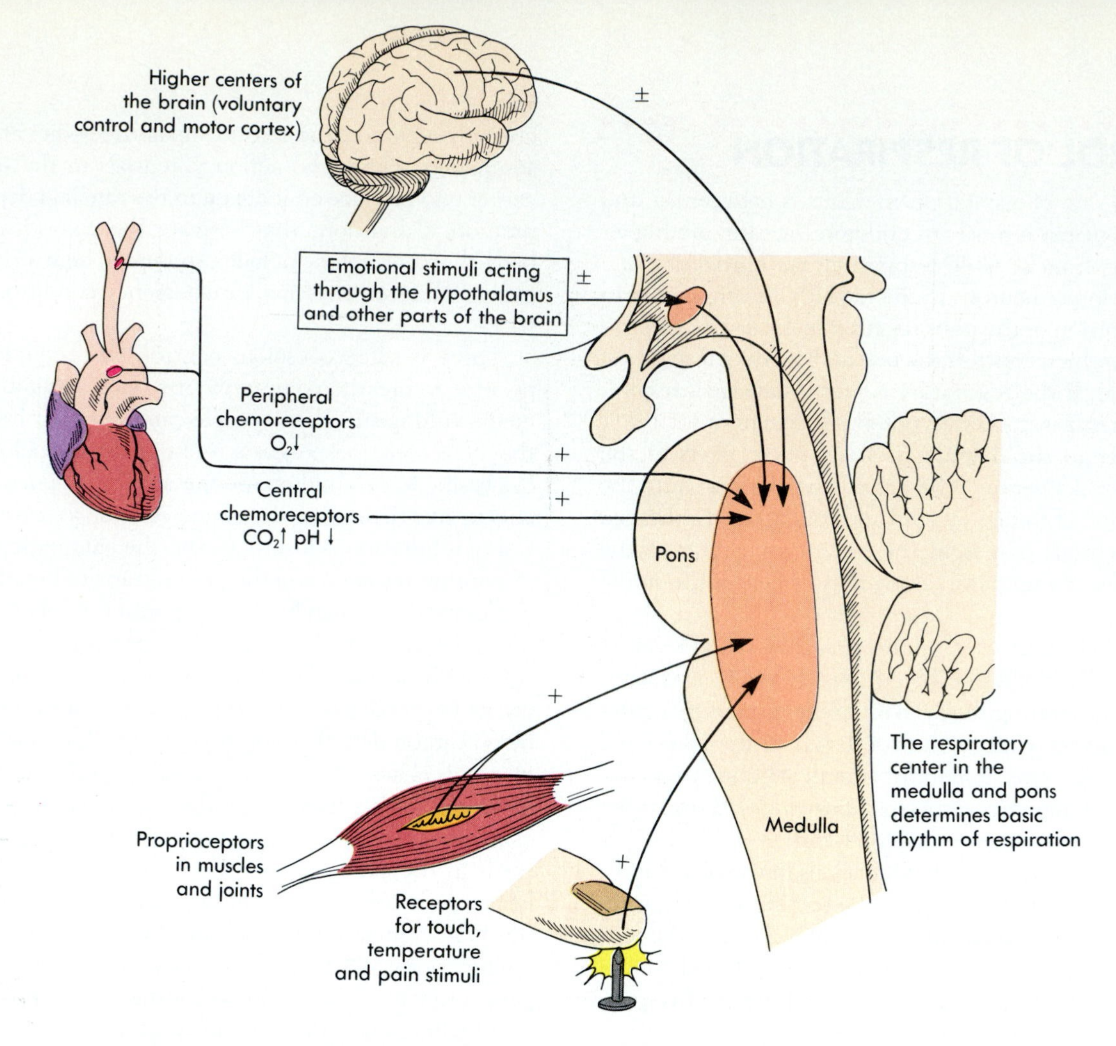

Figure 15-9 ● Regulation of respiration.

The respiratory center determines the basic rate and depth of respiration. Voluntray control, emotions, changes in blood pH, carbon dioxide and oxygen levels, movements of the limbs (proprioception), and stimuli such as touch, temperature, and pain can affect the respiratory center and modify respiration. A plus sign indicates an increase in respiration and a minus sign indicatees a decrease in respiration.

SUMMARY

ANATOMY OF THE RESPIRATORY SYSTEM

The upper respiratory tract consists of the nose, nasal cavity, and pharynx.

The lower respiratory tract consists of the larynx, trachea, bronchi, and lungs.

Nose and nasal cavity

The nasal cavity extends from the nose to the pharynx.

The paranasal sinuses open into the nasal cavity.

Debris is removed from air and the air is warmed and humidified in the nasal cavity.

Pharynx

The pharynx receives air and food. The pharynx connects the nasal and oral cavities to the larynx and esophagus.

Larynx

The larynx contains the vocal cords responsible for voice production.

The epiglottis prevents materials from entering the larynx during swallowing.

Trachea and bronchi

The trachea is held open by cartilage C-rings. It is lined with ciliated epithelium that moves debris trapped in mucus toward the pharynx.

Primary bronchi branch from the trachea and extend to each lung.

Lungs

There are two lungs subdivided into lobes and lobules.

The airway passages of the lungs branch and decrease in size. Bronchi become bronchioles, which give rise to alveoli, the site of gas exchange.

Smooth muscle in bronchioles can regulate air flow.

Pleural cavities

The pleural membranes surround the lungs and provide protection against friction.

VENTILATION

Ventilation is the movement of air into and out of the lungs.

Air moves from an area of higher pressure to an area of lower pressure.

Pressure in the lungs decreases as the volume of the lungs increases, and pressure increases as lung volume decreases.

Inspiration occurs when the diaphragm contracts and the external intercostal muscles lift the rib cage, thus increasing the volume of the thoracic cavity and lungs.

Expiration occurs when the diaphragm relaxes and the internal intercostal muscles depress the rib cage, thus decreasing the volume of the thoracic cavity and lungs.

PULMONARY VOLUMES

There are four pulmonary volumes: tidal volume, inspiratory reserve, expiratory reserve, and residual volume.

Vital capacity is the sum of the inspiratory reserve volume, tidal volume, and expiratory reserve volume.

GAS EXCHANGE

The components of the respiratory membrane include a film of water, the wall of the alveolus, an interstitial space, and the wall of the capillary.

The respiratory membrane is thin and has a large surface area that facilitates gas exchange.

OXYGEN AND CARBON DIOXIDE TRANSPORT IN THE BLOOD

Oxygen diffuses from a higher concentration in the alveoli to a lower concentration in the alveolar capillary.

Oxygen diffuses from a higher concentration in the tissue capillaries to a lower concentration in the tissue spaces.

Oxygen is transported by hemoglobin (97%) and is dissolved in plasma (3%).

Carbon dioxide is transported as bicarbonate ions (72%), in combination with blood proteins (20%), and in solution in plasma (7%).

In tissue capillaries, carbon dioxide combines with water inside the RBCs to form bicarbonate ions and hydrogen ions.

In lung capillaries, bicarbonate ions and hydrogen ions form carbon dioxide that diffuses out of RBCs.

CONTROL OF RESPIRATION

The respiratory center in the medulla oblongata and pons regulates the rate and depth of respiration.

Carbon dioxide is the major chemical regulator of respiration. An increase in carbon dioxide or a decrease in pH of the blood can stimulate chemoreceptors in the medulla oblongata, causing a greater rate and depth of respiration.

Low blood levels of oxygen can stimulate chemoreceptors in the carotid and aortic bodies, which then stimulate the respiratory center.

It is possible to consciously control ventilation.

Input from higher brain centers and from proprioceptors stimulates the respiratory center during exercise.

CONTENT REVIEW

1. What are the functions of the respiratory system?
2. Define respiration.
3. Describe the structure and function of the nasal cavity.
4. Define the pharynx. To what structures is it connected?
5. How do the vocal cords produce sounds of different loudness and pitch?
6. What is the function of the C-shaped cartilages and the ciliated epithelium in the trachea and primary bronchi?
7. Distinguish between the lungs, a lobe of the lung, and a lobule.
8. Explain why breathing becomes more difficult during an asthma attack.
9. Describe the pleura of the lungs. What is their function?
10. How does movement of the diaphragm and ribs affect thoracic volume?
11. Describe the pressure changes that cause air to move into and out of the lungs. What causes these pressure changes?
12. Define tidal volume, inspiratory reserve volume, expiratory reserve volume, residual volume, and vital capacity.
13. List the components of the respiratory membrane. How do changes in thickness or surface area affect the movement of gases across the respiratory membrane?
14. List the ways that oxygen and carbon dioxide are transported in the blood.
15. How can changes in respiration affect blood pH?
16. How does the respiratory center control respiration?
17. What effect does increased blood carbon dioxide, decreased blood pH, and decreased blood oxygen have on respiration? Describe the mechanisms involved in these responses.
18. During exercise, how is respiration regulated?

CONCEPT REVIEW

1. An old technique for artificial respiration is the back-pressure arm-lift method, which is performed with the victim lying face down. The rescuer presses firmly on the base of the scapulae for several seconds, then grasps the arms and lifts them. The sequence is then repeated. Explain why this procedure results in ventilation of the lungs. (Hint: what effect does this type of artificial respiration have on the volume of the thoracic cavity?)
2. If a resting person had a tidal volume of 500 ml and a respiratory rate of 12 respirations per minute, and an exercising person had a tidal volume of 4000 ml and a respiratory rate of 24 respirations per minute, what would be the difference in the total amount of air respired per minute between them?
3. A patient has pneumonia, and fluids accumulate within the alveoli. Explain why this results in an increased rate of respiration and why respiration can be returned to normal by administering oxygen.
4. Patients with diabetes mellitus who are not being treated with insulin therapy rapidly metabolize lipids, and there is an accumulation of acidic byproducts of lipid metabolism in the circulatory system. What effect would this have on respiration? Why is the change in respiration beneficial?

CHAPTER TEST

Answers can be found in Appendix C

MATCHING For each statement in column A select the correct answer in column B (an answer may be used once, more than once, or not at all).

A

1. Increase surface area in the nasal cavity.
2. Common passageway for digestive and respiratory systems.
3. Voice box containing the vocal cords.
4. Connects the larynx to the bronchi.
5. Serous membrane in contact with the lungs.
6. Volume of air inspired during a single normal breath at rest.
7. Determines the basic rhythm of respiration.
8. Location of chemoreceptors most sensitive to small changes in blood carbon dioxide and pH.
9. Location of chemoreceptors most sensitive to low levels of oxygen in the blood.

B

- carotid and aortic bodies
- conchae
- inspiratory reserve volume
- larynx
- parietal pleura
- pharynx
- respiratory center
- tidal volume
- trachea
- visceral pleura

FILL-IN-THE BLANK Complete each statement by providing the missing word or words.

1. The __________ consists of the nose, nasal cavity, and pharynx, whereas the __________ consists of the larynx, trachea, bronchi, and lungs.
2. The __________ and __________ are held open by cartilage C-rings, whereas the __________ have abundant smooth muscle that regulates air flow.
3. The __________ is mostly responsible for thoracic volume changes during respiration.
4. An increase in thoracic volume results in a(n) __________ in air pressure within the lungs.
5. Vital capacity is the sum of __________, __________, and __________.
6. The respiratory membrane consists of a film of water lining the alveolus, the __________, a small interstitial space, and the __________.
7. Most oxygen is transported bound to __________, whereas most carbon dioxide is transported as __________.
8. An increase in respiration can result from a(n) __________ in blood carbon dioxide, a(n) __________ in blood pH, or a(n) __________ in blood oxygen.
9. During exercise, stimulation of the respiratory center by the __________ in the brain or by __________ in joints can result in an increase in respiration.

CHAPTER 16

The Digestive System

KEY TERMS

bile
Fluid secreted from the liver, stored in the gallbladder, and released into the duodenum; consists of bile salts, bile pigments, bicarbonate ions, fats, hormones, and more.

chyme
Semifluid mass of partly digested food passed from the stomach into the duodenum.

colon
Division of the large intestine that extends from the cecum to the rectum.

defecation
Discharge of feces from the rectum.

duodenum
First division of the small intestine; connects to the stomach.

esophagus
The part of the digestive tract between the pharynx and stomach.

lacteal
Lymphatic vessel in the villi of the small intestine; carries lymph from the intestine and absorbs fat.

mastication
The process of chewing.

peristalsis
Waves of contraction and relaxation that propel food along the digestive tube.

villus
Projection of the mucous membrane of the small intestine; functions to increase surface area for absorption and secretion.

FEATURES

OBJECTIVES

After reading this chapter you should be able to:

1. Describe the four layers of the digestive tract.
2. Name the structures of the oral cavity and describe their functions.
3. Describe the pharynx and esophagus and explain how swallowing is accomplished.
4. Outline the stomach's anatomical and physiological characteristics that are most important to its function.
5. Explain how the large surface area of the small intestine is important for its function.
6. List the functions of the liver and pancreas.
7. Describe the peritoneum and the mesenteries.
8. Explain how secretions and movements of the digestive tract are regulated.
9. Describe the digestion of and list the breakdown products of carbohydrates, lipids, and proteins.
10. Describe the absorption of the breakdown products of carbohydrates, lipids, and proteins.

THE DIGESTIVE SYSTEM functions to ingest, digest, and absorb food and liquids. **Ingestion** is the movement of food or liquids into the mouth, **digestion** is the breakdown of the ingested materials into smaller units, and **absorption** is the movement of the digested materials into the body. These processes provide the body with needed nutrients, water, and electrolytes. The digestive system consists of the **digestive tract,** a tube extending from the mouth to the anus, plus the **accessory organs,** which secrete fluids into the digestive tract (Figure 16-1). The term **gastrointestinal (GI) tract** technically refers to the stomach and intestines but is often used as a synonym for the digestive tract.

ANATOMY AND HISTOLOGY OF THE DIGESTIVE SYSTEM

Various portions of the digestive tract are specialized for different functions, but nearly all portions of the digestive tract consist of four layers (Figure 16-2).

1. The innermost layer is a mucous membrane (see Chapter 4) called the **mucosa** (mu-ko´sah). The epithelium of the mucosa is thickened (stratified) in the mouth, esophagus, and rectum to resist abrasion, and is thin (simple) in the stomach and intestine for absorption and secretion.
2. The **submucosa** (sub´mu-ko´sah) is a thick layer of loose connective tissue containing nerves, blood vessels, and small glands.
3. The **muscularis** (mus´ku-la´ris) consists of an inner layer of circular smooth muscle and an outer layer of longitudinal smooth muscle in most parts of the digestive tube. Contractions of the smooth muscle are responsible for **peristalsis** (pĕr´ĭ-stal´sis), wavelike movements that push materials along the digestive tract. Contractions also cause the digestive tract contents to be mixed with digestive secretions. Neurons within the submucosa and muscularis form the **intramural plexus** (in´trah-mu´ral plek´sus), which is important in the control of movement and secretion within the digestive tract.
4. The outermost layer of the digestive tract is a serous membrane (see Chapter 4) called the **serosa** (se-ro´sah), or **visceral peritoneum.**

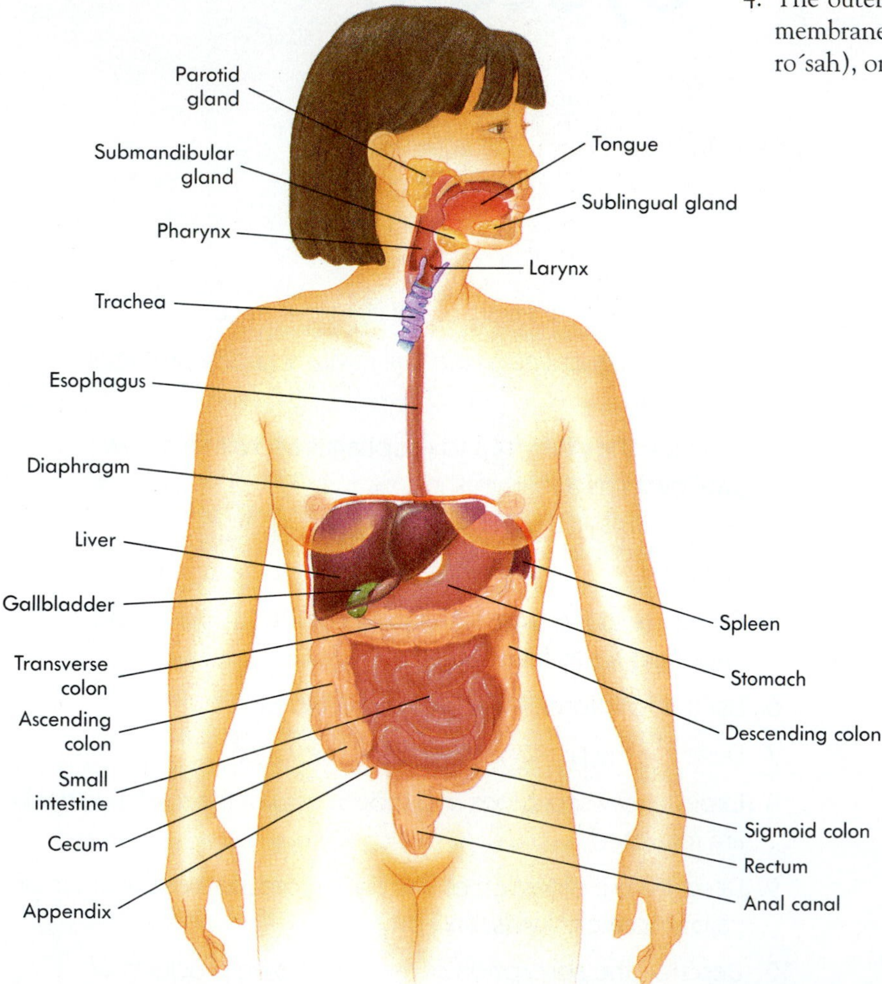

Figure 16-1 ● The digestive system.
The digestive system consists of a tube extending from the mouth to the anus, as well as the accessory organs that empty their secretions into the tube.

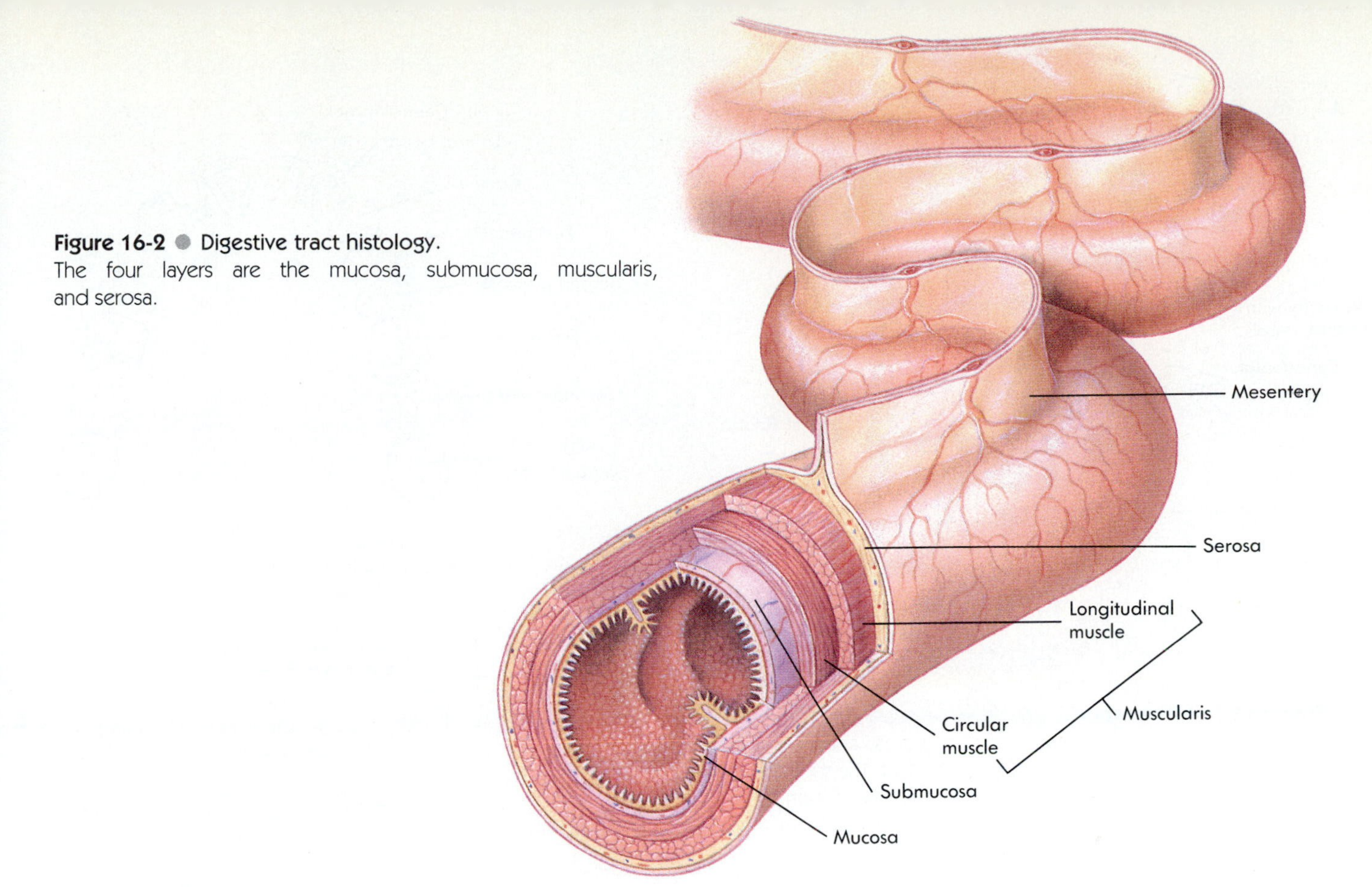

Figure 16-2 ● Digestive tract histology.
The four layers are the mucosa, submucosa, muscularis, and serosa.

Oral cavity

The **oral cavity,** or mouth, is the first portion of the digestive tract. It is bounded by the lips and cheeks, and contains the teeth and tongue. The lips, cheeks, and tongue are important for chewing food and for speech. They help manipulate the food within the mouth and hold the food in place while the teeth crush or tear it. They also help form words during the speech process.

The **tongue** is a large, muscular organ that occupies most of the oral cavity. The tongue plays a major role in the process of swallowing and is a major sensory organ for taste (see Chapter 9).

Teeth

Each **tooth** consists of a **crown** with one or more cusps (points), a **neck,** and a **root** (Figure 16-3, *A*). The center of the tooth is a **pulp cavity,** which is filled with blood vessels, nerves, and connective tissue called **pulp.** The pulp cavity is surrounded by a living, cellular, calcified tissue called **dentin.** The dentin of the tooth crown is covered by an extremely hard, acellular substance called **enamel,** which protects the tooth against abrasion and acids produced by bacteria in the mouth.

The roots of the teeth insert into sockets in the mandible and maxillae and are held in place by **periodontal** (pĕr´e-o-don´tal) **ligaments.** The mandible and maxillae are covered by dense, fibrous connective tissue and moist stratified squamous epithelium, referred to as the **gingiva** (jin´jĭ-vah), or **gums.**

There are 32 teeth in the normal adult mouth, located in the mandible and maxillae (Figure 16-3, *B*). The teeth in the right and left halves of each jaw are roughly mirror images of each other. The jaws can be divided into four parts. Each part contains one **central** and one **lateral incisor;** one **canine;** first and second **premolars;** and first, second, and third **molars.** The third molars are referred to as wisdom teeth because they usually appear when the person is in his late teens or early twenties, when a person is thought to have acquired some degree of wisdom. If there is insufficient room for the wisdom teeth, they may have to be removed.

The teeth of the adult mouth are **permanent,** or **secondary, teeth** (see Figure 16-3, *B*). Most of them are replacements of the 20 **primary,** or **deciduous** (de-sid´u-us), **teeth** (Figure 16-3, *C*) that are lost during childhood. The permanent incisors and canines are replacements for the deciduous incisors and canines. The permanent premolars are replacements for the deciduous molars. The permanent first, second, and third molars are additions to the deciduous teeth.

Food taken into the mouth is chewed, or **masticated** (mas´tĭ-ka´ted) by the teeth. The incisors and the canines primarily cut and tear food, whereas the premolars and molars primarily crush and grind food. Mastication breaks large food particles into many small particles, which have a much larger total surface area than would a few large particles. Because digestive enzymes act on food molecules only at the surface of the particles, mastication increases the efficiency of digestion.

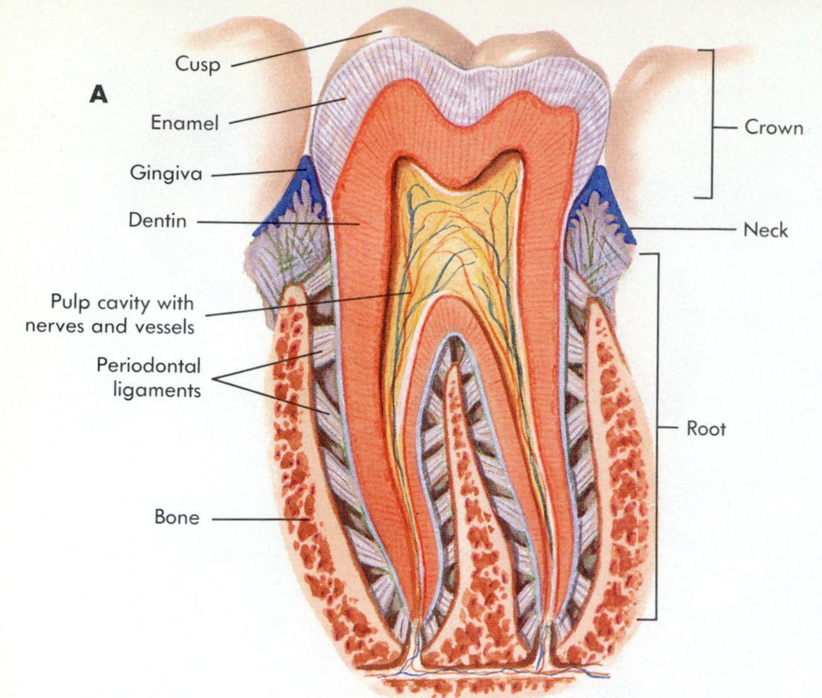

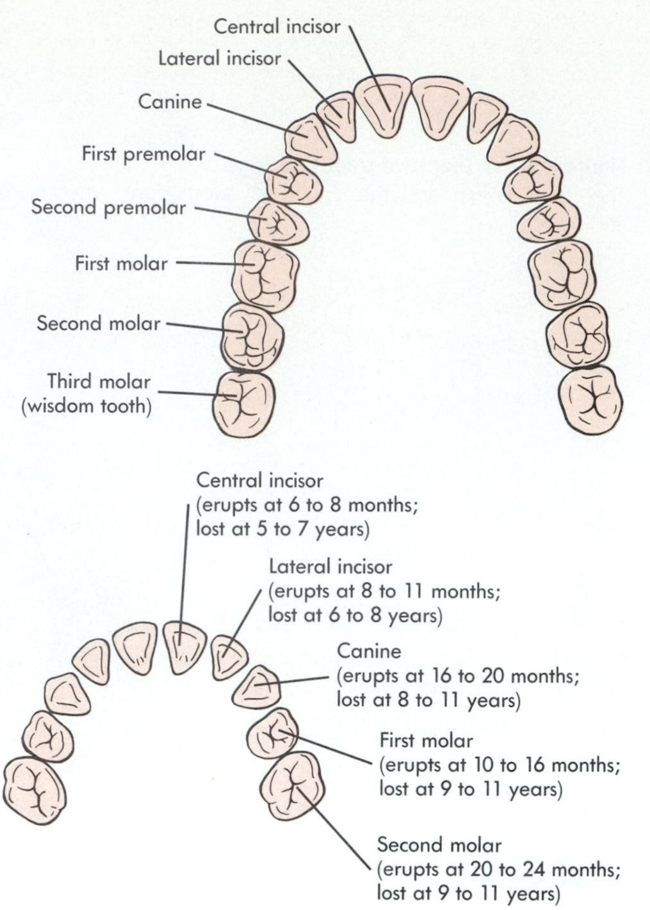

Figure 16-3 The teeth.
A, Molar tooth in place in bone, cut in section to show the pulp. The tooth consists of a crown, a neck, and a root. The tooth is held in the socket by periodontal ligaments. **B,** Permanent teeth. **C,** Deciduous teeth.

Palate

The **palate,** or roof of the oral cavity has two parts. The anterior, bony part is the **hard palate.** The posterior portion, the **soft palate** consists of skeletal muscle and connective tissue. The **uvula** (u´vu-lah) is a projection of the soft palate that can be seen "hanging" in the throat.

Salivary glands

There are three pairs of **salivary glands** (see Figure 16-1). The largest of the salivary glands, the **parotid** (pă-rot´id) glands, are located just anterior to each ear. The mumps is inflammation and swelling of the parotid glands as a result of a viral infection. The **submandibular** glands can be felt as a soft lump along the inferior border of each side of the mandible. In certain people, if the mouth is opened and the tip of the tongue is elevated, saliva may squirt out of the mouth from the ducts of these glands. The **sublingual** glands are the smallest of the three paired salivary glands. They lie immediately below the mucous membrane in the floor of the oral cavity.

The salivary glands produce **saliva,** which is a mixture of mucus and serous (watery) fluids containing digestive enzymes. The mucous part of saliva functions as a lubricant that protects the oral cavity against abrasion. The serous portion of saliva moistens food and contains a digestive enzyme called **salivary amylase** (am´ĭ-lās), which begins the process of digesting carbohydrates. Saliva also prevents bacterial infection in the mouth by washing the oral cavity, and it contains lysozyme, an enzyme that causes some bacterial cells to rupture.

Pharynx and esophagus

The **pharynx** (far´ingks), or throat, connects the oral cavity with the esophagus (see Figure 16-1). The walls of the pharynx have skeletal muscles that constrict and move food into the esophagus. The pharynx is also part of the respiratory system (see Chapter 15).

The **esophagus** (e-sof´ă-gus) is a muscular tube that extends from the pharynx through the thoracic cavity, passes through the diaphragm, and ends at the stomach (see Figure 16-1). The esophagus transports food from the pharynx to the stomach.

Swallowing is partly a voluntary activity and partly an involuntary reflex. During the voluntary phase, a **bolus** (bo´lus), or mass of food, is formed in the mouth and pushed into the pharynx by the tongue. Stimulation of touch receptors in the pharynx initiates reflexes that cause constriction of pharyngeal muscles, and the bolus is pushed into the esophagus. As food passes through the pharynx, the soft palate closes off the openings into the nasal cavity and the **epiglottis** (ep´ĭ-glot´is; see Chapter 15) covers the opening into the larynx.

1 **Why is it important to close off the openings into the nasal cavity and larynx during swallowing? What may happen if a person has an explosive burst of laughter while trying to swallow a liquid? What happens if you try to swallow and speak at the same time?***

Muscular contractions of the esophagus occur in peristaltic waves and the bolus is propelled through the esophagus. Gravity assists the movement of material through the esophagus, especially when liquids are swallowed. However, the peristaltic contractions that move material through the esophagus are sufficiently forceful to allow a person to swallow even while upside down.

Movement of materials between the esophagus and stomach is regulated by a sphincter made of circular smooth muscle that constricts to close the opening between the esophagus and stomach. When a bolus reaches the end of the esophagus, the **esophageal sphincter** relaxes and the bolus enters the stomach. Normally the esophageal sphincter is constricted, which prevents stomach contents from reentering the esophagus. If stomach acid should reflux into the esophagus, the acid can damage the esophagus, producing the unpleasant symptoms of **heartburn.** In severe cases the damage can produce an ulcer, a damaged area of the mucosa that usually is inflamed.

*Answers to predict questions appear in Appendix B at the back of the book.

Stomach

The stomach is an enlarged segment of the digestive tract with three parts (Figure 16-4). The **fundus** (fun´dus) is to the left and superior to the opening of the esophagus, the **body** is the central portion, and the **pylorus** (pi-lōr´us) is the inferior part that connects to the small intestine. The opening from the stomach into the small intestine is surrounded by a relatively thick ring of smooth muscle called the **pyloric** (pi-lōr´ik) **sphincter.**

The muscular layer of the stomach is different from other regions of the GI tract in that it consists of three layers: an outer longitudinal layer, a middle circular layer, and an inner oblique layer. When the stomach is empty, the submucosa and mucosa of the stomach are thrown into large folds called **rugae** (ru´ge). These folds allow the stomach to stretch, and the folds disappear as the stomach is filled.

The stomach is lined with simple columnar epithelium. On the stomach's inner surface, the epithelium produces mucus that coats and protects the stomach lining from stomach acid and digestive enzymes. The epithelium is folded to form many tube-shaped **gastric glands.** The gastric glands produce **gastric juice,** which contains mucus, pepsin (an enzyme that begins protein digestion), **hydrochloric acid** (provides the proper pH environment for pepsin; see Chapter 2), and intrinsic factor (increases the absorption of vitamin B_{12}; see Chapter 11).

The stomach functions primarily as a storage and mixing chamber for ingested food, which typically remains in the

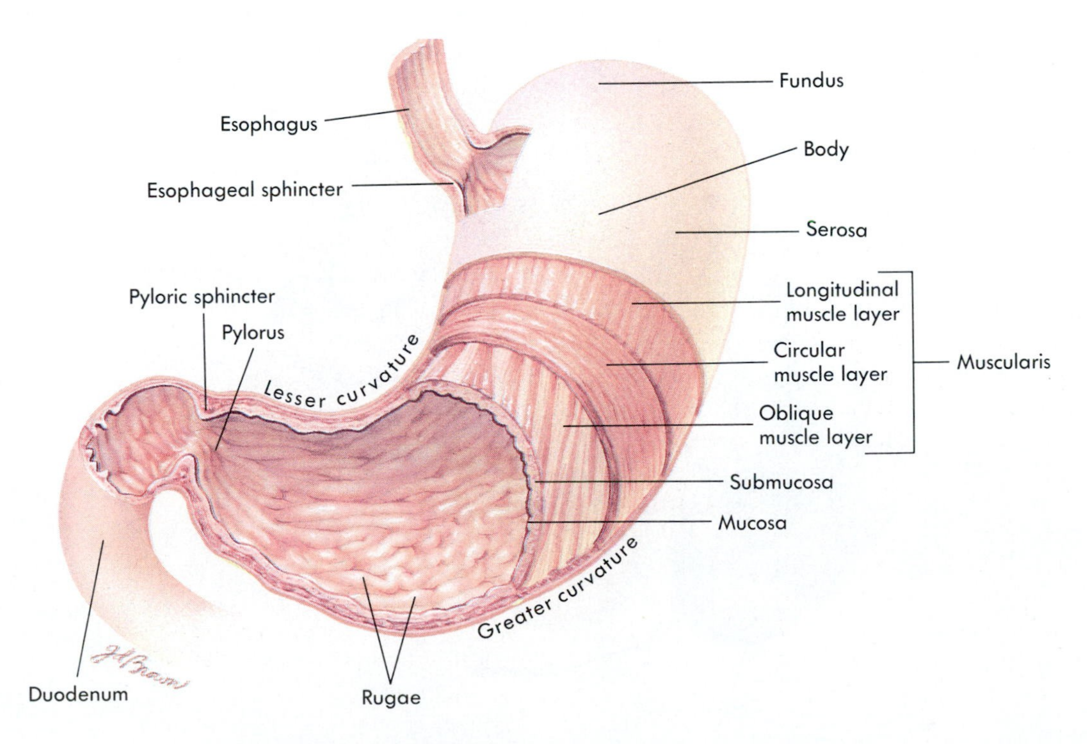

Figure 16-4 ● The stomach.
Cutaway section reveals the layers of the stomach and internal anatomy.

stomach for 1 to 4 hours. Although some digestion and a small amount of absorption occur in the stomach, they are not its principal functions. As food enters the stomach, contractions of the stomach's smooth muscle mixes the food with gastric juice to become a semifluid mixture called **chyme** (kīm). Contractions also force small amounts of chyme past the pyloric sphincter into the small intestine for the next stage of digestion.

Small intestine

The **small intestine** is about 6 meters long and consists of three portions: the **duodenum** (du-o-de´num or du-od´ĕ-num), **jejunum** (jĕ-ju´num), and **ileum** (il´e-um) (Figure 16-5). The duodenum is about 25 cm (1 foot) long, the jejunum is about 2.5 meters (8 feet) long, and the ileum is about 3.5 meters (11.5 feet) long.

The surface of the duodenum is modified to increase its surface area about 600-fold. The increased surface area allows more efficient digestion and absorption of food. The mucosa and submucosa form a series of **circular folds** that run perpendicular to the long axis of the digestive tract (Figure 16-6, *A*). Tiny fingerlike projections of the mucosa form numerous **villi** (vil´i), which are 0.5 to 1.5 mm long (Figure 16-6, *B*). Each villus is covered by simple columnar epithelium and contains a blood capillary network and a lymph capillary called a **lacteal** (lak´te-al) (Figure 16-6, *C*). The blood capillary network and the lacteal function to transport absorbed nutrients (see Chapter 17). Most of the cells composing the surface of the villi have numerous tubelike extensions of the cell membrane, called **microvilli** (mi´kro-vil´i) (Figure 16-6, *D*), which further increase the surface area.

The jejunum and ileum are similar in structure to the duodenum except that there is a gradual decrease in the diameter of the small intestine, in the thickness of the intestinal wall, in the number of circular folds, and in the number of villi as one progresses through the small intestine.

The small intestine, especially the duodenum and jejunum, is the primary site of digestion and absorption of the GI tract. Secretions of the pancreas and liver enter the duodenum and continue the process of digestion begun by saliva and gastric juices. The epithelium lining the small intestine also produces enzymes that complete the digestive process, and the digested food molecules are absorbed through the microvilli.

The mucosa of the small intestine and intestinal glands also produce secretions that contain primarily mucus, electrolytes, and water. Intestinal secretions lubricate and protect the intestinal wall from the acidic chyme and the action of digestive enzymes. They also keep the chyme in the small intestine in a liquid form to facilitate the digestive process.

Contractions of smooth muscle within the wall of the small intestine mix secretions with the chyme and move the chyme to the large intestine. The junction between the small and the large intestine has a ring of smooth muscle,

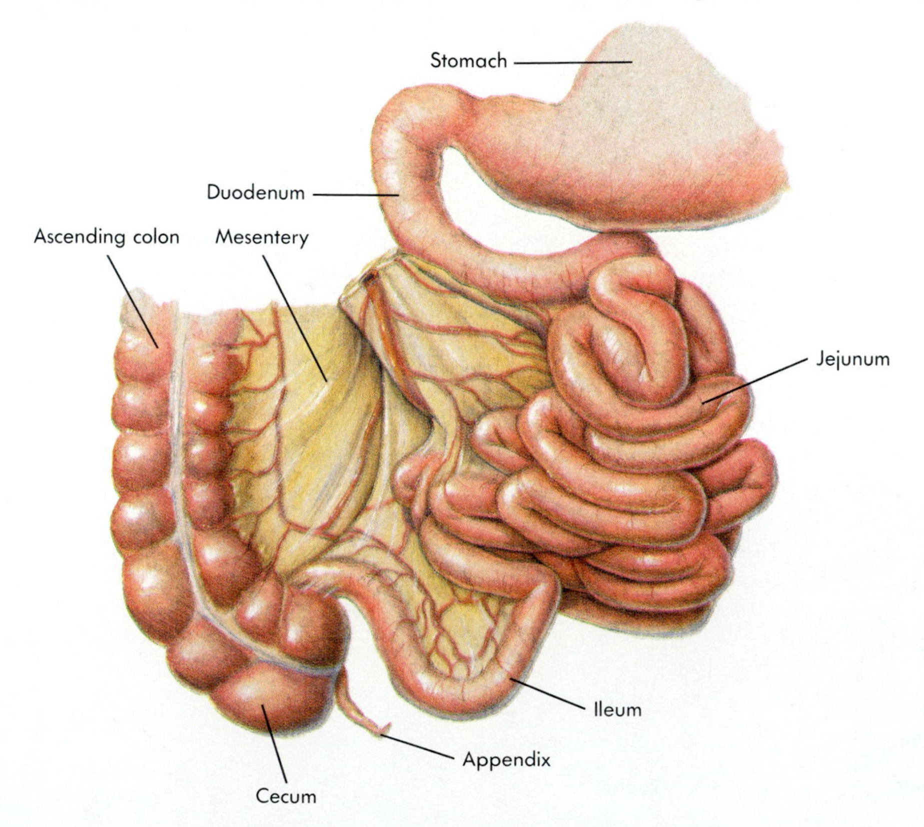

Figure 16-5 ● **The small intestine.**
The duodenum is attached to the stomach and is continuous with the jejunum. The jejunum is continuous with the ileum, which empties into the cecum of the large intestine.

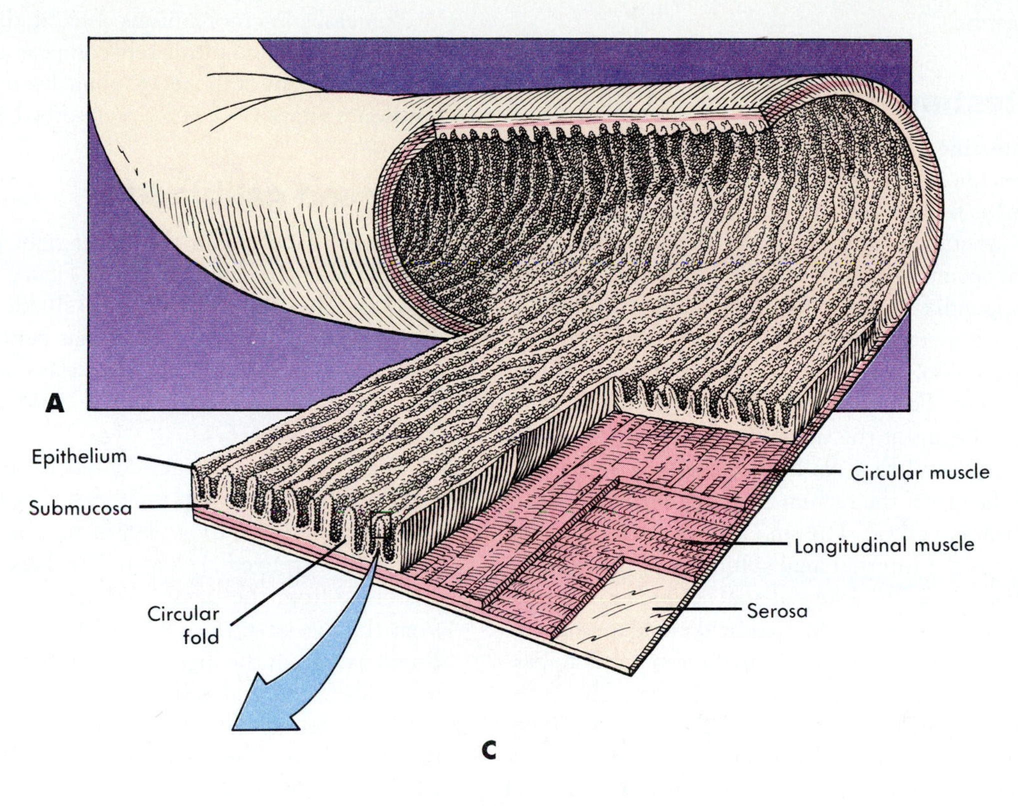

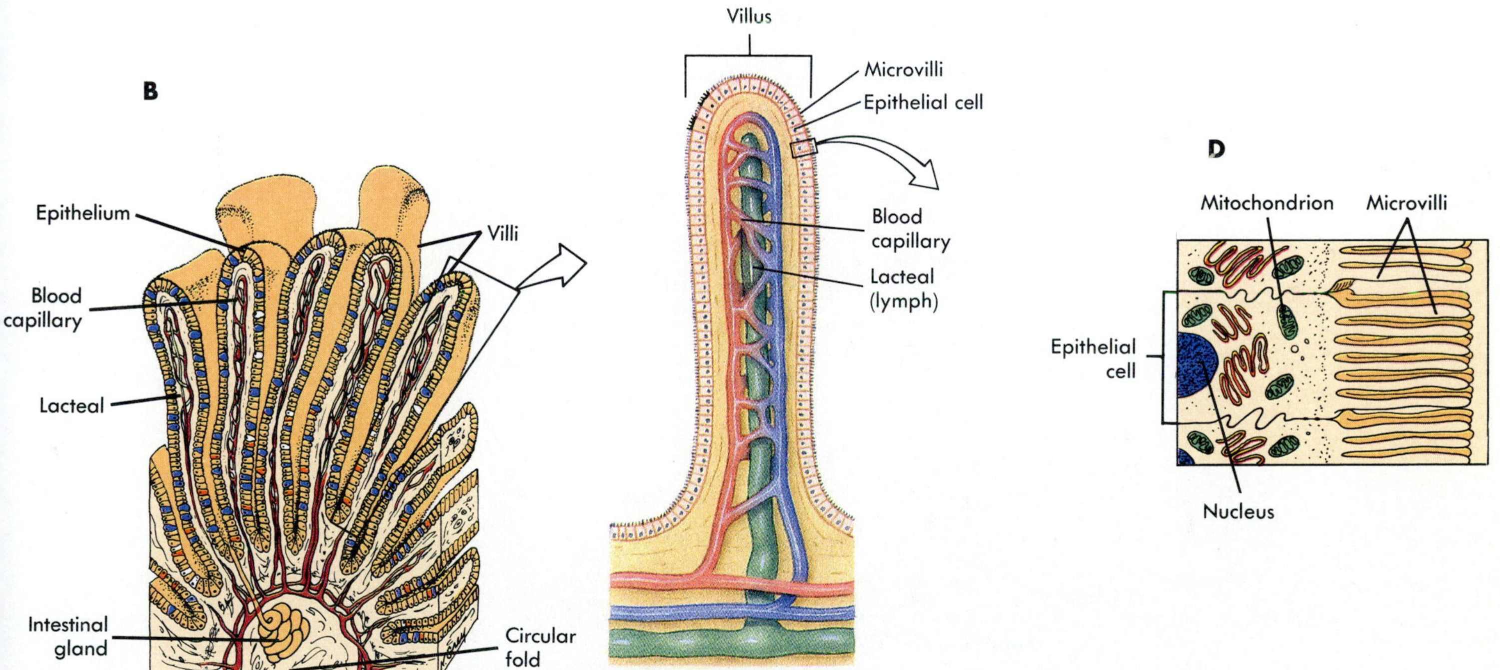

Figure 16-6 ● **Interior view and histology of the duodenum.**
A, The wall of the duodenum has been opened to reveal the circular folds. **B,** The villi on the surface of a circular fold. **C,** A single villus showing the lacteal and capillary. **D,** The microvilli on the surface of a villus cell. The extensive surface area allows more efficient absorption of nutrients.

the **ileocecal** (il´e-o-se´kal) **sphincter,** and a one-way **ileocecal valve.** They control the movement of materials into the large intestine.

Large intestine

The **large intestine** (Figure 16-7 and see Figure 16-1) consists of the cecum, colon, rectum, and anal canal. The **cecum** (se´kum) is a blind sac that extends inferiorly about 6 cm past the junction of the small and large intestine. Attached to the cecum is a small blind tube about 9 cm long called the **appendix** (ă-pen´diks). The **colon** (ko´lon) is about 1.5 to 1.8 meters long and consists of four portions: the ascending colon, transverse colon, descending colon, and sigmoid colon. The **rectum** (rek´tum) is a straight, muscular tube that begins at the sigmoid colon and ends at the anal canal. The last 2 to 3 cm of the digestive tract is the **anal canal.** It begins at the rectum and ends at the **anus,** the external GI tract opening. The smooth muscle layer of the anal canal forms the **internal anal sphincter** at the superior end of the anal canal. The **external anal sphincter** at the inferior end of the anal canal is formed by skeletal muscle.

Normally 18 to 24 hours are required for material to pass through the large intestine in contrast to the 3 to 5 hours required for movement of chyme through the small intestine. While in the colon, chyme is converted to **feces** (fe´sēz). Absorption of water and salts, the secretion of mucus, and extensive action of microorganisms are involved in the formation of feces, which the colon stores until they are eliminated by the process of **defecation** (def-ĕ-ka´shun).

Numerous microorganisms inhabit the colon. They reproduce rapidly and ultimately compose approximately 30% of the dry weight of the feces. Some bacteria in the intestine synthesize vitamin K, which is absorbed in the colon.

Liver and gallbladder

The **liver** is located in the upper-right portion of the abdomen under the diaphragm (see Figure 16-1). It is divided into four lobes and weighs about 1.36 kilograms (3 pounds). The liver is an amazing organ that performs important digestive and excretory functions, stores and processes nutrients, synthesizes new molecules, and detoxifies harmful chemicals (Table 16-1).

The liver has two sources of blood (see Chapter 13) that make possible many of its functions. The **hepatic** (he-pat´ik) **artery** brings to the liver oxygen-rich blood, which supplies liver cells with needed oxygen. The **hepatic portal vein** carries blood from the digestive tract to the liver. The blood from the digestive tract is oxygen poor, because it has released oxygen in the digestive tract, but it is rich in absorbed materials from the digestive tract. Liver cells process nutrients and detoxify harmful substances in the blood. The blood then exits the liver through **hepatic veins** and enters the general circulation.

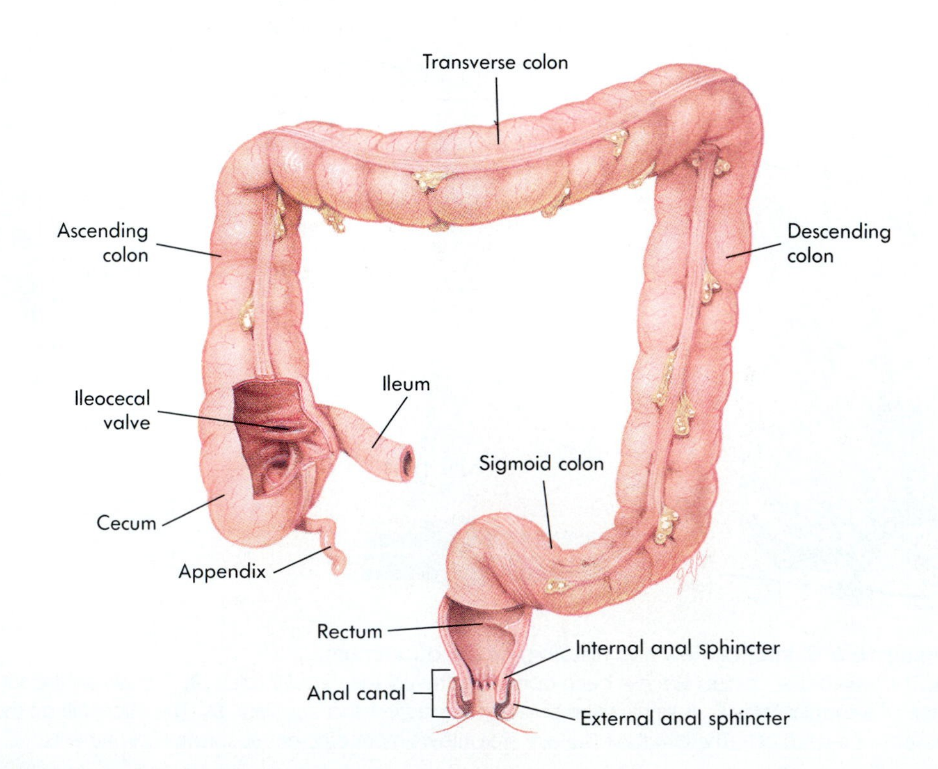

Figure 16-7 ● **The large intestine.**
The large intestine extends from the ileocecal valve to the anus and includes the cecum, colon, rectum, and anal canal.

Table 16-1 • Functions of the liver

Function	Explanation
Digestion	Bile neutralizes stomach acid and emulsifies fats, which facilitates digestion
Excretion	Bile contains excretory products, such as cholesterol, fats, or bile pigments like bilirubin resulting from hemoglobin breakdown
Nutrient storage	Liver cells remove sugar from the blood and store it in the form of glycogen; also store fat, vitamins (A, B_{12}, D, E, and K), copper, and iron
Nutrient conversion	Liver can convert some nutrients into others, for example, amino acids can be converted to lipids or glucose; fats can be converted to phospholipids; vitamin D is converted to its active form
Detoxification of harmful chemicals	Liver cells remove ammonia from the circulation and convert it to urea, which is eliminated in the urine; other substances are detoxified and secreted in the bile or urine
Synthesis of new molecules	Synthesizes blood proteins such as albumin, fibrinogen, globulin, and clotting factors

One major function of the liver is **bile** production. Although bile contains no digestive enzymes, the bicarbonate ions in bile play a role in digestion by diluting and neutralizing stomach acid. In addition, bile salts in bile increase the efficiency of fat digestion and absorption. It also contains excretory products such as bilirubin (see Chapter 11) to be eliminated from the body. Bile is transported from the liver by two ducts that unite to form a single **common hepatic duct.** The common hepatic duct is joined by the **cystic** (sis´tik) **duct** from the gallbladder to form the **common bile duct,** which joins the pancreatic duct to empty into the duodenum (Figure 16-8). The **gallbladder** is a small sac on the inferior surface of the liver that stores bile. Between meals, bile flows from the liver into the common hepatic duct and through the cystic duct into the gallbladder in which it is stored. After a meal, the gallbladder contracts and bile flows out the cystic duct and through the common bile duct to the duodenum.

Pancreas

The **pancreas** (pan´kre-us) (see Figure 16-8) is a complex organ composed of both endocrine and exocrine tissues that perform several functions. The endocrine portion of the pancreas consists of pancreatic islets (islets of Langerhans). The islet cells produce insulin and glucagon, which are important in controlling blood levels of nutrients such as glucose and amino acids (see Chapter 10).

The exocrine portion of the pancreas consists of glands that produce digestive enzymes. Pancreatic secretions also have a high bicarbonate ion content that neutralizes the acidic chyme entering the duodenum. This protects the duodenum from damage by the acidic chyme and provides

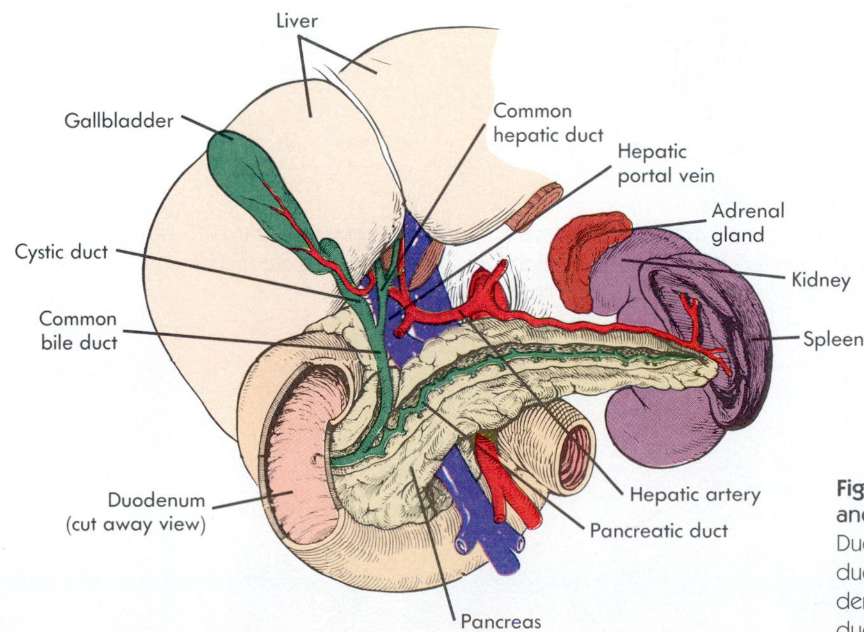

Figure 16-8 • The liver, gallbladder, pancreas, and duct system.
Ducts from the liver join to form the common hepatic duct, which joins the cystic duct from the gallbladder to form the common bile duct. The common bile duct joins the pancreatic duct and enters the duodenum.

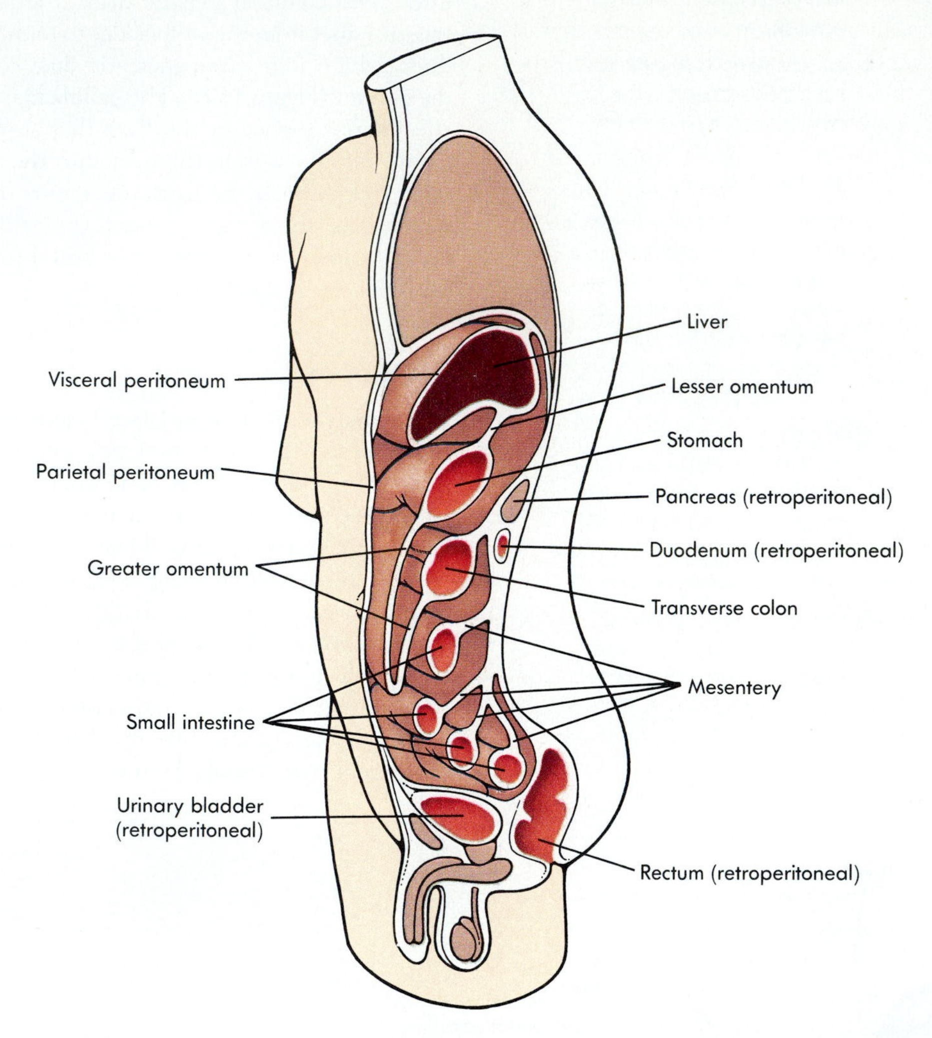

Figure 16-9 ● **Peritoneum and mesenteries.**
The parietal peritoneum lines the abdominal cavity, and the visceral peritoneum covers abdominal organs. Retroperitoneal organs are covered by the parietal peritoneum. The mesenteries are membranes that connect abdominal organs to each other and to the body wall.

the proper pH environment needed for pancreatic and intestinal enzymes to function. The glands are connected by a series of ducts that join to form the **pancreatic duct.** The pancreatic duct joins the common bile duct and empties into the duodenum (see Figure 16-8).

Peritoneum

The body walls and organs of the abdominal cavity are lined with serous membranes (Figure 16-9). The serous membrane that covers the organs is the **visceral peritoneum** (pĕr´ĭ-to-ne´um), and the serous membrane that covers the interior of the body wall is the **parietal peritoneum.**

Many of the organs of the abdominal cavity are held in place by connective tissue sheets called **mesenteries** (mes´en-tĕr´ez). The mesenteries consist of two layers of serous membranes with a thin layer of loose connective tissue in-between (see Figure 16-2). Specific mesenteries are given names. For example, the mesentery connecting the stomach to the liver is the **lesser omentum** (o-men´tum), and the mesentery connecting the stomach to the transverse colon is the **greater omentum.**

Other abdominal organs lie against the abdominal wall, have no mesenteries, and are described as **retroperitoneal** (rĕ´tro-pĕr´ĭ-to-ne´al), or behind the peritoneum. The retroperitoneal organs include the duodenum, pancreas, ascending colon, descending colon, rectum, kidneys, adrenal glands, and urinary bladder.

REGULATION OF THE DIGESTIVE SYSTEM

As food moves through the digestive tract, secretions are added to liquefy and digest the food and to provide lubrication. Each segment of the digestive tract also is specialized to mix the food with secretions and to move its contents to the next segment of the digestive tract. Proper digestion occurs when the movement of materials is coordinated with the secretion of enzymes that break down the materials. For example, if the stomach empties too fast, the efficiency of digestion and absorption is reduced, and the acidic chyme may damage the intestinal wall. If the rate of emptying is too slow, the highly acidic contents of the stomach can damage the stomach wall.

Three types of control mechanisms regulate the activities of the digestive system: the autonomic nervous system, local reflexes, and hormones. The autonomic nervous system has regulatory centers in the medulla oblongata that control the digestive system. The parasympathetic division is more important than the sympathetic division in the day-to-day regulation of the digestive system. The parasympathetic division usually is stimulatory, promoting the secretions and movements necessary for digestion. The sympathetic division usually has an inhibitory effect on the digestive system. For example, during exercise the digestive process slows down.

Autonomic nervous system control is mediated through reflexes. For example, the entry of food into the stomach stimulates stretch or chemical receptors. Action potentials go along nerve fibers to the regulatory centers in the medulla oblongata, which stimulates the stomach to increase its production of gastric juices and increase contractions that result in mixing of the secretions with the ingested food (Figure 16-10, A). Thus the movement of food

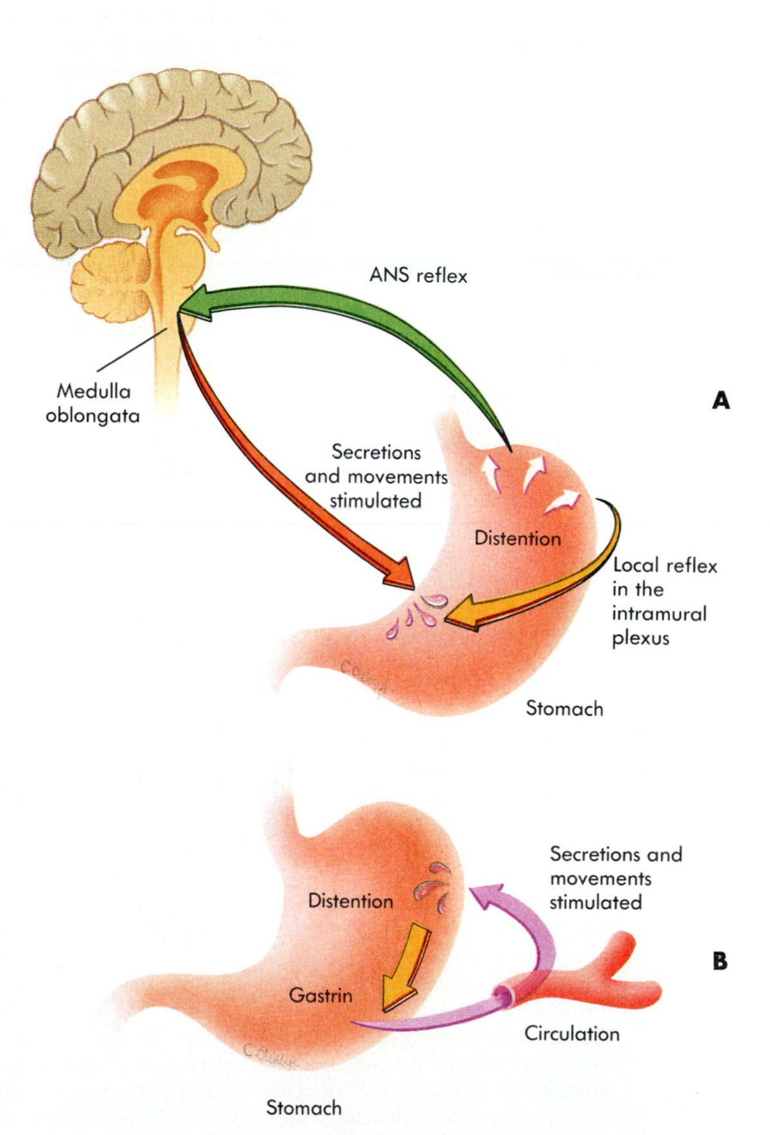

Figure 16-10 ● **Regulation of the digestive tract.**
A, Distention of the stomach stimulates an autonomic reflex that is mediated through the medulla oblongata. Distention also stimulates a local reflex that is mediated through the intramural reflex. The autonomic and local reflexes increase gastric secretions and movements. **B,** Distention of the stomach results in the secretion of gastrin by stomach cells. Gastrin is carried by the circulation back to the stomach where it stimulates gastric juice secretions and movements.

Disorders of the digestive tract

ORAL CAVITY

Formation of **dental caries** (kār-ez), or tooth decay, results from bacteria on the tooth surface. The bacteria convert sucrose and other carbohydrates into acid, which dissolves the tooth enamel. Enamel is nonliving and cannot repair itself. Consequently a dental filling is necessary to prevent further damage. If the caries is not treated, bacteria can enter the dentin and pulp and kill the tooth. A root canal procedure removes the dead pulp.

Periodontal (pĕr´e-o-don´tal) **disease** is an inflammation and degeneration of the periodontal ligaments, gingiva, and bone. This disease is the most common cause of tooth loss in adults.

STOMACH

A **hiatial hernia** (hi-a´tal her-ne´-ah) results when a portion of the stomach protrudes through the diaphragm. There is an opening, or hiatus (hi-a´tus), in the diaphragm through which the esophagus passes to join the stomach. If the opening weakens or becomes abnormally large, part of the stomach can protrude, or herniate, through the diaphragm into the thoracic cavity. Symptoms associated with hiatial hernia can include heartburn, difficulty in swallowing, and ulcer formation.

Vomiting is a reflex resulting from irritation, overdistention, or overexcitation of the stomach or other parts of the digestive tract. Action potentials travel from these areas through nerves to the vomiting center in the medulla oblongata. After the vomiting center is stimulated the following events occur:

1. A deep breath is taken.
2. The hyoid bone and larynx are elevated, opening the upper esophagus.
3. The opening of the larynx is closed.
4. The soft palate is elevated, closing the opening to the nasal cavity.
5. The diaphragm and abdominal muscles are forcefully contracted, strongly compressing the stomach and increasing the pressure within the stomach.
6. The esophageal sphincter is relaxed, and the gastric contents are forcefully expelled.

ULCERS

Peptic (pep´tik) **ulcer** is a condition in which the mucosal lining of the GI tract is damaged and inflamed. The most common site of a peptic ulcer is near the pylorus, usually on the duodenal side (that is, a duodenal ulcer). Ulcers occur less frequently along the lesser curvature of the stomach or at the point where the esophagus enters the stomach (a gastric ulcer).

Most ulcers are probably the result of infection by the bacteria *Helicobacter pylori*. Treatment with antibiotics is effective in eliminating the bacteria and cures the ulcer in approximately 90% of patients. Overproduction of stomach acid resulting from stress or certain drugs such as aspirin also can damage the lining of the digestive tract.

SMALL INTESTINE

Maldigestion is a failure of the chemical process of digestion. As a result, inadequate amounts of nutrients are available for absorption. Common causes of maldigestion are surgical removal of the stomach, diseases that interfere with pancreatic and liver functions, and lactase deficiency. Some individuals, especially blacks, lose the ability to produce the enzyme lactase as adults. Consequently, they cannot digest lactose, the sugar in milk. Lactose and water accumulate in the digestive tract causing diarrhea and abdominal cramps.

Malabsorption is a spectrum of disorders of the small intestine that result in abnormal nutrient absorption. In some people, one type of malabsorption results from the effects of the protein gluten present in oats, rye, barley, and especially wheat. The reaction to gluten can destroy epithelial cells, causing the villi to become blunted and the intestinal surface area to decrease. As a result, the intestinal epithelium is less capable of absorbing nutrients. Another type of malabsorption is apparently caused by bacteria, although no specific bacterium has been identified.

LARGE INTESTINE

Cancer of the digestive tract occurs most commonly in the colon and rectum, being the third most common cause of cancer death in the United States. After age 50, people typically begin to develop polyps (pol´ipz) or localized epithelial growths in the colon and rectum. Usually the polyps are noncancerous. However, over many years cancer can develop in some of the polyps, with most cases occurring in 60 to 70 year olds. Because the polyps sometimes bleed, testing for blood in the feces is a screening test for cancer. A positive test for blood, however, does not mean one has cancer, because there are many causes for blood in the feces, including bleeding from noncancerous polyps. **Colonoscopy** is a procedure that uses a tube through which a doctor can examine the colon and remove the polyps for further testing. In advanced cancers, surgery may be necessary to remove the diseased section of the colon.

Appendicitis is an inflammation of the appendix and usually occurs because of obstruction of the appendix. Secretions from the appendix cannot pass the obstruction and therefore accumulate, causing enlargement and pain. Bacteria in the area cause the appendix to become infected. If the appendix bursts, the infection can spread throughout the peritoneal cavity with life-threatening results. The right inferior quadrant of the abdomen becomes very tender in people with acute appendicitis as a result of pain referred from the inflamed appendix to the body surface.

Constipation is the slow movement of feces through the large intestine. The feces often become dry and hard because of the increased fluid absorption during the extended time they are retained in the large intestine. Constipation often results from irregular defecation patterns that develop after a prolonged time of inhibiting normal defecation reflexes. Spasms of the sigmoid colon resulting from irritation also can result in slow feces movement and constipation.

Diarrhea is an abnormally frequent, watery bowel movement. When the large intestine is irritated and inflamed, such as in patients with enteritis (bacterial infection of the bowel), the intestinal mucosa secretes large amounts of water and electrolytes in addition to mucus. Although diarrhea increases fluid and electrolyte loss, it also moves the infected feces out of the intestine more rapidly and speeds recovery from the disease. Severe cases of diarrhea, however, can result in dehydration and death.

Dysentery (dis´en-tĕr-e) is a severe form of diarrhea in which blood or mucus is present in the feces. Dysentery can be caused by bacteria or protozoa, such as amoebae.

Hemorrhoids (hem´o-roydz) are enlarged or inflamed veins located in the wall of the anal canal.

LIVER

Hepatitis (hep´ă-ti´tis) is an inflammation of the liver that may result from alcohol consumption or viral infection. Chronic intake of large amounts of alcohol causes an accumulation of fat within liver cells as well as other changes that can lead to the death of liver cells. Inflammation occurs in response to the tissue destruction and, as the cells die, the liver is converted into a mass of scar tissue, a condition called **cirrhosis** (sir-ro´sis) of the liver. Because of the loss of liver cells, liver function is impaired and death as a result of liver failure can occur.

Viral hepatitis is the second most frequently reported infectious disease in the United States. Hepatitis A is usually transmitted by poor sanitation practices or from mollusks, such as oysters, living in contaminated waters. Hepatitis B and C are usually transmitted through blood or other body fluids. Hepatitis C is responsible for most cases of hepatitis that develop following transfusions. Symptoms include nausea, diarrhea, loss of appetite, abdominal pain, fever, and chills. Jaundice is seen in about two thirds of the cases, with yellowing of the skin and sclera of the eyes resulting from the accumulation of bile pigments in those tissues.

Cholesterol, secreted by the liver into the bile, may precipitate in the gallbladder to produce **gallstones.** Occasionally a gallstone may pass out of the gallbladder and enter the cystic duct, blocking release of the bile. An accumulation of gallstones or blockage of the cystic duct can cause the gallbladder to become distended, inflamed, and painful. Sometimes the gallbladder must be removed surgically.

INFECTIONS OF THE DIGESTIVE TRACT

Staphylococcal (staf´ĭ-lo-kok´al) **food poisoning** occurs when toxin from the bacteria *Staphylococcus aureus* is ingested. The bacteria usually come from the hands of a person preparing the food. If food is cooked at low temperatures (below 60°C [140°F]) or is allowed to sit for an extended period, the bacteria can reproduce and form toxins. Reheating can eliminate the bacteria but not the toxins. Staphylococcal food poisoning is characterized by nausea, vomiting, and diarrhea from 1 to 6 hours after the contaminated food is ingested.

Salmonellosis (sal´mo-nel-o´sis) is a disease caused by *Salmonella* bacteria. They are ingested with contaminated food (usually meat, poultry, or milk) and grow in the digestive tract. The disease symptoms may not be seen for up to 36 hours after the contaminated food has been consumed. Symptoms include nausea, fever, abdominal pain, and diarrhea. The bacteria are generally destroyed by normal cooking with temperatures greater than 68°C (155°F).

Typhoid (ti´foyd) **fever** is caused by a particularly virulent strain of *Salmonella* bacteria. The bacteria can cross the intestinal wall and invade other tissues. The incubation period is normally about 2 weeks. Symptoms include severe fever and headaches, as well as diarrhea. Poor sanitation practices are the main source of contamination, and typhoid fever is still a leading cause of death in many underdeveloped countries.

Cholera (kol´er-ah) is caused by a bacterium that infects the small intestine. The bacteria produce a toxin that stimulates the secretion of chlorides, bicarbonates, and water from the intestinal tract. The loss of fluid and electrolytes (as much as 12 to 20 liters of fluid loss per day) causes shock, collapse, and even death. Cholera was common in the United States and Europe in the 1800s but is not common in western countries today. Cholera is still a major problem in Asia, however, particularly India.

Giardiasis (je´ar-di´ă-sis) is a disease caused by a protozoan that invades the intestine. Symptoms include nausea, abdominal cramps, weakness, weight loss, and malaise, and may last for several weeks. The disease is carried by humans and wild animals, especially beaver, and commonly affects persons who drink unfiltered water from wilderness streams.

Intestinal parasites are not uncommon in humans, especially under conditions of poor sanitation. **Tapeworms** can infect the digestive tract by way of undercooked beef, pork, or fish. The tapeworms attach to the intestinal wall by suckers and may live in the intestine for 25 years, reaching lengths of 6 meters. There are few symptoms beyond a vague abdominal discomfort. **Pinworms** are common in humans. The tiny worm lives in the digestive tract but migrates out of the anus to lay its eggs. This causes a local itching, and the eggs can be spread by contaminated fingers to numerous surfaces. Eggs resist dehydration and can be picked up from contaminated surfaces by other people. It is common for entire households to be contaminated if one person contracts the disease.

Hookworms attach to the intestinal wall and feed on the blood and tissue of the host, rather than on partially digested food as other parasites do. Infection can cause anemia and lethargy. Because hookworms are spread through fecal contamination of the soil and bare skin contact with contaminated soil, improved sanitation and the practice of wearing shoes has greatly decreased the incidence of hookworm infection.

Ascariasis (as´kă-ri´ă-sis) is caused by a roundworm and is fairly common in the United States. Ingested eggs hatch in the upper intestine into wormlike larvae that pass into the bloodstream and then into the lungs, where they may cause pulmonary symptoms. Extremely large numbers may cause pneumonia. The larvae enter the throat and are swallowed, whereby they return to the intestinal tract. Adults in the intestinal tract cause few symptoms.

Understanding defecation

The large intestine functions to remove water from chyme and convert the chyme into feces. The feces is stored until it can be eliminated from the body, a process that is called **defecation** (def-ĕ-ka´shun), or a **bowel movement**. Defecation begins with the movement of feces into the rectum. Movement of feces into the rectum has a voluntary and involuntary component. Voluntary actions include a large inspiration of air followed by closure of the larynx and forceful contraction of the abdominal muscles. As a consequence, the pressure in the abdominal cavity increases and forces the contents of the colon into the rectum.

The involuntary movement of feces into the rectum results from **mass movements,** which are strong peristaltic contractions of the transverse colon and descending colon that propel their contents toward the rectum. The peristaltic contractions of mass movements are integrated by local reflexes in the intramural plexus of the digestive tract. Stretch of the stomach results in action potentials that travel the length of the digestive tract to the colon in which smooth muscles are stimulated to contract (Figure 16-A). The stretch of the stomach is a signal for the colon to move its contents, which results in room for the new material that has been ingested. Mass movements are most common approximately 15 minutes after breakfast. They usually persist for 10 to 30 minutes and then stop for perhaps half a day.

Distention of the rectal wall by feces acts as a stimulus that initiates the **defecation reflex,** which has local and parasympathetic reflex components (see Figure 16-A). Smooth muscle in the descending colon, sigmoid colon, and rectum contracts. As feces is pushed toward the anus, the internal and external anal sphincters relax. The external anal sphincter, however, is composed of skeletal muscle and is under reflex and voluntary control. At the same time the defecation reflex is activated, action potentials travel along nerve tracts to the brain and initiate the desire to defecate. Voluntary control of the external anal sphincter can block the defecation reflex and thus prevent defecation. At the appropriate time and place, however, the external anal sphincter is voluntarily relaxed and the feces are expelled. Control of the external anal sphincter must be learned by children and can be lost as a result of disease, injury to the nervous system, or old age.

The defecation reflex persists for only a few minutes and quickly disappears. Generally the reflex is reinitiated after a period of time that may be as long as several hours. Mass movements of the colon are usually the reason for the reinitiation of the defecation reflex.

2 An enema is an injection of approximately 1 pint of fluid into the rectum. Explain how an enema stimulates defecation.

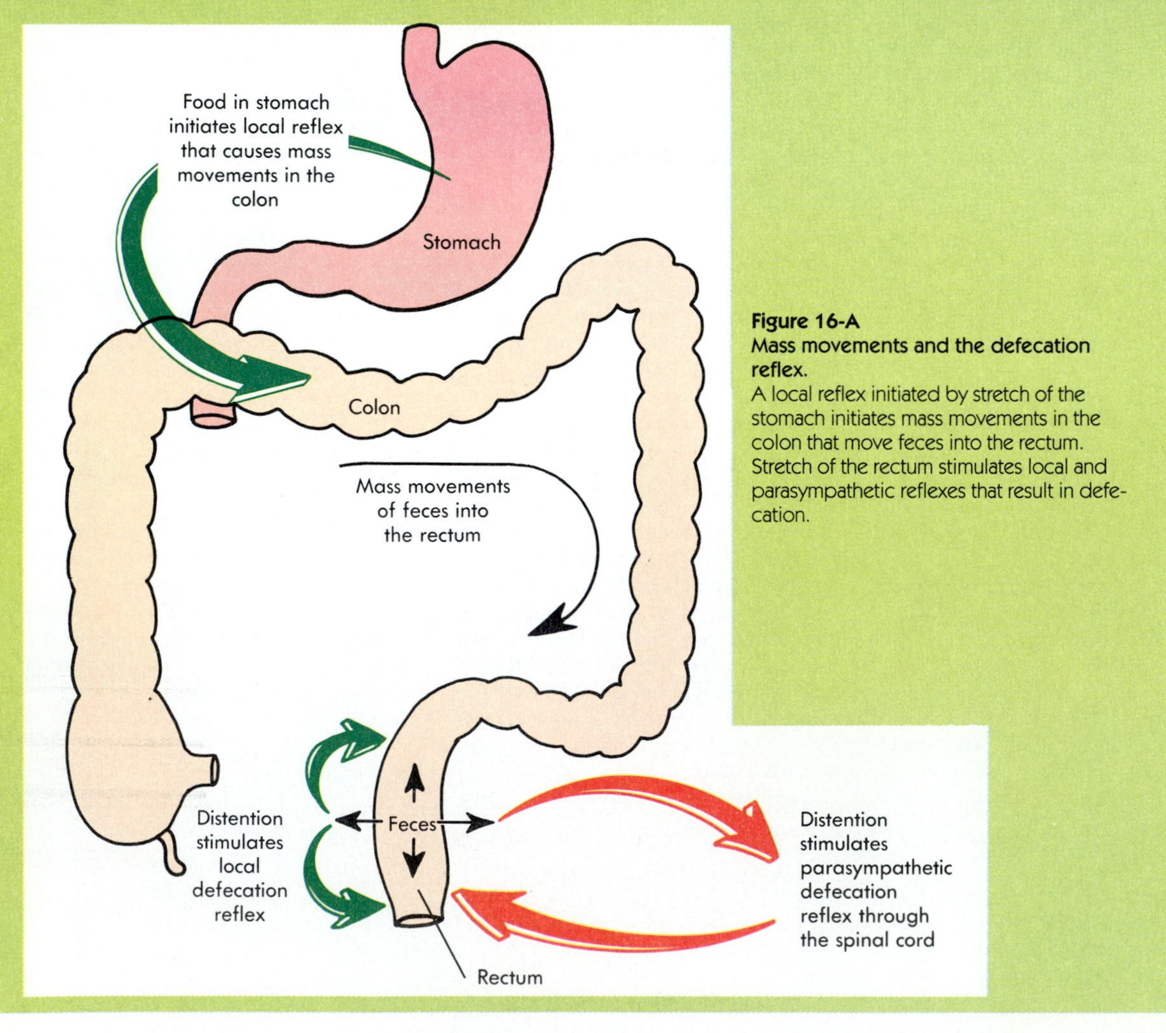

Figure 16-A
Mass movements and the defecation reflex.
A local reflex initiated by stretch of the stomach initiates mass movements in the colon that move feces into the rectum. Stretch of the rectum stimulates local and parasympathetic reflexes that result in defecation.

into the stomach activates the processes that result in the digestion of the food.

Local reflexes are a unique regulatory mechanism found in the digestive tract. Local reflexes are mediated by neurons in the intramural plexus and do not involve the central nervous system. Stimuli such as distention activate stretch receptors within the wall of the digestive tract. Action potentials produced in the neurons of the intramural plexus travel up or down the intramural plexus and produce a response in glands or smooth muscles of the digestive tract. For example, stretch of the stomach wall initiates a local reflex that increases gastric juice secretion and movements (see Figure 16-10, *A*).

Within the epithelium of the stomach and the small intestine are specialized endocrine cells. The hormones produced by these cells can produce a response in glands or smooth muscle of the digestive tract or its accessory organs. For example, the entry of food into the stomach stimulates the release of **gastrin** (gas´trin), which increases secretions and smooth muscle contractions in the stomach (Figure 16-10, *B*).

3 The entry of large amounts of chyme into the small intestine stimulates the release of the hormone secretin (se-kre´tin) from the small intestine. Secretin inhibits gastric secretions and movements, but increases secretions from the pancreas and liver. How are these responses beneficial? (Hint: proper digestion requires that movement of materials be coordinated with the secretion of digestive enzymes.)

DIGESTION AND ABSORPTION

Digestion is the chemical breakdown of organic molecules into smaller component parts that can be absorbed. Digestive enzymes break chemical bonds within organic molecules, digesting carbohydrates into monosaccharides, proteins into amino acids, and fats into fatty acids and glycerol (Table 16-2).

Table 16-2 Functions of digestive secretions

Substance	Source	Function
Mouth		
Saliva	Salivary glands	Moistens and lubricates food
Salivary amylase	Salivary glands	Enzyme that digests starch (polysaccharide)
Stomach		
Pepsin	Gastric glands	Enzyme that digests proteins
Hydrochloric acid	Gastric glands	Provides the proper pH environment for pepsin, kills bacteria
Mucus	Stomach lining and gastric glands	Protects stomach lining from pepsin and hydrochloric acid
Intrinsic factor	Gastric glands	Increases the absorption of vitamin B_{12}
Gastrin	Gastric glands	Hormone that increases gastric secretions and movements
Small intestine and associated organs		
Bile salts	Liver	Emulsifies fats
Bicarbonate ions	Liver and pancreas	Neutralizes stomach acid, proves the proper pH environment for pancreatic and intestinal enzymes
Trypsin	Pancreas	Enzyme that digests proteins
Pancreatic amylase	Pancreas	Enzyme that digests starches (polysaccharides)
Lipase	Pancreas	Enzyme that digests lipids
Mucus	Small intestine	Protects small intestine lining from stomach acid and digestive enzymes
Peptidase	Small intestine	Enzyme that digests proteins
Lipase	Small intestine	Enzyme that digests lipids
Sucrase	Small intestine	Enzyme that digests sucrose (table sugar)
Lactase	Small intestine	Enzyme that digests lactose (milk sugar)
Maltase	Small intestine	Digests maltose (malt sugar)
Secretin	Small intestine	Hormone that inhibits gastric secretions, stimulates pancreatic secretions and bile secretion
Large intestine		
Mucus	Large intestine	Protects stomach lining from pepsin and hydrochloric acid

Absorption is the movement of digested and other materials from the digestive tract into the circulatory or lymphatic system. Absorption begins in the stomach, from which some molecules such as alcohol and aspirin can pass through the stomach epithelium into the circulation. Most absorption, however, occurs in the duodenum and jejunum, although some absorption occurs in the ileum. Some molecules diffuse through the intestinal wall, whereas other molecules are moved by active transport or vesicles (see Chapter 3).

Carbohydrates

Ingested carbohydrates include **polysaccharides** (many sugar molecules joined together), **disaccharides** (two sugar molecules joined together), and **monosaccharides** (one sugar molecule) (see Chapter 2). Examples of polysaccharides are starches and glycogen. Examples of disaccharides are sucrose (table sugar) and lactose (milk sugar), and examples of monosaccharides are glucose and fructose (fruit sugar).

Polysaccharides are broken down by enzymes into disaccharides. **Salivary amylase** (am´ĭ-lās) begins the digestion of polysaccharides in the mouth. The carbohydrates then pass to the stomach, where almost no carbohydrate digestion occurs. In the duodenum, **pancreatic amylase** continues the digestion of the polysaccharides. Disaccharides are broken down into monosaccharides by enzymes that are bound to the microvilli of the intestinal epithelium. Monosaccharides are taken up by the intestinal epithelial cells and are carried by the hepatic portal system to the liver, where different types of monosaccharides are converted to glucose. Glucose is carried by the circulation to cells that take up the glucose and use it as a source of energy (see Chapter 17).

Lipids

Lipids include triglycerides, phospholipids, steroids, and fat-soluble vitamins (see Chapter 2). **Triglycerides** (tri-glis´er-īdz), the most common type of lipid, consist of three fatty acids bound to glycerol. The first step in lipid digestion is **emulsification** (e-mul´sĭ-fi-ka´shun), which is the transformation of large lipid droplets into much smaller droplets. The enzymes that digest lipids are soluble in water and can digest the lipids only by acting at the surface of the droplets. The emulsification process increases the surface area of the lipid exposed to the digestive enzymes by decreasing the droplet size. Emulsification is accomplished by **bile salts** secreted by the liver.

Lipase (li´pās), secreted by the pancreas and intestine, digests lipid molecules. The primary products of this digestive process are fatty acids and glycerol. In the intestine, bile salts aggregate around small droplets of digested lipids to form **micelles** (mi-selz). When a micelle comes into contact with the epithelial cells of the small intestine, the micelle's contents pass through the cell membrane of the epithelial cells by means of simple diffusion.

Within the epithelial cells of the intestinal villi, the digested lipids are packaged inside a protein coat. The packaged lipids leave the epithelial cells and enter the lacteals in the center of the villi. The lacteals are part of the lymphatic system, and lipid-rich lymph is called **chyle** (kil). The lymphatic system carries the lipids to the bloodstream (see Chapter 14). Packaged lipids in the blood may be transported to adipose tissue, where they are stored until an energy source is needed elsewhere in the body. Lipids are also transported by the hepatic portal system to the liver, where they are stored, converted into other molecules, or used as energy.

Cholesterol levels in the blood are of great concern to many older adults. Cholesterol levels of less than 200 mg/dl (milligrams per deciliter) are considered low, which is good. Cholesterol levels of 200 to 239 mg/dl are considered borderline if there are no other risk factors, such as coronary artery disease, cigarette smoking, hypertension, diabetes mellitus, or greater than 30% overweight. Cholesterol levels of 200 to 239 mg/dl are considered high if one or more of these other risk factors are present. Cholesterol levels of greater than 239 mg/dl are considered high in anyone. People with high blood cholesterol levels run a much greater risk of heart disease and stroke than people with low cholesterol levels. People with high levels should reduce intake of foods rich in cholesterol and other fats.

Fats are not soluble in water so they are transported in the blood as lipid-protein complexes, or lipoproteins. For example, **low-density lipoproteins (LDLs)** carry cholesterol to the tissues for use by the cells. When LDLs are in excess, cholesterol is deposited in arterial walls. **High-density lipoproteins (HDLs),** on the other hand, transport cholesterol from the tissues to the liver where cholesterol is removed from the bloodstream, and broken down or excreted in bile. A high HDL/LDL ratio in the bloodstream is related to a lower risk of heart disease. Aerobic exercise is one way to elevate the level of HDL.

Proteins

Pepsin (pep´sin) is an enzyme secreted by the stomach that breaks down proteins, producing smaller polypeptide chains. Approximately 10% to 20% of the total ingested protein is digested by pepsin. After the remaining proteins and polypeptide chains leave the stomach and enter the small intestine, the enzyme **trypsin** (trip´sin), produced by the pancreas, continues the digestive process. Trypsin produces short amino acid chains that are broken down further into amino acids by digestive enzymes bound to the microvilli of the small intestine.

Absorption of individual amino acids occurs through the intestinal epithelial cells by active transport. The amino acids then leave the epithelial cells and enter the hepatic portal system, which transports them to the liver. The amino acids may be modified in the liver, or they may be released into the bloodstream and distributed throughout the body.

Amino acids are actively transported into the various cells of the body. Most amino acids are used as building blocks to form new proteins, but some amino acids may be used for energy. The body cannot store amino acids, so they are partially broken down to their component molecules. Those molecules are used to synthesize glycogen or fat, which can be stored.

4 Achlorhydria (a-klor-hi´dre-ah) is a condition in which the stomach stops producing hydrochloric acid and other secretions. What effect would achlorhydria have on the digestive process?

SUMMARY

ANATOMY AND HISTOLOGY OF THE DIGESTIVE SYSTEM

The gastrointestinal tract is composed of four layers: mucosa, submucosa, muscularis, and serosa.

Oral cavity

The lips, cheeks, and tongue are involved in chewing and speech. The tongue also is involved in taste and swallowing.

Each tooth consists of a crown, neck, and root. There are 32 permanent teeth, including incisors, canines, premolars, and molars.

The roof of the oral cavity is divided into the hard and soft palates.

Salivary glands produce mucous and serous secretions. The three pairs of large salivary glands are the parotid, submandibular, and sublingual glands.

Pharynx and esophagus

Swallowing consists of voluntary movement of a bolus into the pharynx and a reflex that moves the bolus through the pharynx and esophagus.

Stomach

The stomach consists of the fundus, body, and pylorus.

The wall of the stomach consists of three muscle layers: longitudinal, circular, and oblique.

Gastric glands produce mucus, hydrochloric acid, pepsin, and intrinsic factor.

Small intestine

The small intestine is divided into the duodenum, jejunum, and ileum.

Circular folds, villi, and microvilli greatly increase the surface area of the intestinal lining.

Large intestine

The cecum forms a blind sac at the junction of the small and large intestines. The appendix is a blind sac off the cecum.

The colon consists of ascending, transverse, descending, and sigmoid portions.

The rectum is a straight tube that ends at the anal canal.

The anal canal is surrounded by an internal anal sphincter (smooth muscle) and an external anal sphincter (skeletal muscle).

Liver and gallbladder

The hepatic artery brings oxygenated blood to the liver. The hepatic portal vein brings deoxygenated but nutrient-rich blood to the liver.

The cystic duct from the gallbladder joins the common hepatic duct from the liver to form the common bile duct.

The common bile duct joins the pancreatic duct and empties into the duodenum.

Pancreas

The endocrine function of the pancreas is to control blood nutrient levels.

The exocrine function of the pancreas is to produce digestive enzymes and bicarbonate ions.

Peritoneum

The peritoneum is a serous membrane that lines the abdominal cavity and organs.

Mesenteries are peritoneum that extend from the body wall to many of the abdominal organs.

Retroperitoneal organs are located behind the parietal peritoneum.

REGULATION OF THE DIGESTIVE SYSTEM

Autonomic reflexes are controlled through the medulla oblongata.

Local reflexes are mediated through the intramural plexus.

Hormones produced by the digestive tract regulate the digestive tract and accessory organs.

DIGESTION AND ABSORPTION

Digestion is the chemical breakdown of organic molecules into their component parts.

Absorption of digested molecules occurs mainly in the duodenum and jejunum.

Carbohydrates

Polysaccharides are split into disaccharides by salivary and pancreatic amylases.

Disaccharides are broken down to monosaccharides by enzymes on the surface of the intestinal epithelium.

Monosaccharides are absorbed in the blood and are carried to the liver in which they are converted to glucose.

Glucose is carried by the blood to cells.

Lipids

Bile salts emulsify lipids.

Pancreatic lipase breaks down lipids. The breakdown products aggregate with bile salts to form micelles.

Micelles come into contact with the intestinal epithelium, and their contents diffuse into the cells, where they are packaged and released into the lymph and blood.

Lipids are stored in adipose tissue and in the liver, which release the lipids into the blood when energy sources are needed elsewhere in the body.

Proteins

Proteins are split into smaller polypeptides by pepsin in the stomach.

Trypsin from the pancreas splits the polypeptides from the stomach into short amino acid chains.

Enzymes on the surface of intestinal epithelial cells complete the digestive process.

Amino acids are absorbed into the blood and transported to the liver.

Amino acids are used to build new proteins or are used as a source of energy.

CONTENT REVIEW

1. What are the functions of the digestive system?
2. What are the major layers of the digestive tract? What is accomplished by smooth muscle in the digestive tract? What is the intramural plexus?
3. List the functions of the lips, cheeks, and tongue.
4. Describe the parts of a tooth. What are dentin, enamel, and pulp?
5. What are the deciduous and permanent teeth? Name the different kinds of teeth and state their functions.
6. What are the hard and soft palates? What is the function of the palate?
7. Name and give the location of the three largest salivary glands. What is the function of saliva?
8. Describe the events that occur during swallowing.
9. Describe the parts of the stomach. What is chyme?
10. Name and describe the three parts of the small intestine.
11. What are circular folds, villi, and microvilli in the small intestine? What are their functions?
12. Describe the parts of the large intestine. How is chyme converted into feces? Define defecation.
13. Describe the anatomy, location, and functions of the liver and gallbladder. Describe their duct system.
14. Name the two types of tissue in the pancreas and give their functions.
15. What are the peritoneum, mesenteries, and retroperitoneal organs?
16. Describe the three ways in which the digestive system is regulated.
17. State the function of each of the enzymes involved in carbohydrate digestion. Give the location where each enzyme is produced.
18. What enzyme is responsible for lipid digestion? Name two places the enzyme is produced.
19. Describe the role of bile salts in lipid digestion and absorption.
20. State the function of each of the enzymes involved in protein digestion. Give the location where each enzyme is produced.

CONCEPT REVIEW

1. While anesthetized, patients sometimes vomit. Given that the anesthetic eliminates the swallowing reflex, explain why vomiting when anesthetized can be dangerous.
2. Suppose a bolus of food has been swallowed but has not completely moved through the esophagus. What will happen if stretching of the esophagus activates a local reflex that sends action potentials to smooth muscle above the point of distention?
3. Sometimes a gallstone can move into the pancreatic duct and block or impair the flow of pancreatic juices. What symptoms would you expect if this blockage occurred? (Hint: what are the normal functions of pancreatic secretions?)
4. As a result of an automobile accident, the spinal cord of a man is completely severed above the sacral region of the spinal cord. What effect would this have on the ability of the man to control and have bowel movements?

CHAPTER TEST

Answers can be found in Appendix C

MATCHING 1 For each statement in column A select the correct answer in column B (an answer may be used once, more than once, or not at all).

A	B
1. The innermost layer of the digestive tract.	chyme
2. The layer of digestive tract smooth muscle.	dentin
3. Wavelike contractions of the digestive tract.	enamel
4. Hard substance covering the crown of a tooth.	gallbladder
5. Hold teeth in place.	hepatic artery
6. Large folds in an empty stomach.	hepatic portal vein
7. Semifluid mixture from the stomach that enters the duodenum.	hepatic veins
8. Brings nutrients to the liver from the digestive tract.	liver
9. Stores bile.	mucosa
10. Abdominal organs that do not have mesenteries.	muscularis
	periodontal ligaments
	peristalsis
	retroperitoneal
	rugae
	serosa

MATCHING 2 For each statement in column A select the correct answer in column B (an answer may be used once, more than once, or not at all).

A	B
1. Digests starch.	amylase
2. Digests disaccharides such as sucrose.	bile salts
3. Found in saliva and pancreatic juices.	intestinal enzymes
4. Digests lipids.	lipase
5. Emulsifies lipids.	pepsin
6. Stomach enzyme that digests proteins.	trypsin
7. Pancreatic enzyme that digests proteins.	
8. Breaks down short amino acid chains to amino acids.	

FILL-IN-THE BLANK Complete each statement by providing the missing word or words.

1. The types of teeth that function primarily for cutting and tearing of food are ___________ and ___________.
2. There are ___________ (give the number) permanent teeth, which are replacements for or additions to the 20 ___________ (give the type) teeth.
3. The three pairs of salivary glands are the ___________, ___________, and ___________.
4. The ___________ regulates the movement of materials from the stomach to the small intestine.
5. The modifications that increase surface area in the duodenum are the ___________, ___________, and ___________.
6. The ___________ and ___________ regulate the movement of materials between the small intestine and the large intestine.
7. The ___________ duct joins the cystic duct to form the ___________.
8. The digestive system is regulated by ___________ reflexes through the medulla oblongata, ___________ reflexes through the intramural plexus, and ___________ released from the digestive tract.

CHAPTER

17 Nutrition, Metabolism, and Body Temperature

OBJECTIVES

After reading this chapter you should be able to:

1. Define metabolism, anabolism, catabolism, and nutrition.
2. Define a Calorie and list the Calories found in a gram of carbohydrate, lipid, and protein.
3. Describe the dietary sources of carbohydrates, lipids, and proteins and their uses in the body.
4. List several common vitamins and minerals and indicate the function of each.
5. Describe the production of ATP molecules in anaerobic and aerobic respiration.
6. Explain how fats and proteins can be used to produce ATP.
7. Define metabolic rate and describe the uses of metabolic energy.
8. List the ways in which heat is exchanged between the body and the environment.
9. Describe heat production and regulation in the body.

KEY TERMS

aerobic respiration
Breakdown of glucose in the presence of oxygen to produce 6 carbon dioxide, 6 water, and 38 ATP molecules.

anaerobic respiration
Breakdown of glucose in the absence of oxygen to produce two lactic acid and two ATP molecules.

ATP
Energy-storing molecule in cells that provides energy necessary for many chemical reactions.

carbohydrate
Sugar or a molecule formed by chemical-combining sugars.

lipid
Organic substance used for longer-term energy storage; examples include fats, oils, and cholesterol.

metabolic rate
The total amount of energy produced and used by the body per unit of time.

metabolism
Sum of the chemical changes that occur in tissues, consisting of the breakdown of food to produce energy (catabolism) and the buildup of molecules (anabolism).

mineral
Inorganic nutrient necessary for normal metabolic functions.

protein
Organic molecule consisting of many amino acids linked together; source of amino acids (for building new proteins) and energy.

vitamin
One of a group of organic substances, present in small amounts in natural foods, that are essential to normal metabolism.

FEATURES

NUTRITION is the process by which food items (nutrients) are obtained and used by the body. The process includes digestion, absorption, and transport, which are covered in Chapter 16, and cell metabolism, which is a topic of this chapter. Nutrition also can be defined as the evaluation of food and drink requirements for normal body function.

Metabolism (mĕ-tab´o-lizm) is the total of all the chemical changes that occur in the body. It consists of **anabolism** (ah-nab´o-lizm), the energy-requiring process by which small molecules are joined to form larger molecules, and **catabolism** (kah-tab´o-lizm), the energy-releasing process by which large molecules are broken down into smaller molecules. Anabolism occurs in all cells of the body as they divide to form new cells, maintain their own intracellular structure, and produce molecules such as hormones, neurotransmitters, or extracellular matrix molecules for export. Catabolism begins during the process of digestion and is concluded within individual cells in which the energy released by the breaking of covalent bonds is used to produce adenosine triphosphate (ATP) and heat.

The heat resulting from the chemical reactions of metabolism contributes to the maintenance of body temperature within a narrow range, which is essential for normal function. In addition, heat can be gained from or lost to the external environment. A constant body temperature can be maintained by regulating the internal production of heat and the exchange of heat with the external environment.

NUTRITION

Nutrients

Nutrients are the chemicals taken into the body that provide energy and building blocks for new molecules. Some substances in food are not nutrients but provide bulk (fiber) in the diet. Nutrients can be divided into six major classes: carbohydrates, lipids, proteins, vitamins, minerals, and water. Carbohydrates, proteins, and lipids are the major organic nutrients and are broken down by enzymes into their individual subunits during digestion. Subsequently, many of these subunits are broken down further to supply energy. Others are used as building blocks for other molecules. Vitamins, minerals, and water are taken into the body without being digested. They participate in the chemical reactions necessary to maintain life. Some nutrients are required in fairly substantial quantities, and others are required in small amounts (trace elements).

Essential nutrients are nutrients that must be ingested because the body cannot manufacture them or is unable to manufacture adequate amounts of them. The essential nutrients include certain amino acids, linoleic acid (a fatty acid), most vitamins, minerals, water, and a minimal amount of carbohydrate. The term *essential* does not mean, however, that only the essential nutrients are required by the body. Other nutrients are necessary, but, if they are not ingested, they can be synthesized from the essential nutrients. Most of this synthesis takes place in the liver, which has a remarkable ability to transform and manufacture molecules.

Calories

The energy available in foods and released through metabolism is expressed as a measure of heat. A **Calorie** (kal´o-re; Cal) is the amount of heat (energy) required to raise the temperature of 1000 g of water from 14°C to 15°C. The Calories in a serving of food are commonly listed on food packages. For example, one slice of white bread contains approximately 75 Cal, one cup of whole milk contains 150 Cal, a banana contains 100 Cal, a McDonald's Big Mac has 563 Cal, and a soft drink has 145 Cal. For each gram of carbohydrate or protein metabolized by the body, approximately 4 Cal of energy is released. Fats contain more energy per unit of weight than carbohydrates and proteins, and yield approximately 9 Cal per gram.

Carbohydrates

Sources in the diet

Carbohydrates include monosaccharides, disaccharides, and polysaccharides (see Chapter 2). Most of the carbohydrates we ingest come from plants. An exception is lactose (milk sugar), which is found in animal and human milk.

The most common monosaccharides in the diet are glucose and fructose. Plants capture energy in sunlight and use the energy to produce glucose, which can be found in vegetables, fruits, molasses, honey, and syrup. Fructose (fruit sugar) is most often derived from fruits and berries.

The disaccharide sucrose (table sugar) is what most people think of when they use the term *sugar*. Sucrose is a glucose and fructose molecule joined together, and its principal sources are sugar cane and sugar beets. Maltose (malt sugar), derived from germinating cereals, is a combination of two glucose molecules, and lactose (in milk) consists of a glucose molecule and a galactose molecule.

The **complex carbohydrates** are the large polysaccharides starch, glycogen, and cellulose. Starch is an energy storage molecule in plants and is found primarily in vegetables, fruits, and grains. Glycogen is an energy storage molecule in animals and is located in muscle and in the liver. Cellulose forms the cell wall, or outer covering, of plant cells.

Uses in the body

During digestion, polysaccharides and disaccharides are split into monosaccharides that are absorbed into the blood (see Chapter 16). Humans have enzymes that break the bonds between the glucose molecules of starch and glycogen, but do not have enzymes that digest cellulose. Consequently it is important to thoroughly cook or chew plant matter. Breaking down the cell wall exposes the starches inside the cells to enzymes that can digest the starches. The undigested cellulose provides fiber, or "roughage," which increases the bulk of feces and promotes defecation.

Fructose and other monosaccharides absorbed into the blood are converted into glucose by the liver. **Glucose,** whether absorbed from the digestive tract or produced by the liver, is a primary energy source for most cells, which use it to produce ATP molecules (see Cell Metabolism in this chapter). Because the brain relies almost entirely on glucose for its energy, blood glucose levels are carefully regulated (see Chapter 10).

If excess amounts of glucose are present, the glucose is converted into glycogen that is stored in muscle and in the liver. The glycogen can be rapidly converted back to glucose when energy is needed. Because cells can store only a limited amount of glycogen, any additional glucose is converted into fat that is stored in adipose tissue.

In addition to being used as a source of energy, sugars have other functions. They form part of deoxyribonucleic acid (DNA), ribonucleic acid (RNA), and ATP molecules, and they combine with proteins to form glycoprotein receptor molecules on the outer surface of the plasma membrane.

Recommended requirements

It is recommended that 125 to 175 g of carbohydrates be ingested every day. Although a minimum level of carbohydrates is not known, it is assumed that amounts of 100 g or less per day result in over-use of proteins and fats for energy sources. Because muscles are primarily protein, the use of proteins for energy can result in the breakdown of muscle tissue, and the use of fats can result in acidosis (see Chapter 18).

Complex carbohydrates are recommended because starchy foods often contain other valuable nutrients such as vitamins and minerals. Although foods such as soft drinks and candy are rich in carbohydrates, they have little other nutritive value. For example, a typical soft drink is mostly sugar, containing 9 teaspoons of sugar.

Lipids

Sources in the diet

Approximately 95% of the **lipids** in our diets are triglycerides, which consist of three fatty acids attached to a glycerol molecule. Triglycerides often are referred to as fats, which can be divided into saturated and unsaturated fats. **Saturated fats** have fatty acids with only one covalent bond between their carbon atoms (see Chapter 2). Saturated fats are found in the fats of meats (for example, beef and pork), in dairy products (for example, whole milk, cheese, and butter), and in eggs, nuts, coconut oil, and palm oil.

Unsaturated fats have double, or two, covalent bonds between their carbon atoms. **Monounsaturated fats** have one double covalent bond, and **polyunsaturated fats** have more than one double covalent bond. Monounsaturated fats include olive and peanut oils, and polyunsaturated fats are found in fish, safflower, sunflower, and corn oils.

The remaining 5% of lipids include cholesterol and phospholipids. Cholesterol is found in high concentrations in brain, liver, and egg yolks; but it is also present in whole milk, cheese, butter, and meats. Cholesterol is not found in plants. Phospholipids are a major component of cell membranes and they are found in a variety of foods.

Uses in the body

Triglycerides are an important source of energy that can be used to produce ATP molecules, and a gram of triglycerides delivers over twice as many Calories as a gram of carbohydrates. Some cells, for example skeletal muscle cells, derive most of their energy from triglycerides.

After a meal, excess triglycerides that are not immediately used are stored in adipose tissue or the liver. Later, when energy is required, the triglycerides are broken down and their fatty acids are released into the blood, from which they can be taken up and used by various tissues. In addition to storing energy, adipose tissue surrounds and pads organs, and under the skin adipose tissue is an insulator that prevents heat loss.

Cholesterol is an important molecule with many functions in the body. It is obtained in food or it can be manufactured by the liver and most other tissues. Cholesterol is a component of the plasma membrane, and cholesterol can be modified to form other useful molecules such as bile salts and steroid hormones. Bile salts are necessary for fat digestion and absorption. Steroid hormones include the sex hormones estrogen, progesterone, and testosterone, which regulate the reproductive system. Prostaglandins, which are derived from fatty acids, are involved in inflammation, tissue repair, and smooth muscle contraction.

Phospholipids are part of the plasma membrane and are used to construct the myelin sheath around the axons of nerve cells. The phospholipid called lecithin is a major component of plasma membranes, is found in many foods, and is manufactured by the liver.

Recommended requirements

The American Heart Association recommends that fats account for 30% or less of the total Caloric intake. Furthermore, saturated fats should contribute no more than 10% of total fat intake, and cholesterol should be limited to 250 mg (the amount in an egg yolk) or less per day. These guidelines reflect the belief that excess amounts of fats, especially saturated fats and cholesterol, contribute to cardiovascular disease. The typical American diet derives 35% to 45% of its Calories from fats, indicating that most Americans need to reduce fat consumption.

If insufficient amounts of fats are consumed, the body can synthesize fats from carbohydrates and proteins. However, linoleic acid, a fatty acid in many triglycerides, cannot be manufactured by the body. Therefore linoleic acid is an essential fatty acid that must be ingested. It is found in plant oils and milk.

Proteins

Sources in the diet

Proteins are chains of amino acids (see Chapter 2). Twenty amino acids are necessary for good health. The adult human body cannot synthesize eight amino acids, which are called **essential amino acids.** If adequate amounts of the essential amino acids are ingested, they can be used to manufacture the other amino acids, which are called **nonessential amino acids.** A **complete protein** food contains all eight essential amino acids in the correct proportions, whereas an **incomplete protein** food does not. Examples of complete protein foods are meat, fish, poultry, milk, cheese, and eggs. Examples of incomplete proteins are leafy green vegetables, grains, and legumes (peas and beans). If two incomplete proteins such as rice and beans are ingested, each can provide the amino acids lacking in the other. Thus a vegetarian diet can provide all necessary amino acids.

Uses in the body

Proteins perform numerous functions in the human body as the following examples illustrate. Collagen provides structural strength in connective tissue as does keratin in the skin, and the combination of actin and myosin makes muscle contraction possible. Enzymes are responsible for regulating the rate of chemical reactions and protein hormones regulate many physiological processes (see Chapter 10). Proteins in the blood act as buffers to prevent changes in pH, and hemoglobin transports oxygen and carbon dioxide in the blood. Proteins also function as carrier molecules to move materials across plasma membranes, and other proteins in the plasma membrane function as receptor molecules and ion channels. Antibodies, lymphokines, and complement are part of the immune system response that protects us against microorganisms and other foreign substances.

Proteins also can be used as a source of energy, yielding the same amount of energy as carbohydrates. If excess proteins are ingested, the energy in the proteins can be stored by converting their amino acids into glycogen or fats. When protein intake is adequate, the synthesis and breakdown of proteins in a healthy adult occur at the same rate.

Recommended requirements

The recommended daily consumption of protein for an adult is 0.8 g per kg of body weight, or approximately 12% of total Calories. For a 58-kg (128-pound) woman this is 46 g per day, and for a 70-kg man (154-pound) it is 56 g per day. A cup of skim milk contains 8 g protein, 1 ounce of meat contains 7 g protein, and a slice of bread provides 2 g protein.

Vitamins

Vitamins (vi´tah-minz) are organic molecules that exist in minute quantities in food and are essential to normal metabolism. Essential vitamins cannot be produced by the body and must be obtained through the diet. Because no single food item or nutrient class provides all the essential vitamins, it is necessary to maintain a balanced diet by eating a variety of foods. The absence of an essential vitamin in the diet can result in a specific deficiency disease.

There are two major classes of vitamins—**fat soluble** and **water soluble.** Fat-soluble vitamins such as vitamins A, D, E, and K are absorbed from the intestine along with lipids. Some of them can be stored in the body for long periods. Because they can be stored, it is possible to accumulate an overdose of these vitamins in the body to the point of toxicity. Water-soluble vitamins such as the B complex vitamins and vitamin C are absorbed with water from the intestinal tract and remain in the body only a short time before being excreted.

Vitamins are not broken down but are used by the body in their original or slightly modified forms. If the chemical structure of a vitamin is altered, its function usually is lost. The chemical structure of many vitamins is destroyed by heat, such as when food is overcooked. Vitamins function with or as part of enzymes to control chemical reactions in the body (see Chapter 2). The vitamins and their functions are listed in Table 17-1.

1 Predict what would happen if vitamins were broken down during the process of digestion rather than being absorbed intact into the circulation.*

Minerals

A number of inorganic nutrients, **minerals,** are also necessary for normal metabolic functions. They compose approximately 4% of the total body weight and are involved in a number of important functions. Minerals are components of vitamins, enzymes, hemoglobin, and other organic molecules. They add mechanical strength to bone, are necessary for nerve and muscle activity, function as buffers, and are involved with energy transfer process (ATP) and osmosis. Some of the important minerals and their functions are listed in Table 17-2.

Minerals are taken into the body by themselves or in combination with organic molecules. The foods with the highest mineral content include vegetables, legumes, and milk. Foods high in sugar and fats, as well as refined cereals and breads, typically have hardly any minerals. A balanced diet can provide all the necessary minerals, with a few possible exceptions. For example, women who suffer from excessive menstrual bleeding may have to ingest an iron supplement.

*Answers to predict questions appear in Appendix B at the back of the book.

Table 17-1 The principal vitamins

Vitamin	Source	Function	Symptoms of deficiency
Water-soluble vitamins			
B_2 (thiamine)	Yeast, grains, and milk	Carbohydrate and protein metabolism; growth	Beriberi—muscle weakness, neuritis, and paralysis
B_2 (riboflavin)	Green vegetables, liver, wheat germ, milk, and eggs	Energy transport and citric acid cycle	Eye disorders and skin cracking at corners of mouth
Pantothenic acid (part of B_2 complex)	Green vegetables, liver, yeast, grains, and intestinal bacteria	Part of coenzyme A; glucose production	Neuromuscular dysfunction and fatigue
B_3 (niacin)	Fish, liver, red meat, yeast, grains, peas, beans, and nuts	Energy transport, glycolysis, and citric acid cycle	Pellagra—diarrhea, dermatitis, and mental disturbance
B_6 (pyridoxine)	Fish, liver, yeast, tomatoes, and intestinal bacteria	Amino acid metabolism	Dermatitis, retarded growth and nausea
Folic acid	Green vegetables, liver, and intestinal bacteria	Nucleic acid synthesis and blood cell production	Anemia involving enlarged red blood cells
B_{12} (cobalamin)	Liver, red meat, milk, and eggs	Nucleic acid synthesis and blood cell production	Pernicious anemia and nervous system disorders
C (ascorbic acid)	Citrus fruit, tomatoes, and green vegetables	Collagen synthesis and protein metabolism	Scurvy—defective bone growth and poor wound healing
H (biotin)	Liver, yeast, eggs, and intestinal bacteria	Fatty acid and protein synthesis; movement of pyruvic acid into citric acid cycle	Mental and muscle dysfunction; fatigue and nausea
Fat-soluble vitamins			
A (retinol)	From carotene (a provitamin) in vegetables	Vision, skin, bones, and teeth	Night blindness, retarded growth, and skin disorders
D (cholecalciferol)	Fish liver oil, enriched milk; provitamin D converted by sun-light to vitamin D	Calcium and phosphorus absorption; bone and teeth formation	Rickets—poorly developed, weak bones; bone resorption
E (alphatocopherol)	Wheat germ, cottonseed, palm, and rice oils; grain, liver, and lettuce	Prevents catabolism of certain fatty acids	Hemolysis of red blood cells and nerve destruction
K (phylloquinone)	Alfalfa, liver, spinach, vegetable oils, cabbage, and bacteria	Synthesis of several clotting factors	Excessive bleeding resulting from retarded blood clotting

CELL METABOLISM

Metabolism can be divided into the chemical changes that occur during digestion and the chemical processes that occur after the products of digestion are taken up by cells. The chemical processes that occur within cells often are referred to as **cell metabolism.** The digestive products of carbohydrates, proteins, and lipids are taken into cells and they are further metabolized. The energy released in this process is used to form a chemical bond between **adenosine diphosphate (ADP)** and a phosphate group (P), resulting in **adenosine triphosphate,** or **ATP** (Figure 17-1). The chemical bond stores the energy released from the digested molecules.

ATP is often called the energy currency of the cell. When the phosphate group is split from ATP, and the ATP is converted back to ADP, the released energy can be used to drive chemical reactions such as those involved in active transport, muscle contraction, and the synthesis of molecules.

Table 17-2 • Important minerals

Mineral	Function	Symptoms of deficiency
Calcium (Ca)	Bone and teeth formation, blood clotting, and muscle and nerve function	Spontaneous nerve discharge and tetany
Chlorine (Cl)	Blood acid-base balance; HCl production in stomach	Acid-base imbalance
Cobalt (Co)	Part of vitamin B_{12}; erythrocyte production	Anemia
Copper (Cu)	Hemoglobin production and electron-transport chain	Anemia and loss of energy
Fluorine (F)	Extra strength in teeth and prevention of tooth decay	No real pathology
Iodine (I)	Thyroid hormone production and maintenance of normal metabolic rate	Decrease in normal metabolism
Iron (Fe)	Component of hemoglobin; ATP production in electron-transport system	Anemia, decreased oxygen transport, and energy loss
Magnesium (Mg)	Coenzyme constituent; bone formation, and muscle and nerve function	Increased nervous system irritability, vasodilation, and arrhythmias
Manganese (Mn)	Hemoglobin synthesis, growth, and activation of several enzymes	Tremors and convulsions
Phosphorus (P)	Bone and teeth formation, ATP production, and part of nucleic acids	Loss of energy and cellular function
Potassium (K)	Muscle and nerve function	Muscle weakness and abnormal electrocardiogram
Sodium (Na)	Osmosis and nerve and muscle function	Nausea, vomiting, exhaustion, and dizziness
Sulfur (S)	Component of proteins, vitamins, and hormones	Unknown
Zinc (Zn)	Part of several enzymes; carbon dioxide transport; necessary for protein metabolism	Deficient carbon dioxide transport and deficient protein metabolism

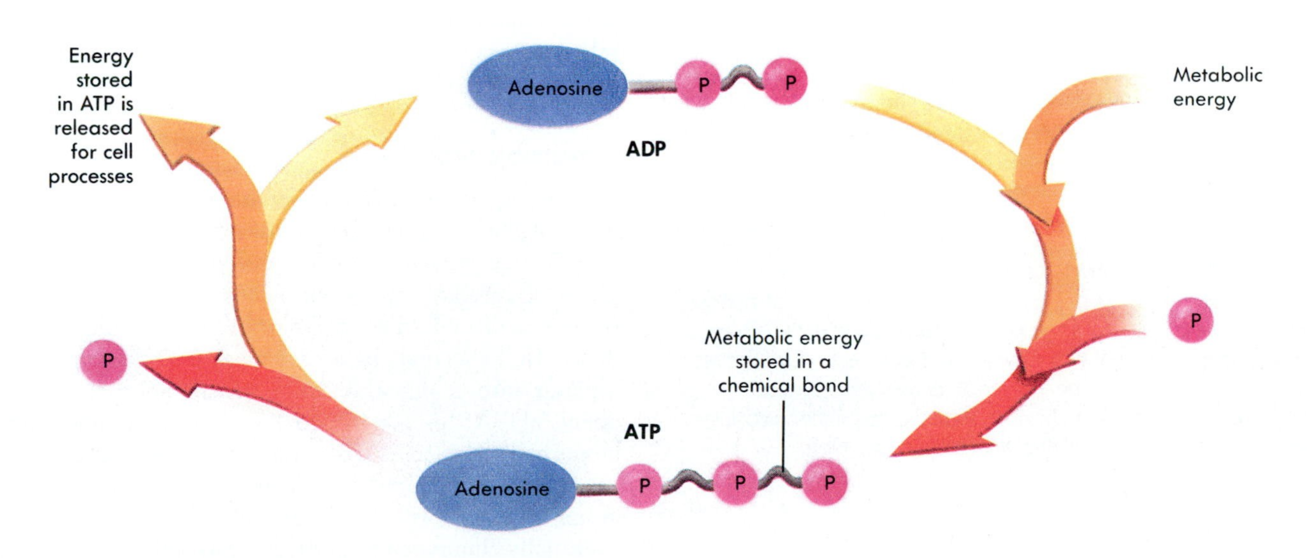

Figure 17-1 • The interconversion of ADP and ATP.
Energy from metabolism and phosphate (P) are required to form ATP from ADP. Energy and a phosphate are given off when ATP is converted back to ADP. The wavy bars represent high-energy bonds.

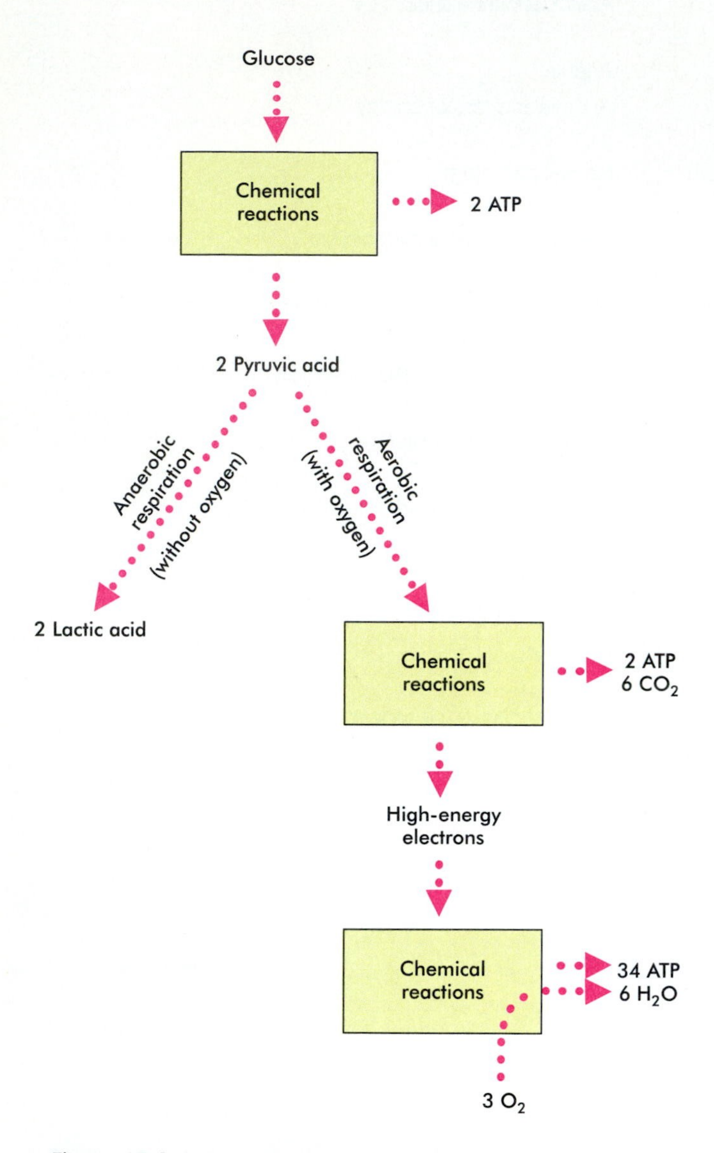

Figure 17-2 ● Carbohydrate metabolism.
Carbohydrate metabolism begins with the breakdown of glucose to pyruvic acid. Anaerobic respiration takes place in the absence of oxygen, converting the pyruvic acid to lactic acid. Aerobic respiration takes place in the presence of oxygen, converting the pyruvic acid into carbon dioxide. Aerobic respiration produces many more ATP molecules than anaerobic respiration.

Carbohydrate metabolism

Glucose is the most important of the monosaccharides as far as cellular metabolism is concerned. Through a series of chemical reactions that occurs in the cytoplasm of most cells, a glucose molecule, which has six carbon atoms, is converted to two **pyruvic** (pi-ru´vik) **acid** molecules, each of which has three carbon atoms. During this process, some of the energy in glucose is used to produce two ATP molecules.

If the cell has adequate amounts of oxygen, the pyruvic acid molecules are used in aerobic respiration. In the absence of oxygen the pyruvic acid molecules are processed by anaerobic respiration.

Anaerobic respiration

Anaerobic (an´ăr-o´bik) **respiration** is the breakdown of glucose in the absence of oxygen to produce two molecules of **lactic** (lak´tik) **acid** and two molecules of ATP. In the first phase of anaerobic respiration, glucose is converted to two pyruvic acid molecules and two ATP molecules (Figure 17-2). In the second phase of anaerobic respiration, the two pyruvic acid molecules are converted into two lactic acid molecules. The ATP produced through anaerobic respiration is a source of energy during activities such as intense exercise when insufficient oxygen is delivered to tissues (see Chapter 7). Although anaerobic respiration can rapidly produce ATP, it can do so only for a short period. Anaerobic respiration does not produce enough ATP to sustain human life.

Lactic acid is released from the cells that produce it and is transported by the blood to the liver. When oxygen becomes available, the lactic acid in the liver can be converted into glucose through a series of chemical reactions, some of which require the input of energy. The glucose then can be released from the liver and transported in the blood to cells that use glucose as an energy source.

Aerobic respiration

Aerobic (ăr-o´bik) **respiration** (Figure 17-3) is the breakdown of glucose in the presence of oxygen to produce 6 carbon dioxide, 6 water, and 38 ATP molecules. In the first phase of aerobic respiration, glucose is converted to two pyruvic acid molecules, and two ATP molecules are produced. In the second phase, pyruvic acid moves from the cytoplasm into a mitochondrion, where another series of chemical reactions takes place. As a result, the three carbon atoms in pyruvic acid are separated from each other and are used to produce carbon dioxide molecules. Thus the carbon atoms that comprise food molecules such as glucose are eventually eliminated from the body as carbon dioxide. We literally breathe out part of the food we eat!

During the conversion of the pyruvic acid molecules to carbon dioxide, two ATP molecules are produced and high-energy electrons are released. These electrons enter a series of reactions in which the energy from the electrons is used to produce 34 ATP molecules. Thus most of the ATP necessary to sustain life is produced at this stage of aerobic respiration. In the last chemical reaction of this process, hydrogen ions combine with oxygen to form water. Without oxygen, these reactions stop. By analogy, if a railroad engine (the last chemical reaction that requires oxygen) stops, then the railroad cars (chemical reactions) behind the railroad engine must also stop. It is necessary for us to get oxygen by breathing because the chemical reactions producing most of our ATP molecules would stop without oxygen.

2 Many poisons function by blocking certain steps in the metabolic pathways. For example, cyanide blocks the use of energy from high energy electrons. Explain why this blockage would cause death.

Fat and protein metabolism

The chemical reactions involved in the breakdown of carbohydrates to yield ATP molecules are the "backbone" of cellular metabolism (see Figure 17-3). The breakdown products of fats and proteins can "feed into" the carbohydrate pathway and be used to produce ATP. Thus excess fats and proteins can be used as sources of energy. On the other hand, many of these reactions are reversible. For example, if inadequate amounts of the nonessential amino acids are ingested, carbohydrates can be used to manufacture them.

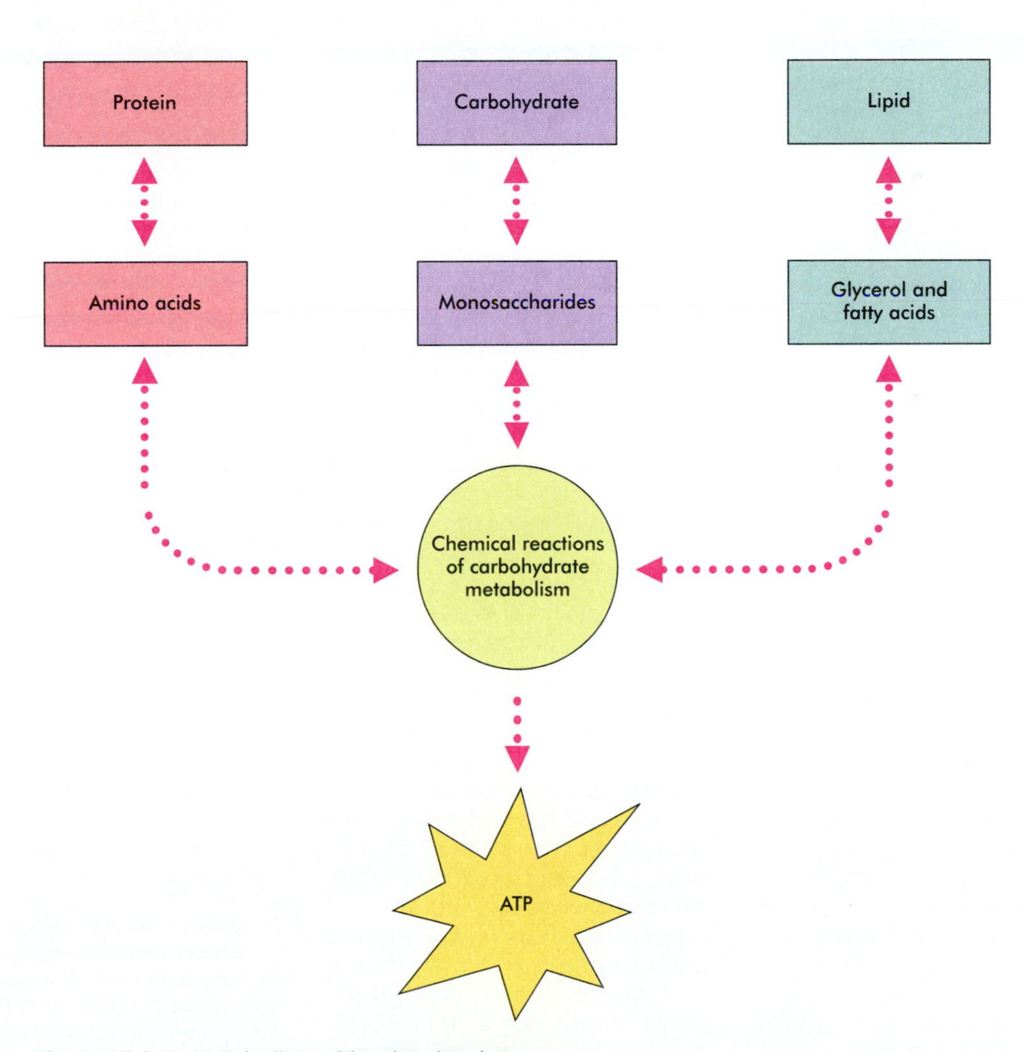

Figure 17-3 ● Metabolism of food molecules.
Carbohydrates are digested into monosaccharides, proteins into amino acids, and fats into fatty acids and glycerol. All of these substances can be used in the chemical reactions of carbohydrate metabolism to produce ATP. In addition, as indicated by the *reverse arrows,* molecules derived from carbohydrates can be used to manufacture proteins and lipids.

Understanding starvation and obesity

STARVATION

Starvation is the inadequate intake of nutrients or the inability to metabolize or absorb nutrients. Starvation can result from a number of causes, such as prolonged fasting, anorexia, deprivation, or disease. No matter what the cause, starvation takes about the same course and consists of three phases. The events of the first two phases occur even during relatively short periods of fasting or dieting. The third phase occurs in prolonged starvation and ends in death.

During the first phase of starvation, blood glucose levels are maintained through the production of glucose from glycogen, proteins, and fats. It is important to maintain blood glucose levels because the brain normally depends on glucose as a source of energy. At first glycogen is broken down into glucose. Enough glycogen is stored in the liver to last only a few hours, however. Thereafter, blood glucose levels are maintained by the breakdown of fats and proteins. Fatty acids derived from the breakdown of fats can be used as a source of energy, especially by skeletal muscle. The decreased use of glucose by tissues other than the brain helps to maintain blood glucose levels, making glucose available for use by the brain. Glycerol, another breakdown product of fats, can be used to make a small amount of glucose, but most of the glucose is formed from the amino acids of proteins.

In the second stage, which can last for several weeks, fats are the primary energy source. The liver metabolizes fatty acids into substances called ketones. An example of a ketone is acetone, the main ingredient in fingernail polish. Ketones can be used as a source of energy. After about a week of fasting, the brain begins to use ketone bodies as well as glucose for energy. This usage decreases the demand for glucose, and the rate of protein breakdown diminishes but does not stop.

The third stage of starvation begins when the fat reserves are nearly depleted, and there is a switch to proteins as the major energy source. Muscles, the largest source of protein in the body, are rapidly depleted. At the end of this stage, proteins essential for cellular functions are broken down, cell function degenerates, and death can occur.

OBESITY

Obesity is the storage of excess fat, and it can be classified according to the number of fat cells and the size of fat cells. The greater the amount of lipids stored in the fat cells, the larger their size. In **hyperplastic obesity,** there is a greater than normal number of fat cells that are also larger than normal. This type of obesity is associated with massive obesity and begins at an early age. In obese and nonobese children, the number of fat cells triples or quadruples between birth and 2 years of age, then remains relatively stable until puberty, when there is a further increase in number. In obese children there is also an increase in the number of fat cells between 2 years of age and puberty. Consequently, obese people end up with more fat cells than nonobese people. **Hypertrophic** (hi′per-trof′ik) **obesity** results from a normal number of fat cells that have increased in size. This type of obesity is more common, is associated with moderate obesity or "overweight," and typically develops in adults. People who were thin or of average weight and quite active when young become less active as they become older. They begin to gain weight at age 20 to 40, and, although they no longer use as many Calories, they still take in the same amount of food as when they were younger. The unused Calories are turned into fat, causing fat cells to increase in size. If all the existing fat cells are filled to capacity with lipids, new fat cells are formed in order to store the excess lipids. Once fat cells are formed, however, dieting and weight loss does not result in a decrease in the number of fat cells—instead, they become smaller in size as their lipid content decreases.

Regulation of body weight is actually a matter of regulating body fat because most changes in body weight reflect changes in the amount of fat in the body. According to one theory of weight control, the body maintains a certain amount of fat called the set point. If the amount of body fat decreases below or increases above this level, mechanisms are activated to return the amount of body fat to its set point.

Some scientists believe that the number of fat cells in the body can also affect appetite. According to this line of reasoning, fat cells maintain their size, and once a "fat plateau" is attained, the body stays at that plateau. Fat cells may accomplish this by effectively taking up triglycerides and converting them to fat. Consequently, there is less energy available for muscle and body organs, and to compensate, appetite increases to provide needed energy. In support of this hypothesis, it is known that obese individuals have an increased amount of the enzyme lipoprotein lipase, which is responsible for the uptake and storage of triglycerides in fat cells. Furthermore, in obese individuals who have lost weight, the levels of lipoprotein lipase increase even more.

It is a common belief that the main cause of obesity is simply overeating. Certainly for obesity to occur, at some time energy intake must have exceeded energy expenditure. A comparison of the Caloric intake of obese and lean individuals at their usual weights, however, reveals that on a per kilogram basis, obese people consume fewer Calories than lean people.

When people lose a large amount of weight their feeding behavior changes. They become hyper-responsive to external food cues, think of food often, and cannot get enough to eat without gaining weight. It is now understood that this behavior is typical of both lean and obese individuals who are below their relative set point for weight. Other changes such as a decrease in basal metabolic rate take place in a person who has lost a large amount of weight. Most of this decrease probably results from a decrease in muscle mass associated with weight loss.

3 Only a small percentage of obese people maintain weight loss on a long-term basis. The typical pattern is one of repeated cycles of dieting and weight loss followed by a rapid regain of the lost weight. Explain why this happens.

METABOLIC RATE

The **metabolic rate** is the total amount of energy produced and used by the body per unit of time. Metabolic rate is usually estimated by measuring the amount of oxygen used per minute. One liter of oxygen consumed by the body is assumed to produce 4.825 Cal of energy.

Metabolic energy can be used in three ways: for basal metabolism, for muscle contraction, and for the assimilation of food (production of digestive enzymes, active transport, and so forth). The **basal metabolic rate (BMR)** is the metabolic rate calculated in expended Calories per square meter of body surface area per hour, and it is measured when a person is awake but restful and has not eaten for 12 hours. A typical BMR for a 70-kilogram (154-pound) man would be 38 Cal/m^2/hr.

BMR is the minimal energy needed to keep the resting body functional. Active transport mechanisms, muscle tone, maintenance of body temperature, beating of the heart, and other activities are supported by basal metabolism. A number of factors can affect the BMR. Males have a greater BMR than females, younger people have a higher BMR than older people, and fever can increase BMR. Greatly reduced Caloric input (for example, during dieting or fasting) depresses BMR.

The daily input of energy should equal the energy demand of metabolism; otherwise, a person will gain or lose weight. For a 23-year-old, 70-kilogram (154-pound) man to maintain his weight, the input should be 2700 Cal per day; for a 58-kilogram (128-pound) woman of the same age, 2000 Cal per day are necessary. For every 3500 Cal above the energy requirement (not necessarily in 1 day), a pound of body fat can be gained, whereas for every 3500 Cal below the requirement (usually over several days), a pound of fat can be lost. Clearly, adjusting Caloric input is an important way to control body weight.

The other way to control weight is through energy expenditure. Physical activity through skeletal muscle movement increases metabolic rate. In the average person, basal metabolism accounts for approximately 60% of energy expenditure, muscular activity 30%, and assimilation of food approximately 10%. A comparison of the number of Calories gained from food and the number of Calories burned during exercise reveals why losing weight can be such a difficult task. For example, it takes 20 minutes of brisk walking to burn off the Calories in one slice of bread (approximately 75 Cal).

BODY TEMPERATURE

Humans can maintain a constant body temperature even though the environmental temperature varies. Normal body temperature is usually considered to be 37°C (98.6°F) when it is measured orally, and 37.6°C (99.7°F) when it is measured rectally. Rectal temperature comes closer to the true body temperature, but an oral temperature is more easily obtained in older children and adults, and therefore is the preferred measure.

Maintenance of a constant body temperature is important to homeostasis because most enzymes are very temperature sensitive and function only within narrow temperature ranges. Body temperature is maintained by balancing heat produced by the body with heat exchanged between the body and the environment. Heat is produced by the body as a byproduct of the chemical reactions of metabolism (see Chapter 2). Approximately 43% of the energy released by metabolism is used to accomplish biological work such as synthesizing molecules, muscular contraction, and other cellular activities. The remaining energy is lost as heat.

Heat may be exchanged with the environment in a number of ways. **Radiation** is the loss of heat as infrared energy, a type of electromagnetic radiation. For example, the coals in a fire give off radiant heat that can be felt some distance away from the fire. **Conduction** is the exchange of heat between objects that are in direct contact with each other (for example, the bottom of the feet and the floor). **Convection** is a transfer of heat between the body and the air. A cool breeze results in movement of air over the body and loss of heat from the body. **Evaporation** is the conversion of water from a liquid to a gaseous state and the loss of the gaseous water from the body; the water carries heat away with it.

The amount of heat exchanged between the environment and the body is determined by the difference in temperature between the body and the environment. The greater the temperature difference, the greater is the rate of heat exchange. Control of the temperature difference can be used to regulate body temperature. If environmental temperature is very cold (for example, on a winter day), there is a large temperature difference between the body and the environment, and there is a large loss of heat. The loss of heat can be decreased by behaviorally selecting a warmer environment (for example, moving inside a heated house) or by insulating the exchange surface (for example, putting on extra clothes). If body temperature begins to drop below normal, heat can be conserved by constriction of blood vessels in the dermis of the skin. The reduced blood flow results in less heat transfer from deeper structures to the skin, and heat loss is reduced (Figure 17-4, *A*).

When environmental temperature is greater than body temperature, vasodilation brings warm blood to the skin, causing an increase in skin temperature. There is a decrease in heat gain from the environment because skin temperature is closer to environmental temperature. At the same time, excess heat can be lost from the skin by evaporation, which prevents heat gain and overheating (Figure 17-4, *B*).

Body temperature regulation is an example of a negative-feedback system (Figure 17-5). Body temperature is maintained around a value called the set point. A small area in the anterior part of the hypothalamus can detect slight increases above the set point through changes in blood temperature. Body temperature is reduced through vasodilation

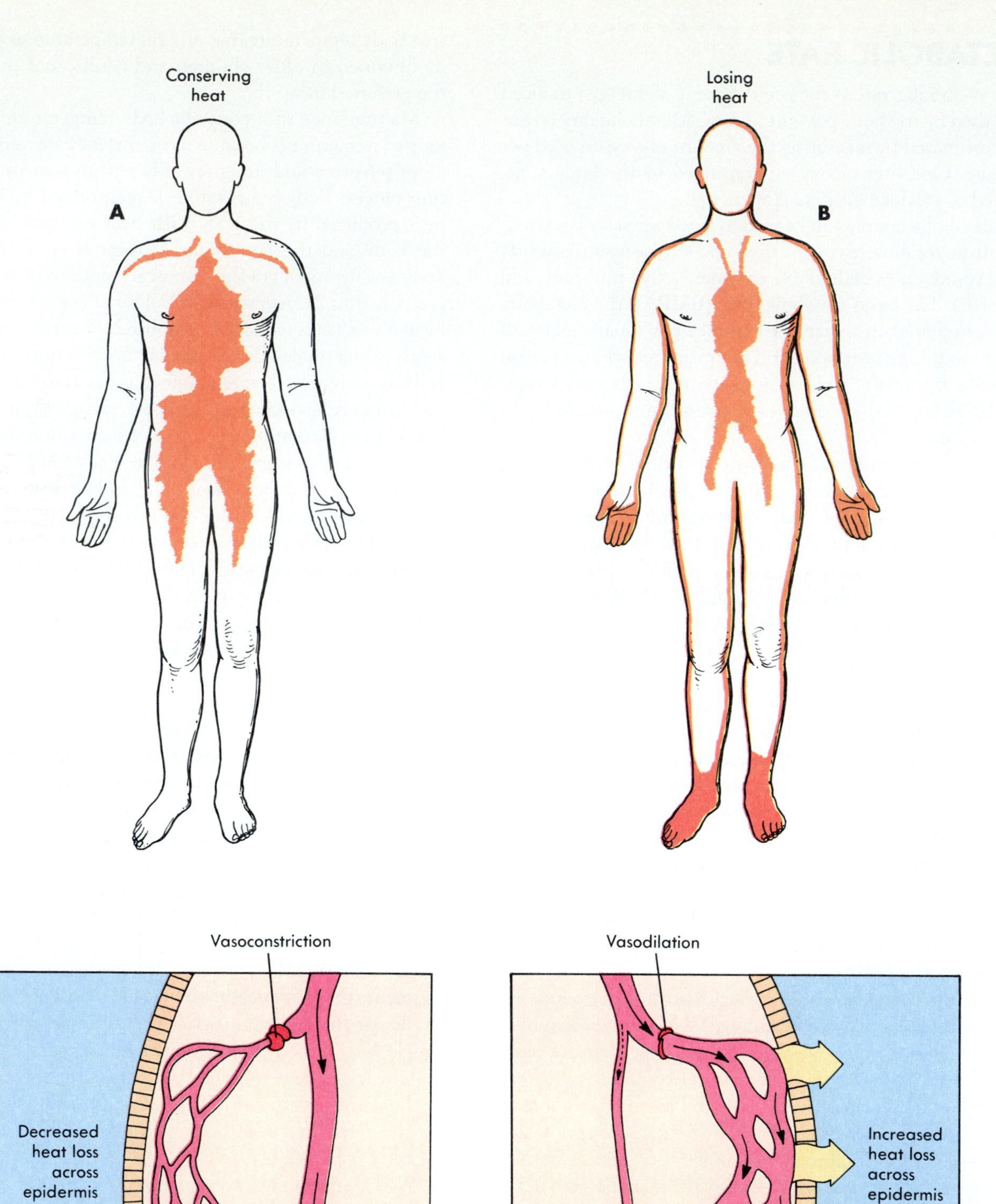

Figure 17-4 ● Heat exchange in the skin.
A, Blood vessels in the dermis of the skin constrict, reducing blood flow to the surface. Consequently less heat is lost from the body. **B,** Blood vessels in the dermis dilate, allowing more blood close to the surface from which heat is lost from the body.

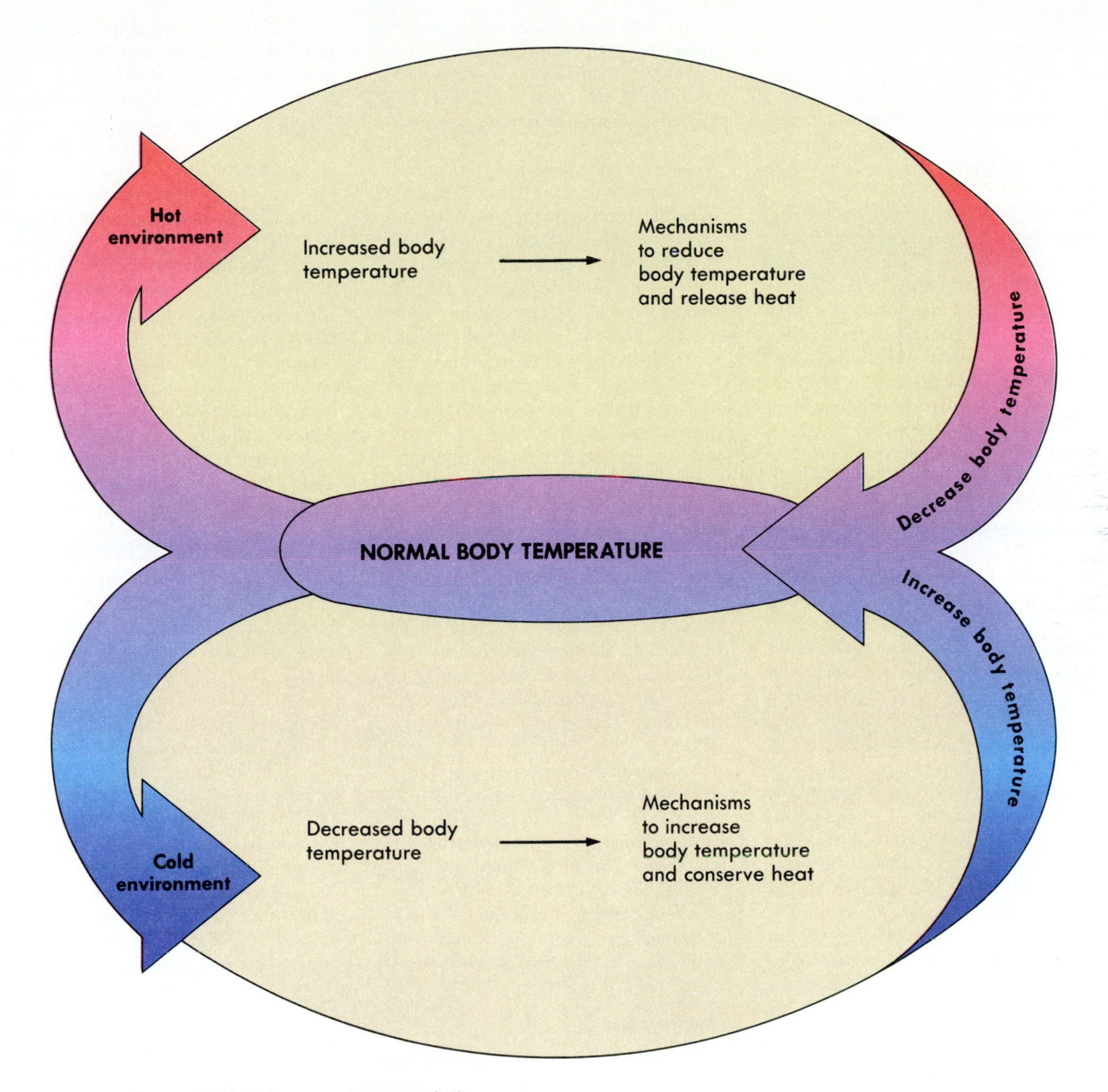

Figure 17-5 ● Temperature regulation.
When environmental factors (*arrows* on left side of Figure) move the body temperature out of its normal range, regulatory mechanisms (*arrows* on right side of Figure) tend to bring the temperature back to its normal range.

and sweating. A small area in the posterior hypothalamus can detect slight decreases in body temperature below the set point. Body temperature is increased by increasing heat production and decreasing heat loss. Heat is produced as a byproduct of the muscular contractions of shivering (see Chapter 7), and heat is conserved by constricting blood vessels in the skin.

Under some conditions, the hypothalamic set point changes. For example, during a fever the set point is raised, causing body temperature to be maintained at a higher-than-normal value. After recovery from the illness that caused the fever, the set point returns to normal.

4 Some diseases result in a high fever. The patient is on the way to recovery when the "crisis is over" and body temperature begins to return to normal. If you were looking for symptoms in a patient who had just passed through the crisis state, would you look for dry, pale skin or for wet, flushed skin? Explain.

Hyperthermia and hypothermia

HYPERTHERMIA

If heat gain exceeds the ability of the body to lose heat, then body temperature increases above normal levels, a condition called **hyperthermia.** Hyperthermia can result from exposure to hot environments, exercise, fever, and anesthesia.

Exposure to a hot environment normally results in the activation of heat loss mechanisms and body temperature is maintained at normal levels. This is an excellent example of a negative-feedback mechanism. Prolonged exposure to a hot environment, however, can result in **heat exhaustion.** The normal negative-feedback mechanisms for controlling body temperature are operating, but they are unable to maintain a normal body temperature. Heavy sweating results in dehydration, decreased blood volume, decreased blood pressure, and increased heart rate. Individuals suffering from heat exhaustion have wet, cool skin because of the heavy sweating. They usually feel weak, dizzy, and nauseated. Treatment includes reducing heat gain by moving to a cooler environment, reducing heat production by muscles by ceasing activity, and restoring blood volume by drinking fluids.

Heat stroke is a breakdown of the normal negative-feedback mechanisms of temperature regulation. If the temperature of the hypothalamus becomes too high, it no longer functions appropriately. Sweating stops and the skin becomes dry and flushed. The person becomes confused, irritable, or even comatose. In addition to the treatment for heat exhaustion, heat loss from the skin should be increased. This can be accomplished by increasing evaporation from the skin by applying wet cloths or by increasing conductive heat loss by immersing the person in a cool bath.

Exercise increases body temperature because of the heat produced as a byproduct of muscle activity (see Chapter 7). Normally vasodilation and increased sweating prevent body temperature increases that are harmful. In a hot, humid environment the evaporation of sweat is decreased and exercise levels have to be reduced to prevent overheating.

Fever is the development of a higher-than-normal body temperature following the invasion of the body by microorganisms or foreign substances. Lymphocytes, neutrophils, and macrophages release chemicals called **pyrogens** (pi´ro-jenz) that raise the temperature set point of the hypothalamus. Consequently body temperature and metabolic rate increase. Fever is believed to be beneficial because it speeds up the chemical reactions of the immune system (see Chapter 14) and inhibits the growth of some microorganisms. Although fever can be beneficial, body temperatures greater than 41°C (106°F) can be harmful. Aspirin lowers body temperature by blocking the effects of pyrogens on the hypothalamus.

Malignant hyperthermia is an inherited muscle disorder. Drugs used to induce general anesthesia for surgery cause sustained, uncoordinated muscle contractions in some individuals. Consequently body temperature increases.

Therapeutic hyperthermia is a local or general body increase in temperature. It is a treatment sometimes used on tumors and infections.

HYPOTHERMIA

If heat loss exceeds the ability of the body to produce heat, body temperature decreases below normal levels, a condition called **hypothermia.** Hypothermia usually results from prolonged exposure to cold environments. At first, normal negative-feedback mechanisms maintain body temperature. Heat loss is decreased by constricting blood vessels in the skin and heat production is increased by shivering. If body temperature decreases despite these mechanisms, then hypothermia develops. The individual's thinking becomes sluggish and movements are uncoordinated. Heart, respiratory, and metabolic rates decline and death results unless body temperature is restored to normal. Rewarming should occur at a rate of a few degrees per hour.

Frostbite is damage to the skin resulting from prolonged exposure to the cold. Normally, blood vessels in the skin constrict to conserve heat in a cold environment. When skin temperatures decrease significantly, however, tissues can be damaged by the cold. Consequently, skin blood vessels dilate, bringing warm blood to the skin. If this protective response is inadequate, frostbite occurs. The fingers, toes, ears, nose, and cheeks are most commonly affected. Damage from frostbite can range from redness and discomfort to loss of the affected part. The best treatment is immersion in a warm water bath. Rubbing the affected area and local, dry heat should be avoided.

Therapeutic hypothermia is sometimes used to slow metabolic rate during surgical procedures such as heart surgery. Because metabolic rate is decreased, tissues do not require as much oxygen as normal and are less likely to be damaged.

SUMMARY

Nutrition is the taking in and the use of food.

Metabolism consists of anabolism and catabolism.
Anabolism is the building up of molecules and requires energy. Catabolism is the breaking down of molecules and gives off energy.

NUTRITION

Nutrients

Nutrients are the chemicals used by the body and consist of carbohydrates, lipids, proteins, vitamins, minerals, and water.

Calories

A Calorie is the heat (energy) necessary to raise the temperature of 1000 g of water from 14°C to 15°C.

A gram of carbohydrate or protein yields 4 Cal, and a gram of fat yields 9 Cal.

Carbohydrates

Carbohydrates are ingested as monosaccharides (glucose, fructose), disaccharides (sucrose, maltose, lactose), and polysaccharides (starch, glycogen, cellulose).

Polysaccharides and disaccharides are converted to glucose. Glucose can be used for energy or stored as glycogen or fats.

Approximately 125 to 175 g of carbohydrates should be ingested each day.

Lipids

Lipids are ingested as triglycerides (95%) or cholesterol and phospholipids (5%).

Triglycerides are used for energy or stored in adipose tissue. Cholesterol forms other molecules such as steroid hormones. Cholesterol and phospholipids are part of the plasma membrane.

The daily diet should include no more than 30% of its Calories from lipids and no more than 250 mg of cholesterol.

Proteins

Proteins are ingested and broken down into amino acids.

Proteins perform many functions: protection (antibodies), regulation (enzymes, hormones), structure (collagen), muscle contraction (actin, myosin), and transport (hemoglobin, carrier molecules, ion channels).

An adult should consume 0.8 g of protein per kg of body weight each day.

Vitamins

Vitamins function with or as part of enzymes.

Most vitamins are not produced by the body and must be obtained in the diet.

Vitamins are classified as either fat soluble or water soluble.

Minerals

Minerals are necessary for normal metabolism. They are components of vitamins, enzymes, and other molecules.

CELL METABOLISM

The energy in carbohydrates, lipids, and proteins is used to produce ATP.

Carbohydrate metabolism

Anaerobic respiration is the breakdown of glucose in the absence of oxygen to two lactic acid molecules and two ATP molecules.

Aerobic respiration is the breakdown of glucose in the presence of oxygen to produce 6 carbon dioxide, 6 water, and 38 ATP molecules.

Fat and protein metabolism

The breakdown products of fat and protein metabolism can enter carbohydrate metabolism reactions to produce ATP.

Carbohydrates can be used to produce amino acids and fats.

METABOLIC RATE

Metabolic rate is the total energy expenditure of the body per unit of time.

Metabolic energy is used for basal metabolism, muscular activity, and the assimilation of food.

BODY TEMPERATURE

Body temperature is a balance between heat gain and heat loss.

Heat is produced through metabolism. Heat is exchanged through radiation, conduction, convection, and evaporation.

The greater the temperature difference between the body and the environment, the greater is the rate of heat exchange.

Body temperature is regulated by a "set point" in the hypothalamus.

CONTENT REVIEW

1. Define metabolism, anabolism, and catabolism.
2. Define a nutrient and list the six major classes of nutrients. What is an essential nutrient?
3. Define a Calorie and state the number of Calories in a gram of carbohydrate, lipid, and protein.
4. List the dietary sources of carbohydrates. After they are converted to glucose what happens to the glucose? What quantities of carbohydrate should be ingested daily?
5. List the dietary sources of lipids, explain how triglycerides, cholesterol, and phospholipids are used in the body, and describe the recommended dietary intake of lipids.
6. List the dietary sources of complete and incomplete protein foods. Describe some of the functions performed by proteins in the body. What is the recommended daily consumption of proteins?
7. What are vitamins? Name the water-soluble vitamins and the fat-soluble vitamins. List some of the functions of vitamins.
8. List some of the minerals and give their functions.
9. Define anaerobic respiration. How many ATP molecules are produced? What happens to the lactic acid produced?
10. Define aerobic respiration and list the products it produces.
11. How are fats and proteins used to produce ATP molecules?
12. What is meant by metabolic rate? Describe its three component parts.
13. Describe how heat is produced by and lost from the body. How is body temperature regulated?

CONCEPT REVIEW

1. Why does a vegetarian usually have to be more careful about his or her diet than a person who includes meat in the diet?
2. Lotta Bulk, a muscle builder, wanted to increase her muscle mass. Knowing that proteins are the main components of muscle, she consumed large amounts of protein daily (high protein diet), along with small amounts of lipid and carbohydrate. Explain why this strategy will or will not work.
3. Why can some people lose weight on a low Calorie per day diet, and other people cannot?
4. Some people claim that fasting occasionally for short times may be beneficial. How can fasts be damaging?

CHAPTER TEST

Answers can be found in Appendix C

MATCHING 1 For each statement in column A select the correct answer in column B (an answer may be used once, more than once, or not at all).

A

1. Type of nutrient that must be ingested because the body cannot manufacture it.
2. A measure of the energy in food.
3. Polysaccharide that is undigestible by humans; provides bulk.
4. Polysaccharide used for energy storage in animals.
5. Table sugar; disaccharide of glucose and fructose.
6. Lipid with only single covalent bonds between carbon atoms.
7. Group to which vitamins A, D, E, and K belong.

B

Calorie
cellulose
essential
fat-soluble vitamins
glycogen
maltose
saturated fat
starch
sucrose
unsaturated fat
water-soluble vitamins

MATCHING 2 For each statement in column A select the correct answer in column B (an answer may be used once, more than once, or not at all).

A

1. Energy expended at rest.
2. Accounts for 30% of energy expenditure.
3. Accounts for 10% of energy expenditure.
4. Heat transfer between the body and air.
5. Heat lost from the body by gaseous water.
6. Change in blood vessels that results in heat loss from the body.
7. Area of the brain in which body temperature is regulated.

B

assimilation of food
basal metabolic rate
cerebrum
conduction
convection
evaporation
hypothalamus
muscular activity
radiation
vasoconstriction
vasodilation

FILL-IN-THE BLANK Complete each statement by providing the missing word or words.

1. Phosphate, __________, and energy are used to produce ATP.
2. The breakdown of glucose in the absence of oxygen is called __________.
3. The last molecule produced in anaerobic respiration is __________.
4. When oxygen is available, lactic acid can be converted into __________.
5. For each glucose molecule, anaerobic respiration yields __________ ATP molecules; whereas aerobic respiration produces __________ ATP molecules.
6. The carbon atoms in glucose are used to produce __________ in aerobic respiration.
7. The last step of aerobic respiration is the formation of __________.
8. In addition to carbohydrates, __________ and __________ can enter the reactions of carbohydrate metabolism to produce ATP molecules.

KEY TERMS

aldosterone
Hormone produced by the adrenal cortex that acts on the kidneys causing sodium reabsorption from the urine, and potassium and hydrogen ion excretion.

antidiuretic hormone (ADH)
Hormone secreted from the posterior pituitary that acts on the kidney to reduce the output of urine.

filtration
The movement of water, ions, and small molecules (but not blood cells) across a membrane from the blood in to Bowman's capsule; occurs because of a pressure difference.

kidney
One of a pair of bean-shaped organs, which function to cleanse blood and produce urine.

micturition reflex
Contraction of the urinary bladder stimulated by stretching of the urinary bladder wall; results in emptying of the urinary bladder.

nephron
Functional unit of the kidney, consisting of Bowman's capsule, the proximal convoluted tubule, the loop of Henle, and the distal convoluted tubule.

tubular reabsorption
Movement of materials, by means of diffusion or active transport, from the filtrate back into the blood.

tubular secretion
Movement of materials, by means of active transport, from the blood into the filtrate.

urine
Liquid waste excreted from the kidneys; consists of mostly water, excess ions, and organic waste products such as urea.

FEATURES

CHAPTER 18

Urinary System and Fluid Balance

OBJECTIVES

After reading this chapter you should be able to:

1. List the structures that make up the urinary system and describe the functions of the urinary system.
2. Describe the location and anatomy of the kidneys.
3. Describe the structure of the nephron.
4. Define filtration, reabsorption, and secretion.
5. Describe the ureters, urinary bladder, and urethra.
6. Explain how antidiuretic hormone, aldosterone, and atrial natriuretic hormone influence the volume and concentration of urine.
7. Describe the micturition reflex.
8. Describe how sodium ions and potassium ions are regulated in the extracellular fluid.
9. Explain how the pH of body fluids is regulated.

THE URINARY SYSTEM consists of the kidneys, ureters, urinary bladder, and urethra (Figure 18-1). The kidneys remove waste products from the blood; help control the volume, ion concentration, and pH of the blood; and control red blood cell (RBC) production (see Chapter 11). Although the kidneys are the major organs of excretion, the skin, liver, intestines, and lungs also eliminate small amounts of waste. However, if the kidneys fail to function, the other structures cannot compensate.

The skin secretes water, salts, and small amounts of nitrogenous waste products, mainly in perspiration. Water evaporates from the surface of the skin leaving a salty residue that is more noticeable following heavy perspiration in hot weather or following exercise.

The liver excretes bile, containing a number of organic waste products, into the intestines. Some salts and small amounts of water also enter the intestine across its wall and are eliminated from the body in the feces.

Carbon dioxide and a small amount of water are eliminated through the lungs. These substances pass out of the lungs as gaseous carbon dioxide and as water vapor during expiration.

This chapter describes the anatomy of the urinary system, the means by which waste products are removed from the blood, mechanisms that regulate the volume and ion concentration in the extracellular fluid, and control of acid base balance.

URINARY SYSTEM

Kidneys

The **kidneys** are bean-shaped organs and each is about the size of a tightly clenched fist. They are located on the posterior abdominal wall to either side of the vertebral column (Figure 18-2). A tough connective tissue capsule surrounds each kidney, and there is a thick layer of fat around the connective tissue capsule, which protects the kidney from mechanical shock.

On the medial side of each kidney is the **hilum** (hi´lum), where the renal artery and nerves enter and the renal vein and ureter exit the kidney. The hilum opens into a cavity called the **renal sinus,** which is filled with fat and other connective tissue (Figure 18-3). Within the renal sinus is the **renal pelvis.** Several funnel-shaped portions of the renal pelvis called **calyces** (kal´ĭ-sēz; singular **calyx,** ka´liks) extend from the renal pelvis to the kidney tissue. The kidney tissue produces urine, which flows into the calyces. At the hilum, the renal pelvis narrows to form the **ureter** (u-re´ter).

In a longitudinal section of the kidney, the outer **cortex** and an inner **medulla** can be identified with the unaided eye (see Figure 18-3). The medulla consists of a number of cone-shaped **renal pyramids.** The base of each renal pyramid extends toward the cortex, and the tip of each pyramid projects toward the medulla, where it is surrounded by a calyx.

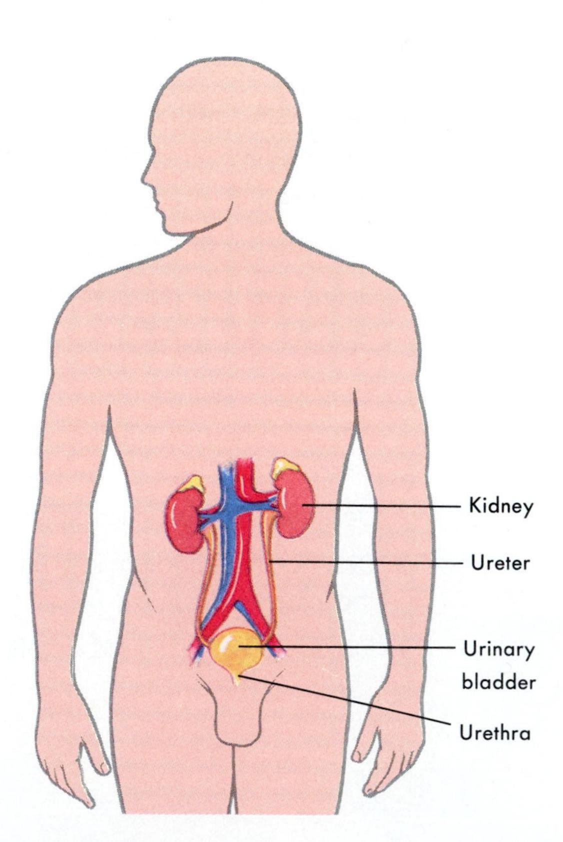

Figure 18-1 ● The urinary system.
The urinary system consists of two kidneys, two ureters, a urinary bladder, and a urethra.

A

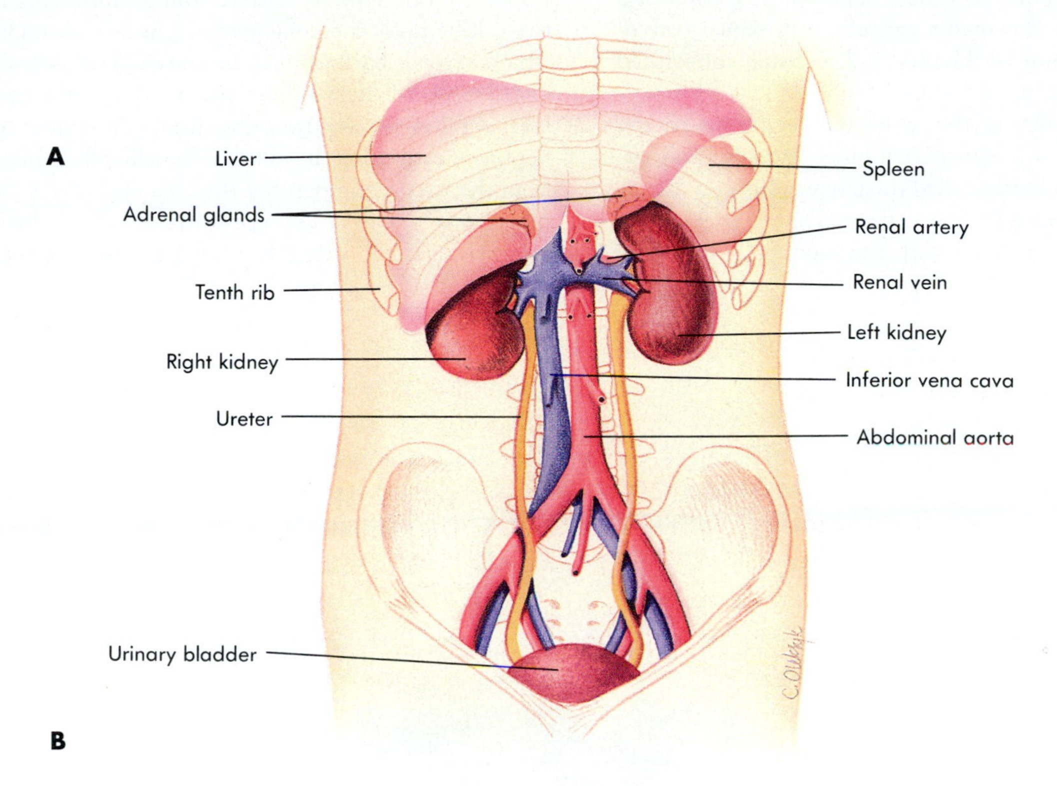

B

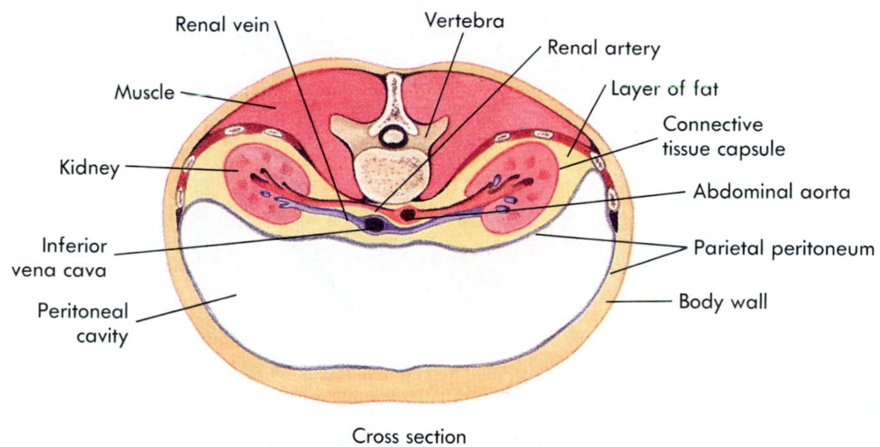

Figure 18-2 ● Anatomy of the kidney.
A, The kidneys are located in the abdominal cavity, with the right kidney just below the liver and the left kidney below the spleen. The ureters extend from the kidneys to the urinary bladder within the pelvic cavity. An adrenal gland is located at the superior pole of each kidney. **B,** The kidneys are located behind the parietal peritoneum. A layer of fat surrounds each kidney. The renal arteries extend from the abdominal aorta to each kidney, and the renal veins extend from the kidneys to the inferior vena cava.

The nephron

The functional unit of the kidney is the **nephron** (nef´ron) (Figure 18-4). Each nephron is capable of producing urine, and about one third of the 1.3 million nephrons in each kidney must be functional to ensure survival. The parts of a nephron include a **Bowman's capsule,** a **proximal convoluted tubule,** a **loop of Henle,** and a **distal convoluted tubule.**

Bowman's capsule is the enlarged beginning of the nephron. The side of Bowman's capsule is indented and surrounds a tuft of capillaries called the **glomerulus** (glo-mĕr´u-lus), which resembles a ball of yarn (see Figures 18-4). Bowman's capsule is continuous with the proximal convoluted tubule.

Fluid called **filtrate** is filtered from the glomerulus into Bowman's capsule. After entering Bowman's capsule, the filtrate flows into the proximal convoluted tubule (see Figure 18-4, *B*). From there, filtrate flows into the loop of Henle. Each loop of Henle has a **descending limb,** which extends toward the tip of the renal pyramid, and an **ascending limb,** which extends back toward the cortex. The ascending limb of the loop of Henle gives rise to the distal convoluted tubule. Filtrate passes from the distal convoluted tubules of nephrons into collecting ducts. The **collecting ducts** extend from the cortex through the medulla and empty their contents into a calyx. As the filtrate passes through the nephrons and collecting ducts of the kidney, it is processed into a fluid called **urine.**

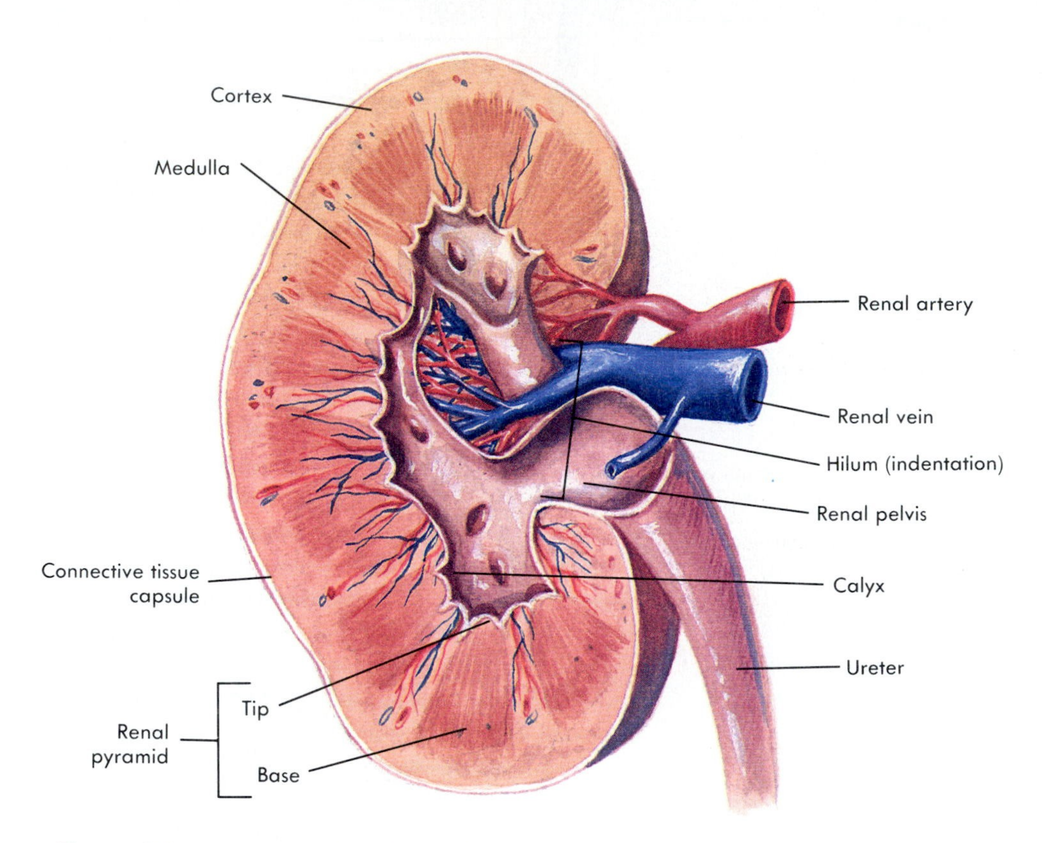

Figure 18-3 ● Longitudinal section of the kidney.
The cortex forms the outer part of the kidney, and the medulla forms the inner part. In the medulla are renal pyramids. The tip of each renal pyramid is surrounded by a calyx, which extends to the renal pelvis. Urine flows from the tip of the renal pyramid through the calyx and renal pelvis into the ureter.

Figure 18-4 ● The nephron.
A, A longitudinal section of a kidney showing the location of a nephron and collecting duct in the kidney. **B,** Each nephron consists of a Bowman's capsule, a proximal convoluted tubule, a loop of Henle, and a distal convoluted tubule. The distal convoluted tubules join the collecting duct that extends to the tip of the renal pyramid. An afferent arteriole carries blood to the glomerulus. An efferent arteriole carries blood from the glomerulus and gives rise to peritubular capillaries, which surround the nephron.

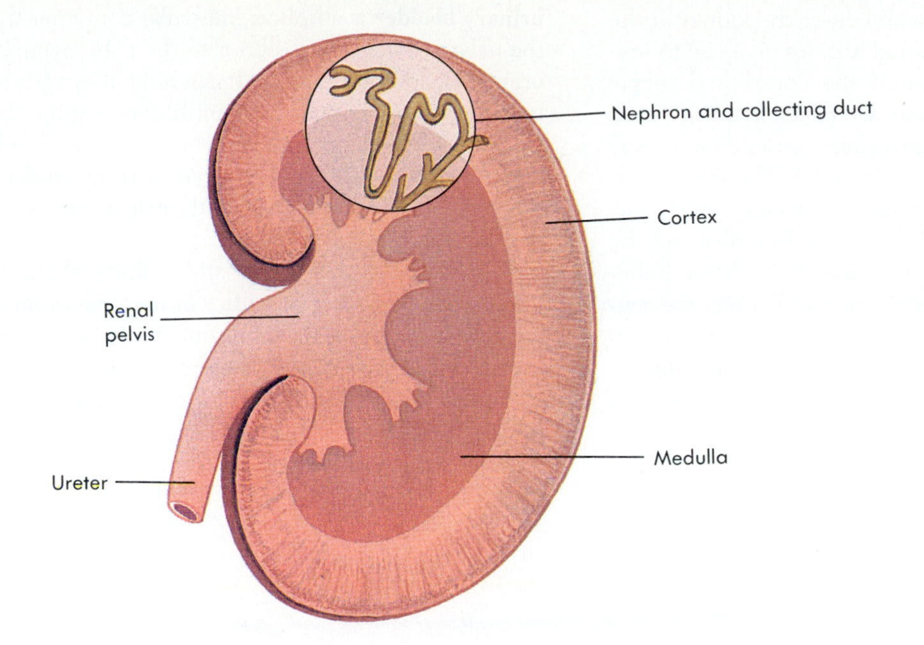
A
Nephron and collecting duct
Cortex
Renal
pelvis
Medulla
Ureter

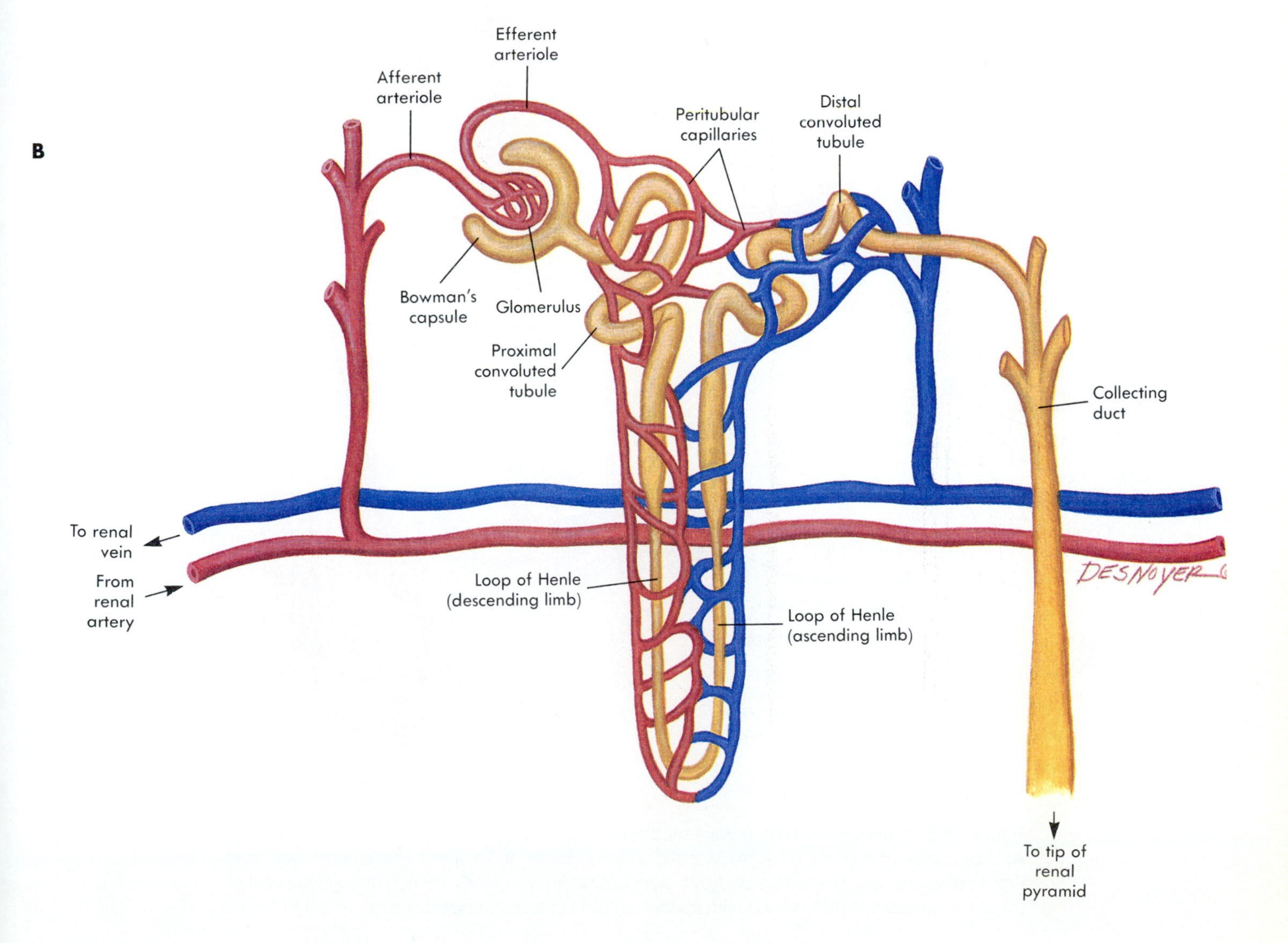
B
Efferent
arteriole
Afferent
arteriole
Peritubular
capillaries
Distal
convoluted
tubule
Bowman's
capsule
Glomerulus
Proximal
convoluted
tubule
Collecting
duct
To renal
vein
From
renal
artery
Loop of Henle
(descending limb)
Loop of Henle
(ascending limb)
To tip of
renal
pyramid
DESNOYER

Arteries and veins

Approximately 20% of the blood pumped by the heart each minute flows through the kidneys. The **renal arteries** branch off the abdominal aorta and enter the kidney at the hilum (see Figure 18-3). The renal arteries give rise to several branches that extend toward the cortex and supply blood to the kidney. An **afferent arteriole** extends to each glomerulus, and an **efferent arteriole** carries blood away from each glomerulus (see Figure 18-4, *B*). The efferent arterioles extend to the **peritubular capillaries,** which surround the proximal and distal convoluted tubules and the loop of Henle (see Figure 18-4, *B*). Blood from the peritubular capillaries returns to the renal veins and enters the general circulation.

Water, ions, and molecules pass from the filtrate in the nephrons into the peritubular capillaries. Although some of the filtrate becomes urine, most of it is reabsorbed from the nephron.

Ureters, urinary bladder, and urethra

The **ureters** (u-re´turz) are small tubes that carry urine from the renal pelvis to the urinary bladder (Figure 18-5). The **urinary bladder** is a hollow, muscular container that lies in the pelvic cavity just posterior to the pubic symphysis. The urinary bladder stores urine. It can hold from a few milliliters to a maximum of about 1000 milliliters of urine. Normally, the urinary bladder is emptied when it reaches a volume of a few hundred milliliters. The **urethra** (u-re´thrah) is a tube that exits the urinary bladder and carries urine to the outside of the body.

The walls of the ureter and urinary bladder contain smooth muscle cells. Peristaltic contractions of smooth muscle cells force urine to flow from the kidneys through the ureters to the urinary bladder. Contractions of smooth muscle in the urinary bladder wall force urine to flow from the urinary bladder through the urethra to exit the body.

Where the urethra exits the urinary bladder, the smooth

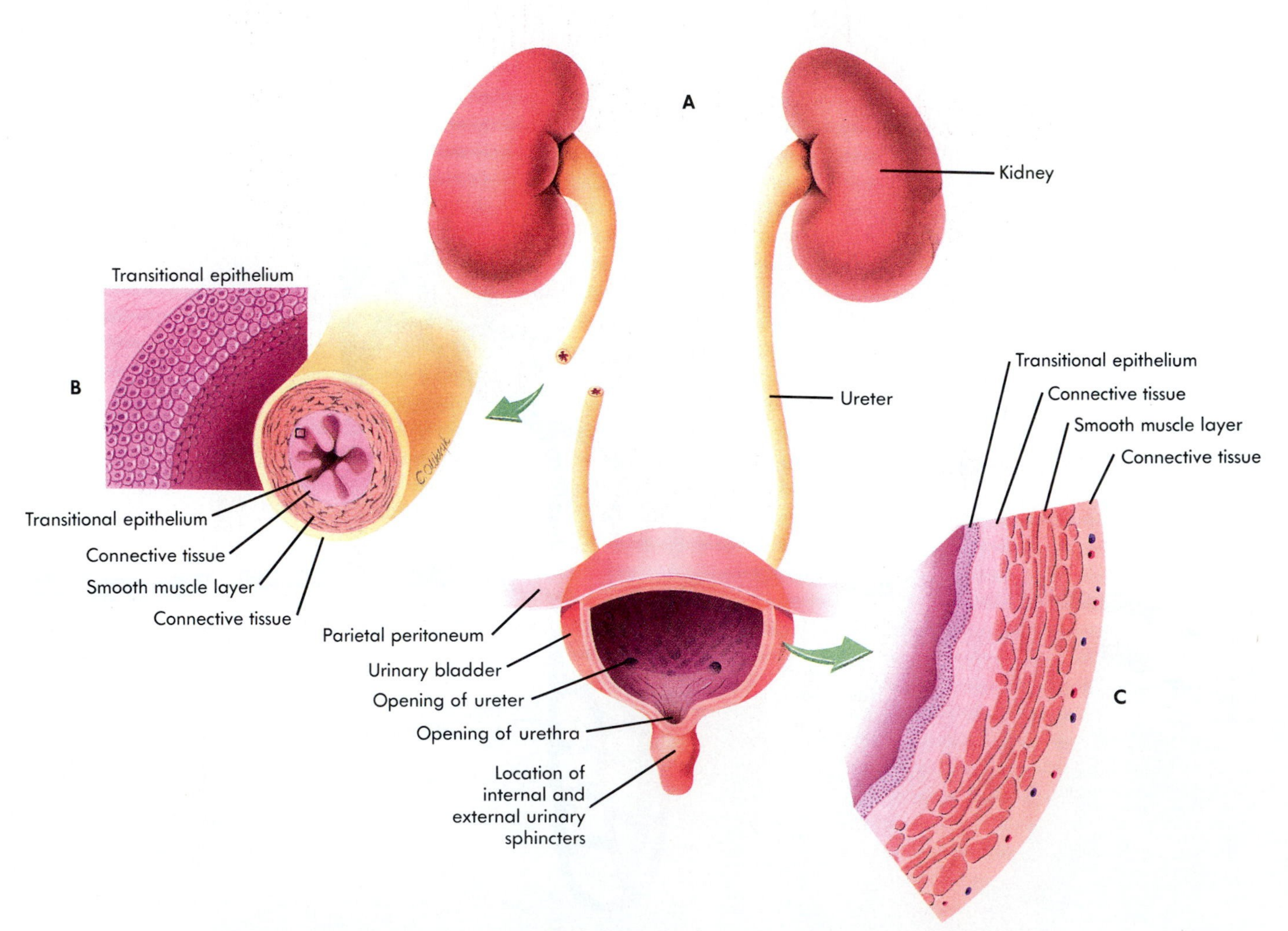

Figure 18-5 ● Ureters and the urinary bladder.
A, Ureters extend from the renal pelvis to the urinary bladder. **B,** The walls of the ureters are lined with epithelium, which is surrounded by a layer of smooth muscle. **C,** Section through the wall of the urinary bladder, which is lined with epithelium and contains abundant smooth muscle.

muscle of the bladder wall forms the **internal urinary sphincter.** The **external urinary sphincter** is skeletal muscle that surrounds the urethra near the urinary bladder. The sphincters regulate the flow of urine through the urethra.

In the male, the urethra extends to the end of the penis, where it opens to the outside. The female urethra is much shorter than the male urethra and opens into a space anterior to the vaginal opening.

1 Urinary bladder infection (cystitis) often occurs when bacteria from outside the body enter the bladder through the urethra. Who is more prone to urinary bladder infection, males or females? Explain.*

URINE PRODUCTION

Urine is mostly water and contains organic waste products such as urea, uric acid, and creatinine; and excess ions such as sodium, potassium, chloride, bicarbonate, and hydrogen.

*Answers to predict questions appear in Appendix B at the back of the book.

Urea is a by-product of protein metabolism. When amino acids are used as a source of energy, they are split to yield **ammonia** (NH_3) and molecules that enter carbohydrate metabolism (see Chapter 17). Ammonia is toxic to cells, and it is converted by the liver into **urea,** which is carried by the blood to the kidneys, where it is eliminated.

The three processes critical to the formation of urine are filtration, reabsorption, and secretion. Urine produced by the nephrons consists of the substances that are filtered and secreted into the nephron minus those substances that are reabsorbed (Figure 18-6).

Substances smaller than blood cells and proteins are filtered into Bowman's capsule of the nephron. Most of those small molecules and water are reabsorbed and reenter the circulatory system through the peritubular capillaries. A small volume of water, waste products, and excess ions remain in the nephron. Secretion results in the movement of some additional waste products and ions into the nephron. Consequently, the nephrons produce urine, which consists of a small volume of water containing mostly waste products and excess ions. Blood cells, proteins, and beneficial molecules remain within the circulatory system.

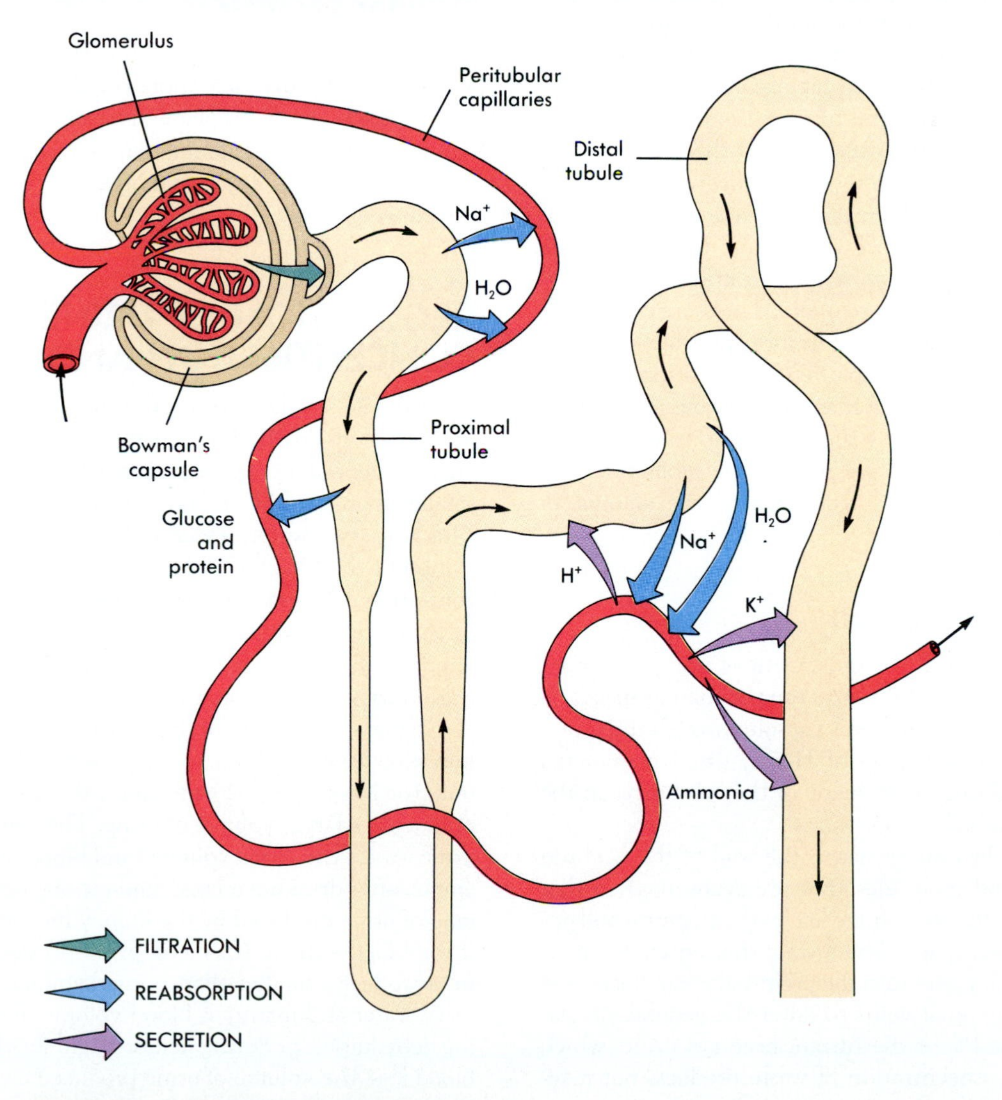

Figure 18-6 ● Overview of urine production.
The three essential processes that result in urine formation are filtration, reabsorption, and secretion. Filtration results in the movement of materials from the glomerulus into Bowman's capsule. Reabsorption is the movement of materials from the nephron into the peritubular capillary. Secretion is the movement of materials from the peritubular capillary into the nephron .

Filtration

Filtration is the movement of fluid from the blood across the wall of the glomerulus and into Bowman's capsule of the nephron (see Chapter 3). Together, the wall of the glomerulus and the wall of Bowman's capsule are called the **filtration membrane.** Of the plasma that flows through each glomerulus, about 19% passes into Bowman's capsule. The portion of the plasma crossing the filtration membrane to enter Bowman's capsule becomes the **filtrate.** In both kidneys, about 180 liters of filtrate are produced each day. Once the filtrate is produced, it flows from Bowman's capsule through the other portions of the nephron.

Water and ions and molecules of small molecular diameter readily pass from the glomerulus across the filtration membrane into Bowman's capsule. However, blood cells and large molecules such as proteins are too large and cannot cross through the filtration membrane. Consequently, the filtrate contains no cells and very little protein.

The formation of filtrate depends on a pressure difference called the **filtration pressure** between the glomerulus and Bowman's capsule. Blood pressure in the glomerular capillary is higher than the pressure inside Bowman's capsule. Consequently, the filtration pressure forces fluid from the glomerulus, across the filtration membrane, into Bowman's capsule. In general, when the filtration pressure increases, the volume of the filtrate increases, and the urine volume increases. When the filtration pressure decreases, the volume of filtrate and the volume of urine produced decrease. If the filtration pressure falls below a certain value, no filtrate crosses the filtration membrane and no urine is produced. For example, a sudden decrease in blood pressure during cardiovascular shock reduces the filtration pressure to nearly zero. In addition, strong sympathetic stimulation of blood vessels during periods of excitement or vigorous physical activity result in constriction of the afferent arteriole. As a result blood flow and blood pressure in the glomerulus decrease, filtration pressure is reduced, and the amount of urine produced decreases.

Tubular reabsorption

Tubular reabsorption is the movement of water and other molecules out of the nephron into the peritubular space. As the filtrate flows from Bowman's capsule through the proximal convoluted tubule, loop of Henle, distal convoluted tubule, and collecting ducts, many of the substances in the filtrate are reabsorbed.

Water moves by osmosis across the wall of the nephron with the ions and molecules that are reabsorbed so that about 99% of the filtrate volume leaves the nephron and enters the peritubular space. The filtrate that enters the peritubular space then passes into the peritubular capillaries and flows through the renal veins to enter the general circulation. Only about 1% of the filtrate becomes urine, which contains a high concentration of waste products not reabsorbed from the filtrate.

Substances reabsorbed from the nephron are useful, and reabsorption helps prevent them from being lost in the urine. For example, inorganic salts such as sodium and chloride ions and organic molecules such as glucose and protein molecules are reabsorbed. Protein and glucose are found in urine only in very small amounts unless these substances pass across the filtration membrane in abnormally high concentrations.

2 People who suffer from untreated diabetes mellitus can experience very high blood levels of glucose (blood sugar). The glucose can easily cross the filtration membrane into Bowman's capsule. Although some glucose can be reabsorbed across the wall of the nephron, if the concentration of glucose in the nephron is too high, not all of the glucose is reabsorbed. How will the volume of urine produced by a person with untreated diabetes mellitus differ from that of a normal person? (Hint: note that water passes across the wall of the nephron by osmosis and that a high level of glucose in the filtrate will influence water movement.)

Tubular secretion

Movement of substances across the wall of the nephron into the filtrate is **tubular secretion.** Some substances, including many toxic by-products of metabolism, are secreted into the nephron. For example, ammonia, excess hydrogen ions, and excess potassium ions are among those substances secreted into the nephron.

REGULATION OF URINE CONCENTRATION AND VOLUME

The volume and composition of urine changes, depending on conditions in the body. If the blood concentration increases (the water content decreases), the kidneys produce a smaller-than-normal amount of very concentrated urine. This conserves water, which helps prevent further concentration of the blood. On the other hand, if the blood concentration decreases (the water content increases), the kidneys produce a larger volume of dilute urine, which helps eliminate the excess water and increases the blood concentration back toward normal.

Changes in the rate of urine production also help maintain blood volume (and blood pressure). When blood volume (or blood pressure) increases, the kidneys respond by producing a larger volume of urine. The loss of water in the urine lowers the blood volume (and blood pressure). For example, after drinking a large amount of a beverage, the volume of urine produced by the kidney increases. Conversely, if the blood volume (or blood pressure) decreases, the kidneys produce a smaller-than-normal volume of urine to conserve water and maintain blood volume. For example, during dehydration or during hemorrhagic shock resulting from blood loss, the volume of urine produced by the kidneys decreases.

HORMONAL MECHANISMS

Hormones play important roles in controlling the functions of the kidneys. Hormones influence urine volume and urine concentration as well as help regulate the concentration of ions, such as sodium, potassium, calcium, and hydrogen ions, in the blood. The major hormones that influence the functions of the kidneys are antidiuretic hormone, aldosterone, and atrial natriuretic hormone.

Antidiuretic hormone

Antidiuretic (an´tĭ-di-u-ret´ik) **hormone** (ADH), secreted by the posterior pituitary gland, passes through the circulatory system to the kidneys. Certain cells in the hypothalamus are sensitive to changes in the concentration of the blood. If the concentration of the blood increases, these cells cause ADH to be released from the posterior pituitary. When ADH levels in the blood increase, the permeability of the distal convoluted tubules and collecting ducts to water increases, and more water moves by osmosis from the nephron to the peritubular space. Consequently, an increase in ADH secretion results in the production of a smaller volume of more concentrated urine. This helps prevent the loss of water from the body in the form of urine and helps prevent a further concentration of the blood.

On the other hand, a reduced concentration of blood inhibits ADH secretion. When ADH levels decrease, the distal convoluted tubules and collecting ducts become less permeable to water. As a result, less water is reabsorbed, and a larger volume of dilute urine is produced (Figure 18-7). The production of a large volume of dilute urine helps increase the concentration of the blood toward its normal range of values.

Diabetes insipidus (in-sip´ĭ-dus) results when the posterior pituitary fails to secrete ADH, or when the kidney tubules cannot respond to ADH. Without a normal response to ADH, much of the filtrate entering the nephron becomes urine. People with diabetes insipidis produce a large volume of dilute urine (as much as 20 to 30 liters) each day. Because they lose so much water, they are continually in danger of severe dehydration and ionic imbalances.

Aldosterone

Aldosterone (al-dos´ter-ōn), secreted by the adrenal glands, increases the rate of sodium and chloride ion reabsorption in the distal portion of the nephron and in the collecting ducts. The secretion of aldosterone by the adrenal glands is controlled by a series of events that starts with the release of an enzyme called **renin** (ren´in) from the kidney. Renin is released from specialized structures in the kidney called **juxtaglomerular** (juks´tă-glo-mĕr´u-lar) **complexes.** A juxtaglomerular complex is found where the distal convoluted tubule of a nephron passes between the afferent and efferent arterioles. The juxtaglomerular complex consists of specialized cells in the wall of the afferent arteriole that are in contact with specialized cells of the distal convoluted tubule.

Renin causes a protein in the blood called angiotensinogen to be converted to angiotensin I, which is rapidly converted to angiotensin II. Angiotensin II stimulates aldosterone secretion from the adrenal gland (see Chapter 13, Figure 13-13, *B*).

Renin secretion by the kidney increases when blood pressure decreases (see Figure 13-13, *B*). The resultant increase in aldosterone causes sodium ions and water to be reabsorbed from the nephron. Thus the volume of water lost as urine declines, and the conservation of water helps prevent a further decline in blood pressure.

Renin secretion by the kidney decreases as the blood pressure increases, resulting in a decline in aldosterone secretion. Consequently, reabsorption of sodium and water from the nephron slows and more water and sodium are lost in the urine. The loss of water decreases blood volume and blood pressure.

3 One result of Addison's disease is an abnormally low level of aldosterone secretion. Would you expect a person with Addison's disease to have abnormally high or low blood pressure? Explain.

Atrial natriuretic hormone

Atrial natriuretic (na´tre-u-ret´ik) **hormone** is secreted by cells in the wall of the atria of the heart. It is released from the atria when they are stretched. Atrial natriuretic hormone travels in the blood to the kidney where it increases the volume of urine produced and the sodium ion concentration of the urine. Atrial natriuretic hormone appears to be important after the ingestion of a large volume of fluid containing salt. In response to the increased atrial natriuretic hormone, a larger volume of urine containing a substantial amount of salt is produced until the blood volume decreases to its normal concentration range.

Diuretics (di´u-ret´ikz) are substances that increase urine volume. There are several mechanisms by which diuretics increase urine volume. Alcohol and caffeine are examples of common diuretics. Alcohol inhibits the secretion of ADH from the posterior pituitary. Consequently, alcoholic beverages result in the formation of a large volume of dilute urine. The volume of urine easily can exceed the volume of water consumed with the alcohol. Several hours after drinking alcoholic beverages (such as the morning after an intense celebration), some dehydration and intense thirst may exist. Caffeine and related substances increase urine volume by increasing blood pressure in the glomerular capillaries, thus increasing the filtration pressure.

4 Some diuretics inhibit the reabsorption of sodium ions in the nephron. Explain how these diuretic drugs could cause an increase in urine volume.

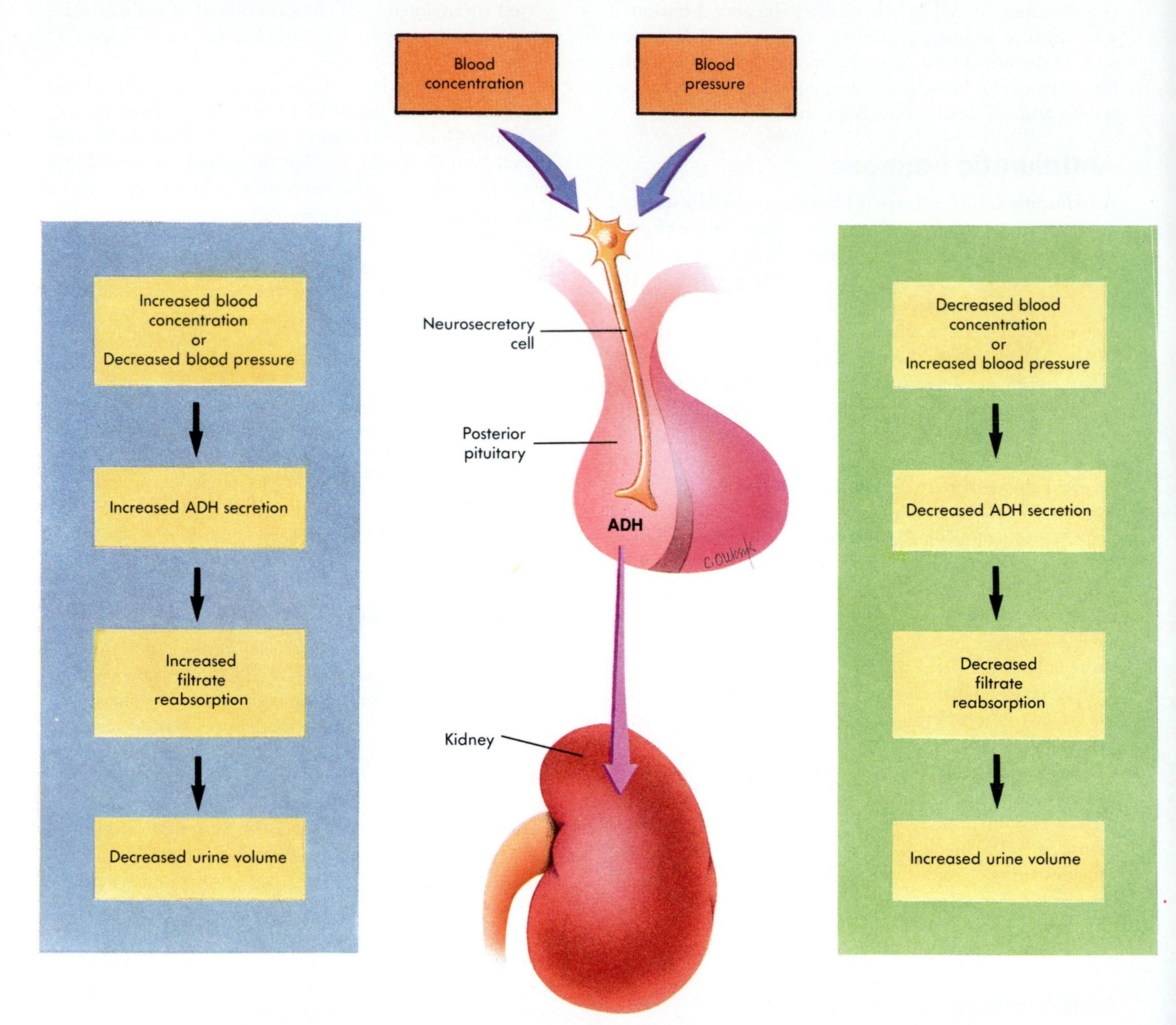

Figure 18-7 ● ADH secretion.
Increased blood concentration or decreased blood pressure results in increased ADH secretion from the posterior pituitary. The increased ADH acts on the distal convoluted tubule and collecting duct, causing an increased reabsorption of water from the nephron and the production of a smaller-than-normal volume of concentrated urine. Decreased blood concentration or increased blood pressure results in decreased ADH secretion from the posterior pituitary. The decreased ADH causes a decreased reabsorption of water from the nephron and the production of a large volume of dilute urine.

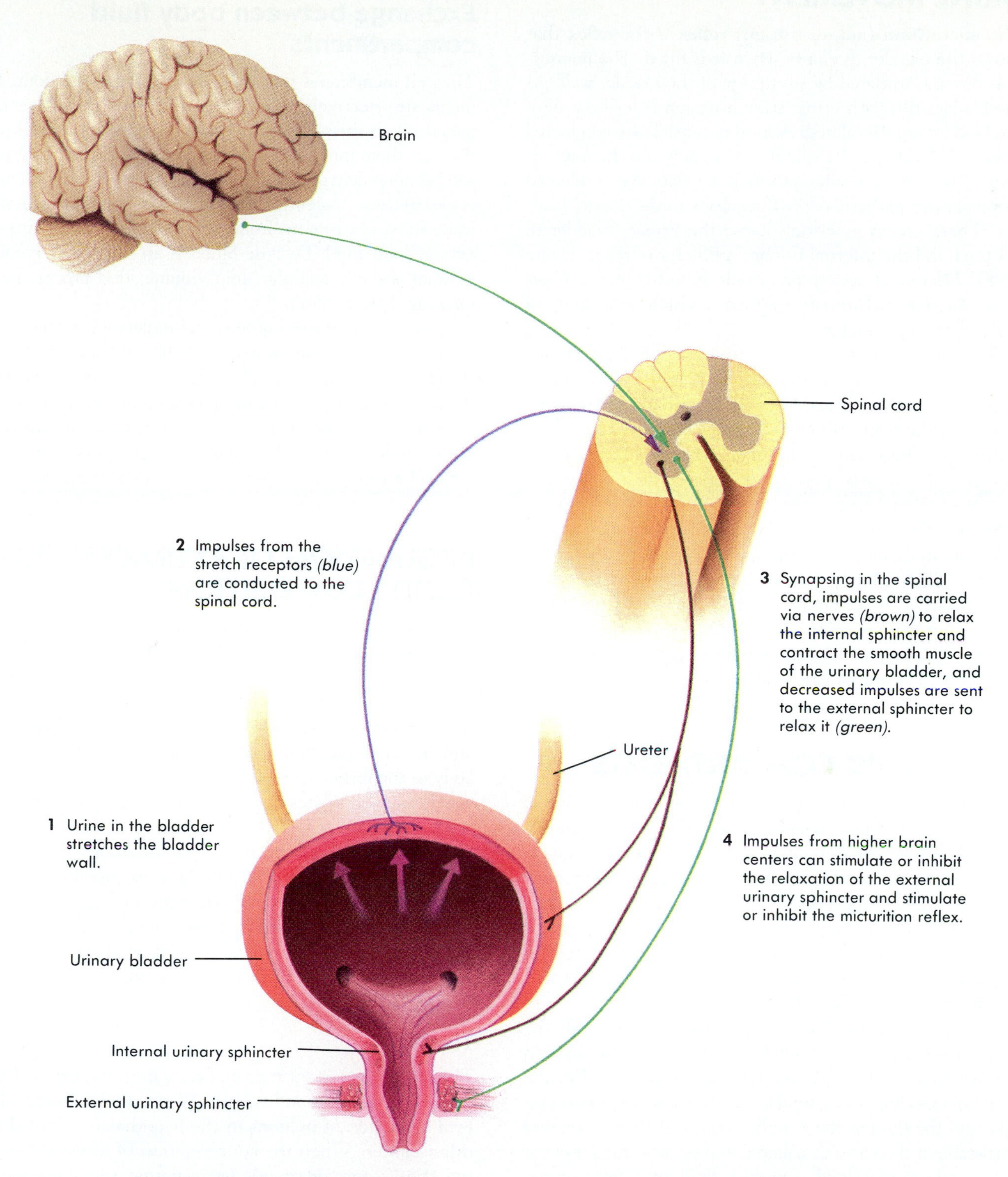

Figure 18-8 ● The micturition reflex.
An increased volume of urine in the bladder stretches the bladder wall, sending impulses *(blue)* along the spinal nerves to the spinal cord. The action potentials result in increased parasympathetic action potentials *(brown)*, which return to the bladder causing the bladder wall to contract and the internal urinary sphincter to relax. Decreased action potientials are sent to the external urinary sphincter to cause it to relax. Action potentials are also sent to the brain from the spinal cord stimulating the desire to urinate. When convenient, the brain sends action potentials to the spinal cord that stimulate the micturition reflex. The bladder wall contracts, the urinary sphincters relax, and urine is expelled.

URINE MOVEMENT

The **micturition** (mik-tu-rish´un) **reflex** is the reflex that causes the bladder to empty when it is filled. The micturition reflex is initiated by stretching of the bladder wall. As the bladder fills with urine, stretch receptors in the wall of the bladder are stimulated. Action potentials are conducted from the bladder to the spinal cord to activate the micturition reflex. As a result, action potentials are conducted along parasympathetic nerve fibers back to the urinary bladder. These action potentials cause the urinary bladder to contract and the internal urinary sphincter to relax (Figure 18-8). Decreased action potentials in other nerve fibers cause the external urinary sphincter, which is made up of skeletal muscle, to relax.

The micturition reflex is an automatic spinal cord reflex that occurs when the urinary bladder is stretched. Whether or not micturition occurs, however, is influenced by higher centers in the brain. When the urinary bladder is stretched, action potentials sent to the brain stimulate the desire to urinate. If the time is appropriate, action potentials are sent from the brain to the spinal cord to facilitate the micturition reflex. On the other hand, if the time is not appropriate, action potentials are sent to the spinal cord that inhibit micturition. Thus voluntary control of micturition is possible.

In addition to stretch, irritation of the urinary bladder or the urethra such as occurs during bacterial infections of the urinary bladder, also initiate the urge to urinate, even though the bladder may be nearly empty.

BODY FLUID COMPARTMENTS

For a 70-kilogram adult, approximately 40 liters, or 57% of the total body weight, consists of water. The water and its electrolytes are distributed in two major body compartments. **Electrolytes** include all of the ions dissolved in water such as sodium, potassium, calcium, and chloride ions. The **intracellular fluid compartment** includes all the water and electrolytes inside the cells of the body. Approximately 25 of the 40 liters of body fluid, or 63% of the total body water, is contained within the cells.

The **extracellular fluid compartment** includes all the fluid outside the cells. About 37% of the total body water is extracellular fluid, which includes the water found in the tissue spaces (interstitial fluid), the plasma within blood vessels, and the fluid in the lymph vessels. Smaller volumes of extracellular fluid are the aqueous and vitreous humor of the eye, cerebrospinal fluid, synovial fluid in joint cavities, serous fluid in the body cavities, and fluid secreted by glands.

Intracellular fluid has a similar composition from cell to cell. Like intracellular fluid, the extracellular fluid has a fairly constant composition. However, the concentrations of sodium, chloride, and bicarbonate ions are higher, and the concentrations of protein, potassium, calcium, magnesium, phosphate, and sulfate ions are lower in the extracellular fluid when compared with intracellular fluid.

Exchange between body fluid compartments

The cell membranes that separate the body fluid compartments are selectively permeable (see Chapter 3). Water can pass through them easily, but most ions and molecules pass through them much more slowly. For example, when a person becomes dehydrated, the concentration of ions and molecules in the extracellular fluid increases. As a consequence, water moves by osmosis from the intracellular fluid into the extracellular fluid. Because blood is an important component of the extracellular fluid volume, this process helps maintain blood volume.

If the concentration of ions and molecules in the extracellular fluid decreases, water moves by osmosis from the extracellular fluid into the cells. The water movement causes the cells to swell. Movement of water between the intracellular and extracellular fluid compartments occurs continuously. Under most conditions, the movement is maintained within limits consistent with survival of the individual.

REGULATION OF EXTRACELLULAR FLUID COMPOSITION

Homeostasis requires that the intake of substances such as water and electrolytes is equal to their elimination. Ingestion of water and electrolytes adds them to the body, whereas they are excreted primarily by the kidneys. Some water and electrolytes are eliminated through the liver, skin, and digestive tract. Large amounts of water are lost from the body in the form of perspiration (sweat) on warm days, and water and electrolytes can be lost in varying amounts in feces. Over a long period, the total amount of water and electrolytes entering the body is equal to the amount eliminated from the body unless the individual is growing, gaining weight, or losing weight. The regulation of water and electrolytes involves the participation of several organ systems, but the most important organ regulating the loss of water and electrolytes from the body is the kidney.

Thirst

Thirst is an important means of regulating the extracellular fluid concentration and volume. Water intake is controlled by a collection of neurons in the hypothalamus called the **thirst center.** When the concentration of blood increases, the thirst center responds by initiating the sensation of thirst. When water or some other dilute solution is consumed, the concentration of the blood decreases and so does the sensation of thirst.

When blood pressure decreases, such as during shock, the thirst center is also activated and the sensation of thirst is triggered. Consumption of water increases the blood volume and allows the blood pressure to increase toward its normal value.

Stimuli other than blood concentration and blood pressure can also temporarily trigger the sensation of thirst. For example, if the mouth becomes dry, the thirst center is activated.

Ions

Keeping the extracellular fluid composition within a normal range of values is required to sustain life. If the concentration of ions in the extracellular fluid deviate from their normal range, cells cannot function normally and cell death can occur.

The concentrations of positively charged ions such as sodium, potassium, and calcium ions in the body fluids are particularly important. A normal range of concentrations is necessary for the conduction of action potentials, contraction of muscles, and maintenance of normal cell membrane permeability. Important control mechanisms regulate the concentration of these ions in the body.

In general, negatively charged ions are attracted to positively charged ions. There are fewer control mechanisms that regulate the negatively charged ions. Also, there are fewer conditions in which the concentrations of negatively charged ions are abnormal.

Sodium ions

Sodium ions are one of the dominant types of extracellular ions. The recommended intake of sodium is 2.4 grams per day. However, most people in the United States consume 20 to 30 times more than the amount needed. The kidneys provide the major route by which the excess sodium ions are excreted. The primary mechanisms that regulate the sodium ion concentration in the extracellular fluid are the same mechanisms that regulate blood pressure and changes in the concentration of the extracellular fluid.

Potassium ions

Electrically excitable tissues, such as muscle and nerve, are highly sensitive to slight changes in the extracellular potassium concentration. The extracellular concentration of **potassium ions** must be maintained within a narrow range for these tissues to function normally.

Aldosterone plays a major role in regulating the concentration of potassium ions in extracellular fluid. Circulatory system shock resulting from plasma loss or dehydration increases the concentration of potassium in the extracellular fluid. Tissue damage, for example in burn patients, results in the release of potassium from cells into the extracellular fluid.

Increased blood levels of potassium strongly stimulate aldosterone secretion. Aldosterone, in addition to increasing the rate at which sodium ions are reabsorbed from the nephron, stimulates the rate of potassium secretion into the nephron. Consequently, potassium ions are secreted into the nephron and the concentration of potassium ions in the blood declines (Figure 18-9).

Calcium ions

The extracellular concentration of **calcium ions,** like that of potassium ions, is maintained within a narrow range. Increases and decreases in the extracellular concentration of calcium ions have dramatic effects on the electrical properties of excitable tissues. The mechanisms that control calcium levels in the blood are outlined in Chapter 10.

REGULATION OF ACID-BASE BALANCE

The concentration of hydrogen ions in the body fluids is reported as the pH of the body fluids (see Chapter 2). When the hydrogen ion concentration increases, pH decreases, and when the hydrogen ion concentration decreases, pH increases. The pH of body fluids is controlled by buffers in those fluids, the respiratory system, and the kidneys. The pH of the body fluids is maintained between 7.35 and 7.45. Any deviation from that range is life-threatening. Consequently the mechanisms that regulate body pH are critical for survival.

Buffers

Buffers are chemicals that resist a change in the pH of a solution when either acids or bases are added to the solution (see Chapter 2). The buffers found in the body fluids combine with hydrogen ions when excess hydrogen ions are added to a solution, or release hydrogen ions when bases are added to a solution. Because of these characteristics, buffers tend to keep the hydrogen ion concentration and thus the pH within a narrow range of values. The three principal classes of buffers in the body fluids are the proteins, the phosphate buffer system, and the bicarbonate buffer system. Proteins and phosphate ions are in high concentrations in the body fluids and are important components of the buffer system of the body.

The bicarbonate buffer system is unable to combine with as many hydrogen ions as the protein and phosphate buffers, but it is critical because it can be regulated by the respiratory system and kidneys. Carbon dioxide in the body fluids (CO_2) combines with water (H_2O) to form hydrogen ions (H^+) and bicarbonate ions (HCO_3^-):

$$CO_2 + H_2O \longleftrightarrow H^+ + HCO_3^-$$

A change in carbon dioxide levels can alter the pH of the body fluids. For example, an increase in the concentration of carbon dioxide causes a greater number of hydrogen ions and bicarbonate ions to be formed. The increase in hydrogen ions results in a decreased pH. On the other hand, the reaction is reversible. If carbon dioxide levels decline, hydrogen and bicarbonate ions combine to form carbon dioxide and water. The decrease in hydrogen ions results in an increase in pH.

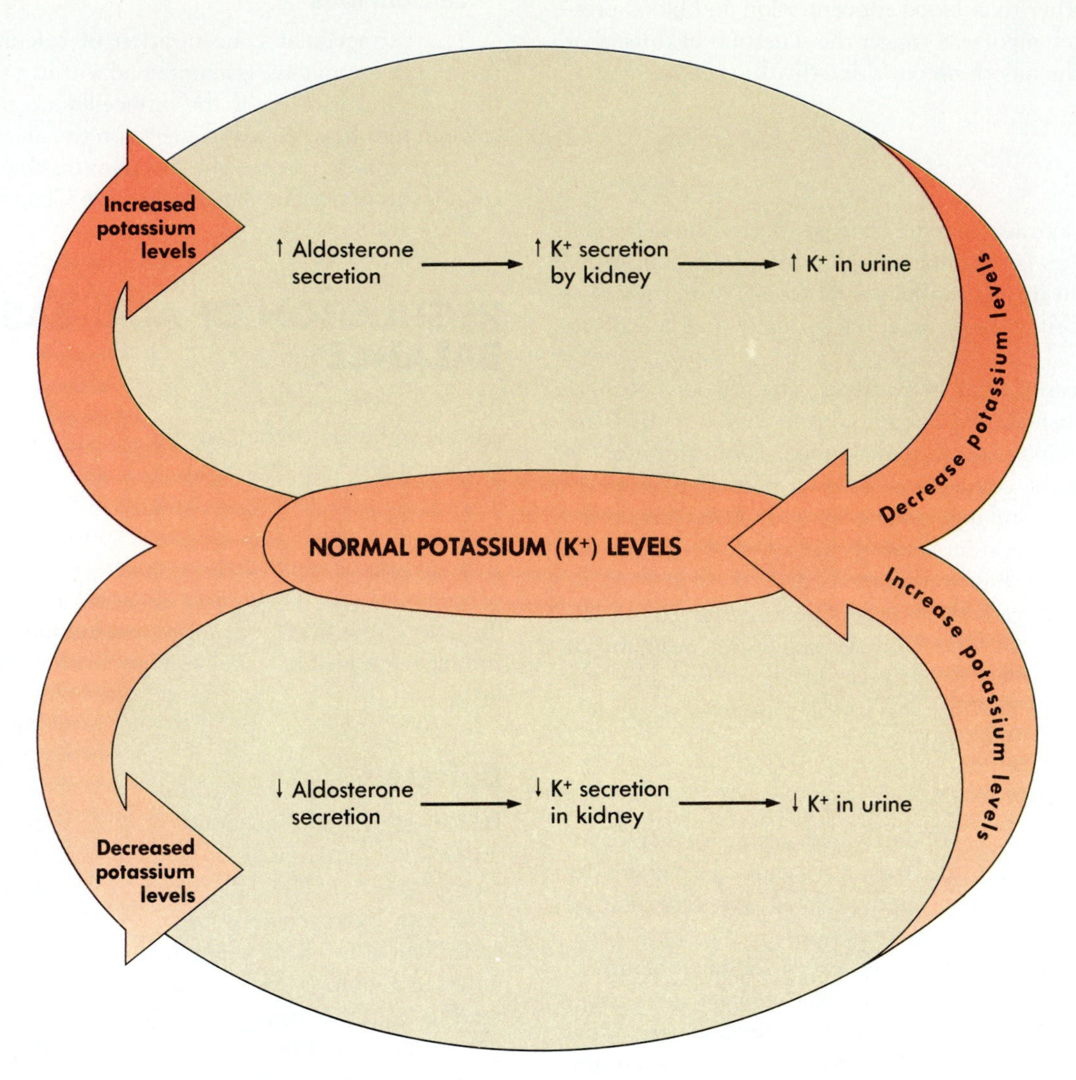

Figure 18-9 ● Regulation of potassium.
Increased extracellular potassium stimulates aldosterone secretion. Aldosterone acts on the kidneys, resulting in increased potassium secretion into the nephron, which lowers the extracellular potassium concentration. Decreased extracellular potassium inhibits aldosterone secretion, resulting in a reduced rate of potassium secretion into the nephron. The result is an increase in extracellular potassium.

Respiratory system

The **respiratory system** responds rapidly to changes in pH and functions to bring the pH of body fluids back toward its normal range. If the metabolic activity increases, such as during exercise, or if respiration is not adequate, carbon dioxide accumulates in the body fluids. Increasing carbon dioxide levels and a corresponding decreasing body fluid pH stimulate neurons in the respiratory center of the brain and cause the rate and depth of respiration to increase. As a result, carbon dioxide is eliminated from the body through the lungs at a greater rate, the concentration of carbon dioxide in the body fluids decreases, and the pH of body fluids increases.

If the blood levels of carbon dioxide decrease, resulting in a corresponding increase in the blood pH, or if the pH of the body fluids increases for other reasons, the rate of respiration decreases in response. The lower rate of respiration allows carbon dioxide to accumulate in the circulatory system and the blood pH to increase toward its normal value (Figure 18-10).

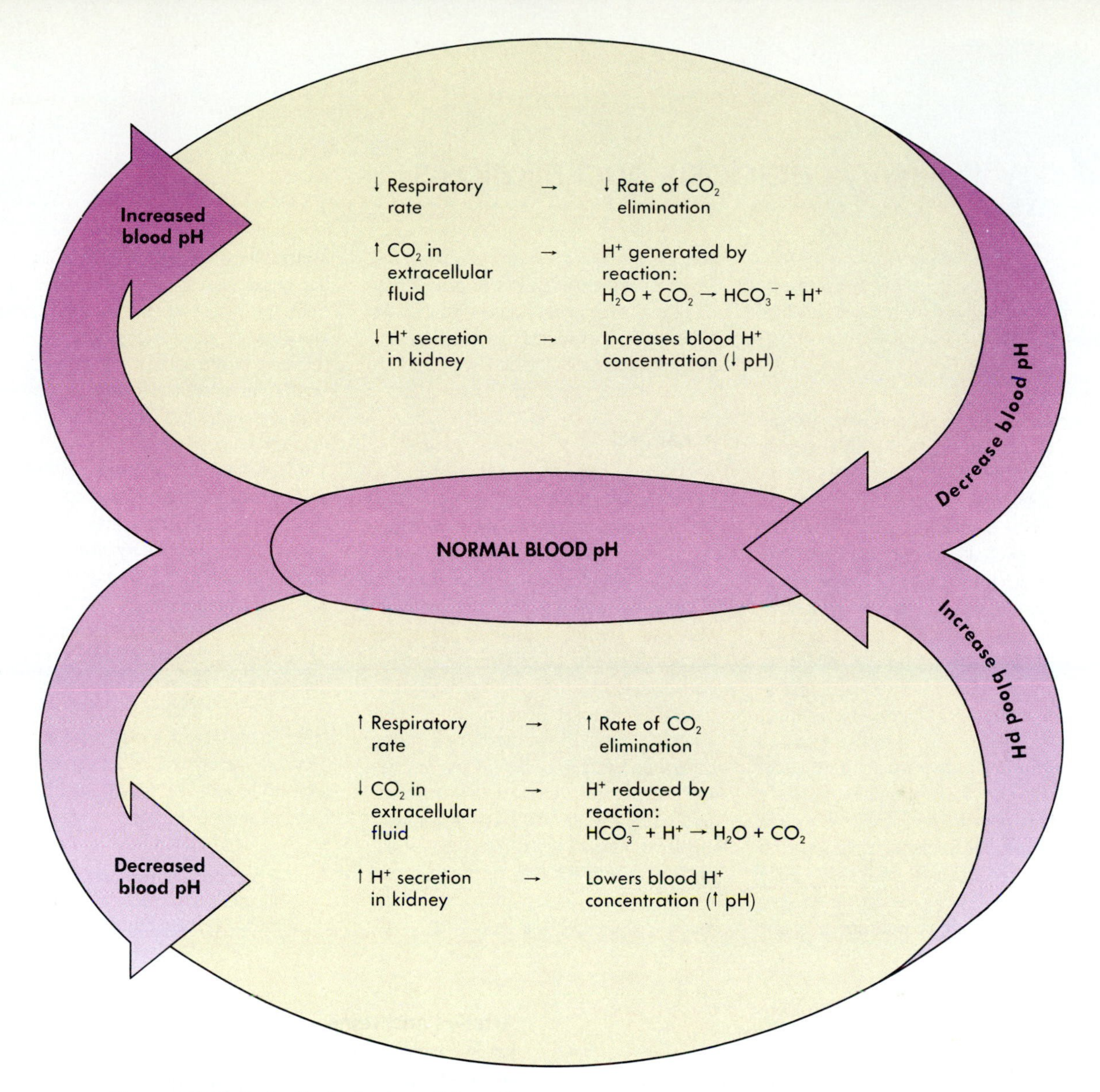

Figure 18-10 ● Control of extracellular pH.
When blood pH increases, the rate of respiration slows, resulting in the accumulation of carbon dioxide in the circulatory system. The carbon dioxide reacts with water to form hydrogen ions and bicarbonate ions, resulting in a decrease in pH. In addition, the rate of hydrogen ion secretion by the kidneys declines. When blood pH decreases, the rate of respiration increases, resulting in the elimination of carbon dioxide from the circulatory system. Because carbon dioxide is eliminated, the pH of the extracellular fluid increases. In addition, the rate of hydrogren ion secretion by the kidneys increases.

5 Under stressful conditions, some people hyperventilate. Predict the effect of the rapid rate of ventilation on the pH of body fluids. In addition, explain why a person who is hyperventilating may benefit from breathing into a paper bag.

Kidneys

The kidney is a powerful regulator of pH, but it takes several hours to respond, whereas the respiratory system begins to respond almost immediately. Cells in the walls of the tubules of nephrons secrete hydrogen ions into the filtrate and therefore can regulate pH of the body fluids directly. As the pH of the body fluids decreases below normal, the rate at which the kidneys secrete hydrogen ions into the urine increases. The removal of hydrogen ions from the body fluids causes the pH of the body fluids to increase toward normal.

On the other hand, as the pH of the body fluids increases above normal, the rate of hydrogen ion secretion by the kidneys declines. Consequently, the concentration of hydrogen ions in the body fluids increases and the pH of the body fluids decreases toward its normal value (see Figure 18-10).

Understanding acidosis and alkalosis

Failure of the buffer systems, the respiratory system, or the urinary system to maintain normal pH levels can result in acidosis or alkalosis.

ACIDOSIS

Acidosis (as-ĭ-do´sis) occurs when the pH of the blood falls below 7.35. The central nervous system malfunctions, and the individual becomes disoriented and, as the condition worsens, possibly comatose. There are two categories of acidosis. **Respiratory acidosis** results when the respiratory system is unable to eliminate adequate carbon dioxide. Carbon dioxide accumulates in the circulatory system resulting in an increased formation of hydrogen ions in the body fluids, which causes the pH to decline. For example, injury to the respiratory system resulting in a collapsed lung or blockage of the respiratory passageways are causes of respiratory acidosis. **Metabolic acidosis** results from the excessive production of acidic substances because of increased metabolism or because of a decreased ability of the kidneys to eliminate hydrogen ions in the urine. For example, excessive production of acidic substances occurs in people suffering from diabetes mellitus, and kidney failure reduces the ability of the kidney to eliminate hydrogen ions.

ALKALOSIS

Alkalosis (al´kah-lo´sis) occurs when the pH of the blood increases above 7.45. A major effect of alkalosis is hyperexcitability of the nervous system. Peripheral nerves are affected first, resulting in spontaneous nervous stimulation of muscles. Muscular spasms, sustained muscular contractions, and extreme nervousness or convulsions may occur. Sustained contractions of respiratory muscles can cause death because normal breathing requires contraction and relaxation of the respiratory muscles (see Chapter 15). There are two categories of alkalosis. **Respiratory alkalosis** results from hyperventilation, such as may occur in response to stress. Hyperventilation causes the loss of too much carbon dioxide from the lungs. **Metabolic alkalosis** usually results from the rapid elimination of hydrogen ions from the body, such as during severe vomiting, or when excess aldosterone is secreted by the adrenal glands (excess aldosterone causes an increase of hydrogen ion secretion into the filtrate of the nephron).

In the case of respiratory alkalosis or acidosis, the kidneys help compensate for the change in pH. The kidneys reduce the severity of the respiratory alkalosis or acidosis. For example, if a person is suffering from respiratory acidosis, due to a condition such as severe emphysema, the kidney increases its rate of hydrogen ion secretion into the filtrate to reduce the hydrogen ion concentration. On the other hand, in the case of metabolic acidosis or alkalosis, the respiratory system compensates to reduce the severity of the metabolic alkalosis or acidosis.

6 If a person is suffering from metabolic acidosis, how does the respiratory system compensate?

SUMMARY

URINARY SYSTEM

The urinary system consists of the kidneys, ureters, urinary bladder, and urethra.

The urinary system removes waste products from the blood; helps control blood volume, ion concentration, and pH; and controls RBC production.

Kidneys

Each kidney is surrounded by a connective tissue and a layer of fat.

The ureter expands to form the renal pelvis within the renal sinus, and the renal pelvis has extensions called calyces.

The kidney is divided into an outer cortex and an inner medulla.

Each renal pyramid in the medulla has a base that ex tends into the cortex. The tip of each renal pyramid projects to a calyx.

The functional unit of the kidney is the nephron. The parts of a nephron are Bowman's capsule, the proximal convoluted tubule, the loop of Henle, and the distal convoluted tubule.

The filtrate passes from glomerular capillaries into Bowman's capsule.

Arteries and veins

Renal arteries give rise to afferent arterioles.

Afferent arterioles supply the glomeruli, and efferent arterioles carry blood from glomeruli to peritubular capillaries.

Blood from peritubular capillaries flows to renal veins.

Ureters, urinary bladder, and urethra

Ureters carry urine from the renal pelvis to the urinary bladder. The urethra carries urine from the urinary bladder to the outside of the body.

The ureters and urinary bladder are lined with epithelium and have smooth muscle in their walls.

The internal and external urinary sphincter muscles regulate the flow of urine through the urethra.

URINE PRODUCTION

Urine is produced by the processes of filtration, reabsorption, and secretion.

Filtration

The filtration membrane consists of the wall of the glomerulus and the wall of Bowman's capsule.

The filtrate passes through the filtration membrane and contains no blood cells and few blood proteins.

Filtration pressure forces fluid through the filtration membrane into Bowman's capsule.

Disorders of the urinary system

Glomerular nephritis (ne-fri´tis) is the inflammation of the wall of the glomerulus and the wall of Bowman's capsule. The permeability of the wall of the glomerulus and Bowman's capsule increases, and plasma proteins and white blood cells (WBCs) enter the filtrate. The plasma proteins cause the urine volume to increase because they increase the concentration of the filtrate.

Acute glomerular nephritis often occurs 1 to 3 weeks after a severe bacterial infection, such as streptococcal sore throat or scarlet fever. Antigen-antibody complexes (see Chapter 14) associated with the disease become deposited in the wall of the glomerulus and the wall of Bowman's capsule and cause inflammation. Acute glomerular nephritis normally subsides after several days.

Chronic glomerular nephritis is long term and usually progressive. The wall of the glomerulus and Bowman's capsule thickens and eventually is replaced by connective tissue. In its early stages, chronic glomerular nephritis resembles the acute form. In the advanced stages, many Bowman's capsules and glomeruli are replaced by connective tissue, and the kidney eventually becomes nonfunctional.

Cystitis (sis-ti´tis) is inflammation of the urinary bladder, usually resulting from bacterial infection. The inflammation causes diffuse pain in the lower back and a burning sensation during urination. The irritation of the urinary bladder results in a frequent urge to urinate.

Kidney stones are precipitates of substances such as calcium salts and usually form in the renal pelvis. They can cause irritation and increase the chance of kidney infection, and small parts can break off and pass through the ureter. Passage of a kidney stone through the ureter usually is very painful. Ultrasound techniques have been developed to pulverize kidney stones without the use of surgery.

Renal failure can result from any condition that interferes with kidney function. **Acute renal failure** occurs when damage to the kidney is rapid and extensive. It leads to the accumulation of urea and other metabolites in the blood and to acidosis. If renal failure is complete, death can occur in 1 to 2 weeks. Acute renal failure can result from acute glomerular nephritis, or it can be caused by damage to, or blockage of, the renal tubules. Lack of blood supply or exposure to certain toxic substances can also cause damage to the epithelial cells of the nephron and lead to acute renal failure.

Chronic renal failure is the result of permanent damage to so many nephrons that the remaining nephrons are inadequate for normal kidney function. Chronic renal failure can result from chronic glomerular nephritis, trauma to the kidneys, tumors, urinary tract obstruction by kidney stones, or severe lack of blood supply resulting from atherosclerosis. Chronic renal failure leads to the inability to eliminate toxic metabolic by-products. Water retention and edema result from the accumulation of ions and molecules in the body fluids. Potassium levels become elevated and acidosis develops. The toxic effects of accumulated metabolic waste products are mental confusion, coma, and finally death when chronic renal failure is severe.

Dialysis (di-al´ĭ-sis) is used when a person is suffering from severe acute or chronic kidney failure. The procedure substitutes for the excretory functions of the kidney. Renal dialysis is based on blood flow through tubes composed of a selectively permeable membrane. Blood is usually taken from an artery, passed through the tubes of the dialysis machine, and then returned to a vein. On the outside of the dialysis tubes is a fluid that contains the same concentration of ions and molecules as the plasma except for the metabolic waste products. As a consequence, the metabolic wastes diffuse from the blood to the dialysis fluid. The dialysis membrane has pores that are too small to allow blood cells and plasma proteins to pass through them. Smaller beneficial ions and molecules in the plasma do not diffuse across the dialysis membrane because the dialysis fluid has the same concentration of these substances as the plasma.

Kidney transplants are performed on people who suffer from severe renal failure. A transplantation requires a donor kidney from an individual who has an immune system similar to that of the recipient. Usually the donor has suffered an accidental death and has granted permission to have his or her kidneys used for transplantation. In most cases, the transplanted kidney functions well, although the recipient does have to take medication to prevent his or her immune system from rejecting the transplanted kidney. The major cause of kidney transplant failure is rejection by the immune system of the recipient.

Tubular reabsorption

About 99% of the filtrate volume is reabsorbed; 1% becomes urine.

Proteins, amino acids, glucose, and ions, such as sodium, potassium, calcium, bicarbonate, and chloride ions, are among the substances reabsorbed.

Tubular secretion

Some by-products of metabolism are actively secreted into the nephron.

REGULATION OF URINE CONCENTRATION AND VOLUME

Hormonal mechanisms

ADH is secreted from the posterior pituitary when the blood concentration increases or when blood volume (pressure) decreases. ADH increases the permeability to water of the distal convoluted tubule and collecting duct. It increases water reabsorption by the kidney.

Aldosterone, secreted in response to angiotensin II which, in turn is produced in response to renin secretion from the kidney when the blood pressure decreases, causes the rate of sodium chloride reabsorption to increase. It also increases potassium and hydrogen ion secretion.

Atrial natriuretic hormone, secreted from the right atrium in response to increases in blood volume, acts on the kidney to increase sodium and water loss in the urine.

URINE MOVEMENT

Increased volume in the urinary bladder stretches its wall and activates the micturition reflex.

Contraction of the urinary bladder and relaxation of the internal and external urinary sphincters occurs in response.

Higher brain centers can inhibit or facilitate the micturition reflex.

BODY FLUID COMPARTMENTS

Approximately 63% of the total body water is found within cells (intracellular fluid).

Approximately 37% of the total body water is found outside cells (extracellular fluid), mainly in tissue spaces, plasma of blood, and lymph.

The concentrations of sodium, chloride, and bicarbonate ions are higher, and the concentrations of protein, potassium, calcium, magnesium, phosphate, and sulfate ions are lower in the extracellular fluid when compared with intracellular fluid.

Exchange between body fluid compartments

Water moves between body fluid compartments more readily than ions and molecules. Water moves between compartments by osmosis.

REGULATION OF EXTRACELLULAR FLUID COMPOSITION

The total amount of water and electrolytes in the body does not change unless the person is growing, gaining weight, or losing weight.

Thirst

Sensation of thirst increases if blood pressure decreases or the extracellular fluid becomes more concentrated.

Ions

The primary mechanisms controlling the sodium ion concentration in the extracellular fluid are the same as those that control blood concentration and blood pressure.

Aldosterone increases potassium secretion in the urine. Increased blood levels of potassium stimulate aldosterone secretion and decreased blood levels of potassium inhibit aldosterone secretion.

The extracellular concentration of calcium ions has a dramatic effect on electrical properties of excitable tissues.

REGULATION OF ACID-BASE BALANCE

Buffers

Protein, phosphate, and bicarbonate buffers resist changes in the pH of the extracellular fluid.

An increase in carbon dioxide results in a decrease in pH, whereas a decrease in carbon dioxide results in an increase in pH.

Respiratory system

The respiratory system functions to regulate pH. It responds rapidly. An increased respiratory rate raises the pH because the rate of carbon dioxide elimination is increased, and a reduced respiratory rate reduces the pH because the rate of carbon dioxide elimination is reduced.

Kidneys

The kidneys excrete excess hydrogen ions in response to a decreasing blood pH, and they reduce hydrogen ion excretion in response to an increasing blood pH.

CONTENT REVIEW

1. Name the structures that make up the urinary system. List the functions of the urinary system.
2. Describe the relationships of the renal pyramids, calyces, renal pelvis, and ureter.
3. What is the functional unit of the kidney? Name its parts.
4. Describe the blood supply of the kidney.
5. What are the functions of the ureters, urinary bladder, and urethra? Describe their structure.
6. Name the three general processes that are involved in the production of urine.
7. Describe the filtration membrane. What substances do, and do not, pass through it?
8. How do changes in filtration pressure in the glomerulus affect the volume of filtrate produced?
9. What substances are reabsorbed in the nephron? What happens to most of the filtrate volume that enters the nephron?
10. What substances are secreted into the nephron?
11. What effect does ADH have on urine volume? What causes an increase in ADH secretion?
12. Where is aldosterone produced and what effect does it have on urine volume? What stimulates aldosterone secretion?
13. Where is atrial natriuretic hormone produced, and what effect does it have on urine production?
14. Describe the micturition reflex. How is voluntary control of micturition accomplished?
15. What stimuli result in an increased sensation of thirst?
16. How are sodium levels regulated in the body fluids?
17. How are potassium levels regulated in the body fluids?
18. Describe the role of buffers in the extracellular fluid.
19. Explain how the respiratory system and the kidneys respond to changes in the pH of body fluids.

CONCEPT REVIEW

1. Mucho McPhee did an experiment after reading the urinary system chapter in his favorite anatomy and physiology textbook. He drank 2 liters of water in 15 minutes and then monitored the rate of urine production and urine concentration over the next 2 hours. Explain what he observed.
2. A man ate a full bag of salty (sodium chloride) potato chips but drank no liquids. What effect did this have on urine concentration and the rate of urine production over the next 2 hours? Explain the mechanisms involved.
3. During severe exertion in a hot environment, a person can lose up to 4 liters of sweat per hour (sweat is less concentrated than extracellular fluid in the body). What effect would this loss have on urine concentration and rate of production? Explain the mechanisms involved.

4. Swifty Trotts has an infection that produces severe vomiting. Vomiting causes the loss of hydrogen ions from the stomach. What would this vomiting do to his blood pH and urine pH?

5. When Spanky and his mother went to a grocery store, Spanky eyed some candy he decided he wanted. His mother refused to buy it, so Spanky became angry. He held his breath for 2 minutes. What effect did this have on his body fluid pH? After 2 minutes, what mechanism was most important in reestablishing the normal body fluid pH?

CHAPTER TEST

Answers can be found in Appendix C

MATCHING For each statement in column A select the correct answer in column B (an answer may be used once, more than once, or not at all).

A

1. The outer portion of the kidney.
2. The functional unit of the kidney.
3. A capillary that forms part of the filtration membrane.
4. A blood vessel that carries blood to the glomerulus.
5. A blood vessel that carries blood away from the glomerulus.
6. The capillary that surrounds the nephron.
7. The duct that carries urine from the kidney to the urinary bladder.
8. The hormone released from the posterior pituitary gland that makes the distal nephron more permeable to water.
9. The hormone that increases the rate of sodium ion reabsorption, and potassium ion secretion, in the nephron.
10. The hormone released from the atria of the heart.
11. The fluid compartment made up of all of the fluid outside of cells.
12. The condition that results when the pH of the body fluids declines below 7.35.
13. The duct that carries urine from the urinary bladder to the exterior.

B

acidosis
afferent arteriole
aldosterone
alkalosis
antidiuretic hormone
atrial natriuretic hormone
Bowman's capsule
cortex
efferent arteriole
extracellular fluid
glomerulus
intracellular fluid
medulla
nephron
peritubular capillary
ureter
urethra

FILL-IN-THE BLANK Complete each statement by providing the missing word or words.

1. The urinary system is made up of the ___________, ___________, ___________, and urethra.
2. On the medial side of each kidney is the ___________, where the renal artery and nerves enter and the renal vein and ureter exit the kidney.
3. The medulla of the kidney is made up of several cone-shaped ___________.
4. The three main processes that are responsible for the formation of urine are ___________, ___________, and ___________.
5. The portion of the plasma entering the nephron becomes the ___________.
6. ___________ is the movement of water and other molecules out of the nephron across its wall.
7. ___________ increases the reabsorption of water from the nephron.
8. The ___________ reflex is initiated by stretching of the urinary bladder wall.
9. ___________ are chemicals that resist a change in the pH of a solution when either acids or bases are added.

CHAPTER

19 The Reproductive System

OBJECTIVES

After reading this chapter you should be able to:

1. Describe the structure and function of the testes.
2. Describe the process of spermatogenesis and the route sperm cells follow from the site of their production to the outside of the body.
3. Describe the structure of the penis.
4. List the hormones that influence the male reproductive system and describe their functions.
5. Describe the structure and function of the organs of the female reproductive system.
6. Discuss the development of the follicle and the oocyte.
7. Describe the processes of ovulation and fertilization.
8. Describe the changes that occur in the ovary and uterus during the menstrual cycle.
9. List the hormones of the female reproductive system and explain how their secretion is regulated.
10. Define menopause and describe the changes that occur as a result.

KEY TERMS

estrogen
Hormone secreted primarily by the ovaries; involved in maintenance and development of female reproductive organs, secondary sexual characteristics, and the menstrual cycle.

menopause
Permanent cessation of the menstrual cycle.

menstrual cycle
Cyclical series of changes that occur in sexually mature, nonpregnant females resulting in shedding of the uterine lining.

oocyte
Female gamete or sex cell; contains the genetic material transmitted from the female.

ovary
One of two female reproductive glands located in the pelvic cavity; produces oocytes, estrogen, and progesterone.

ovulation
Release of an oocyte from the ovary.

progesterone
Hormone secreted by the ovaries and by the placenta; necessary for uterine and mammary gland development and function in pregnant and nonpregnant women.

sperm cell
Male gamete or sex cell, composed of a head, midpiece, and tail; contains the genetic information transmitted by the male.

testis, pl. testes
One of two male reproductive glands located in the scrotum; produces testosterone and sperm cells.

testosterone
Hormone secreted primarily by the testes; aids in spermatogenesis, controls maintenance and development of male reproductive organs and secondary sexual characteristics, and influences sexual behavior.

FEATURES

THE MAJOR FUNCTION of the male and female reproductive systems is to produce offspring. Although a functional reproductive system is not necessary for the survival of the individual, it is obvious that some people must produce offspring to keep the species from becoming extinct.

This chapter describes the structure and function of the male and female reproductive organs. The reproductive organs produce the sex cells (oocytes in the female and sperm cells in the male), sustain them, and transport them to the site where fertilization may occur. In the female, the developing offspring is nurtured before and, for a time, after birth.

Reproductive organs produce hormones that play important roles in the development and maintenance of the reproductive system. These hormones help determine sexual characteristics, influence sexual behavior, and play a major role in regulating the physiology of the reproductive system.

FORMATION OF SEX CELLS

The formation of sex cells in males and females occurs by a special type of cell division called **meiosis** (mi-o´sis). For both males and females, meiosis begins in cells that contain 23 chromosome pairs (a total of 46 chromosomes) (see Chapter 3). Before the beginning of cell division, the genetic material is duplicated. During meiosis two cell divisions occur (Figure 19-1). In the male, for each cell that begins the process four sex cells, called **sperm cells,** or

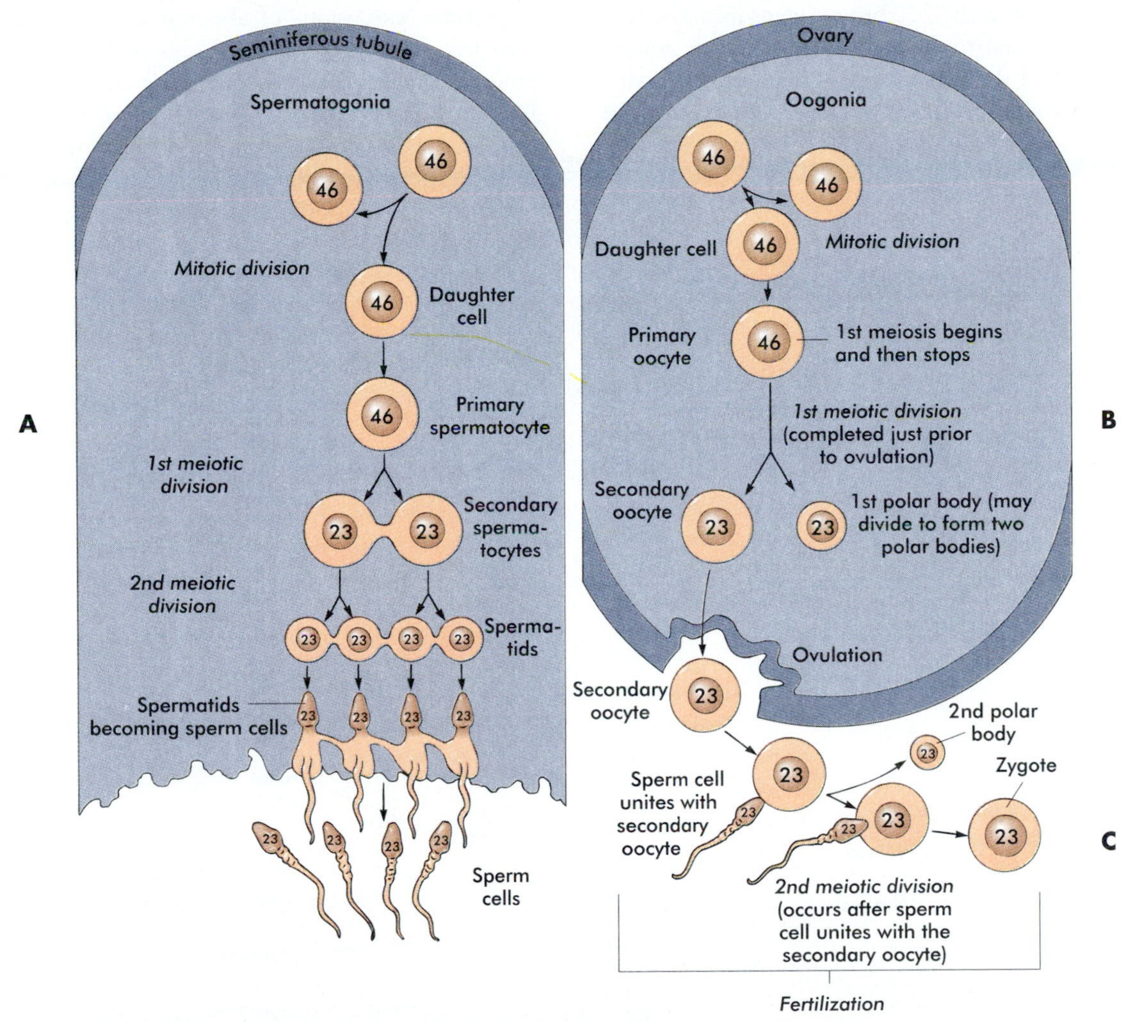

Figure 19-1 ● Meiosis and fertilization in males and females.
A, Meiosis in the male occurs continuously in the seminiferous tubules after puberty. From each primary spermatocyte that begins meiosis, 4 sperm cells are produced. The primary spermatocytes have 46 chromosomes (23 pairs) and the sperm cells that are produced each have 23 chromosomes (1 from each pair). **B,** Meiosis in the female starts before birth but stops early in the first meiotic division. For each primary oocyte that begins meiosis, 4 cells are produced. However, because the divisions of the cytoplasm are not equal, only 1 is capable of being fertilized. Each primary oocyte has 46 chromosomes (23 pairs), and each secondary oocyte has 23 chromosomes (1 from each pair). **C,** Upon fertilization 23 chromosomes from the male and 23 chromosomes from the female unite to make 46 chromosomes (23 pairs) in the zygote.

spermatozoa (sper´mă-to-zo´ah), are produced. Each sperm cell contains 23 chromosomes with one chromosome coming from each of the original 23 pairs of chromosomes. In the female, during each of the meiotic divisions, distribution of the cytoplasm among the sex cells is unequal. Most of the cytoplasm remains with one of the resulting cells, which is called a secondary **oocyte.** Like sperm cells, the secondary oocyte contains 23 chromosomes, with one chromosome coming from each of the original 23 pairs of chromosomes. The cells receiving little cytoplasm are called **polar bodies,** and they are not functional as sex cells.

The secondary oocyte is released from the ovary and unites with a single sperm cell before the completion of meiosis. After the sperm cell unites with the secondary oocyte, meiosis is completed. Then the 23 chromosomes from the sperm cell and the 23 chromosomes from the secondary oocyte combine to form 23 pairs of chromosomes. The combination of the chromosomes of the oocyte with the chromosomes of the sperm cell is called **fertilization.** All of the cells of the body contain 23 pairs of chromosomes except for the sex cells.

MALE REPRODUCTIVE SYSTEM

The male reproductive system consists of the primary and secondary reproductive organs. The **primary reproductive organs** are the testes, and the **secondary reproductive organs** include the scrotum, epididymis, ductus deferens, seminal vesicles, urethra, prostate gland, bulbourethral glands, and penis (see Figure 19-1, A). Some of the secondary reproductive organs form a duct system that carries the sperm cells to the outside of the body, and other secondary reproductive organs are accessory glands that produce secretions. The **male external genitalia** (jen´ĭ-ta´le-ah) are the scrotum and the penis, which are visible externally.

Scrotum

The testes are located in the **scrotum** (skro´tum), which is divided into two compartments by a connective tissue septum. Externally the scrotum consists of skin. Beneath the skin is a layer of smooth muscle (Figure 19-2).

In cold temperatures, the smooth muscle of the scrotum contracts, causing the skin of the scrotum to become firm and wrinkled, and reducing the scrotum's overall size. At the same time, extensions of abdominal muscles that project into the scrotum contract. Consequently, the testes are pulled nearer to the body and their temperature is raised. During warm weather or exercise, these muscles relax, the skin of the scrotum becomes loose and thinner, and the testes descend away from the body, which lowers their temperature. Control of the temperature in the testes is important. If the testes become too warm or too cold, normal development of sperm cells does not occur.

Testes

The **testes** (tes´tez) are oval organs within the scrotum, each about 4 to 5 cm long (see Figure 19-2). The outer portion of each testis consists of a thick, white connective tissue capsule. Extensions of the capsule project into the interior of

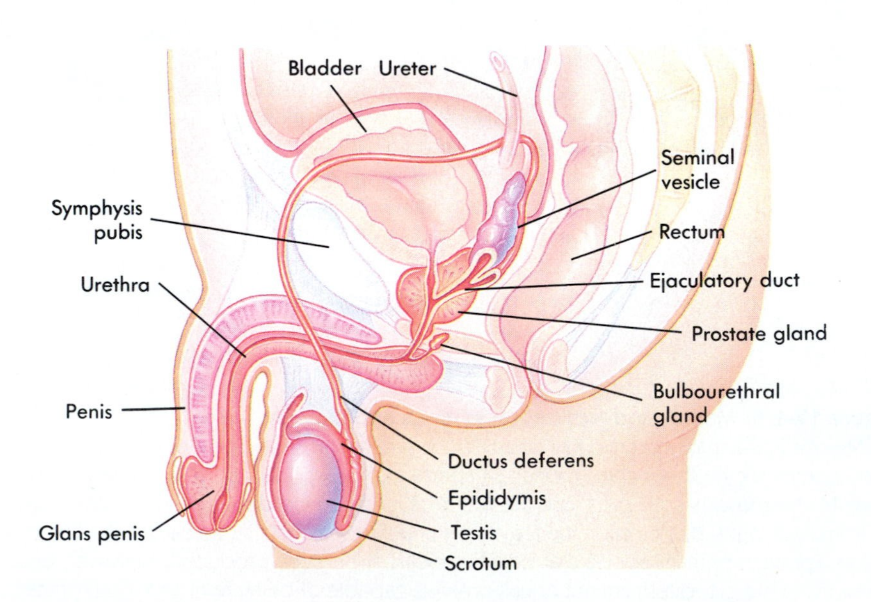

Figure 19-2 ● Sagittal view of the male pelvis.
Sagittal section of the male pelvis showing the male reproductive structures.

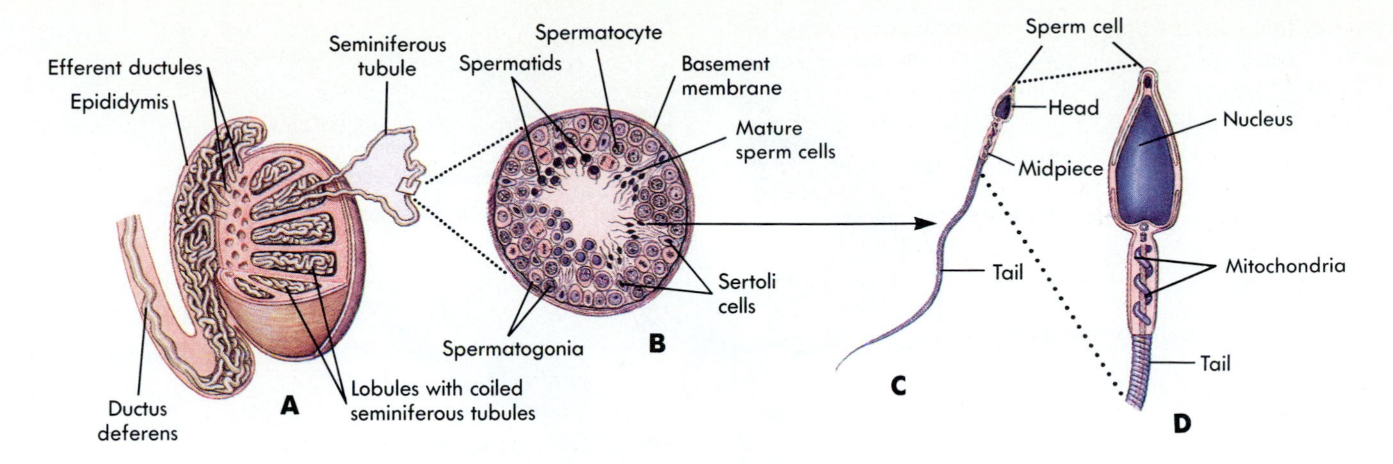

Figure 19-3 ● Structure of the testis.
A, Gross anatomy of the testis with a section cut away to reveal internal structures. **B,** Cross section of a seminiferous tubule. Spermatogonia are near the periphery, and mature sperm cells are near the lumen of the seminiferous tubule. **C,** The head, midpiece, and tail of a sperm cell. **D,** The head of a sperm cell contains the nucleus.

the testis and divide each testis into **lobules** (Figure 19-3, *A*). The lobules contain **seminiferous** (sem′ĭ-nif′er-us) **tubules,** in which sperm cells develop. Delicate connective tissue surrounding the tubules contains clusters of endocrine cells called **interstitial cells,** or **cells of Leydig,** which secrete testosterone.

The seminiferous tubules empty into a tubular network in the posterior portion of the testis, which gives rise to approximately 20 small tubules called **efferent ductules.** The efferent ductules exit the testis and enter the epididymis (see Figure 19-3, *A*).

The testes develop in the abdominopelvic cavity. They move from the abdominopelvic cavity to the scrotum, through the **inguinal canals,** which are passageways through the abdominal wall. The descent of the testes occurs either before birth or, sometimes, shortly after birth. Failure of the testes to descend into the scrotum causes sterility because of the inhibiting effect of normal body temperature on the development of sperm cells. After the testes descend, the inguinal canals narrow permanently, but they represent a weak spot in the abdominal wall. If an inguinal canal does not narrow or if it ruptures, this can cause an **inguinal hernia** through which a loop of intestine can protrude. This herniation can be quite painful and even dangerous, especially if the inguinal canal compresses the intestine and cuts off its blood supply. Fortunately, inguinal hernias can be repaired surgically.

Spermatogenesis

Spermatogenesis (sper′mă-to-jen′ĕ-sis) is the formation of sperm cells. Before puberty, the testes remain relatively simple and unchanged from the time of their initial development. The interstitial cells are small and few in number, and the seminiferous tubules are small and not yet functional. At the time of puberty, the interstitial cells increase in size and number. The seminiferous tubules enlarge, and spermatogenesis begins.

The seminiferous tubules contain **germ cells** and **Sertoli** (ser-to′le) **cells** (Figure 19-3, *B*). Sertoli cells are large cells that extend from the outside toward the center of the seminiferous tubule. They nourish and support the germ cells.

Germ cells are scattered between the Sertoli cells. The most peripheral cells are **spermatogonia** (sper′mă-to-go′ne-ah), which divide through mitosis. Some cells produced from these mitotic divisions remain as spermatogonia and continue to divide by mitosis. Other cells form **primary spermatocytes** (sper′mă-to′sītz), which divide by meiosis.

Each primary spermatocyte (with 23 pairs of chromosomes) passes through the first meiotic division to produce two **secondary spermatocytes.** Each secondary spermatocyte undergoes a second meiotic division to produce two smaller sex cells called **spermatids** (sper′mă-tidz), each having 23 chromosomes. After meiosis is complete, each spermatid develops a head, midpiece, and flagellum (tail) and becomes a sperm cell (see Figure 19-3, C and *D*).

Ducts

After sperm cells are produced in the seminiferous tubules, the sperm cells leave the testes and pass through a series of ducts to reach the exterior of the body.

Epididymis

The efferent ductules lead from the testis to the **epididymis** (ep-ĭ-did′ĭ-mis), which is a collection of coiled tubules outside of the testis (see Figures 19-2 and 19-3, *A*). Sperm cells become capable of fertilization after spending some time in the epididymis.

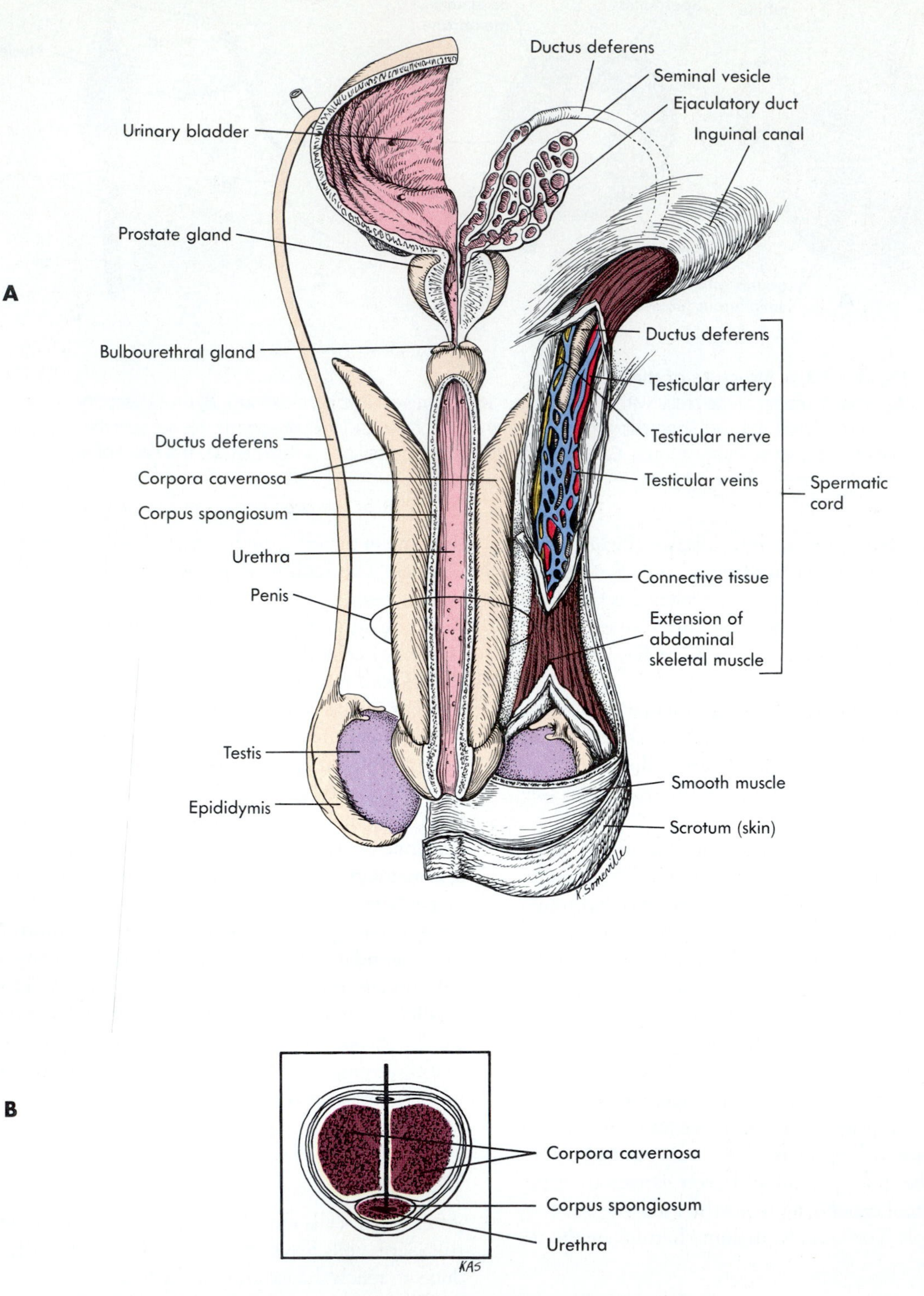

Figure 19-4 ● **Frontal view of the male reproductive organs.**
A, The testis and epididymis are in the scrotal sac. The ductus deferens extends from the epididymis through the inguinal canal to the prostate gland. Note that the ductus deferens joins the artery, vein, and nerves that supply the testes to form the spermatic cord. **B,** Cross section of the penis.

Ductus deferens

The **ductus deferens** (duk´tus def´er-enz), or **vas deferens,** extends from the epididymis through the inguinal canal. After passing through the inguinal canal, the ductus deferens extends to the prostate gland (see Figures 19-2 and 19-4, A). Contractions of smooth muscle in the wall of the ductus deferens move the sperm cells from the epididymis through the ductus deferens.

As the ductus deferens emerges from the epididymis it becomes associated with the blood vessels and nerves that supply the testis. Together these structures form the **spermatic cord,** which can be palpated through the wall of the scrotum (see Figure 19-4).

Ejaculatory duct

A sac-shaped gland called the **seminal** (sem´ĭ-nal) **vesicle** is adjacent to the ductus deferens where it approaches the prostate gland. A short duct from the seminal vesicle, and the ductus deferens, join together to form the **ejaculatory duct.** Each ejaculatory duct extends through the prostate gland and ends by opening into the urethra within the prostate gland (see Figure 19-4).

Urethra

The male **urethra** (u-re´thrah) extends from the urinary bladder to the distal end of the penis (see Figures 19-3 and 19-4). The urethra is a passageway for both urine and male reproductive fluids.

Penis

The penis contains three columns of erectile tissue (see Figure 19-4). Engorgement of the erectile tissue with blood causes the penis to enlarge and become firm, a process called **erection.** The penis is the male organ of copulation and functions in the transfer of sperm cells from the male to the female. Two of the erectile columns form the dorsal portion and the sides of the penis and are called the **corpora cavernosa** (kor´por-ah kav´er-no-sah). The third and smaller erectile column occupies the ventral portion of the penis and is called the **corpus spongiosum** (kor´pus spun´je-o´sum). The corpus spongiosum expands over the distal end of the penis to form a cap, the **glans penis.** The urethra passes through the corpus spongiosum, penetrates the glans penis, and opens to the exterior.

The skin of the penis, especially the glans penis, is well supplied with sensory receptors. A loose fold of skin formed at the base of the glans penis, called the **prepuce** (pre´pūs), or **foreskin,** lies over the glans penis. **Circumcision** is the surgical removal of the prepuce.

Glands and secretions

Semen (se´men) is a mixture of sperm cells and secretions from the male reproductive glands. **Emission** is the discharge of semen into the urethra, and **ejaculation** is the forceful expulsion of semen from the urethra. Only about 5% of the semen volume is produced by the testes.

The seminal vesicles (see Figures 19-2 and 19-4) produce about 60% of the semen volume. The thick, mucous-like secretion of the seminal vesicles contains nutrients that nourish the sperm cells.

The **prostate** (pros´tāt) **gland** is about the size and shape of a walnut, and it surrounds the urethra and the two ejaculatory ducts (see Figures 19-2 and 19-4). There are several short ducts that carry secretions of the prostate gland to the urethra. Approximately 30% of the semen volume is produced by the prostate gland. Its thin, milky secretion helps neutralize the acidic urethra and the more acidic secretions of the testes, seminal vesicles, and vagina. The increased pH is important for normal sperm cell function.

1 The prostate gland can enlarge for several reasons, including infections and tumors. Noncancerous enlargement of the prostate occurs in many elderly men. Cancer of the prostate is the second most common cause of male death from cancer in the United States (less than lung cancer and more than colon cancer). The detection of enlargement or changes in the prostate is important. Suggest a way that the prostate gland can be examined by palpation for any abnormal changes. (Hint: see Figures 19-2 and 19-4.)*

The **bulbourethral** (bul´bo-u-re´thral) **glands** are a pair of small mucus-producing glands located near the base of the penis (see Figures 19-2 and 19-4). A single duct from each gland enters the urethra. The bulbourethral glands produce about 5% of the semen volume. The secretions are produced several minutes before ejaculation, and the mucus lubricates the urethra, neutralizes the contents of the acidic urethra and vagina, and provides a small amount of lubrication during intercourse.

Before ejaculation, both ductus deferens begin to contract rhythmically, propelling sperm cells and testicular fluid from each epididymis through the ductus deferens. Contractions of the ductus deferens, seminal vesicles, and ejaculatory ducts cause the sperm cells, testicular secretions, and seminal fluid to move into the urethra, in which they mix with prostatic secretions released as a result of contraction of the prostate.

The normal volume of semen is 2 to 3 ml. The normal sperm cell count is about 100 million sperm cells per milliliter of semen. If the sperm cell count falls below about 20 million sperm cells per milliliter of semen, sterility usually results, even though millions of sperm cells present appear to be normal.

*Answers to predict questions appear in Appendix B at the back of the book.

PHYSIOLOGY OF MALE REPRODUCTION

The male reproductive system depends on both hormonal and neural mechanisms to function normally. Hormones are responsible for the development and maintenance of reproductive structures, the development of secondary sexual characteristics, and the control of spermatogenesis. They also influence sexual behavior. Nervous mechanisms control the sexual act and the expression of sexual behavior.

Regulation of sex hormone secretion

Gonadotropin-releasing hormone (GnRH) is released from the hypothalamus and passes to the anterior pituitary gland. In response to GnRH, cells in the anterior pituitary gland secrete two hormones into the blood, **luteinizing** (lu´te-ĭ-nīz-ing) **hormone (LH)** and **follicle-stimulating hormone (FSH).** LH and FSH are named for their functions in females, but they also are important in males.

LH causes the interstitial cells in the testes to secrete testosterone, and FSH stimulates the function of Sertoli cells in the seminiferous tubules and promotes spermatogenesis.

Puberty

Puberty is the sequence of events by which a child is transformed into a young adult. The reproductive system matures and assumes its adult functions, and the structural differences between adult males and females become more apparent. In boys, puberty commonly begins at about age 11 and is largely completed by age 18. At puberty, the hypothalamus and anterior pituitary gland increase the rate of GnRH, LH, and FSH secretion. Elevated LH levels cause the interstitial cells to secrete larger amounts of testosterone. Elevated FSH levels stimulate the function of Sertoli cells and promote spermatogenesis.

Effects of testosterone

Testosterone is the major male hormone secreted by the testes. It is responsible for the differentiation and enlargement of the male genitals, and for the development of the reproductive duct system. It is necessary for spermatogenesis as well. Testosterone is also responsible for the **secondary sexual characteristics**—those structural and behavioral changes, other than reproductive organs, that develop at puberty and that distinguish males from females (Table 19-1). After puberty, testosterone maintains the adult structure of the male genitals, reproductive ducts, and secondary sexual characteristics. It also is important in maintaining the male sexual drive.

2 Predict the effect on secondary sexual characteristics, external genitalia, and sexual behavior, if the testes failed to produce normal amounts of testosterone at puberty.

A few athletes, especially those who depend on muscle strength, take synthetic androgens (testosterone is a type of androgen) in an attempt to increase muscle mass. These synthetic androgens are commonly called anabolic steroids, or simply **steroids.** The effect of synthetic androgens on muscle is much greater than their effect on the reproductive organs. However, they are often taken in large amounts, and they can influence the reproductive system. Large doses of synthetic androgens reduce GnRH, LH, and FSH secretion.

Table 19-1 Effects of testosterone on target tissues

Target tissue	Response
Penis and scrotum	Enlargement and differentiation
Hair follicles	Hair growth and coarser hair: pubic area, legs, chest, axillary region, the face, and occasionally the back; male pattern baldness on the head if the person has the appropriate genetic makeup
Skin	Coarser texture of skin; increased rate of secretion of sebaceous glands, frequently resulting in acne at the time of puberty; increased secretion of sweat glands in axillary regions
Larynx	Enlargement of larynx and deeper masculine voice
Most tissues	Increased rate of metabolism
Red blood cells	Increased rate of red blood cell production; red blood cell count increased by about 20% as a result of increased erythropoietin secretion
Kidney	Retention of sodium and water to a small degree, resulting in increased extracellular fluid volume
Skeletal muscle	Skeletal muscle mass increases at puberty; the average is greater in men than in women
Bone	Rapid bone growth resulting in increased rate of growth and in early cessation of growth; males who mature sexually at a later age do not exhibit a rapid period of growth, but they grow for a longer period and may become taller than men who mature earlier

As a result, the testes can atrophy, and sterility can develop. Other side effects of large doses of synthetic androgens include kidney and liver damage, heart attack, and stroke. Taking synthetic androgens is highly discouraged by the medical profession and is a violation of the rules for most athletic organizations.

Male sexual behavior and the male sexual act

The male sexual act is a complex series of reflexes that result in erection of the penis, secretion of mucus into the urethra, emission, and ejaculation. Sensations that are normally interpreted as pleasurable occur during the male sexual act and result in a **male climax,** or **orgasm,** associated with ejaculation. After ejaculation, a phase called **resolution** generally occurs in which the penis becomes flaccid, an overall feeling of satisfaction exists, and the male is unable to achieve erection and a second ejaculation.

Erection is the first major component of the male sexual act. The arteries that supply blood to the erectile tissues dilate. Blood then fills small venous sinuses in the erectile tissue of the penis. The increased blood pressure in the sinuses causes the erectile tissue to become inflated and rigid.

Emission occurs as a result of nervous stimulation of the reproductive ducts and glands. As a result, the epididymis, seminal vesicles, and prostate gland release their contents into the urethra to form the semen. Sperm cells and the secretion of the epididymis make up only a small portion of the semen. Most of the volume of the semen consists of prostate and seminal vesicle secretions. Skeletal muscles that surround the base of the penis produce rhythmic contractions that force the semen out of the urethra resulting in ejaculation.

FEMALE REPRODUCTIVE SYSTEM

The female reproductive organs consist of the ovaries, uterine tubes, uterus, vagina, external genitalia, and mammary glands. The internal reproductive organs of the female (Figures 19-5 and 19-6) are located within the pelvic cavity. The uterus and the vagina are in the midline with an ovary on each side of the uterus.

Ovaries

The **ovaries** are suspended in the pelvic cavity by ligaments (see Figure 19-6). The ovarian arteries, veins, and nerves course through the ligaments to enter the ovaries. A layer of visceral peritoneum covers the surface of each ovary. The outer portion of each ovary is made up of dense connective tissue and contains **ovarian follicles.** Each of the ovarian follicles contains an **oocyte** (o´o-sīt). The inner portion of the ovary, which consists of loose connective tissue, is where blood vessels, lymph vessels, and nerves enter the ovary.

Follicle and oocyte development

In the ovaries of female fetuses, cells called **oogonia** (o´o-go´ne-ah), which give rise to oocytes, divide by mitosis until several million oogonia are produced. The oogonia, which have 23 pairs of chromosomes, begin the process of meiosis. At this stage the cells are called **primary oocytes.** However, development is arrested before the cells complete the first meiotic division. Just before ovulation the first meiotic division is completed to form a **secondary oocyte** and a **polar body.** The secondary oocyte is then released from the ovary during ovulation. The second meiotic division is not complete until the secondary oocyte unites with a sperm cell.

Each female has about 2 million ovarian follicles by the time she is born. The primary oocyte is surrounded by a layer of cells and the entire structure is called a **primary follicle** (Figures 19-7 and 19-8). From birth to puberty, the number of primary follicles declines to about 400,000; of these only about 450 continue to develop and ovulate. Follicles that do not ovulate degenerate.

Beginning at puberty, approximately every 28 days, hormonal changes cause a few ovarian follicles to continue to develop (Figure 19-8). FSH from the anterior pituitary stimulates several follicles to develop during each menstrual cycle. The cells surrounding the oocyte multiply and form an increasing number of layers. The center of the follicle becomes a fluid-filled chamber called the **antrum** (an´trum). After antrum formation has started, the follicle is called a **secondary follicle.** The secondary follicle continues to enlarge and forms a lump on the surface of the ovary. The follicle is then called a **mature follicle.** Usually only one of the

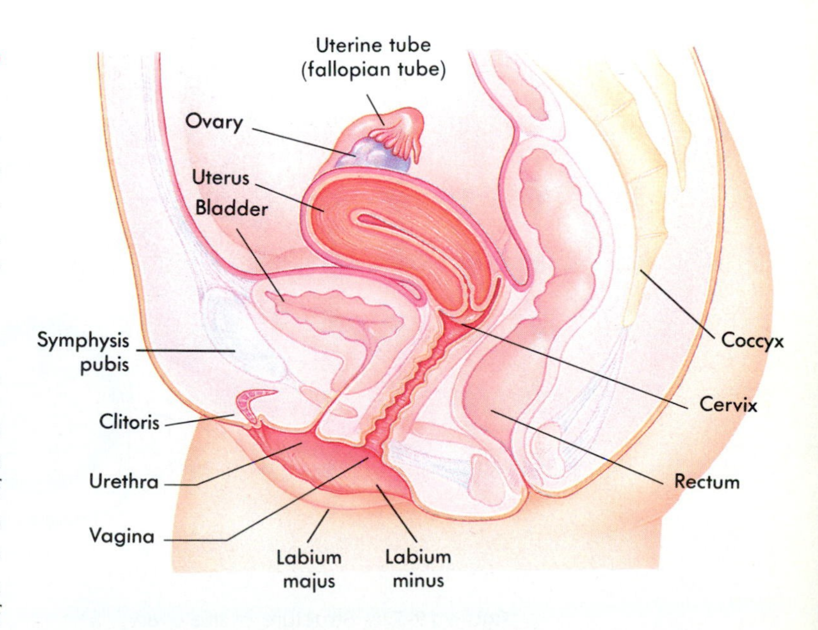

Figure 19-5 ● Sagittal view of the female pelvis.
Anatomy of the female reproductive organs.

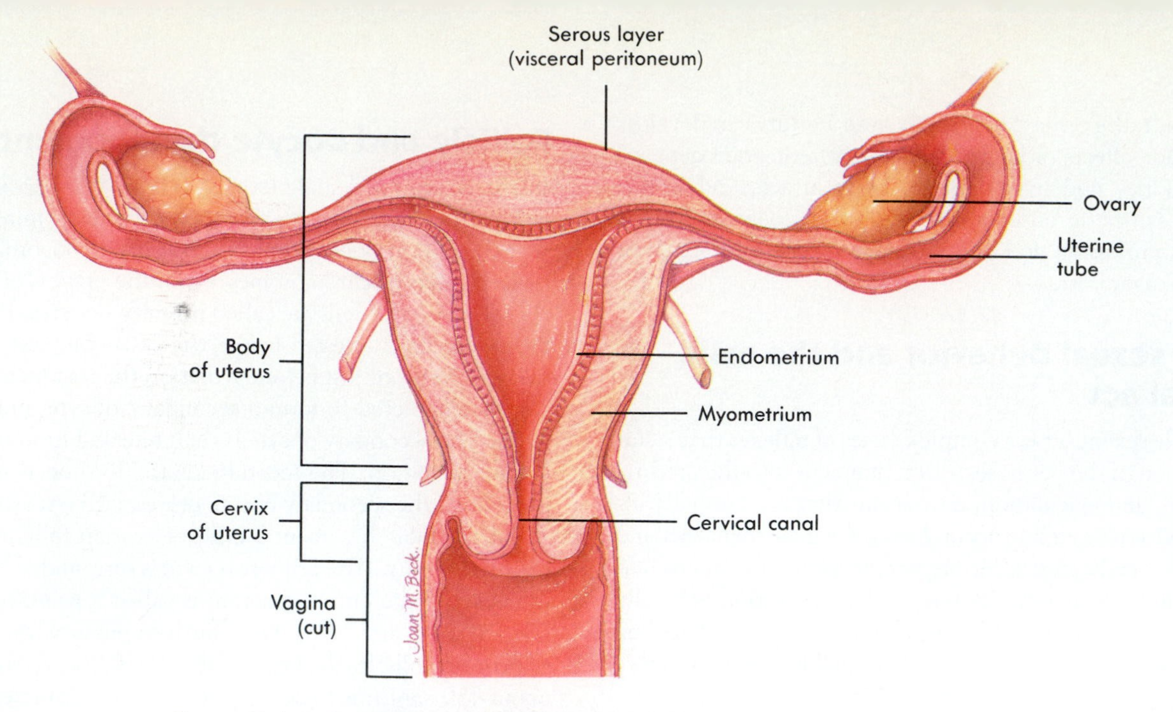

Figure 19-6 ● Frontal view of female reproductive organs.
The uterus, uterine tubes, and vagina are cut in section to show the internal anatomy.

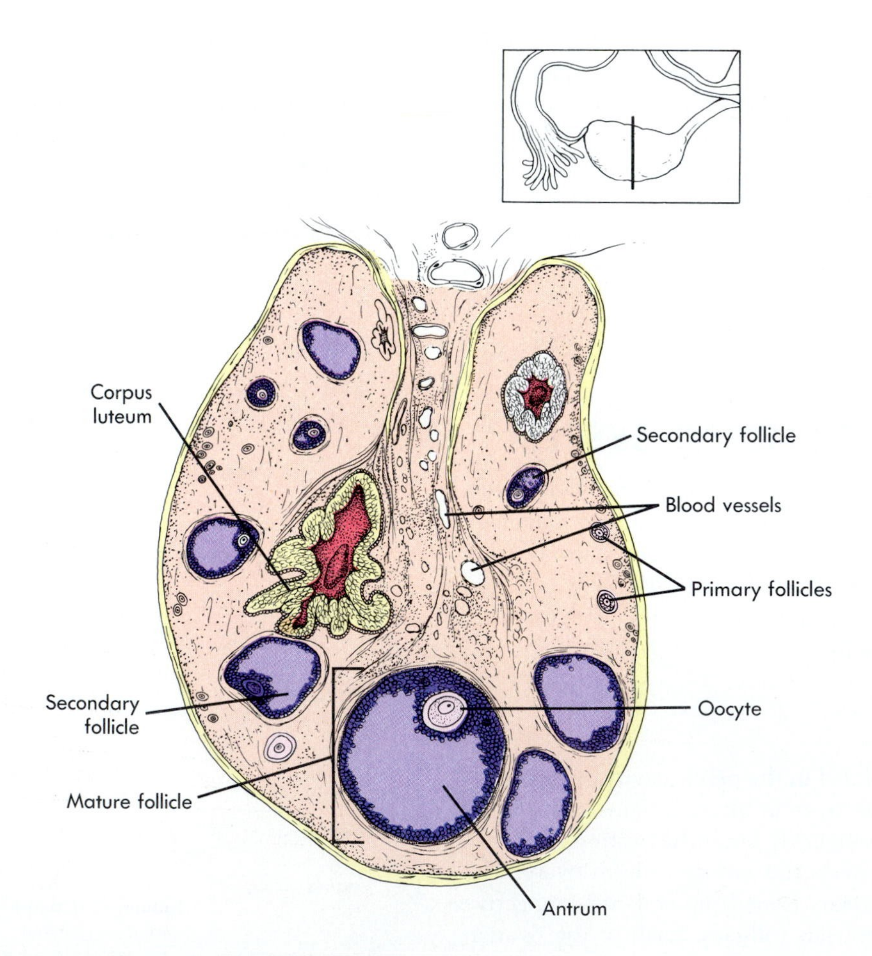

Figure 19-7 ● Structure of the ovary.
The ovary is sectioned to illustrate its internal structure (*inset* shows plane of section). Ovarian follicles from each major stage of development are present, and a corpus luteum is present.

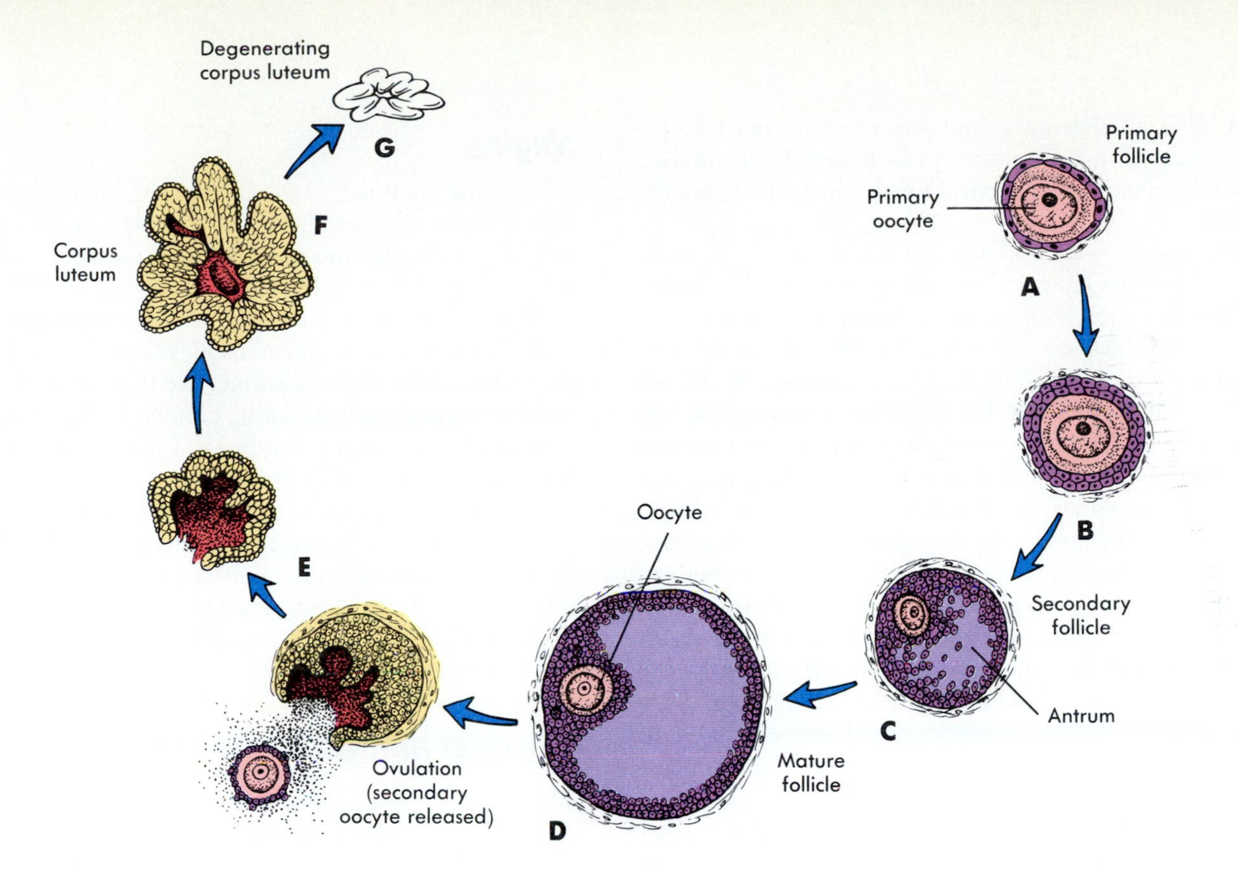

Figure 19-8 ● Maturation of the follicle, oocyte, and corpus luteum.
A, A primary oocyte begins to mature one or two menstrual cycles before it is ovulated. Several follicles begin to mature at the same time, but only one reaches the final stage of development and undergoes ovulation. **B,** The primary follicle enlarges. **C,** An antrum begins to form and to fill with fluid to form a secondary follicle. **D,** When a follicle becomes mature, it enlarges to its maximum size, and a large antrum is present. **E,** During ovulation the oocyte is released from the follicle along with some surrounding cells. **F,** Subsequently, the cells of the follicle divide rapidly and enlarge to form the corpus luteum. **G,** The corpus luteum degenerates if fertilization does not occur.

follicles that started to develop becomes a mature follicle. The other developing follicles degenerate.

Developing follicles secrete a hormone called **estrogen.** Estrogen plays an important role in coordinating the menstrual cycle and preparing the uterus to receive the fertilized oocyte.

Ovulation

The mature follicle can be seen on the surface of the ovary as a blister. **Ovulation** occurs when the follicle ruptures forcing the secondary oocyte out of the follicle into the peritoneal cavity. Ovulation occurs in response to LH secreted by the anterior pituitary gland.

After ovulation, the cells of the ruptured follicle become transformed into a structure called the **corpus luteum** (lu´te-um) (see Figures 19-7 and 19-8). Cells of the corpus luteum secrete **progesterone** and smaller amounts of estrogen.

If pregnancy occurs, the corpus luteum enlarges. Maintenance of pregnancy depends on progesterone secreted by the corpus luteum for about the first trimester (first third or 12 weeks) of pregnancy. After the first trimester of pregnancy, progesterone is produced by the placenta, and the corpus luteum is no longer essential. If pregnancy does not occur, the corpus luteum begins to degenerate after about 10 to 12 days.

Uterine tubes

There are two **uterine tubes,** also called **fallopian tubes,** or **oviducts.** Each uterine tube extends from the area of one ovary to the uterus. The uterine tubes open directly into the peritoneal cavity near the ovary (see Figure 19-6). As soon as the secondary oocyte is ovulated, it enters the uterine tube. Fertilization usually occurs in the portion of the uterine tube near the ovary. The uterine tube transports the secondary oocyte from the ovary to the uterus.

Uterus

The **uterus** (u´ter-us) is the size of a medium-sized pear (see Figures 19-6). It is slightly flattened and is oriented in the pelvic cavity with the larger, rounded portion called the body directed superiorly, and the narrower portion, the **cervix** (ser´viks), directed inferiorly. Internally, the **uterine cavity** continues as the **cervical canal,** which opens into the vagina.

The uterine wall is composed of three layers: a serous layer, a muscular layer, and the endometrium (see Figure 19-6). The thin outer layer, or **serous layer,** of the uterus is the visceral peritoneum. The thick middle layer is the **myometrium** (mi´o-me´tre-um), which consists of smooth muscle. The innermost layer of the uterus is the **endometrium** (en´do-me´tre-um). The endometrium consists of simple, columnar epithelial cells and a connective tissue layer. Uterine glands are formed by folds of the epithelium.

Cancer of the uterine cervix is a relatively common type of cancer. A Pap smear is a simple diagnostic test that can detect the presence of cancer of the uterine cervix. A sample of epithelial cells is taken from the area of the cervix by inserting a swab through the vagina. The cells are examined microscopically to determine whether any of them show signs of being cancerous. Cells that are cancerous appear to be more immature than the characteristic epithelial cells of the cervix. The more immature the cells appear, the more severe the cancer. When cancer of the cervix is detected before it spreads to other organs, the chances of survival are good.

Vagina

The **vagina** (vă-ji´nah) extends from the uterus to the outside of the body (see Figures 19-5 and 19-6). The vagina functions to receive the penis during intercourse. It also allows menstrual flow and childbirth.

The wall of the vagina consists of an outer muscular layer and an inner mucous membrane. The muscular layer is smooth muscle and elastic connective tissue that allows the vagina to stretch greatly during childbirth. The inner mucous membrane is moist stratified squamous epithelium that forms a protective surface layer.

In young females, the vaginal opening is partially covered by a thin mucous membrane called the **hymen.** The hymen usually is perforated by one or several holes that allow menstrual fluid to pass through. The openings in the hymen can be enlarged during the first sexual intercourse or the hymen can be torn earlier.

External genitalia

The external female genitalia, also called the **vulva,** consist of the vestibule and its surrounding structures (Figure 19-9). The **vestibule** (ves´tĭ-būl) is the space into which the

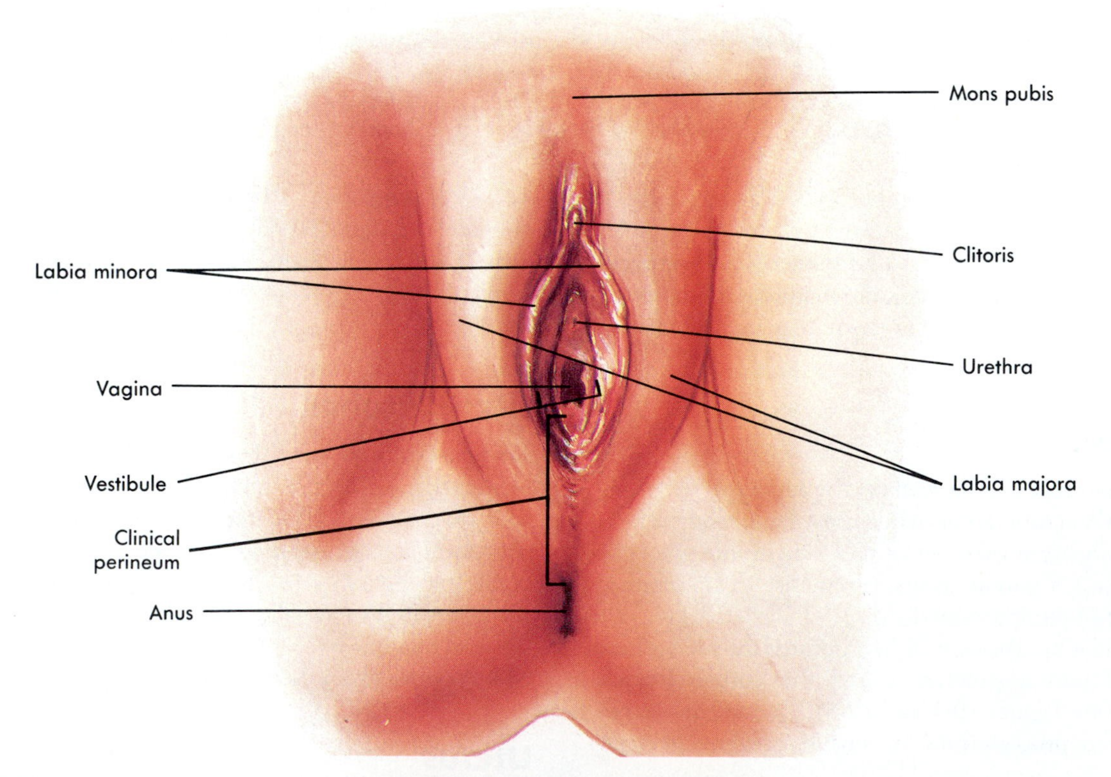

Figure 19-9 ● Female external genitalia.

vagina and urethra open. The urethra opens just anterior to the vagina. The vagina is bordered by a pair of thin, longitudinal skin folds called the **labia minora** (la´be-ah mi-no´rah). A small erectile structure called the **clitoris** (klit´o-ris) is located in the anterior margin of the vestibule. The clitoris contains sensory neurons that are stimulated during sexual intercourse. Between the vaginal opening and the labia minora are openings of the **vestibular glands.** The vestibular glands produce a lubricating fluid that helps maintain the moistness of the vestibule.

Lateral to the labia minora are two prominent, rounded folds of skin called the **labia majora** (la´be-ah ma-jo´rah). The two labia majora unite anteriorly in an elevation over the pubic symphysis called the **mons pubis.** The lateral surfaces of the labia majora, as well as the surface of the mons pubis, are covered with coarse hair. Most of the time, the labia majora are in contact with each other across the midline, concealing the deeper structures within the vestibule.

The region between the vagina and the anus is the **clinical perineum** (per´ĭ-ne-um). The skin and muscle of this region may tear during childbirth. To prevent such tearing, an incision called an **episiotomy** (e-piz-e-ot´o-me) is sometimes made in the clinical perineum. Alternatively, allowing the perineum to stretch slowly during childbirth may prevent tearing, making an episiotomy unnecessary.

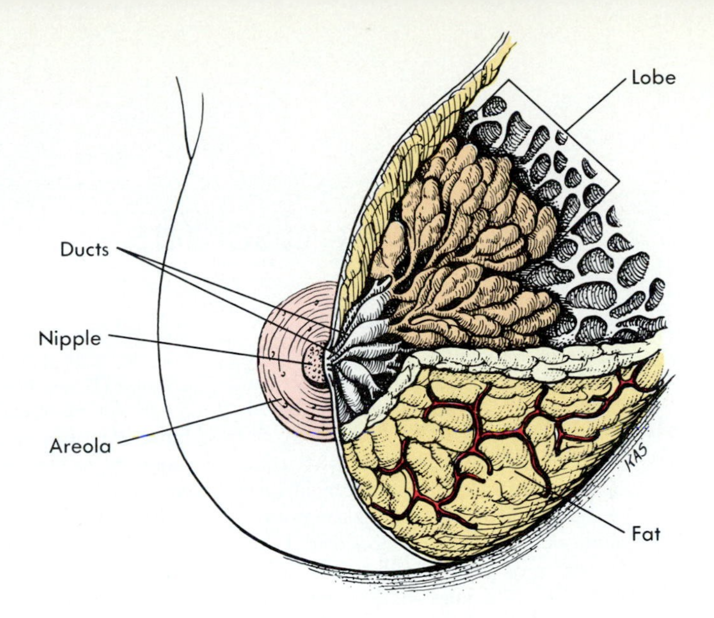

Figure 19-10 ● Anatomy of the breast.
The section illustrates the duct system, secretory units, and adipose tissue of the breast.

Mammary glands

The **mammary glands** are the organs of milk production located within the **breasts.** Externally, the breast of both males and females has a raised **nipple** surrounded by a circular, pigmented **areola** (ă-re´o-lah) (Figure 19-10).

Each adult female breast consists of 15 to 20 lobes of the mammary gland covered by a considerable amount of adipose tissue. The superficial fat gives the breast most of its form. Each lobe possesses a single duct that opens independently on the surface of the nipple. The duct of each lobe subdivides into smaller ducts. In the milk-producing breast, the ends of these small ducts expand to form secretory sacs called **alveoli.**

The nipples are sensitive to tactile stimulation and contain smooth muscle that can contract, causing the nipple to become erect in response to stimulation. These smooth-muscle fibers respond similarly to general sexual arousal.

Lactation is the production of milk by the breasts. It occurs in women after **parturition** (par-tu-rish´un, childbirth) and may continue for 2 or 3 years, provided suckling occurs often and regularly. For the first few days after parturition, the mammary glands secrete **colostrum** (ko-los´trum), which contains little fat and less lactose than milk. Eventually, more nutritious milk is produced. Colostrum and milk provide nutrition and antibodies that help protect the nursing baby from infection.

During pregnancy, the placenta produces estrogen and progesterone. These hormones cause additional fat deposition, and the breasts increase in size. These hormones also stimulate development of the duct system and the alveoli of the breast. At the same time, estrogen and progesterone prevent the secretion of milk during pregnancy.

After parturition, the placenta is no longer present to produce estrogen and progesterone. In the absence of estrogen and progesterone, **prolactin,** produced by the anterior pituitary, stimulates milk production. Suckling stimulates action potentials that are sent from the nipple to the brain and result in the release of prolactin. However, if nursing is

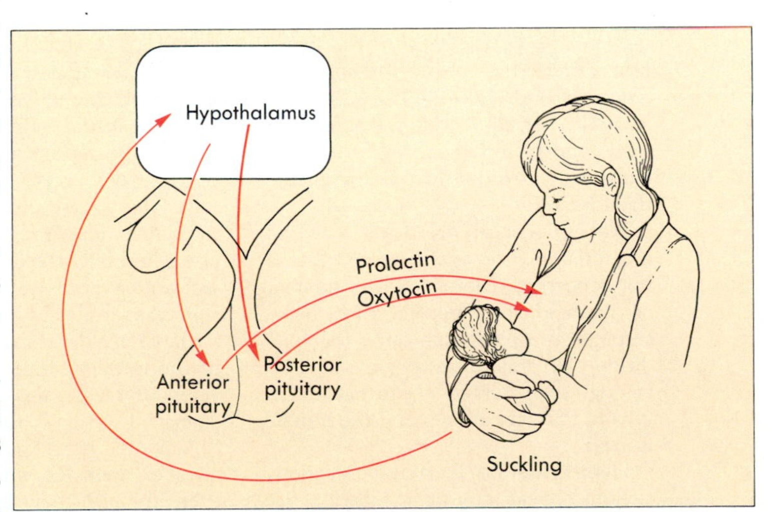

Figure 19-11 ● Hormonal control of lactation.
Suckling stimulates action potentials that pass along nerve fibers to the mothers's brain. The brain signals anterior pituitary to release prolactin and the posterior pituitary to release oxytocin. The prolactin stimulates milk production. The oxytocin stimulates alveolar cells to contract, causing milk to flow from the breasts.

Reproductive disorders

SEXUALLY TRANSMITTED DISEASES

Sexually transmitted diseases (STDs) are infectious diseases spread by intimate sexual contact between individuals.

Trichomonas is a protozoan commonly found in the vagina of females and the urethra of males. If the normal acidity of the vagina is disturbed *Trichomonas* can grow rapidly. It is more common in females than in males. Inflammation and a greenish-yellow discharge with a foul odor are common.

Gonorrhea is caused by *Neisseria gonorrhoeae*. The organisms attach to the epithelial cells of the vagina or the male urethra. The invasion of bacteria establishes an inflammatory response in which pus is formed. Males become aware of a gonorrheal infection by painful urination and the discharge of pus-containing material from the urethra. Symptoms appear within a few days to a week. Recovery can eventually occur without complication, but when complications do occur, they can be serious. The urethra can become partially blocked or sterility can result from blockage of reproductive ducts with scar tissue. In some cases, other organ systems such as the heart, meninges of the brain, or joints may become infected. In females, the early stages of infection may not be noticeable, but the infection can lead to pelvic inflammatory disease. Gonorrheal eye infections can occur in newborn children of women with gonorrheal infections. Antibiotics are usually effective in treating gonorrheal infections.

Nongonococcal urethritis refers to any inflammation of the urethra that is not gonorrhea. Factors such as trauma, or passage of a nonsterile catheter through the urethra can cause this condition, but many cases are acquired through sexual contact. Antibiotics are usually effective in treating the condition.

Genital herpes is a viral infection. Lesions appear after an incubation period of about 1 week and cause a burning sensation. After this, blisterlike areas of inflammation appear. In males and females, urination can be painful, and walking or sitting may be unpleasant, depending on the location of the lesions. The viruses can exist in a latent condition in infected tissues and may initiate periods of inflammation in response to factors such as menstruation, emotional stress, or illness. If active lesions are present in the mother's vagina or external genitalia, a caesarean delivery is performed to prevent newborns from becoming infected with the herpes virus. Because genital herpes is caused by a virus, there is no effective cure for the condition.

Genital warts result from a viral infection and are quite contagious. Genital warts are common, and their frequecy is increasing. Genital warts can also be transmitted from infected mothers to their infants. Genital warts vary from separate, small warty growths to large cauliflower-like clusters. The lesions are usually not painful, but they may cause painful intercourse, and they bleed easily. For women who have genital warts, there is an increased risk of developing cervical cancer. Treatments for genital warts include topical agents, freezing techiques, or surgical methods.

Syphilis is caused by the bacterium *Treponema pallidum*, which can be spread by sexual contact of all kinds. Typically symptoms do not appear for a period of from 2 weeks to several months. The disease progresses through several recognized stages. In the primary stage, the initial symptom is a small sore that usually appears at the site of infection. Several weeks after the primary stage, the disease enters the secondary stage, characterized mainly by skin rashes and mild fever. The symptoms of secondary syphilis usually subside after a few weeks. In less than half the cases, a tertiary stage develops after many years. In the tertiary stage, many lesions develop that can cause extensive tissue damage. Damage to the nervous system is among the more serious consequences. Females who have syphilis, even though they may not exhibit symptoms, are likely to have babies who are infected. Antibiotics are used to treat syphilis, although some strains are very resistant to certain antibiotics.

Acquired immune deficiency syndrome (AIDS) is caused by infection with the human immunodeficiency virus (HIV), which infects cells of the immune system. After the initial infection there may be a symptom-free period lasting up to several years. Eventually, the virus becomes activated and destroys cell types that are essential to the maintenance of the immune system. Victims do not die directly from HIV. The destruction of the immune system by AIDS makes the individual vulnerable to other infections. These infections ultimately cause the death of the AIDS victim. There is no cure for AIDS at this time. Current treatments only prolong the life of the AIDS victim.

Preventative measures against HIV infection provide the only real protection against AIDS at this time. The most common mechanisms of transmission of the virus are through sexual contact with a person infected with HIV and through sharing needles with an infected person during the administration of illicit drugs. Screening techniques now implemented make the transmission of HIV through blood transfusions very rare. Some documented cases of transmission of HIV through accidental needle sticks in hospitals and other health care facilities exist, but the frequency is rare. There is no evidence that casual contact with a person who has AIDS or who is infected with HIV will result in transmission of the disease. Transmission appears to require exposure to body fluids of an infected person in a way that allows HIV into the interior of another person.

OTHER INFECTIOUS DISEASES

Pelvic inflammatory disease (PID) is a bacterial infection of the pelvic organs. For example, a vaginal or uterine infection may spread throughout the pelvis. Early symptoms of PID include increased vaginal discharge and pelvic pain. Early treatment with antibiotics can stop the spread of PID, but lack of treatment results in a life-threatening infection.

stopped, within a few days the ability of the breast to respond to prolactin is lost and milk production stops (Figure 19-11).

At the time of nursing, milk contained in the alveoli and ducts of the breast is forced out of the breast by the contracting walls of these structures. In addition to stimulating the prolactin release, action potentials sent from the nipple cause the release of **oxytocin** from the posterior pituitary. Oxytocin stimulates cells surrounding the alveoli to contract; milk then flows from the breasts, a process that is called **milk letdown.**

Cancer of the breast is a serious, often fatal disease in women. The use of mammography and regular self-examination of the breast can lead to early detection of breast cancer and effective treatment. **Mammography** utilizes low-intensity x-rays to detect tumors in the soft tissue of the breast. With modern techniques, tumors often can be identified before they can be detected by palpation. Once a tumor is identified, a small tissue sample can be taken and examined to determine whether the tumor is benign or malignant. Most tumors of the mammary glands are benign, but those that are malignant have the potential to spread to other areas of the body.

PHYSIOLOGY OF FEMALE REPRODUCTION

As in the male, female reproduction is under hormonal and neural regulation.

Puberty

Puberty in females is marked by the first episode of menstrual bleeding, which is called **menarche** (mĕ-nar´ke). Puberty commonly begins about the age of 10 in females and is largely complete by age 16. During puberty, the vagina, uterus, uterine tubes, and external genitalia begin to enlarge. Fat is deposited in the breasts and around the hips, causing them to enlarge and assume an adult form. Pubic and axillary hair begins to grow, and sexual drive begins to develop.

Before puberty, GnRH from the hypothalamus, and LH and FSH from the anterior pituitary are secreted in small amounts. Only small amounts of estrogen and progesterone are secreted by the ovaries. After the onset of puberty, the normal cyclic pattern of reproductive hormone secretion that occurs during the menstrual cycle becomes established. The changes associated with puberty are primarily the result of the elevated rate of estrogen and progesterone secretion by the ovaries.

Menstrual cycle

The **menstrual cycle** is the series of changes that occur in sexually mature, nonpregnant females that result in menses. **Menses** (men´sēz) is a period of mild hemorrhage during which the endometrium is sloughed and expelled from the uterus. Typically, the menstrual cycle is about 28 days long, although it may be as short as 18 days or as long as 40 days (Figure 19-12 and Table 19-2).

The first day of menstrual bleeding, called the **menstrual phase,** or **menses,** is considered to be day 1, and menses typically lasts 4 or 5 days. Ovulation occurs on about day 14 of the menstrual cycle, although the timing of ovulation varies from individual to individual and can vary within an individual from one menstrual cycle to the next.

The time between the ending of menses and ovulation is called the **proliferative phase.** During this phase, FSH from the anterior pituitary stimulates the follicles in the ovary. As the follicles mature, they secrete increasing amounts of estrogen, which acts on the uterus and causes the cells of the endometrium to divide and rapidly proliferate. The endometrium thickens and tubular glands form.

The sustained increase in the amount of estrogen secreted by the developing follicles stimulates GnRH secretion from the hypothalamus. The GnRH triggers LH and FSH secretion from the anterior pituitary gland, which causes the follicle to develop further and secrete more estrogen. The maximum rate of LH and FSH secretion occurs approximately on about day 14 of the menstrual cycle.

On approximately day 14 of the menstrual cycle, a large increase in LH secretion from the anterior pituitary gland causes a mature follicle to undergo ovulation. After ovulation, the corpus luteum in the ovary secretes increasing amounts of progesterone and some estrogen. The progesterone acts on the uterus causing the cells of the endometrium to become larger and to secrete a small amount of fluid.

The time between ovulation and the next menses is called the **secretory phase** of the menstrual cycle because of the small amount of fluid secreted by the cells of the endometrium. During the secretory phase, the lining of the uterus reaches its maximum thickness. After ovulation, the combination of progesterone and estrogen from the corpus luteum inhibits LH and FSH secretion, and LH and FSH secretion decline to low levels in the blood.

3 Predict the effect of administering a relatively large amount of estrogen and progesterone just before the increase in LH that precedes ovulation.

If the ovulated oocyte unites with a sperm cell in the oviduct (see Chapter 20), it forms a **zygote,** the zygote begins to undergo cell divisions to form a structure called a **blastocyst** (see Chapter 20). By 7 or 8 days following ovulation (day 21 or 22 of the menstrual cycle), the endometrium is prepared to receive the developing blastocyst. The developing blastocyst becomes implanted in the thick endometrium where development proceeds. If the oocyte is not united with a sperm cell, the corpus luteum begins to produce less progesterone by day 24 or 25 of the menstrual cycle. By day 28 the declining progesterone causes the endometrium to slough away to begin menses and the next menstrual cycle.

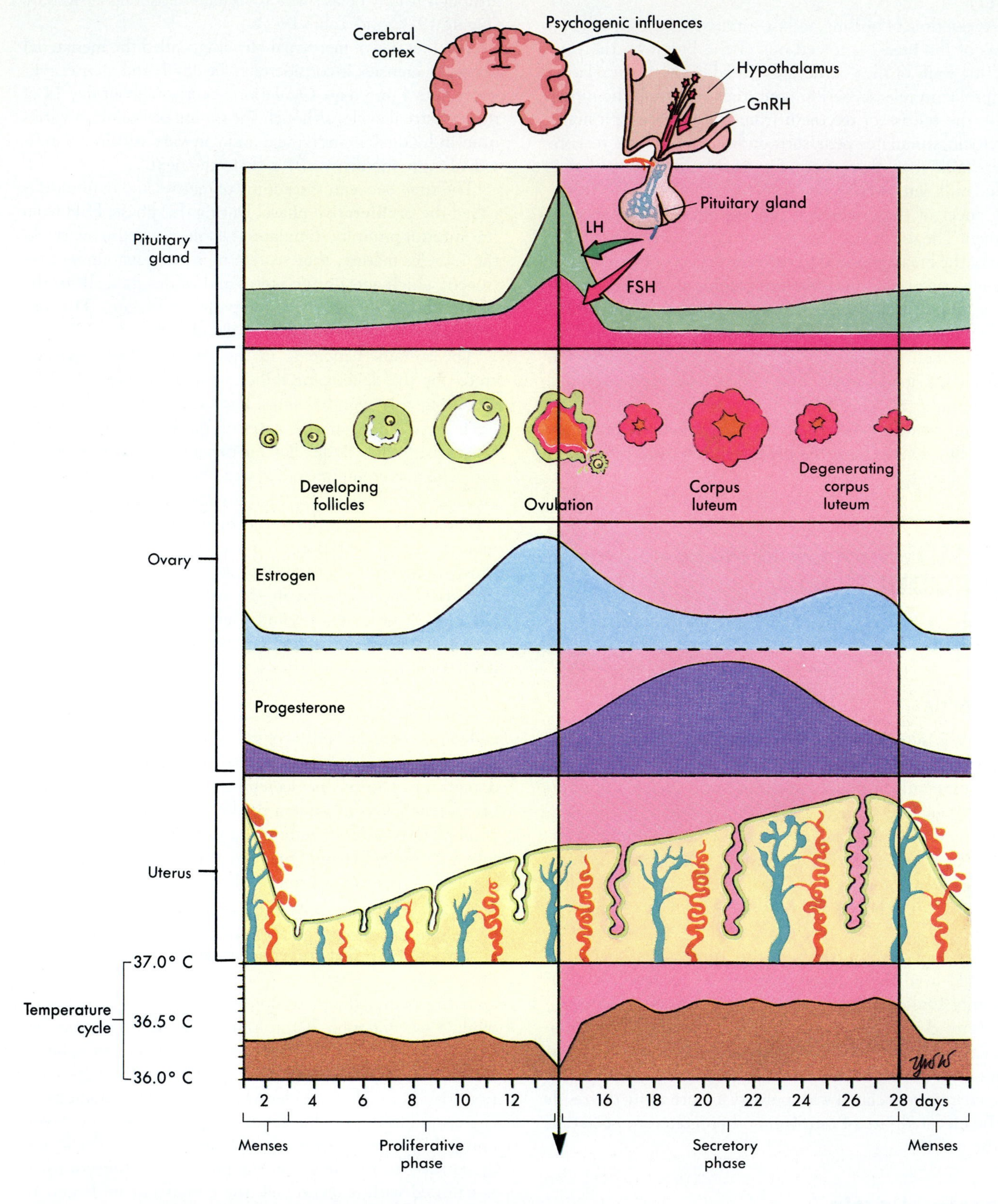

Figure 19-12 ● The menstrual cycle.
Hormonal changes that occur during the menstrual cycle, changes that occur in the ovary, and changes that occur in the endometrium of the uterus arranged relative to each other.

Table 19-2 • Events during the menstrual cycle

• Menses (day 1 to day 4 or 5 of the menstrual cycle)	
Pituitary gland	The rate of LH and FSH secretion is low.
Ovary	The rate of estrogen and progesterone secretion is low after degeneration of the corpus luteum produced during the previous menstrual cycle.
Uterus	In response to declining progesterone levels, the endometrial lining of the uterus sloughs off, resulting in menses followed by repair of the endometrium.
• Proliferative phase (from day 4 or 5 until ovulation on about day 14)	
Pituitary gland	The rate of FSH and LH secretion is low during most of the proliferative phase; FSH and LH secretion increase near the end of the proliferative phase in response to increasing estrogen secretion from the ovaries.
Ovary	Developing follicles secrete increasing amounts of estrogen, especially near the end of the proliferative phase; increasing FSH and LH cause additional estrogen secretion from the ovaries near the end of the proliferative phase.
Uterus	Estrogen causes endometrial cells of the uterus to divide. The endometrium of the uterus thickens and tubular glands form.
• Ovulation (about day 14)	
Pituitary gland	The rate of FSH and LH secretion increases rapidly just before ovulation in response to increasing estrogen levels. Increasing FSH and LH levels stimulate estrogen secretion resulting in a positive-feedback cycle.
Ovary	LH causes final maturation of a mature follicle and initiates the process of ovulation. FSH acts on immature follicles and causes several of them to begin to enlarge.
Uterus	Cells of the endometrium continue to divide in response to estrogen.
• Secretory phase (from about day 14 to day 28)	
Pituitary gland	Estrogen and progesterone reach levels high enough to inhibit LH and FSH secretion from the pituitary gland.
Ovary	After ovulation the follicle is converted to the corpus luteum; the corpus luteum secretes large amounts of progesterone and smaller amounts of estrogen from shortly after ovulation until about day 24 or 25. If fertilization does not occur, the corpus luteum degenerates after about day 25 and the rate of progesterone secretion rapidly declines to low levels.
Uterus	In response to progesterone the endometrial cells enlarge, the endometrial layer thickens, and the glands of the endometrium reach their greatest degree of development; the endometrial cells secrete a small amount of fluid. After progesterone levels decline, the endometrium begins to degenerate.
• Menses (day 1 to day 4 or 5 of the next menstrual cycle)	
Pituitary gland	The rate of LH and FSH secretion is low.
Ovary	The rate of estrogen and progesterone secretion is low.
Uterus	In response to declining progesterone levels, the endometrial lining of the uterus sloughs off, resulting in menses followed by repair of the endometrium.

Menstrual cramps result from strong smooth muscle contraction of the uterus just before and during menses. The cramps result, in part, from excessive production of inflammatory substances such as prostaglandins. Inflammatory substances are produced when the endometrium degenerates and is sloughed. In some women, menstrual cramps are extremely uncomfortable.

Some women suffer from more or less severe changes in mood that can result in depression, irritability, and other socially undesirable behaviors just before menses. This condition is called **premenstrual syndrome (PMS).** The fluctuations in estrogen and progesterone associated with the menstrual cycle may trigger these mood changes, although the precise cause of the condition is unknown.

Understanding the prevention of pregnancy

This chapter describes how the reproductive system functions and how development begins. In our society, methods that prevent reproduction are very important. For example, many couples elect to limit the number of children in their families. Some women elect to pursue a profession, or experience economic pressures that play a role in the decision to delay having children or to not have children at all. Several methods are available to prevent pregnancy. Most, but not all, of them prevent pregnancy by acting on the female reproductive system. It is unfortunate that, in our society, the burden for the prevention of pregnancy generally is the responsibility of women. The pregnancy prevention techniques most commonly and successfully used are those that women elect to use.

Several methods are available to prevent or terminate pregnancy (Figure 19-A). These include the prevention of fertilization, called **contraception,** and the removal of the implanted embryo, called **abortion.** Many methods of contraception are quite effective when done properly and used consistently. For example abstinence, when practiced consistently, is a sure way to prevent pregnancy, but it is not an effective method when used occasionally. Although many of the methods that prevent pregnancy are not permanent, some do result in the permanent loss of the ability to reproduce, or **sterility.** Methods that prevent reproduction can be categorized as behavioral methods, barrier methods, chemical methods, surgical methods, or methods that prevent implantation.

BEHAVIORAL METHODS

Coitus interruptus (ko′ĭ-tus in′ter-rup′tis) is removal of the penis from the vagina just before ejaculation so that sperm cells are not deposited in the vagina. This is a very unreliable method of preventing pregnancy, since it requires perfect awareness and willingness to withdraw the penis at the correct time.

The **rhythm method** requires abstaining from sexual intercourse near the time of ovulation. A major factor in the success of this method is the ability to predict accurately the time of ovulation. Although the rhythm method provides some protection against becoming pregnant, it has a relatively high rate of failure resulting from both the inability to predict the time of ovulation and the failure to abstain during the period of fertility.

BARRIER METHODS

A **condom** is a sheath of animal membrane, latex (rubber), or plastic that is placed over the erect penis. It is a barrier device, since the semen is collected within the condom instead of within the vagina. Latex condoms also provide protection against sexually transmitted diseases.

A condom used by females and placed in the vagina prior to sexual activity also has been developed. Like the condom placed over the penis, it functions as a barrier device.

Methods to prevent sperm cells from reaching the oocyte once they are in the vagina include use of a diaphragm, spermicidal agents, and the vaginal sponge. A **diaphragm** is a flexible plastic or rubber dome that is placed over the cervix, where it prevents passage of sperm cells from the vagina through the cervical canal of the uterus. The most commonly used spermicidal agents are foams or creams that kill sperm cells; they are inserted into the vagina before intercourse. A device to flood or rinse the vagina with a spermicidal agent is called a **spermicidal douche.** Spermicidal douches used alone and after intercourse are not very effective. **The sponge,**—a soft, disposable polyurethane device containing a spermicidal agent—is placed over the cervix where it acts as a barrier *and* kills the sperm cells. Note that a combination of devices—for example, the sponge used with a condom or condoms used with foams or creams—improves effectiveness.

CHEMICAL METHODS

Synthetic estrogen and progesterone in **oral contraceptives,** or **birth control pills,** effectively suppress fertility in females. These substances can have more than one action, but they reduce LH release from the anterior pituitary. The reduced LH prevents ovulation. Over the years, the dose of estrogen and progesterone in birth control pills has been reduced. The current lower dose birth control pills have fewer side effects than earlier dosages. There is an increased risk of heart attack or stroke in women using oral contraceptives and who smoke, or those who have a history of hypertension or coagulation disorders. For most women, the pill is effective and has a minimum frequency of complications until women exceed 35 years of age.

An implant has been developed that can stop the menstrual cycle for several months. The implants are approximately the diameter of a pencil lead, contain a progesterone-like substance, and are placed beneath the skin. Implants have the advantage of preventing pregnancy without a woman having to take a pill each day.

Chemical methods of reducing sperm cell production are being developed. Some techniques appear to be promising, but none are currently available.

Lactation prevents the menstrual cycle for a few months after childbirth. Action potentials sent to the hypothalamus that cause the release of oxytocin and prolactin also inhibit GnRH release from the hypothalamus. Reduced GnRH reduces LH, which prevents ovulation. Despite continual lactation, the ovarian and uterine cycles eventually resume. Because ovulation normally precedes menstruation, relying on lactation to prevent pregnancy is not consistently effective.

SURGICAL METHODS

Vasectomy (vă-sek′to-me) is a common method used to render males permanently incapable of fertilization without affecting the performance of the sexual act. Vasectomy is a surgical procedure used to cut and tie the ductus deferens within the scrotum, preventing sperm cells from becoming part of the semen. Since such a small volume of semen comes from the testis and epididymis, vasectomy has little effect on the volume of

Figure 19-A
Contraceptive devices and techniques.
A, Condom. **B,** Diaphragm with spermicidal jelly. **C,** The vaginal sponge. **D,** Spermicidal foam. **E,** Oral contraceptives. **F,** Vasectomy. **G,** Tubal ligation.

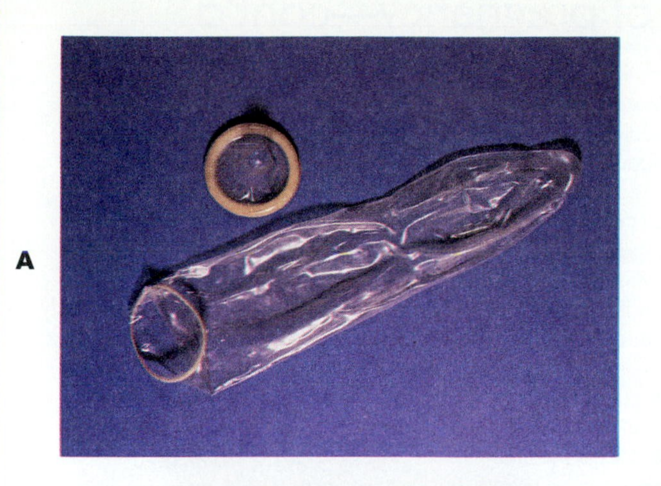

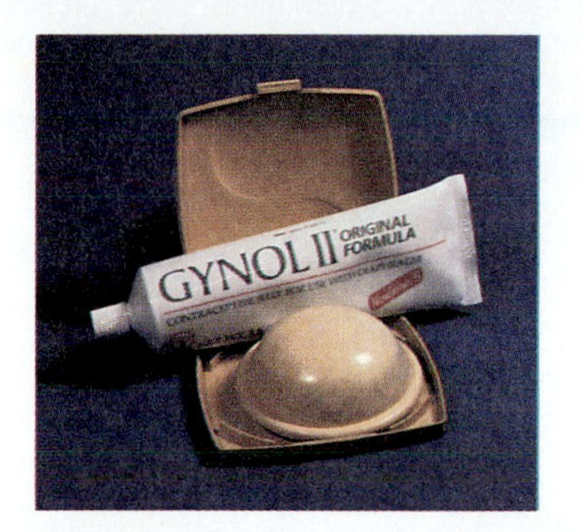

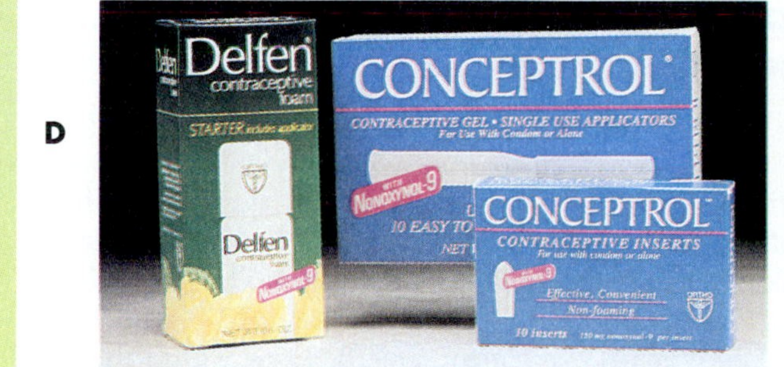

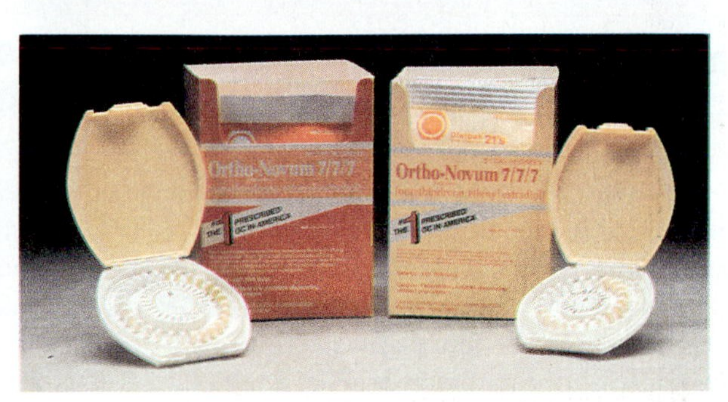

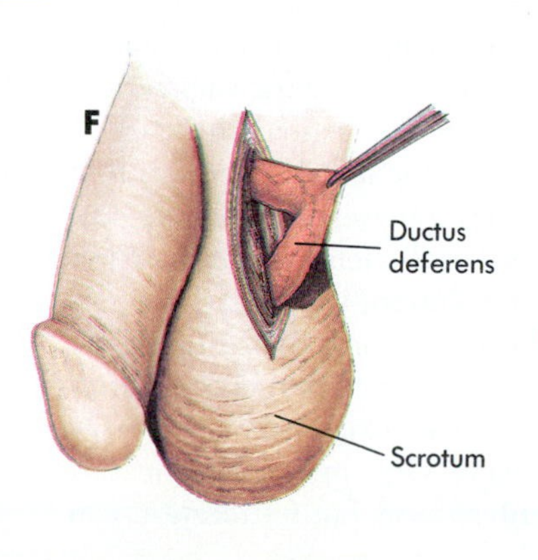

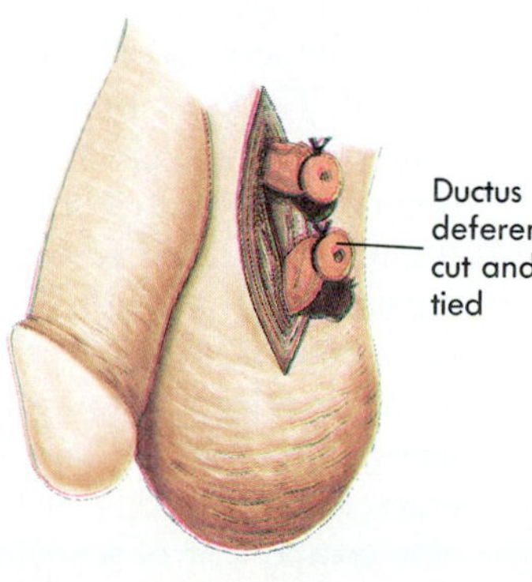

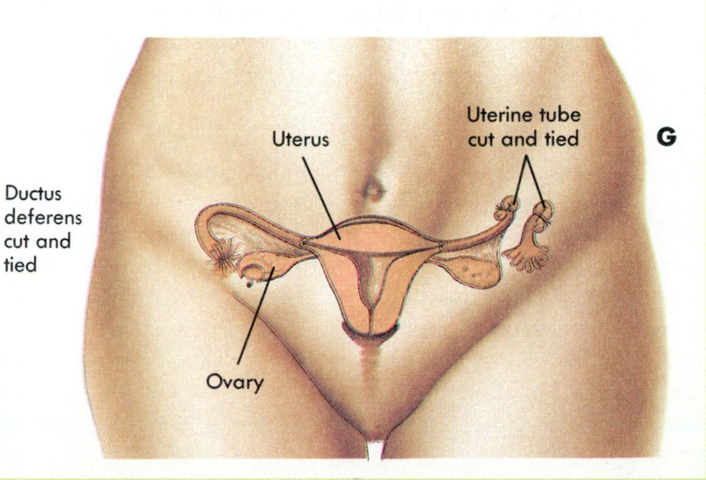

Continued.

Understanding the prevention of pregnancy—cont'd

the semen. The sperm cells are reabsorbed in the epididymis.

A common method of permanent birth control in females is **tubal ligation,** a procedure in which the uterine tubes are tied and cut or clamped through an incision made through the wall of the abdomen. This procedure closes off the pathway between the sperm cells and the oocyte. **Laparoscopy** (lap-ă-ros′ko-pe), in which a special instrument is inserted into the abdomen through a small incision, is commonly used so that only small openings are required to perform the operation.

In some cases, pregnancies are terminated by surgical procedures called **abortions.** The most common method for performing abortions is the insertion of an instrument through the cervix into the uterus. The instrument scrapes the endometrial surface, and at the same time, a strong suction is applied. The endometrium and the embedded embryo are disrupted and sucked out of the uterus. This technique is normally used in pregnancies that have progressed less than 3 months.

A new drug, RU486, is currently available in some European countries and may be available soon in the United States. RU486 blocks the action of progesterone causing the endometrium of the uterus to slough off as it does at the time of menstruation. Therefore it can be used to induce menstruation and reduce the possibility of implantation when sexual intercourse has occurred near the time of ovulation. It can also be used to terminate pregnancies.

PREVENTION OF IMPLANTATION

Intrauterine devices (IUDs) are inserted into the uterus through the cervix, and they prevent normal implantation of the developing embryo within the endometrium. Some early IUD designs produced serious side effects such as perforation of the uterus and, as a result, many IUDs have been removed from the market. Data indicate, however, that IUDs are effective in preventing pregnancy.

4 Of the methods that are available to prevent pregnancy, which methods result in sterility and which of those contraceptive methods do not have a permanent effect on fertility?

Menopause

When a woman is 40 to 50 years old, the menstrual cycles become less regular, and ovulation does not consistently occur during each cycle. The time when menstrual cycles no longer occur is called **menopause.**

Menopause is the result of age-related changes in the ovary. Just prior to menopause, the number of follicles remaining in the ovaries of menopausal women is small, and they are less sensitive to stimulation by FSH and LH. Consequently, fewer mature follicles and corpora lutea are produced. Gradual changes occur in women in response to the reduced amount of estrogen and progesterone produced by the ovaries (Table 19-3).

Prior to menopause, some women experience "hot flashes," irritability, fatigue, anxiety, and occasionally, severe emotional disturbances. Many of these symptoms can be treated successfully by administering small amounts of estrogen and then gradually decreasing the dosage, or by providing psychological counseling. Some, but not all, data suggest that a side effect of estrogen can be a slightly increased possibility of breast and uterine cancer but a reduced risk of heart attack and strokes.

Female sexual behavior and the female sexual act

Sexual drive in females, like sexual drive in males, is dependent on hormones. Testosterone-like hormones and possibly estrogens affect sexual behavior. Testosterone-like hormones are produced primarily in the adrenal gland. The female neural pathways involved in controlling sexual responses are similar to those found in the male.

During sexual excitement, erectile tissue within the clitoris and around the vaginal opening become engorged with blood. The mucous glands within the vestibule, especially the vestibular glands, secrete small amounts of mucus. Larger amounts of mucus-like fluid pass into the vagina from its wall. These secretions lubricate the vagina during intercourse. The tactile stimulation of the female's genitals that occurs during sexual intercourse, as well as psychological stimuli, can trigger an **orgasm,** or the **female climax.** After orgasm, there is a period of **resolution,** which is characterized by an overall sense of satisfaction and relaxation. However, females are more receptive to further immediate stimulation than males and they can experience successive orgasms more quickly than males. Orgasm can be a pleasurable component of sexual intercourse, but fertilization can occur in females without it.

Table 19-3 Possible changes in post-menopausal women caused by decreased ovarian hormone secretion

Changes	
Menstrual cycle	5-7 years before menopause the cycle becomes irregular; the number of cycles in which ovulation does not occur and corpora lutea do not develop increases.
Uterus	Irregular menstruation gradually is followed by no menstruation; the endometrium and the uterus become smaller.
Vagina and external genitalia	Epithelial lining becomes thinner; external genitalia become thinner and less elastic; labia majora become smaller; pubic hair decreases; reduced secretion leads to dryness; the vagina is more easily inflamed and infected.
Skin	Epidermis becomes thinner; melanin synthesis increases and skin becomes darker.
Cardiovascular system	Hypertension and atherosclerosis occur more frequently.
Vasomotor instability	Hot flashes and increased sweating are correlated with vasodilation of cutaneous blood vessels; hot flashes are related to decreased estrogen levels.
Libido	Temporary changes, usually a decrease, in sex drive are associated with the onset of menopause.
Fertility	Fertility begins to decline about 10 years before the onset of menopause; by age 50 almost all oocytes and follicles have degenerated.
Pituitary function	The low levels of estrogen and progesterone produced by the ovary cause the pituitary gland to secrete larger-than-normal amounts of LH and FSH. The increased levels of these hormones have little effect on the postmenopausal ovary.

SUMMARY

The reproductive organs produce the sex cells, sustain them, and transport them to the site where fertilization occurs. The female reproductive system nurtures the developing child.

FORMATION OF SEX CELLS

The formation of male and female sex cells occurs by meiosis.

Prior to meiosis each cell contains 46 chromosomes (23 pairs), and after meiosis each sex cell contains 23 chromosomes (1 from each of the 23 pairs).

After fertilization the fertilized cell contains 46 chromosomes (23 pairs).

MALE REPRODUCTIVE SYSTEM

Scrotum

The scrotum is a two-chambered sac that contains the testes.

The temperature in the scrotum is slightly lower than the body temperature, which is required for sperm cell development.

Testes

The testes are divided into lobules containing the seminiferous tubules (produce sperm cells) and interstitial cells (produce testosterone).

The seminiferous tubules join a tubular network that opens into the efferent ductules, which enter the epididymis.

During development the testes pass from the abdominal cavity through the inguinal canal to the scrotum.

Spermatogenesis

Spermatogenesis, the production of sperm cells, begins at the time of puberty.

Sertoli cells nourish and support the germ cells.

Spermatogonia give rise to spermatocytes, which divide by means of meiosis to form spermatids.

Spermatids develop a head, midpiece, and flagellum to become sperm cells.

Ducts

The epididymis is the site of sperm maturation.

The ductus deferens passes from the epididymis into the abdominal cavity.

The ejaculatory duct is formed by the joining of the ductus deferens and the duct from the seminal vesicle.

The ejaculatory ducts join the urethra in the prostate gland.

The urethra extends from the urinary bladder through the penis to the outside of the body.

Penis

The penis consists of erectile tissue, the two corpora cavernosa, and the corpus spongiosum.

The prepuce covers the glans penis.

Glands and secretions

The seminal vesicles empty into the ejaculatory duct. Their secretions nourish the sperm cells.

The prostate gland empties into the urethra. The secretions neutralize acid.

The bulbourethral glands empty into the urethra. Their secretions neutralize acid.

PHYSIOLOGY OF MALE REPRODUCTION

Regulation of sex hormone secretion

Gonadotropin-releasing hormone (GnRH) is produced in the hypothalamus.

GnRH stimulates release of luteinizing hormone (LH) and follicle-stimulating hormone (FSH) from the anterior pituitary.

LH stimulates the interstitial cells to produce testosterone.

FSH stimulates spermatogenesis.

Puberty

During puberty there is an increased secretion of GnRH, FSH, LH, and testosterone.

Effects of testosterone

Testosterone causes enlargement of the genitals and is necessary for spermatogenesis.

Testosterone is responsible for the development of secondary sexual characteristics.

Male sexual behavior and the male sexual act

The male sexual act is a complex series of reflexes that result in erection, emission, and ejaculation.

A climax sensation called orgasm is associated with ejaculation.

FEMALE REPRODUCTIVE SYSTEM

Ovaries

The ovaries are located in the pelvic cavity.

Follicles containing oocytes are located in the ovary.

Follicle and oocyte development

As follicles mature in response to FSH they enlarge, antrums form, and the oocytes increase in size.

Ovulation is the release of the oocyte from the ovary. Ovulation occurs in response to LH. After ovulation the follicle becomes the corpus luteum. If fertilization occurs, the corpus luteum persists. If there is no fertilization, it degenerates.

Uterine tubes

The uterine tube transports the oocyte from the ovary to the uterus.

Fertilization usually occurs in the upper portion of the uterine tube.

Uterus

The uterus is a pear-shaped organ. The uterine cavity and the cervical canal are the spaces formed in the uterus.

The wall of the uterus consists of the serous layer, myometrium (smooth muscle), and endometrium.

Vagina

The vagina connects the uterus (cervix) to the vestibule.

The vagina consists of a layer of smooth muscle and a mucous membrane.

The hymen covers the opening of the vagina.

External genitalia

The vestibule is a space into which the vagina and the urethra open.

The clitoris is composed of erectile tissue and is important in detecting sexual stimuli.

The labia minora are folds lateral to the vestibule.

The vestibular glands produce a mucous fluid, and lubricating fluid is produced by the wall of the vagina.

The labia majora usually cover the labia minora.

The mons pubis is an elevated area superior to the labia majora.

Mammary glands

The mammary glands are located in the breasts.

The lobes of the mammary gland connect to the nipple through ducts. The nipple is surrounded by the areola.

The mammary glands develop and enlarge in response to estrogen and progesterone during pregnancy.

After parturition the mammary glands produce milk. Milk is secreted in response to prolactin. Milk is ejected from the alveoli in response to oxytocin.

PHYSIOLOGY OF FEMALE REPRODUCTION

Puberty

Puberty begins with the first menstrual bleeding (menarche). Puberty begins when GnRH, LH, and FSH levels increase.

Menstrual cycle

The cyclical changes in the uterus are controlled by estrogen and progesterone produced by the ovary.

Cyclic changes in the uterus:

A. Menses (day 1 to days 4 or 5). Menses is composed of sloughed cells, secretions, and blood.
B. Proliferative phase (day 5 to day 14). Epithelial cells multiply and form glands in response to estrogen.
C. Secretory phase (day 15 to day 28). The endometrium becomes thicker, and endometrial glands secrete in response to progesterone. The uterus is prepared for implantation of the blastocyst by day 21.

FSH initiates the development of the follicles.

LH stimulates ovulation and formation of the corpus luteum.

Menopause

The cessation of the menstrual cycle is called menopause.

Female sexual behavior and the female sexual act

Female sexual drive is partially influenced by testosterone-like hormones (produced by the adrenal gland) and possibly estrogens produced by the ovary.

Nervous stimulation causes erectile tissue to become engorged with blood, the vestibular glands to secrete mucus, and the vagina to produce a lubricating fluid.

CONTENT REVIEW

1. What is the scrotum? Explain how the temperature of the testis is regulated.
2. Where are sperm cells produced in the testes? Describe the process of spermatogenesis.
3. Name the ducts the sperm cells travel through to go from their site of production to the outside of the body.

4. Describe the erectile tissue of the penis. State where the seminal vesicles, prostate gland, and bulbourethral glands empty into the male reproductive duct system.
5. Define emission and ejaculation of semen.
6. Where are GnRH, FSH, LH, and testosterone produced?
7. Describe the effects of testosterone during puberty and on the adult male.
8. Describe the male sexual act.
9. Describe the process of follicle formation and ovulation.
10. What is the corpus luteum? What happens to the corpus luteum if fertilization occurs? If fertilization does not occur?
11. Describe the normal pathway followed by the oocyte after ovulation. Where does fertilization usually take place?
12. Describe the structure of the uterus and vagina.
13. Describe the parts of the external genitalia.
14. What effect does estrogen, progesterone, prolactin, and oxytocin have on the mammary glands?
15. List the stages of the menstrual cycle and describe the changes that occur in the ovary and uterus in each stage.
16. Define menopause and list some of the changes that occur in women as a result.
17. Describe the female sexual act.

CONCEPT REVIEW

1. If the pituitary gland in an adult male stopped secreting LH and FSH, what effect would that have on the function of the testes?
2. If the pituitary gland in an adult female stopped secreting LH and FSH, what effect would that have on the function of the ovaries?
3. Birth control pills for women contain synthetic estrogen and progesterone compounds. Explain how these hormones can prevent pregnancy.
4. Predict what would happen if progesterone levels in the blood remained elevated rather than decreasing just prior to menses.
5. Predict what would happen if estrogen levels did not increase during the secretory phase of the menstrual cycle?
6. Illustrated below are the steps necessary to produce a new individual. At each step indicate the method(s) of birth control that would prevent the next step from occurring.

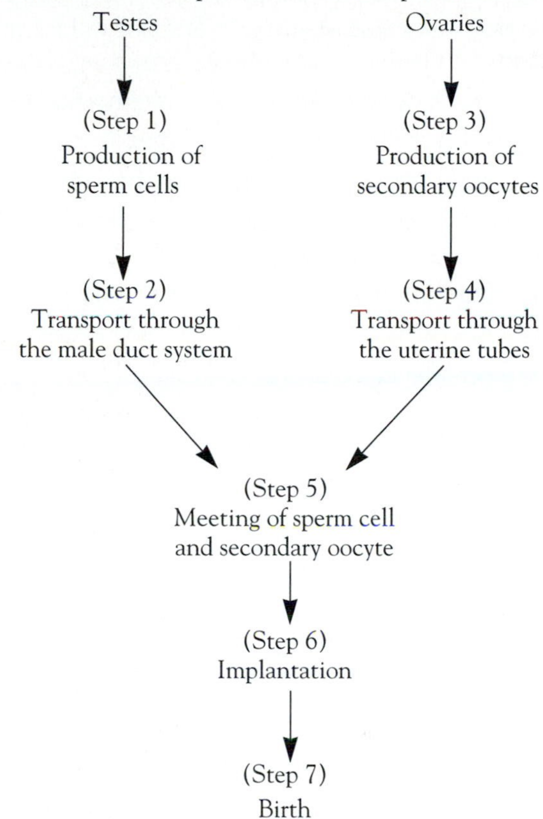

CHAPTER TEST

Answers can be found in Appendix C

MATCHING For each statement in column A select the correct answer in column B (an answer may be used once, more than once, or not at all).

A

1. Germ cells located in the seminiferous tubule that divide by mitosis.
2. The duct that carries sperm cells from the epididymis toward the prostate gland.
3. The process by which the prepuce is removed from the penis.
4. The discharge of semen into the urethra.
5. The forceful expulsion of semen from the urethra.
6. The major hormone responsible for the secondary sexual characteristics in males.
7. The major hormone secreted by developing follicles.
8. The major hormone secreted by the corpus luteum.
9. The inner layer of the uterus.
10. The space into which the urethra and the vagina open.
11. The time during which the endometrium is sloughed and eliminated from the uterus.
12. The cessation of menstrual cycles in a female after about 50 years of age.
13. The hormone responsible for milk production in the mammary gland after parturition.

B

circumcision
ductus deferens
ejaculation
ejaculatory duct
emission
endometrium
estrogen
follicular phase
menopaurse
menses
myometrium
ovum
progesterone
prolactin
secretory phase
sperm cells
spermatogonia
testosterone
vestibule

FILL-IN-THE BLANK Complete each statement by providing the missing word or words.

1. __________ are large cells in the seminiferous tubule that support and nourish the germ cells.
2. Cells located in delicate connective tissue around the seminiferous tubules that secrete testosterone are __________.
3. The gland that produces about 60% of the semen volume is called the __________.
4. The gland through which the male urethra passes is called the __________.
5. The two erectile bodies in the dorsal portion of the penis are called the __________ __________.
6. The pituitary hormone that stimulates follicle development is __________.
7. Fertilization of the oocyte normally occurs in the __________.
8. The phase of the menstrual cycle during which progesterone is secreted from the corpus luteum is the __________ phase.
9. The average number of days following the beginning of menses that ovulation occurs is __________
10. The hormone that is responsible for ovulation is __________.

CHAPTER 20

Development and Heredity

KEY TERMS

blastocyst
Early stage of development consisting of a hollow ball of cells with an inner cell mass; stage at which implantation in the uterus occurs.

embryo
In prenatal development, the developing human between the end of the second week and the end of the eighth week.

fetus
In prenatal development, the developing human between the end of the eighth week and birth.

gene
The functional unit of heredity; a group of nucleotides in DNA containing the instructions necessary for making a protein.

genotype
Genetic makeup of the individual.

heterozygous
Having two different genes for a given trait.

homozygous
Having two identical genes for a given trait.

parturition
Childbirth; the delivery of a baby at the end of pregnancy.

phenotype
Characteristic observed in the individual due to expression of the genotype.

zygote
The single-celled product of fertilization, resulting from the union of a sperm cell and a secondary oocyte.

OBJECTIVES

After reading this chapter you should be able to:

1. Describe fertilization and the formation of the blastocyst.
2. List the three germ layers, describe their formation, and list the adult derivatives of each layer.
3. Define embryo and fetus and describe the major events that occur during their development.
4. Define implantation and give the function of the placenta.
5. Describe the events that occur during parturition.
6. Define genetics and explain how chromosomes are related to genetics.
7. Define a gene and explain how genes control cell functions.

FEATURES

THE HUMAN BODY consists of trillions of specialized cells organized to form tissues, organs, and organ systems. This complex organism begins as a single cell that divides and gives rise to the cells of the body. **Development** is the process by which cell division and cell specialization results in the formation and continued existence of an individual. From before birth until death we are continually developing. This chapter considers those aspects of development that occur before birth.

The ability of a single cell to produce the trillions of cells of our bodies during development and the characteristics of our complex bodies are partly determined by our genetic makeup, that is, our deoxyribonucleic acid (DNA) (see Chapter 3). **Genetics** is the science that studies how these abilities and characteristics are transmitted from one generation to the next. Genetics also considers the effects of abnormalities in our genetic makeup.

DEVELOPMENT

Usually the life span of a person is considered to be the time from birth to death. However, the events that occur during the 9 months before birth, called the **prenatal period,** have profound effects on the rest of a person's life.

Early cell divisions

The prenatal period begins with **fertilization,** the union of a sperm cell with a secondary oocyte to form a single cell called the **zygote** (zi´gōt). Approximately 18 to 36 hours after fertilization, the zygote divides to form two cells. Those two cells divide to form four cells, which divide to form eight, and so on. Approximately 3 or 4 days after fertilization a hollow ball of cells called a **blastocyst** (blas´to-sist) has formed (Figure 20-1). At one end of the blastocyst is a collection of cells called the **inner cell mass.** Some of the inner cell mass cells give rise to the developing organism. The single layer of cells forming the wall of the blastocyst becomes part of the placenta (see Implantation and the Placenta in this chapter).

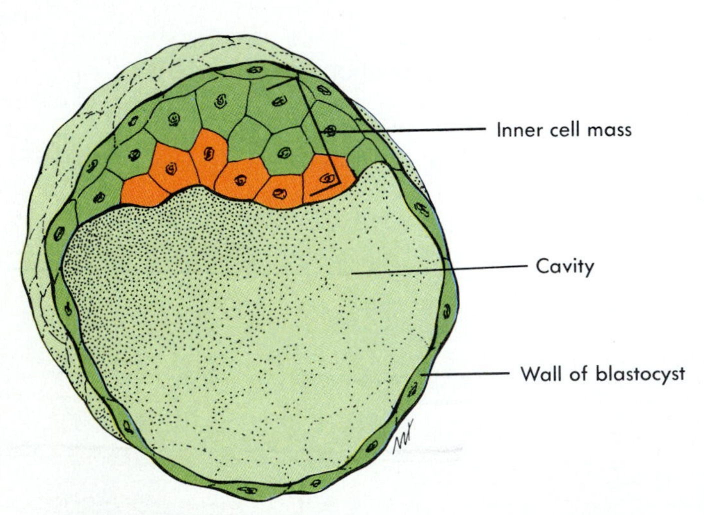

Figure 20-1 ● Blastocyst.
The orange cells of the inner cell mass become the embryo. All the green cells become the amniotic cavity, yolk sac, or part of the placenta.

Twins are two individuals born at the same time from the same mother. Twins occur in approximately 1% of all pregnancies. **Identical twins** look alike because they are genetically identical (have identical DNA). Identical twins result from the fertilization of an oocyte by a sperm cell, followed by a separation of the developing cells into two individuals. **Fraternal twins** have a different appearance and can even be a different sex because they are not genetically identical (have different DNA). Two oocytes, each fertilized by a different sperm cell, develop into fraternal twins. Fraternal twins are no more or less alike than any pair of siblings, except that they result from a single pregnancy.

Embryo and fetus

Within the inner cell mass of the blastocyst two cavities form, the **yolk sac** and the **amniotic** (am´ne-ot´ik) **cavity** (Figure 20-2). In some animals, such as birds, the yolk sac contains nutrients called yolk, which the developing animal uses as a source of energy. In humans, the placenta (see below) provides nutrients and the yolk sac has other functions such as producing blood cells. The amniotic cavity enlarges as a fluid-filled amniotic sac into which the developing organism will grow. The amniotic fluid within the cavity forms a protective cushion around the developing organism.

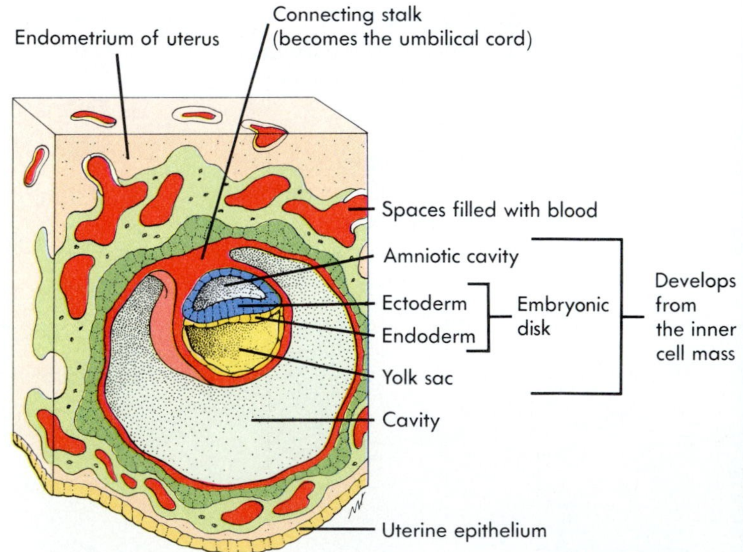

Figure 20-2 ● Embryonic disk.
The blastocyst is embedded within the wall of the uterus. The amniotic cavity and the yolk sac have formed within the inner cell mass of the blastocyst. The embryonic disk consists of ectoderm and endoderm. The connecting stalk, which attaches the embryo to the uterus, will become part of the umbilical cord.

Between the yolk sac and the amniotic cavity is a flat disk of tissue called the **embryonic disk** (see Figure 20-2). The embryonic disk is composed of two layers of cells—an **ectoderm** (ek´to-derm) and an **endoderm** (en´do-derm). Proliferating cells of the ectoderm migrate toward the center of the disk, forming a thickened line called the **primitive streak.** Some ectoderm cells migrate through the primitive streak and emerge between the ectoderm and endoderm as a new layer, the **mesoderm** (mez´o-derm) (Figure 20-3). These three **germ layers,** ectoderm, mesoderm, and endoderm, are the beginning of the **embryo** (em´bre-o). The developing organism is considered an embryo from the end of the second week to the end of the eighth week of development.

1 Approximately 30% of identical twins result from the development of two separate blastocysts from a single zygote. The other 70% of identical twins result from a separation of the inner cell mass of a single blastocyst into two groups of cells that develop independently of each other. Two primitive streaks are formed, resulting in two embryos. Rarely, the primitive streaks remain in contact with each other. What would be the result of this rare event?*

During the embryonic period the germ layers give rise to all the tissues and organs of the body. Although there are exceptions, the following generalizations can be made:

1. Ectoderm gives rise to the epidermis and glands of the skin; the brain, spinal cord, and nerves; the pituitary gland; and the adrenal medulla.
2. Endoderm gives rise to the lining of the digestive tract, glands associated with the digestive tract, and the lining of the respiratory and urinary tracts.
3. Mesoderm gives rise to most muscle tissue, most connective tissues, serous membranes, blood vessels, reproductive organs, and the adrenal cortex.

*Answers to predict questions appear in Appendix B at the back of the book.

Figure 20-4, *A* shows a 35-day-old embryo. The head and eyes can be seen, a beating heart is present, and limb buds are being formed. Because the embryonic period is the time of organ formation, serious, and even fatal, malformations of the organs can occur. For example, the drug thalidomide was once used to treat morning sickness in pregnant women. Unfortunately it also interfered with normal limb development and some children were born with only partially formed limbs.

During the last 7 months of the prenatal period the developing organism is called a **fetus** (fe´tus) (Figure 20-4, *B*). During this time the organ systems grow and become more mature as the fetus greatly increases in size and weight. At the end of 8 weeks the embryo is approximately 1.25 inches in length and weighs approximately 1 ounce. At birth the fetus is approximately 20 inches in length and weighs approximately 7.3 pounds.

Implantation and the placenta

Fertilization normally occurs in the uterine tube. As the blastocyst forms, the dividing cells move from the uterine tube to the uterus. Approximately 7 days after ovulation (day 21 of the menstrual cycle), the uterus is prepared for **implantation,** which is the attachment and subsequent embedding of the blastocyst in the uterine wall. The single

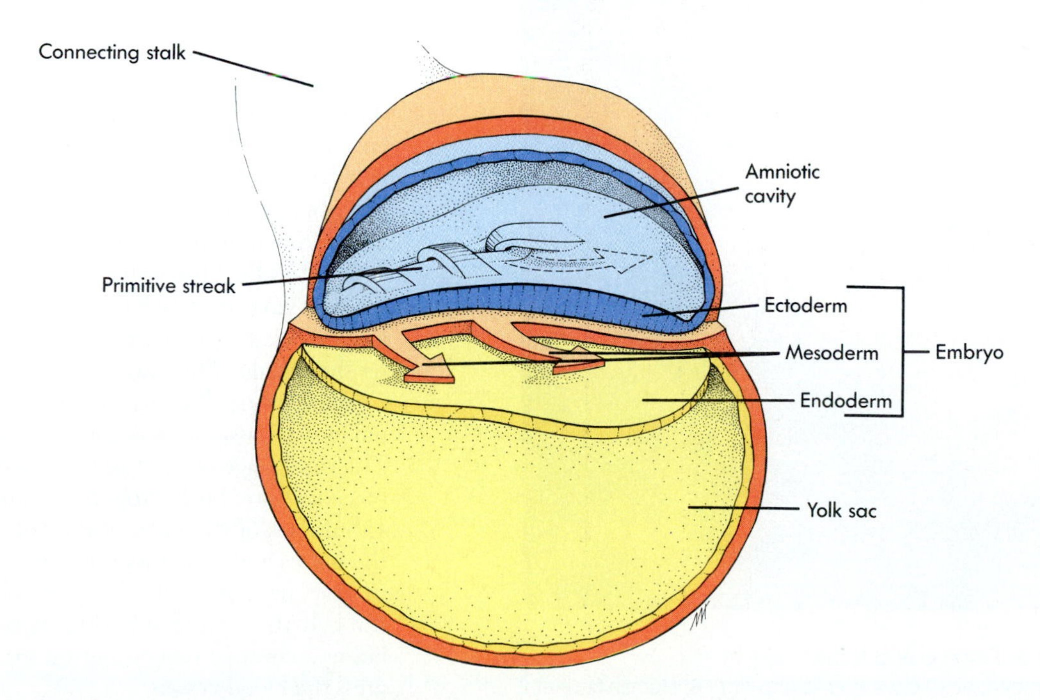

Figure 20-3 ● Formation of the embryo.
Ectoderm migrates into the primitive streak to become mesoderm, located between the ectoderm and endoderm. The embryo consists of ectoderm, mesoderm, and endoderm.

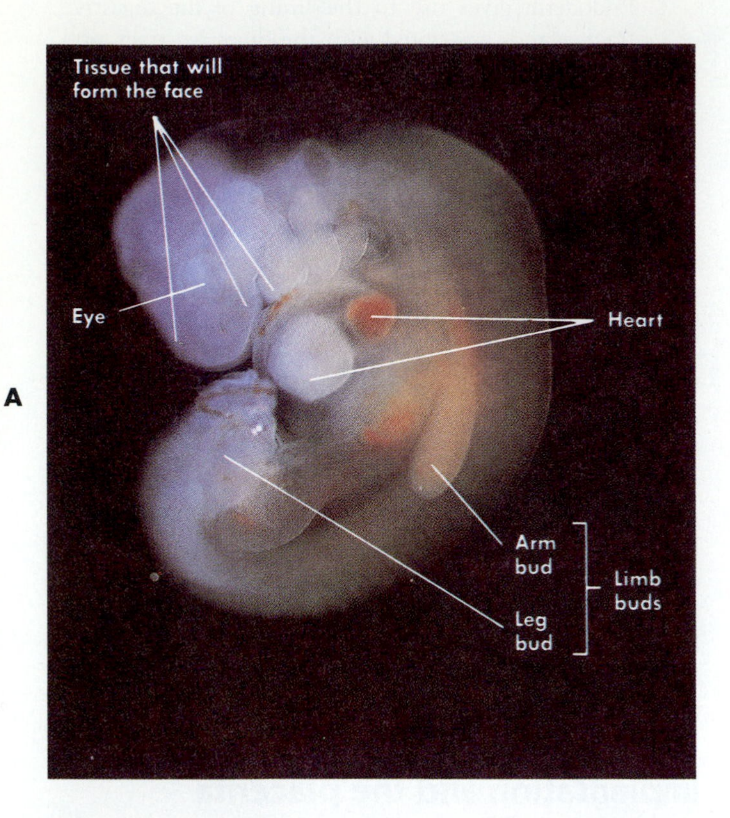

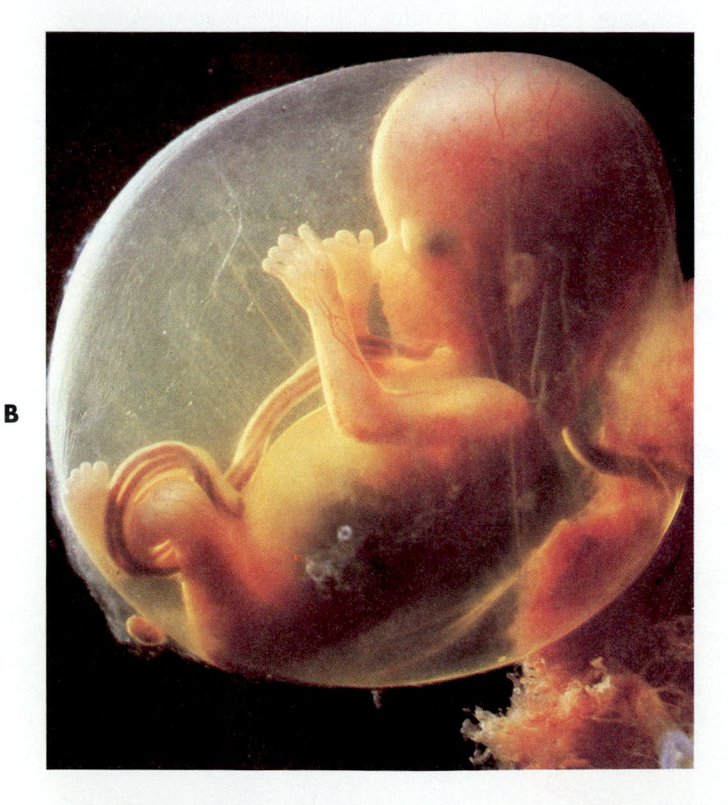

Figure 20-4 ● Embryo and fetus.
A, Human embryo at 35 days after fertilization. **B,** Human fetus at 3 months of development.

layer of cells forming the wall of the blastocyst eventually becomes the **chorion** (ko´re-on). The chorion and the lining of the uterus form the **placenta** (pla-sen´tah), which functions to exchange nutrients, gases, and waste products between the mother and the embryo.

Initially the embryo is attached to the placenta by a connecting stalk (see Figures 20-2 and 20-3). As the embryo matures and becomes a fetus, the connecting stalk elongates and is known as the **umbilical cord** (Figure 20-5). Blood vessels within the umbilical cord carry blood between the placenta and fetus. Within the placenta there is no mixing of maternal and fetal blood. Fetal blood vessels are inside fingerlike projections called **chorionic villi** that extend into cavities containing pools of maternal blood within the uterine wall.

The chorion secretes **human chorionic gonadotropin (HCG),** which is transported in the blood to the maternal ovary. HCG stimulates the corpus luteum in the ovary to secrete estrogen and progesterone, which are essential for the maintenance of the endometrium of the uterus for the first 3 months of pregnancy (see Chapter 19). Most pregnancy tests are designed to detect HCG in either urine or blood.

PARTURITION

Parturition (par-tu-rish´un) refers to the process by which the baby is born. It also is called **labor** or **delivery.** Near the end of pregnancy, the uterus has occasional contractions that become stronger and more frequent prior to parturition. At parturition the contractions become regular and stronger. Although parturition can differ greatly from woman to woman, it usually is divided into three stages.

1. The first stage begins with the onset of regular uterine contractions and extends until the cervix dilates to a diameter approximately the size of the fetus' head (10 cm). This stage takes approximately 24 hours, but it may be as short as a few minutes in some women who have had more than one child. Contractions of the uterus can rupture the amniotic sac, which is often called "breaking the water."
2. The second stage of labor lasts from the time of maximum cervical dilation until the time that the baby exits the vagina. This stage may last from 1 minute to 1 hour. During this stage, contraction of the abdominal muscles assists the uterine contractions.
3. The third stage of labor involves the expulsion of the placenta (after birth) from the uterus. It occurs within 15 minutes of the birth of the baby. Contractions of the uterus and abdominal muscles cause the placenta to tear away from the wall of the uterus and exit through the vagina. Bleeding is normally restricted because contractions of uterine smooth muscle compress the blood vessels.

The precise signal that triggers parturition is not known, but many factors that support parturition have been identi-

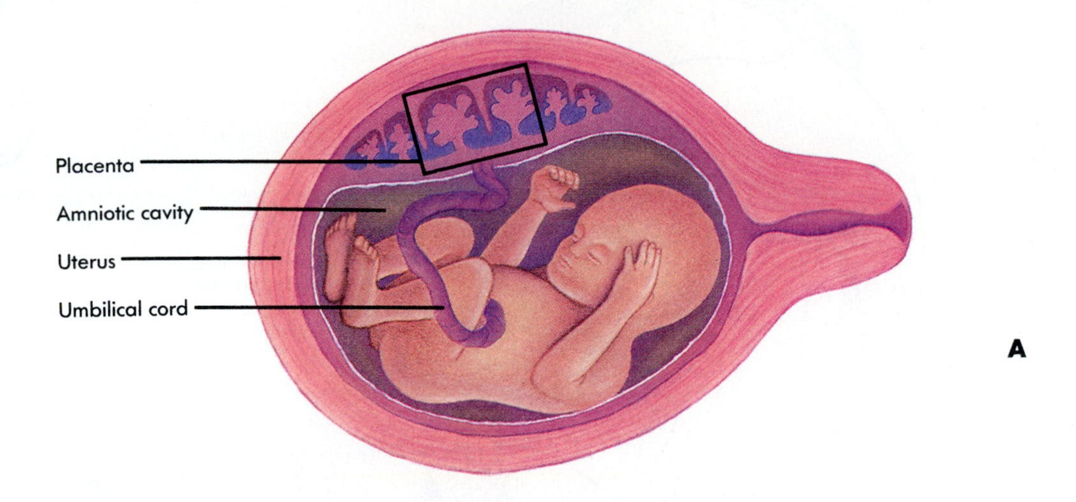

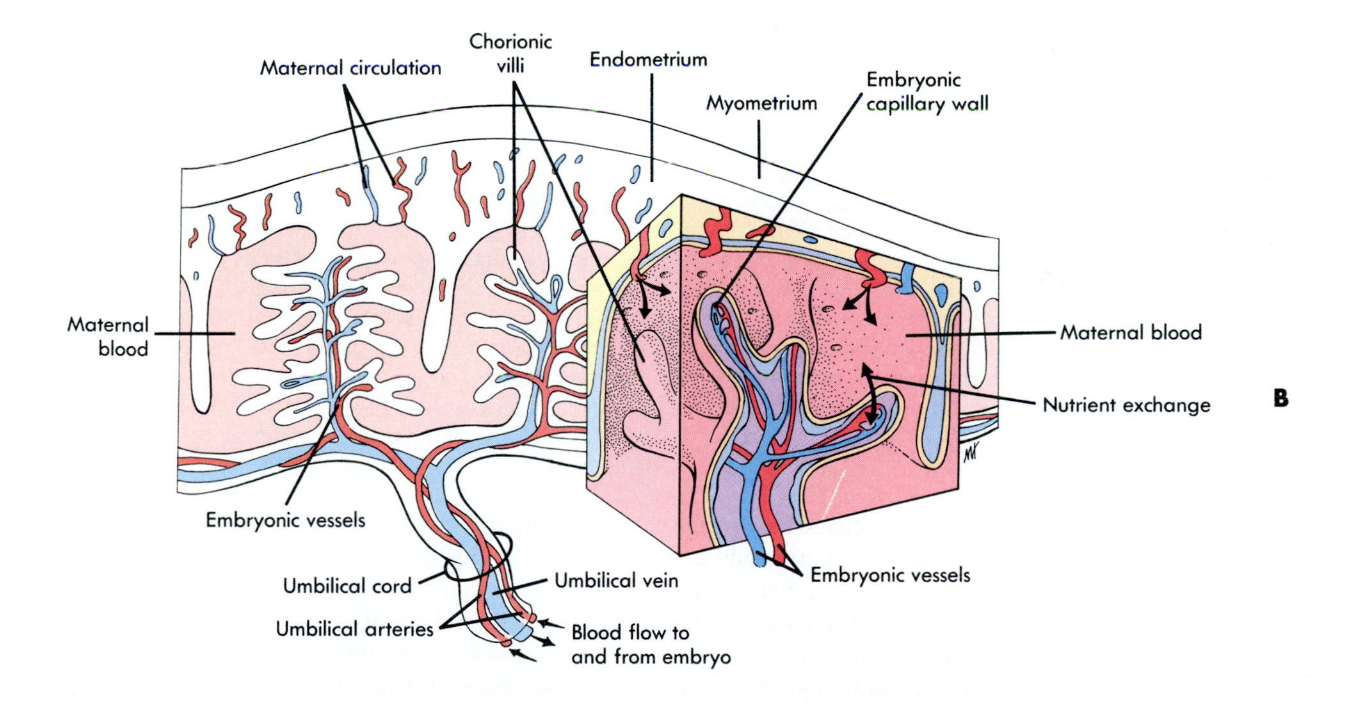

Figure 20-5 ● The placenta.

A, Location of the placenta and umbilical cord. **B,** Blood vessels from the fetus pass to the placenta through the umbilical cord. Blood vessels of the placenta are within chorionic villi, which extend into spaces filled with maternal blood. Exchange of nutrients and gases takes place without mixing of maternal and fetal blood.

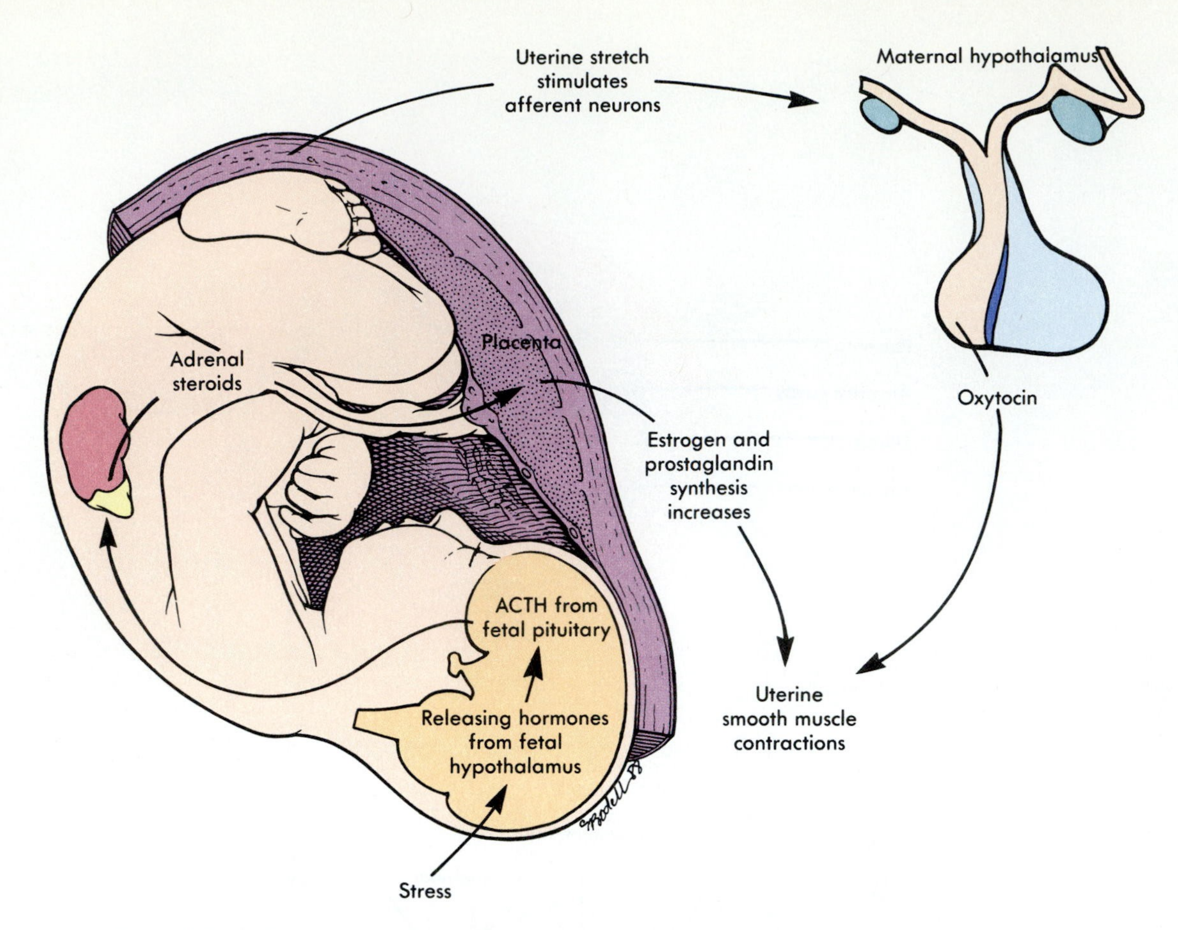

Figure 20-6 ● **Factors that influence parturition.**
As the fetus increases in size, becoming more and more confined within the uterus, it becomes stressed. This stress stimulates releasing hormones in the hypothalamus, which stimulate the fetal pituitary gland. The fetal pituitary gland secretes adrenocorticotropic hormone (ACTH) in greater amounts near parturition. ACTH causes the fetal adrenal gland to secrete greater quantities of adrenal cortical steroids, which travel in the umbilical blood to the placenta. There the adrenal steroids cause estrogen and prostaglandin synthesis to increase, making the uterus more irritable. Stretch of the uterus results in action potentials that go to the brain and stimulate the secretion of oxytocin. Oxytocin also causes the uterine smooth muscle to contract.

fied (Figure 20-6). Stress caused by the increased size of the fetus within the limited space of the uterus results in adrenocorticotropic hormone (ACTH) secretion from the fetal pituitary gland (see Chapter 10). ACTH stimulates the release of adrenal steroids that cause the placenta to increase secretions of estrogen and prostaglandins, both of which increase uterine contractility. In addition, stretch of the uterus stimulates the release of oxytocin from the mother's pituitary gland. The oxytocin also increases uterine contractions.

GENETICS

Genetics is the study of heredity, that is, those characteristics inherited by children from their parents. Although environment plays a role in the expression of genetic makeup, your genetic makeup is primarily responsible for your physical characteristics and many of your abilities. Your genetic makeup explains much about you. For example, your appearance is similar to that of your parents and your more distant ancestors. Many of your abilities, your susceptibility to disease, and even your life span are influenced by heredity. Because many of the diseases caused by microorganisms are now preventable or treatable, diseases that have a genetic basis are receiving more attention.

Chromosomes

Deoxyribonucleic (de-ox´sĭ-ri´bo-nu-kle´ik) **acid (DNA)** is the hereditary material of cells and is responsible for controlling cell activities. DNA and associated protein molecules become visible as densely stained bodies, called **chromosomes,** during cell division (see Figure 3-8). Somatic

Disorders of pregnancy

ECTOPIC PREGNANCY

The term *ectopic* means out of place, and an **ectopic** (ek-top´ik) **pregnancy** is one that occurs outside the uterus. The most common site of an ectopic pregnancy is the uterine tube (tubal pregnancy). The blastocyst does not reach the uterine cavity and instead implants into the wall of the uterine tube. The uterine tube cannot expand enough to accommodate the growing fetus and if the fetus is not removed, the tube will eventually rupture. The ruptured uterine tube causes life-threatening internal bleeding.

MISCARRIAGE

It is estimated that as many as 50% of all zygotes are lost before delivery. Most are lost before implantation. Approximately 15% of all pregnancies end in **miscarriage,** or **spontaneous abortion.** Before approximately 24 weeks a miscarriage results in the death of the fetus because the fetus cannot live outside the uterus. If delivery occurs between 24 and 37 weeks, the baby is referred to as **premature.**

Although there is a higher incidence of birth defects among aborted fetuses, the vast majority of the fetuses appear to be normal. There are many factors that may cause a miscarriage, many of which do not directly involve the fetus, and many of which are unknown. One common cause of miscarriage is improper implantation of the blastocyst in the uterus. In most cases, the blastocyst implants in the upper part of the uterus, but occasionally a blastocyst may implant near the opening into the cervical canal, a condition called **placenta previa** (pre´ve-ah). As the fetus grows and the uterus stretches, the previa placenta may tear away from the uterine wall, a condition called **placental abruption.** When this occurs, the fetus often dies. The associated hemorrhaging may be life threatening to the mother as well.

PREGNANCY-INDUCED HYPERTENSION

One reason the mother's weight is carefully monitored during pregnancy is that a sudden weight gain associated with edema may be a sign of **pregnancy-induced hypertension** (toxemia of pregnancy). The cause of the disorder is unknown, but, it can result in convulsions, kidney failure, and death of both the mother and the fetus.

TERATOGENS

Teratogens (tĕr´ă-to-jenz) are drugs that can cross the placenta and cause birth defects in the developing embryo. **Fetal alcohol syndrome (FAS)** is seen in children of women who consumed substantial amounts of alcohol during the pregnancy. This syndrome consists of brain dysfunction, growth retardation, and facial peculiarities. It has been estimated that FAS may occur as often as 1 in 350 births, and may account for as much as 33% of all cases of mental retardation. **Fetal alcohol effect** includes brain dysfunction without the facial characteristics of FAS, and may be three times as common as FAS.

Cocaine addiction in the newborn can occur in infants whose mothers are cocaine users. A fetus can also suffer strokelike symptoms if the mother ingests cocaine during the latter part of pregnancy.

cells contain 23 pairs of chromosomes (46 chromosomes), and gametes contain 23 chromosomes. **Somatic** (so-mat´ik) **cells** are all the body's cells except for the **gametes** (gam´ētz) or sex cells. Examples of somatic cells are epithelial cells, muscle cells, neurons, fibroblasts, lymphocytes, and macrophages. In the male, the gametes are sperm cells, and in the female, the gametes are secondary oocytes (see Chapter 19).

A **karyotype** (kar´e-o-tīp), or display of the chromosomes in a somatic cell, can be produced by photographing the chromosomes through a microscope, cutting the pictures of the chromosomes out of the photograph, and arranging the chromosomes in pairs (Figure 20-7). The 23 pairs of chromosomes are divided into two groups. There are 22 pairs of **autosomal** (aw´to-so´mal) **chromosomes** and one pair of **sex chromosomes.** For convenience, the autosomes are numbered in pairs from 1 through 22, and sex chromosomes are denoted as X or Y chromosomes. A normal female has two X chromosomes (XX) in each somatic cell, whereas a normal male has one X and one Y chromosome (XY) in each somatic cell.

Gametes are derived from somatic cells by **meiosis** (mi-o´sis). In this process the somatic cells divide twice and the chromosomes from the somatic cells are distributed to the gametes. Meiosis is called reduction division because the

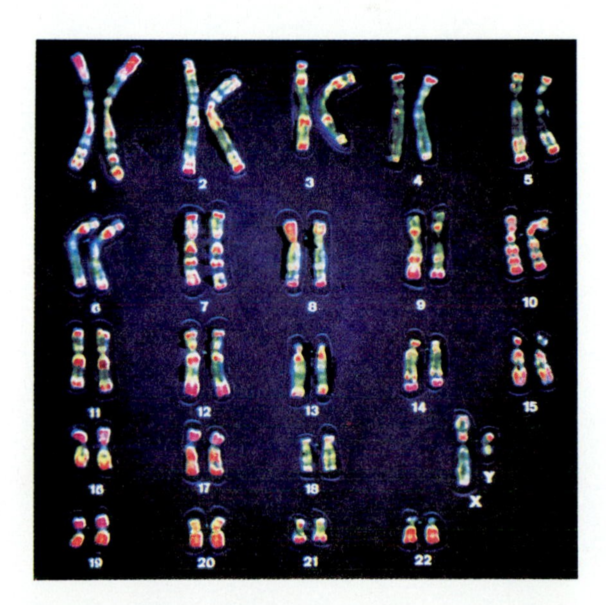

Figure 20-7 ● **Human karyotype.**
The 23 pairs of chromosomes in humans consists of 22 pairs of autosomal chromosomes (numbered 1 to 22) and 1 pair of sex chromosomes. This karyotype is of a male and has an X and a Y sex chromosome. A female karyotype would have two X chromosomes.

number of chromosomes in the gametes is half the number in the somatic cells. Reduction of the number of chromosomes in sperm cells and oocytes is important. When a sperm cell and an oocyte fuse during fertilization, each contributes one half of the chromosomes necessary to produce new somatic cells. Therefore half of an individual's genetic makeup comes from the father and half comes from the mother.

During meiosis, the chromosomes are distributed in such a way that each gamete receives only one chromosome from each pair of chromosomes. The inheritance of sex illustrates how chromosomes are distributed during gamete formation and fertilization. During the formation of gametes, the pair of sex chromosomes separates and an oocyte has only an X chromosome, whereas a sperm cell has only an X chromosome or only a Y chromosome (Figure 20-8). When a sperm cell fertilizes an oocyte to form a single cell, the sex of the individual is determined. If the oocyte is fertilized by a sperm cell with a Y chromosome, a male results, but if the oocyte is fertilized by a sperm cell with an X chromosome, a female results. When all the possible combinations of sperm cells with oocytes are considered, half the individuals should be female and the other half should be male.

Genes

The functional unit of heredity is the gene. Each **gene** consists of a certain portion of a DNA molecule but not necessarily a continuous stretch of DNA (see Figure 3-7). The information in a gene can be used to produce an mRNA molecule that leaves the nucleus of a cell and moves to a ribosome where a protein is synthesized (see Figure 3-8). Proteins form structural components of cells or function as enzymes that regulate the chemical activities of cells (see Chapter 2). By determining the structure of proteins, genes are responsible for the characteristics of cells and, therefore, the characteristics of the entire organism.

The importance of genes is dramatically illustrated in situations in which the alteration of a single gene results in a genetic disorder. For example, in **phenylketonuria** (fen´il-ke´to-nu´re-ah) **(PKU)** the gene responsible for producing an enzyme that converts the amino acid phenylalanine to the amino acid tyrosine is defective. Therefore phenylalanine accumulates in the blood and is eventually converted to harmful substances that can cause mental retardation.

Each chromosome contains thousands of genes, and each gene occupies a specific position on the chromosome. For example, on chromosome 11 there is a gene responsible for the production of an enzyme necessary for the synthesis of melanin. Melanin is the pigment responsible for skin, hair, and eye color (see Chapter 5). A person with normal coloration has the melanin gene that produces the functional enzyme. An **albino** is a person who has defective melanin genes and is unable to produce melanin. Therefore the albino lacks normal skin, hair, and eye color. Instead, albino coloration consists of shades of pink, blue, and yellow. The pink and blue colors result from blood (see Chapter 5) and the yellow color from the natural accumulation of ingested yellow plant pigments in the skin.

Because chromosomes are paired, each chromosome of a

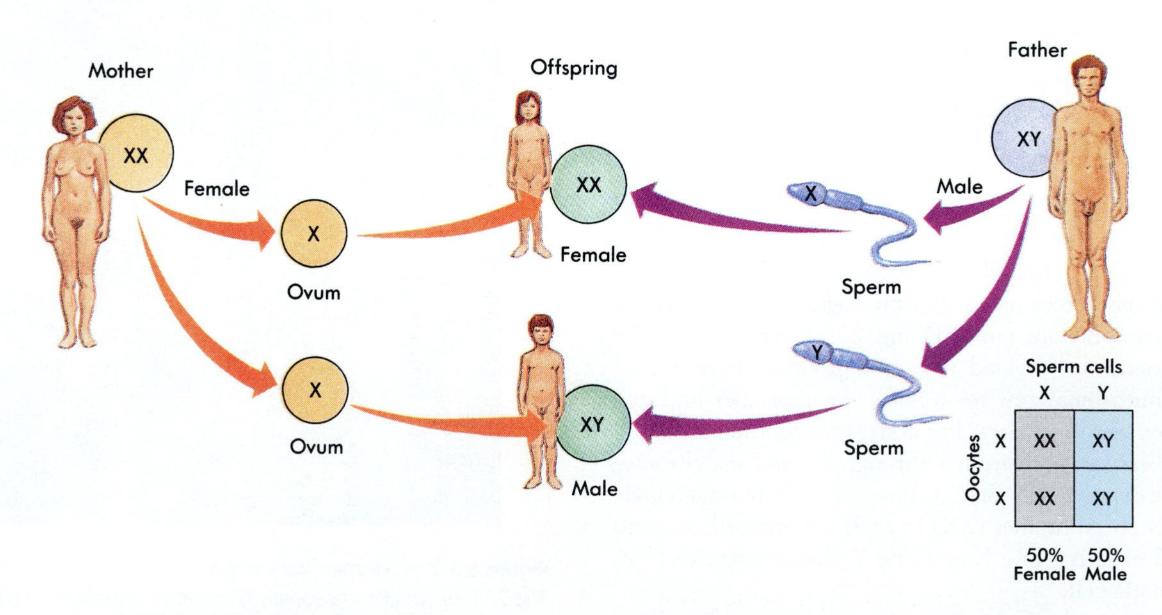

Figure 20-8 ● **Inheritance of sex.**
The female produces oocytes containing one X chromosome, whereas the male produces sperm cells with either an X or a Y chromosome. There are four possible combinations of an oocyte with a sperm cell, half of which produce females and half of which produce males.

given pair contains similar genes for a particular trait. For example, there are two chromosomes 11 and each chromosome 11 has a melanin gene. If the two genes for a trait are identical, the person is **homozygous** (ho-mo-zi´gus) for the trait. If the two genes for a trait are different, the person is **heterozygous** (het´er-o-zi´gus) for the trait.

Dominant and recessive genes

For many genetic traits, the effects of one gene for a trait can mask the effect of another gene for that same trait. For example, a person who is heterozygous for the melanin genes has the normal gene for melanin production on one chromosome 11 and the defective gene for melanin production on the other chromosome 11. Such a person produces melanin and appears normal. The gene whose trait is expressed is the **dominant** gene, whereas the gene whose trait is masked by the dominant gene is the **recessive** gene. Thus normal pigmentation is a dominant trait and albinism is a recessive trait. By convention, dominant traits are indicated by upper case letters and recessive traits are indicated by lower case letters. For example, the letter "A" designates the dominant normal condition and the letter "a" designates the recessive albino condition.

The possible combinations of dominant and recessive genes for melanin are AA (homozygous dominant), Aa (heterozygous), and aa (homozygous recessive). The actual genes that a person possesses is his **genotype** (jen´o-tīp). The expression of those genes is his **phenotype** (fe´no-tīp). A person with the genotype AA or Aa would have the phenotype of normal pigmentation, whereas a person with the genotype aa would have the phenotype of albinism. Note that a recessive trait is expressed when it is not masked by the dominant trait.

2 Polydactyly (pol-e-dak´tĭ-le) is a condition in which a person has extra fingers and/or toes. Given that polydactyly is a dominant trait, list all the possible genotypes and phenotypes for polydactyly. Use the letters "D" and "d" for the genotypes.

The inheritance of dominant and recessive traits can be determined if the genotypes of the parents are known. For example, if a heterozygous normal person (Aa) mates with a heterozygous normal person (Aa), the probability is that one fourth of their children will be albino (aa), and three fourths will be normal (AA or Aa) (Figure 20-9). A **carrier** is a heterozygous person with an abnormal recessive gene, but with a normal phenotype because he has a normal dominant gene. As this example shows, it is possible for normal carriers to have children with recessive traits.

3 If a homozygous normal person mates with a carrier for albinism, what is the likelihood that any of their children will be albinos? Explain.

Sex-linked traits

Traits resulting from genes on the sex chromosomes are called **sex-linked traits.** An example of a sex-linked trait is hemophilia A (classic hemophilia) in which there is an inability to produce a clotting factor (see Chapter 11). Consequently, clotting is impaired and persistent bleeding can occur either spontaneously or as a result of an injury.

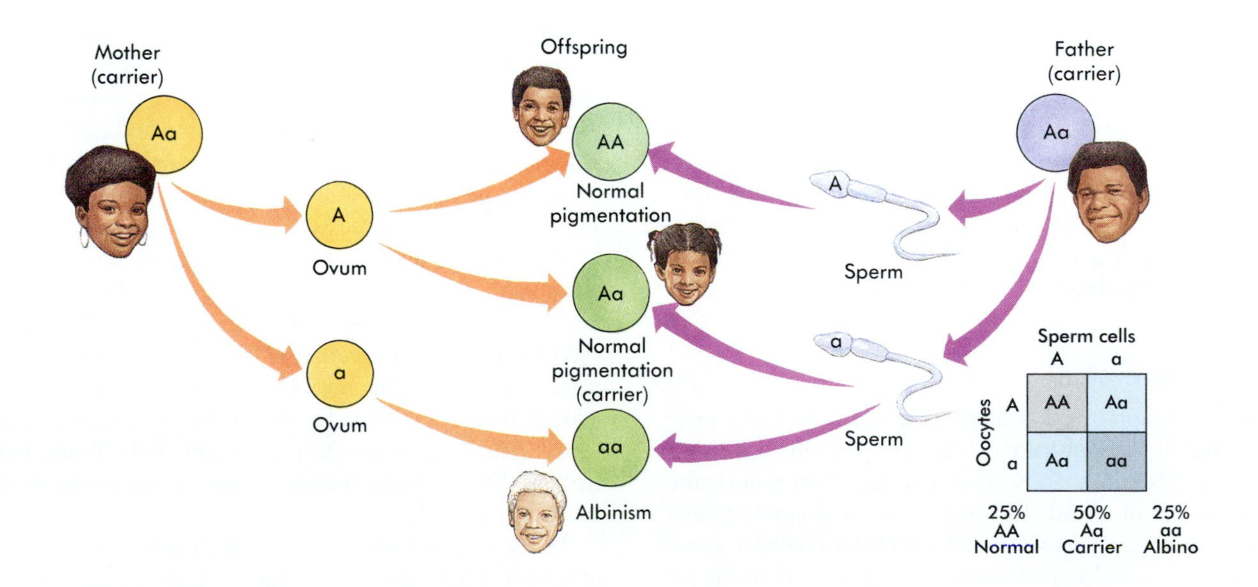

Figure 20-9 ● **Inheritance of albinism.**
Albinism is a recessive trait. The symbol *A* represents the normal, pigmented condition and the symbol *a* represents the recessive unpigmented condition. The cross is between two carriers (heterozygous). The gametes each produces and the offspring produced by the possible gamete combinations are illustrated.

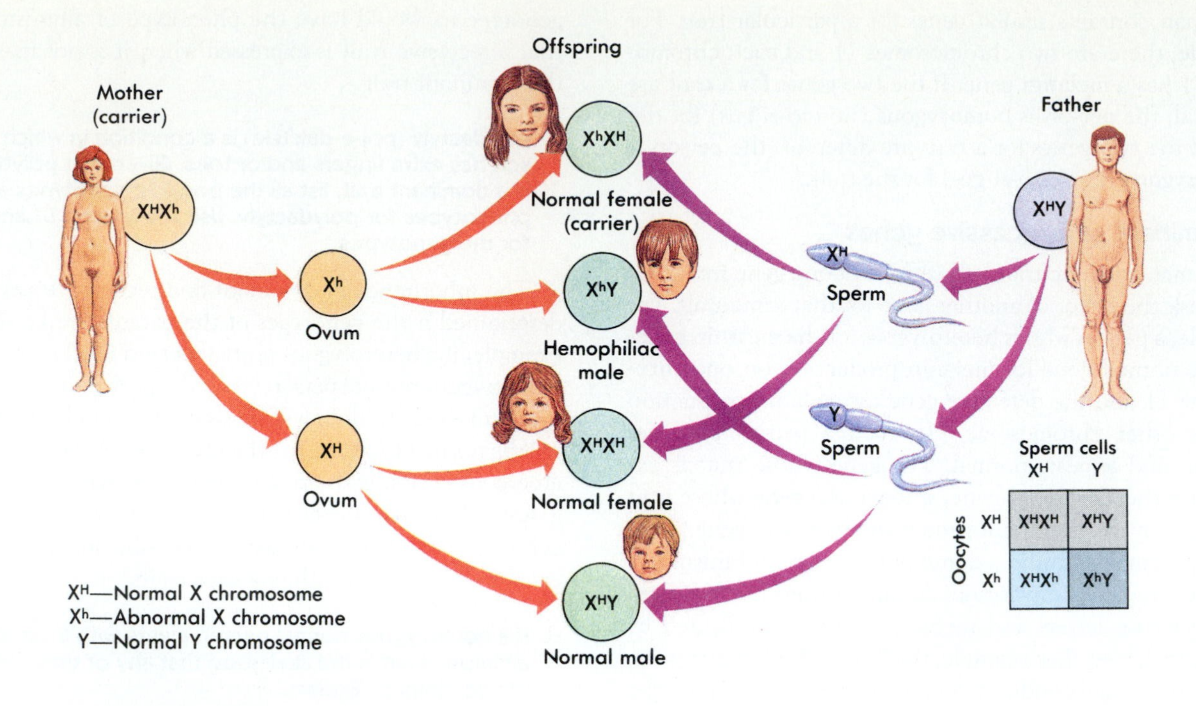

Figure 20-10 ● Inheritance of hemophilia.
Hemophilia is a sex-linked trait in which X^H represents the normal X chromosome condition with all clotting factors, and X^h represents the X chromosome lacking a gene for one clotting factor. The illustration represents a mating between a normal male and a normal carrier female.

Hemophilia A is a recessive trait located on the X chromosome. Therefore the possible genotypes and phenotypes are X^HX^H (normal homozygous female), X^HX^h (normal heterozygous female), X^hX^h (hemophiliac homozygous female), X^HY (normal male), and X^hY (hemophiliac male). Note that a female must have both recessive genes to be a hemophiliac, whereas a male, because he has only one X chromosome, is a hemophiliac if he has only one of the recessive genes. An example of the inheritance of hemophilia is illustrated in Figure 20-10. If a woman who is a carrier for hemophilia mates with a man who does not have hemophilia, none of their daughters but half of their sons will have hemophilia.

Gene expression

The expression of a dominant gene over a recessive gene is the simplest manner in which genes determine a person's phenotype. There are many other ways in which genes cause the expression of a trait. In some cases the dominant gene does not completely mask the effects of the recessive gene. For example, in **sickle cell anemia** the gene responsible for producing hemoglobin in red blood cells (RBCs) is abnormal. Consequently, the hemoglobin produced by the gene is abnormal. The result is RBCs that are stretched into an elongated sickle shape. These RBCs tend to stick in capillaries, blocking blood flow to tissues. In addition, the sickle-shaped cells tend to rupture. The normal gene (S) for producing normal hemoglobin is dominant over the sickle cell gene (s) responsible for producing the abnormal hemoglobin. A normal person has genes SS and has normal hemoglobin. A person with sickle cell anemia has genes ss and has abnormal hemoglobin. A person who is heterozygous has genes Ss and has half normal hemoglobin and half abnormal hemoglobin. This condition is called **sickle-cell trait** and normally only a few RBCs are sickle-shaped. Usually a person with sickle-cell trait exhibits no adverse symptoms.

In another method of gene expression, different genes can combine to produce an effect without either of them being dominant or recessive. For example, people with type AB blood have A antigens and B antigens on the surface of their RBCs (see Chapter 14). The antigens result from a gene that causes the production of the A antigen and a different gene that causes the production of the B antigen. In this case A and B are neither dominant nor recessive in relation to each other.

Many traits are determined by the expression of multiple genes on different chromosomes. Examples are a person's height, intelligence, eye color, and skin color. Multiple gene traits typically are characterized by having a great amount of variability. For example, there are many different shades of eye color. Because of the many genes involved, it is difficult to predict how a multigenic trait will be passed from one generation to the next.

Understanding the human genome

The **human genome** (je´nom) is all of the genes found in human beings. It is estimated that humans have 50,000 to 100,000 genes. A **genomic map** is a description of the DNA nucleotide sequence of genes and their location on chromosomes. To date, approximately 2300 genes have been mapped at least to their location on chromosomes. By the year 2005, the U.S. Human Genome Project hopes to have a complete genomic map of all the genes.

Armed with the knowledge of a person's genome and what effects the genome has on a person's physical, mental, and behavioral abilities, medicine and society will be transformed in many ways. Medicine, for example, will shift emphasis from the curative to the preventative. The potential disorders or diseases a person is likely to develop can be prevented or their severity lessened. When prevention is not possible, knowledge of the enzymes involved will result in new drugs and techniques that can compensate for the genetic disorder. Knowledge of the genes involved may result in gene therapy that repairs or replaces defective genes, resulting in cures of genetic disorders.

Despite the great promise of benefits from the Human Genome Project, the knowledge that will be produced has raised a number of ethical and legal questions for society. Should a person's genome be public knowledge? Should a person with a genome that predisposes him to cancer or behavioral disorders be barred from certain types of employment or be refused medical insurance because he is a high risk? Can prospective mates demand to know a person's genome? Should parents know the genome of their fetus and be allowed to make decisions regarding abortion based on this knowledge? Should the same gene therapy techniques that allow alteration of the genome to cure genetic disorders be used to create genomes that are deemed to be superior? Such questions raise the specter of genetic discrimination.

4 It is estimated that the Human Genome Project will cost approximately 3 billion dollars. Explain why you believe this will be money well spent or why you believe the project should be stopped.

Genetic disorders

Genetic disorders are caused by abnormalities in genetic makeup, that is, in DNA. Genetic disorders are often confused with **congenital disorders. Congenital** means "present at birth" and is commonly referred to as a birth defect. Approximately 15% of all congenital disorders have a known genetic cause, and approximately 70% of all birth defects are of unknown cause. In some cases, however, the birth defect results from damage to the fetus during development (for example, fetal alcohol syndrome; see p. ###).

One cause of genetic disorders is a **mutation,** a change in a gene that usually involves a change in the number or kinds of nucleotides composing DNA (see Chapter 3). Mutations are known to occur by chance (randomly without known cause) or be caused by chemicals, radiation, or viruses. In most cases, a specific cause of a mutation cannot be determined. Once a mutation has occurred, however, the abnormal trait can be passed from one generation to the next.

Chromosomal disorders result from an abnormal distribution of chromosomes or parts of chromosomes during gamete formation. Sometimes a pair of chromosomes does not separate during meiosis and a gamete with both chromosomes of a pair results. For example, an oocyte could have two chromosomes 21 instead of the usual one chromosome 21. If an oocyte with two chromosomes 21 is fertilized by a sperm cell with one chromosome 21, then an individual with three chromosomes 21 is produced. This condition is called **Down syndrome,** and can result in mental retardation. Table 20-1 gives some examples of genetic disorders caused by the improper distribution of chromosomes.

Cancer is a tumor resulting from uncontrolled cell divisions. **Oncogenes** (ong´ko´jenz) are genes associated with cancer. Many oncogenes apparently function in the regulation of normal cell divisions. A change in an oncogene or in the regulation of an oncogene can result in uncontrolled cell divisions and the development of cancerous tumors. It is believed that certain chemicals called **carcinogens** (kar´sin´o-jenz) can induce such changes and thereby initiate the development of cancer. For example, chemicals in cigarette smoke are known to cause cancer. Thus, even though everyone has oncogenes, an outside agent may be necessary for the cancer to begin.

A change in somatic cells that results in cancer is not inheritable. Nonetheless, there may be a genetic basis that allows cancer development, especially under the right environmental conditions. In this sense, the inheritance of cancer and other abnormalities has been described as **genetic susceptibility** or **genetic predisposition.** For example, if a woman's close relatives, such as her mother or sister, have breast cancer, she has a greater-than-average risk of developing breast cancer. Similar genetic susceptibilities have been found for diabetes mellitus, schizophrenia, and other disorders.

Treatment of genetic disorders usually involves treatment of symptoms, but does not result in a cure because the

Table 20-1 • Genetic disorders

Disorder	Description
• Dominant traits	
Achondroplasia	Dwarfism characterized by shortening of the upper and lower limbs
Huntington's disease	Severe degeneration of the basal ganglia and frontal cerebral cortex; chracterized by purposeless movements and mental deterioration; onset is usually between 40 to 50 years of age
Hypercholesterolemia	Elevated blood cholesterol levels that contribute to atherosclerosis and cardiovascular disease
Marfan syndrome	Abnormal connective tissue results in increased height, elongated digits, and weakness in the aortic wall
Neurofibromatosis	Small pigmented lesions (café-au-lait spots) in the skin and disfiguring tumors (noncancerous) caused by proliferation of neuroglia cells along nerves; best known as the disease of the "Elephant Man"
Osteogenesis imperfecta	Abnormal phosphate metabolism results in brittle bones that repeatedly break
• Recessive traits	
Albinism	Lack of the enzyme necessary to produce the pigment melanin; characterized by lack of skin, hair, and eye coloration
Cystic fibrosis	Impaired transport of chloride ions across cell membranes; results in excessive production of thick mucus that blocks the respiratory and gastrointestinal tract; the most common fatal genetic disorder
Phenylketonuria	Lack of the enzyme necessary to convert the amino acid phenylalanine to the amino acid tyrosine; an accumulation of phenylalanine leads to mental retardation
Severe combined immune deficiency	Inability to form the white blood cells (B cells, T cells, and phagocytes) necessary for an immune system response
Sickle cell anemia	Inability to produce normal hemoglobin; results in abnormally shaped red blood cells that clog capillaries or rupture
Tay-Sachs disease	Lack of the enzyme necessary to break down certain fatty substances; an accumulation of fatty substances impairs action potential propagation, resulting in deterioration of mental and physical functions and death by 3 to 4 years of age
Thalassemia	Decreased rate of hemoglobin synthesis; results in anemia, enlargment of the spleen, increased cell numbers in red bone marrow, and congestive heart failure
• Sex-linked traits	
Hemophilia	Most commonly, a recessive gene causes a failure to produce blood clotting factors; results in prolonged bleeding
Red-green color blindness	Most commonly, a recessive gene causes a deficiency of functional green-sensitive cones; inability to distinguish between red and green colors
• Chromosomal disorders	
Down syndrome	Caused by having three chromosomes 21; results in mental retardation, short stature, and poor muscle tone
Duchenne muscular dystrophy	Caused by deletion or alteration of part of the X chromosome; results in progressive weakness and wasting of muscles
Klinefelter's syndrome	Caused by two or more X chromosomes in a male (XXY); results in small testes, sterility, and development of femalelike breasts
Turner's syndrome	Caused by having only one X chromosome; results in immature uterus, lack of ovaries, and short stature

basic genetic material is unchanged by the treatment. For example, treating the problems associated with mucus buildup in cystic fibrosis does not cure the disorder. Research is underway, however, to actually affect a person's genetic makeup. Ultimately it may be possible to insert a normal gene into cells to replace an abnormal or missing gene. Then if the normal gene functions, the disorder will be cured.

Genetic counseling

Genetic counseling includes predicting the possible results of matings involving carriers of harmful genes and talking to parents or prospective parents about the possible outcomes and treatments of a genetic disorder. With this knowledge, prospective parents can make informed decisions about having children.

A first step in genetic counseling is to determine the genotype of the individuals involved. For example, a couple may suspect they are carriers for a genetic disorder. A family tree or **pedigree** provides historical information about family members. Sometimes by knowing the phenotypes of relatives it is possible to determine a person's genotype. Direct means of obtaining genetic information are also available. A karyotype can be taken from white blood cells (WBCs) or the epithelial cells lining the inside of the cheek. Or, the amount of a substance produced by a carrier can be tested. Sometimes carriers produce slightly more or less of a substance because they are heterozygous and have only one dominant gene for the normal trait. For example, carriers for cystic fibrosis produce more salt in their sweat than is normal.

Sometimes it is suspected that a fetus may have a genetic abnormality. Fetal cells can be tested by **amniocentesis** (am´ne-o-sen-te´sis), which takes cells floating in the amniotic fluid, or **chorionic villus sampling,** which takes cells from the fetal side of the placenta.

SUMMARY

DEVELOPMENT

Early cell divisions

Fertilization, the union of the oocyte and a sperm cell, results in a zygote.

The zygote undergoes divisions until it becomes a hollow ball of cells, the blastocyst.

Embryo and fetus

The yolk sac and amniotic cavity form within the inner cell mass of the blastocyst.

The embryonic disk consists of ectoderm and endoderm.

Mesoderm forms between the ectoderm and endoderm.

The embryo is the developing organism from the end of the second week to the end of the eighth week. Ectoderm, mesoderm, and endoderm give rise to the organs.

The fetus is the developing organism from the end of the eighth week of development until birth. The organs grow and mature.

Implantation and the placenta

The blastocyst implants into the uterus approximately 7 days after fertilization.

The outer wall of the blastocyst becomes the chorion, which with the lining of the uterus forms the placenta.

Maternal and fetal blood do not mix in the placenta.

PARTURITION

Uterine contractions force the baby out of the uterus during parturition.

ACTH from the fetal adrenal gland and increased estrogen and prostaglandin from the placenta initiate parturition.

Stretching of the uterus stimulates oxytocin secretion, which stimulates uterine contractions.

GENETICS

Chromosomes

Humans have 46 chromosomes in 23 pairs; 22 pairs of autosomes and 1 pair of sex chromosomes.

Males have the sex chromosomes XY and females have the sex chromosomes XX.

During gamete formation, the chromosomes of each pair of chromosomes separate; therefore half of a person's genetic makeup comes from his father and half from his mother.

Genes

A gene is a portion of a DNA molecule. Genes determine the proteins in a cell.

Genes are paired (located on the paired chromosomes).

Dominant genes mask the effects of recessive genes.

Sex-linked traits result from genes on the sex chromosomes.

A heterozygous person can have traits that result from both the dominant and the recessive gene.

Some genes are neither dominant nor recessive in relationship to each other.

Many traits result from the expression of multiple genes.

Genetic disorders

A mutation is a change in the number or kinds of nucleotides in DNA.

Some genetic disorders result from an abnormal distribution of chromosomes during gamete formation.

Oncogenes are genes associated with cancer.

Genetic predisposition makes it more likely a person will develop a disorder.

Genetic counseling

A pedigree (family history) can be used to determine the risk of having children with a genetic disorder.

Specific chemical tests or examination of a person's karyotype can be used to determine a person's genotype.

CONTENT REVIEW

1. How is a zygote produced? What is a blastocyst?
2. How is ectoderm, endoderm, and mesoderm formed? What adult structures are derived from each of these germ layers?
3. What major events distinguish embryonic and fetal development?
4. What is the difference between identical and fraternal twins? At what stages of development are identical twins formed?
5. What is the function of the placenta? Does maternal and fetal blood mix within the placenta?
6. What events occur during each stage of parturition? What hormonal changes are involved with parturition?
7. What is the number and type of chromosomes in the karyotype of a human somatic cell. How do the chromosomes of a male and female differ from each other?
8. How do the chromosomes in somatic cells and gametes differ from each other?
9. What is a gene and how are genes responsible for the structure and function of cells?
10. Define homozygous dominant, heterozygous, and homozygous recessive.
11. What is the difference between genotype and phenotype?
12. What is a sex-linked trait? Give an example.
13. What kinds of gene expression are responsible for sickle cell anemia, type AB blood, and a person's height?
14. What is a mutation?
15. What is the cause of the genetic disorder called Down syndrome?
16. What are oncogenes and carcinogens?
17. What is genetic susceptibility?
18. How are pedigrees, karyotypes, chemical tests, amniocentesis, and chorionic villus sampling used in genetic counseling?

CONCEPT REVIEW

1. If a woman contracts rubella (German measles) while pregnant, the virus can cross the placenta and infect the embryo or fetus. Such an infection can result in a spontaneous abortion or congenital disorders such as malformation of the heart, deafness, and cataracts. If the mother is infected in the first month of pregnancy versus the third month, what effect would this time difference have on the likelihood of a spontaneous abortion or of congenital disorders?
2. Dimpled cheeks are inherited as a dominant trait. If two parents, each of whom is heterozygous for this trait, have children, is it possible for them to have a child that does not have dimpled cheeks? Explain.
3. The ability to roll the tongue to form a "tube" is due to a dominant gene. Suppose that a woman and her son can both roll their tongues, but her husband cannot. Is it possible to determine if the husband is the father of her son? Explain.
4. A woman who does not have hemophilia marries a man who has the disorder. Determine the genotype of both parents if half of their children have hemophilia.

CHAPTER TEST

Answers can be found in Appendix C

MATCHING 1 For each statement in column A select the correct answer in column B (an answer may be used once, more than once, or not at all).

A	B
1. Results from the union of a sperm cell and an oocyte.	amniotic cavity
2. Hollow ball of cells.	blastocyst
3. Part of blastocyst that becomes the individual.	chorion
4. Cavity into which the developing organism grows.	embryo
5. Ectoderm, mesoderm, and endoderm.	fetus
6. The developing organism from the end of the second week to the end of the eighth week.	germ layers
7. Organ formation takes place in this stage of development.	implantation
8. Attachment of the blastocyst to the uterus.	inner cell mass
9. Becomes part of the placenta and produces HCG.	parturition
10. Delivery or labor.	yolk sac
	zygote

MATCHING 2 For each statement in column A select the correct answer in column B (an answer may be used once, more than once, or not at all).

A	B
1. Display of chromosomes in a cell.	autosomal chromosomes
2. X and Y chromosomes.	carcinogen
3. The functional unit of heredity.	dominant
4. Having two identical genes.	gene
5. A gene that masks the effects of another gene.	genotype
6. The expression of a person's genes.	heterozygous
7. A change in the makeup of a gene.	homozygous
8. Gene associated with cancer.	karyotype
9. Substance that causes cancer.	mutation
	oncogene
	phenotype
	recessive
	sex chromosomes

FILL-IN-THE BLANK Complete each statement by providing the missing word or words.

1. All cells of the body, except reproductive cells, are called ___________. Reproductive cells are called ___________.
2. Dark staining bodies that contain the cell's DNA and become visible during cell division are ___________.
3. The process by which gametes are derived from somatic cells is ___________.
4. A male has sex chromosomes ___________, whereas a female has sex chromosomes ___________.
5. A ___________ controls the production of proteins within cells.
6. Characteristics resulting from genes on the X chromosomes are called ___________.
7. The tendency to develop cancer or other disorders under certain environmental conditions is called ___________.
8. A ___________ is a history (family tree) of the occurrence of a genetic disorder within a person's relatives.

CHAPTER

21

Infectious Diseases

OBJECTIVES

After reading this chapter you should be able to:

1. Define and use common terminology associated with diseases.
2. Describe the structures of a bacterial cell and a virus.
3. Describe asexual reproduction in bacteria.
4. Describe the basic process by which bacteria are identified in the laboratory.
5. List the steps that occur in viral reproduction.
6. State and give examples of the basic principle used to design treatment for parasitic diseases.
7. Summarize the characteristics of protozoans, fungi, and parasitic worms.
8. Define a reservoir and give the ways that pathogens are transmitted into humans to cause disease.
9. Describe the techniques used to prevent the transmission of disease in the medical setting and for public health.

KEY TERMS

asepsis
To keep microorganisms away from an object or person.

carrier
An infected person with no symptoms who can transmit the infectious agent to others.

disease
The interference with or cessation of normal body functions.

endospore
A dormant structure formed inside a bacterial cell that is resistant to heat, chemicals, and lack of water.

opportunistic pathogen
A pathogen already present in the body that causes disease when transferred to another part of the body, or when some condition of the host changes (for example, reduced resistance to disease).

pathogen
Any organism or substance that causes disease.

portal
The place where a pathogen enters or exits the body.

reservoir
Living or nonliving material in or on which a disease-causing organism can live and reproduce; the source of infections.

sterilization
The complete destruction of all microorganisms in a given area.

vector
An invertebrate animal (for example, mosquito, louse, or tick) capable of transmitting a disease to a vertebrate (for example, human).

FEATURES

A PARASITE is an organism that lives on or in another organism, called the **host.** The parasite derives its nourishment from the host and in the process the host is harmed. For example, human lice attach to hair, puncture the skin with their needlelike mouthparts, and suck the host's blood for nourishment. Some parasites cause their host little harm, whereas many can cause disease and even death. This chapter is an introduction to **pathogens** (path´o-jens), parasites that cause disease or death. The emphasis is on pathogenic microorganisms. **Microorganisms** are organisms so small they can be seen only with the aid of a microscope. Microorganisms also are called **microbes,** or **germs,** and include bacteria, viruses, fungi, and protozoans. Humans are host to several hundred different kinds of parasites, some of which are described in Appendix A.

Before considering the nature of microorganisms and their role in causing disease, it should be noted that many microorganisms are either harmless, or are beneficial. Microorganisms are responsible for the decomposition of dead organisms, allowing the recycling of nutrients. Some bacteria fix nitrogen in the air, converting it into a form usable by plants. Many valuable products, such as antibiotics, yogurt, cheese, beer, and wine, are produced through the activities of microorganisms. Finally, the use of microorganisms in genetic engineering is opening a new era in the treatment and prevention of disease.

DISEASE TERMINOLOGY

Disease is defined as an interference with normal body functions such that homeostasis is not maintained. In addition to pathogens, disease can be caused by chemicals (for example, poisoning), physical agents (for example, excessive heat or injury), malnutrition, birth defects, tumors, or degeneration (for example, old age).

Disease produces symptoms and signs. A **symptom** is any departure from normal function, appearance, or sensation experienced by the patient. A **sign** is any departure from normal that the health care provider can objectively observe. For example, a statement by a patient that she "feels hot" is a symptom, whereas a body temperature 3 degrees above normal is a sign. Often the terms *symptom* and *sign* are used interchangeably even though they have different meanings. Symptoms and signs are used to identify or make a **diagnosis** of the disease, because each pathogen damages the body in a relatively unique way. For example, the measles cause a particular kind of rash in the skin, whereas the mumps cause the salivary glands to swell. After a diagnosis is made, a course of treatment or **therapy** can be determined and a **prognosis** (prog-no´sis) or estimation of the outcome of the disease can be made.

When a pathogen invades and multiplies within the body, an **infection** occurs. The period between entry and the first appearance of symptoms and signs is the **incubation period.** During this time the pathogen becomes established in the body. Depending on the characteristics of the pathogen, the numbers of the pathogen introduced, and the condition of the host, the incubation period varies in length. Therefore the incubation period usually is reported as a range. The incubation period is 1 to 2 days for the common cold, 3 to 5 days for gonorrhea, 14 to 16 days for chickenpox, and 14 to 21 days for mumps. The incubation period ends when the pathogen causes enough damage to produce the symptoms and signs of disease.

Diseases can be classified according to the length and severity of their symptoms and signs, as well as according to their location in the body. An **acute** disease lasts a short time and usually has severe symptoms. The common cold is an example. A **chronic** disease lasts for a long time and the symptoms can slowly become more severe—for example, tuberculosis. A **subacute** disease is in-between an acute and chronic disease. Subacute bacterial endocarditis, for example, is a continual infection of the heart that at first produces no symptoms. If pathogens stay in one area of the body, a **local** infection results (for example, a pimple), whereas a **systemic** infection occurs when pathogens spread throughout the body (for example, a bacterial infection of the blood).

Diseases also can be classified according to their distribution in the population. The **incidence** of a disease is the number of new cases that appear in a given period, and the **prevalence** of the disease is the number of old and new cases at a given time. If the incidence of a disease greatly increases in a given area, the disease is said to be **epidemic.** A sudden outbreak of influenza (the flu) is an example. Many diseases, however, are **endemic,** meaning they have a low prevalence but are continuously present in a population. For example, at any give time a small number of children in the United States have the measles.

DISEASE-CAUSING ORGANISMS

Bacteria

Bacteria are single-celled organisms that occur almost everywhere. Some, like plants, are capable of photosynthesis, but most must take in nourishment in some way. Interestingly, bacteria vary in their need for oxygen. Like the cells in the human body, some bacteria are **aerobic** (ăr-o´bik) and use oxygen. Others, however, are **anaerobic** (an´ăr-o-bik) and survive in the absence of oxygen. For example, the bacteria that cause botulism (bot´u-lizm) food poisoning grow without oxygen inside the sealed containers used to store preserved foods. Some bacteria can function aerobically if oxygen is present, and anaerobically if oxygen is absent.

Classification by shape and arrangement

Most bacteria exist in one of three basic shapes: a rod-shaped **bacillus** (bă-sil´us; pl. bacilli, bă-sil´i), a spherical-shaped **coccus** (kok´us; pl. cocci, kok´si), or a **spiral shape**

(Figure 21-1). Bacteria can exist as individual cells with no particular relationship to each other. This is most typical of the bacilli and spiral bacteria. The cocci, however, are often attached to each other. For example, **streptococci** (strep´to-kok´si) are joined together to form long chains, like a string of beads, and **staphylococci** (staf´ĭ-lo-kok´si) are arranged in clusters.

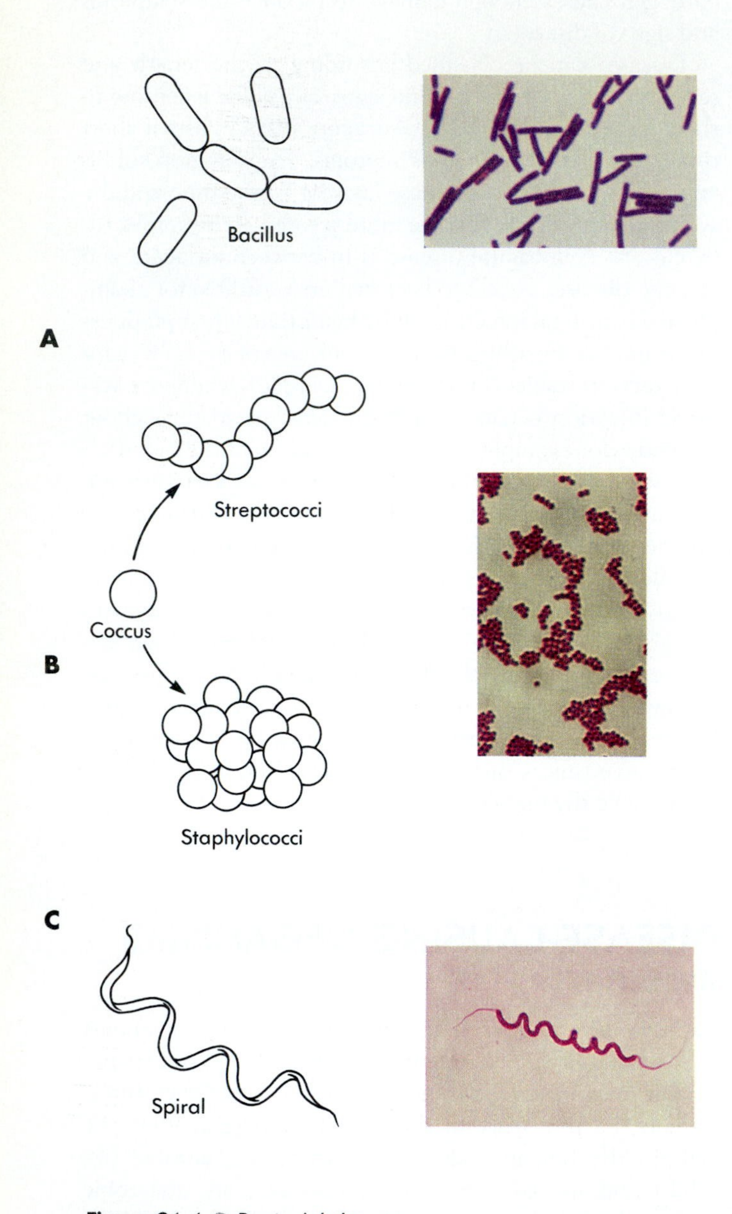

Figure 21-1 ● Bacterial shapes.
A, Bacillus or rod-shaped. **B,** Coccus or spherical-shaped; streptococci are cocci arranged in chains; staphylococci are cocci arranged in clusters. **C,** Spiral-shaped.

Structure

Figure 21-2 shows the structure of a typical bacterial cell. Understanding the structure of bacteria is important because structure and function are related. Bacterial structures not only enable bacteria to perform the functions necessary for life, but are also responsible for the ability of bacteria to cause diseases in humans. Knowledge of bacterial structures and functions makes it possible to treat diseases by attacking specific parts of the bacteria.

Some, but not all, bacteria secrete a gelatinous, sticky material that surrounds the bacteria. This **capsule** helps the bacteria stick to surfaces. For example, the capsule of certain bacteria sticks to the surface of teeth. When these bacteria ingest and break down sugar for an energy source, an acid by-product is produced that causes tooth decay. Brushing and flossing the teeth helps remove the bacteria and prevent tooth decay.

A **pilus** (pi´lus) is a short, hollow tube that appears as a hairlike appendage on some bacteria. One type of pilus functions to attach the bacteria to cell surfaces. Another type of pilus is involved with the transfer of genetic material between bacterial cells (see box, p. 356).

A **flagellum** (flă-jel´um) is a long, semirigid structure shaped like a corkscrew. Rotation of the flagellum results in movement of the bacterial cell. Bacteria with flagella are able to control the direction in which they move.

The **cell wall** is a nonliving, rigid structure responsible for the shape of bacteria. The molecules making up the cell wall make it structurally strong and resistant to rupture. One of these is **peptidoglycan** (pep´tĭ-do-gly´can), a combination of sugars and amino acids. This molecule is important because humans do not have peptidoglycan. Therefore chemicals that destroy peptidoglycan kill bacteria but do not harm humans. **Lysozyme** (li´so-zīm) is an enzyme found in saliva, tears, and phagocytes that destroys peptidoglycan. Once this happens, the bacteria are no longer surrounded by a rigid container, they swell with water, and lyse (burst).

Another component of cell walls in some bacteria is the **lipopolysaccharide** (lip´o-pol-e-sak´ă-rīd), or **LPS, layer.** The presence or absence of an LPS layer can be determined using the **Gram stain.** Bacteria having an LPS layer stain red and are called **gram negative,** whereas bacteria without an LPS layer stain purple and are **gram positive.** When gram negative bacteria die in the body, the cell wall breaks up and the pieces of the LPS layer are called **endotoxins.** Endotoxins produce symptoms of fever, aches, weakness, and sometimes shock.

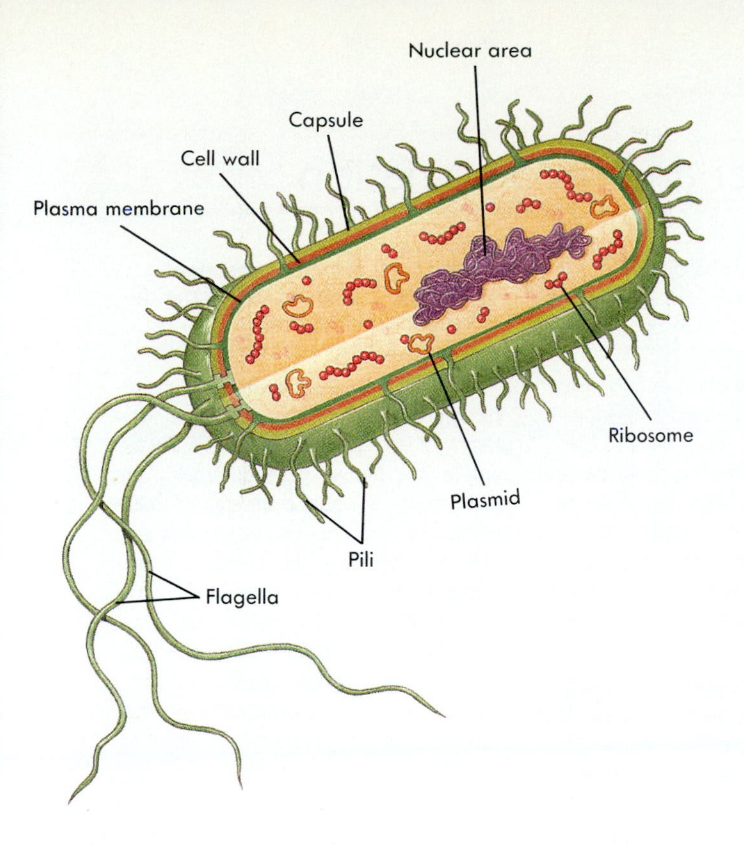

Figure 21-2 ● **Bacterial structure.**
A composite of typical bacterial structures. Not all bacteria have every structure shown.

Immediately inside the cell wall, the **plasma membrane,** or **cell membrane,** is found. It is the outer, living boundary of a bacterial cell. Like the cell membrane in human cells, it consists of phospholipids and proteins that regulate the movement of materials into and out of the cell.

Inside the plasma membrane, the differences between human and bacterial cells become apparent. Bacteria do not have mitochondria, endoplasmic reticulum, or a nucleus. The **nuclear area** of a bacterial cell consists of a single, long, circular molecule of deoxyribonucleic acid (DNA), the bacterial chromosome, which functions to control cell activities (see Chapter 3). Bacteria also have short, circular molecules of DNA, called **plasmids** (plaz´midz), which are not normally required for cell growth and survival. However, plasmids are advantageous under certain conditions. For example, some plasmids carry genes responsible for making bacteria resistant to antibiotics.

Bacteria have ribosomes (see Chapter 3). Although the ribosomes are structurally different from those found in humans, the ribosomes are sites of protein synthesis and produce proteins necessary for bacterial structure and function. In addition, proteins involved in disease are produced at the ribosomes. Some bacteria release enzymes that kill human cells responsible for fighting infections (see Chapter 14). Other enzymes break down tissues, allowing the bacteria to spread throughout the body. **Exotoxins** are proteins released from bacteria that are responsible for the symptoms of many diseases. For example, the bacteria that cause botulism food poisoning release an exotoxin that contaminates food. After the contaminated food is ingested, the exotoxin inhibits the release of acetylcholine from neurons (see Chapter 8). Consequently the person with botulism food poisoning develops double vision, dizziness, and difficulty in swallowing and breathing. Death can result from failure of respiratory muscles to contract. Other examples of diseases involving exotoxins are diphtheria (dif-the´re-ah), cholera (kol´er-ah), and tetanus (tet´ă-nus; lockjaw) (see Appendix A).

Some bacteria have the ability to form **endospores,** a dormant form of the bacteria that is resistant to destruction. An endospore consists of a **spore coat** surrounding the bacterial chromosome and a small amount of cytoplasm. The spore coat protects the chromosome and cytoplasm against extreme conditions of heat, dehydration, and exposure to harsh chemicals. Endospores can remain dormant for long periods, up to hundreds of years. When conditions become favorable, the spore coat breaks open and a bacterial cell emerges. This single bacterial cell is capable of reproducing and starting a whole new population of bacteria. Because endospores are so resistant to destruction, special care must be taken to be sure they have all been destroyed. The bacteria that cause tetanus, botulism food poisoning, and gas gangrene (gang´grēn) can form endospores (see Appendix A).

Reproduction and growth

Bacteria reproduce by **asexual reproduction** in which one cell divides to form two "daughter" cells that are genetically identical to the original "parent" cell. This type of reproduction is similar to the cell division (mitosis) in humans that is responsible for growth and repair (see Chapter 3). **Growth** in bacteria does not usually refer to an increase in the size of the bacterial cell. Instead it means there is an increase in the number of bacteria. Bacteria can reproduce and increase their numbers rapidly by asexual reproduction.

Identification

A patient with pneumonia goes to the doctor. A specimen of the patient's sputum (mucus and cells coughed up from the lungs) is sent to the laboratory where various tests are performed to identify the microorganism causing the pneumonia. This is necessary because different treatments are not equally effective against all microorganisms. For example, some bacteria are resistant to the antibiotic penicillin but are susceptible to the antibiotic tetracycline. Once the microorganism causing a disease is known, the best treatment can be selected.

The laboratory has many ways to identify microorganisms. A microscopic examination of the sample after a Gram stain has been done reveals the shape, organization, and staining characteristics of the bacteria. However, many different types of bacteria have the same shape, organization, and staining characteristics, so additional tests are used to more precisely identify the bacteria. **Metabolic tests** determine nutritional requirements. Different types of bacteria

Understanding genetic transfer and recombination

Genetic material, or DNA, controls the activities of cells (see Chapter 3). If a bacterial cell receives DNA that it does not already have, then the bacterial cell may acquire the ability to do new activities. Transfer of DNA between bacteria can occur in several ways. One method uses the **sex pilus,** a hollow tube that connects two bacterial cells (Figure 21-A). The donor bacterial cell replicates a piece of DNA such as part of a chromosome or plasmid. The DNA is then transferred through the sex pilus to the recipient bacterial cell. This process is sometimes called sexual reproduction, which is somewhat inappropriate because the process starts and ends with two bacterial cells. Another method of transfer involves the uptake of DNA across the bacterial cell membrane. Following the death and rupture of a bacterial cell, DNA leaks out of the dead cell and is taken up by other bacteria.

Genetic recombination occurs when a piece of DNA is inserted into another piece of DNA. For example, part of a chromosome transferred from a donor bacterial cell to a recipient bacterial cell can recombine, or be incorporated, into the chromosome of the recipient bacterial cell. After recombination, the transferred DNA functions to give the recipient bacterial cell new capabilities.

1 A teenager who takes the antibiotic tetracycline to treat acne develops a bacterial infection that causes diarrhea. Tetracycline usually is an effective treatment against the bacteria causing the diarrhea, but in the case of the teenager it does not work. Explain.*

Recombinant DNA technology, or **genetic engineering,** is the deliberate combining of DNA (genes) by humans. For example, the human gene controlling the production of insulin is transferred in the laboratory from a human cell to a bacterial cell. The bacterial cell asexually reproduces, forming large numbers of bacteria. The bacteria produce insulin that is extracted from the bacterial culture and purified to treat diabetes. Genetic engineering has been used to produce hormones, vaccines, antibodies, antibiotics, vitamins, and other useful substances.

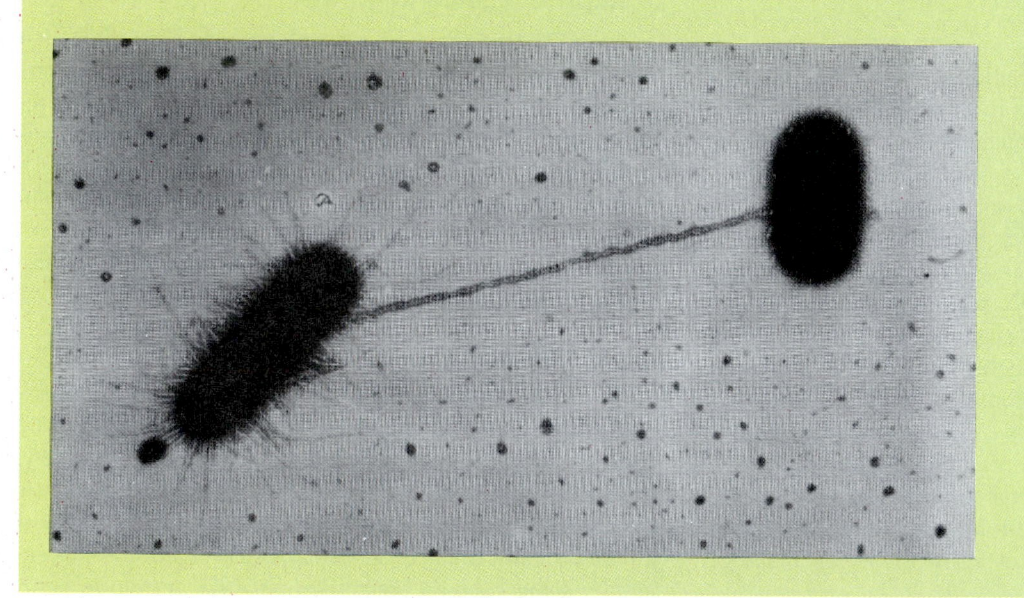

Figure 21-A Transfer of genetic material.
Two bacterial cells are connected by a pilus through which genetic material is passed. After transfer the genetic material may give the receiving bacteria new capabilities such as the ability to resist antibiotics.

have different nutritional requirements. For example, some can use glucose for an energy source and others cannot. By testing the nutritional needs of the bacteria in the sample, it is possible to make a "metabolic fingerprint" that fits the identity of only one type of bacteria. **Antibody tests** use proteins called antibodies (see Chapter 14). Each antibody binds to only one type of bacteria. For example, *Streptococcus pneumoniae* (nu-mo´ne-i) is one of several bacteria that can cause pneumonia. If the antibodies for this bacteria bind to the bacteria in the sputum sample, then the cause of the pneumonia is known. However, if the *Streptococcus pneumoniae* antibodies do not bind to the bacteria in the sputum sample, then some other microorganism must be causing the disease.

Treatment strategies

The basic treatment strategy is to kill the bacteria and do as little harm as possible to healthy human cells. One way to do this is to attack parts of the bacteria that are either not found in humans or are different from human cells.

2 An effective drug for treating bacterial infections in humans would act on which of the following structures: the cell wall, cell membrane, endoplasmic reticulum, or ribosomes? Explain. (Hint: consider the differences between bacterial and human cells.)

Viruses

Most bacteria are **extracellular,** meaning they live outside other cells. Even though there is a bacterial infection in the body, the bacteria are not inside the individual cells of the body. All viruses (and some bacteria) are **intracellular,** meaning they must at some time exist inside a living cell. Viruses are very small and cannot be seen with a light microscope. Instead, the more powerful electron microscope is used. The smallest viruses are the size of large molecules and the largest viruses are approximately the size of the smallest bacterial cells. Outside of a living cell viruses are inert particles, but once inside a cell they reproduce.

Structure

A **virus** is very simple, consisting of nucleic acid (DNA or ribonucleic acide [RNA]) surrounded by a protein coat called a **capsid** (kap´sid). The capsid of many animal viruses is surrounded by a lipoprotein **envelope** that sometimes has projections called **spikes** (Figure 21-3). The nucleic acid is responsible for controlling the reproduction of the virus, and the capsid or envelope functions to protect the nucleic acid and to attach the virus to cells. Although there are many different viral shapes, many viruses are roughly spherical in shape.

Reproduction

The general pattern of reproduction of all viruses is similar, although the details vary. Only an overview of the reproduction of typical animal viruses is considered here. There are five basic steps (Figure 21-4).

1. *Attachment*. The envelope or capsid of the virus attaches to the cell membrane of the cell to be infected.
2. *Entry*. The virus is taken into the cell by endocytosis (see Chapter 3).
3. *Uncoating*. The envelope and capsid are broken down and the viral nucleic acid is released. Uncoating is a poorly understood process that varies with the type of virus.
4. *Production*. Just as the cell's DNA controls the activities of the cell (see Chapter 3), the released viral nucleic acid can direct the cell's chemical activities. The cell makes copies of the viral nucleic acid and manufactures viral proteins. The viral proteins form a capsid around the duplicated nucleic acids to produce new viruses. There can be many thousands of viruses produced inside a single cell.
5. *Release*. The new viruses are released from the infected cell. As they pass through the cell membrane some types of viruses are covered by an envelope formed from viral proteins and parts of the cell membrane. Viruses without envelopes are usually released when the cell lyses (ruptures).

Once the new viruses are released they can attach to and infect other cells. In this way a viral infection can spread rapidly. The infected cells usually die or they don't function properly. This cell destruction and loss of function produces the symptoms of viral infections.

Treatment strategies

Viruses are responsible for a number of diseases in humans such as the common cold, influenza, chickenpox, measles,

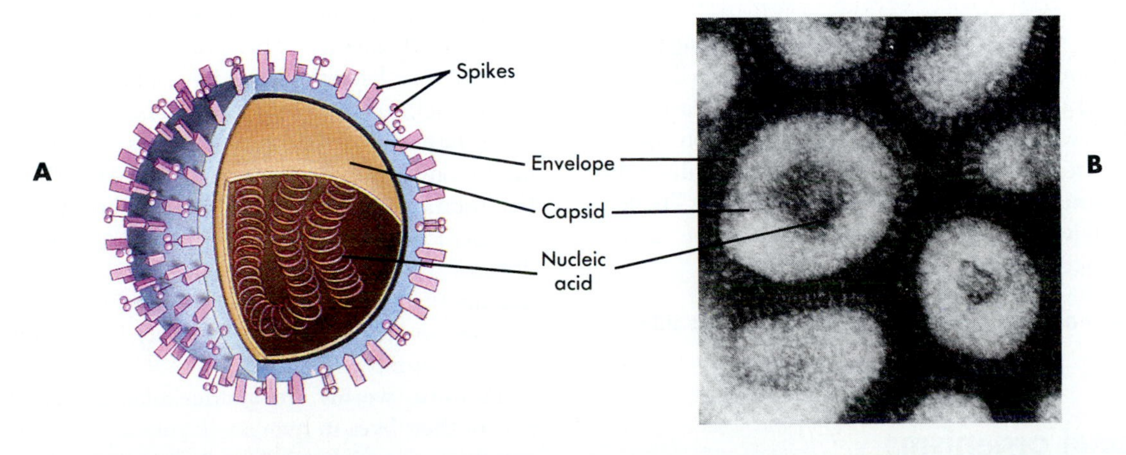

Figure 21-3 ● **Viral structure.**
A, Diagram showing inner core of nucleic acid surrounded by a capsid. In some viruses the capsid is surrounded by an envelope that can have spikes. **B,** Electron micrograph of the influenza virus.

A

Envelope
Capsid
Nucleic acid
Attachment
Cell membrane
Entry by endocytosis
Vesicle containing the virus
Cell nucleus
Uncoating
Viral nucleic acid
Production
Copies of viral nucleic acid
Capsid forms around viral nucleic acid
Release by exocytosis
Envelope
Capsid
Nucleic acid

B

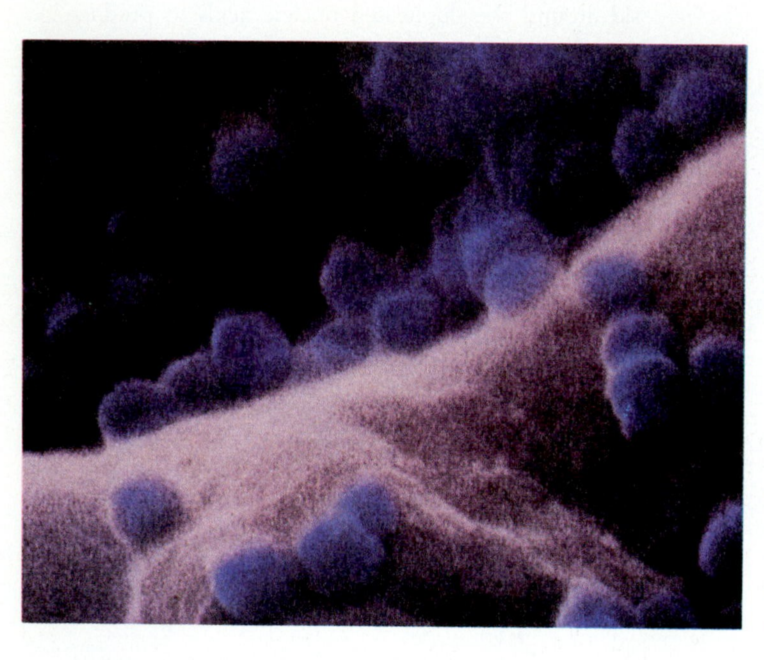

Figure 21-4 ● Viral reproduction.
A, The virus attaches to a cell and enters the cell by endocytosis. The envelope and capsid are removed (uncoating) and the viral nucleic acid directs the production of new viral nucleic acid molecules. Viral proteins produced by the cell form capsids around the viral nucleic acid molecules, resulting in the production of viruses. The viruses are then released from the cell. **B,** AIDS viruses *(blue)* are released from the surface of a white blood cell.

mumps, polio, rabies, herpes, and acquired immune deficiency syndrome (AIDS). Once a person is infected with a virus, treatment is not usually very effective. Outside of cells, viruses are inert and resistant to destruction. Inside cells, viruses are in a protected environment and killing the infected cells may be necessary to kill the viruses. The few effective antiviral drugs act inside cells to prevent nucleic acids from functioning.

3

Explain why antiviral drugs often produce serious side effects.

Additional organisms

Organisms other than bacteria and viruses cause disease in humans. The different types will be briefly mentioned here. For more details see Appendix A.

Protozoans (pro-to-zo´anz) are single-celled organisms that are larger than bacteria (Figure 21-5, *A*). Structurally they are like animal cells and have a nucleus and other typical organelles. Protozoans cause certain types of diarrhea, dysentery (dis´en-tĕr-e), trichomoniasis (trik´o-mo-ni´ă-sis), toxoplasmosis (tok´so-plaz-mo´sis), malaria, and trypanosomiasis (trĭ-pan´o-so-mi´ă-sis; African sleeping sickness).

Fungi (fun´ji) are plantlike single-celled or multicellular organisms (Figure 21-5, *B*). They are larger and more complicated than bacteria. Fungi have a nucleus and other typical organelles, but do not have chlorophyll, the green pigment responsible for photosynthesis. Yeasts, molds, and mushrooms are examples of fungi. Some fungi are beneficial. For instance, yeasts are used in making bread and the antibiotic penicillin is derived from a fungus. A few fungi cause diseases in humans such as ringworm, athlete's foot, candidiasis (kan-dĭ-di´ă-sis; thrush), coccidioidomycosis (kok-sid-e-oy´do-mi-ko´sis; valley fever), and histoplasmosis (his´to-plaz-mo´sis).

Parasitic worms are multicellular animals that spend part of their lives in humans (Figure 21-5, C). Examples of parasitic worm infections include intestinal roundworms, tapeworms, flukes, schistosomiasis (skis-to-so-mi´ă-sis), pinworms, hookworms, filariasis (fil-ă-ri´ă-sis), and trichinosis (trik-ĭ-no´sis).

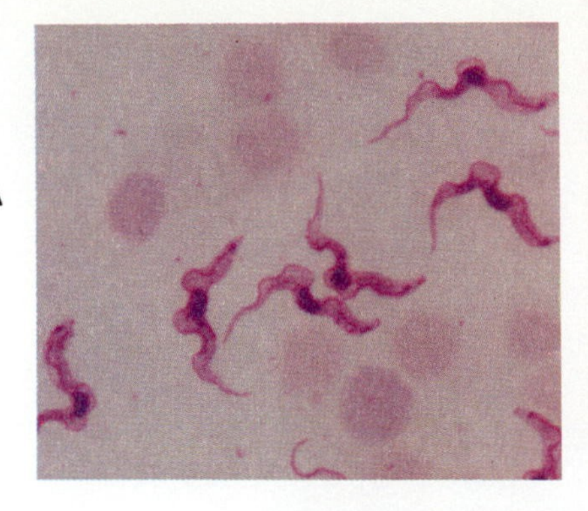

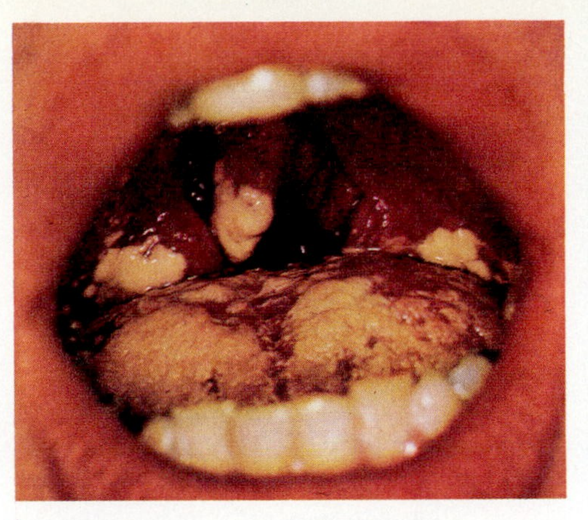

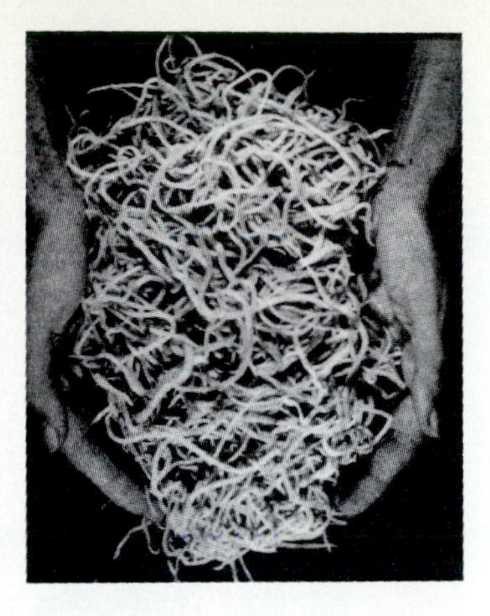

Figure 21-5 ● **Examples of protozoans, fungi, and worms.**
A, The protozoan that causes trypanosomiasis (African sleeping sickness). The dark-staining nucleus is visible within the cell. RBCs are in the background. **B,** The fungus that causes candidiasis (thrush) is growing on the surface of the tongue and oral cavity. **C,** Intestinal roundworms taken from the intestine of a deceased 2-year-old African child.

THE SPREAD OF PATHOGENS

Reservoirs

A **reservoir** is the source of a disease-producing organism. Examples of nonliving reservoirs are water, food, and soil. Living reservoirs include humans and animals (Figure 21-6). The most important reservoir for human infectious diseases is other human beings. When a person is sick and has the symptoms of a disease, he or she usually releases pathogens that can infect others. However, a carrier can spread the disease even when there are no symptoms. A **carrier** is usually an infected person who releases pathogens before the symptoms of disease appear or for a short time following recovery from the disease. In some cases, however, a carrier can have a continual, low level infection. Although such a carrier appears healthy, transmission of the disease can occur.

Wild and domestic animals can also serve as reservoirs for human diseases. Examples of diseases and their reservoirs include psittacosis (sit-ă-ko´sis) in parakeets and parrots; tularemia (tu´lă-re´me-ah) in wild rabbits; trichinosis in pigs; tapeworm in cattle; rabies in bats, skunks, and dogs; and toxoplasmosis in cats (see Appendix A).

Transmission

Transmission of pathogens from a reservoir to a noninfected human can occur by direct or indirect contact (see Figure 21-6). As the term **direct contact** implies, there is a close association between the reservoir and the human to be infected. Examples of direct contact and the diseases they transmit include sexual intercourse, which transmits venereal diseases; kissing, which transmits mononucleosis; handshaking, which transmits colds; droplets of water released during talking or sneezing, which transmit influenza, streptococcal sore throat, and diphtheria; placental transmission, which transmits German measles; blood transfusions, which transmit hepatitis and AIDS; and animal bites, which transmit rabies.

In **indirect contact,** the pathogen can be transmitted by contaminated objects, contaminated food or water, and vectors. An infected person can shed large numbers of pathogens that can contaminate clothing, bedding, eating and drinking utensils, medical instruments, or other objects. Pathogens also can be transmitted through the air on dust particles or on small, dried secretions from the respiratory tract. If a noninfected person comes into contact with these objects while the pathogen is still alive, that person can become infected.

Just as objects can become contaminated, so can food and water. Many people are carriers for the bacteria *Staphylococcus aureus* (ă´re-us), which is found in the nasal passages and from there it can be transferred to the hands. Handling food can result in contamination. The bacteria grow in the food and produce an exotoxin that is eaten with the food. The result is food poisoning with the symptoms of nausea, vomiting, and diarrhea. Other examples of diseases transmitted from contaminated food or water are botulism, *Salmonella* (sal´mo-nel´ah) food infection, typhoid fever, cholera, hepatitis, traveler's diarrhea, giardiasis (je´ar-di´ă-sis; backpacker's diarrhea), amoebic (ă-me´bik) dysentery, and polio (see Appendix A).

A **vector** is an invertebrate (without a backbone) animal that transmits a disease. For example, the black plague is caused by the bacteria *Yersinia pestis* (yer-sin´e-ah pes´tis). The normal reservoir for the bacteria is rats, and the disease is transmitted from one rat to another by fleas. However, fleas can bite humans and infect them. Other examples of vectors and the diseases they transmit are mosquitoes, which transmit malaria, yellow fever, and encephalitis; lice, which transmit typhus fever; ticks, which transmit Lyme disease

Transmission

Direct contact:
Body contact
Droplets
Across placenta
Blood transfusion
Animal bite

Indirect contact:
Contaminated objects
Contaminated food/water
Vectors

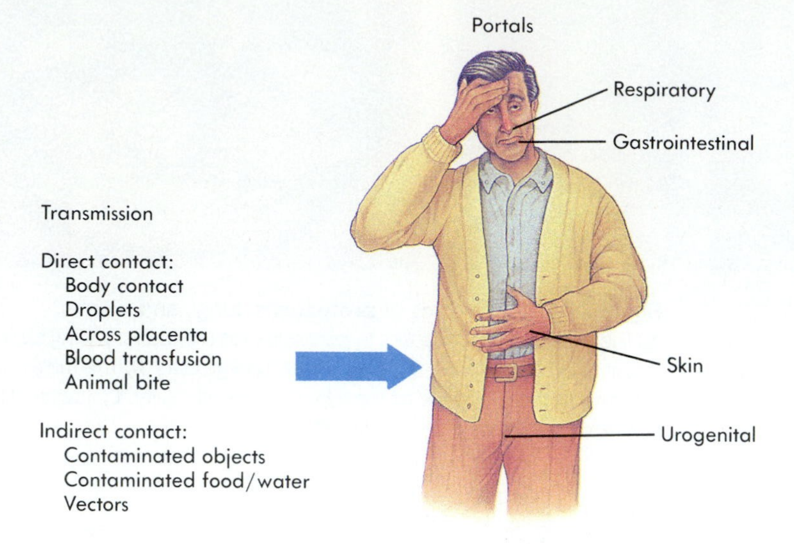

Figure 21-6 ● The spread of pathogens.
Pathogens are found in reservoirs. Pathogens are directly or indirectly transmitted to a noninfected person where they can enter a portal.

and Rocky Mountain spotted fever; and the tsetse (tset´se) fly, which transmits trypanosomiasis (see Appendix A).

Portals

Once a pathogen has been directly or indirectly transmitted from a reservoir to a human, it can enter the human. A **portal** is the point where a pathogenic organism enters or exits the body. The four portals are the respiratory tract, gastrointestinal tract, urogenital tract, and the skin (see Figure 21-6). Often the portal of entry is also the portal of exit from the body. For instance, the bacteria that cause pneumonia enter and exit through the respiratory tract. The chickenpox virus, however, enters through the respiratory tract and can exit through the skin. The virus first infects the respiratory tract, and then is carried by the blood to the skin where it causes small red bumps and blisters. The virus can be released from the ruptured blisters or from the respiratory tract.

4 List as many ways as you can that the chickenpox virus could be transmitted from an infected to a noninfected child.

Opportunistic pathogens

Following transmission of a pathogen from its reservoir to a noninfected person, the pathogen can incubate and cause a disease. Some pathogens, called **opportunistic pathogens,** enter and take up residence in the body, but do not cause disease until conditions become favorable. For example, a microorganism in one location in the body does not cause disease, but when transferred to another location it does. *Escherichia coli* (esh-er-ik´e-ah ko´le) is normally a harmless bacteria in the large intestine. Improper hygiene can result in transfer of the bacteria to the urinary tract, resulting in a bladder infection.

Opportunistic pathogens can also flourish following a change in some condition within the host. For example, weakening due to one infection (for example, influenza) makes it possible for opportunistic pathogens to start another infection (for example, bacterial pneumonia).

Destruction of the normal, nonpathogenic microorganisms of the body also can result in growth of opportunistic pathogens. Normally the nonpathogenic microorganisms compete with opportunistic pathogens for space and nutrients. As a result, the number of opportunistic pathogens is kept small. If nonpathogenic microorganisms are killed by an antibiotic, but opportunistic pathogens are not, then treatment with the antibiotic can reduce competition. The opportunistic pathogens can increase in number and cause disease. The development of the fungal infection called thrush following antibiotic treatment is an example.

Nosocomial infections

Although medical therapy is intended to cure, it often happens that diagnostic procedures and treatments actually cause disease. **Nosocomial** (nos´o-ko´me-al) infections are those that are acquired in the hospital, nursing home, or other health-related facility. In other words, the patient did not have the infection prior to entering the hospital, but developed it as a result of something that happened in the hospital. Approximately 90% of nosocomial infections are caused by bacteria and the rest are mainly caused by viruses.

Several factors interact to produce nosocomial infections. First, patients are often in a weakened condition, which makes them more susceptible to infections. Other illnesses can weaken the patient's resistance, and cancer therapy often involves the deliberate suppression of the immune system. Many of the nosocomial pathogens are either opportunistic pathogens or pathogens that would not ordinarily cause disease in a healthy individual. Some hospitals do not allow surgical or burn patients to receive flowers or plants because, in the patient's weakened condition, bacteria on the flowers or plants can cause nosocomial infections.

Second, already infected patients, personnel and visitors who are carriers, and contaminated inanimate objects or food all combine to make the hospital environment a reservoir for many pathogens. As a result of constant exposure, some of these pathogens become resistant to antibiotics and disinfectants and are particularly difficult to control. Some pathogens actually thrive within disinfectant solutions!

Third, the opportunities for transmission of pathogens is greatly increased. Direct contact transmission from patient to patient and from health-care personnel to patients can occur. Many of the procedures performed in hospitals are invasive and provide an opportunity for pathogens to enter the body. Most nosocomial infections develop in the urinary tract, usually as a result of a urinary catheter (a tube used to drain urine from the urinary bladder). Other nosocomial infections include surgical wounds, pneumonia (related to respiratory equipment that aids in breathing or administering drugs), and bacteremia (bacteria in the blood resulting from intravenous catheters used to pass fluids, nutrients, or drugs).

It is estimated that 5% to 15% of hospital patients develop nosocomial infections and approximately 20,000 people per year die from them. The number of nosocomial infections can be lowered by reducing the exposure of patients to pathogens. Proper sterilization and disinfection of inanimate objects, careful attention to aseptic technique, and conscientious handwashing in-between patients are all vital. Most hospitals have an infection control committee (or person) whose job is to monitor and identify problems that can result in nosocomial infections.

PREVENTING DISEASE

The immune system, chemotherapy, and preventing exposure to pathogens are three basic ways we can deal with disease. Two of these ways, the immune system and chemotherapy, deal with pathogens after they have entered the body. The immune system consists of phagocytes and white blood cells (WBCs) that destroy pathogens. The details of the operation of the immune system are covered in Chapter 14. **Chemotherapy** uses drugs to kill or suppress pathogens. For example, penicillin kills bacteria by inhibiting peptidoglycan synthesis in the cell wall, and AZT (azidothymidine) inhibits viral replication by inhibiting nucleic acid synthesis.

Preventing exposure to pathogens is an effective way to deal with diseases. Diseases do not occur if pathogens are not allowed to infect the body. Medical techniques and public health measures are designed to kill pathogens before they enter the body, or they keep pathogens from coming into contact with the body.

Medical techniques

Standard medical techniques are designed to prevent pathogens from entering the body. **Sterilization** is the complete destruction of all organisms. A number of techniques are used to achieve sterilization including autoclaving, baking, burning, filtration, radiation, and treatment with ethylene oxide (Table 21-1). **Disinfection** is the destruction of pathogens on inanimate objects, whereas **antisepsis** (an´tĭ-sep´sis) is the destruction of pathogens on the skin or on tissues. Disinfection and antisepsis do not kill all the microorganisms present, and different methods are not equally effective in killing pathogens. Usually chemicals are used for disinfection and antisepsis (Table 21-1).

Surgical asepsis (ă-sep´sis) keeps all microorganisms away from an object or person, whereas **medical asepsis** keeps away pathogens associated with communicable diseases. **Quarantine,** an example of medical asepsis, is the isolation of an infected patient to keep the pathogen from coming into contact with noninfected people.

5 Suppose a surgical team needed to achieve surgical asepsis for a patient who was having her appendix removed. Using Table 21-1 suggest some of the methods they could use. Explain how each method would result in surgical asepsis.

Public health

Routine public health measures have been adopted by most developed societies and are the best method of preventing the outbreak of epidemics.

Table 21-1 • Methods used to control microorganisms

Method	Description	Use
Heat		
Autoclaving	Steam under 15 pounds of pressure per square inch reaches a temperature of 121°C; application is for 15 minutes	The most used method of sterilization: any item that can withstand the moisture, pressure, and temperature can be autoclaved (e.g., glassware, instruments, dressings, etc.)
Canning	Various techniques utilize water under pressure	If properly done, canned items are sterile; an effective method for preserving food
Baking	Heated air at a temperature of 170°C for 2 hours	An effective sterilization technique used for items that can't be wet (e.g., powders) and for glassware, instruments, etc.
Burning	Destruction by fire	Disposal of paper products, dressings, and tissues; effective sterilization of inoculating loops
Boiling	Heating water to 100°C for 10 to 30 minutes	Boiling kills most, but not all pathogens; it should be used only when no better method is available
Pasteurization	Enough heat (72°C for 15 seconds for milk) is applied to kill most microorganisms, but not enough heat to ruin the substance being pasteurized	Milk and other dairy products are pasteurized to kill tuberculosis bacteria and other pathogens; beer and wine are pasteurized to prevent spoilage
Refrigeration and freezing	Refrigeration occurs at low temperatures (approximately 4°C) and freezing occurs below 0°C	Refrigeration slows microbial growth and retards spoilage; freezing stops microbial growth and prolonged freezing below 10°C kills some pathogens
Filtration	A filter allows passage of liquid or air but not microorganisms	Sterilization of solutions (e.g., drugs) that would be damaged by heat; surgical masks; air entering operating rooms
Radiation	Destroys DNA	Gamma rays have good penetration power and are used industrially to sterilize disposable plastic products (e.g., syringes and petri dishes); ultraviolet light has little penetration power and is used to kill microorganisms in the air or on the surface of objects
Ultrasound	High frequency sound waves produce tiny bubbles in liquids that collapse and produce a suctioning type of action that removes soil from objects	A very effective means of cleaning but not a reliable method of destroying pathogens
Chemicals		
Ethylene oxide	A gas that effectively penetrates paper and plastic wrappings; it causes proteins and nucleic acids to become nonfunctional	Used to sterilize items affected by high temperatures (e.g., catheters, sutures, disposable plastic supplies, heart-lung machines, etc.)
Phenol derivatives	Damages the cell membrane and inactivates enzymes	Cresols are the main ingredients in the disinfectant Lysol and hexylresorcinol is used in throat lozenges
Chlorine	Chlorine combines with water to produce a reactive form of oxygen that damages microorganisms	Used to disinfect drinking water, swimming pools, and sewage; also found in bleach
Iodine	Interferes with enzyme activity	Commonly used for wound antisepsis, preoperative skin antisepsis, and disinfection of thermometers; examples are tincture of iodine and Betadine; also used to disinfect drinking water
Alcohol	Changes the shape of proteins; an effective cleaner	Isopropyl alcohol is widely used, but ethyl alcohol (in liquor) is also effective; used as an antiseptic (e.g., prior to an injection)
Soap	Breaks up oils and debris so they can be removed by water	Soaps are useful for cleaning the skin, but have little ability to kill most skin microorganisms; some soaps, however, have added chemicals that kill microorganisms

Food safety

Food safety includes measures that prevent the transmission of disease through food. For example, proper canning of food prevents food poisoning as well as food spoilage, and pasteurization stops tuberculosis and other diseases from being spread in milk and other dairy products (Table 21-1). Cooking meat adequately can eliminate trichinosis and tapeworms. Trichinosis is caused by the roundworm *Trichinella spiralis* (trik´ĭ-nel´ah spi-rah´lis). In animal populations, the worm is found in cysts (sacs) within muscle. When the infected animal is eaten by another animal, the worms are taken into the digestive tract where they reproduce. The worms then enter the blood and are carried to muscle where they form new cysts. Symptoms of trichinosis include abdominal pain, diarrhea, fever, muscle aches, and swelling around the eyes. Normally these worms are found in rodents, pigs, and bears. If a human eats meat (for example, pork) that has the encysted worms, he or she can become infected. Once common in the United States, trichinosis is now rare because it is known that high temperatures (137° C.) kill the worm. Garbage fed to pigs is cooked and the public is urged to cook pork thoroughly.

Food-handling establishments must meet certain sanitary requirements that are designed to prevent food contamination. If they fail to do so, diseases such as *Salmonella* food infection may result. The main reservoir for the *Salmonella* bacteria is the intestinal tract of poultry, but humans can become carriers. During butchering, chicken or turkey meat can become contaminated, or laying hens can contaminate eggs. Improper processing, cooking, or refrigeration of poultry and egg products, or poor sanitary practices by human carriers, can result in live *Salmonella* bacteria in food. When ingested, the bacteria multiply in the intestinal tract and release endotoxins. Symptoms include a moderate fever, abdominal pain, diarrhea, and nausea.

Water safety

Water safety includes measures that prevent the transmission of disease through water. This is accomplished by filtration, disinfection of water supplies with chlorine, and constant monitoring of the water to ensure noncontamination.

Sewage disposal

Many pathogens are shed from the body in feces or urine. Proper plumbing and sewage disposal prevent exposure to pathogens and kill pathogens. Sewage treatment separates sewage into **effluent** (liquid wastes) and **sludge** (solid waste). Unfavorable oxygen levels, decomposition by microorganisms, and disinfection with chlorine kills pathogens. The sludge is used as fertilizer, burned, or dumped. The effluent is used for irrigation, is dumped, or, in some treatment facilities, is reclaimed as drinking water.

Animal control

Animal control includes measures that prevent the transmission of diseases from animals to humans. The likelihood that humans will contract diseases from animals is reduced if the diseases are prevented in the animals. The vaccination of pet dogs to prevent rabies in humans is an example. Dogs can get rabies by being bitten by an animal such as a skunk that has the disease. However, if dogs are routinely vaccinated against rabies they are immune to the disease. Vaccinated dogs do not develop rabies and therefore do not transmit the disease to humans.

Destroying or eliminating contact with vectors is also effective in disease prevention. For example, insecticides can kill vectors such as mosquitoes, fleas, and ticks, and good personal hygiene can eliminate lice.

Decreasing the size of animal populations that are reservoirs for human diseases reduces the likelihood that humans will come into contact with a diseased animal. For example, garbage is a source of food for many animals such as rats and mice. Proper garbage disposal greatly decreases the number of animals that are reservoirs for or transmit disease.

SUMMARY

Parasites derive nourishment from a host. Pathogens cause disease or death of the host.

Microorganisms include bacteria, viruses, fungi, and protozoans.

DISEASE TERMINOLOGY

Disease is an interference or cessation of normal body functions.

Disease produces symptoms and signs.

The incubation period precedes the appearance of symptoms and signs.

Diseases can be classified several ways: by severity and length (for example, acute, chronic, and subacute); by location in the body (for example, local and systemic); and by distribution in the population (for example, incidence, prevalence, epidemic, and endemic).

DISEASE-CAUSING ORGANISMS

Bacteria

Bacterial shapes are bacillus, coccus, and spiral shaped.

Bacterial structures and their functions include the capsule and pili (for attachment), flagella (for movement), cell wall (provides shape and prevents lysis), plasma membrane (regulates the movement of materials into and out of the cell), chromosome (regulates cell functions), and ribosomes (protein production).

Endotoxins (LPS layer of cell wall) and exotoxins (proteins secreted by the cell) produce disease symptoms.

Endospores allow bacteria to resist harsh environmental conditions.

Bacteria reproduce asexually. Bacterial growth is an increase in bacterial numbers as a result of asexual reproduction.

Bacteria are identified based on shape, staining characteristics, metabolic requirements, and antibody binding.

The best treatment against bacteria acts on parts of the bacteria not found in humans.

Viruses

Viruses must live inside a cell at some time.

A virus is nucleic acid (DNA or RNA) surrounded by a capsid. Sometimes an envelope surrounds the capsid.

Viruses reproduce by causing a cell to produce viral nucleic acid and capsids.

Additional organisms

Protozoans, fungi, and parasitic worms can also cause disease.

THE SPREAD OF PATHOGENS

Reservoirs

A reservoir is a source of a pathogen. A carrier is a human reservoir that has no disease symptoms.

Humans and other animals are reservoirs for human diseases.

Transmission

Direct contact involves a close association between the reservoir and the person to be infected.

Indirect contact involves contaminated objects, food, or water, or involves vectors (invertebrate animals).

Portals

A portal is the place of entry of pathogens, for example, respiratory tract, gastrointestinal tract, urogenital tract, or skin.

Usually the place of exit of pathogens is the same as the place of entry.

Opportunistic pathogens

Following a change in body location or change in host condition, opportunistic pathogens can cause disease.

PREVENTING DISEASE

Medical techniques

Sterilization, disinfection, and antisepsis kill pathogens.

Asepsis keeps microorganisms away from objects or persons.

Public health

Food and water safety measures and proper sewage disposal prevent transmission of diseases through the gastrointestinal tract.

Animal control regulates animal reservoirs and prevents transmission by vectors.

CONTENT REVIEW

1. Explain the difference between the following terms:
   a. Parasite and host
   b. Disease and infection
   c. Symptom and signs
   d. Diagnosis, prognosis, and therapy
   e. Acute, chronic, and subacute diseases
   f. Local and systemic diseases
   g. Incidence and prevalence
   h. Epidemic and endemic
   i. Aerobic and anaerobic bacteria
2. How are bacteria classified according to shape and arrangement?
3. Give the location and function of the following bacterial structures: capsule, pili, flagella, cell wall, cell membrane, nuclear area, chromosome, and ribosomes.
4. What is the difference between gram negative and gram positive bacteria? Contrast endotoxins and exotoxins. What are they responsible for?
5. What are endospores? Why is it important to be aware of endospores?
6. Describe asexual reproduction in bacteria. What is meant by bacterial growth?
7. Why is it important to identify bacteria taken from a sick person? How is this done?
8. How can knowledge of bacterial structures be used in designing treatment for bacterial diseases?
9. Describe the structures that make up a virus. List the five steps of viral reproduction.
10. Briefly describe protozoans, fungi, and parasitic worms.
11. Explain the difference between the following terms:
    A. Reservoir, carrier, and portal
    B. Animal reservoir and vector
    C. Direct and indirect contact
12. What is an opportunistic pathogen?
13. Explain the differences between the following terms:
    A. Sterilization, disinfection, and antisepsis
    B. Surgical asepsis, medical asepsis, and quarantine
14. List ways that the safety of food and water can be ensured.
15. Why is it important to properly dispose of sewage?
16. How can animal control result in the control of some diseases?

CONCEPT REVIEW

1. When bacterial cells and red blood cells (RBCs) are placed in a solution of pure water, the RBCs lyse (rupture) but the bacterial cells do not. Explain.
2. Ima Saver canned some vegetables from her garden. She placed the raw vegetables in glass jars, sealed the lids shut, and placed the jars in boiling water for 30 minutes. The boiling cooked the vegetables and Ima thought the boiling also safely preserved the vegetables. Later, however, Ima's family was poisoned when they ate the vegetables. What was the cause of the food poisoning? Why didn't Ima's method work? What method should she have used? (Hint: see Table 21-1.)
3. Folic acid is a vitamin necessary for bacterial and human cell function. Folic acid must be manufactured by many bacteria, whereas folic acid used by human cells is obtained from ingested food. Sulfa drugs inhibit the formation of folic acid. How can sulfa drugs be effective against bacteria and yet not stop cell function in humans?
4. Why is a knowledge of portals important in controlling disease? Give an example of such control for each of the body's portals.
5. Bacterial endocarditis is an infection of the valves in the heart. People who for some reason already have damaged heart valves are more likely to develop the disease than people with normal heart valves. The condition can destroy the heart valves and can be fatal. Prior to having her teeth cleaned a patient was advised to take penicillin. Explain why this was done.

CHAPTER TEST

Answers can be found in Appendix C

MATCHING 1 For each statement in column A select the correct answer in column B (an answer may be used once, more than once, or not at all).

A	B
1. Any departure from normal observed by a health care provider.	bacillus
2. Estimated outcome of a disease.	coccus
3. The number of new cases of a disease that appear in a given time.	endemic
4. Disease with a low, but continuous prevalence in a population.	epidemic
5. Rod-shaped bacteria.	incidence
	prevalence
	prognosis
	sign
	symptom
	therapy

MATCHING 2 For each statement in column A select the correct answer in column B (an answer may be used once, more than once, or not at all).

A	B
1. The source of a disease-producing organism.	antisepsis
2. A person who has no disease symptoms but can infect others.	asepsis
3. An invertebrate animal that transmits a disease.	carrier
4. Point where pathogenic organisms enter or exit the body.	disinfection
5. Pathogen in the body that does not cause disease unless conditions become favorable.	opportunistic pathogen
6. Complete destruction of all organisms.	portal
7. Destruction of pathogens on the skin.	reservoir
8. To keep microorganisms away from an object or person.	sterilization
	vector

FILL-IN-THE BLANK Complete each statement by providing the missing word or words.

1. The ___________ and ___________ of bacteria help them stick to surfaces.
2. A substance found in bacterial cell walls, but not human cells, that makes the cell wall structurally strong and resistant to rupture is ___________.
3. The ___________ of bacterial cell walls makes some bacteria gram ___________; it can also be released from dead bacteria as an ___________ that causes disease.
4. DNA is found in a single, long, circular chromosome located in the ___________ of bacterial cells. DNA is also found in short, circular structures called ___________.
5. Proteins secreted from living bacterial cells that cause disease are ___________.
6. Some bacteria can form ___________, which enables the bacteria to resist destruction by unfavorable environmental conditions.
7. Asexual reproduction in bacteria results in an increase in the number of bacteria, which is referred to as bacterial ___________.
8. A virus is nucleic acid (DNA or RNA) surrounded by a protein coat called the ___________. Some animal viruses also have a lipoprotein ___________.
9. List in sequential order the five steps of viral replication.

APPENDIX A

Some Common Disorders Caused by Pathogens

Disorder	Causative agent and description of the disorder
Integumentary system	
Acne	*Propionibacterium acnes* and other bacteria. Bacteria within a hair follicle cause an inflammatory response. Accumulating pus results in a pimple. A blackhead is an accumulation of sebum and sloughed skin cells within a hair follicle.
Anthrax	*Bacillus anthracis* (bacteria). Endospores of the bacteria, found on infected animals (for example, cattle, sheep) or their products, enter the skin to cause a localized pustular infection. Less commonly the bacteria can cause pneumonia or septicemia.
Boil	*Staphylococcus aureus* and other bacteria. A painful nodule consisting of an area of inflammation surrounding a central core of pus where bacteria have infected a hair follicle.
Chickenpox	Chickenpox virus. The virus infects the respiratory system and is carried by the blood to the skin, where it causes a blisterlike rash.
Cold sore	Herpes simplex virus type. The initial infection produces no symptoms or can cause inflammation of the gums. The virus becomes latent and causes recurring blisterlike lesions in the skin and mucous membranes.
Impetigo	*Staphylococcus aureus* (bacteria). A superficial skin infection with small, round elevations containing pus that rupture. Bacteria in the released fluid can spread to other skin areas or to other individuals.
Lice	*Pediculus humanus* (insect). Lice are transmitted on clothing or by physical contact. Bites of the feeding lice cause intense itching of the scalp and skin.
Lyme disease	*Borrelia burgdorferi* (bacteria). The bacteria are transmitted from mice to humans by ticks. Symptoms include fever and a ring-shaped rash. Months later arthritis, heart inflammation, or neurological abnormalities can develop.
Measles	Measles (rubeola) virus. The virus infects the respiratory system causing a sore throat and Koplik spots (tiny red patches with central white specks). The virus spreads through the blood to the skin to produce a rash consisting of small, solid elevations of the skin. Complications such as pneumonia, middle ear infections, and encephalitis can occur.
	German measles (rubella) virus. The virus infects the respiratory system and is carried by the blood to the skin, where it causes a rash consisting of nonelevated discolored patches. Infection during pregnancy can damage the fetus (for example, deafness, heart defects, and mental retardation).
Ringworm	Several species of fungus cause blisters and scaling of the skin. The fungi are transmitted from contaminated objects or from dogs and cats. It can occur in the scalp (ringworm of the scalp), the feet (athlete's foot), and the groin (jock itch).
Rocky Mountain spotted fever	*Rickettsia rickettssii* (bacteria). Ticks transmit the bacteria to humans from dogs, rabbits, spotted fever and squirrels. Symptoms include fever and a rash.
Scabies	*Sarcoptes scabiei* (mite). Transmitted by skin-to-skin contact or by contaminated clothing or bedding. The female mite tunnels through the skin laying eggs. Symptoms include burrows in the skin and intense itching.
Shingles	Chickenpox virus. After contracting chickenpox, the virus remains latent in nerve cells. Illness or stress activates the virus, which produces a blisterlike rash along the pathway of the nerve.
Typhus	*Rickettsia* species (bacteria). Epidemic typhus is transmitted from person to person by the human louse. Symptoms include fever and a rash. Murine (relating to rodents) typhus is transmitted from rats by the rat flea, and is a milder form of typhus.

Wart	Wart virus. A wart is a usually benign (noncancerous) growth of the skin that is spread by contact-transfer of the virus. Warts often disappear spontaneously or can be removed chemically, surgically, or by cryotherapy (liquid nitrogen).
Wound	Damage to the skin can introduce bacteria or allow opportunistic pathogens to cause alocalized infection.

• Skeletal system

Infectious arthritis	Inflammation of the joint is caused by bacteria (for example, gonorrhea, tuberculosis, syphilis, Lyme disease), viruses (for example, mumps, measles, hepatitis, mononucleosis), and fungi (for example, coccidioidomycosis, histoplasmosis). Usually the pathogens spread to the bone through the blood from another site of infection in the body. Note that most arthritis, such as osteoarthritis and rheumatoid arthritis, does not result from an infection.
Osteomyelitis	Inflammation of the bone, usually caused by bacteria. Most commonly the bacteria are transported from another site of infection by the blood.

• Muscular System

Gas gangrene	*Clostridium perfringens* (bacteria). Found in the soil (endospores) or the intestinal tract of animals and humans, the bacteria are introduced in a wound., The bacteria destroy muscle and connective tissue. They also produce a gas that causes the tissue to swell.

• Nervous system

Encephalitis	Encephalitis is an infection of the brain by bacteria, viruses, or fungi. Mosquitoes can transmit a virus from birds or horses to humans. Symptoms include fever, headache, coma, and death.
Leprosy	*Mycobacterium leprae* (bacteria). The bacteria live inside Schwann cells and skin cells, causing loss of sensation, disfiguring nodules, and tissue death. The bacteria are spread from the discharge of lesions to small cuts in the skin. However, the disease is not very contagious.
Meningitis	*Neisseria meningitidis* (bacteria), other bacteria, viruses, and fungi. Bacterial meningitis usually begins with an infection of the throat, producing no symptoms or the symptoms of a cold. The bacteria travel by blood to the meninges causing headache, vomiting, confusion, and a stiff neck. Death can occur rapidly.
Poliomyelitis	Polio virus. Ingestion of the virus in water contaminated with human feces is the usual mode of infection. The virus infects the throat and small intestine, causing sore throat, fever, and nausea. In 1% to 2% of cases the virus spreads through the blood to the nervous system, kills motor neurons, and causes paralysis.
Rabies	Rabies virus. A skin wound inflicted by an infected animal is the usual mode of infection. At first the virus multiplies in skeletal muscle and connective tissue, but then travels along nerves to the brain and spinal cord. Destruction of nervous tissue results in paralysis and respiratory failure.
Subacute sclerosing panencephalitis	Measles virus. Five to 12 years after contracting measles the virus causes sclerosis (hardening) of brain tissue and death.
Tetanus	Clostridium tetani (bacteria). Found in soil and the intestinal tract of animals, the bacteria are introduced in a wound. The bacteria produce a toxin that is carried to the nervous system. The results are spastic contractions of skeletal muscles. Lockjaw (muscles hold jaw closed) and death from spasms of the respiratory muscles can occur.
Toxoplasmosis	Toxoplasma gondii (protozoan). Infection occurs from eating undercooked meat or inhaling dried cat feces while cleaning a litter box. Most cases produce no symptoms or an undefined mild illness. The protozoan can cross the placenta and cause brain damage in the fetus.
Trypanosomiasis	*Trypanosoma brucei* (protozoan). Found in cattle and antelope, the protozoan is transmitted to humans by the bite of the tsetse fly. Symptoms are decreased physical and mental activity, coma, and death. Also called sleeping sickness.

• Special Senses

Conjunctivitis	Different bacteria cause an inflammation of the conjunctiva. Haemophilus aegyptius (bacteria) causes pinkeye.
Herpetic keratitis	Herpes simplex virus type. The virus causes corneal ulcers. The virus can become latent and then cause repeated ulcer development at later times.
Ophthalmia neonatorum	*Neisseria gonorrhoeae* (bacteria). During delivery , an infected mother can transmit the bacteria from her reproductive tract to the eyes of the newborn. Blindness can result.
Otitis externa	*Pseudomonas aeruginosa* (bacteria). Infection of the skin in the ear canal. The bacteria are highly resistant to chlorine and can infect users of swimming pools.

Otitis media — Bacterial species. Infection of the middle ear can be a complication of nose or throat infections. Pus and fluid accumulate, producing pressure that causes pain and interferes with hearing.

Trachoma — *Chlamydia trachomatis* (bacteria). An infection of the cornea that causes blindness. The bacteria are transmitted by towels, fingers, etc.

Cardiovascular System

Endocarditis — Different bacterial species cause an inflammation of the endocardium and heart valves. The bacteria are carried by the blood from another location in the body (for example, from a tooth extraction site or a localized infection).

Malaria — *Plasmodium* species (protozoans). Transmitted by a mosquito, the protozoans infect the liver, undergo developmental changes, and then infect erythrocytes. Symptoms include recurring cycles of fever and chills.

Pericarditis — Inflammation of the pericardium can result from bacterial, viral, or fungal infections. Symptoms include chest pain that worsens with breathing movements.

Rheumatic fever — *Streptococcus pyogenes* (bacteria). The disease begins with a streptococcal throat infection. Weeks later an autoimmune reaction develops. Antibodies that destroyed the bacteria also destroy heart valves and joints (arthritis).

Septicemia — Different bacterial species cause septicemia or blood poisoning. The bacteria enter the blood from a localized infection in the body. Endotoxins cause fever, a decrease in blood pressure, and shock.

Lymphatic System and Immunity

AIDS — Acquired immune deficiency syndrome (AIDS) is caused by the HIV (human immunodeficiency virus). Transmission occurs through sexual intercourse, blood products, contaminated needles, and across the placenta. The virus infects T cells, causing suppression of the immune system. Death usually results from secondary infections.

Brucellosis — *Brucella* species (bacteria). The bacteria enter the body through abrasions in the skin (while handling swine carcasses) or through the digestive system (from drinking unpasteurized milk). The bacteria live inside phagocytes in lymphatic tissue. Symptoms include fever, weakness, and aching.

Mononucleosis — Epstein-Barr virus. Transmission is by direct (for example, kissing) or indirect (for example, drinking from the same glass) transfer of saliva. The virus infects B lymphocytes. Most people have been infected, but only a few individuals develop the symptoms of fever, sore throat, enlarged lymph nodes, and enlarged spleen.

Plague — *Yersinia pestis* (bacteria). Rat fleas transmit the bacteria from rats to humans. Symptoms include swollen lymph nodes, fever, and a blackish skin (due to hemorrhage).

Tularemia — *Francisella tularensis* (bacteria). Contracted from contact with infected animals (for example, rabbits, squirrels) or by arthropod bites. Tularemia results in enlarged lymph nodes filled with pus, fever, and a localized ulcer in the skin.

Respiratory System

Coccidioidomycosis — *Coccidioides immitis* (fungus). The fungus is transmitted by air from the soil. Usually no symptoms or symptoms of fever, coughing, and chest pain develop. In rare cases, the disease progresses to a tuberculosis-like disease.

Cold — Many different cold viruses. Symptoms include sneezing, excessive nasal secretions, and congestion. The viruses are transmitted through respiratory secretions.

Diphtheria — *Corynebacterium diphtheriae* (bacteria). Transmitted by water droplets in the air, the bacteria infect the throat causing a sore throat and fever. A membrane of dead human and bacterial cells can form in the throat and sometimes blocks air movement into the lungs. The bacterial cells also produce an exotoxin that causes paralysis and cardiac arrest.

Histoplasmosis — *Histoplasma capsulatum* (fungus). The fungus is transmitted by air from the soil. Usually no symptoms or a minor respiratory illness results. In severe cases, a chronic pulmonary disease like tuberculosis develops.

Influenza — Influenza virus. Characterized by fever, chills, headache, and muscular aches, influenza usually occurs as an annual epidemic. Death is rare and often is due to a secondary bacterial pneumonia.

Legionnaire's disease — *Legionella pneumophila* (bacteria). A type of pneumonia. The bacteria are transmitted by air from environmental sources such as the water in air conditioning systems.

Pneumonia — *Streptococcus pneumoniae* causes most bacterial pneumonias. Symptoms are fever, chest pain, and difficulty in breathing. Air sacs (alveoli) fill with fluid. The bacteria are transmitted in nasal secretions, or lowered resistance (for example, other illness such as viral infections of the respiratory system) allows bacteria already present to multiply (opportunistic pathogen). Viral pneumonias are caused by a number of viruses and can occur as a complication of influenza. *Pneumocystis carinii* (protozoan) causes pneumonia in patients with a depressed immune system, such as AIDS patients.

Psittacosis	*Chlamydia psittaci* (bacteria). A type of pneumonia contracted by inhaling the bacteria on dried particles from bird droppings.
Scarlet fever	*Streptococcus pyogenes* (bacteria). The bacteria causing streptococcal sore throat can release a toxin that causes a pink-red skin rash.
Streptococcal sore throat	*Streptococcus pyogenes* (bacteria). Commonly transmitted through respiratory secretions, the symptoms include fever and sore throat.
Tuberculosis	*Mycobacterium tuberculosis* (bacteria). The bacteria are usually acquired by inhalation. In the lungs a tubercle is formed, which walls off and contains the bacteria. The bacteria can survive indefinitely inside the tubercle, or they can escape to infect other areas of the lung or body.
Whooping cough	*Bordetella pertussis* (bacteria). The bacteria infect the trachea and bronchi causing the production of large amounts of mucus that restricts air movement. A "whoop" sound is produced as the victim gasps for air between coughs.

Digestive System

Botulism	*Clostridium botulinum* (bacteria). The bacteria are found in soil. They form endospores that contaminate food items. The endospores survive improper techniques used to preserve the food. Bacteria from the endospores produce an exotoxin which, when ingested, causes flaccid paralysis. Double vision, difficulty in swallowing, and respiratory or cardiac failure can result.
Cholera	*Vibrio cholerae* (bacteria). These noninvasive bacteria produce an exotoxin that increases the permeability of the intestinal wall to water. Large amounts of fluids and electrolytes are lost into the intestine, causing diarrhea, shock, and possibly death. Transmission involves ingestion of food or water contaminated by human feces.
Diarrhea	Many different species of bacteria, viruses, and protozoans cause diarrhea. Escherichia coli (bacteria) is a primary cause of infant's diarrhea and traveler's diarrhea. See also giardiasis.
Dysentery	*Shigella* species (bacteria) and *Entamoeba histolytica* (protozoan). Dysentery is diarrhea with a bloody, mucous feces. Transmission is through food or water that has been contaminated by human feces. Flies and cockroaches also transmit the pathogens from feces to food.
Food poisoning	Many species of bacteria cause food poisoning. For specific examples see botulism, salmonellosis, and staphylococcal food poisoning.
Giardiasis	*Giardia lamblia* (protozoan). Ingestion of fecally contaminated (by humans or animals) water or food results in a prolonged diarrheal disease.
Hepatitis	An inflammation of the liver caused by the hepatitis virus, chemicals, and allergic reactions. Symptoms include fever, chills, nausea, and jaundice. Infectious hepatitis, caused by the hepatitis A virus, is spread by the oral-intestinal route. Serum hepatitis, caused by hepatitis viruses B, C, or D, is usually acquired through transfusions.
Hookworm	*Necator* species and other roundworms. The eggs of hookworms are passed with human feces into the soil. Larvae from the eggs can penetrate the skin, causing inflammation and itching. Carried by the blood to the lungs, the larvae penetrate the alveoli, ascend the respiratory tract, and are swallowed. In the intestines the larvae become adults that suck blood from the intestinal wall. Symptoms of anemia, abdominal pain, and diarrhea can develop.
Mumps	Mumps virus. Primarily a childhood disease, the virus is transmitted in saliva and respiratory droplets, and infects the respiratory system. The virus multiplies and is carried by the blood to the parotid salivary glands, which become swollen. In males past puberty, the virus can reach the testes and cause sterility.
Pinworm	*Enterobius vermicularis* (roundworm). The female worms lay eggs on and around the anus, causing itching. Infection results when the eggs are ingested.
Salmonellosis	*Salmonella* species (flatworm). The bacteria are found in the intestines of animals, and infection occurs by ingestion of contaminated meat or poultry products. The bacteria multiply in the intestine, releasing an endotoxin that causes nausea, abdominal pain, and diarrhea.
Schistosomiasis	*Schistosoma* species (flatworm). The eggs of the flatworm are released in human wastes into water. Larvae hatch from the eggs and infect snails. After further development the larvae leave the snails and penetrate the skin of a person who contacts the water. The larvae are carried by the blood to the liver or urinary bladder and mature into an adult. The disease usually causes liver damage.
Staphylococcal food poisoning	*Staphylococcus aureus* (bacteria). Human carriers of the bacteria contaminate food. The bacteria grow in the food and produce a toxin. The ingested toxin causes nausea, vomiting, and diarrhea.

Tapeworm — A number of flatworm species cause tapeworm. The larvae of the worms encyst in the muscles of host animals. When humans ingest undercooked meat (for example, beef, pork, fish) the larvae develop into an adult tapeworm that attaches to the intestinal wall. Often no symptoms are present, or abdominal pain and anemia can develop.

Typhoid fever — *Salmonella typhi* (bacteria). Released in feces from humans, the bacteria contaminate water and food. The ingested bacteria cross the intestinal wall causing fever and malaise that lasts about 2 weeks.

Trichinosis — *Trichinella spiralis* (roundworm). The larvae of the worm encyst in the muscles of animals. Human ingestion of undercooked meat (usually pork) results in infection. The larvae become adults, which produce a new generation of larvae. These larvae burrow into tissues causing damage until they encyst.

Yellow fever — Yellow fever virus. A mosquito transmits the virus between humans, or from monkeys to humans. Symptoms include jaundice (due to liver damage) and fever.

Urinary System

Cystitis — Cystitis is an infection of the urinary bladder. Common causes are infection by *Escherichia coli* (bacteria) from the intestinal tract (for example, poor hygiene) or *Trichomonas vaginalis* (protozoan) from the reproductive tract.

Glomerulonephritis — *Streptococcus pyogenes* (bacteria). An autoimmune disease in which antibodies against the bacteria cause kidney damage.

Pyelonephritis — *Escherichia coli* or other bacteria cause a diffuse inflammation of the kidneys. The bacteria are carried to the kidneys by the blood from another infection site, or are a complication of lower urinary tract infections.

Reproductive System

Candidiasis — *Candida albicans* (fungus). An opportunistic pathogen that grows on the mucous membranes of the genitourinary tract and oral cavity.

Crabs — *Phthirus pubis* (insect). These lice are usually transmitted sexually. Bites of the feeding lice cause itching of the skin in the pubic region.

Chancroid — *Haemophilus ducreyi* (bacteria). A sexually transmitted disease that produces a genital ulcer.

Genital wart — Human papillomaviruses. A sexually transmitted disease. A few forms of the virus are linked to cervical cancer and to cancer of the penis.

Gonorrhea — *Neisseria gonorrhoeae* (bacteria). A sexually transmitted disease. In males symptoms include painful urination and a pus-containing discharge from the urethra. In females there are usually no symptoms.

Herpes (genital) — Usually caused by herpes simplex type 2 virus. A sexually transmitted disease. Symptoms include genital itching and a rash, followed by clusters of blisterlike lesions that break and form sores. The virus can remain latent in nerve cells and cause recurrent outbreaks of lesions.

Nongonococcal urethritis — *Chlamydia trachomatis* (bacteria) that are sexually transmitted are responsible for most cases. Nongonococcal urethritis is any inflammation of the urethra not caused by the gonorrhea bacteria. In males there are no symptoms, or a watery discharge and urethral itching. For females, see pelvic inflammatory disease.

Pelvic inflammatory disease — *Chlamydia trachomatis* and *Neisseria gonorrhoeae* (bacteria). Usually a sexually transmitted disease. The uterus, uterine tubes, and peritoneum become infected. Symptoms include bleeding, fever, and painful urination.

Puerperal fever — *Streptococcus pyogenes* (bacteria). As a result of childbirth or abortion the uterus is infected. The bacteria can spread to the abdominal cavity causing peritonitis, septicemia, and death.

Syphilis — *Treponema pallidum* (bacteria). A sexually transmitted disease. Initial symptom is a hard sore (chancre) at the site of infection. Weeks later a skin rash develops. Years later soft, gummy lesions develop in various organs.

Thrush — See candidiasis.

Trichomoniasis — *Trichomonas vaginalis* (protozoan). A sexually transmitted disease. Males usually have no symptoms, whereas females have a foul-smelling vaginal discharge.

APPENDIX B

Answers to Predict Questions

CHAPTER 1

1 p. 7 Donating a pint of blood results in a decrease in blood pressure. Negative-feedback mechanisms such as an increase in heart rate return blood pressure toward a normal value. When a negative-feedback mechanism fails to return a value to its normal level, the value can continue to deviate from its normal range. Homeostasis is not maintained in this situation and the person's health can be threatened.

2 p. 8 The thirst sensation is associated with a decrease in body fluid levels. The thirst mechanism causes the person to drink fluids, which returns the fluid level to normal. Thirst therefore is a sensation involved in negative-feedback control of body fluids.

3 p. 11 When a man is standing on his head, his nose is superior to his mouth. Remember that directional terms refer to a person in the anatomical position and do not refer to the body's current position.

CHAPTER 2

1 p. 18 Because atoms are electrically neutral, the iron (Fe) atom has the same number of protons and electrons. The loss of three electrons would mean the iron atom has three more protons than electrons and therefore a charge of plus three. The correct symbol would be Fe^{3+}.

2 p. 19 During exercise, muscle contractions increase. This requires the release of stored energy in ATP. Some of the energy is used to drive muscle contractions and some of it is released as heat. Because the rate of these reactions increases during exercise, more heat is produced than when at rest, and body temperature increases.

3 p. 20 Adding a base to a solution would make the solution more basic, and the pH would increase. If the buffer removed the base from the solution, the increase in pH would be resisted.

4 p. 22 An increase in respiration rate results in a decrease in blood carbon dioxide. Hydrogen ions combine with bicarbonate ions to form more carbon dioxide. Consequently, hydrogen ion levels decrease, resulting in an increase in blood pH, or alkalosis.

CHAPTER 3

1 p. 33 Urea is produced continually by cells and diffuses from the cells into the extracellular spaces and from the extracellular spaces into the blood. If the kidneys stop eliminating urea, it begins to accumulate in the blood. Because the concentrations increase in the blood, urea can not diffuse from the extracellular spaces. Because urea accumulates in the extracellular spaces, it cannot diffuse from the cells. The urea finally reaches concentrations high enough to be toxic to cells, causing cell damage followed by cell death.

2 p. 34 Normally blood is isotonic to the fluid inside cells, that is the cells neither swell nor shrink. If glucose accumulates in the blood, it has more dissolved substances and proportionately less water than before the glucose accumulated. Water moves from the cells where it is in higher concentration into the blood where it is in lower concentration. The cells shrink because of the loss of water.

3 p. 35 *(A)* Cells highly specialized to perform active transport would have a large surface area (many microvilli) exposed to the fluid from which substances are actively transported. Numerous mitochondria would be present near the membrane across which active transport occurs because the mitochondria produce ATP necessary for active transport. *(B)* Cells highly specialized to phagocytize foreign substances would have numerous lysosomes in their cytoplasm and evidence of phagocytic vesicles.

CHAPTER 4

1 p. 48 Capillaries are where gas and nutrients diffuse across the wall of the capillary. The walls of capillaries consist of simple squamous epithelium. It is thin and well adapted for allowing substances to diffuse across it. Because the ducts of sweat glands secrete sweat, it is reasonable to predict that they are lined with either simple cuboidal or simple columnar epithelial cells. Actually the ducts of sweat glands are lined with simple cuboidal epithelial cells. These cells contain the organelles needed to secrete sweat.

2 p. 48 In tendons, tension is applied along the length of the tendon. Collagen fibers are arranged so that they are parallel to the length of the tendon and therefore are able to resist the effect of the tension. In skin, tension is applied in many directions. Fiber arrangement is in many directions and the skin is well adapted to resisting tension from many directions.

3 p. 56 With only the information given, it may not be possible to tell for sure. However, it is not likely that cancer would arise in more than one location so soon after surgery in the woman. It is more likely that cancer cells had spread (metastasized) to new locations before the surgery, but had not yet formed tumors large enough to be detected. They are likely to have spread through the lymphatic vessels, which are abundant in the breast.

CHAPTER 5

1 p. 62 Because the permeability barrier is composed mainly of lipids surrounding the epidermal cells, substances that are lipid soluble could diffuse through the barrier easily. This fact is used as a basis for administering some medications through the skin. On the other hand, water-soluble substances would have difficulty diffusing through the skin. The skin's lipid barrier prevents water loss from the body.

2 p. 63 *(A)* The posterior surface of the forearm appears darker because of the tanning effect of ultraviolet light from the sun. The posterior surface of the forearm usually is exposed to more sunlight than the anterior surface of the forearm. *(B)* The lips are pinker or redder than the palms of the hand. Several explanations for this are possible and any of the following explanations are reasonable. Often more than one explanation is possible and further investigation is needed to arrive at an answer. Some possible explanations for lips being pinker or redder than the palms are that there could be more blood vessels in the lips, there could be increased blood flow in the lips, or the blood vessels could be easier to see through the epidermis of the lips. The last possibility actually explains most of the difference in color between the lips and palms. The epidermis of the lips is thinner and not as heavily keratinized as that of the palms, which are more resistant to abrasion. In addition, the papillae containing the blood vessels in the lips are "high" and closer to the surface.

3 p. 63 It is not reasonable to believe the story. Hair color is due to melanin that is added to the hair in the hair bulb as the hair grows. The hair itself is dead. To turn white, the hair must grow out without the addition of melanin. This, of course, takes more time than 1 night.

4 p. 64 Because the hair pulled out easily, the hair follicle has been destroyed. Since the hair follicle extends deep into the dermis, this would indicate the epidermis and dermis have been destroyed. It is probably a third-degree burn.

CHAPTER 6

1 p. 71 If all the mineral is removed, the bone consists mainly of collagen. The bone becomes so flexible and weak that it cannot support weight. Loss of mineral can produce bowed bones in children who do not ingest or absorb adequate amounts of calcium. If all the collagen is removed, the bone consists mainly of minerals and becomes very brittle and easily broken. This is a problem with many older people whose bones are fragile and easily broken.

2 p. 74 If cartilage growth and bone formation failed to occur at the epiphyseal plate, the bone would be normal in diameter (or even greater in diameter than normal) but would be much shorter than normal. This is the condition seen in one type of dwarfism, where the head and trunk are more normal in size, but the long bones of the limbs are very short.

3 p. 80 If only the muscles attached to the spinous processes contracted, the vertebral column would bend backward. If only the muscles attached to the transverse processes on one side contracted, the vertebral column would bend toward the side on which the muscles were contracting.

4 p. 91 Abduction of the arm and flexion of the forearm.

CHAPTER 7

1 p. 98 Cross bridge movement causes the Z lines of sarcomeres to come closer together. As each sarcomere shortens, the myofibrils, which consist of many sarcomeres, also shorten. When the myofibrils in muscle fibers shorten, the muscle fibers shorten, causing muscle bundles to shorten. As the muscle bundles shorten, the muscle shortens.

2 p. 101 During a 1-mile run, aerobic respiration is the primary source of ATP production for muscle contraction. Anaerobic respiration provides the short burst of energy for the sprint at the finish. After the race, aerobic respiration is elevated for a time to repay the oxygen debt, causing the heavy breathing after the race.

3 p. 105 Raising eyebrows—occipitofrontalis muscle; winking—orbicularis oculi muscle; whistling—orbicularis oris and buccinator muscles; smiling—zygomaticus muscles; frowning—corrugator supercilii and depressor anguli oris muscles; sneering—levator labii superioris muscles.

4 p. 105 Contraction of the right sternocleidomastoid muscle rotates the head to the left, and contraction of the left sternocleidomastoid muscle rotates the head to the right. If the right sternocleidomastoid were damaged it would not exert its normal force of contraction on the head. Consequently the left sternocleidomastoid muscle would be less "opposed" and would cause rotation of the head to the right.

5 p. 106 Postural muscles, which must contract continuously for long periods without fatiguing would have a higher percentage of slow-twitch muscle fibers. The arm muscles, which are capable of rapid movements but which are more readily fatigued, would have a higher percentage of fast-twitch muscle fibers.

6 p. 106 The man should use very heavy weights, resulting in only a few repetitions of each exercise. The intense, anaerobic exercise that results would affect primarily fast-twitch muscle fibers to produce large, strong muscles. The woman should use light weights with many repetitions of each exercise. The aerobic exercise that results would affect primarily slow-twitch muscle fibers. Her muscle would not greatly enlarge, although the definition and endurance of the muscles would increase.

CHAPTER 8

1 p. 122 Active transport is the movement of ions or molecules from areas of lower concentration to areas of higher concentration. This movement requires the expenditure of energy in the form of ATP (see Chapter 3). The movement of Na^+ ions out of the cell is by active transport, because movement is from a lower concentration inside the cell to a higher concentration outside the cell. The movement of K^+ ions is also active transport because movement is from a lower concentration outside the cell to a higher concentration inside the cell.

2 p. 124 Inferior movement of the brainstem can compress nervous tissue as the brainstem is forced through the foramen magnum. Specifically, the inferior part of the brainstem, the medulla oblongata, is most likely to be damaged. Damage to nuclei within the medulla oblongata can result in loss of the functions controlled by the nuclei. If the nuclei involved with controlling heart rate, blood vessel diameter, or breathing stopped functioning, death could result.

3 p. 126 The prefrontal area plans and initiates the movements of the head. Action potentials from the prefrontal area go to the motor area. In the motor area, nerve pathways going to the neck muscles are activated. Action potentials go to the skeletal muscles, causing contraction and turning of the head. The image (light) from the person sitting next to you enters the eye and causes the production of action potentials that travel up sensory nerve tracts to the visual area of the brain.

4 p. 130 As the trunk moves to one side (say the right side) it causes stretch of skeletal muscles on the opposite side (the left side). Sensory receptors detect the stretch, and action potentials go to the spinal cord and cause the activation of motor neurons. Action potentials are sent along motor neurons to the stretched skeletal muscles. They contract slightly and the trunk straightens to its original position.

5 p. 137 Because the phrenic nerves originate in the cervical region of the spinal cord, damage to the spinal cord in the thoracic region would not affect the diaphragm, which is largely responsible for breathing. Damage to the upper cervical region, however, would cut the nerve tract between the brain and the motor neurons in the spinal cord that supply the diaphragm. This would eliminate phrenic nerve function and greatly interfere with breathing. Death could result.

6 p. 139 *(A)* A person who is extremely angry has activated the sympathetic nervous system and the expected responses might include increased heart rate, blood pressure, respiration, and perspiration. *(B)* In a person who has just finished eating and is now relaxing, the activities of the parasympathetic nervous system would predominate. The responses might be decreased heart rate and respiration. There would be an increase in digestive activities, such as secretion and motility in the digestive tract.

CHAPTER 9

1 p. 146 Much of taste is based on olfactory function. A cold may include a stuffy nose, which may interfere with olfaction, and thus with taste.

2 p. 147 Medication placed into the eyes may pass through the nasolacrimal duct into the nasal cavity, where its odor can be detected. Because much of our taste sensation is based on olfactory function, the medication is perceived to have a taste that is detected.

3 p. 148 The action potentials from the pain receptors are traveling at different speeds. The nerves for the sharp pain are myelinated and the nerves for the diffuse pain are unmyelinated (see Chapter 8).

4 p. 149 Pain from visceral organs is not well localized because of the absence of touch receptors and is normally perceived as a diffuse pain. The pain from the colon would probably be referred to the anterior, inferior surface of the abdomen (see Figure 9-A).

5 p. 150 In dim light, cones, which are involved in color perception but are less sensitive to dim light, tend to quit functioning, leaving only the rods, which can function in dim light but cannot perceive color.

CHAPTER 10

1 p. 163 Following a meal, blood levels of glucose would increase to much higher-than-normal levels if insulin was not secreted.

2 p. 168 The TSH-like chemical binds to the same receptors as TSH resulting in an increase in the synthesis and secretion of thyroid hormones. In addition, the TSH-like chemical is not sensitive to the inhibitory effect (negative feedback) of thyroid hormones on TSH secretion. Consequently, the rate of thyroid hormone secretion and the blood levels of thyroid hormone become elevated. A chemical similar to TSH is the cause of many cases of oversecretion of the thyroid gland (hyperthyroidism). Symptoms associated with hypersecretion of thyroid hormone would become obvious. The thyroid gland also enlarges (goiter). In addition, connective tissue is deposited behind the eyeballs causing them to protrude and appear larger than normal, a condition called exophthalamos. This type of hyperthyroidism is called Grave's disease.

3 p. 168 Insufficient vitamin D results in insufficient calcium absorbed by the intestine. As a result, blood levels of calcium begin to decrease. In response to the low blood calcium levels, parathyroid hormone is secreted from the parathyroid glands. Parathyroid hormone acts primarily on bone, causing bone to be broken down and calcium to be released into the blood to maintain blood levels of calcium within the normal range. Eventually so much calcium is removed from bones that they become soft, fragile, and easily broken. The condition is most obvious in children, where the bones are bent and deformed, and it is called rickets.

4 p. 172 In the absence of ACTH the entire adrenal gland degenerates. As a consequence the hormones of the adrenal cortex, which includes cortisol, aldosterone, and androgens, are secreted in much smaller amounts. Hormones of the adrenal medulla, epinephrine and norepinephrine, are also secreted in smaller amounts.

5 p. 174 After a large meal, glucose enters the blood from the intestine. The increasing blood glucose stimulates insulin secretion and decreases glucagon secretion. The insulin acts on most cells of the body and causes them to take up glucose and either use it as an energy source or store it as glycogen or fat. After 12 hours without eating, blood glucose levels would tend to decrease. Decreasing blood glucose levels would result in a decreased rate of insulin secretion and would stimulate glucagon secretion. The glucagon acts on the liver to cause it to release glucose into the blood.

6 p. 175 Joe was suffering from insulin shock. Blood glucose levels fell to very low levels and he began to suffer from disorientation, which would lead to convulsions and loss of consciousness. The insulin shock occurred because he failed to eat after taking his insulin injection. Joe could treat the symptoms by eating something with a high sugar content. The sugar would increase the blood levels of glucose. If he became very disoriented or began to suffer from convulsions or was unconscious, glucose would have to be given intravenously. After blood glucose levels were increased to their normal levels, the symptoms of insulin shock would disappear.

CHAPTER 11

1 p. 180 Carbon monoxide binds to the iron of hemoglobin and prevents the transport of oxygen. The decreased oxygen stimulates the release of erythropoietin, which increases red blood cell production in red bone marrow. Consequently, the number of red blood cells in the blood increases.

2 p. 185 The blood should be type B negative. Transfused type B blood would not cause a transfusion reaction in the ABO blood group because it matches the baby's ABO blood type. Rh negative blood has no antigens to react to any Rh antibodies in the baby. Keep in mind that erythroblastosis fetalis results from Rh antibodies reacting against the Rh antigens of the baby's blood. If the transfused blood was Rh positive it could make the condition worse. Even though the baby receives Rh negative blood, the baby's blood type is not changed, because blood type is genetically determined. After the transfused Rh negative blood cells die, the baby will produce its own Rh positive blood cells as replacements.

3 p. 186 People with type AB blood were called universal recipients because they could receive type A, B, AB, or O blood with little likelihood of a transfusion reaction. Type AB blood does not have antibodies against type A or B antigens. Therefore transfusion of these antigens in type A, B, or AB blood does not cause a transfusion reaction in a person with type AB blood. The term is misleading, however, for two reasons. First, other blood groups can cause a transfusion reaction. Second, antibodies in the donor's blood can cause a transfusion reaction. For example, type O blood contains A and B antibodies that can react against the A and B antigens in type AB blood.

4 p. 188 An increase in the white blood cell count often indicates a bacterial infection. A white blood cell differential count with an abnormally high neutrophil percentage would help confirm the diagnosis.

CHAPTER 12

1 p. 195 Complete blockage of the left coronary artery would stop blood flow to most of the left side of the heart including the left ventricle. The left ventricle would soon lose its ability to contract and to pump blood. Since the left ventricle pumps blood to all portions of the body except the lungs, a positive feedback cycle occurs, which results in a continuous decline in the left ventricle's ability to pump blood. The person would probably die.

2 p. 198 If the normal blood supply is reduced to the small area of the heart through which the left bundle branch passes, conduction of action potentials through the left side of the heart is reduced or blocked. As a consequence, the left side of the heart does not contract at its normal rate. The right side of the heart functions more normally. The reduced rate of contraction of the left ven-ventricle reduces the pumping effectiveness of the left ventricle.

3 p. 199 An increase in the number of P waves will also increase the number of QRS complexes, up to a point. However, if there are many ectopic action potentials arising in the atria, each of them giving rise to a P wave, not all of the P waves will give rise to QRS complexes. Consequently, the heart rate will be increased somewhat and there will be several P waves for each QRS complex.

4 p. 199 Most of ventricular contraction occurs between the first and second heart sounds of the same beat. Therefore, between the first and second heart sounds, blood is ejected from the ventricles into the pulmonary trunk and the aorta. Between the second heart sound of one beat and the first heart sound of the next beat, the ventricles are relaxing. No blood passes from the ventricles into the aorta or pulmonary trunk during that period.

5 p. 201 The *swish* sound made after a heart sound is created by the backward flow of blood after closure of a leaky or incompetent valve. A swishing sound immediately after the second heart sound (lubb-dupp*swish*) represents a leaky aortic semilunar or pulmonary semilunar valve. The *swish* sound before a heart sound is created by blood being forced through a narrowed or stenosed valve just before the valve closes. The lubb-*swish*dupp suggests that there is a swishing sound immediately before the second heart sound, so a stenosed aortic or pulmonary semilunar valve is indicated.

6 p. 203 In response to severe hemorrhage, blood pressure decreases, which is detected by baroreceptors. A reduced frequency of action potentials is sent from the baroreceptors to the medulla oblongata. This causes the cardioregulatory center to increase sympathetic stimulation of the heart and to increase the heart rate. Sympathetic stimulation of the heart also increases stroke volume as long as the volume of blood returned to the heart is adequate. Following hemorrhage, however, the blood volume in the body and the volume of blood returned to the heart from the body are reduced. Because the heart is contracting rapidly there is less time for the heart to fill with blood. Also, because there is less blood returning to the heart, the heart does not fill as rapidly. As a consequence, the volume of blood in the heart is lower than normal when the heart contracts, resulting in a reduced stroke volume. The reduced stroke volume is an illustration of Starling's law of the heart. The heart rate can be very high and the stroke volume can be quite low in people who are suffering from severe hemorrhage.

CHAPTER 13

1 p. 212 Gravity acts on blood in the veins of the lower limbs and tends to force it to flow toward the feet (when the feet are below the level of the heart). The valves in the veins prevent blood from flowing toward the feet, and periodic compression of the veins by skeletal muscles in the lower limbs causes blood to flow toward the heart. For people who have varicose veins, the valves are incompetent. Thus blood tends to pool in the veins of the lower limbs, especially when people with varicose veins are in standing positions. Elevating the lower limbs assists blood flow from the veins in the lower limbs downhill toward the heart.

2 p. 212 Atherosclerosis slowly reduces blood flow through the carotid arteries and therefore the amount of blood flowing to the brain. In advanced stages of atherosclerosis, the resistance to blood flow increases so much that the blood flow to the brain through the carotid arteries is reduced significantly. As a consequence, there is a reduced ability of the brain to carry out normal functions, which results in symptoms such as confusion and a loss of memory.

3 p. 220 When a blood vessel is blocked, oxygen and nutrients are depleted and waste products accumulate in tissues supplied by the blocked blood vessel. The reduced supply of oxygen and nutrients and the accumulated waste products all cause relaxation of the precapillary sphincters. Consequently, blood flow through the area increases greatly after the block has been removed.

4 p. 220 In a long distance runner blood vessels in contracting skeletal muscles dilate allowing blood flow to exercising muscles to increase greatly. Dilation of the skeletal muscle blood vessels decreases peripheral resistance. However, blood vessels in the viscera constrict. Therefore the total resistance to blood flow does not decrease dramatically. Also, the increased blood flow through the exercising muscles results in an increased volume of blood returning to the heart. The increased volume of blood returning to the heart, because of Starling's law of the heart, results in an increased cardiac output. Increased sympathetic stimulation of the heart also results in a cardiac output. Blood pressure will actually increase because of the large increase in cardiac output (due to the increased heart rate and stroke volume) even though the overall peripheral resistance decreases.

5 p. 221 Doing a headstand increased blood pressure in the carotid sinuses and aortic arch because of the effect of gravity on blood. The increased pressure activates the baroreceptor reflex, resulting in decreased sympathetic stimulation of the heart. The heart would respond with a decreased heart rate. The increased pressure detected by the baroreceptors would also reduce sympathetic stimulation of blood vessels (reduce vasomotor tone), allowing them to dilate in an attempt to reduce the blood pressure.

CHAPTER 14

1 p. 234 Cutting and tying off the lymph vessels would prevent the movement of fluid from the affected tissue. The result would be edema.

2 p. 238 Antigens are too small to be seen with the unaided eye. By joining many antigens together they form a large enough clump to be seen. For example, in blood typing, the antibodies join the antigens on the surface of different red blood cells together, producing an agglutination reaction, which is easily identified.

3 p. 240 The first exposure to the disease-causing agent (antigen) would evoke a primary immune system response. Gradually, however, the antibodies would degrade, and memory cells would die. If, before all the memory cells were eliminated, a second exposure to the same antigen occurred, a secondary immune system response would result. The memory cells produced could provide immunity until the next exposure to the antigen.

4 p. 240 The booster shot stimulates a memory (secondary) response, resulting in the formation of large amounts of antibodies and memory cells. Consequently there is better, longer-lasting immunity.

CHAPTER 15

1 p. 247 When you sleep with your mouth open, air does not pass through the nasal passages. This is especially true when nasal passages are plugged as a result of having a cold. As a consequence air is not humidified and warmed. The dry air dries the throat and the trachea, thus irritating them.

2 p. 250 During respiratory movements, the parietal and visceral pleura slide over the surface of each other. When the pleural membranes are inflamed, they rub against each other creating an intense pain. The pain is worse when a person takes a deep breath because the movement of the membranes is greater than during normal breaths.

3 p. 252 In respiratory distress syndrome the lungs tend to collapse because of the high surface tension of the water lining the alveoli. Adding surfactant to inspired air reduces the surface tension and the tendency of the alveoli to collapse. Administering air under pressure inflates the alveoli by increasing the pressure within them. The pressurized air pushes against the alveolar walls and helps prevent them from collapsing.

4 p. 255 A rapid rate of respiration increases the blood pH because carbon dioxide is eliminated from the blood more rapidly during rapid respiration. As carbon dioxide is lost, hydrogen ions and bicarbonate ions combine to form carbon dioxide and water. The decrease in hydrogen ions causes an increase in blood pH.

Holding one's breath results in a decrease in pH because carbon dioxide accumulates in the blood. The carbon dioxide combines with water to form hydrogen ions and bicarbonate ions. The increase in hydrogen ions causes a decrease in blood pH.

5 p. 257 When a person breathes rapidly and deeply for several seconds, the carbon dioxide levels decrease and blood pH increases. Carbon dioxide is an important regulator of respiratory movements. A decrease in blood carbon dioxide and an increase in blood pH result in a reduced stimulus to the respiratory center. As a consequence, respiratory movements stop until blood carbon dioxide levels build up again in the body fluid. This normally takes only a short time.

CHAPTER 16

1 p. 265 It is important to close off the nasopharynx during swallowing so that food, and especially liquid, doesn't pass into the nasal cavity. If a person has an explosive burst of laughter while trying to swallow a liquid, the liquid may be explosively expelled from the mouth and even from the nose. Speaking requires that the epiglottis be elevated so that air can pass out of the larynx. If you do this while you are swallowing, the food, and especially liquid, may pass into the larynx, causing you to choke.

2 p. 274 Introducing fluid into the rectum by way of an enema causes distention of the rectum. Distention stimulates the defecation reflex.

3 p. 275 Large amounts of chyme in the small intestine is an indication that the digestive activities of the small intestine need to increase. This is accomplished by increasing the secretions of the pancreas and liver. At the same time, secretin inhibits the activities of the stomach. Consequently, the entry of chyme into the small intestine decreases, allowing the small intestine time to complete its digestive functions.

4 p. 277 Achlorhydria would eliminate pepsin digestion of proteins in the stomach, but trypsin in the duodenum would still function and digest proteins so that little effect would be seen on protein digestion.

CHAPTER 17

1 p. 283 If vitamins were broken down during the process of digestion, their structures would be destroyed, and as a result, their ability to function would be lost.

2 p. 287 If the transfer of energy from high energy electrons is blocked, then ATP is not produced aerobically, and the patient dies because too little energy is available for the body to maintain vital functions. Anaerobic metabolism can provide energy for only very short periods and cannot sustain life very long.

3 p. 288 A person who has lost a large amount of weight typically has an increased appetite and a decreased ability to expend energy because basal metabolic rate is reduced. Consequently, overeating frequently occurs and the excess Calories are not expended, but are stored as fat.

4 p. 291 During fever production the hypothalamic set point increases and mechanisms that increase body temperature are activated. The body produces heat by shivering. The body also conserves heat by constriction of blood vessels in the skin (producing a pale skin) and by reduction in sweat loss (producing a dry skin). When the fever breaks (that is, the "crisis is over"), the hypothalamic set point is lowered back to normal and heat is lost from the body to lower body temperature. This is accomplished by dilation of blood vessels in the skin (producing a flushed skin) and increased sweat production (producing a wet skin).

CHAPTER 18

1 p. 301 Because the urethra of females is much shorter than the urethra of males, the female urinary bladder is more accessible to bacteria from the exterior. This accessibility is one of the reasons that urinary bladder infection is more common in women than in men.

2 p. 302 If large amounts of glucose enter the nephron and are not reabsorbed, the glucose causes the concentration of the filtrate to increase. The glucose molecules attract water and, because they are trapped in the nephron, the amount of water that remains in the nephron is increased. A large volume of urine that contains glucose is a symptom of diabetes mellitus.

3 p. 303 Aldosterone increases the reabsorption of sodium and chloride ions from the filtrate in the nephron. As the sodium and chloride are reabsorbed, water follows the sodium and chloride by osmosis. As a result, the volume of the extracellular fluid in the body increases and so does the blood pressure.

4 p. 303 Diuretics that inhibit the reabsorption of sodium and chloride ions in the nephron cause the concentration of sodium and chloride ions to increase in the filtrate. Because the concentration of these ions increases in the urine, the volume of water that is reabsorbed across the wall of the nephron decreases. The water remains in the nephron with the sodium and chloride ions due to osmosis. Therefore the volume of urine increases and the urine will have a relatively high concentration of sodium and chloride ions in it.

5 p. 309 Hyperventilation results in a greater-than-normal rate of carbon dioxide loss from the circulatory system. As carbon dioxide is lost from the circulatory system, hydrogen ions combine with bicarbonate ions to form carbon dioxide. As the hydrogen ion concentration decreases, the pH of body fluids increases. A hyperventilating person benefits from breathing into a paper bag, because this causes the person to rebreathe air that has a higher concentration of carbon dioxide. The result is an increase in carbon dioxide in the body. Consequently the hydrogen ion concentration increases and pH decreases toward normal levels.

6 p. 311 The respiratory system compensates by increasing the rate of respiration. The increased rate of respiration increases the rate of carbon dioxide loss from the blood. As a result, the concentration of hydrogen ions decreases in the blood because the hydrogen ions combine with bicarbonate ions to form carbon dioxide and water. The loss of hydrogen ions from the blood increases pH, which counteracts the effect of metabolic acidosis.

CHAPTER 19

1 p. 319 The prostate gland is located just anterior to the rectum. A physician can insert fingers into the rectum and palpate the prostate through the wall of the rectum. The procedure does not require surgical procedures and involves relatively minor discomfort.

2 p. 320 The secondary sexual characteristics, external genitalia, and sexual behavior develop in response to testosterone. If the testes failed to produce normal amounts of testosterone at puberty, secondary sexual characteristics and external genitalia would remain juvenile and normal adult sexual behavior would not develop.

3 p. 327 Prior to ovulation a sustained increase in estrogen stimulates GnRH, LH, and FSH secretion. However, in combination with progesterone, and when progesterone is present in high concentrations with estrogen, they inhibit GnRH, LH, and FSH secretion. Thus, administration of a large amount of estrogen and progesterone just before the preovulatory LH surge would inhibit the release of GnRH, LH, and FSH. Consequently ovulation could not occur.

4 p. 332 *(A)* Methods resulting in sterility: vasectomy, tubal ligation (in some cases these methods can be surgically reversed). *(B)* Effective methods that do not have a permanent effect: rhythm method (marginally effective), condoms, diaphragm and sponge (most effective when used with spermicidal agents), oral contraceptives, IUDs.

CHAPTER 20

1 p. 339 Two primitive streaks forming in one embryonic disk can result in twins. If the two streaks are touching, the twins are attached to each other. The attachment can be fairly simple, and the twins can be separated fairly easily by surgery, or the attachment can be extensive involving internal organs and cannot be corrected easily. Twins joined to one another are called conjoined, or Siamese, twins.

2 p. 345 Genotype DD (homozygous dominant) would have the polydactyly phenotype, genotype Dd (heterozygous) would have the polydactyly phenotype, and genotype dd (homozygous recessive) would have the normal phenotype.

3 p. 345 The genotype of the carriers is Aa (heterozygous). Each can produce gametes with an A gene or an a gene. The possible combinations of these gametes are AA, Aa, Aa and aa. Out of the four possibilities, one (aa) results in albinism. Therefore the likelihood is that one fourth of their children will be albinos.

4 p. 347 The answer to this question depends on your values and/or a weighing of the possible benefits against the possible harm that could result from a complete knowledge of the human genome. You might also consider whether this is a function of the government or if the money could be better spent on other projects. Discuss the issue with others.

CHAPTER 21

1 p. 356 The teenager has bacteria in her digestive tract that are resistant to the tetracycline. The bacteria could have a plasmid with genes that make the bacteria resistant to tetracycline. Transfer of the plasmid to the bacteria causing the diarrhea has made the disease-causing bacteria resistant to the tetracycline. Resistance to antibiotics as a result of DNA transfer is a serious problem. Not only can resistance be transferred between bacterial cells of the same species, but it also can be transferred between different species of bacteria.

2 p. 357 The best drugs would act on bacterial structures but not human structures. The cell wall is a good choice because it has peptidoglycans, which are not found in humans. Penicillin is an antibiotic that inhibits the synthesis of peptidoglycans. Consequently, the bacteria produce weak cell walls that rupture, resulting in the death of the bacteria. Ribosomes would also be a good choice because the ribosomes of human and bacterial cells are different from each other. Tetracycline is an antibiotic that kills bacteria by stopping protein synthesis at bacterial ribosomes, but not at human ribosomes. The cell membranes of human and bacterial cells are similar and would not be an ideal choice. Human cells have endoplasmic reticulum but bacterial cells do not, so this would also be a poor choice.

3 p. 358 Antiviral drugs not only inhibit viral nucleic acid synthesis, but also the synthesis of nucleic acids in human cells. As a result human cells no longer function properly.

4 p. 360 The virus could be spread by indirect contact. Ruptured blisters could release the virus, which could be carried by dust particles in the air to the respiratory tract. The virus from the respiratory tract of the infected child also could be carried in dried secretions by the air to the respiratory tract of the noninfected child. The virus also could contaminate bedding, clothing, towels, etc. If the noninfected child touched these objects and then rubbed his nose, the virus could reach the respiratory tract. The virus also could be spread by direct contact; for example, by water droplets released during talking or sneezing, or by directly touching the blisters.

5 p. 361 Surgical asepsis keeps all microorganisms away from an object or person. All objects that come into contact with the patient should be sterile because sterile objects have no living microorganisms that can be transferred to the patient. For example, instruments, gowns, gloves, masks, and dressings could be sterilized by autoclaving, baking, radiation, or ethylene oxide treatment. It is not possible to sterilize the surgical team or patient. However, the surgical team could use scrubbing, soap, and antiseptics (for example, Betadine) to reduce the number of microorganisms on their hands and on the patient's skin. The operating room should be disinfected (for example, phenol derivatives) to kill pathogens on the walls, floors, and other surfaces. The air entering the operating room should be filtered and exposed to ultraviolet light to remove or kill microorganisms in the air.

APPENDIX C

Answers to Chapter Test Questions

CHAPTER 1

Matching

1. organelle
2. tissue
3. organ
4. superior
5. posterior
6. lateral
7. sagittal
8. transverse

Fill-in-the Blank

1. homeostasis
2. physiology
3. Negative-feedback
4. anatomical position
5. distal (or inferior)
6. medial
7. thoracic
8. mediastinum
9. pelvic cavity

CHAPTER 2

Matching

1. matter
2. element
3. electron
4. molecule
5. ion
6. base
7. acid
8. organic
9. monosaccharide
10. fat
11. amino acid
12. enzyme
13. DNA

Fill-in-the Blank

1. chemistry
2. atom
3. nucleus
4. ionic
5. acidic
6. buffer
7. dissociate
8. glycerol and fatty acids
9. lock and key model
10. Denaturation

CHAPTER 3

Matching 1

1. plasma membrane
2. extracellular
3. nucleus
4. rough endoplasmic reticulum
5. smooth endoplasmic reticulum
6. Golgi apparatus
7. lysosome
8. mitochondria
9. cilia
10. microvilli

Matching 2

1. gene
2. transcription
3. mRNA
4. tRNA
5. mitosis
6. interphase
7. chromatin
8. chromosome

Fill-in-the Blank

1. selectively permeable
2. diffusion, osmosis, and filtration
3. membrane channels
4. water
5. filtration
6. active transport
7. Phagocytosis
8. exocytosis

CHAPTER 4

Matching

1. basement membrane
2. free surface
3. simple squamous epithelium
4. simple columnar epithelium
5. simple cuboidal epithelium
6. stratified squamous epithelium
7. dense connective tissue
8. adipose tissue
9. skeletal muscle
10. serous membrane

Fill-in-the Blank

1. simple cuboidal, simple columnar
2. Simple squamous epithelium
3. Stratified squamous epithelium
4. Transitional epithelium
5. exocrine
6. Collagen
7. Areolar or loose connective tissue
8. Cartilage
9. Smooth muscle

CHAPTER 5

Matching

1. striae
2. papillae
3. stratum basale
4. keratin
5. melanin
6. albinism
7. cyanosis
8. hair bulb
9. sebaceous gland
10. nail matrix

Fill-in-the Blank

1. dermis
2. hypodermis
3. lipids
4. melanocytes
5. hair follicle
6. arrector pili
7. watery, organic
8. vitamin D
9. dilate
10. more

CHAPTER 6

Matching 1

1. tendon
2. diaphysis
3. red marrow
4. osteoblast
5. osteoclast
6. axial skeleton
7. pelvic girdle

Matching 2

1. plane joint
2. hinge joint
3. ellipsoid joint
4. pivot joint
5. ellipsoid joint
6. ball-and-socket joint
7. ellipsoid and saddle joint
8. abduction
9. pronation
10. rotation

Fill-in-the Blank

1. spongy bone
2. Haversian canals, canalic-uli
3. fontanels
4. epiphyseal plate
5. callus
6. suture
7. intervertebral disk, costal cartilage, or pubic symphysis
8. articular cartilage, joint capsule, synovial fluid

CHAPTER 7

Matching 1

1. myofibril
2. sarcomere
3. cross bridges
4. isotonic
5. aerobic
6. anaerobic
7. psychological fatigue
8. origin

Matching 2

1. sternocleidomastoid
2. erector spinae
3. trapezius
4. latissimus dorsi
5. triceps brachii

6. anterior forearm muscles
7. gluteus maximus
8. iliopsoas (also part of the quadriceps femoris)
9. quadriceps femoris
10. tibialis anterior

Fill-in-the Blank

1. movement, posture, and heat production
2. sliding filament mechanism
3. recruitment
4. Anaerobic
5. oxygen debt

CHAPTER 8

Matching 1

1. sensory division
2. axon
3. neuroglia
4. neuroglia
5. node of Ranvier
6. synaptic cleft
7. nucleus
8. nerve

Matching 2

1. medulla oblongata
2. pons
3. midbrain
4. thalamus
5. hypothalamus
6. Broca's area
7. basal ganglia
8. cerebellum

Matching 3

1. dura mater
2. pia mater
3. choroid plexus
4. arachnoid granulation
5. cranial nerves
6. sympathetic division
7. sympathetic division

Fill-in-the Blank

1. myelinated
2. medulla oblongata, pons, and midbrain
3. occipital
4. temporal
5. frontal
6. dorsal, ventral
7. sensory receptor, sensory neuron, association neuron, motor neuron, effector organ
8. ventricles, subarachnoid space

CHAPTER 9

Matching 1

1. conjunctiva
2. lacrimal gland
3. sclera
4. cornea
5. iris
6. ciliary body
7. cones
8. anterior compartment

Matching 2

1. tympanic membrane
2. auditory ossicles
3. auditory tube
4. organ of Corti
5. round window
6. crista ampullaris

Fill-in-the Blank

1. olfactory nerves, olfactory tract
2. sour, salty, bitter, and sweet
3. optic nerves, visual areas
4. malleus, incus, and stapes
5. maculae

CHAPTER 10

Matching

1. endocrine glands
2. hypothalamus
3. anterior pituitary
4. target tissue
5. adrenocorticotropic hormone (ACTH)
6. thyroid-stimulating hormone (TSH)
7. antidiuretic hormone (ADH)
8. parafollicular cells
9. parathyroid hormone
10. epinephrine
11. aldosterone
12. insulin

Fill-in-the Blank

1. hypothalamic-pituitary portal system
2. receptor
3. Growth hormone
4. Adrenocorticotropic hormone
5. anterior pituitary
6. iodine
7. Parathyroid hormone
8. Cortisol
9. glucagon
10. Prostaglandin

CHAPTER 11

Matching

1. plasma
2. red blood cell
3. platelet
4. hemoglobin
5. erythropoietin
6. bilirubin
7. jaundice
8. iron
9. vitamin B_{12}
10. vitamin K

Fill-in-the Blank

1. neutrophils and monocytes
2. lymphocytes
3. prothrombin, thrombin
4. fibrin, thrombin
5. anticoagulants
6. plasmin
7. antigens, antibodies
8. agglutination
9. B, A

CHAPTER 12

Matching

1. epicardium
2. myocardium
3. bicuspid valve
4. intercalated disk
5. SA node
6. P wave
7. cardiac cycle
8. pulmonary trunk
9. stethoscope
10. murmur
11. Starling's law of the heart

Fill-in-the Blank

1. right atrium
2. tricuspid valve
3. mitochondria
4. aortic semilunar, aorta
5. left coronary, right coronary
6. infarct
7. cardiac output
8. baroreceptor

CHAPTER 13

Matching

1. artery
2. tunica intima
3. venule
4. artherosclerosis
5. aortic arch
6. internal carotid arteries
7. radial artery
8. common iliac arteries
9. superior vena cava
10. systolic pressure
11. baroreceptor reflex
12. renin-angiotensin-aldosterone mechanism

Fill-in-the Blank

1. hepatic portal system
2. pulmonary
3. muscular
4. common carotid, subclavian
5. celiac, superior mesenteric, inferior mesenteric
6. hypertension
7. precapillary sphincters

CHAPTER 14

Matching 1

1. lymph
2. chyle
3. lymph vessel
4. lymph node
5. germinal center
6. spleen
7. thymus

Matching 2

1. complement
2. interferon
3. antigen
4. B cell
5. plasma cell
6. cytotoxic T cell
7. active artificial immunity
8. passive artificial immunity

Fill-in-the Blank

1. fluid, fats, defense
2. right lymphatic duct
3. contraction of skeletal muscle, contraction of smooth muscle in the lymph vessel wall, and pressure changes in the thorax during respiration
4. specific immunity
5. neutrophils and macrophages

6. basophils and mast cells, eosinophils
7. inflammatory response

CHAPTER 15

Matching

1. conchae
2. pharynx
3. larynx
4. trachea
5. visceral pleura
6. tidal volume
7. respiratory center
8. respiratory center
9. carotid and aortic bodies

Fill-in-the Blank

1. upper respiratory tract, lower respiratory tract
2. trachea, bronchi, bronchioles
3. diaphragm
4. decrease
5. inspiratory reserve, tidal volume, and expiratory reserve
6. alveolar wall, capillary wall
7. hemoglobin, bicarbonate ion
8. increase, decrease, decrease
9. motor cortex, proprioceptors

CHAPTER 16

Matching 1

1. mucosa
2. muscularis
3. peristalsis
4. enamel
5. periodontal ligaments
6. rugae
7. chyme
8. hepatic portal vein
9. gallbladder
10. retroperitoneal

Matching 2

1. amylase
2. intestinal enzymes
3. amylase
4. lipase
5. bile salts
6. pepsin
7. trypsin
8. intestinal enzymes

Fill-in-the Blank

1. incisors and canines
2. 32, deciduous
3. parotid, submandibular, sublingual
4. pyloric sphincter
5. circular folds, villi, microvilli
6. ileocecal sphincter, ileocecal valve
7. common hepatic duct, common bile duct
8. parasympathetic, local, hormones

CHAPTER 17

Matching 1

1. essential
2. Calorie
3. cellulose
4. glycogen
5. sucrose
6. saturated fat
7. fat-soluble vitamins

Matching 2

1. basal metabolic rate
2. muscular activity
3. assimilation of food
4. convection
5. evaporation
6. vasodilation
7. hypothalamus

Fill-in-the Blank

1. adenosine diphosphate (ADP)
2. anaerobic respiration
3. lactic acid
4. glucose
5. 2, 38
6. carbon dioxide
7. water
8. fats, proteins

CHAPTER 18

Matching

1. cortex
2. nephron
3. glomerulus
4. afferent arteriole
5. efferent arteriole
6. peritubular capillary
7. ureter
8. antidiuretic hormone
9. aldosterone
10. atrial natriuretic hormone
11. extracellular fluid
12. acidosis
13. urethra

Fill-in-the Blank

1. kidneys, ureters, urinary bladder
2. hilum
3. renal pyramids
4. filtration, reabsorption, secretion
5. filtrate
6. Reabsorption
7. Antidiuretic hormone (ADH)
8. micturition
9. Buffers

CHAPTER 19

Matching

1. spermatogonia
2. ductus deferens
3. circumcision
4. emission
5. ejaculation
6. testosterone
7. estrogen
8. progesterone
9. endometrium
10. vestibule
11. menses
12. menopause
13. prolactin

Fill-in-the Blank

1. Sertoli cells
2. interstitial cells
3. seminal vesicles
4. prostate gland
5. corpora cavernosa
6. FSH
7. uterine tube
8. secretory
9. 14
10. LH

CHAPTER 20

Matching 1

1. zygote
2. blastocyst
3. inner cell mass
4. amniotic cavity
5. germ layers
6. embryo
7. uterine tube
8. implantation
9. chorion
10. parturition

Matching 2

1. karyotype
2. sex chromosomes
3. gene
4. homozygous
5. dominant
6. phenotype
7. mutation
8. oncogene
9. carcinogen

Fill-in-the Blank

1. somatic, gametes or sex cells
2. chromosomes
3. meiosis
4. XY, XX
5. gene
6. sex-linked traits
7. genetic susceptibility or genetic predisposition
8. pedigree

CHAPTER 21

Matching 1

1. sign
2. prognosis
3. incidence
4. endemic
5. bacillus

Matching 2

1. reservoir
2. carrier
3. vector
4. portal
5. opportunistic pathogen
6. sterilization
7. antisepsis
8. asepsis

Fill-in-the Blank

1. capsule, pili
2. peptidoglycan
3. lipopolysaccharide layer, negative, endotoxin
4. nuclear area, plasmids
5. exotoxins
6. endospores
7. growth
8. capsid, envelope
9. attachment, entry, uncoating, production, and release

Glossary

A

abdominal cavity Space bounded by the diaphragm, the abdominal wall, and the pelvis.

abdominopelvic cavity The abdominal and pelvic cavities considered together.

abduction [L. *abductio,* take away] Movement away from the midline.

accommodation Increase in thickness and convexity (roundness) of the lens to focus an object closer than twenty feet onto the retina.

acetylcholine (as-ĕ-til-ko′lēn) Neurotransmitter substance released from motor neurons, all parasympathetic neurons, some sympathetic neurons, and some CNS neurons.

acetylcholinesterase (as′ĕ-til-ko-lin-es′ter-ās) Enzyme found in the synaptic cleft that causes the breakdown of acetylcholine.

acid Any substance that is a proton donor; any substance that releases hydrogen ions.

acidic solution Solution with more hydrogen ions than hydroxide ions; has a pH of less than 7.

acidosis (as-ĭ-do′sis) A decrease in blood pH below 7.35.

acromegaly (ak′ro-meg′al-e) [Gr. *acro* + *megas,* large] Disorder marked by progressive enlargement of the bones of the head, face, hands, feet, and thorax as a result of excessive secretion of growth hormone by the anterior pituitary.

actin myofilament (ak′tin) Thin myofilament within the sarcomere.

action potential Brief reversal of the charge difference across a cell membrane; depolarization and repolarization.

active artificial immunity Deliberate exposure to an antigen that results in a specific immune system response.

active natural immunity Natural exposure to an antigen that results in a specific immune system response.

active transport Movement of materials across the plasma membrane that requires a carrier molecule and ATP; can move substances from an area of lower concentration to an area of higher concentration.

adduction [L. *adductus,* to bring forward] Movement toward the midline.

adenosine triphosphate Energy-storing molecule in cells that provides the energy necessary for many chemical reactions.

adipose tissue (ad′ĭ-pōs) Fat.

adrenal cortex The outer portion of the adrenal gland, which secretes the following hormones: glucocorticoids, mainly cortisol; mineralocorticoids, mainly aldosterone; and androgens.

adrenal gland (a-dre′nal) [L. *ad,* to + *ren,* kidney] One of two endocrine glands located on the superior pole of each kidney; secretes the hormones epinephrine, norepinephrine, aldosterone, cortisol, and androgens.

adrenal medulla The inner portion of the adrenal gland, which secretes mainly epinephrine but also small amounts of norepinephrine.

adrenaline (ă-dren′ă-lin) Synonym for epinephrine.

adrenocorticotropic hormone (ACTH) (ă-dre′no-kor′tĭ-ko-tro′pik) Hormone of the anterior pituitary that stimulates the adrenal cortex to secrete cortisol.

aerobic respiration (ăr-o′bik) Breakdown of glucose in the presence of oxygen to produce 6 carbon dioxide, 6 water, and 38 ATP molecules.

afferent arteriole A small artery in the cortex of the kidney that supplies blood to the glomerulus.

agglutination (ă-glu′tĭ-na′shun) [L. *ad,* to + *gluten,* glue] The process by which cells stick together to form clumps.

albinism (al′bĭ-nizm) [L. *albus,* white] Inability to produce melanin, resulting in unpigmented skin and hair; a recessive genetic trait.

aldosterone (al-dos′ter-ōn) Hormone produced by the adrenal cortex that acts on the kidneys causing sodium reabsorption from the urine and potassium and hydrogen ion secretion.

alkaline solution (al′kah-līn) See basic solution.

alkalosis (al′kah-lo′sis) An increase in blood pH above 7.45.

all-or-none response When a stimulus is applied to a cell, an action potential is either produced or not.

alveolus, pl. **alveoli** (al′ve-o′lus, al′ve-o′li) [L. cavity] The terminal saclike endings of the respiratory system, where gas exchange occurs.

amino acid (ah-me′no) Class of organic acids containing an amine group (NH_2) that makeup the building blocks of proteins.

amniocentesis (am′ne-o-sen-te′sis) Removal of fluid and cells from the contents of the amniotic sac; used to determine the fetus′ genotype for traits that could cause genetic abnormalities.

amniotic cavity (am′ne-ot′ik) Space formed within the inner cell mass of the blastocyst; provides a space for embryonic and fetal growth; fluid within the cavity forms a protective cushion around the embryo or fetus.

amylase (am′ĭ-lās) One of a group of starch-splitting enzymes that cleave starch, glycogen, and related polysaccharides; produced by the salivary glands and the pancreas.

anabolism (ah-nab′o-lizm) The chemical reactions of the body in which smaller molecules join together to form larger molecules; requires the input of energy.

anaerobic respiration (an′ăr-o′bik) Breakdown of glucose in the absence of oxygen to produce two lactic acid and two ATP molecules.

anal canal Terminal portion of the digestive tract.

anaphase (an′ă-fāz) Time during cell division when duplicated chromosomes separate from each other and move to opposite poles of the cell.

anatomical position Position in which a person is standing erect with the feet facing forward, arms hanging to the sides, and the palms of the hands facing forward.

anatomy (ă-nat´o-me) [Gr. *ana,* up + *tome,* a cutting] Scientific discipline that investigates the structure of the body.

androgen (an´dro-jen) General term for male sex hormones, for example testosterone; secreted by the testes and, in small amounts, from the adrenal cortex.

anemia (ă-ne´me-ah) [Gr. *an,* without + *haima,* blood] Any condition that results in lower than normal hemoglobin in the blood or lower than normal number of erythrocytes.

angioplasty (an´je-o-plas´te) Technique used to dilate the coronary arteries by threading a small balloon-like device into a partially blocked coronary artery, then inflating the balloon to enlarge the diameter of the vessel.

angiotensin I (an´je-o-tin-sin) A peptide derived when renin acts on angiotensinogen; angiotensin I is subsequently converted to angiotensin II.

angiotensin II (an´je-o-tin-sin) A peptide derived when a converting enzyme acts on angiotensin I. Angiotensin II is a potent vasoconstrictor, and it stimulates the secretion of aldosterone from the adrenal cortex.

angiotensinogen (an´je-o-tin´sin-o-jen) A protein produced by the liver; renin acts on angiotensinogen to produce angiotensin.

antagonist (an-tag´o-nist) A muscle acting in opposition to another muscle.

anterior [L. to go before] That which goes first. In humans, toward the belly or front; synonymous with ventral.

anterior pituitary Portion of the pituitary that secretes TSH, ACTH, prolactin, LH, and FSH.

antibody Protein found in blood plasma that binds to a specific, usually foreign or toxic, substance.

antibody-mediated immunity Immunity resulting from B cells and the production of antibodies.

anticoagulant (an´tĭ-ko-ag´u-lant) Chemical that prevents coagulation or blood clotting.

antidiuretic hormone (ADH) (an´tĭ-di-u-ret´ik) Hormone secreted from the posterior pituitary that acts on the kidney to reduce the output of urine; also called vasopressin.

antigen (an´tĭ-jen) Any substance that induces a state of sensitivity and/or resistance to infection or toxic substances after a latent period; substance that stimulates the specific immune system. Self antigens are produced by the body and foreign antigens are introduced into the body.

antisepsis (an´tĭ-sep´sis) Destruction of pathogens on the skin or on tissues.

antiserum Serum with antibodies that provide passive artificial immunity.

anus Lower opening of the digestive tract through which fecal matter is extruded.

aorta (a-or´tah) [Gr. *aorte,* from *aeiro,* to lift up] A large elastic artery that is the main trunk of the systemic arterial system; carries blood from the left ventricle of the heart and passes through the thorax and abdomen.

aortic body Small organ near the aorta that detects changes in blood oxygen, carbon dioxide, and pH.

apex (a´peks) [L. tip] The extremity of a conical or pyramidal structure; the apex of the heart is the rounded tip directed anteriorly and slightly inferiorly.

aphasia (ă-fa´ze-ah) [Gr. *a,* unable + *phasis,* speech] Impaired or absent ability to communicate by speech or writing or an inability to comprehend spoken or written language; results from damage or disease of the brain.

apocrine (ăp´o-krin) See sweat gland.

appendicular skeleton (ap´pen-dik´u-lar) [L. *appendo,* to hang something on] The limbs and the bones (girdles) that attach the limbs to the trunk.

appendix (ă-pen´diks) An appendage, specifically the blind-ending sac that attaches to the cecum.

aqueous humor Watery, clear solution that fills the anterior compartment of the eye.

arachnoid (ar-ak´noyd) [Gr. *arachne,* spider, cobweb] Thin, cobweb-appearing meningeal layer surrounding the brain; the middle of three such layers.

arachnoid granulation Extension from the arachnoid layer into the dural sinus; site of return of cerebrospinal fluid to blood.

arrector pili (ah-rek´tor pi´li) [L. that which raises hair] Smooth muscle attached to the hair follicle and dermis that raises the hair when it contracts.

arteriole (ar-te´re-ōl) The smallest diameter artery; consists of three thin tunics with only a few layers of smooth muscle in the tunica media; blood flows from arterioles through capillaries.

arteriosclerosis (ar-te´re-o-sklĕ-ro´sis) [L. *arterio-* + Gr. *sklerosis,* hardness] Calcified lesions in the arteries that narrows the lumen, or passage, and makes the walls of the arteries less elastic; commonly called hardening of the arteries.

artery (ar´ter-e) Blood vessel that carries blood away from the heart.

articular cartilage Hyaline cartilage covering the ends of bones within a synovial joint.

articulation A place where two bones come together; a joint.

asepsis (ă-sep´sis) To keep microorganisms away from an object or person.

association neuron Neuron connecting one neuron to another neuron.

atherosclerosis (ath´er-o-sklĕ-ro´sis) Lipid deposits in the tunica intima of large and medium-sized arteries.

atom (at´om) [Gr. *atomos,* indivisible, uncut] Smallest particle of an element that retains the properties of that element; composed of neutrons, protons, and electrons.

atomic number The number of protons in an atom; each element has a unique atomic number.

ATP See adenosine triphosphate.

atrial natriuretic hormone (na´tre-u-ret´ik) Hormone released from cells in the atrial wall of the heart in response to increased blood volume; acts to lower blood pressure by increasing the rate of urine production.

atrioventricular bundle (a´tre-o-ven´trik´u-lar) A bundle of modified cardiac muscle fibers that projects from the AV node through the interventricular septum; conducts action potentials from the AV node rapidly through the interventricular septum; also called the bundle of His.

atrioventricular (AV) node Small collection of specialized cardiac muscle fibers located in the lower portion of the right atrium and which gives rise to the atrioventricular bundle; briefly delays the conduction of action potentials to the atrioventricular bundle.

atrium pl. **atria** (a´tre-um, a´tre-ah) [L. entrance chamber] One of the two upper chambers of the heart; collects blood from the body or lungs during ventricular contraction and pumps blood into the ventricles during ventricular relaxation.

atrophy (at´ro-fe) Wasting of tissues resulting from decreased cell size or number; decrease in the size of a muscle.

auditory tube Air-filled passage between the middle ear and pharynx;

functions to equalize pressure between the middle ear and the external environment.

auricle (aw′rĭ-kl) [L. *auris,* ear] Part of the external ear that protrudes from the side of the head; also called the pinna. Small pouch projecting from each atrium of the heart.

autoimmune disease Disorder resulting from a specific immune system reaction against self antigens.

autonomic nervous system Composed of nerve fibers that send action potentials from the central nervous system to smooth muscle, cardiac muscle, and glands.

autosomal chromosome (aw′to-so′mal) Any of the 22 paired chromosomes other than the sex chromosomes.

axial skeleton Skull, vertebral column, and thoracic cage.

axon (ak′son) [Gr. axis] Main central process of a neuron that normally conducts nerve impulses away from the neuron cell body.

B

B cell Lymphocyte responsible for antibody-mediated immunity.

bacillus, pl. **bacilli** (bă-sil′us, bă-sil′i) [L. rod] Any rod-shaped bacterium.

ball-and-socket joint A ball (head) of one bone inserted into a socket in another bone; capable of a wide range of movements; e.g. the hip joint.

baroreceptor (bar′o-re-sep′tor) Sensory nerve endings in the walls of the aorta and internal carotid arteries; sensitive to stretching of the wall caused by increased blood pressure.

baroreceptor reflex A reflex in which baroreceptors detect changes in blood pressure and produce changes in heart rate, force of heart contraction, and blood vessel diameter that return blood pressure toward normal levels.

basal ganglia Nuclei at the base of the cerebrum involved in controlling motor functions.

basal metabolic rate Metabolic rate of an awake but restful person who has not eaten for 12 hours; expressed as heat produced per unit of surface area over a specified time period.

base Any substance that is a proton acceptor; any substance that binds to hydrogen ions.

basement membrane The structure that attaches most epithelia to underlying tissue; consists of carbohydrates and proteins secreted by the epithelia and the underlying connective tissue.

basic solution Solution with less hydrogen ions than hydroxide ions; has a pH greater than 7; an alkaline solution.

basophil (ba′so-fil) [Gr. *basis,* base + *phileo,* to love] White blood cell with granules that stain purple with basic dyes; promotes inflammation.

benign (be-nīn′) A type of tumor that increases in size, but does not spread.

bicuspid valve (bi-kus′ped) Valve consisting of two cusps of tissue; located between the left atrium and left ventricle of the heart, also called the mitral valve.

bile Fluid secreted from the liver, stored in the gallbladder, and released into the duodenum; consists of bile salts, bile pigments (e.g., bilirubin), bicarbonate ions, cholesterol, fats, hormones, and more.

bile salt Organic salt secreted by the liver that functions as an emulsification agent.

blastocyst (blas′to-sist) [Gr. *blastos,* germ + *kystis,* bladder] Early stage of development consisting of a hollow ball of cells with an inner cell mass; stage at which implantation in the uterus occurs.

blood group A category of red blood cells based on the type of antigen on the surface of the red blood cell; the ABO blood group is involved with transfusion reactions.

blood pressure [L. *pressus,* to press] The force blood exerts against the blood vessel walls; expressed relative to atmospheric pressure and reported in the form of mm Hg pressure.

blood-brain barrier Permeability barrier controlling the passage of substances from the blood into brain tissue; primarily due to the structure of the capillary walls; neuroglia cells influence the structure and function of the capillaries.

bolus (bo′lus) Food that has been chewed, mixed with saliva, and is swallowed.

Bowman's capsule The enlarged end of the nephron; Bowman's capsule and the glomerulus make up the renal corpuscle.

brain case The part of the skull that encloses the brain.

brainstem Portion of the brain consisting of the medulla oblongata, pons, and midbrain; connects the superior part of the brain to the spinal cord and the cerebellum.

breast The anterior surface of the thorax; hemispherical projections of variable size located on each side of the chest, consisting of skin, fat, connective tissue, and mammary glands; rudimentary in males, organs of milk production in females.

Broca's area Portion of the motor area of the cerebrum that controls the skeletal muscles necessary for speech; found in the left cerebral hemisphere of most individuals.

bronchiole (brong′ke-ol) One of the finer subdivisions of the bronchial tubes, less than 1 mm in diameter, and having no cartilage in its wall, but relatively more smooth muscle and elastic fibers.

bronchus, pl. **bronchi** (brong′kus, brong′ki) [Gr. *bronchos,* windpipe] Any one of the tubes conducting air from the trachea to the bronchioles. A bronchus has cartilage rings or plates in its wall, and it varies in diameter from approximately 1 cm in the primary bronchi to approximately 1 mm in the smallest (tertiary) bronchi.

buffer A chemical that resists changes in pH when either an acid or base is added to a solution containing the buffer.

bulbourethral gland (bul′bo-u-re′thral) One of two small glands that produce a mucous secretion; located near the base of the penis, its secretions are released into the penile urethra.

bundle of His See atrioventricular bundle.

bursa, pl. **bursae** (bur′sah, bur′se) [L. purse] Closed sac or pocket containing synovial fluid, usually found where friction occurs, e.g., between bone and muscle or between bone and skin.

C

calcitonin (kal-sĭ-to′nin) Hormone released from cells of the thyroid gland that acts on tissues, especially bone, to cause a decrease in blood levels of calcium ions.

callus (kal′us) [L. hard skin] Thickening of the stratum corneum of skin in response to friction. The hard bonelike substance that develops at the site of a broken bone.

Calorie The amount of heat (energy) required to raise the temperature of 1000 grams of water from 14° to 15°

C; used as a measure of the amount of energy in foods.

calyx, pl. **calyces** (ka´liks, kal´ĭ-sēz) The small containers into which urine flows as it leaves the collecting ducts at the tip of the renal pyramids; the calyces come together to form the renal pelvis.

canaliculus, pl. **canaliculi** (kan-ă-lik´u-lus, kan-ă-lik´u-li) [L. *canalis,* canal] Little canal, e.g., in bone containing osteocyte cell processes.

capillary (kap´ĭ-ler-e) Tiny blood vessel consisting of only simple squamous epithelium and a basement membrane; major site for the exchange of substances between the blood and tissues.

capsid (kap´sid) A protein coat that surrounds and protects the nucleic acid core of a virus; can also function to attach the virus to a cell.

capsule A layer of slime consisting of polysaccharides, polypeptides, or both that surrounds some bacteria; functions to attach bacteria to surfaces.

carbohydrate Simple sugar (monosaccharide) or the molecules made by chemically binding simple sugars together, e.g., starch.

carbonic anhydrase Enzyme inside red blood cells that increases the chemical reaction between carbon dioxide and water.

carcinogen (kar-sin´o-jen) Any cancer-producing substance.

carcinoma (kar-sĭ-no´mah) [Gr. *karkinoma,* cancer] A malignant tumor derived from epithelial tissue.

cardiac cycle Complete round of cardiac diastole and systole.

cardiac muscle Muscle of the heart responsible for pumping blood; consists of striated, cylindrical, branching cells connected by intercalated disks, usually with one centrally-located nucleus.

cardiac output The volume of blood pumped by each ventricle of the heart per minute; about 5 liters/min for the heart of a healthy adult at rest.

cardioregulatory center A specialized area within the medulla oblongata of the brain that receives sensory input and functions to control parasympathetic and sympathetic innervation to the heart.

carotid body (kă-rot´id) Small organ near the carotid sinuses that detects changes in blood oxygen, carbon dioxide, and pH.

carrier A heterozygous person with an abnormal recessive gene; appears normal because of the expression of the dominant normal gene. An infected person with no symptoms who can transmit the infectious agent to others.

carrier molecule Protein that binds to substances and moves them across the plasma membrane.

cartilage (kar´tĭ-lij) [L. cartilage, gristle] Firm, smooth, resilient, nonvascular connective tissue.

cartilaginous joint Bones connected by cartilage; examples include intervertebral disks, the costal cartilages, and the pubic symphysis.

catabolism (kah-tab´o-lizm) Chemical reactions of the body in which larger molecules are broken apart to form smaller molecules; releases energy.

cecum (se´kum) Blind ending sac that forms the beginning of the large intestine; joined to the small intestine.

cell [L. *cella,* chamber] Basic living unit of all plants and animals.

cell membrane See plasma membrane

cell metabolism The chemical processes that take place within cells.

cell-mediated immunity Immunity resulting from the actions of T cell.

central nervous system (CNS) Major subdivision of the nervous system consisting of the brain and spinal cord.

centriole (sen´tre-ōl) Usually paired organelle shaped like a small cylinder; during cell division each centriole moves to the opposite pole (end) of a cell.

cerebellum (sĕr´e-bel´um) [L. little brain] A part of the brain attached to the brainstem and important in maintaining muscle tone, balance, and coordination of movements.

cerebrum (sĕr´e-brum) [L. brain] The largest part of the brain, consisting of two hemispheres and including the cortex, nerve tracts, and basal ganglia.

cerumen (sĕ-roo´men) A specific type of sebum produced in the external auditory meatus; earwax.

cervical canal Passageway extending from the uterine cavity to the opening of the uterus into the vagina.

cervix (ser´viks) [L. neck] Lower part of the uterus extending into the vagina; contains the cervical canal.

chemical bond The transfer or sharing of one or more electrons between atoms.

chemical formula Symbols of atoms used to indicate the composition of a molecule.

chemical reaction Process by which atoms or molecules interact to form or break chemical bonds.

chemistry [Gr. *chemeia,* alchemy] Scientific study of the composition of substances and the reactions they undergo.

chemoreceptor Sensory cell that is stimulated by a change in the concentration of chemicals to produce action potentials. Examples include taste receptor, olfactory receptor, and receptors in blood vessels or the medulla oblongata that respond to changes in oxygen, carbon dioxide, or pH.

chemoreceptor reflex Process in which chemoreceptors detect changes in oxygen levels, carbon dioxide levels, and pH in the blood and produce changes in heart rate, force of heart contraction, and blood vessel diameter that return these values toward their normal levels.

chordae tendineae (kor´de ten´dĭ-ne-e) [L. cord, heart strings] Tendinous strands running from the papillary muscles to the free margin of the cusps that make up the tricuspid and bicuspid valves; prevent the cusps of these valves from extending up into the atria during ventricular contraction.

chorion (ko´re-on) Membrane forming part of the placenta; derived from the wall of the blastocyst.

chorionic villus (ko´re-on´ik vil´us) Fingerlike projection, containing embryonic or fetal blood vessels, that extends into spaces containing maternal blood; site of nutrient and gas exchange between the embryo or fetus and the mother, but no mixing of embryonic or fetal blood with the mother´s blood occurs.

chorionic villus sampling Removal of cells from the fetal side of the placenta; used to determine the fetus´ genotype for traits that could cause genetic abnormalities.

choroid (ko´royd) Middle layer of the eye that covers the sclera within the posterior compartment of the eye; black in color in order to absorb light.

choroid plexus (plek´sus) Vascular membrane lining the ventricles that produces cerebrospinal fluid.

chromatin (kro´mah-tin) Unraveled DNA and associated proteins that is dispersed throughout the nucleus during interphase.

chromosome (kro´mo-sōm) [Gr. *chroma,* color + Gr. *soma,* body] One of the bodies (normally 46 in humans) in the cell nucleus that carry the cell´s genetic information; becomes visible during prophase of cell division.

chyle (kīl) [Gr. *chylos,* juice] Milky colored lymph with a high fat content.

chyme (kīm) [Gr. *chymos,* juice] Semifluid mass of partly digested food passed from the stomach into the duodenum.

ciliary body (sil′e-ăr-e) Structure continuous with the choroid layer of the eye at its anterior margin that contains smooth muscle cells; attached to the lens by suspensory ligaments; regulates thickness of the lens.

cilium, pl. **cilia** (sil′e-um, sil′e-ah) Extension of the cell surface that moves; functions to move materials along the surface of the cell.

circular fold One of many folds of the mucosa and submucosa of the small intestine that increase the surface area of the intestinal lining.

circumcision [L. *circumcido,* to cut around] Operation in which part or all of the prepuce is removed in males; usually at the time of birth.

circumduction [L. around + *ductus,* to draw] Movement in a circular motion.

cleft palate An opening in the hard palate resulting from failure of the bones of the hard palate to join together during development.

clitoris (klit′o-ris, kli-to-ris) Small cylindrical, erectile body, situated at the most anterior portion of the vestibule and projecting beneath the prepuce.

clot A network of protein fibers, called fibrin, in which blood cells, platelets, and fluid becomes trapped; functions to stop bleeding.

clotting factor One of many proteins found in the blood in an inactivate state; activated in a series of chemical reactions that result in the formation of a blood clot.

coagulation (ko-ag′u-la-shun) Process of changing from a liquid to a solid, especially of blood; blood clotting.

coccus, pl. **cocci** (kok′us, kok′si) [Gr. kokos, berry] A round or oval bacterium.

coccyx (kok′siks) The four or five fused coccygeal vertebrae on the inferior end of the vertebral column.

cochlea (kok′le-ah) [Gr. *kochlias,* snail] The portion of the inner ear involved in hearing; shaped like a snail shell.

collagen (kol′lă-jen) [Gr. *koila,* glue + gen, producing] Ropelike protein of the extracellular matrix. Provides support and resists stretching.

collecting duct Straight tubule that extends from the cortex of the kidney to the tip of the renal pyramid; filtrate from the distal convoluted tubules enters the collecting duct and is carried to the calyces.

colon (ko′lon) Division of the large intestine that extends from the cecum to the rectum; consists of the ascending, transverse, descending, and sigmoid colons.

common bile duct Duct formed by the union of the cystic duct and the common hepatic duct; it empties into the duodenum of the small intestine.

common hepatic duct Duct formed by the joining of the right and left hepatic ducts; joins the cystic duct to form the common bile duct.

compact bone Bone that is more dense and has fewer spaces than spongy bone; consists of layers of matrix surrounding haversian canals.

complement Group of plasma proteins that stimulates phagocytosis, inflammation, and lysis of cells.

complex carbohydrate Large polysaccharides such as starch, glycogen, and cellulose.

compound Two or more different types of atoms joined by a chemical bond. See molecule.

concave A structure with a surface that is depressed inward.

conduction Transfer of energy such as heat from one point to another without evident movement in the conducting body; heat exchange between objects that are in direct contact with each other.

condyle (kon′dīl) [Gr. *kondylos,* knuckle] Rounded articulating surface of a joint.

condyloid joint See ellipsoid joint.

cone Photoreceptor cell in the retina with cone-shaped processes; functions in visual acuity, color vision, and vision in bright light.

congenital A characteristic that is present at birth; cause is unknown, genetic disorder, or damage from disease, chemicals, etc.

conjunctiva (kon-junk-ti′vah) Mucous membrane covering the anterior surface of the eye (except for the cornea) and the inner lining of the eyelids.

connective tissue One of the four major tissue types; typically consists of cells surrounded by large amounts of extracellular material; holds other tissues together and provides a supporting framework for the body.

convection Transfer of heat in liquids or gases by movement of the heated particles.

convex A structure with a surface that bulges outward.

corn [L. *cornu,* horn] Thickening of the stratum corneum of the skin over a bony projection in response to friction or pressure.

cornea (kor′ne-ah) Transparent, anterior portion of the eye; refracts light entering the eye.

coronal section See frontal plane.

coronary artery (kor′o-năr-e) Artery that carries blood to the muscle of the heart; arises from the base of the aorta.

coronary bypass surgery Surgery in which a vein from some other part of the body is grafted to a coronary artery in such a way as to bypass a blocked coronary artery.

coronary sinus A short trunk that receives most of the veins of the cardiac muscle and empties into the right atrium.

coronary thrombosis Clot formation within a coronary artery.

corpora cavernosa (kor′por-ah kav′er-no′sah) The two parallel columns of erectile tissue forming the dorsal part of the body of the penis.

corpus (kor′pus) [L. body] Any body or mass; the main part of an organ.

corpus callosum (kah-lo′sum) Largest nerve tract connecting the left and right cerebral hemispheres.

corpus luteum (kor′pus lu′te-um) Yellow endocrine body formed in the ovary at the site of a ruptured follicle immediately after ovulation; secretes progesterone and smaller amounts of estrogen.

corpus spongiosum (kor′pus spun′je-o′sum) Median column of erectile tissue located between and ventral to the corpora cavernosa in the penis; it terminates as the glans penis; the urethra runs through it.

cortex The outer portion of an organ such as the kidney or adrenal gland.

cortisol (kor′tĭ-sol) Steroid hormone released by the adrenal cortex; increases blood glucose and inhibits inflammation; it is a glucocorticoid.

covalent bond Chemical bond that is formed when two atoms share one or more pairs of electrons.

coxa, pl. **coxae** (kok′sah, kok′se) Hip bone formed by the fusion of the ilium, ischium, and pubis.

cranial nerve Nerve that originates from a nucleus within the brain; there are 12 pairs of cranial nerves.

crenation (kre-na´shun) [L. *crena,* notched] Denoting the outline of a shrunken cell, i.e., a cell that has lost water.

cretinism (kre´tĕ-nizm) A condition resulting from reduced thyroid hormone secretion during development or infancy; results in short stature, abnormally formed body, and mental retardation.

crista ampullaris (kris´tah am-pul-lar´is) Structure at the base of the semicircular canals consisting of hair cells with hairs embedded in a gelatinous mass; functions to detect movements of the head (kinetic equilibrium).

cross bridge The connection between actin and myosin myofilaments; movement of cross bridges, powered by energy from ATP, is responsible for muscular contraction.

cutaneous membrane The skin.

cuticle (ku´ti-kl) [L. *cutis,* skin] Outer thin layer, usually horny; the growth of the stratum corneum onto the nail.

cyanosis (si-ă-no´sis) [Gr. dark blue color] Blue coloration of the skin and mucous membranes caused by insufficient oxygenation of blood.

cystic duct (sis´tik) Duct from the gallbladder that joins the common hepatic duct to form the common bile duct.

cytoplasm (si´to-plazm) Cellular material surrounding the nucleus.

cytotoxic T cell Type of lymphocyte that responds to an antigen by releasing chemicals that promote inflammation and phagocytosis; also lyses cells; responsible for cell-mediated immunity.

D

deciduous tooth (de-sid´u-us) One of the 20 teeth belonging to the first set of teeth that are replaced by the permanent teeth; primary tooth.

deep [O.E. *deop,* deep] Away from the surface, internal.

defecation (def-ĕ-ka´shun) [L. *defaeco,* to purify] Discharge of feces from the rectum.

defecation reflex Combination of a local reflex and parasympathetic reflex initiated by distention of the rectum and resulting in movement of feces out of the lower colon.

denaturation A change in molecular shape, resulting from heating or changes in pH, that makes molecules such as enzymes nonfunctional.

dendrite (den´drīt) [Gr. tree] Short, tree-like cell process of a neuron; usually receives stimuli.

deoxyribonucleic acid (de-ox´sĭ-ri´bo-nu-kle´ik) Type of nucleic acid containing the sugar deoxyribose; the genetic material of cells, which directs the production of proteins.

depolarization Change in the electrical charge difference across the cell membrane of a resting cell; typically results in a reversal of the charge difference, with the inside of the cell becoming positive relative to the outside.

dermis (der´mis) [Gr. *derma,* skin] Dense connective tissue that forms the deep layer of the skin; responsible for the structural strength of the skin.

diagnosis Determining the nature of a disease.

diaphragm (di´a-fram) Muscular partition between the abdominal and thoracic cavities; its movement produces most of the changes in thoracic volume responsible for respiration.

diaphysis pl. **diaphyses** (di-af´ĭ-sis, di-af´ĭ-sez) [Gr. growing between] Shaft of a long bone.

diastole (di-as´to-le) [Gr. *diastole,* dilation] Relaxation of the heart chambers during which they fill with blood; usually refers to ventricular relaxation.

diastolic pressure (di-ă-stol´-ik) [Gr. *diastole,* dilation] The lowest pressure in large arteries; it occurs during diastole of the heart.

diencephalon (di-en-sef´ă-lon) [Gr. *dia,* through + *enkephalos,* brain] Central portion of the brain consisting of the thalamus and hypothalamus; covered by the cerebrum.

differentiation The process by which cells develop specialized structures and functions.

diffusion [L. *diffundo,* to pour in different directions] Tendency for molecules to move from an area of higher concentration to an area of lower concentration in solution; the product of the constant random motion of all atoms, ions, or molecules in a solution.

disaccharide (di-sak´ă-rīd) Two monosaccharides chemically bound together; sucrose is glucose and fructose chemically joined.

disease The interference with or cessation of normal body functions.

disinfection The destruction of pathogens on inanimate objects.

dissociate [L. *dis-* + *socio,* to disjoin, separate] To separate a molecule into positive and negative ions by surrounding the ions with H_2O molecules.

distal [L. *di-* + *sto,* to be distant] Farther from the point of attachment to the body than another structure.

distal convoluted tubule Convoluted tubule of the nephron that extends from the ascending limb of the loop of Henle and ends in a collecting duct.

DNA See deoxyribonucleic acid.

dorsal [L. *dorsum,* back] See posterior.

dorsal root Sensory root of a spinal nerve.

dorsal root ganglion Collection of sensory neuron cell bodies within the dorsal root of a spinal nerve.

ductus arteriosus A short artery that connects the pulmonary trunk with the aorta in a fetus; allows blood to bypass the lungs.

ductus deferens (duk´tus def´er-enz) Duct of the testicle, running from the epididymis to the ejaculatory duct; also called the vas deferens.

duodenum (du-o-de´num, du-od´ĕ-num) [L. *duodeni,* twelve] First division of the small intestine; connects to the stomach.

dura mater (du´rah ma´ter) [L. hard mother] Tough, fibrous membrane forming the outer covering of the brain and spinal cord.

dural sinus Space within the dura mater that contains blood; receives blood and cerebrospinal fluid from the brain.

E

eardrum See tympanic membrane.

ectoderm (ek´to-derm) Outermost of the three germ layers of the embryo.

edema (e-de´mah) [Gr. *oidema,* a swelling] Excessive accumulation of fluid, usually causing swelling.

effector organ A peripheral organ that receives and responds to action potentials.

efferent arteriole Vessel that carries blood from the glomerulus to the peritubular capillaries.

ejaculation Reflexive expulsion of semen from the penis.

ejaculatory duct Duct formed by the union of the ductus deferens and the excretory duct of the seminal vesicle, which opens into the urethra.

electrocardiogram (ECG) (e-lek´tro-kar´de-o-gram) Graphic record of the heart´s electrical currents obtained with an electronic recording instrument.

electroencephalogram (EEG) (e-lek′tro-en-sef′ă-lo-gram) Graphic recording of the brain′s electrical activities.

electrolyte (e-lek-tro-līt) [Gr. *electro-* + *lytos,* soluble] Positive and negative ions in solution that conduct electricity.

electron Negatively charged particle found in the orbitals of atoms.

element [L. *elementum,* a rudiment] Substance composed of atoms of only one kind.

ellipsoid joint Football shaped surface of one bone inserted into the concave surface of another bone; allows movement in two planes; an example is the joint between the occipital condyles and the first cervical vertebrae.

embolus (em′bo-lus) [Gr. *embolos,* a plug] A detached clot or other foreign body that occludes a blood vessel.

embryo (em′bre-o) In prenatal development, the developing human between the end of the second week and the end of the eighth week.

embryonic disk Layer of ectoderm and endoderm located between the yolk sac and the amniotic cavity.

emission [L. *emissio,* to send out] Discharge; accumulation of secretions of reproductive glands and the testes in the ejaculatory duct and penis prior to ejaculation.

emulsification (e-mul′sĭ-fi-ka′shun) To form an emulsion.

emulsion (e-mul′shun) Two liquids with one liquid dispersed through the other liquid or in very fine globules; e.g., droplets of lipid suspended in an aqueous solution in the intestine.

endocarditis (en′do-kar-di′tis) Inflammation of the endocardium.

endocardium (en′do-kar′de-um) Innermost layer of the heart, including the stratified squamous epithelium of the endothelium.

endocrine gland (en′do-krin) [Gr. *endon,* inside + *krino,* to separate] Ductless gland that secretes internally, usually into the circulatory system.

endocrine system All of the endocrine glands of the body and the hormones they secrete.

endoderm (en′do-derm) Innermost of the three germ layers of the embryo.

endometrium (en′do-me′tre-um) Mucous membrane comprising the inner layer of the uterine wall; consists of simple columnar epithelium and an underlying layer of connective tissue; contains simple tubular uterine glands.

endoplasmic reticulum (ER) (en′do-plaz′mik re-tik′u-lum) Membranous network inside the cytoplasm; see rough and smooth ER.

endospore A dormant structure formed inside a bacterial cell that is resistant to heat, chemicals, and lack of water.

endotoxin Lipopolysaccharide in the bacterial cell wall that is released when the cell dies; produces the symptoms of fever, aches, weakness, and sometimes shock.

envelope A lipoprotein coat that surrounds the capsid in many animal viruses; protects the inner nucleic acid and helps attach the virus to cells.

enzyme (en′zīm) [Gr. *en,* in + *zyme,* leaven] A protein molecule that increases the rate of a chemical reaction without being permanently altered.

eosinophil (e-o-sin′o-fil) [Gr. *eos,* dawn + *phileo,* to love] White blood cell with granules that stain red with acidic dyes; inhibits inflammation.

epicardium (ep′ĭ-kar′de-um) [Gr. *epi-* + *kardia,* heart] The serous membrane covering the surface of the heart; also called the visceral pericardium.

epicondyle (ep′ĭ-kon′dīl) [Gr. *epi,* on + *kondylos,* a knuckle] Projection on (usually to the side of) a condyle; an attachment site for muscles.

epidermis (ep′ĭ-der′mis) [Gr. *epi,* upon + *derma,* skin] Outer portion of the skin formed of epithelial tissue that rests on the dermis; resists abrasion and forms a permeability barrier.

epididymis (ep-ĭ-did′ĭ-mis) [Gr. *epi,* upon + *didymos,* twin] Elongated structure made up of tubules connected to the posterior surface of the testis, site of storage and maturation of the sperm cells.

epiglottis (ep-ĭ-glot′is) Plate of elastic cartilage covered with mucous membrane; serves to cover the opening to the larynx during swallowing.

epinephrine (ep′ĭ-nef′rin) Hormone released from the adrenal medulla; increases cardiac output and blood glucose levels; also called adrenaline.

epiphyseal plate (e-pif′ĭ-se-al) Cartilage located between the epiphysis and diaphysis of a long bone; the site of bone growth in length.

epiphysis, pl. **epiphyses** (e-pif′ĭ-sis, e-pif′ĭ-sez) The end of a bone; consists of spongy bone with an outer covering of compact bone.

epithelial tissue (ep-ĭ-the′le-al) One of the four major tissue types; consists of cells with a basement membrane, little extracellular material, and no blood vessels; covers the free surfaces of the body and forms glands.

epithelium, pl. **epithelia** (ep-ĭ-the′le-um, ep-ĭ-the′le-ah) See epithelial tissue.

erection [L. *erectio,* to set up] Condition of erectile tissue when filled with blood, causing the erectile tissue to become hard and unyielding; especially referring to the penis.

erythroblastosis fetalis (ĕ-rith′ro-blast-to′sis fe-tă′lis) [erythroblast + -osis, condition] Destruction of red blood cells in the fetus or newborn caused by antibodies produce in the Rh negative mother acting on the Rh positive blood of the fetus or newborn.

erythrocyte (ĕ-rith′ro-sīt) [Gr. *erythro,* red + *kytos,* cell] See red blood cell.

erythropoietin (ĕ-rith′ro-poy′ĕ-tin) Protein hormone that stimulates RBC formation in red bone marrow.

esophagus (e-sof′ă-gus) [Gr. *oisophagos,* gullet] Part of the digestive tract between the pharynx and stomach.

essential nutrient Nutrient that must be ingested because the body cannot manufacture them or is unable to manufacture adequate amounts of them.

estrogen (es′tro-jen) Hormone secreted primarily by the ovaries; involved in the maintenance and development of female reproductive organs, secondary sexual characteristics, and the menstrual cycle.

Eustachian tube (u-sta′she-an) See auditory tube.

evaporation Change from liquid to vapor form; water evaporation results in heat loss from the body.

exocrine gland (eks′so-krin) [Gr. *exo-,* outside + *krino,* to separate] Gland that secretes to a surface or outward through a duct.

exocytosis (eks-o-si-to′sis) Elimination of materials from a cell through the formation of a vesicle.

exophthalmos (eks-ŏf-thal′mos) [Gr. *ex,* out, + *ophthalmos,* eye] Protrusion of the eyeballs associated with hyperthyroidism.

exotoxin Protein toxins released by bacteria and responsible for many disease symptoms.

expiration The movement of air out of the lungs.

expiratory reserve volume The amount of air that can be expired forcefully after expiration of the normal tidal volume.

extension [L. *extensio,* to stretch out] To straighten a joint.

external anal sphincter Ring of skeletal muscle around the inferior end of the anal canal.

external auditory meatus (me-a´tus) Short canal that opens to the exterior and ends at the eardrum.

extracellular Outside of the cell.

extrinsic muscle Muscle located outside the structure being moved.

F

fallopian tube See uterine tube.

false rib A rib that attaches to the costal cartilage of the seventh rib or does not have an anterior attachment.

fascia (fash´e-ah) Dense connective tissue that encloses and separates muscles.

fast-twitch muscle fiber Muscle fiber that contracts rapidly and relies primarily upon anaerobic respiration to produce ATP.

fat Greasy, soft-solid lipid found in animal tissues and many plants, composed of glycerol and fatty acids; triglycerides are the most common type in humans.

fat-soluble vitamin Vitamin such as A, D, E, and K that dissolves in lipids and is absorbed from the intestine along with lipids.

fatty acid Straight chain of carbon atoms with a carboxyl group (— COOH) attached at one end.

fertilization Union of a sperm cell and an oocyte to form a zygote.

fetus (fe´tus) In prenatal development, the developing human between the end of the eighth week and birth.

fibrin (fi´brin) [L. *fibra,* fiber] A threadlike protein fiber derived from fibrinogen by the action of thrombin; forms a clot, i.e., a network of fibers that traps blood cells, platelets, and fluid, which stops bleeding.

fibrous joint Bones held together by fibrous connective tissue; exhibits little or no movement; e.g. a suture.

filtrate The fluid that enters the nephron by passing through the filtration membrane of the renal corpuscle.

filtration The movement, resulting from a pressure difference, of a liquid through a partition that prevents some or all of the substances in the liquid from passing; water, ions, and small molecules (but not blood cells) move by filtration from the blood into Boman´s capsule.

filtration membrane The membrane formed by the glomerular capillary endothelium and the wall of Bowman's capsule.

filtration pressure The pressure difference across the filtration membrane that is responsible for the movement of materials into Bowman´s capsule.

first-degree burn See partial-thickness burn.

flagellum, pl. **flagella** (flă-jel´um, flă-jel´ah) [L. a whip] A long filamentous extension of the plasma membrane that moves cells such as sperm cells or certain bacteria.

flexion [L. *flectus,* to bend] The act of flexing or bending a joint.

floating rib A rib that is only attached to a vertebra.

follicle (fol´ĭ-kl) Structure in the outer portion of the ovary. Contains an oocyte which is surrounded by cells of the follicle; the cells of the follicle give rise to the corpus luteum following ovulation.

follicle-stimulating hormone (FSH) Hormone of the anterior pituitary; in the female, stimulates the follicles of the ovary, assists in maturation of the follicle, and causes secretion of estrogen from the follicle; in the male, stimulates the epithelium of the seminiferous tubules and is partially responsible for inducing spermatogenesis.

fontanel (fon´tă-nel) Membranous area between the skull bones of an infant.

foramen, pl. **foramina** (fo-ra´men, fo-ram´ĭ-nah) A hole; referring to a hole or opening in a bone.

foramen ovale (o-val´e) In the fetal heart, the oval opening in the interatrial septum with a valve that allows blood to low from the right to the left atrium but not in the opposite direction; allows blood to bypass the lungs in the fetus; closes at birth.

formed element One of the cells (e.g., red blood cell or white blood cell) or cell fragments (e.g., platelet) in blood.

fraternal twin One of two individuals that have different genetic makeups that are born at approximately the same time; formed from two different oocytes, each of which is fertilized by a different sperm cell.

frontal plane Plane running vertically through the body and separating it into anterior and posterior portions; coronal section.

full-thickness burn Burn that destroys the epidermis and the dermis and sometimes the underlying tissue as well; sometimes called a third degree burn.

fundus (fun´dus) [L. bottom] The rounded end of a hollow organ, e.g., the fundus of the stomach or uterus.

G

gallbladder Pear-shaped receptacle on the inferior surface of the liver; serves as a storage reservoir for bile.

gamete (gam´ēt) A reproductive cell; a sperm cell in males or an oocyte in females; contains 23 chromosomes.

ganglion, pl. **ganglia** (gan´gle-on, gan´gle-ah) [Gr. knot] A group of nerve cell bodies in the peripheral nervous system.

gastrin (gas´trin) Hormone released by the stomach that stimulates the stomach to increase gastric secretions and movements.

gastrointestinal tract The stomach, small intestine, and large intestine.

gene The functional unit of heredity; a group of nucleotides in DNA containing the instructions necessary for making a protein.

general sense A sensation that is detected by receptors located throughout the body; includes touch, temperature, pain, and proprioception.

genetics The science that studies the transmission of traits from one generation to the next; considers the effects of genetic abnormalities.

genotype (jen´o-tīp) Genetic makeup of the individual.

germ layer Ectoderm, mesoderm, or endoderm; gives rise to all the organs of the body.

gingiva (jin´jĭ-vah) Dense fibrous connective tissue, covered by mucous membrane, that covers the bone around the teeth; the gums.

girdle A bony ring or belt that attaches a limb to the body; see pectoral and pelvic girdle.

gland A single cell or a multicellular structure that secretes substance into the blood, into a cavity, or onto a surface.

glans penis [L. acorn] Conical expansion of the corpus spongiosum that forms the head of the penis.

gliding joint See plane joint.

glomerulus (glo-mĕr´u-lus) [L. *glomus,*

ball of yarn] Mass of capillary loops at the beginning of each nephron, nearly surrounded by Bowman's capsule.

glucagon (glu′kă-gon) Hormone secreted from the pancreatic islets of the pancreas that acts primarily on the liver to release glucose into the circulatory system.

glucocorticoid (glu′ko-kor′tĭ-koyd) A term for a group of hormones that increase blood glucose and inhibit inflammation; examples are cortisol, secreted from the adrenal cortex, and cortisone.

glucose Simple sugar that is a primary energy source for most cells.

glycerol (glis′er-ol) A three-carbon molecule with a hydroxyl group attached to each carbon.

goiter (goy′ter) [L. *guttur,* throat] An enlargement of the thyroid gland, not due to a tumor, usually caused by a lack of iodine in the diet.

Golgi apparatus (gol′je) Stacks of flattened sacks, formed by membranes, that concentrate and package materials for secretion from the cell.

gonadotropin (gon′ă-do-tro′pin) Hormone capable of promoting gonadal growth and function; two major gonadotropins are luteinizing hormone (LH) and follicle stimulating hormone (FSH).

gonadotropin-releasing hormone (GnRH) Hypothalamic hormone that stimulates the secretion of LH and FSH from the anterior pituitary.

gray matter Collections of nerve cell bodies, their dendrites, and associated neuroglia cells within the central nervous system.

growth hormone (GH) Protein hormone of the anterior pituitary; it promotes body growth, fat mobilization, and inhibition of glucose utilization.

H

hair A threadlike outgrowth of the skin consisting of columns of dead epithelial cells filled with keratin.

hair follicle Deep, narrow pit in the skin; contains the root of the hair.

hard palate Floor of the nasal cavity formed by the maxillary and palatine bones; separates the nasal cavity from the oral cavity.

haversian canal (hă-ver′shan) Canal running parallel to the long axis of a bone; contains blood vessels, nerves, and loose connective tissue.

heart rate The number of complete cardiac cycles (heart beats) per minute.

heart sound The normal sounds that result from closure of the valves of the heart. The first heart sound results from the closure of the tricuspid and bicuspid valves and the second heart sound is caused by closure of the aortic and pulmonary semilunar valves.

hematocrit (hem′ă-to-krit) [Gr. *hemato-,* blood + *krino,* to separate] The percentage of total blood volume composed of red blood cells.

hemoglobin (he′mo-glo′bin) A substance in red blood cells consisting of four globin proteins, each with an iron-containing red pigment heme; transports oxygen and carbon dioxide.

hemolysis (he-mol′ĭ-sis) [Gr. *hemo-,* blood + *lysis,* destruction] The rupture of red blood cells.

hepatic duct (he-pat′ik) One of two ducts (left and right) that drain bile from the liver and join to form the common hepatic duct.

hepatic portal system Blood flow through the veins in the small intestine, spleen, and stomach that carry blood to the liver.

herniated disk Rupture of the anulus fibrosus and protrusion of the nucleus pulposus, which can press on the spinal cord or spinal nerves; sometimes incorrectly called a slipped disk.

heterozygous (het′er-o-zi′gus) Having two different genes for a given trait.

hilum (hi′lum) [L. a small amount or trifle] Part of an organ where the nerves and vessels enter and leave.

hinge joint A convex cylinder of bone inserted into a concave surface of another bone; permits movement in one plane, the elbow and knee joints are examples.

histology (his-tol′o-je) Scientific discipline that studies the structure of cells, tissues, and organs and their relation to function.

homeostasis (ho′me-o-sta′sis) [Gr. *homoio,* like + *stasis,* a standing] Existence and maintenance of a relatively constant environment within the body with respect to functions and the composition of fluids and tissues.

homozygous (ho-mo-zi′gus) Having two identical genes for a given trait.

hormone (hor′mōn) [Gr. *hormon,* to set into motion] A substance secreted by endocrine tissues into the blood that acts on a target tissue to produce a specific response.

host [L. *hospes,* a host] The organism on or in which a parasite lives.

hydrocephalus (hi′dro-sef′ă-lus) [Gr. *hydor,* water + *kephale,* head] Excessive fluid within the ventricles or subarachnoid space of the brain resulting in swelling and compression of brain tissue.

hydrochloric acid Acid of gastric juices; provides the necessary pH environment for pepsin activity.

hymen [Gr. membrane] Thin, membranous fold partly blocking the vaginal orifice; normally disrupted by sexual intercourse or other mechanical phenomena.

hypertension High blood pressure; for most adults hypertension exists when the systolic blood pressure exceeds 150 mm Hg and the diastolic pressure exceeds 90 mm Hg.

hyperthyroidism A condition resulting from over-secretion of thyroid hormone; results in nervousness, elevated rate of metabolism, weight loss, diarrhea, irregular heart beat, rapid heart beat, and elevated body temperature.

hypertonic (hi′per-ton′ik) Solution that causes cells to shrink.

hypertrophy (hi-per′tro-fe) Increase in bulk or size due to an increase in cell size; increase in the size of a muscle.

hypodermis (hy′po-der′mis) [Gr. *hypo,* under + *dermis,* skin] Loose connective tissue under the dermis that attaches the skin to muscle and bone.

hypophysis (hi-pof′ĭ-sis) Synonym for the pituitary gland.

hypothalamic-pituitary portal system A series of blood vessels that carry blood from the hypothalamus to the anterior pituitary; they originate from capillary beds in the hypothalamus and terminate as a capillary bed in the anterior pituitary.

hypothalamus (hi′po-thal′ă-mus) [Gr. *hypo,* under, below + *thalamus,* bedroom] Important autonomic and endocrine control center of the brain located beneath the thalamus.

hypothyroidism A condition resulting from reduced thyroid hormone secretion; results in lethargy, reduced rate of metabolism, sluggish mental activity, weight gain, and constipation.

hypotonic (hi′po-ton′ik) Solution that causes cells to swell.

I

identical twin One of two individuals

that have the same genetic makeup; formed from a single sperm cell and an oocyte.

ileocecal sphincter (il′e-o-se′kal) Thickening of the circular smooth muscle between the ileum and the cecum.

ileocecal valve Valve formed between the ileum and the cecum.

ileum (il′e-um) Third portion of the small intestine, extending from the jejunum to the large intestine.

immunity Ability to resist damage from foreign substances such as microorganisms and harmful chemicals.

implantation Attachment of the blastocyst to the wall of the uterus.

incubation period The period of time between the entry of a pathogen into the body and the first appearance of symptoms.

infarct (in′farkt) An area of necrosis resulting from a sudden insufficiency of arterial blood supply.

infection The multiplication of an organism within a reservoir, e.g., the human body.

inferior [L. lower] Down, or lower, with reference to the anatomical position.

inferior vena cava (ve′nah ca′vah) A large vein that receives blood from the lower limbs and the greater part of the pelvic and abdominal organs and empties into the right atrium of the heart.

inflammation Complex sequence of events involving chemicals and immune system cells that results in the isolation and destruction of foreign substances such as bacteria; symptoms include redness, heat, swelling, pain, and disturbance of function.

infundibulum (in-fun-dib′u-lum) The stalk that connects the pituitary gland to the hypothalamus of the brain.

inguinal canal (ing′gwĭ-nal) Passage through the lower abdominal wall through which the spermatic cord passes in the male.

inner cell mass Thickened area of several layers of cells in the blastocyst; gives rise to the embryo, yolk sac, and amniotic cavity.

inorganic Substances that do not contain carbon atoms; see organic.

insertion The more movable attachment point of a muscle.

inspiration The movement of air into the lungs; inhalation.

inspiratory reserve volume The amount of air that can be inspired forcefully after inspiration of the normal tidal volume.

insulin (in′su-lin) Protein hormone secreted from the pancreas that increases the uptake of glucose and amino acids by most tissues.

interatrial septum Cardiac muscle partition separating the right and left atria.

intercalated (in-ter′kă-la-ted) **disks** Special cell-to-cell attachments that allow action potentials to pass from one cardiac muscle cell to another.

interferon (in′ter-fēr′on) A protein that prevents viral replication.

internal anal sphincter Smooth muscle ring at the upper end of the anal canal.

interphase Time between cell divisions.

interstitial cell (in-ter-stish′al) Cell between the seminiferous tubules of the testes; secretes testosterone; also called cell of Leydig.

interventricular septum The cardiac muscle partition separating the right and left ventricles.

intracellular Inside of the cell.

intramural plexus (in′trah-mu′ral plek′sus) [L. within the wall] A nerve plexus within the walls of the gastrointestinal tract; functions to coordinate secretion and muscular movements of the gastrointestinal tract.

ion (i′on) Atom or group of atoms carrying an electrical charge due to loss or gain of one or more electrons.

ionic bond (i′on-ik) Chemical bond that is formed when one atom loses an electron and another atom accepts that electron.

iris (i′ris) The "colored" part of the eye that can be seen through the cornea; consists of smooth muscles that regulate the amount of light entering the eye through the pupil.

isometric contraction (i′so-met′rik) [Gr. *isos,* equal + *metron,* measure] Muscle contraction in which the length of the muscle does not change, but the amount of tension increases.

isotonic (i′so-ton′ik) Solution that causes cells to neither shrink nor swell.

isotonic contraction [Gr. *isos,* equal + *tonos,* a stretching] Muscle contraction in which the amount of tension is constant, and the muscle shortens.

isotope (i-so-tōp) Either of two or more forms of the same elements that have the same number of protons and electrons, but different numbers of neutrons.

J

jaundice (jawn′dis) [Fr. *jaune,* yellow] Yellowish staining of the skin, sclerae, and deeper tissues and excretions with bile pigments.

jejunum (jĕ-ju′num) Second portion of the small intestine located between the duodenum and ileum.

joint See articulation

juxtaglomerular complex (juks′tă-glo-mĕr′u-lar) Cells of the afferent arteriole and cells of the distal convoluted tubule where the distal convoluted tubules passes close to the afferent arteriole near the Bowman′s capsule; produces the hormone renin.

K

karyotype (kar′e-o-tīp) A display of chromosomes arranged by pairs.

keratin (kĕr′ah-tin) [Gr. *keras,* horn] Fibrous protein complex found in the stratum corneum, hair, and nails; provides structural strength.

kinetic equilibrium (kĭ-net′ik) The sense of balance that monitors the rate of change of head movements.

kidney One of a pair of bean-shaped organs, which function to cleanse blood and produce urine.

kyphosis (ki-fo′sis) [Gr. humpback] Abnormal posterior curvature of the thoracic region of the vertebral column.

L

labia majora (la′be-ah ma-jo′rah) One of two rounded folds of skin surrounding the labia minora and vestibule; anteriorly they unite to form the mons pubis.

labia minora (la′be-ah mi-no′rah) One of two narrow longitudinal folds of mucous membrane enclosed by the labia majora; anteriorly they unite to form the prepuce over the clitoris.

lacrimal gland (lak′rĭ-mal) Tear-producing gland in the superolateral corner of the orbit.

lacteal (lak′te-al) Lymphatic vessel in the wall of the small intestine; carries chyle from the intestine and absorbs fats.

lacuna pl. **lacunae** (lă-ku′nah, lă-ku′ne) [L. a pit] A small space or cavity.

large intestine Portion of the digestive tract extending from the small intestine to the anus; consists of the cecum, appendix, colon, rectum, and anal canal.

larynx (lăr′ingks) Organ of voice production located between the pharynx and the trachea; consists of a framework of cartilages and elastic ligaments containing the vocal cords and the muscles that control the position and tension of the vocal cords.

lateral [L. *latus,* side] Away from the middle or midline of the body.

lens Biconcave structure capable of being flattened or thickened to adjust the focus of light entering the eye.

leukemia (lu-ke′me-ah) [Gr. *leukos,* white + *haima,* blood] A tumor of the red bone marrow that results in the production of large numbers of abnormal white blood cells; often accompanied by decreased production of red blood cells and platelets.

leukocyte (lu′ko-sīt) [Gr. *leukos,* white + *kytos,* cell] See white blood cell.

leukocytosis (lu-ko-si-to′sis) [*leukocyte* + Gr. *-osis,* a condition] A higher than normal number of white blood cells.

leukopenia (lu-ko-pe′ne-ah) [leukocyte + Gr. *penia,* poverty] A lower than normal number of white blood cells.

Leydig cell See interstitial cell.

ligament A band of dense connective tissue that connects two or more bones, cartilages, or other structures.

lipase (li′pās) In general, any fat-splitting enzyme.

lipid [Gr. *lipos,* fat] Substance composed principally of carbon, oxygen, and hydrogen; generally soluble in noncharged (nonpolar) solvents such as acetone; used for longer-term energy storage; fats, oils, phospholipids, cholesterol, and steroids are examples.

lipopolysaccharide (lip′o-pol-e-sak′ă-rīd) A lipid polysaccharide complex; a component of some bacterial cell walls; bacteria with the LPS layer stain Gram negative.

lipoprotein A lipid protein complex; low density lipoproteins carry cholesterol to tissues and high density lipoproteins carry cholesterol from tissues to the liver.

liver Largest gland of the body, lying in the upper-right portion of the abdomen just inferior to the diaphragm; secretes bile and is of great importance in carbohydrate and protein metabolism and in detoxifying chemicals.

local reflex Reflex of the intramural plexus that does not involve the brain or spinal cord.

lock and key model The shapes of an enzyme and the reactants allow them to fit together so that the reactants undergo a chemical reaction.

loop of Henle U-shaped part of the nephron extending from the proximal to the distal convoluted tubule and consisting of descending and ascending limbs; many of the loops of Henle extend into the renal pyramids.

lordosis (lor-do′sis) [L. bending backward] Abnormal anterior curvature (swayback) of lumbar region of the vertebral column.

lower respiratory tract The larynx, trachea, bronchi, and lungs.

lung One of a pair of organs in the thoracic cavity; the principle organs of respiration.

lunula (lu′nu-lah) [L. *luna,* moon] White, crescent-shaped portion of the nail matrix visible through the proximal end of the nail.

luteinizing hormone (LH) (lu′te-ĭ-nīz-ing) Hormone of the anterior pituitary that, in the female, initiates the final maturation of the follicle, its rupture to release the oocyte, the conversion of the ruptured follicle into the corpus luteum, and the secretion of progesterone; in the male, stimulates the secretion of testosterone in the testes, and is sometimes referred to as interstitial cell-stimulating hormone (ICSH).

lymph (limf) [L. *lympha,* clear spring water] Clear or yellowish fluid derived from tissue fluid and found in lymph vessels.

lymph capillary Tiny, close-ended vessel consisting of simple squamous epithelium; fluid from tissues moves into lymph capillaries to form lymph.

lymph node Encapsulated mass of lymph tissue found along lymph vessels; functions to filter lymph and produce lymphocytes.

lymph vessel Vessel that carries lymph from lymph capillaries to the blood; usually connected to a lymph node.

lymphatic tissue Connective tissue with very fine collagen fibers and several types of cells including lymphocytes and macrophages.

lymphocyte (lim′fo-sīt) Nongranular WBC; B or T cell; functions include antibody production, allergic reactions, graft rejections, tumor control, and regulation of the immune system.

lysis (li′sis) Process by which a cell swells and ruptures.

lysosome (li′so-sōm) Membrane-bound vesicle containing intracellular digestive enzymes.

M

macrophage (mak′ro-faj) [Gr. *makros,* large + *phagein,* to eat] Any large mononuclear phagocytic cell.

macula (mak′u-lah) One of the sensory structures in the vestibule, consisting of hair cells and a gelatinous mass; responds to gravity (static equilibrium).

malignant (mă-lig′nant) A type of tumor that increases in size and spreads to surrounding and to distant places in the body.

mammary gland Lobed gland within the breasts that produces milk; each lobe has a duct that empties milk onto the nipple of the breast.

mass movement Strong peristaltic contraction of the large intestine that is the result of a local reflex initiated by distention of the stomach.

mast cell Connective tissue cell that releases histamine and other chemicals that promote inflammation.

mastication (mas′tĭ-ka′shun) [L. *mastico,* to chew] The process of chewing.

matrix (ma′triks) The extracellular substance of a tissue.

matter Anything that occupies space.

medial [L. *medialis,* middle] Toward the middle or midline of the body.

mediastinum (me′de-as-tin′num) [L. middle septum] The middle wall of the thorax consisting of the trachea, esophagus, thymus, heart, and other structures.

medulla (mĕ-dul′ah) Any soft marrow-like structure, especially in the center of an organ such as the kidney or adrenal gland.

medulla oblongata (ob′long-gah′tah) Inferior portion of the brainstem that connects the spinal cord to the brain; contains nuclei controlling heart rate, respiration, and swallowing; contains ascending and descending nerve tracts.

medullary cavity Large, marrow-filled cavity in the diaphysis of a long bone.

meiosis (mi-o′sis) Type of cell division in the testes and ovaries that produces sex cells, each having half the number of chromosomes as the parent cells.

melanin (mel′ah-nin) [Gr. *melas,* black] Brown to black pigment responsible for skin, hair color, and eye color.

melanocyte (mel′ă-no-sīt) [Gr. *melas,* black + *kytos,* cell] Cells producing the brown or black pigment melanin; found mainly in the stratum basale of skin.

melanocyte-stimulating hormone (MSH) Hormone secreted by the anterior pituitary; increases melanin production by melanocytes, making the skin darker in color.

melanoma (mel′ă-no′mah) [Gr. *melas,* black + *oma,* tumor] A malignant tumor derived from melanocytes.

membrane channel Protein that extends across the plasma membrane and through which ions and molecules can diffuse.

memory cell Lymphocytes derived from B or T cells after they have been exposed to an antigen; when exposed to the same antigen responsible for their formation, memory cells rapidly respond to provide immunity.

memory response Immune response that occurs when the immune system is exposed to an antigen against which it has already had a primary response; results in the production of large amounts of antibodies and memory cells. Also called a memory response.

menarche (mĕ-nar′ke) The first onset of menstrual cycle.

meninx, pl. **meninges** (me′ningks, mĕ-nin′jez) [Gr. membrane] Connective tissue membranes that surround and protect the brain and spinal cord; the dura mater, arachnoid, and pia mater.

menopause [Gr. *mensis,* month + *pausis,* cessation] Permanent cessation of the menstrual cycle.

menses (men′sēz) [L. *mensis,* month] Loss of blood and tissue as the endometrium of the uterus sloughs away at the end of the menstrual cycle; occurring at approximately 28-day intervals in the nonpregnant female of reproductive age.

menstrual cycle Cyclic series of changes that occur in sexually mature, nonpregnant females resulting in shedding of the uterine lining (menses); specifically includes the cyclical changes that occur in the uterus and ovary.

merocrine (mer′o-krin) See sweat gland.

mesentery (mes′en-tĕr-e) Serous membranes and connective tissue that extend from the abdominal wall to organs; holds the organs in place.

mesoderm (mez′o-derm) Middle of the three germ layers of the embryo.

messenger RNA Type of RNA that is a copy of the information contained in DNA; moves out of the nucleus to ribosomes where the information is used to produce a protein.

metabolic rate The total amount of energy produced and used by the body per unit of time.

metabolism (mĕ-tab′o-lizm) [Gr. *metabole,* change] Sum of the chemical changes that occur in tissues, consisting of the breakdown of food to produce energy (catabolism) and the buildup of molecules (anabolism, which requires energy).

metaphase (met′ă-fāz) Time during cell division when the chromosomes line up along the center of the cell.

metastasis (mĕ-tas′tă-sis) [Gr. *meta,* a removing + *stasis,* a placing] Movement of a disease from one location in the body to another; tumor cells travel in the circulatory or lymphatic system to distant places in the body.

micelle (mi-sel′) [L. *micella,* a small morsel] Droplet of lipid surrounded by bile salts in the small intestine.

microorganism An organism so small that it can be seen only with a microscope; includes bacteria, viruses, fungi, and protozoans.

microvillus, pl. **microvilli** (mi′kro-vil′us, mi′kro-vil′i) Minute projection of the plasma membrane that greatly increases the surface area.

micturition reflex (mik-tu-rish-un) Contraction of the urinary bladder stimulated by stretching of the urinary bladder wall; results in emptying of the urinary bladder.

midbrain Superior part of the brainstem; involved with auditory and visual reflexes; contains ascending and descending nerve tracts.

midsagittal (mid′saj′ĭ-tal) Plane running vertically through the body and dividing it into equal right and left parts.

mineral Inorganic nutrient necessary for normal metabolic functions.

mineralocorticoid (min′er-al-o-kor′tĭ-koyd) Hormone released from the adrenal cortex; acts on the kidney to increase the rate of sodium reabsorption from the nephron and potassium and hydrogen ion secretion into the nephron; an example is aldosterone.

mitochondrion, pl. **mitochondria** (mi′to-kon′dre-on, mi′to-kon′dre-ah) [Gr. *mitos,* thread + *chandros,* granule] Small, bean-shaped or rod-shaped structure in the cytoplasm that is the site of ATP production.

mitosis (mi-to′sis) [Gr. thread] Division of the nucleus. Process of cell division that results in two daughter cells with the same number and type of chromosomes as the parent cell.

mitral valve (mi′tral) See bicuspid valve.

mixed nerve Nerve with sensory and motor functions.

molecule Two or more atoms of the same or different type joined by a chemical bond. See compound.

monocyte (mon′o-sīt) A type of white blood cell that transforms to become a macrophage.

monosaccharide (mon-o-sak′ă-rīd) The basic building block from which more complex carbohydrates are constructed; glucose and fructose.

mons pubis [L. mountain] Prominence formed by a pad of fatty tissue over the pubic symphysis in the female.

motor division Part of the peripheral nervous system that sends action potentials from the central nervous system to the periphery.

motor neuron Neuron that carries action potential from the central nervous system to skeletal muscle, cardiac muscle, smooth muscle, or glands.

mRNA See messenger RNA.

mucosa (mu-ko′sah) Mucous membrane consisting of epithelium and connective tissue. In the digestive tract there is also a layer of smooth muscle.

mucous membrane (mu′kus) Thin sheet consisting of epithelial and connective tissue that lines cavities opening to the outside of the body; many contain mucous glands, which secrete mucus.

murmur An abnormal sound produced within the heart.

muscle bundle A group of muscle fibers surrounded by connective tissue within a muscle.

muscle fatigue Fatigue due to depletion of ATP within the muscle fiber.

muscle fiber Muscle cell. Shortens forcefully. Muscle fibers are responsible for most bodily movements.

muscle tissue One of the four major tissue types; consists of cells with the ability to contract; skeletal, cardiac, and smooth muscle.

muscle tone Constant tension produced by muscles as a result of isometric contractions.

muscularis (mus′ku-la′ris) The muscular layer of an organ, i.e., the smooth muscle in the digestive tract.

mutation A change in the number or

kinds of nucleotides in the DNA of a gene.

myelin sheath (mi′ĕ-lin) Cell membrane and myelin (a lipoprotein) wrapped many times around an axon.

myelinated fiber Axon surrounded by myelin sheath produced by neuroglia; transmits action potentials more rapidly than unmyelinated fibers.

myocarditis (mi′o-kar-di′tis) Inflammation of cardiac muscle.

myocardium (mi′o-kar′de-um) [Gr. *mys,* muscle + *kardia,* heart] The middle layer of the heart, consisting of cardiac muscle.

myofibril A fine longitudinal fibril of skeletal muscle consisting of sarcomeres placed end to end.

myofilament An ultramicroscopic protein thread that helps form myofibrils in skeletal muscle. Thin myofilaments are composed of actin and thick myofilaments are composed of myosin.

myoglobin Protein within muscle fibers that stores and transports oxygen.

myometrium (mi′o-me′tre-um) Muscular wall of the uterus, composed of smooth muscle.

myosin myofilament (mi′o-sin) Thick myofilament within the sarcomere.

myxedema (miks-ĕ-de′mah) [Gr. *myxa,* mucus + Gr. *oidema,* swelling] Hypothyroidism characterized by edema or swelling of subcutaneous connective tissue due to a change in its structure.

N

nail A thin, horny plate at the ends of the fingers and toes, consisting of several layers of dead epithelial cells containing a hard keratin.

nail matrix Portion of the nail bed from which the nail is formed.

nasal cavity Cavity divided by the nasal septum, and extending from the nostrils anteriorly to the pharynx posteriorly, and bounded inferiorly by the hard palate.

nasal septum Bony partition that separates the nasal cavity into left and right parts; composed of the vomer, the perpendicular plate of the ethmoid, and hyaline cartilage.

nasolacrimal duct Duct that carries tears from the medial corner of the eye to the nasal cavity.

negative feedback Mechanism by which any deviation from an ideal normal value is made smaller or is resisted; returns a parameter to its normal range and thereby maintains homeostasis.

nephron (nef′ron) [Gr. *nephros,* kidney] Functional unit of the kidney, consisting of Bowman's capsule, the proximal convoluted tubule, the loop of Henle, and the distal convoluted tubule.

nerve Bundle of nerve fibers with their associated sheaths in the peripheral nervous system; there are 12 pairs of cranial nerves and 31 pairs of spinal nerves.

nerve cell See neuron.

nerve fiber Cell process of a neuron; a dendrite or axon.

nerve impulse An action potential that moves along the cell membrane of a nerve fiber.

nerve tract Bundle of nerve fibers with their associated sheaths in the central nervous system.

nervous tissue One of the four major tissue types; consists of neurons, which have the ability to conduct actions potentials, and neuroglia, which are support cells.

neuroglia (nu-rog′le-ah) [Gr. *neuro,* nerve + *glia,* glue] Cells of the nervous system other than neurons; play a support role in the nervous system; include astrocytes, oligodendrocytes, and Schwann cells; also called glia.

neuron (nu′ron) [Gr. nerve] The functional unit of the nervous system that is specialized to receive and transmit action potentials; consists of the cell body, dendrites, and axon; a nerve cell.

neurotransmitter [Gr. *neuro,* nerve + L. *transmitto,* to send across] A chemical that a neuron releases into the synapse; serves to transmit information to another neuron or effector cell.

neutral solution Solution with equal numbers of hydrogen and hydroxide ions; has a pH of 7.

neutron [L. *neuter,* neither] Electrically neutral particle found in the nucleus of atoms.

neutrophil (nu′tro-fil) [L. *neuter,* neither + Gr. *phileo,* to love] White blood cell with granules that stain with neither basic or acidic dyes; small phagocytic white blood cell.

node of Ranvier (ron′ve-a) Bare area of a nerve fiber between wrappings of the myelin sheath.

nonspecific resistance Immune system response that is the same upon each exposure to an antigen; there is no ability to remember a previous exposure to the antigen.

noradrenaline (nor′a-dren′ă-lin) Synonym for norepinephrine.

norepinephrine (nor′ep′ĭ-nef′rin) A neurotransmitter released from many nerve fiber endings of the sympathetic division of the autonomic nervous system and the central nervous system; a hormone released from the adrenal medulla that increases cardiac output and blood glucose; also called noradrenaline.

nosocomial (nos′o-ko′me-al) [Gr. hospital] Infections that are acquired in a hospital, nursing home, or health-related facility.

nuclear pore Opening in the membranes that form the outer boundary of the nucleus.

nucleic acid Molecule consisting of many nucleotides chemically bound together; deoxyribonucleic acid and ribonucleic acid.

nucleolus, pl. **nucleoli** (nu-kle′o-lus, nu-kle′o-li) Somewhat rounded, dense, well-defined nuclear body with no surrounding membrane; the site of production of ribosomal subunits.

nucleotide (nu′kle-o-tīd) Basic building block of nucleic acids consisting of sugar (either ribose or deoxyribose), one of several types of organic bases, and a phosphate group.

nucleus, pl. **nuclei** (nu′kle-us, nu′kle-i) [L. inside of a thing] Cell organelle containing most of the cell's genetic material (DNA) and controlling the activities of the cell; collection of nerve cell bodies within the central nervous system; center of atom consisting of protons and neutrons.

nutrient Chemical that provides the body with energy and/or building blocks to synthesize new molecules.

nutrition Process by which nutrients are obtained and used in the body.

O

olfaction The sense of smell.

olfactory bulb Ganglion-like enlargement that receives the olfactory nerves and sends nerve fibers to the olfactory area.

oncogene (ong′ko-jēn) A gene that can change or be activated to cause cancer.

oncology (ong-kol′o-je) The study of tumors.

oocyte (o′o-sīt) [Gr. *oon,* egg + *kytos,*

cell] The female gamete or sex cell; contains the genetic material transmitted from the female.

opportunistic pathogen A pathogen already present in the body that causes disease when transferred to another part of the body, or when some condition of the host changes, e.g., reduced resistance to disease.

optic disc Point at which axons from the retina come together to form the optic nerve.

optic nerve Nerve carrying action potentials from the retina to the optic chiasma.

orbital The region around the nucleus of an atom in which electrons are found.

organ [Gr. *organon,* tool] Part of the body composed of two or more tissue types and performing one or more specific functions.

organ of Corti (kor′te) Specialized region of the cochlea consisting of hair cells with hairs embedded in a rigid gelatinous shelf; functions to detect sound vibrations.

organ system Group of organs classified as a unit because of a common function or set of functions.

organelle Specialized part of a cell performing one or more specific functions.

organic Substances that contain a carbon atom (carbon dioxide is an exception); see inorganic.

organism Any living thing considered as a whole, whether composed of one cell or many.

orgasm [Gr. *orgao,* to swell, be excited] Climax of sexual arousal, associated with a pleasurable sensation.

origin The less movable attachment point of a muscle.

osmosis (os-mo′sis) [Gr. *osmos,* thrusting or an impulsion] Diffusion of water through a selectively permeable membrane from a less concentrated solution (less dissolved substances, more water) to a more concentrated solution (more dissolved substances, less water).

ossicle (os′ĭ-kl) Bone of the middle ear; the malleus, incus, or stapes.

osteoblast (os′te-o-blast) [Gr. *osteon,* bone + *blastos,* germ] Bone-forming cell.

osteoclast (os′te-o-klast) [Gr. *osteon,* bone + *klastos,* broken] Large multinucleated cell that breaks down bone.

osteocyte (os′te-o-sīt) [Gr. *osteon,* bone + *kytos,* cell] Mature bone cell surrounded by bone matrix.

oval window Opening to which the stapes attaches; transmits vibrations to the inner ear.

ovary One of two female reproductive glands located in the pelvic cavity; produces the oocyte, estrogen, and progesterone.

ovulation Release of an oocyte from the ovary.

oxygen debt The amount of oxygen required to convert the lactic acid produced during anaerobic respiration to glucose.

oxytocin (ok-sĭ-to′sin) Peptide hormone secreted by the posterior pituitary that increases uterine contraction and stimulates milk ejection from the mammary glands.

P

P wave The first wave of the electrocardiogram which is the result of depolarization of the atria.

pancreas (pan′kre-us) Abdominal gland that secretes pancreatic juice into the duodenum of the small intestine and insulin and glucagon from the pancreatic islets into the blood.

pancreatic duct Excretory duct of the pancreas that extends through the gland to the duodenum of the small intestine.

pancreatic islets Cellular mass in the tissue of the pancreas; composed of different cell types that comprise the endocrine portion of the pancreas and are the source of insulin and glucagon.

papilla, pl. **papillae** (pă-pil′ah, pă-pil′e) [L. nipple] A small nipplelike process. Projection of the dermis, containing blood vessels and nerves, into the epidermis. Projections on the surface of the tongue.

papillary muscle (pap′ĭ-lĕr′e) A raised area of cardiac muscle in the ventricle to which the chordae tendineae of the tricuspid and bicuspid valves attach.

parafollicular cell (păr′ah-fŏ-lik′u-lar) Cell of the thyroid gland located between the thyroid follicles and which secretes calcitonin.

paranasal sinus Air-filled cavity within certain skull bones that connects to the nasal cavity; the four sets of paranasal sinuses are the frontal, maxillary, sphenoidal, and ethmoidal.

parasite [Gr. *parasitos,* a growth] An organism that lives on or in another organism from which it derives nourishment and which it harms.

parasympathetic division Subdivision of the autonomic nervous system involved in vegetative functions such as digestion, defecation, and urination.

parathyroid gland (păr-ă-thi′royd) One of four small glands embedded in the posterior surface of the thyroid gland; secretes parathyroid hormone.

parathyroid hormone (PTH) Peptide hormone consisting of a series of amino acids produced by the parathyroid gland that increases bone breakdown and blood calcium levels.

parietal peritoneum (pĕr′ĭ-to-ne′um) [L. wall] Portion of the serous membranes of the abdominal cavity lining the inner surface of the body wall.

partial-thickness burn Burn that damages only the epidermis (first degree burn) or the epidermis and part of the dermis (second degree burn).

parturition (par-tu-rish′un) [L. *parturio,* to be in labor] Childbirth; the delivery of a baby at the end of pregnancy.

passive artificial immunity Transfer of antibodies to a person from an animal or another person.

passive natural immunity Transfer of antibodies from a mother to her fetus (across the placenta) or baby (in mother′s milk).

pathogen (path′o-jen) Any organism or substance that causes disease.

pectoral girdle (pek′to-ral) Site of attachment of the upper limb to the trunk; consists of the scapula and the clavicle.

pelvic cavity Space completely surrounded by the pelvic bones.

pelvic girdle Site of attachment of the lower limb to the trunk; ring of bone formed by the sacrum and the coxae.

pepsin (pep′sin) Principle digestive enzyme of gastric juice; digest proteins into smaller polypeptide chains.

peptidoglycan (pep′tĭ-do-gly′can) A component of bacterial cell walls consisting of sugars and amino acids; makes the cell wall structurally strong.

pericardial fluid (pĕr′ĭ-kar′de-al) The serous fluid found within the pericardial cavity.

pericardial membrane A serous membrane; thin sheet consisting of epithelium and connective tissue that lines the pericardial cavity around the heart.

pericardial cavity The space between the visceral and parietal pericardium, filled with pericardial fluid; a cavity that surrounds the heart.

pericardial sac See pericardium.

pericarditis (pĕr-ĭ-kar-di′tis) Inflammation of the pericardium.

pericardium (pĕr′ĭ-kar′de-um) [Gr. *pericardion,* the membrane around the heart] The membrane consisting of the epicardium and parietal pericardium (of the serous layers) and the outer fibrous pericardium that forms the pericardial sac.

perineum (per′ĭ-ne′um) Area inferior to the pelvic diaphragm between the thighs; extends from the coccyx to the pubis. The clinical perineum extends from the vaginal opening to the anus.

periodontal ligament (pĕr′e-o-don′tal) [Gr. *peri,* around + *odous,* tooth] Connective tissue that surrounds the tooth root and attaches it to its bony socket.

periosteum (pĕr′e-os′te-um) [Gr. *peri,* around + *osteon,* bone] Connective tissue sheath covering the entire surface of a bone except the articular surface, which is covered with cartilage.

peripheral circulation Blood flow through the systemic circulation and the pulmonary circulation; includes all blood flow except that through the heart tissue itself.

peripheral nervous system (PNS) Major subdivision of the nervous system consisting of nerves and ganglia.

peristalsis (pĕr′ĭ-stal′sis) Waves of contraction and relaxation moving along a tube; propels food along the digestive tube.

peritoneal membrane (pĕr′ĭ-to-ne′al) A serous membrane; thin sheet consisting of epithelium and connective tissue that lines the peritoneal cavity, which is located inside the abdominopelvic cavity.

peritubular capillary The capillary network located in the cortex of the kidney; associated with the distal and proximal convoluted tubules.

permanent tooth One of the 32 teeth belonging to the second set of teeth; secondary tooth.

pH scale A measure of the hydrogen ion concentration of a solution. The scale extends from 0 to 14 with a pH of 7 being neutral, a pH of less than 7 acidic, and a pH of greater than 7 basic.

phagocytosis (fag′o-si-to′sis) [Gr. *phagein,* to eat + *kytos,* cell + *osis,* condition] Cell eating; process of ingestion by a cell of solid substances such as other cells, bacteria, bits of dead tissue, and foreign particles.

phantom pain Pain perceived in a body part that is no longer present.

pharynx (făr′ingks) [Gr. throat] Upper expanded portion of the digestive tube located between the esophagus and the oral and nasal cavities.

phenotype (fe′no-tīp) Characteristic observed in the individual due to expression of the genotype.

phlebitis (fle-bi′tis) Inflammation of the veins.

phospholipid Lipid with phosphorus resulting in a molecule with a charged (polar) end and a noncharged (nonpolar) end; main component of cell membranes.

physiology (fiz′e-ol′o-je) [Gr. *physis,* nature + *logos,* study] Scientific discipline that deals with the processes or functions of living things.

pia mater (pe′ah ma′ter) [L. tender mother] Delicate membrane forming the inner covering of the brain and spinal cord.

pilus, pl. **pili** (pi′lus, pi′li) [L. a hair] A hollow protein tube on the surface of bacteria; functions for attachment or for sexual reproduction.

pinna (pin′ah) See auricle.

pinocytosis (pin′o-si-to′sis) [Gr. *pineo,* to drink + *kytos,* cell + *osis,* condition] Cell drinking; uptake of liquid by a cell.

pituitary (pit-u′ĭ-tĕr-e) An endocrine gland attached to the hypothalamus; secretes hormones that influence the function of several other endocrine glands and tissues.

pituitary dwarf A condition resulting from too little growth hormone secreted in young people; results in short stature but may be of normal intelligence if it results from only a lack of growth hormone.

pivot joint Joint consisting of a cylindrical bony process that rotates within a ring composed partly of bone and partly of ligament; restricts movement to rotation around a single axis; the joint between the first and second cervical vertebrae is an example.

placenta (pla-sen′tah) The organ that exchanges gases, nutrients, waste products, and other substances between the embryo or fetus and the mother; derived from the chorion and the lining of the uterus.

plane joint Two flat bony surfaces that slide over each other in many different directions; the joint between articular processes of vertebrae is an example.

plasma (plaz′mah) Fluid portion of blood; blood minus the cells and fragments.

plasma membrane (plaz′mah) Outermost component of the cell, surrounding and binding the rest of the cell contents; regulates the movement of materials into and out of the cell; the cell membrane.

plasmid (plaz′mid) Extrachromosomal DNA; controls characteristics of bacteria that are not normally required for cell growth and survival, but which can be advantageous under certain circumstances, i.e., resistance to antibiotics.

platelet (plāt′let) A cell fragment involved in platelet plug and clot formation; also called a thrombocyte.

platelet plug Accumulation of platelets that stick to connective tissue and to each other to prevent blood loss from damaged blood vessels.

pleural cavity (ploor′al) Space between the parietal and visceral layers of the pleura, normally filled with pleural fluid.

pleural membrane A serous membrane; thin sheet consisting of epithelium and connective tissue that lines the pleural cavities around the lungs; the pleura.

plexus (plek′sus) [L. braid] Intertwining network of nerves or blood vessels.

pneumothorax (nu-mo-tho′raks) The introduction of air into the plural cavity resulting from equalization of pressure in the plural cavity and atmospheric air.

polycythemia (pol′e-si-the′me-ah) [Gr. *polys,* many + *kytos,* cell] Increase in red blood cell numbers above the normal value.

polysaccharide (pol-e-sak-ă-rīd) Many monosaccharides chemically bound together; glycogen and starch.

pons [L. bridge] That portion of the brainstem between the medulla oblongata and midbrain; connects to the cerebellum.

portal [L. *portalis,* pertaining to a gate] The place where a pathogen enters or exits the body.

positive feedback Mechanism by which any deviation from an ideal normal value is made greater.

posterior [L. *posterus,* following] That which follows. In humans, toward the back and synonymous with dorsal.

posterior pituitary The posterior portion of the pituitary gland, which consists of processes of nerve cells that have their cell bodies located in the hypothalamus; secretes oxytocin and antidiuretic hormone.

precapillary sphincter The smooth muscle sphincter that regulates the flow of blood from an arteriole into a capillary.

prepuce (pre´pus) In males, a free fold of skin that almost completely covers the glans penis; the foreskin. In females, the external fold of the labia majora that covers the clitoris.

primary curvature Normal bending of the vertebral column present at birth; curved posteriorly.

primary response Immune response that occurs as a result of the first exposure to an antigen; results in the production of antibodies and memory cells.

prime mover Muscle that plays a major role in accomplishing a movement.

primitive streak A thickened line in the ectodermal surface of the embryonic disk; cells migrating through the streak become mesoderm.

product Substance produced in a chemical reaction.

progesterone (pro-jes´ter-on) Hormone secreted by the ovaries and by the placenta; necessary for uterine and mammary gland development and function in pregnant and nonpregnant women.

prognosis (prog-no´sis) Estimating the probable outcome of a disease.

prolactin (pro-lak´tin) Hormone of the anterior pituitary that stimulates the secretion of milk.

pronation (pro-na´shun) [L. *pronare,* to bend forward] In the anatomical position, rotation of the forearm so the anterior surface faces posteriorly; rotation of the forearm (when the forearm is parallel to the ground) so that the anterior surface faces down.

prophase (pro-fāz) First stage of mitosis in which chromatin strands condense to form chromosomes and the nuclear membrane and nucleolus disappear.

prostaglandin (pros´tă-glan´din) Class of hormones present in many tissues; effects include vasodilation, stimulation and contraction of uterine smooth muscle, and promotion of inflammation and pain.

prostate gland (pros´tat) [Gr. *prostates,* one standing before] A gland that surrounds the beginning of the urethra in the male; the secretion of the gland is a milky fluid that is discharged into the urethra as part of the semen.

protein [Gr. *proteios,* primary] Organic molecule consisting of many amino acids linked by chemical bonds; source of amino acids (for building new proteins) and energy.

proton [Gr. *protos,* first] Positively charged particle found in the nucleus of atoms.

proximal [L. *proximus,* nearest] Closer to the point of attachment to the body than another structure.

proximal convoluted tubule Convoluted portion of the nephron that extends from Bowman´s capsule to the descending limb of the loop of Henle.

psychological fatigue Fatigue caused by the central nervous system.

puberty [L. *pubertas,* grown up] Series of events that transform a child into a sexually mature adult; involves an increase in the secretion of all reproductive hormones.

pubic symphysis The fibrocartilage joint between each coxa.

pulmonary circulation (pul´mo-nĕr-e) Blood flow from the right ventricle of the heart to the lungs and back to the left atrium.

pulmonary semilunar valve The valve found at the base of the pulmonary trunk where it exits from the right ventricle.

pulmonary trunk The large elastic artery that carries blood from the right ventricle of the heart to the right and left pulmonary arteries.

pulmonary vein One of the four veins that carry blood from the lungs to the left atrium.

pulse The dilation of an artery produced by the increased volume of blood pumped into arteries by the contraction of the heart. The pulse can be detected in large and medium sized arteries close to the surface of the skin.

pupil Opening in the iris of the eye through which light passes.

Purkinje fibers (pur-kin´je) Interlacing fibers formed from modified cardiac muscle cells found beneath the endocardium of the ventricles. Specialized to conduct action potentials.

pyloric sphincter (pi-lor´ik) Thickening of the circular layer of gastric musculature encircling the junction between the stomach and duodenum.

pyrojen (pi´ro-jen) Chemical released by lymphocytes, neutrophils, and macrophages; stimulates fever production by acting on the hypothalamus.

Q

QRS complex The main deflection in the electrocardiogram representing depolarization of the ventricles.

R

radiation Radiant heat such as infrared energy.

reactant Substance taking part in a chemical reaction.

receptor A molecule in the membrane, cytoplasm, or nucleus of cells to which a chemical such as a hormone or neurotransmitter can bind. The binding initiates a response in the cell.

recruitment An increase in the number of muscle fibers stimulated by the nervous system, resulting in gradually increasing force of contraction of the muscle.

rectum (rek´tum) [L. *rectus,* straight] Portion of the digestive tract that extends from the sigmoid colon to the anal canal.

red blood cell (RBC) Biconcave disk-shaped cell that contains hemoglobin but does not have a nucleus; transports oxygen and carbon dioxide; also called an erythrocyte.

red marrow Connective tissue within the cavities of bone; the site of blood cell production.

referred pain Pain perceived in a body part that is not the source of the painful stimulus.

reflex Automatic response to a stimulus that occurs without conscious thought; produced by a reflex arc.

reflex arc Smallest portion of the nervous system that is capable of receiving a stimulus and producing a response; composed of a receptor, sensory neuron, association neurons, motor neuron, and effector organ.

releasing hormone A hormone that is released from neurons of the hypothalamus and flows through the hypothalamic-pituitary portal system to the anterior pituitary; regulates the secretion of hormones from the anterior pituitary gland.

renal pelvis The expanded end of the

ureter within the renal sinus which receives urine from the calyces.

renal pyramid Cone-shaped structure that extends from the renal sinus, where the apex is located, into the cortex of the kidney, where the base is located.

renal sinus The cavity central to the medulla of the kidney that is filled with adipose tissue and contains the renal pelvis.

renin (ren´in) Enzyme secreted by the kidney that converts angiotensinogen to angiotensin I.

repolarization Change in the electrical charge difference across the cell membrane following depolarization; return of the charge difference to the resting condition, that is, the cell membrane is positively charged on the outside compared to the inside.

reservoir Living or nonliving material in or on which a disease-causing organism can live and reproduce; the source of infections.

residual volume The volume of air still remaining in the respiratory passages and lungs after a maximum expiration.

respiratory center Neurons in the medulla oblongata and pons of the brain that control inspiration and expiration.

respiratory membrane Membrane in the lungs across which gas exchange occurs with blood; consists of a thin layer of fluid, the alveolar wall, interstitial space, and the capillary wall.

resting membrane potential Charge difference across the cell membrane of an unstimulated cell; the outside of the cell membrane is positively charged compared to the inside.

reticular activating system Scattered group of nuclei within the brain stem and extending into the thalamus; responsible for arousal and maintaining consciousness.

retina (ret´ĭ-nah) The inner, light-sensitive layer of the eye; contains rods and cones.

retroperitoneal (rĕ´tro-pĕr´ĭ-to-ne´al) Organs behind (covered by) the peritoneum such as the duodenum, pancreas, ascending colon, descending colon, rectum, kidneys, adrenal glands, and urinary bladder.

rheumatic heart disease Inflammation of the endocardium and the valves of the heart as a result of an immune response to a streptococcal infection.

ribonucleic acid (ri´bo-nu-kle´ik) Type of nucleic acid containing the sugar ribose; involved in protein synthesis.

ribosome (ri´bo-sōm) Small, spherical, cytoplasmic organelle where protein synthesis occurs.

right lymphatic duct Lymphatic duct that empties into the right subclavian vein; drains the right side of the head and neck, the right upper thorax, and the right limb.

RNA See ribonucleic acid.

rod Photoreceptor cell in the retina with rod-shaped processes; functions in low-light vision.

rotation Movement of a structure about its axis.

rough endoplasmic reticulum Endoplasmic reticulum with attached ribosomes; a site of protein synthesis.

round window Membranous structure separating the inner and middle ear; functions to absorb and dampen sound vibrations within the inner ear.

ruga, pl. **rugae** (ru´gah, ru´ge) [L. a wrinkle] Fold or ridge; fold of the mucous membrane of the stomach; transverse ridge in the mucous membrane of the vagina.

S

sacrum (sa´krum) The single bone formed by the fusion of (usually) five sacral vertebrae.

saddle joint Two saddle-shaped articulating surfaces oriented at right angles to each other; the joint between the metacarpal and carpal at the base of the thumb.

sagittal (saj´ĭ-tal) [L. *sagitta,* the flight of an arrow] Plane running vertically through the body and dividing it into right and left parts.

salivary gland Gland that produces and secretes saliva into the oral cavity. The three major pairs of salivary glands are the parotid, submandibular, and sublingual glands.

salt Molecule consisting of a positively charged ion other than hydrogen, and a negatively charged ion other than hydroxide; formed by a reaction between an acid and a base.

sarcomere (sar´ko-mer) [Gr. *sarco,* flesh, means muscle + *meros,* part] The structural and functional unit of a muscle; the part of a myofibril formed of actin and myosin myofilaments, extending from Z line to Z line.

saturated fat Fatty acid in which there are only single covalent bonds between carbon atoms.

sclera (skler´ah) [L. *skleros,* hard] The dense, white, opaque outer layer of the eye; white of the eye.

scoliosis (sko-le-o´sis) [Gr. *skoliosis,* a crookedness] Abnormal lateral curvature of the spine.

scrotum (skro´tum) Musculocutaneous sac containing the testes.

sebaceous gland (se-ba´shus) [L. *sebum,* tallow] Gland of the skin that produces sebum; usually associated with a hair follicle.

sebum (se´bum) [L. tallow] Oily, white, fatty substance produced by the sebaceous glands; oils hair and the surface of the skin.

second-degree burn See partial-thickness burn.

secondary curvature Anterior bending of the spine; develops in the cervical region when a baby lifts his head and in the lumbar region when he walks.

secondary response See memory response.

secretin (se-kre´tin) Hormone released by the small intestine that inhibits gastric juice secretions and movements but stimulates increased secretions from the pancreas and liver.

secretory vesicle Membrane-bound sac that pinches off from the Golgi apparatus, moves to and fuses with the plasma membrane, and releases its contents to the outside of the cell.

selectively permeable Allowing some substances to pass through but not others.

semen (se´men) [L. seed] Penile ejaculate; viscous fluid containing sperm cells and secretions of the testes, seminal vesicles, prostate gland, and bulbourethral glands.

semicircular canal One of three canals in each temporal bone involved in the detection of motion of the head.

semilunar valve One of two valves in the heart composed of three semilunar-shaped cusps that prevent flow of blood back into the ventricles following ejection; located at the beginning of the aorta and pulmonary trunk.

seminal vesicle (sem´ĭ-nal) One of two glandular structures that empty into the ejaculatory ducts; its secretion is one of the components of semen.

seminiferous tubule (sem´ĭ-nif´er-us) Tubule in the testis in which sperm cells develop.

sensory division Part of the peripheral nervous system that sends action potentials from the periphery to the central nervous system.

sensory neuron Neuron that carries action potentials to the central nervous system.

septal defect A defect (e.g., an opening) in the septum that separates either the atria or ventricles of the heart.

serosa (se-ro´sah) The outermost layer of an organ that lies in a body cavity; consists of epithelium resting on connective tissue; serous membrane.

serous membrane (sēr´us) Thin sheet consisting of epithelial and connective tissue that lines cavities not opening to the outside of the body; does not contain glands but does secrete serous fluid.

Sertoli cell (ser-to´le) Cell in the wall of the seminiferous tubules to which spermatogonia and spermatids are attached.

serum Fluid portion of blood after the removal of fibrin and formed elements.

sex chromosome An X or a Y chromosome; determines sex in that an individual that is XX is female and XY is male.

sex-linked trait Characteristic resulting from the expression of a gene on a sex chromosome.

shoulder girdle See pectoral girdle.

sign An objective departure from normal function or appearance that can be recorded.

sinoatrial (SA) node (si´no-a´tre-al) Collection of specialized cardiac muscle fibers that acts as the "pacemaker" of the cardiac conduction system.

skeletal muscle Muscle attached to the skeleton or skin and responsible for body movements; consists of long, cylindrical, striated cells with several peripherally located nuclei.

sliding filament mechanism The mechanism by which muscle contraction occurs wherein actin myofilaments slide past myosin myofilaments.

slow-twitch muscle fiber Muscle fiber that contracts slowly and relies primarily upon aerobic respiration to produce ATP. Well supplied with blood vessels and contains myoglobin.

small intestine Portion of the digestive tract between the stomach and the cecum; consists of the duodenum, jejunum, and ileum.

smooth endoplasmic reticulum Endoplasmic reticulum without ribosomes; site of lipid synthesis.

smooth muscle Muscle in the wall of hollow organs, pupil of the eye, skin, and glands; consists of nonstriated spindle-shaped cells with a single, centrally-located nucleus.

sodium-potassium exchange pump Transport mechanism that uses energy derived from ATP to move sodium and potassium ions from areas of lower to higher concentrations; as sodium moves across the cell membrane in one direction, potassium moves across in the opposite direction.

soft palate Posterior muscular portion of the palate that separates the oropharynx and nasopharynx; prevents materials from entering the nasal cavity during swallowing.

somatic cell (so-mat´ik) Any body cells except for a reproductive cell; contains 23 pairs of chromosomes.

somatic nervous system Part of the peripheral nervous system that sends action potentials to skeletal muscles.

special sense A sensation that is detected by receptors located in highly specialized organs; includes taste, smell, sight, hearing, and balance.

specific resistance Immune system response in which there is an ability to recognize, remember, and destroy a specific antigen.

sperm cell [Gr. *sperma,* seed] Male gamete or sex cell, composed of a head, midpiece, and tail; contains the genetic information transmitted by the male; also called a spermatozoon.

spermatic cord Cord formed by the ductus deferens, testicular artery and veins, and testicular nerves; ensheathed by connective tissue. Extends from the scrotum through the inguinal canal.

spermatid (sper´mă-tid) A male reproductive cell that results from meiosis in the seminiferous tubule; undergoes changes in its structure to produce a sperm cell.

spermatocyte (sper´mă-to-sīt) Cell arising from a spermatogonium and destined to give rise to sperm cells.

spermatogenesis (sper´mă-to-jen´ĭ-sis) Formation and development of sperm cells.

spermatogonia (sper´mă-to-go´ne-ah) [Gr. *sperma,* seed + *gone,* generation] Cells that divide by mitosis to form primary spermatocytes.

spermatozoon, pl. **spermatozoa** (sper´mă-to-zo´on, sper´mă-to-zo´ah) [Gr. *sperma,* seed; zoon, animal] See sperm cell.

spinal nerve One of 31 pairs of nerves formed by the joining of the dorsal and ventral roots that arise from the spinal cord.

spindle fibers Protein fibers that form during mitosis; help to pull chromosomes to opposite poles of the cell.

spine See vertebral column.

spirometer (spi-rom´ĕ-ter) Device used to measure pulmonary volumes and vital capacity.

spirometry (spi-rom´ĕ-tre) The process of measuring the volumes of air that move into and out of the respiratory system.

spleen (splēn) Large lymphatic organ in the upper part of the abdominal cavity on the left side, between the stomach and diaphragm; responds to microorganisms and foreign substances in the blood, destroys worn out red blood cells, and is a reservoir for blood.

spongy bone Bone with a latticelike appearance, having spaces filled with marrow.

Starling's law of the heart The volume of blood pumped from the heart during each contraction of the heart is proportional to the degree to which the ventricles are filled with blood (the greater the amount of blood flowing into the atria, the greater the amount of blood pumped from the ventricles).

static equilibrium The sense of balance that evaluates the position of the head relative to gravity.

stenosis (stĕ-no´sis) Narrowing of a valve so that its opening is partially closed.

sterilization The complete destruction of all living organisms in a given area.

sternum The breastbone; attachment site for the true ribs; formed by the fusion of the manubrium, body, and xiphoid process.

steroid Large family of lipids, including some hormones, vitamins, and cholesterol.

stethoscope (steth´o-skop) [Gr. *stetho-,* chest + *skopeo,* to view] An instrument originally devised for aid in hearing the respiratory and cardiac sounds in the chest and now used in hearing other sounds in the body as well.

stratum, pl. **strata** ([L. bed cover, layer] Layer of tissue.

stratum basale (ba-să´le) [L. layer + basal] Deepest layer of the epidermis;

new cells produced here move to the surface of the epidermis.

stratum corneum (kor′ne-um) [L. layer + *corneus,* horny] Most superficial layer of the epidermis consisting of flat, dead cells filled with keratin.

stria, pl. **striae** (stri′ah, stri′e) [L. channel] Line or streak in the skin that is a different texture or color from the surrounding skin; a stretch mark resulting from damage to the dermis of the skin.

stroke volume The volume of blood ejected from either the right or left ventricle during each heart beat.

subarachnoid space (sub-ă-rak′noyd) Space between the arachnoid layer and the pia mater that is filled with cerebrospinal fluid.

subcutaneous (sub′ku-ta′ne-us) [L. *sub,* under + *cutis,* skin] Under the skin; also known as the hypodermis.

submucosa (sub′mu-ko′sah) A layer of tissue beneath a mucous membrane.

superficial [L. *superficialis,* surface] Toward or on the surface.

superior [L. higher] Up, or higher, with reference to the anatomical position.

superior vena cava (ve′nah ka′vah) The large vein that returns blood from the head and neck, upper limbs, and thorax to the right atrium.

supination (su′pĭ-na′shun) [L. *supino,* to bend backward, place on back] Rotation of the forearm to the anatomical position; rotation of the forearm (when the forearm is parallel to the ground) so that the anterior surface is facing up.

surface tension A measure of the attraction of water molecules for each other at a boundary between water and some other substance.

surfactant (sur-fak′tant) Lipoproteins forming a layer over the film of water in the alveoli; reduces surface tension and the tendency for alveoli to collapse.

suspensory ligament Attachment between the lens and the ciliary body; holds the lens in place.

suture (su′chur) [L. *sutura,* a seam] Junction between the flat bones of the skull.

sweat gland Secretory organ (merocrine sweat gland) that usually produces a watery secretion called sweat that is released onto the surface of the skin; some sweat glands (apocrine sweat glands), however, produce an organic secretion.

sympathetic chain ganglion Collection of sympathetic neurons that are connected to each other to form a chain along both sides of the spinal cord.

sympathetic division Subdivision of the autonomic nervous system generally involved in preparing the body for immediate physical activity.

symphysis pubis See pubic symphysis.

symptom Any departure from normal function, appearance, or sensation experienced by a person.

synapse (sin′aps) [Gr. *syn,* together + *haptein,* to clasp] Junction between a neuron and some other cell.

synaptic cleft Space between the axon ending and another cell such as a neuron, muscle cell, or gland cell.

synergist (sin′er-jist) A muscle that works with another muscle to cause a movement.

synovial joint (sĭ-no′ve-al) Bone joint in which the ends of the bones are covered with articular cartilage but are separated by a joint cavity filled with synovial fluid. The bones are held together by the joint capsule.

synovial membrane Connective tissue membrane lining the joint cavity (except for articular cartilage) or bursa; produces synovial fluid.

systemic circulation Blood flow from the left ventricle of the heart to the tissues of the body and back to the right atrium.

systole (sis′to-le) [Gr. *systole,* a contracting] Contraction of the heart chambers during which blood leaves the chambers; usually refers to ventricular contraction.

systolic pressure (sis′tol′ik) [Gr. *systole,* a contracting] The maximum pressure achieved in the aorta, which is close to the maximum pressure achieved in the left ventricle during systole.

T

T cell Lymphocyte responsible for cell-mediated immunity; also regulates specific immunity.

T wave Deflection in the electrocardiogram following the QRS complex, representing repolarization of the ventricles.

target tissue Tissue upon which a hormone acts.

taste bud Sensory structure found mostly on the tongue; contains taste cells with receptors that respond to dissolved chemicals.

taste cell Sensory cell in a taste bud; hairs of the taste cell have receptors that detect chemicals involved with taste.

telophase (tel′o-fāz) Last phase of mitosis; the chromosomes have reached the opposite poles (ends) of the cell and the nuclear membrane and nucleolus reappear.

tendon Band of dense connective tissue that connects a muscle to a bone or other structure.

testis, pl. **testes** (tes′tis, tes′tēz) One of two male reproductive glands located in the scrotum; produces testosterone and sperm cells.

testosterone (tes-tos′tĕ-rōn) Hormone secreted primarily by the testes; aids in spermatogenesis, controls maintenance and development of male reproductive organs and secondary sexual characteristics, and influences sexual behavior.

thalamus (thal′ă-mus) A collection of nuclei that forms most of the diencephalon; major relay center for sensory nerve tracts going to the cerebrum.

thick myofilament Filament composed of myosin within the sarcomere.

thin myofilament Filament composed of actin within the sarcomere.

third-degree burn See full-thickness burn.

thoracic cage The thoracic vertebrae, the ribs with their associated cartilages, and the sternum; protects the organs of the thoracic cavity; the rib cage.

thoracic cavity Space bounded by the neck, the thoracic wall, and the diaphragm.

thoracic duct Largest lymph vessel in the body; empties into the left subclavian vein; drains the left side of the head and neck, the left upper thorax, the left upper limb, and the inferior half of the body.

threshold Stimulus that is just large enough to elicit a response.

thrombocyte (throm′bo-sīt) [Gr. *thrombos,* clot + *kytos,* cell] See platelet.

thrombus (throm′bus) [Gr. *thrombos,* a clot] Clot in the cardiovascular system formed from coagulated components of blood; can block a blood vessel or attach to the vessel wall without completely blocking the vessel.

thymus (thi′mus) Bilobed lymphatic organ located in the inferior neck and superior mediastinum; involved with the maturation of T cells.

thyroid gland (thi'royd) [Gr. *thyreoeides,* shield] Endocrine gland located inferior to the larynx and consisting of two lobes connected by a narrow band; secretes the thyroid hormones.

thyroid cartilage Largest and most superior cartilage of the larynx; protrudes to form the Adam's apple.

thyroid hormone Hormone secreted by the thyroid gland; especially those such as thyroxine that contain iodine and function to regulate metabolism and maturation of tissues.

thyroid-stimulating hormone (TSH) Hormone released from the anterior pituitary that stimulates thyroid hormone secretion from the anterior pituitary.

thyroxine (thi-rok'sin) One of the thyroid hormones that contain iodine atoms.

tidal volume The volume of air inspired or expired in a single breath during regular breathing; usually measured at rest but can refer to air movement during exercise.

tissue [L. *texo,* to weave] A collection of cells with similar structure and function, and the substances between the cells.

tissue regeneration Substitution of viable cells for damaged or dead cells by new cells of the same type that were damaged.

tissue repair Substitution of viable cells for damaged or dead cells by tissue regeneration or tissue replacement.

tissue replacement Substitution of viable cells for damaged or dead cells by new cells of a different type, usually connective tissue cells resulting in a scar.

tonsil Any collection of lymphoid tissue; usually refers to large collections of lymphoid tissue beneath mucous membranes of the oral cavity and pharynx; lingual, pharyngeal, and palatine tonsils.

tooth, pl. **teeth** One of the hard conical structures set in the bone of the upper and lower jaws; consists of a crown, neck, and root; used in mastication and speech.

trachea (tra'ke-ah) Air tube extending from the larynx into the thorax, where it divides to form the two primary bronchi; has 16 to 20 C-shaped rings of cartilage in its walls; functions to conduct air.

transcription Process of forming mRNA from DNA.

transfer RNA Type of RNA that attaches to amino acids and transports them to ribosomes where the amino acids are connected to form proteins.

translation Synthesis of proteins at the ribosomes in response to information contained in mRNA.

transmission The transfer of a pathogen from a reservoir to a susceptible host. It can be direct contact (e.g., body contact, respiratory droplets) or indirect contact (e.g., contaminated objects, contaminated food/water, or vectors).

transverse plane Plane running horizontally through the body and dividing it into superior and inferior parts; a horizontal cross section.

tricuspid valve (tri-kus'ped) Valve consisting of three cusps of tissue; located between the right atrium and right ventricle of the heart.

triglyceride (tri-glis'er-īd) A fat consisting of a glycerol molecule with three attached fatty acids.

tRNA See transfer RNA.

true rib Rib that directly attaches to the sternum by costal cartilage; rib pairs one through seven.

trypsin (trip'sin) Digestive enzyme in pancreatic juice; digest proteins and polypeptide chains into short amino acid chains.

tubular reabsorption Movement of materials, by means of diffusion or active transport, from the filtrate of a nephron into the blood.

tubular secretion Movement of materials, by means of active transport, from the blood into the filtrate of a nephron.

tumor Abnormal tissue growth resulting from cellular divisions that continue after normal cell division of the tissue has stopped or slowed.

tunica externa The outermost fibrous coat of a vessel or an organ that is derived from the surrounding connective tissue.

tunica intima The innermost layer of a blood or lymphatic vessel; consists of endothelium and a small amount of connective tissue.

tunica media The middle, usually muscular, coat of an artery or other tubular structure.

tympanic membrane (tim-pan'ik) Thin partition separating the external ear from the middle ear; vibrates in response to sound waves.

U

umbilical artery One of the two arteries that supply blood to the placenta from the fetal circulation; the umbilical arteries originate from the iliac arteries of the fetus.

umbilical cord Connection between the embryo or fetus and the placenta; carries blood between the embryo or fetus and the placenta.

umbilical vein The vein in the umbilical cord of the fetus by which the fetus receives nourishment from the maternal system; blood flows from the placenta to the fetus.

unmyelinated fiber Nerve fibers that do not have myelin sheaths formed by the wrappings of neuroglia.

unsaturated fat Fatty acid chain with one (monounsaturated) or more (polyunsaturated) double covalent bonds between its carbon atoms.

upper respiratory tract The nose, nasal cavity, and pharynx.

ureter (u-re'ter) [Gr. *oureter,* urinary canal] Tube that conducts urine from the kidney to the urinary bladder.

urethra (u-re'thrah) Tube that carries urine from the urinary bladder to the exterior of the body.

urinary bladder Container located in the pelvic cavity that stores urine until it is eliminated by the process of micturition.

urine Liquid waste excreted from the kidneys; consists of mostly water and organic waste products, such as urea, and excess ions.

uterine tube One of the tubes leading on either side from the uterus to the ovary; also called the fallopian tube.

uterus (u'ter-us) Hollow muscular organ in which the fertilized oocyte develops into a fetus.

uvula (u'vu-lah) [L. *uva,* grape] Small grapelike appendage at the posterior margin of the soft palate.

V

vaccine Preparation of killed microorganisms, altered microorganisms, or derivatives of microorganisms intended to produce immunity. Usually administered by injection, but sometimes ingestion is preferred.

vagina (vă-ji'nah) [L. sheath] Genital canal in the female, extending from the uterus to the vestibule.

vas deferens See ductus deferens.

vasomotor center (va´so-mo´tor) The center within the medulla oblongata and lower pons of the brain that regulates peripheral resistance by controlling the diameter of blood vessels through the sympathetic innervation of the blood vessels.

vasomotor tone Partial constriction of blood vessels resulting from relatively constant sympathetic stimulation.

vasopressin (va-zo-pres´in) See antidiuretic hormone.

vector [L. a carrier] An invertebrate animal (e.g., mosquito, louse, or tick) capable of transmitting a disease to a vertebrate (e.g., human).

vein Blood vessel that carries blood toward the heart.

ventilation The movement of air into and out of the lungs.

ventral [L. *ventr,* belly] See anterior.

ventral root Motor root of a spinal nerve.

ventricle (ven´tri̮-kul) [L. *venter,* belly] One of four cerebrospinal-filled cavities within the brain; chamber of the heart that pumps blood into arteries (i.e., the left and right ventricles).

venule (ven´ūl) The smallest diameter veins; have a structure similar to capillaries but with a larger diameter; blood flows from capillaries into venules.

vertebral column The 26 vertebrae considered together; bears the weight of the trunk, protects the spinal cord, is the site of exit of the spinal nerves, and provides attachment sites for muscles.

vesicle (ves´ĭ-kl) [L. *vesica,* bladder] Small sac containing a liquid or gas, e.g., a blister in the skin or an intracellular, membrane-bound sac such as a secretory vesicle or a lysosome.

vestibule (ves´tĭ-būl) [L. antechamber] Middle region of the inner ear containing the maculae; responsible for static equilibrium.

villus, pl. **villi** (vil´us, vil´i) [L. shaggy hair (of beasts)] Projection of the mucous membrane of the small intestine; functions to increase surface area for absorption and secretion.

visceral peritoneum (pĕr´ĭ-to-ne´um) [L. organ] That part of the serous membrane in the abdominal cavity covering the surface of some abdominal organs.

vital capacity The sum of the inspiratory reserve volume, tidal volume, and the expiratory reserve volume.

vitamin (vi´tah-min) [L. *vita,* life + *amine,* from ammonia] One of a group of organic substances, present in minute amounts in natural foods, that are essential to normal metabolism.

vitamin D Fat-soluble vitamin produced from precursor molecule in skin exposed to ultraviolet light; increases calcium and phosphate uptake in the intestine.

vitreous humor (vit´re-us) Transparent jellylike material that fills the posterior compartment of the eye (space between the lens and the retina).

vocal cord One of two folds of elastic ligaments covered by mucous membrane that extends across the superior end of the larynx; vibrations of the vocal cords are responsible for voice production.

W

water A molecule consisting of one atom of oxygen joined by covalent bonds to two atoms of hydrogen.

water-soluble vitamin Vitamin such as one of the B complex vitamins or vitamin C that dissolves in water and is absorbed with water from the intestinal tract.

white blood cell (WBC) Round, nucleated cell involved in immunity. The five types of white blood cells are neutrophils, eosinophils, basophils, lymphocytes, and monocytes; also called a leukocyte.

white matter Bundles of nerve fibers and their associated sheaths.

Y

yellow marrow Adipose (fat) tissue within the cavities of bones.

yolk sac Cavity formed within the inner cell mass of the blastocyst; in humans gives rise to blood cells; in other animals provides nutrients (yolk).

Z

Z line Delicate membranelike structure found at either end of a sarcomere to which the actin myofilaments attach.

zygote (zi´gōt) [Gr. *zygotos,* yoked] The single-celled product of fertilization, resulting from the union of a sperm cell and a secondary oocyte.

Credits

CHAPTER 1
1-1 (art), Christine Oleksyk; **1-1** (photo), Pat Watson; **Table 1-1,** Cynthia Turner Alexander/Terry Cockerham, Synapse Media Production; **1-2, 1-3,** Rolin Graphics; **1-4, 1-5a, 1-6, 1-K,** Terry Cockerham, Synapse Media Production; **1-5(b,c,d),** courtesy of R.M.H. McMinn and R.T. Hutchings (from *Color Atlas of Human Anatomy,* ed. 2, 1988, Year Book Medical Publishers); **1-7, 1-8,** Nadine Sokol.
CHAPTER 2
2-1, 2-2b, 2-5, 2-6, Ron Ervin; **2-2c,** Pat Watson.
CHAPTER 3
3-1, 3-2, 3-5, 3-6, 3-7, 3-8, Christine Oleksyk; **3-3, 3-4,** Rolin Graphics; **3-9,** Ronald J. Ervin; **3-A,** Barbara Cousins.
CHAPTER 4
4-1a, 4-4b, 4-7, Rolin Graphics; **4-1b, 4-5,** Christine Oleksyk; **4-2a** (location), G. David Brown; **4-2b** (location), Stackhouse; **4-2** (c,d,e,f—locations), **4-4** (a,c,d,e,f—locations), Eileen Draper; **4-2** (a,b,d,f,g—photos), **4-4** (b,c—photos), Ed Reschke; **4-2c** (photo), **4-4** (a,e—photos), Trent Stephens; **4-2e** (photo), **4-4** (d,f—photos), Carolina Biological Supply; **4-3,** Michael P. Schenk; **4-6,** Scott Bodell.
CHAPTER 5
5-1, 5-3, Christine Oleksyk; **5-2,** John Cunningham/Visuals Unlimited; **5-A,** Ronald J. Ervin; **5-B,** courtesy of Thomas P. Habif (from *Clinical Dermatology,* ed. 2, 1990, Mosby-Year Book, Inc.).
CHAPTER 6
6-1a, 6-2, 6-3, 6-4, 6-5, 6-7, 6-8, 6-9, 6-10, 6-11 (2,3), 6-13, 6-14, 6-15, 6-17, 6-A (c,d,e), David J. Mascaro & Associates; **6-1 (b,c,d),** John V. Hagen; **6-6,** Karen Waldo; **6-12, 6-16, 6-19,** Terry Cockerham/Synapse Media Production; **6-18,** Rusty Jones; **6-A (a,b,d), 6-11 (1,4)** Rolin Graphics; **6-B (a,b),** Christine Oleksyk; **6-C,** Paul R. Manske.
CHAPTER 7
7-1 (a,b), Barbara Cousins; **7-1c, 7-2,** William Ober; **7-4, 7-7** (inset), Rolin Graphics; **7-5 (a,b), 7-6, 7-7, 7-8 (a,b), 7-9, 7-10,** John V. Hagen; **7-8c, 7-11,** Terry Cockerham/Synapse Media Production.
CHAPTER 8
8-1, Pagecrafters; **8-2, 8-3c, 8-4, 8-6, 8-7, 8-11, 8-12a, 8-13a,** Scott Bodell; **8-3 (a,b),** Joan Beck; **8-5, 8-8a,** courtesy of R.M.H. McMinn and R.T. Hutchings (from *Color Atlas of Human Anatomy,* ed. 2, 1988, Year Book Medical Publishers); **8-8b,** William Ober; **8-9,** Christine Oleksyk; **8-10, 8-12b, 8-13b, 8-16,** Barbara Cousins; **8-14,** Rolin Graphics.
CHAPTER 9
9-1, Janis K. Atlee/Michael P. Schenk; **9-2, 9-3, 9-4, 9-6, 9-9, 9-10a, 9-11a, 9-12a,** Marcia Dohrmann; **9-5,** Hagen; **9-7,** Terry Cockerham/Synapse Media Production; **9-8, 9-11 (b,c), 9-12 (b,c,d), 9-A, 9-B,** G. David Brown; **9-10b,** Rolin Graphics.
CHAPTER 10
10-1, Joan Beck; **10-2 (a,b),** Barbara Stackhouse; **10-2c,** Rolin Graphics; **10-3 (a,b),** Nadine Sokol; **10-3c,** Trent Stephens; **10-4, 10-6, 10-7,** Christine Oleksyk; **10-5, 10-8,** Pagecrafters.
CHAPTER 11
11-1, William Ober; **11-2, 11-3,** Christine Oleksyk; **11-5, 11-6, 11-A,** Molly Babich/John Daugherty; **11-7,** Trent Stephens.
CHAPTER 12
12-1, 12-A, Rusty Jones; **12-2,** Rolin Graphics; **12-3, 12-5, 12-8,** Christine Oleksyk; **12-4,** Barbara Cousins; **12-6,** Ronald J. Ervin; **12-7,** Trent Stephens.
CHAPTER 13
13-1, 13-2, William Ober; **13-3, 13-12, 13-13,** Christine Oleksyk; **13-4, 13-14,** John Daugherty; **13-5,** David J. Mascaro & Associates; **13-6, 13-9,** Karen Waldo; **13-7, 13-8,** George J. Wassilchenko; **13-10,** Joan Beck/Donna Odle; **13-11,** G. David Brown.
CHAPTER 14
14-1, Ronald J. Ervin; **14-2,** G. David Brown; **14-3,** David J. Mascaro & Associates; **14-5,** John Daugherty; **14-6,** Rolin Graphics; **14-A,** Michael Schenk.
CHAPTER 15
15-1, Joan Beck; **15-2, 15-8,** Christine Oleksyk; **15-3, 15-7,** Jody L. Fulks; **15-4,** courtesy of Branislav Vidic; **15-5, 15-6,** John Daugherty; **15-9,** Barbara Cousins.
CHAPTER 16
16-1, Lisa Shoemaker; **16-2, 16-6c,** William Ober; **16-3a,** K. Born; **16-3 (b,c), 16-8,** David J. Mascaro & Associates; **16-4, 16-7,** G. David Brown; **16-5,** John Daugherty; **16-6 (a,b,d),** Kathy Mitchell Grey; **16-9,** Michael P. Schenk; **16-10,** Christine Oleksyk; **16-A,** Joan Beck.
CHAPTER 17
17-1, Christine Oleksyk; **17-4 (a,b),** Barbara Stackhouse; **17-4** (insets), Ronald J. Ervin.
CHAPTER 18
18-1, Joan Beck; **18-2, 18-5, 18-7, 18-8,** Christine Oleksyk; **18-3,** David J. Mascaro & Associates; **18-4,** courtesy of Branislov Vidic; Barbara Stackhouse; **18-6,** Rolin Graphics.
CHAPTER 19
19-1, Rolin Graphics; **19-2, 19-5,** Barbara Cousins; **19-3,** William Ober; **19-4, 19-10,** Kevin Sommerville; **19-6,** Ronald J. Ervin; **19-6,** Joan Beck; **19-7,** Kevin Somerville/Kathy Mitchell Grey; **19-8,** Kevin Somerville/Kathy Mitchell Grey/Scott Bodell; **19-9,** David J. Mascaro & Associates; **19-12,** Yvonne Wylie Walston; **19-A (a,b,c)** Laura J. Edwards; **19-A (d,e),** courtesy of Ortho Pharmaceutical Corporation; **19-A (f,g),** Ronald J. Ervin.
CHAPTER 20
20-1, 20-2, 20-3, 20-5, Marcia Hartsock; **20-4a,** Lennart Nilsson (copyright *A Child Is Born,* Dell Publishing Company); **20-4b,** Lennart Nilsson (copyright *Behold Man,* Little, Brown, and Company); **20-6,** Scott Bodell; **20-7,** CNRI/Science Photo Library; **20-8, 20-9, 20-10,** Rolin Graphics.
CHAPTER 21
21-1 (drawings), **21-2, 21-4a, 21-6,** Rolin Graphics; **21-1** (photos), **21-5a,** Ed Reschke; **21-3a,** William Ober; **21-3b,** Visuals Unlimited; **21-4b,** Lennart Nilsson (copyright Boehringer Ingelheim International GmbH, *The Incredible Machine,* National Geographic Society); **21-5b,** courtesy of R.T.D. Emond and H.A.K. Rowland (from *A Colour Atlas of Infectious Diseases,* ed. 2, 1987 Wolfe Medical Publications Ltd.); **21-A,** Charles C. Brinton/Judith Carnahan.

Index

N

O